PREHISTORIC

DK

PREHISTORIC

LONDON, NEW YORK, MELBOURNE,
MUNICH, AND DELHI

DORLING KINDERSLEY

Senior Art Editor
Ina Stradins

Project Art Editors
Yen Mai Tsang, Francis Wong,
Steve Woosnam-Savage

Designers
Sonia Barbate, Alison Gardner,
Helen McTeer, Simon Murrell

Design Assistants
Riccie Janus, Rebecca Tennant

Production Editor
Phil Sergeant

Creative Technical Support
Adam Brackenbury, John Goldsmid

Jacket Designer
Duncan Turner

Illustrators
Antbits Ltd (Richard Tibbitts), Dotnamestudios
(Andrew Kerr), Francisco Gascó, Peter Minister,
pixel-shack.com (Jon Hughes and
Russel Gooday)

Special Photography
Gary Ombler

Cartographer
Paul Eames

Senior Managing Art Editor
Philip Ormerod

Art Director
Bryn Walls

Senior Editors
Angeles Gavira Guerrero, Peter Frances

Project Editors
Cressida Malins, Ruth O'Rourke, Gill Pitts,
David Summers, Victoria Wiggins

Editors
Jamie Ambrose, Daniel Gilpin, Salima Hirani,
Tom Jackson, Nathan Joyce, Lara Maiklem,
Emma Marriott, Claire Nottage, Miezan van Zyl

Editorial Assistants
Elizabeth Munsey, Jaime Tenreiro

Researcher
Graeme Lloyd

Indexer
Hilary Bird

Production
Rita Sinha

Picture Researcher
Myriam Mégharbi

Managing Editor
Sarah Larter

Publishing Manager
Liz Wheeler

Reference Publisher
Jonathan Metcalf

DK INDIA

Art Director
Shefali Upadhyay

Designers
Mahua Mandal, Neerja Rawat,
Arijit Ganguly

Head of General Books
Aparna Sharma

Editor
Garima Sharma

DTP Designers
Harish Aggarwal,
Dheeraj Arora,
Preetam Singh,
Jagtar Singh

First published in Great Britain in 2009
by Dorling Kindersley Limited
80 Strand, London WC2R 0RL

A Penguin Company

2 4 6 8 10 9 7 5 3 1

A CIP catalogue record for this book is available from the
British Library.

ISBN: 978-1-4053-4758-7

Colour reproduction by Media Development Printing Ltd, UK
Printed and bound in Singapore by Star Standard

See our complete catalogue at
www.dk.com

Authors

YOUNG EARTH
Dr Douglas Palmer – science writer and lecturer based in Cambridge, UK

MICROSCOPIC LIFE
Dr Martin Brasier – Professor of Palaeobiology, University of Oxford, UK

PLANTS
David Burnie Bsc – award-winning natural history writer

Chris Cleal – Head of Vegetarian History, Amgueddfa Cymru (National Museum Wales)

Professor Sir Peter Crane FRS – Director of the Royal Botanic Gardens, Kew until 2006, Professor at the Department of Geophysical Sciences, University of Chicago, USA

Professor Barry A. Thomas – Honorary Professor, Aberystwyth University, Wales

INVERTEBRATES
Dr Caroline Buttler, Professor John C.W. Cope, and Dr Robert M. Owens – Department of Geology, Amgueddfa Cymru (National Museum Wales), Cardiff, Wales

VERTEBRATES
Dr Jason Anderson – vertebrate palaeontologist and Assistant Professor of Veterinary Anatomy, University of Calgary, Canada

Dr Roger Benson – palaeontologist, University of Cambridge, UK

Stephen Brusatte – vertebrate palaeontologist, American Museum of Natural History, New York, USA

Professor Jennifer Clack – Professor and Curator of Vertebrate Palaeontology, University Museum of Zoology, Cambridge, UK

Dr Kim Dennis-Bryan – palaeontologist, writer, and lecturer, Open University, London, UK

Dr Christopher Duffin – palaeontologist and teacher, London, UK

Dr David Hone – vertebrate palaeontologist, Beijing Institute of Vertebrate Palaeontology and Palaeoanthropology, Beijing, China

Dr Zerina Johanson – curator of fossil fish, Natural History Museum, London, UK

Dr Andrew Milner – vertebrate palaeontologist and research associate, Natural History Museum, London, UK

Dr Darren Naish – science writer and honorary research associate at the University of Portsmouth, UK

Dr Katie Parsons – biologist and natural history writer, London, UK

Dr Donald Prothero – Professor of Geology, Occidental College and lecturer at California Institute of Technology, Los Angeles, USA

Professor Xu Xing – palaeontologist, Beijing Institute of Vertebrate Palaeontology and Palaeoanthropology, Beijing, China

PERIOD INTRODUCTIONS
Dr Ken McNamara – palaeobiologist, Department of Earth Sciences, University of Cambridge, UK

THE RISE OF HUMANS
Dr Fiona Coward – research fellow, Royal Holloway University of London, UK

GLOSSARY
Richard Beatty

Consultants

YOUNG EARTH
Dr Simon Lamb – lecturer, Department of Earth Sciences, University of Oxford, UK

Felicity Maxwell Bsc – freelance environmental consultant and science editor

PLANTS
Professor Sir Peter Crane FRS – see above

Dr Paul Kenrick – researcher, Palaeontology Department, Natural History Museum, London, UK

INVERTEBRATES
Professor Euan N.K. Clarkson – Professor Emeritus of Palaeontology, University of Edinburgh, Scotland

VERTEBRATES
Professor Michael J. Benton – Professor of Vertebrate Palaeontology, University of Bristol, UK

Dr Stephen Brusatte – see above

Dr Robert Carr – lecturer in comparative anatomy and general biology, Ohio University, USA

Professor Jennifer Clack – see above

Dr Gareth Dyke – palaeontologist and lecturer, University College Dublin, Ireland

Professor Christine Janis – mammalian palaeontologist and lecturer, Brown University, Providence, USA

THE RISE OF HUMANS
Dr Katerina Harvati – palaeoanthropologist, Max Planck Institute, Germany

YOUNG EARTH

After the Earth's formation 4.56 billion years ago, a unique series of events and geological processes transformed it into a planet capable of sustaining life.

Scales and sizes

Most of the profiles in Life on Earth and The Rise of Humans include a scale drawing to indicate the size (usually the maximum) of the animals or plants being described. In the plant entries, scale drawings are only included where there is reliable and easily accessible reference for reconstruction of the whole plant.

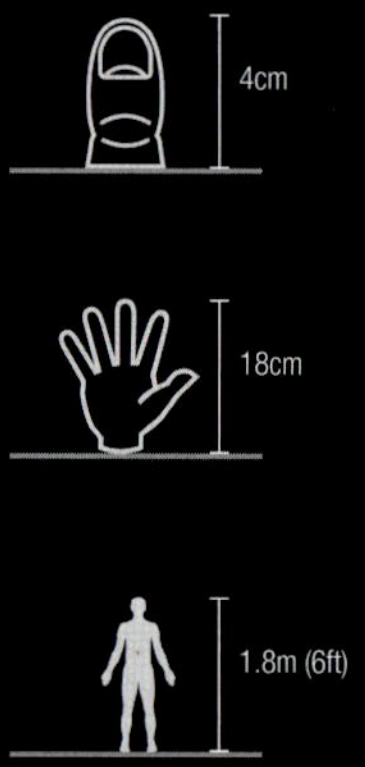

LIFE ON EARTH

The earliest-known form of life on Earth appeared about 3.5 billion years ago. From trilobites to tree ferns, hydrosaurs to hominids, this catalogue profiles hundreds of fascinating species, as they evolved.

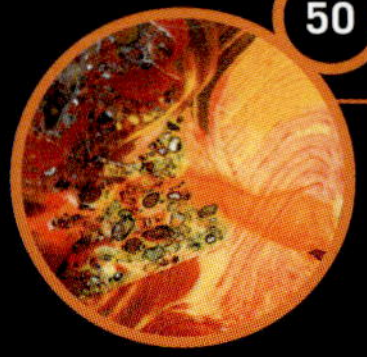

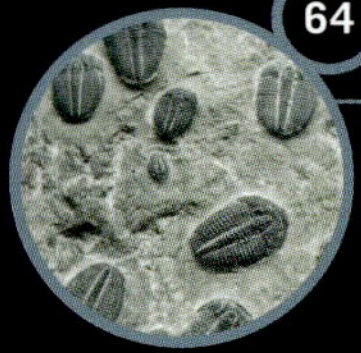

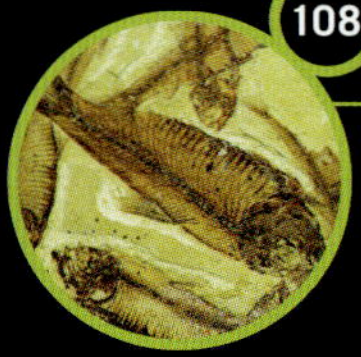

contents

THE RISE OF THE HUMANS

The story of human evolution stretches back seven million years and features an array of species of which our own, *Homo sapiens*, is the sole survivor.

century, he divined correctly that an ancient sea had once covered the area. So began the science of palaeontology – the interpretation of the life of the ancient past. At a time when other philosophers explained such fossils as artificial "sports of nature" or the lunchtime rubbish of Roman legionaries, Leonardo applied modern scientific methods of deduction.

Fossils can be beautiful and amazing. Anyone who has stood beneath the skeleton of a dinosaur, or who has cracked open a rock to reveal the bright mother-of-pearl colours of a seashell that died over 100 million years ago, will know the sense of wonder. That the dinosaur skeleton was once covered with muscle and skin, and the shellfish once lived on the sea floor, filtering food particles from the water, shows how fossils allow us access to different times.

By 1800, most naturalists accepted that the Earth was very ancient, although they could not measure its exact age; that many plants and animals found as fossils were actually extinct; and that there was some kind of succession of rocks and fossils through time. It took another two centuries to improve methods for dating rocks, to be able to read ancient environments from the sediments, and to understand the diets and habits of ancient life.

The modern palaeobiologist makes observations in the field, but also does experiments in the chemistry laboratory and on supercomputers. Through trial and error it is now possible to look beneath the surface and test ideas about the diet of a particular dinosaur or the scale and effects of a particular mass extinction. Palaeobiology rests on the wonder of the fossils, but it is a powerful science, where investigators piece together the clues like Sherlock Holmes to reveal the wonder of the life of the past.

This book is written by active professional palaeontologists from all over the world, and it translates the current science into a rich narrative of the astonishing story of the evolution of life on Earth.

MICHAEL J. BENTON

YOUNG

EA

RTH

Exploring Earth's Past

More than two hundred years of scientific investigation have revealed a great deal about Earth's dynamic 4.5-billion-year history and development. Geological time has been measured and divided, and its environments and life reconstructed. Every year, new discoveries and new techniques of investigation generate new ideas about the depths and processes of the planet's geological past, but there is still a great deal to be explored and explained by the relatively young science of geology.

The present and the past

The geological processes that transformed Earth from its origins over 4 billion years ago have left records in rocks and minerals. Most of what we know about the planet's history has been recovered from the study of rocks and fossils found at or near the planet's surface. Additionally, some of these rocks preserve information about the environments in which living things evolved, along with their remains and traces. Reading this rock and fossil record, interpreting the information in it, and reconstructing a history of Earth and its living things have taken hundreds of years – and the task is still far from complete.

UNIFORMITARIANISM
A section of a modern sand dune (left) reveals a similar pattern of layers, called cross-bedding, to those in 200-million-year-old sandstones (above), suggesting formation by the same processes.

BIOGRAPHY

CHARLES LYELL

Trained as a barrister, the Scottish geologist Charles Lyell (1797–1875) used his skills of dealing with evidence and presenting arguments to examine the mass of emerging geological information in order to understand and reconstruct Earth's geological history. Lyell popularized an idea called the principle of uniformitarianism, which states that geological processes have occurred in the past at the same rate and intensity as they do today. This contradicted the prevailing view of an Earth shaped by a short history of catastrophic events, such the biblical flood.

BROKEN RECORD
In the eighteenth century, Scottish farmer and naturalist James Hutton noticed "unconformities" or breaks in layers of grey shale and red sandstone at Siccar Point, Scotland (above). These gaps in the record, he reasoned, meant the Earth had a protracted history, with earlier episodes of mountain-building, erosion, and sedimentation.

Earth's structure

Different types of evidence suggest that Earth has a layered internal structure. By studying vibrations triggered by earthquakes, scientists have discovered that, like an onion, our planet has many layers. By combining this evidence with the study of volcanic rocks and meteorites, various layers with distinct compositions or physical properties can be recognized, from a hot, partly liquid core to a cold surface crust.

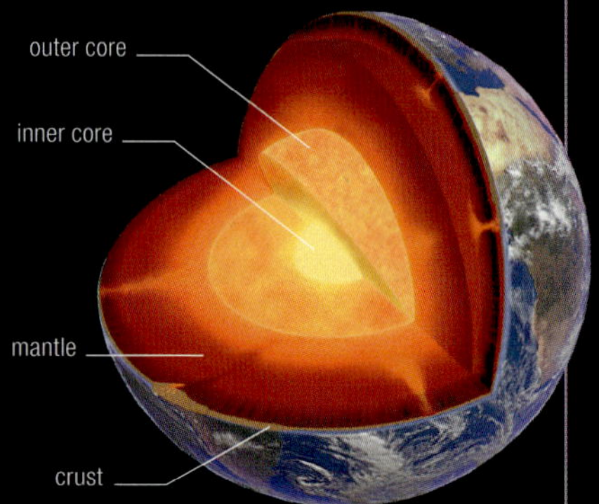

CORE TO CRUST
Earth's intensely hot iron and nickel core is divided into solid inner and liquid outer parts. Next is the mantle, made of dense silicate rock. Although almost completely solid, the hot, ductile rock flows. The outermost layer is the cool, thin crust with a variety of igneous, metamorphic, and sedimentary rocks.

SEDIMENTARY ROCKS
The "layer-cake" appearance of Arizona's Painted Desert is actually a succession of sediment deposits laid down over 200 million years ago that contain both plant and dinosaur fossils.

Rock and fossil records

Most rocks can be assigned to one of three types, depending on how they formed. Igneous rocks have cooled and crystallized from a molten state; they range from quick–cooled, fine-grained, volcanic lava erupted at Earth's surface to coarse-grained rocks such as granites that have cooled slowly at depths of more than 5km (3 miles). Weathering and erosion of rocks generate sediment that is transported to areas such as inland basins or the oceans, where it is laid down as successive layers. Over time, with the application of pressure and heat, these layers are transformed into sedimentary rocks. Such rocks may contain fossils. Where either igneous or sedimentary rocks are subjected to extremely high temperatures and pressures, they are transformed or "morphed" into metamorphic rocks.

IGNEOUS ROCK
Melting of rocks deep in Earth's continental crust led to the formation of magma that slowly cooled and crystallized into this coarse-grained granite.

METAMORPHIC ROCK
When rocks are pushed deep into the crust, heat and pressure cause them to flow and recrystallize into metamorphic rocks such as this gneiss.

REVEALED FOSSIL
Fossils can be used to date rocks. For example, the presence of the fossilized jaw bone of a large, crocodile-like amphibian called *Rhinesuchus* in this rock layer in the Karoo region of South Africa shows that the rock is between 260 and 265 million years old.

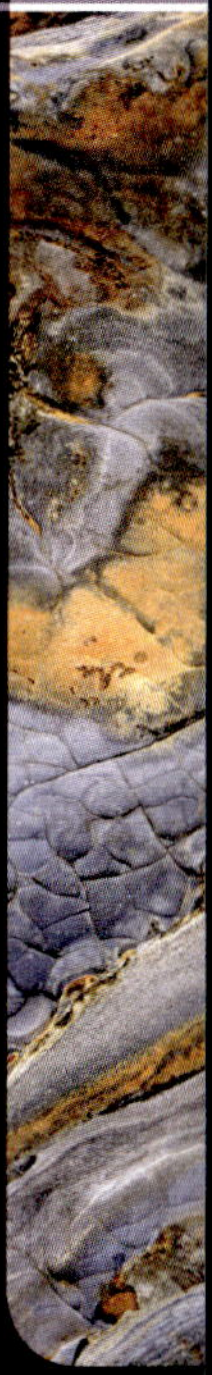

Dating techniques

The first dating of Earth's rocks consisted of using common fossils to order recognized divisions of sedimentary layers. The types of fossils change over time because species evolve. Although there were many early attempts to calculate a chronological age for Earth and its formation, none were accurate until the phenomenon of radioactivity was discovered at the end of the nineteenth century. Knowledge of the fact that radioactive elements decay and have isotopes (see below) allowed scientists to date the formation of certain minerals within igneous rocks, from which it has become possible to establish a chronology of Earth's history.

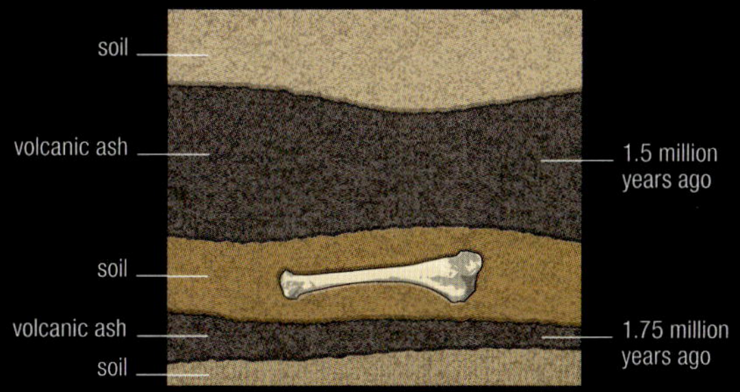

DATING FOSSILS
The radiometric dating of fossils depends upon dating the nearest appropriate igneous rocks, such as volcanic lava and ash deposited as part of the strata. Dating lava above and below a fossil gives its maximum and minimum age.

RADIOMETRIC ROCK DATING
Radioactive elements such as uranium decay from the moment of their formation by shedding electrons (negatively charged particles present in each atom). This loss occurs over time at a regular rate, and results in the creation of a series of "daughter atoms" known as isotopes. By measuring the relative proportions of isotopes, the time that has passed since their formation can be calculated.

KEY
Uranium-235 atom
Lead-207 atom

RATIO
32 uranium-235:
0 lead-207

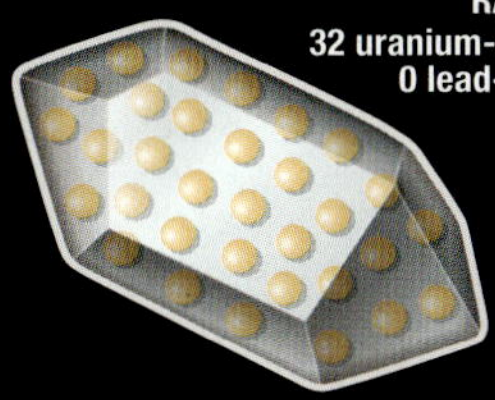

YEAR OF FORMATION
A mineral crystallizes from molten rock and contains radioactive uranium-235, which will decay to the isotope lead-207.

RATIO
1 uranium-235:
1 lead-207

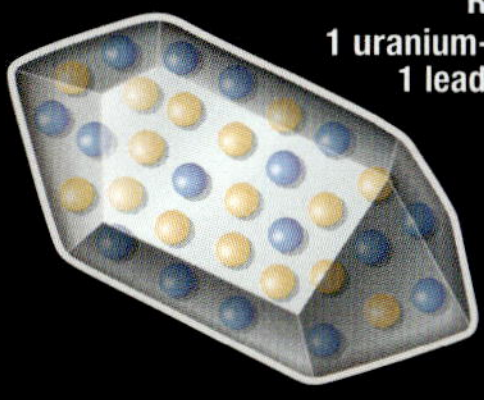

700 MILLION YEARS LATER
Fifty per cent (half) of the uranium atoms are now atoms of lead-207. Thus, the "half-life" of uranium-235 is 700 million years.

RATIO
1 uranium-235:
3 lead-207

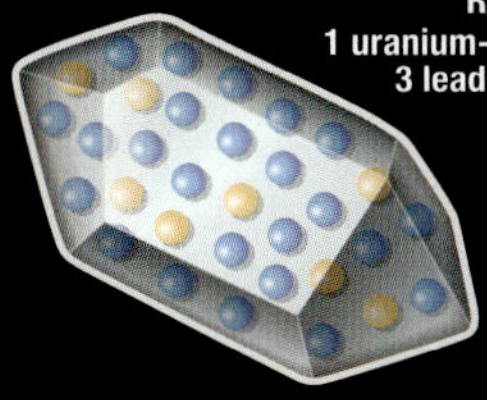

1,400 MILLION YEARS LATER
By this time, another 50 per cent of the uranium-235 atoms has decayed into lead 207. The ratio of uranium to lead atoms is now 1:3.

RATIO
1 uranium-235:
7 lead-207

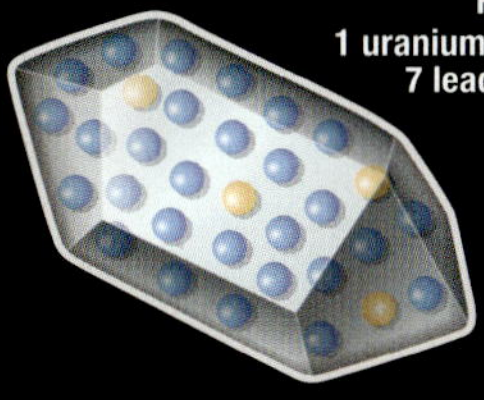

2,100 MILLION YEARS LATER
A geologist measures the ratio remaining in the rock (now 1:7) and dates the rock to three half-lives, that is, 2,100 million years old.

The Origin of the Earth

Born of gas and dust about 4.56 billion years ago, Earth has had a dramatic history. Scientists have attempted to reconstruct its earliest stages from evidence preserved in meteorites and Earth itself, as well as direct observations of distant stars and nebulae. But our knowledge of events in this remotest part of our planet's past remains incomplete.

The origin of the Solar System

The Solar System began to form about 4.56 billion years ago, when an immense cloud of gas and dust, the solar nebula, started to collapse under gravity. As it collapsed, the cloud flattened into an ever-faster spinning disc, with a bulging centre that heated and condensed to form the Sun. The orbiting debris formed the four inner rocky planets. In the cooler outer disc, the four gas giants formed, then the small dwarf planets (including Pluto), and finally, a vast cloud of comets. Altogether, the Solar System extends about 6,000 billion km (3,700 billion miles) from the Sun.

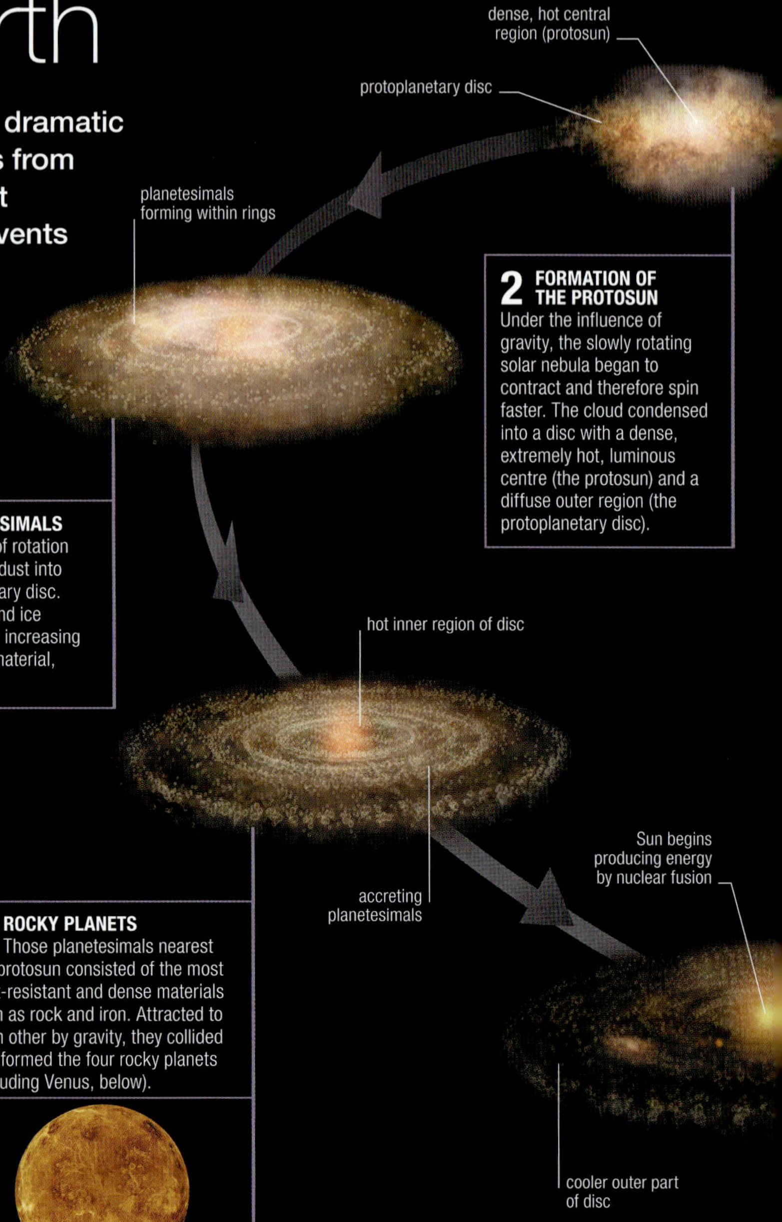

2 FORMATION OF THE PROTOSUN Under the influence of gravity, the slowly rotating solar nebula began to contract and therefore spin faster. The cloud condensed into a disc with a dense, extremely hot, luminous centre (the protosun) and a diffuse outer region (the protoplanetary disc).

3 RINGS AND PLANETESIMALS The increasing speed of rotation condensed the icy gas and dust into rings within the protoplanetary disc. Colliding particles of dust and ice clumped together, and their increasing gravity attracted yet more material, forming planetesimals.

4 ROCKY PLANETS Those planetesimals nearest the protosun consisted of the most heat-resistant and dense materials such as rock and iron. Attracted to each other by gravity, they collided and formed the four rocky planets (including Venus, below).

STAR-FORMING NEBULA
The first stars formed in the clouds of hydrogen and helium that dominated the early Universe. Their birth (above) and death (right) generated new elements such as carbon, oxygen, silicon, and iron.

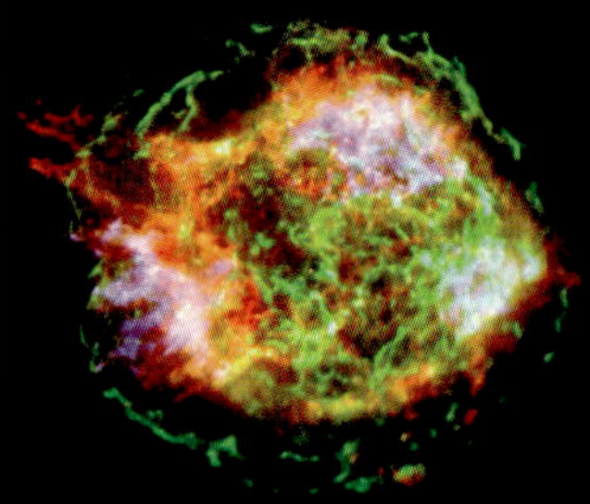

SUPERNOVA EXPLOSION

The formation of the Earth

According to the most widely accepted theory of how the Solar System formed, known as the nebular hypothesis, the rocks and ice that shared the same orbit around the developing Sun coalesced under gravity, in a process called cold accretion. The largest bodies in each ring attracted the most material and formed planetesimals – loose collections of rock and ice with a uniform structure. As a planetesimal grew larger, its gravitational pull increased. It became more tightly held together, and it drew in rocks from its immediate surroundings with greater force, leading to a period of intense bombardment and growth. Earth, and the three other rocky planets of the inner Solar System, were formed in this way about 4.56 billion years ago.

IMPACT CRATERS ON MERCURY
Mercury, the smallest of the rocky planets, has, like part of Earth's Moon, a highly cratered surface interspersed with dark lava fields. These impact craters all resulted from the same phase of intense meteorite bombardment that was suffered by Earth and which lasted until about 3.5 billion years ago.

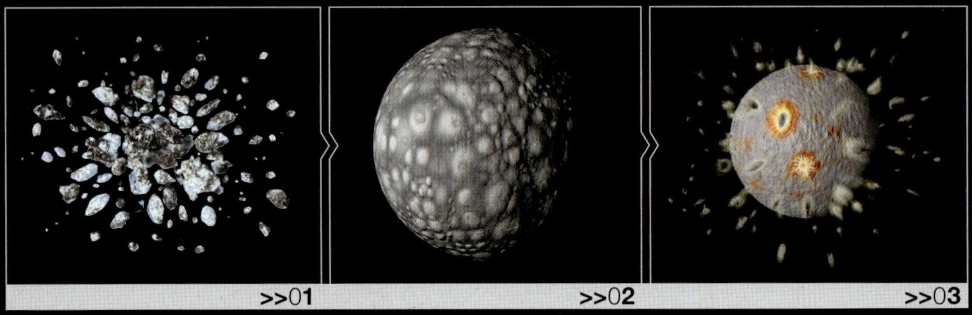

BIRTH OF A PLANET
>>01 Pieces of rock and ice sharing the same orbit around the Sun were pulled together by gravitational attraction, forming ever-larger clumps, known as planetesimals. **>>02** As they grew larger in size and mass, so did their gravitational field, pulling them into more spherical shapes. **>>03** Rocks were pulled towards the primordial planet at increasing speeds, producing high-energy impacts that generated localized heating and melting. Heavy bombardment continued until most of the local planetesimals were accreted and the gravitational field was so strong that the planet contracted.

1 THE SOLAR NEBULA FORMS
The solar nebula was initially a vast but dense cloud of cold gas and dust, several times larger than the present Solar System. The gas and dust is thought to have originated from the death of even older stars, and was effectively recycled.

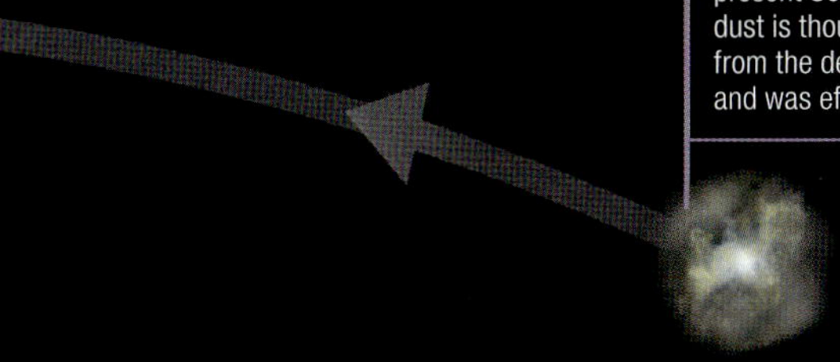

THE NEBULAR HYPOTHESIS
The nebular hypothesis was developed by the French mathematician Pierre Laplace and the German philosopher Immanuel Kant in the late eighteenth century. The six stages shown here explain many of the known facts about the Solar System, such as why the orbits of most of the planets lie roughly in the same plane and why the planets all orbit the Sun in the same direction.

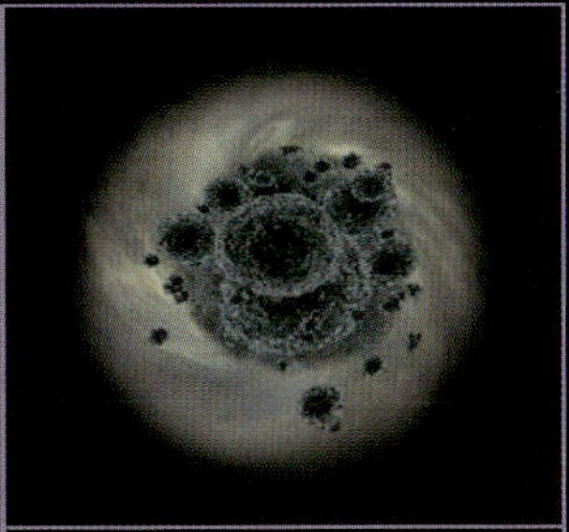

5 GAS GIANTS
In the cool outermost rings of the protoplanetary disc, beyond the asteroid belt, ice and gas could survive. Here, planetesimals made of rock and ice grew large enough to attract, and be enveloped by, deep clouds of gas (above). The four gas giants were formed, and shortly afterwards the protosun became a fully fledged star.

6 REMAINING DEBRIS
Following the formation of the planets, some gas and other unaccreted material still remained in the protoplanetary disc. Most was blown away by radiation generated by nuclear fusion in the Sun. The remaining planetesimals formed the vast and distant Oort Cloud of comets at the edge of the Solar System.

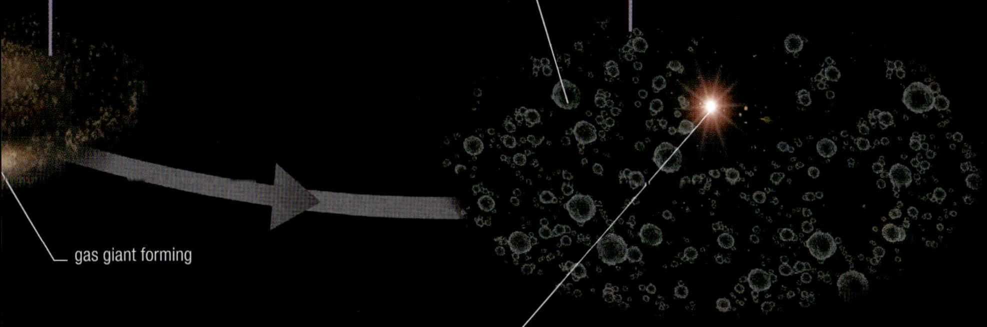

The Earth in the Solar System

Earth's size, orbit, and position in the Solar System have all promoted the evolution of life on this blue planet. The third planet from the Sun, Earth is the only one to lie in the habitable zone, where conditions are favourable to life. Earth's distance from the Sun and its mass, gravity, and internal heat have allowed the development and retention of an oxygen-rich atmosphere and abundant surface water. In contrast, Earth's small, lifeless Moon has little internal heat, an insubstantial atmosphere, and no surface water. Earth's orbit is elliptical, but its rotation and near-circular path mean that seasonal variation in exposure to solar radiation is not so extreme as to extinguish life.

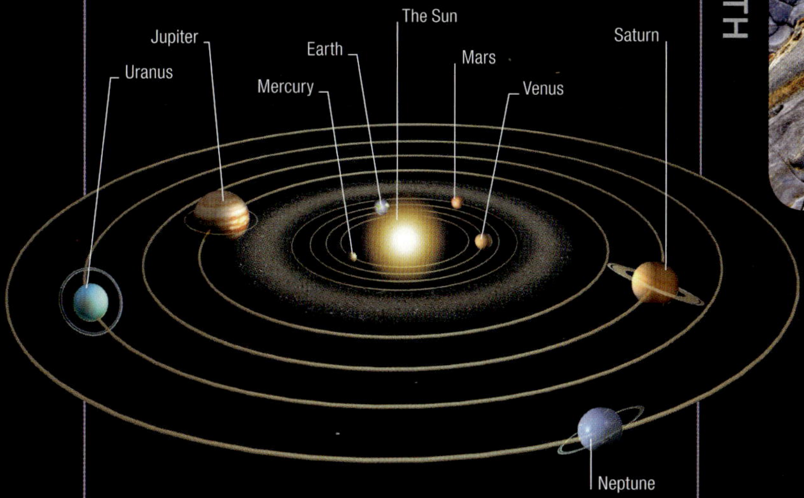

THE PLANETS

Planet	Average Distance from the Sun
Mercury	57.9 million km (36 million miles)
Venus	108.2 million km (67.2 million miles)
Earth	149.6 million km (93 million miles)
Mars	227.9 million km (141.6 million miles)
Jupiter	778.3 million km (483.6 million miles)
Saturn	1.43 billion km (888 million miles)
Uranus	2.87 billion km (1.78 billion miles)
Neptune	4.5 billion km (2.8 billion miles)

The formation of the Moon

The earliest-known Moon rocks have been reliably dated at about 4.5 billion years old, indicating that Earth's satellite was formed not long after Earth itself. Most astronomers agree with the giant-impact theory, which proposes that the Moon originated when a massive asteroid collided with the young Earth and tore away a huge amount of its surface (see below). Continuous heavy meteorite bombardment over the following billion years left the Moon's rocky surface severely cratered. A period of volcanic activity then followed, and lava oozed out of cracks in the crust to fill low-lying craters. The lava solidified, forming the Moon's vast dark maria (seas), which are still visible from Earth today.

MOON ROCK
Over 380kg (838lb) of rock were recovered from the Moon by the Apollo missions. The rocks are mostly igneous, meaning that they formed from the solidification of cooling molten material (magma). Although closely related to Earth rocks, they are depleted in the volatile elements sodium and potassium.

>>01

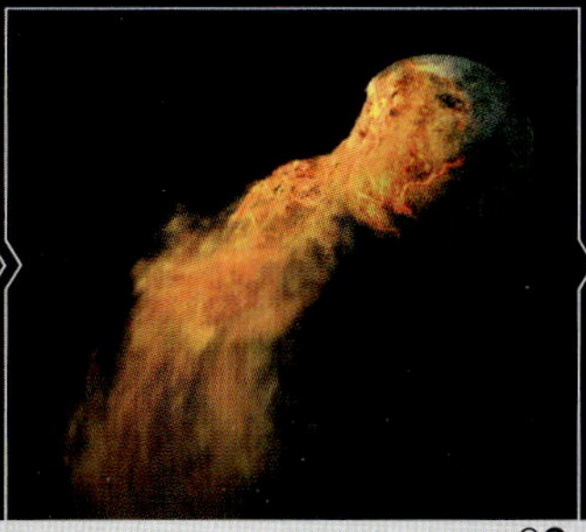
>>02

>>03

>>04

EARTH'S MOON
>>01 About 4.5 billion years ago, a Mars-sized asteroid collided with Earth and tore away a huge amount of silicate rock material from its surface. >>02 A vast cloud of gas and fragmented rock debris was ejected into space and rapidly cooled. >>03 Earth's gravitational pull held the ejecta in a circular orbit, forming a dense ring. >>04 The rocks collided, growing into ever larger clumps until eventually the Moon, a single satellite body over 3,400km (2,100 miles) wide, circled the Earth.

The First 500 Million Years

Earth's early history was violent and dramatic. Its mass began to take shape slightly over 4.5 billion years ago. Within 50 million years, its core had formed and, in turn, generated a magnetic field. However, it was not until the atmosphere and surface of the crust were relatively stable, about 3.8 billion years ago, that life had a good chance to evolve and thrive.

Earth's core and mantle

Soon after it formed, most of Earth's mineral material separated from a uniform ball into the intensely hot, metallic core and the cooler, rocky mantle. The iron-nickel composition of the core is indicated by measurements of density, the chemistry of iron meteorites, and Earth's magnetic field. The magnetic field also reveals that part of the core must be liquid and circulate electrically conductive molten iron that generates magnetism. Analysis of earthquake waves shows the outer core is liquid, while the inner core is solid. As iron changes from solid to liquid at the boundary, energy is released, driving convection in the outer core. Within the mantle, gravity, which acts on differences in density between hot and cold rock, causes the mantle to flow in a pattern of convection. Colder, denser material sinks deep into the mantle, especially in subduction zones. This downward flow is balanced by the upward rise of hot and less dense mantle, either as plumes beneath hotspots, or upwelling beneath mid-ocean spreading ridges (see pp.20–21).

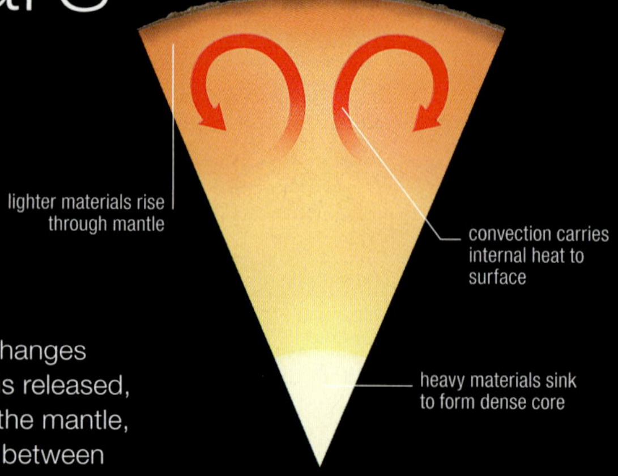

CONVECTION AND DIFFERENTIATION
The differentiation of Earth's mineral material into the metallic core and rocky mantle resulted in the establishment of the magnetic field and the transfer of heat energy away from the core, as hot rock of relatively low density rose up through the mantle.

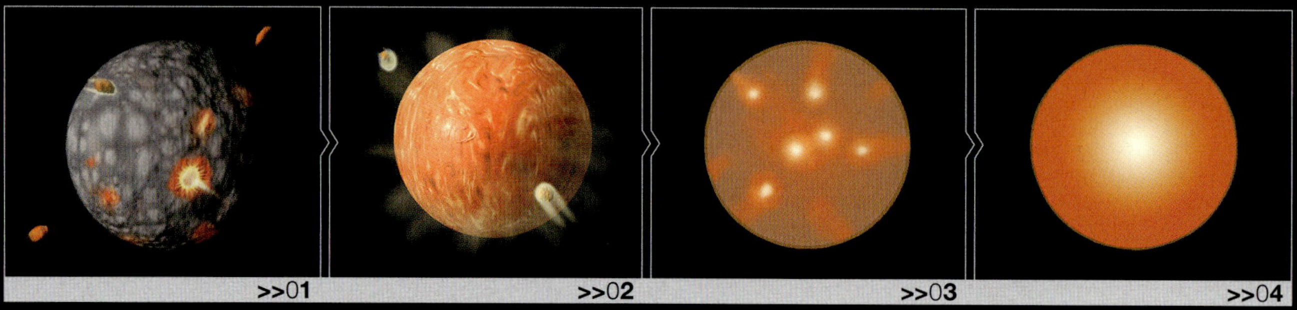

THE IRON CATASTROPHE
>>01 Around 4.56 billion years ago, gravity caused relatively cool nebular material to coalesce. >>02 As this continued, Earth contracted under the strength of its own gravity, heating up in the process. >>03 Eventually, it was weak enough for dense iron and nickel to sink deep into the interior, starting the formation of the core. >>04. The sinking and aggregation of this material released even more heat, possibly causing meltdown in the mantle.

IRON-OXIDE DEPOSTS AND OXYGEN IN THE ATMOSPHERE
Laid down in water 2.5 billion years ago, these deposits hold clues about when the atmosphere and oceans were oxygenated. Iron oxide is insoluble, so the iron must have been transported in a non-oxidised form. This could happen only if there was almost no oxygen in the atmosphere and ocean.

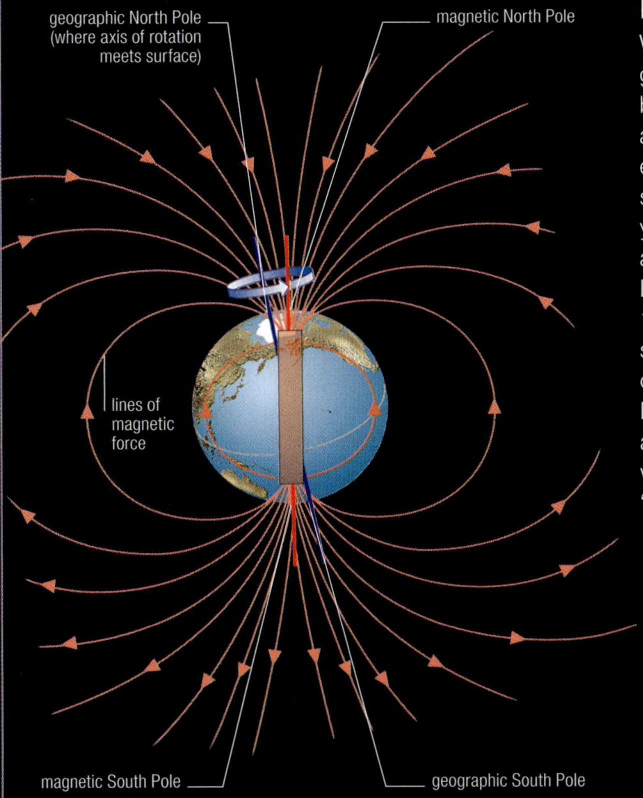

EARTH'S MAGNETIC FIELD
Earth's magnetic field is doughnut-shaped (toroidal), like a huge bar magnet whose long axis is close to, but not aligned with, Earth's axis of rotation. The magnetic poles lie some way from the geographic poles.

Earth's magnetism

With its two opposite poles, the Earth's magnetic field corresponds to that generated by a bar magnet, but it is formed by electrical currents generated by the fluid motion of the outer core. The mechanism may work like that of an electrical dynamo, which converts mechanical energy into electromagnetic energy. On average, the magnetic field switches its polarity or direction about every 500,000 years, but the last reversal was some 780,000 years ago. The axis of polarity is also aligned differently to Earth's axis of rotation (at present, they are separated by 11 degrees). The intensity of the field fluctuates, but is sufficient to align tiny, iron-rich particles as if they were compass needles within certain rocks formed at the Earth's surface. Because of this, some solidified lavas and other rocks provide a record of the field's polarity when they were originally formed. Measurement of these "fossil" or palaeomagnetic fields has revealed a chronological history of Earth's polarity reversals.

MAGNETIC SHARK?
A sensory organ in this great white shark's snout detects weak electromagnetic fields, such as that produced by prey. We now know that sharks also navigate by sensing the Earth's magnetic field, positioning themselves relative to lines of magnetic force.

AURORAE
Luminous aurorae appear in the polar night skies when Earth's magnetic field traps charged particles carried from the Sun by the solar wind. Atmospheric gas particles produce a spectrum of colours.

The crust

Earth's outer crust is a thin layer that varies from 5 to 11km (3 to 7 miles) thick as oceanic crust and 25 to 90km (15 to 56 miles) thick as continental crust. The oldest remaining fragments of existing crust date from about 4.0 billion years ago, but it is likely that primitive oceanic crust formed earlier than this, soon after Earth itself formed. As the oceanic crust formed, subduction (see p.20) drew as much again back into the mantle. Volcanic and tectonic (crust-shifting) processes exposed mantle-derived rocks to weathering, sedimentation, and subduction. The latter dragged watery sediments deep into the mantle, which triggered melting and generated molten rock that, because it was of relatively low density, rose into the crust above, leading to the formation of new continental material (see p.19). Slow melting at the base of the growing continental crust added silica-rich magma that cooled and hardened into crystalline rock.

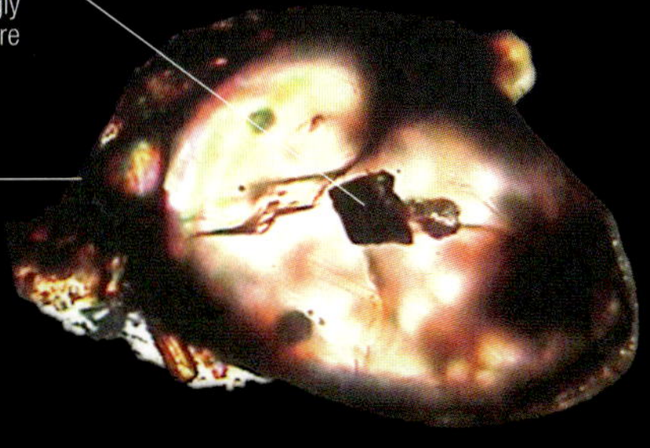

OLDEST TERRESTRIAL DIAMOND
These "microdiamonds" trapped inside a zircon crystal from the Jack Hills region of Western Australia are more than 4 billion years old – the oldest identified fragments of Earth's crust. It has been suggested that the diamonds crystallized under high pressure within 300 million years of Earth's formation

The formation of the atmosphere

Earth's present oxygen-rich atmosphere differs greatly from its original atmosphere, which consisted of the light gases hydrogen and helium and other volatile gases. However, in the latter stages of the Sun's formation, this first atmosphere was blasted away by a surge of the solar wind – the continuous stream of atomic particles given off by the Sun – only to be replaced by a second, more stable, atmosphere as Earth continued to evolve and develop. Intense volcanic activity expelled vast amounts of volatile gases. Known as outgassing, this process released abundant nitrogen, carbon dioxide, and water vapour as well as ammonia, methane, and smaller amounts of other gases. It is believed the amount of oxygen in the atmosphere slowly increased as micro-organisms converted carbon dioxide into oxygen via photosynthesis. Clouds of water vapour condensed and precipitated, forming surface water and the first oceans.

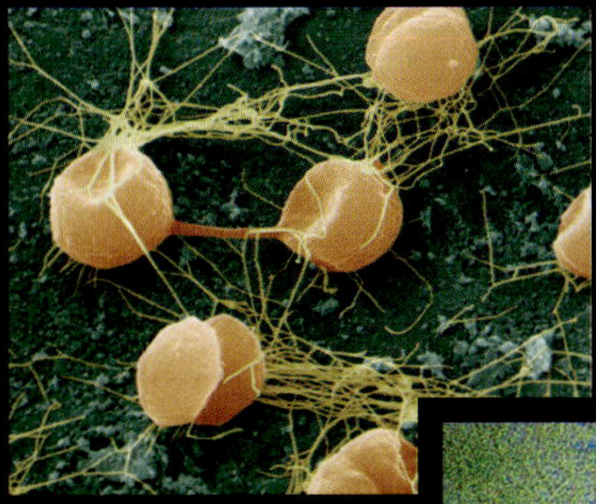

ARCHAEA
Archaea are organisms similar to bacteria, yet most are anaerobic – they require no oxygen. *Pyrococcus furiosus*, shown here, thrives on sulphur in near-boiling sea water. Anaerobes such as these were probably among the first life forms on Earth.

CYANOBACTERIA
Earth's early seas became full of microscopic organisms similar to this blue-green algae or cyanobacteria. These organisms played an important part in oxygenating Earth's atmosphere

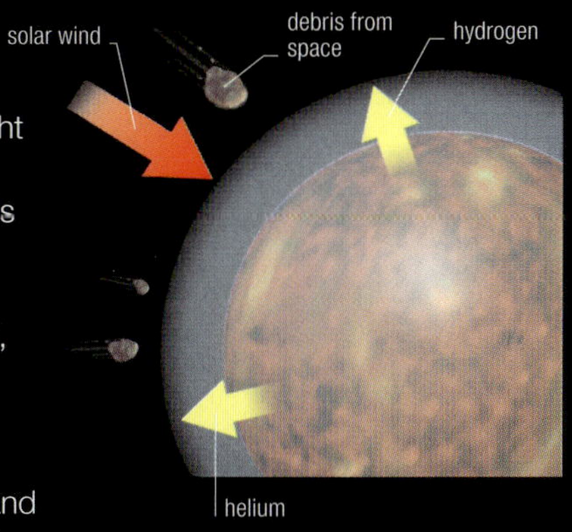

FIRST ATMOSPHERE
During Earth's accretion within the solar nebula, the first atmosphere was formed from the light gases hydrogen and helium, only to be stripped away by an intense blast of solar wind.

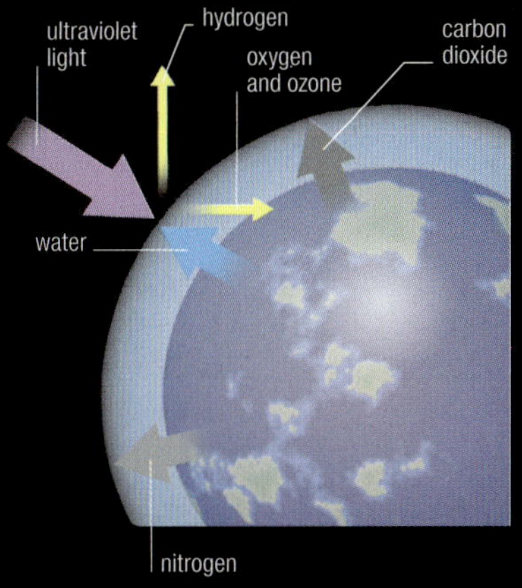

SECOND ATMOSPHERE
Earth's second atmosphere of water vapour, carbon dioxide and nitrogen probably formed as a result of volcanic outgassing. Ultraviolet light split the water into hydrogen, oxygen, and ozone.

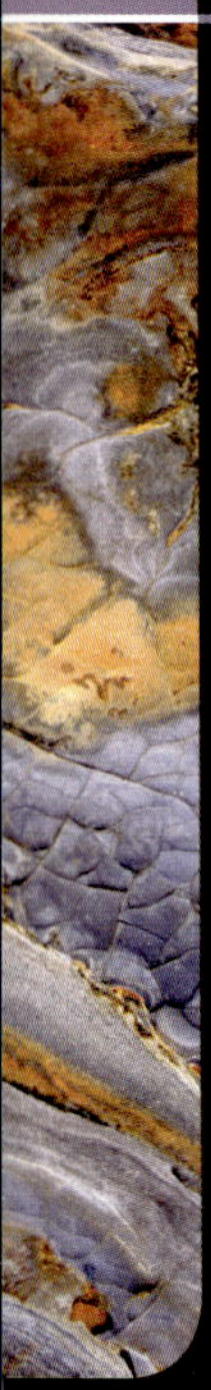

The formation of the oceans

Earth is unique among the planets of the Solar System in having abundant surface water that is being constantly recycled between its atmosphere and terrestrial water bodies such as seas, lakes, and oceans. Today, around two-thirds of Earth's surface is covered with sea water, and interactions between the oceans and atmosphere are vital for maintaining the planet's climate and life. Ocean formation probably began during the first 500 million years of Earth history, when the planet first cooled sufficiently to allow water molecules to condense, fall onto the surface, and persist as free-standing water bodies. Zircon mineral grains laid down by water have been dated to over 4 billion years old, indicating that some surface water existed at that time. Some of Earth's oldest rocks are "pillow lavas" from western Greenland, many of which are up to 3.8 billion years old and were formed by underwater eruption. The early ocean waters reacted with carbon dioxide from the atmosphere to deposit calcium and magnesium carbonates as limestones. Weathering of rocks on the first continental landmasses also leached soluble salts into sea water. Australian limestone formations known as stromatolites, which were formed by microscopic blue-green algae or cyanobacteria, indicate that fully saline oceans existed around 3.5 billion years ago.

480 kph **The speed at which tidal waves, known as tidal bores, crashed into early Earth's shores**

LIVING REEF
Coral reefs are present-day biodiversity hot spots – the ocean's equivalent of tropical rainforests. The largest living structures on Earth, even the skeletons and shells of their inhabitants build up the sea bed, altering the underwater environment both biologically and physically.

TIDAL FRICTION
The Moon's gravitational pull generated tides in Earth's first oceans which in turn produced bores that ripped across coasts at speeds of 480 kph (280 mph). Such strong currents stripped, weathered, and eroded rock materials from the land, dumping the sediment at sea.

SCIENCE

CHANGING DAY LENGTH

Many living organisms record daily, monthly, and seasonal growth cycles by the changing rates of growth in their shells and skeletons. Coral, for example, deposits a new layer of limestone every day, and it is particularly influenced by lunar monthly growth cycles. By studying fossil corals from the early Devonian period (400 million years ago), such as the one shown here, we know from growth-line evidence that there were 410 days in a year during this part of Earth's history. Since Earth's orbit around the Sun has remained constant, the Devonian day must have been shorter – just 21 hours long – and Earth must have been spinning faster.

Early plate tectonics

Convection in the mantle, driven by gravity, has led to the formation of a number of tectonic plates, made up of Earth's cold, strong outer layer (called the lithosphere), which are in constant motion. As the early plates moved apart, there was an outpouring of new crustal material from the mantle. Since Earth's mass is not expanding, as much crust must return to the mantle as is newly generated, with the result that one plate is often forced below another along what are called subduction zones. Such plate movement began when Earth cooled sufficiently to form an outer crust over 4 billion years ago. The oldest rocks associated with this process are 3.8 billion years old, but this crustal recycling has destroyed all of the early oceanic crust and the oldest known existing oceanic crust beneath the present oceans is 180 million years old.

THE CHANGING FACE OF EARTH
Using various lines of evidence, geologists can reconstruct the past distribution of Earth's oceans and continents. This map shows them as they probably looked about 650 million years ago. Before this date, not enough evidence has survived for accurate maps to be made.

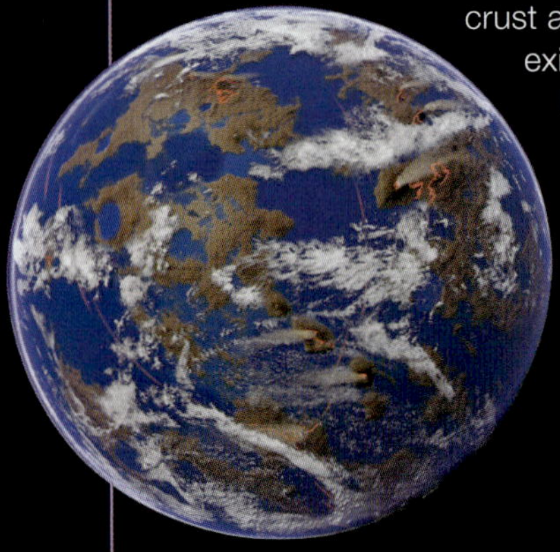

EARLY TECTONICS
Early tectonic processes probably differed from those operating today. It is likely that during Earth's early history, convection in the mantle was more vigorous than it is now, leading to faster crustal movements, smaller plates, and more volcanic activity.

OPHIOLITE
An ophiolite is a part of the ocean floor that has been thrust above sea level by the action of plate tectonics – in this instance, when two plates collided. This brown rock formation in Oman, now weathered into rugged hills, is a rare and well-preserved example.

The first continents

Today, continents make up about one-third of Earth's surface but contain the oldest rocks on the planet, over 3.8 billion years old. Analysis of these rocks reveals even older zircon minerals that formed over 4 billion years ago. Geochemical investigation of the zircons and smaller fragments within them shows they formed at relatively low pressures and temperatures in molten material rich in water and silica at convergent plate boundaries, such as volcanic island arcs. This suggests plate movement and subduction were active and liquid water and continental crust were present before 4 billion years ago. Subduction of the primitive crust rocks led to selective melting with increasing heat at depth. Preferential melting of silicate minerals with the lowest melting points and relatively lower density formed magmas that rose into the crust and solidified, forming granitic rock bodies near the surface. These initial island arcs, microcontinents, and their granitic bodies grew further as they converged and joined together.

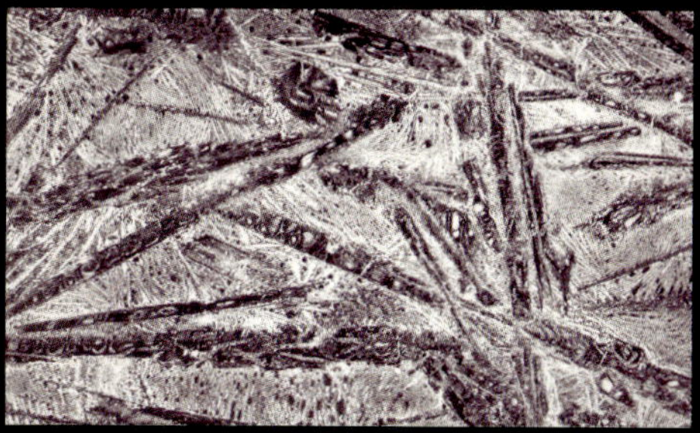

BARBERTON GREENSTONE BELT
The 3.5-billion-year-old pillow lavas and sediments of South Africa's Barberton Greenstone Belt (left) preserve remnants of volcanic island-arc rocks involved in the early formation of continental crust. Magnesium-rich volcanic rocks, called komatiites, demonstrate the presence of liquid water at this time. The lava erupted on the early ocean floor and quenched to give rise to needle-shaped crystals (above). Eruption of this lava suggests the mantle was hotter, wetter, and more fluid than it is now.

THE DEVELOPMENT OF CONTINENTAL CRUST
It is likely that the first continental crust formed after a primitive crust had already developed and convection had started in the mantle. Continental crust is formed when rocks in the mantle melt and later solidify, in the process becoming differentiated from the mantle. The process was probably particularly rapid above sinking flows in the mantle and slower above rising flows, where the continuous supply of mantle rocks slowed the rate of differentiation.

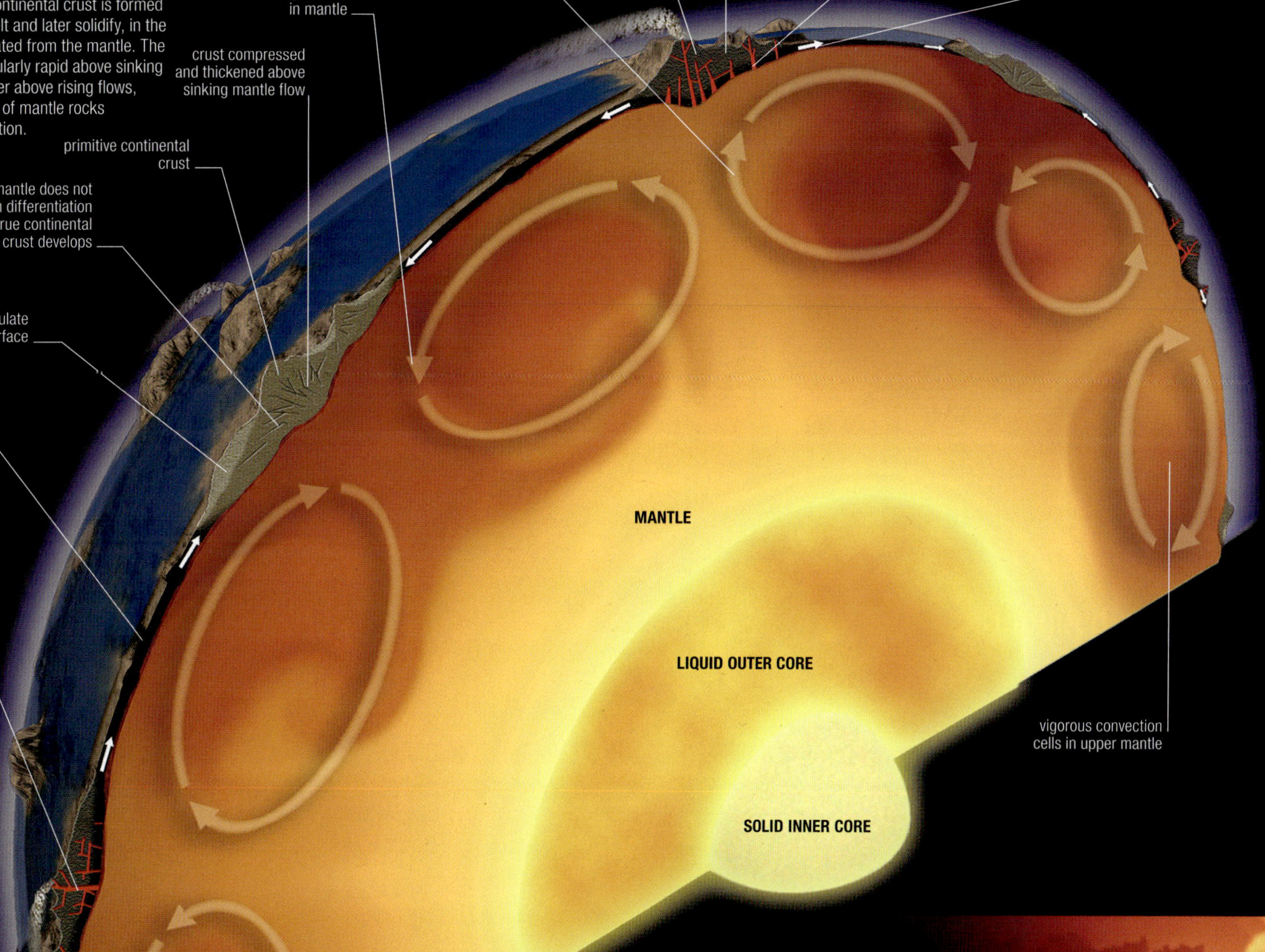

LAVA
Volcanic activity has shaped Earth since its origins. Lava is the term for a magma formed of molten upper-mantle rock that rises through the crust and erupts at the surface. It also refers to the solidified rock that forms when a molten lava flow cools.

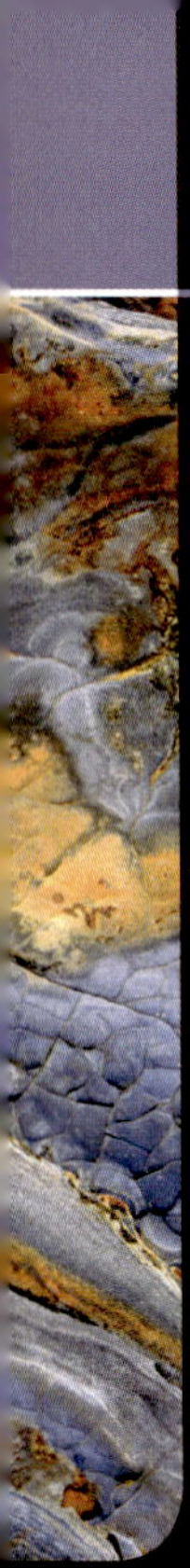

Plate Tectonics

Earth's thin outer crust and upper mantle, down to a depth of about 100–300km (60–200 miles), are divided into continent-sized plates that jostle against one another. As the plates move, oceans are created and later disappear, and volcanoes and mountain chains are formed.

Divergent boundaries

Oceanic plates move under gravity because they are colder and denser than the mantle below. As they do so, the mantle wells up, and the crust bulges, ruptures along weak points called faults, and eventually rifts apart. Pressure release allows the hot crust to melt, forming magma that erupts as lava through the crest of a lifted ridge. As this process continues, the splitting plate spreads apart or diverges, forming ridges and valleys on either side. As they slowly cool and shrink, the ridge flanks subside and their surface is smoothed out by the deposition of blankets of sediment.

RED SEA
As Africa moves away from Arabia, the rifted valley between the two plates has been flooded by the Red Sea and will eventually become a wide ocean.

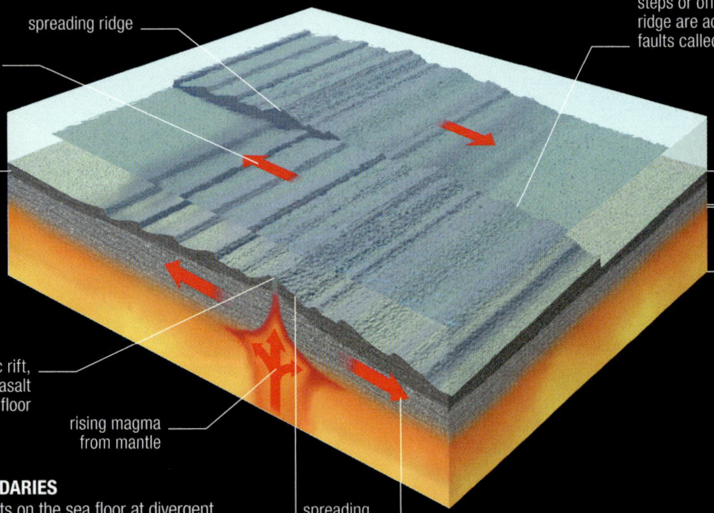

DIVERGENT BOUNDARIES
As molten rock erupts on the sea floor at divergent boundaries, it solidifies on contact with ocean water and forms new oceanic crust. As the plates continue to diverge (a process called sea-floor spreading), new oceans open up over millions of years.

Convergent boundaries

New crust is created at spreading ridges, but since Earth is not expanding, divergence in one place results in convergence in another. On average, the crust is less dense than the mantle, and oceanic plates are denser than continental plates because they contain a thinner crust. This means that where oceanic and continental plates collide, the heavier oceanic plate is overridden by the continental plate and descends (is subducted) into the mantle, melting and releasing magma, which erupts at the surface.

OCEANIC–OCEANIC
Where two oceanic plates meet, the older, cooler, and slightly denser plate is subducted and drawn deep into the mantle. Magma erupts at the surface to form chains of volcanic islands, like the one shown here.

OCEANIC–CONTINENTAL
Where plates of different densities converge, the denser plate is subducted below the other, descending hundreds of kilometres into the mantle. This usually generates earthquakes, as well as volcanic activity as water rising from the subducted slab leads to melting of the overlying mantle.

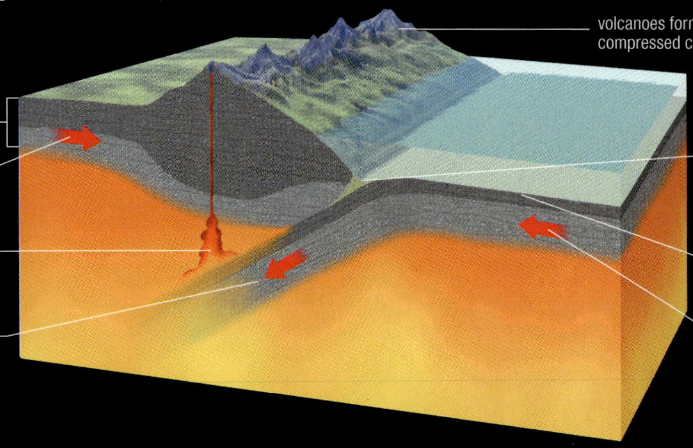

CONTINENTAL–CONTINENTAL
When two continental plates converge, neither is subducted. Instead, convergence leads to mountain formation as the crust folds and thrusts upwards – which is how the Himalayas, above, were formed.

MID-ATLANTIC RIDGE, ICELAND
The typical volcanic rocks and structures of an active spreading ridge can be seen on Iceland, where the Mid-Atlantic Ridge has been lifted above sea level. During eruptions, lava flows from long fissures, such as the one seen here.

Volcanoes and earthquakes

Volcanoes and earthquakes are violent expressions of Earth's internal dynamic forces. The vast majority occur at plate boundaries, and they are intimately connected to plate interaction. Diverging plates stretch and break, generating shallow earthquakes and volcanic eruptions, most of which occur at spreading ridges in ocean depths and produce magma made up mainly of basalt. Converging plates, however, generate earthquakes as far down as 700km (400 miles). Magma rises through the crust, assimilating rock materials and changing composition as it goes before erupting explosively through surface volcanoes, some of which form volcanic islands.

OCEANIC VOLCANOES
Volcanoes can also rise from the ocean floor, away from plate boundaries above hot spots in the mantle. An example is Kilauea, which exudes free-running lava.

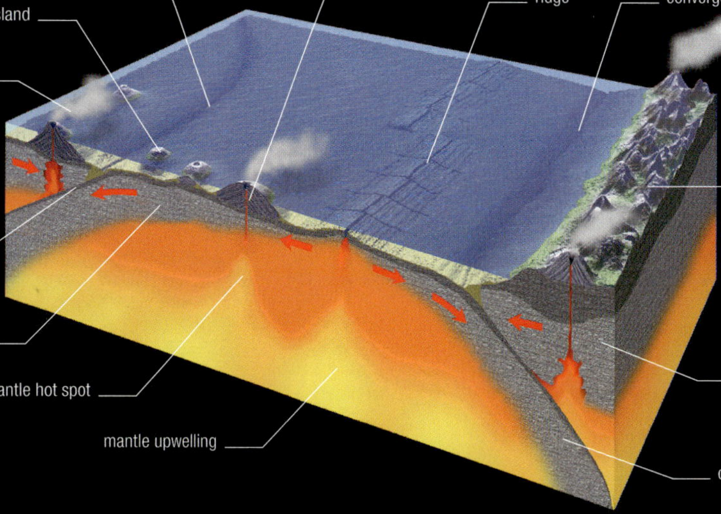

VOLCANOES AND PLATES
The vast majority of Earth's active volcanoes are situated along plate boundaries, and most of these lie within ocean-floor spreading zones. More evident are the volcanoes associated with subduction zones. But some eruptions also occur over hot spots in the middle of plates, where plumes of molten rock rise from deep in the mantle.

SAN ANDREAS FAULT
The continuing interaction between the Pacific and North American plates is most apparent at the 1,300km- (808-mile-) long San Andreas transform fault zone in coastal California. Sideways movement along the fault occurs at 35mm (1½in) per year and led to the catastrophic 1906 San Francisco earthquake.

Tectonics and life

Plate movement has had a significant impact on the evolution and distribution of life. Convergence brings different organisms together in competition, while divergence separates species groups, which then evolve in different conditions. An example is the supercontinent of Gondwana, formed around 542–488 million years ago. Evolving life forms spread throughout this enlarged land mass, leaving a record of themselves behind as fossils. Thus, fossils of the same species have been recovered in rocks from what are now widely separated continents. Only when Gondwana began to split apart, isolating these creatures from one another, did groups begin to evolve in different ways.

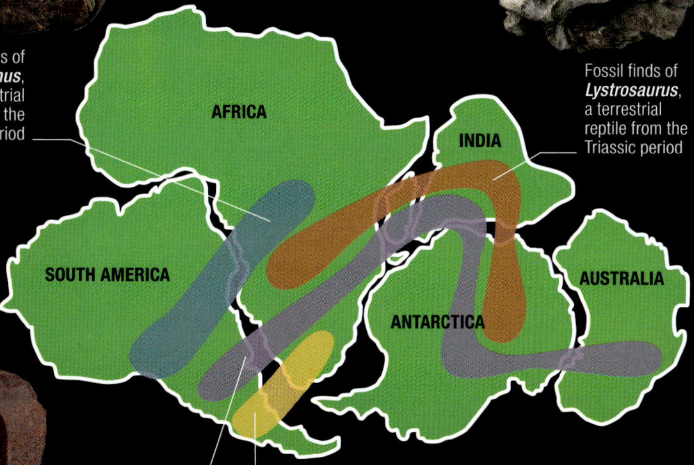

Fossil finds of *Cynognathus*, a terrestrial reptile from the Triassic period

Fossil finds of *Lystrosaurus*, a terrestrial reptile from the Triassic period

Fossil finds of *Glossopteris*, an extinct plant from the Permian period

Fossil finds of *Mesosaurus*, a freshwater reptile from the Permian period

WEGENER MAP
As long ago as 1915, German meteorologist Alfred Wegener collected geological and fossil evidence to show that the southern continents once joined together to form a supercontinent, but at the time he had no way of explaining how this had occurred.

WALLACE'S LINE
Distinctive Asian broadbills on the island of Borneo and Australian cockatoos on the island of Sulawesi, first noticed by Alfred Russel Wallace, were originally brought together by plate movements. The boundary is now called Wallace's line.

Changing Climates

There is clear evidence that Earth's climates have changed over time, notably in the form of cold periods (ice ages) when there were long-lived ice sheets at the poles and warmer greenhouse periods with largely ice-free poles. Long-term changes are probably due to plate movements, volcanic activity, and ocean circulation changes. Within ice ages, change is driven by cycles in Earth's orbit around the Sun and greenhouse-gas fluctuations.

Evidence in the rock record

Since the early nineteenth century, evidence from rocks and fossils has pointed to repeated, often dramatic changes in past climates. Ancient coral-reef limestones, rainforest and tropical-swamp coal deposits, desert sandstones, and glacial deposits are all climatic indicators in the rock record. However, movement of tectonic plates across climate zones must be taken into account before a global change can be proven. The presence of successive glacial and interglacial deposits beyond the poles is one of the best indicators of climate change in the rock record. Glacial events have been identified in many stages of the geologic timescale (see pp.44–45); evidence for Precambrian ice ages has even been found in low latitudes, leading to the controversial idea that Earth may have been almost entirely covered in snow and ice.

DROPSTONE
Some 635 million years ago, an 8cm- (3¾in-) diameter rock dropped from melting sea ice into sea-bed sediment now exposed in northwest Namibia. Plate-tectonic reconstruction indicates Namibia was located at the Equator at the time – meaning there were icebergs in low latitudes.

CARBONIFEROUS LIMESTONE CLIFFS
The limestone cliffs of the Gower Peninsula in Wales are actually deposits of tropical seas, as shown by their coral fossils. A tropical climate had not "reached Wales", however; instead, plate tectonics had carried southern Britain into tropical latitudes.

GLACIATED LANDSCAPE
The U-shaped Lauterbrunnen Valley in Switzerland features almost vertical rock walls and hanging valleys. Massive amounts of glacial ice were required to sculpt out the valley by removing vast amounts of rock.

Evidence in the fossil record

Eighteenth-century discoveries of fossilized elephants in North America and Europe stirred debates about climate change and the biblical flood. By the next century, scientists realized these fossils were actually the remains of animals adapted to cold conditions. They had lived in an ice age that extended across northern Europe and North America – as had Neanderthals, our extinct human relatives. Finding fossilized tropical plants and animals at high latitudes and ancient glacial deposits in low latitudes also pointed to widespread past climate change, yet many (but not all) discrepancies were explained by the theory that plate tectonics had pushed continents through different climate zones at different times. Nonetheless, there is good evidence that global climates have repeatedly swung from cold to warm conditions throughout Earth's history.

DRYAS OCTOPETALA
The mountain aven, *Dryas octopetala*, is a tundra and alpine plant. Its pollen is found throughout 12,000-year-old glacially related deposits in Europe.

FOSSIL FERN
About 100 million years ago, Antarctica was close to its present position, but without an ice cap. Dinosaurs roamed wooded polar landscapes and ate plants such as this now-extinct fern.

Oxygen isotopes

An isotope is an atom of an element whose nucleus contains the same number of positively charged particles (protons) as the "parent" element, but a different number of neutral particles (neutrons). Oxygen is present in water molecules as two isotopes: oxygen-16 and oxygen-18. Oxygen-16 is lighter than oxygen-18 and evaporates more quickly. During glacial periods, oxygen-16-enriched water vapour falls as ice and snow, depleting the ocean of this isotope and enriching it with oxygen-18 (see below). Some marine organisms secrete calcium-carbonate skeletons derived from sea water, and the oxygen content of these shells reflects the ratio of light to heavy isotopes in the water at the time, as well as the temperature of the water. Knowing the ratios from fossil foraminiferan shells allows scientists to calculate the oxygen composition of the sea water and work out the climate of the time.

FORAMINIFERAN FOSSIL
Millimetre-sized, coiled, and chambered shells made of calcium carbonate are secreted by single-celled micro-organisms called foraminiferans, which are common in ocean waters.

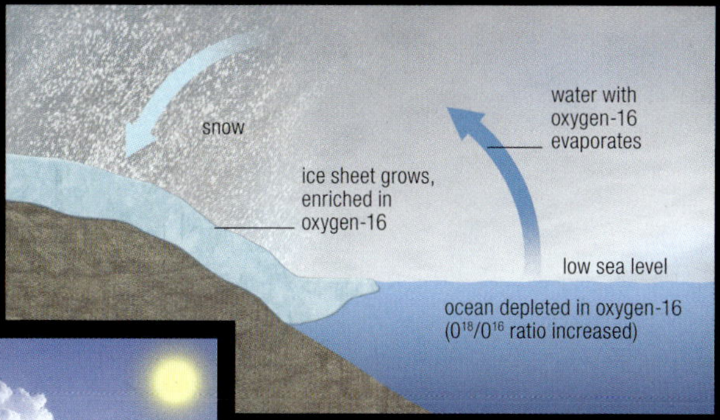

GLACIAL PERIOD
Preferential evaporation of the light oxygen-16 isotope in atmospheric water vapour and its precipitation as snow and ice lead to depletion of oxygen-16 relative to oxygen-18 in ocean waters during glacial phases.

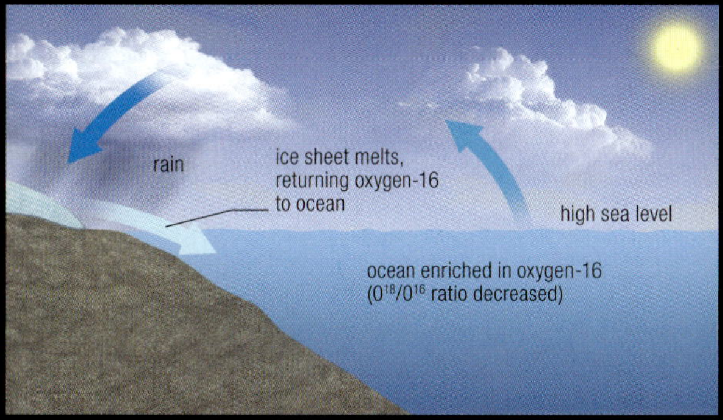

INTERGLACIAL PERIOD
During interglacial phases, the melting of glaciers and ice sheets returns huge volumes of oxygen-16-rich fresh water to the oceans, thus altering the ratio of oxygen isotopes in sea water.

Orbital cycles

Over time, many factors control the amount of solar radiation that reaches Earth's surface and influences its climate, resulting in what are called Milankovich cycles (see panel, right). The shape of Earth's orbit varies from nearly circular to slightly elliptical, in cycles lasting 90,000–120,000 and about 413,000 years. In addition, the axis of rotation shifts between 21.8 and 24.4° over 41,000 years. At its maximum tilt, summers receive more solar radiation and are warmer, while winters receive less radiation and are colder. The axis of rotation also wobbles (or precesses) due to gravitational forces exerted by the Sun and Moon as well as by Jupiter and Saturn. This results in a variation in solar radiation and length of the seasons in the northern and southern hemispheres every 26,000 years.

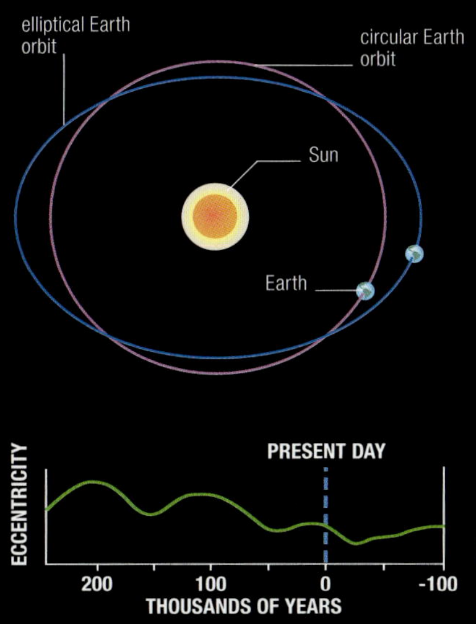

ORBITAL ECCENTRICITY
Earth's orbit around the Sun varies because of the gravitational influences of Jupiter and Saturn. The result is a variation in the amount of solar radiation as well as in the length of the seasons.

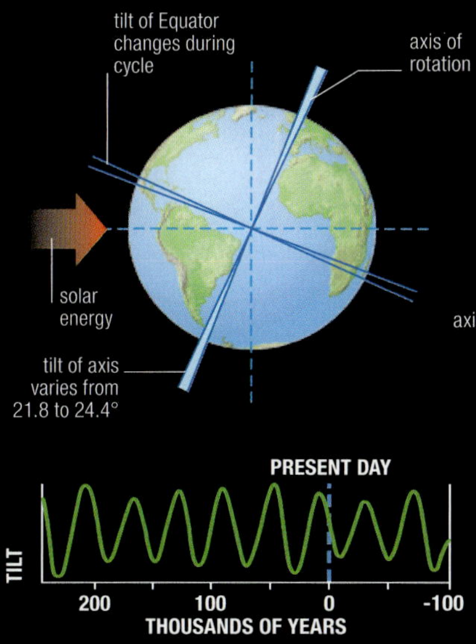

TILT
The tilt of Earth's rotation axis (obliquity) varies periodically during a 41,000-year cycle, resulting in colder winters and warmer summers when the tilt reaches its maximum of 24.4°.

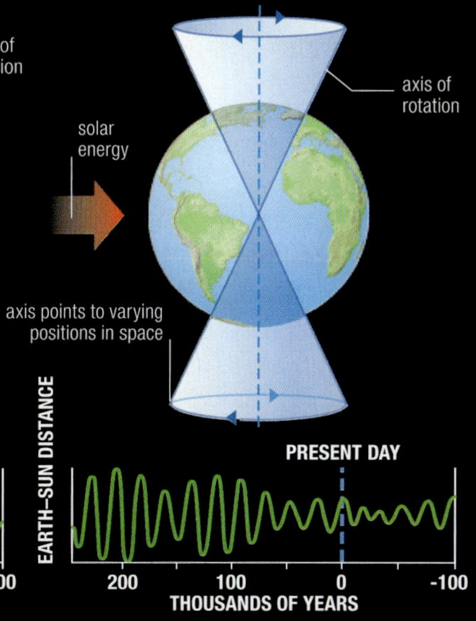

PRECESSION
A wobble (precession) in Earth's rotation axis relative to fixed stars means that the length and intensity of seasons vary between northern and southern hemispheres over a 26,000-year cycle.

BIOGRAPHY

MILUTIN MILANKOVICH

Serbian mathematician Milutin Milankovich (1879–1958) helped scientists understand the link between astronomical cycles, solar radiation, and climate change. Scottish scientist James Croll (1821–90) first linked Earth's orbital eccentricity to solar variation, which influences surface temperature and climates. Milankovich built on this theory, thanks to more accurate data about Earth's orbital variations. In 1920, he published an abstract on the curve of insolation (solar variation) at Earth's surface and a mathematical theory of periodic thermal effects of solar radiation, which established his international reputation. Milankovich went on to develop ideas about the motions of Earth's poles, linking them to the theory of glacial periods now known as Milankovich cycles.

The composition of the atmosphere

Milankovich cycles are an external cause of climate change, but change can also be caused by various factors on Earth itself. One of these is the composition of the atmosphere, particularly the levels of gases such as carbon dioxide, methane, and water vapour. Known as greenhouse gases, these absorb infrared radiation emitted from Earth's surface and re-radiate it back to the surface. As their level rises, the atmosphere tends to warm up. Important causes of change in greenhouse-gas concentrations include biological activity, volcanic eruptions, and weathering of rocks (see below). Organisms exchange carbon dioxide with the atmosphere as they respire and photosynthesise; carbon dioxide is released into the atmosphere by volcanic eruptions; and many types of rock react with carbon dioxide and water to form new minerals, removing carbon dioxide from the atmosphere.

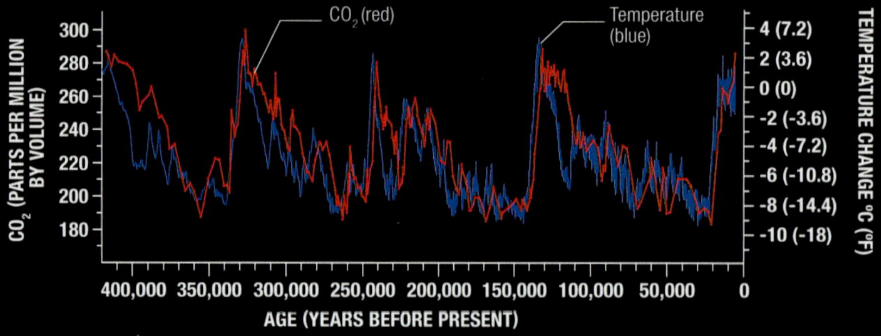

CARBON DIOXIDE AND TEMPERATURE
Changing levels of carbon dioxide can be measured in air bubbles trapped in ice. This graph, taken from an Antarctic ice core, shows the close link between carbon dioxide and temperature. Four glacial cycles can be seen on the graph.

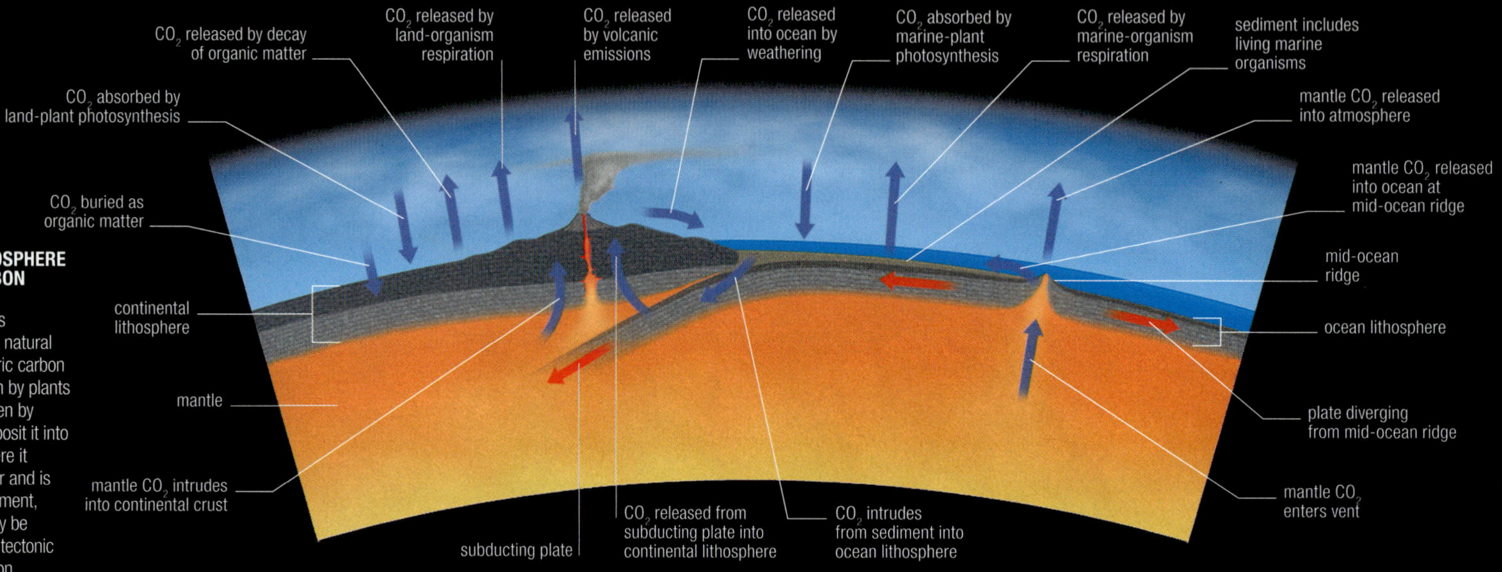

EARTH'S ATMOSPHERE AND THE CARBON CYCLE
Carbon circulates continuously in a natural cycle. Atmospheric carbon dioxide is taken in by plants that are then eaten by animals; they deposit it into ecosystems, where it dissolves in water and is deposited in sediment, from where it may be recycled through tectonic activity and erosion.

Changing geography

The global circulation of oceanic water redistributes heat and humidity to the atmosphere, affecting the climate. Tropical surface waters, heated by solar radiation, are carried towards the poles, warming the land and atmosphere in high latitudes. For example, coastal regions of northwest Europe are warmed by the Gulf Stream and warm air associated with it in winter. If the Gulf Stream were to cease or change direction, northwest Europe would have much colder winters. Since the circulation pattern is partly directed by the arrangement of ocean basins and continents, any change in that configuration affects regional climates. This means that plate movements can bring about climate change (see illustration below, for example).

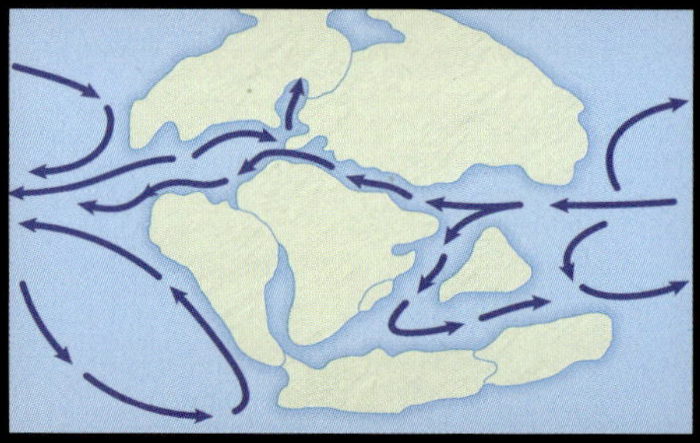

MESOZOIC CURRENTS
One hundred million years ago, tropical waters flowed east to west through the Tethys Ocean, which stretched from southeast Asia through what is now the Mediterranean to the Pacific between North and South America.

TODAY'S CIRCULATION
As a result of crustal plate movement, Africa has joined to Asia while North and South America are linked, leading to a much more fragmented global circulation.

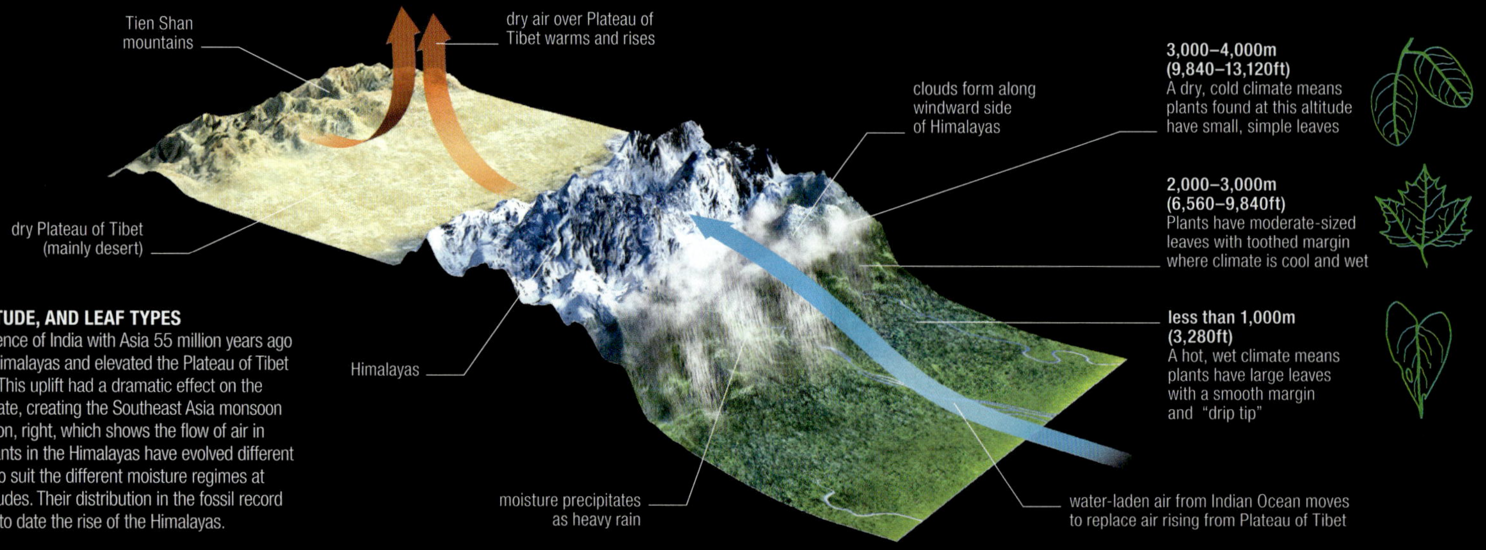

RAIN, ALTITUDE, AND LEAF TYPES
The convergence of India with Asia 55 million years ago formed the Himalayas and elevated the Plateau of Tibet to the north. This uplift had a dramatic effect on the region's climate, creating the Southeast Asia monsoon (see illustration, right, which shows the flow of air in summer). Plants in the Himalayas have evolved different leaf shapes to suit the different moisture regimes at different altitudes. Their distribution in the fossil record can be used to date the rise of the Himalayas.

DROWNED COAST
Eroded limestone in Ha Long Bay, Vietnam, shows one effect of the current warmer climate. Rising sea levels, caused by the melting of icesheets and glaciers, have reduced the distinctive topography to a series of rocky islands.

Warm phases

"Greenhouse" phases with little or no polar ice and high sea levels have occurred throughout Earth's history, linked to significant increases in the concentration of greenhouse gases in the atmosphere. For example, fossil evidence shows that during the Cretaceous Period lush forests grew in Antarctica and northern Alaska, providing food for dinosaurs. Warm polar regions tend to promote warming at lower latitudes too, creating seasonally arid zones near the Equator, with huge, shallow seas.

HIGH AND DRY
Australia's marsupial animals, such as the wombat (above) evolved in a continent that was cooler and wetter than it is now. As a result, they are continually having to adapt to drier conditions.

MISSOULA FLOODS
Dry Falls in Washington, USA, was once the world's largest waterfall. In the last ice age, floodwaters 120m (400ft) deep poured over its edge, with a force about ten times greater than that of all the world's rivers combined.

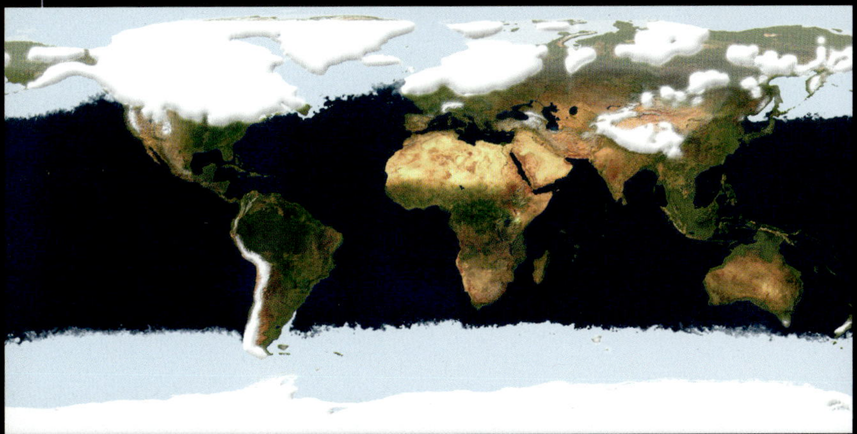

FOSSIL MAMMOTH TOOTH
Found in glacial deposits throughout northern Eurasia and North America, fossil teeth of cold-adapted mammoths reflect the extent of frigid glacial climates.

THE LAST ICE AGE
Glacial erosion and deposition from the last glaciation have left traces on landscapes and sediments. Geologists use these to reconstruct the original extent of glaciers and icesheets.

Cold phases

Repeatedly, Earth's climates have entered glacial phases lasting millions of years at a time. Known as ice ages, they consist of alternating colder and warmer phases. Geological evidence points to at least two major Precambrian ice ages, known as "snowball events", that even reached low-latitude areas. The most recent ice age began about 35 million years ago, with the last glacial event ending just 11,000 years ago. Fossils show how rapid climate change associated with ice ages had a drastic impact upon life, sea levels, and terrestrial environments on a global scale.

Life and Evolution

Genetic, biological, and fossil evidence shows that, for over 3.5 billion years, life has evolved, diversified, and changed from tiny, ocean-dwelling microbes to the millions of species that colonize all Earth's habitable environments. This astonishing feat has been achieved by evolution.

What is life?

Life can be defined as a condition in which organic, animate matter is distinguished from inorganic, inanimate matter by its ability to renew its complex, highly ordered structure. This includes a capacity for change, growth, and reproduction, with functionality maintained until death, when its constituents disperse back into the environment. In order to maintain itself, life must be able to get energy and raw materials from the environment as well as manufacture everything necessary for growth, repair, and replication. The only element that is known to form living structures is carbon, which can combine with itself and other elements, notably nitrogen, oxygen, and hydrogen, to form molecules of great diversity and complexity. Four main groups of organic carbon compounds are found in living organisms: carbohydrates and fats, which supply energy; proteins, built from amino acids, which form structural tissue; and nucleic acids, the basic components of genes.

SEXUAL REPRODUCTION
Sexual reproduction, as seen in this image of two dragonflies mating, creates a new individual via the fusing of two sex cells or gametes – a male sperm and a female egg.

PHOTOSYNTHESIS
Plants use a pigment called chlorophyll to convert sunlight into food. Light energy captured by chlorophyll converts water, carbon dioxide, and minerals into "chemical energy" (sugars the plant uses as food) and oxygen, which benefits other life forms.

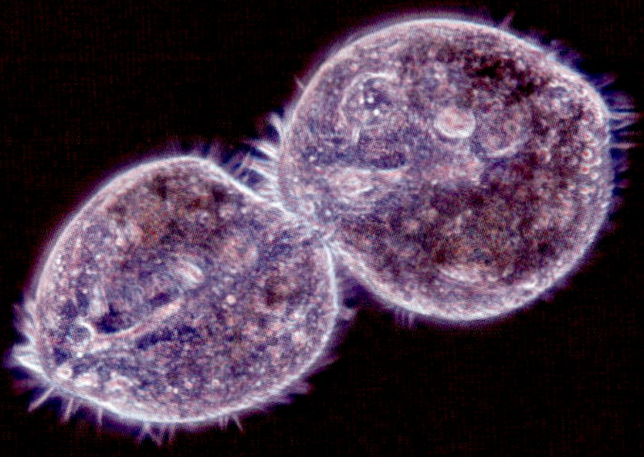

ASEXUAL REPRODUCTION
Some organisms, such as this ciliate protozoan, reproduce without sex cells (gametes). Here, the protozoan uses binary fission: it duplicates its DNA, then divides into two parts. Each new protozoan holds a copy of the "parent's" DNA.

LIFE IN THE PAST
Numerous species of plants and animals have evolved and become extinct, preserved only as fossils. The first dinosaurs evolved in the Triassic Period, and the group became extinct at the end of the Cretaceous.

What is evolution?

Evolution is a scientific theory which states that plants and animals change genetically over time, modifying and adapting over generations in response to the demands of Earth's changing environments. This biological process includes reproduction, diversification, and adaptation. Charles Darwin (see p.26), who formulated his theory of natural selection in 1837–39, called the process "descent with modification": the descent from a common ancestor with the modification of biological characteristics over time. Natural selection is a key evolutionary process, brought about by the survival of organisms best suited to their local environments: the "survival of the fittest".

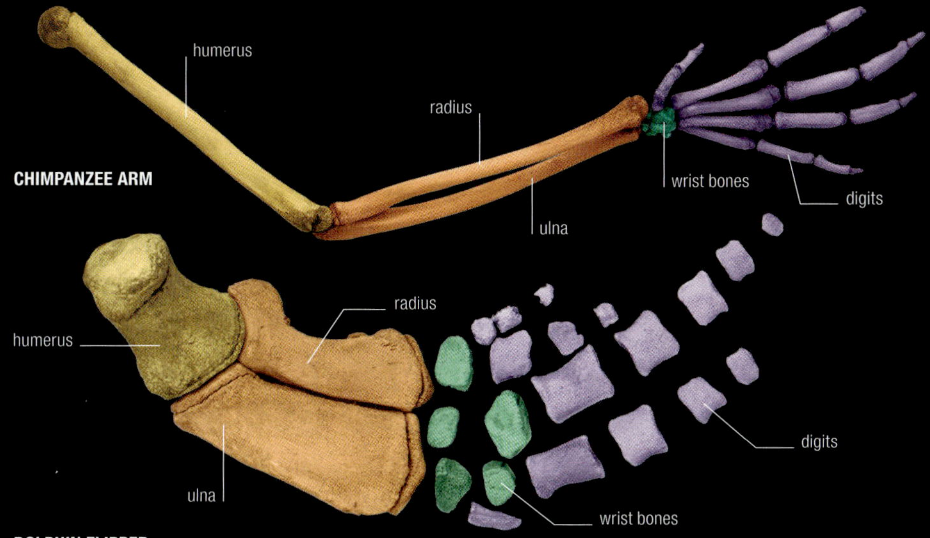

ARMS AND FLIPPERS
Anatomy often reveals a shared ancestry between species. For instance, a chimpanzee's arm looks very different to a dolphin's flipper, but the sequence from the shoulder of a single humerus bone followed by a paired radius and ulna and a series of wrist and finger bones can be identified in both animals.

Genes and DNA

Almost every cell of every living organism, from single-celled amoebas to whales, has its own set of molecular instructions. The function of each cell is encoded in chromosomes: the thread-like part of a cell that carries hereditary information in the form of genes. Each human cell contains between 20,000 and 25,000 genes, each with its own set of instructions responsible for particular characteristics. The code is recorded mainly in the form of a molecule called DNA, which includes chemicals called bases arranged in pairs. A gene is encoded by a specific sequence of base pairs.

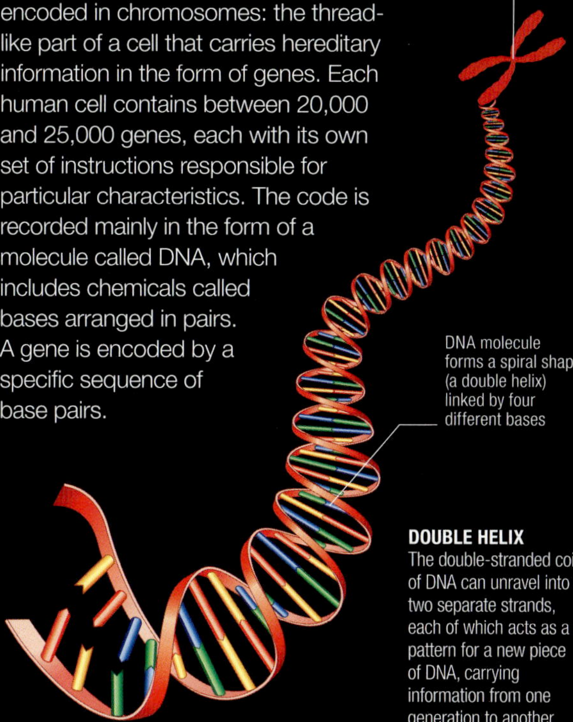

DOUBLE HELIX
The double-stranded coil of DNA can unravel into two separate strands, each of which acts as a pattern for a new piece of DNA, carrying information from one generation to another.

What drives evolution?

Evolution is fuelled mainly by the processes of selection and competition. These act upon species, which respond by having offspring that contain inherited variations. Most species tend to produce more offspring than can survive. Natural selection favours and promotes the "survival of the fittest": i.e. those best adapted to the physical and biological environment, such as a particular climate or escaping a predator. These survivors select other, similarly adapted mates to produce offspring that survive in greater numbers to breed new generations.

PEPPERED MOTHS
Genetic variation is demonstrated in the pale and dark forms of the peppered moth. The latter survive where industrial pollution and blackened trees give them a camouflage advantage over their pale relatives.

SEXUAL RIVALRY
Male Hercules beetles are armed with giant, pincer-like horns with which they battle over sexual access to females. The horns are powerful enough to fracture the loser's tough exoskeleton (outer shell).

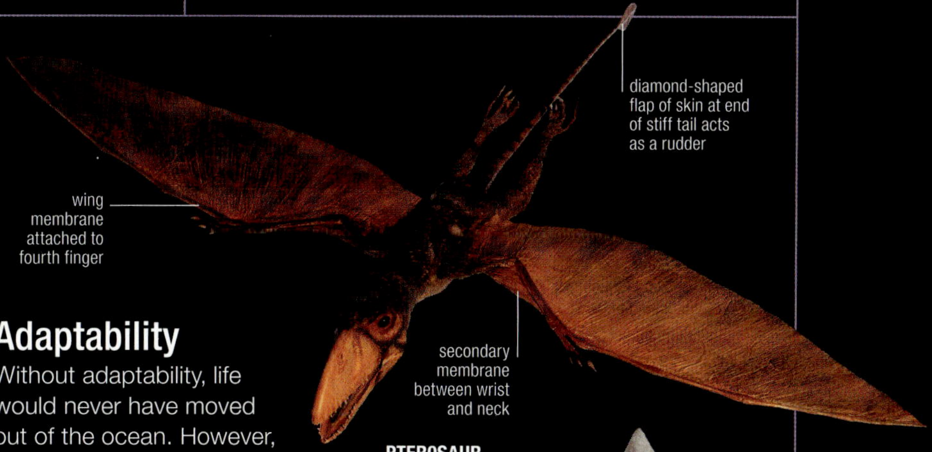

Adaptability

Without adaptability, life would never have moved out of the ocean. However, even adaptability is largely a result of cause and effect. Life cannot predict a future need. Instead, new traits are produced in individuals by genetic variation and mutation (when random changes in genetic code are introduced, for example by errors in copying). If these are favourable, natural selection promotes their transmission to future generations. For instance, flight in pterosaurs and bats, based on a membrane stretched between or from fingers, may have originated as an adaptation to glide away from predators, which themselves later became adapted to powered flight.

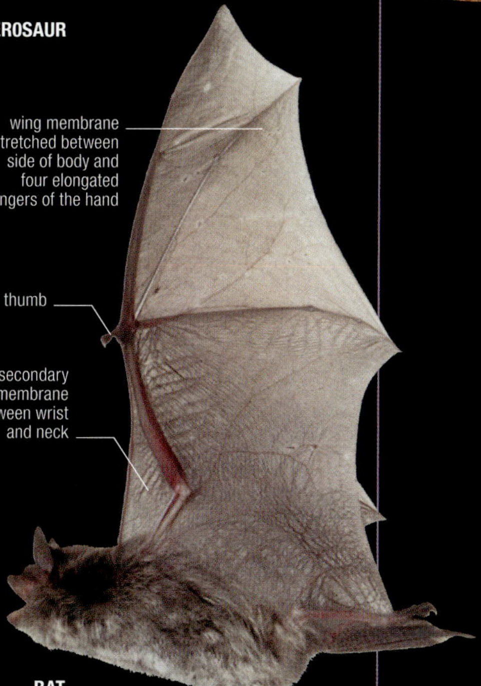

FLIGHT DEVELOPMENT
Adaptation for flight has been achieved many times by animals as diverse as extinct pterosaurs, such as the reptilian Jurassic-age *Dimorphodon* (above), and in mammals such as Daubenton's bat (right).

Speciation and extinction

The fossil record shows that, throughout Earth history, species have evolved and died out. The majority of species that have evolved are now extinct, but their genes survive in living descendants. Speciation is the process by which new species evolve from an ancestral species, and it can be brought about in a number of ways. For example, environmental change can fragment populations, leading to geographical and genetic isolation of different groups. Over time, genetic variations and adaptation to new circumstances alter the gene pool of these groups to such an extent that even if they did come together again, members of different groups would not be able to breed and produce fertile young because they would have evolved into separate species. Additionally, various regional and global environmental catastrophes have impacted upon life and resulted in mass extinctions.The largest of these, at the end of Permian times, caused 96 per cent of all species on Earth to die out – and yet life recovered.

TAPIRS
Today, the Brazilian tapir (above) and Malaysian tapir (right) are both surviving descendants of an original pig-like plant-eater, yet they are now geographically and genetically distinct. Such "living fossils" are relics of a much more widespread ancestral group of primitive mammals that lived 35 million years ago. Their survival has been due to their shy natures and retiring lifestyles in dense forests.

BIOGRAPHY

CHARLES DARWIN

The son of a wealthy provincial English doctor, Charles Darwin (1809–82) trained as a doctor, then as a clergyman, but he was mainly interested in natural history. The chance to join *HMS Beagle*'s survey voyage in 1831–36 gave him a unique opportunity to advance his knowledge, and his observations and specimen collections provided a wealth of ideas and questions about the nature of life, its origin, and history. In 1858, Alfred Russel Wallace's (see p.21) similar ideas prompted the co-publication of an outline theory of evolution, but it was Darwin who wrote down and expanded their theory into a book, *On the Origin of Species*, in 1859.

Co-evolution

Where two or more species interact over a long period, the possibility of mutually beneficial adaptation or co-evolution arises. For example, flowering plants develop a flower with coloured petals to attract insects, which then cross-pollinate and fertilize the plants, so that the plants produce seeds. The form of different flower species have co-evolved to accommodate specific insects and, less frequently, birds. In addition, the evolution of various attractive, nutritious, fleshy berries and fruits has led different birds and fruit-eating mammals to develop feeding habits that utilize such resources. The benefit to the plant is that its seeds pass through the gut of the consumer unharmed and are distributed in droppings of natural fertilizer.

BEES
Attracted by a flower, a bee is rewarded with nectar for transmitting pollen on its hairy body to another flower of the same species, which it will fertilize.

BILLS AND FLOWERS
Hummingbirds have co-evolved with certain flowering plants. As flowers have developed deeper trumpet shapes, hummingbird bills and tongues have lengthened to reach more distant nectar sources. Viewed in detail, the shape of the birds' bills have also adapted to the shape of flowers of particular species.

Reversals in evolution

Animals as diverse as whales, snakes, and ostriches are all essentially tetrapods – four-limbed creatures – descended from a land-living ancestor that used its limbs for walking. Over time, they have all adapted to very different lifestyles and habitats, with radical modifications of their appendages and bodies. The whale's forelimbs have become paddles for swimming and steering, while the hind limbs have been lost altogether, although vestiges of the pelvic girdle remain. By losing all four limbs, snakes have adapted to a different mode of movement: throwing the body into S-shaped curves and using the friction of scales to move forwards. Finally, the flightless ostrich is descended from a flying bird, which had forelimbs modified into wings. Such loss of original limbs is termed a reversal in evolution (although evolution is a process of constant adaptation rather than of steady progress).

LOSS OF FLIGHT
Having flown onto the islands of Réunion and Mauritius in the Indian Ocean, the Dodo, a relative of the pigeon, underwent a reversal of evolution and lost its ability to fly. With no natural predators, this plant-eater thrived – until humans arrived in the 1590s, hunting it to extinction in less than 100 years.

RETURN TO THE SEA
Like all seals and walruses, the Australian sea lion belongs to a group of mammals that evolved on land. It has undergone an evolutionary reversal: over time, its limbs have been modified from ancestral legs used in walking into highly efficient flippers for fast swimming in order to catch its fish prey.

Macroevolution and microevolution

Genetic changes that lead to the creation of a new species are known as microevolution. Changes in higher-level taxonomic groups (such as families) and their patterns of evolution are referred to as macroevolution. For instance, the small-scale changes separating house-sparrow populations (see right) or the herring gull (*Larus argentatus argentatus*) from the lesser black-backed gull (*Larus fuscus graellsi*), which share a common ancestor, are microevolutionary. The changes that occurred between mammals such as sea lions and their remote, egg-laying ancestors, however, are macroevolutionary. Sea lions are aquatic mammals, but they give birth to live young on land and suckle them with milk. The platypus is a monotreme, the most ancient order of mammal; it has a duck-like bill and lays eggs yet also suckles its young with milk.

SHELLED EGG
Primitive tetrapods laid unprotected eggs in water to be fertilized later, a process called external fertilization. Internal fertilization (in which sperm is transferred from the male directly into the female's body) and the laying of protected eggs allowed reproduction to occur outside water. These were major innovations that defined tetrapods such as crocodiles and birds.

PLATYPUS
The platypus is a curious primitive mammal that lays eggs. When the young hatch, however, the mother suckles them with milk – just like all modern mammals which give birth to live young.

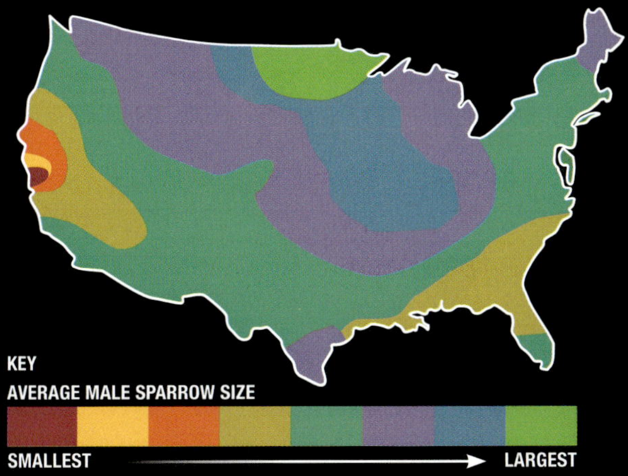

HOUSE SPARROW SIZE
Since their 19th-century introduction to North America, house sparrow populations have evolved different traits, such as body size, in different locations as they spread across the continent.

Evolution through geological time

Over 99.99 per cent of all species that ever lived are extinct because life has evolved for over 3.5 billion years, and species rarely survive beyond a few tens of millions of years. For 3 billion years, "life" meant microscopic marine organisms; it only moved into fresh water and onto land 470 million years ago, and not until 300 million years ago were plants able to colonize dry uplands. Whole groups of organisms have evolved and become extinct, some in mass-extinction events. Fossils reveal a pattern of rapid bursts of evolution, dispersal, and extinction, sometimes with waves of evolving populations replacing one another as well as gradual background evolutionary change. Evolutionary innovations have allowed new niches to be exploited; and extinction has emptied niches and allowed newly evolving groups to diversify into them. Additionally, climate change, changing sea levels, extensive volcanism, and plate movements have all affected evolution at different times in different places.

SCIENCE

PUNCTUATED EQUILIBRIUM

Darwin assumed that evolution required a great deal of time to accommodate all the macroevolutionary changes from the emergence of microbial life in the oceans. However, the late American palaeontologist S.J. Gould (right) claimed that rapid bursts of evolution are followed by inactive periods of little or no change – in other words, stasis. Called punctuated equilibrium, this idea has been difficult to prove, but detailed microevolutionary studies of large samples of continuously evolving related species such as trilobites reveal that it can happen.

Classification

With many millions of species estimated to be alive today, and many in danger of extinction, there is a need to make sense of the remarkable diversity of life on Earth. Beyond a mere key to identification, modern classification methods seek to reveal the evolutionary connections and the descent of all life from a common ancestor or ancestors.

Linnaean and phylogenetic classification

Scientific classification of life was founded by the 18th-century Swedish botanist Carl Linnaeus, who attempted to group like with like, based on species, in a hierarchy designed to illustrate a God-given order. Adopted by evolutionary biologists, it attempted to reflect natural groups and evolutionary relationships. Thus, feathered birds in the class Aves were placed at the same level, or rank, as the class Reptilia. But fossils have shown that birds evolved from reptilian dinosaurs – their "class" nests within the "class" of reptiles. Phylogenetic classification tries to remove such contradictions by distinguishing a series of descendent groups with common ancestry that is based on unique shared characteristics.

LIVING AND EXTINCT FISH GROUPS
Fossils show that the Linnaean class Pisces is not a true taxon, but it is a useful informal grouping. This phylogenetic chart shows the evolutionary relationship and descent of living and extinct groups of fish. The groups branch off from each other when a new unique characteristic arises to distinguish one from another.

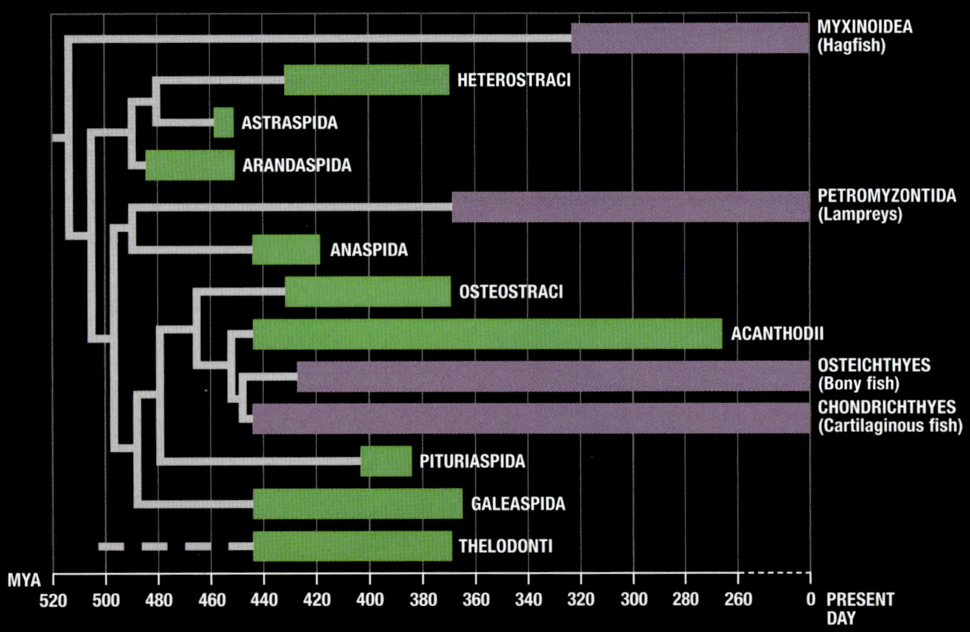

LINNAEAN HIERARCHY

The Linnaean hierarchy begins with a biological species, identified by its Latin species and genus names; there can be one or more closely related species in a genus. Genera are grouped in ascending order into families, orders, classes, phyla, kingdoms, and domains, with the aim of reflecting a meaningful relatedness in evolutionary terms.

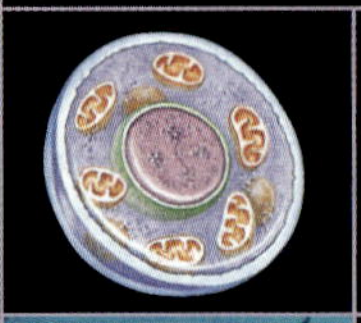

DOMAIN Eukarya
The highest currently recognized taxonomic division of life, introduced in 1990, with three domains: Archaea, Bacteria, and Eukarya. Eukarya are distinguished by a membrane-bound cell nucleus.

KINGDOM Animalia
Today, there are at least 28 kingdoms whereas there used to be just two: Animalia and Plantae. Animals are mainly multicelled eukaryotes that consume other organisms as primary food sources.

PHYLUM Chordata
A major subdivision of the animal kingdom, made up of classes. Phylum Chordata, for example, comprises animals that at some point in their life cycle possess a notochord, which is the precursor of a backbone.

CLASS Mammalia
This taxonomic level is made up of orders and their respective subgroups. Mammals, for example, are distinguished from other chordates by having a single jaw bone (the dentary), hair, and mammary glands.

ORDER Cetacea
Each order contains one or more families and their subgroups. Cetaceans (whales and dolphins), for example, are aquatic mammals that have lost their hind limbs and developed a fluked tail.

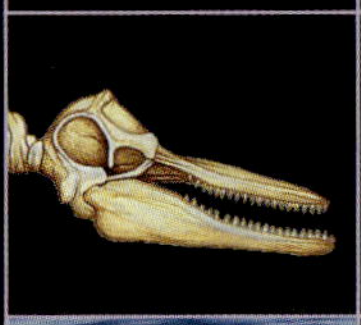

FAMILY Delphinidae
A family comprises one or more genera and their subgroups. Family Delphinidae, for example, includes all dolphins (a subgroup of toothed cetaceans) that have "beaked" jaws.

GENUS *Delphinus*
A genus is a subdivision of a family. Dolphins show considerable variation in body shape and colour, but genetic analysis suggests there are just two or three species in the *Delphinus* genus.

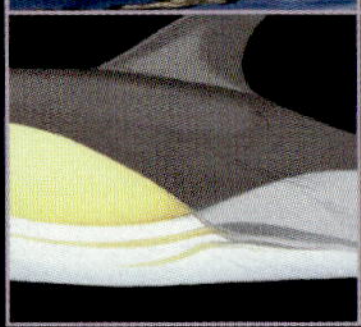

SPECIES *Delphinus delphis*
This group comprises similar individuals able to interbreed in the wild. *Delphinus delphis*, or the short-beaked common dolphin, is distinguished by its black-and-white colour pattern and beak size.

Molecular clocks

The idea of a molecular clock is that evolutionary changes occur at regular time intervals. It is assumed that the rate of genetic change (mutation) in an organism's DNA has not changed over time, or at least can be averaged. The molecular genetic difference or "distance" between two species must be measured and their rate of genetic change estimated. The latter is derived from radiometric dating (see p.13) of the fossils of a known ancestral species. From these, it is possible to calculate when the two species diverged from their common ancestor.

50 MILLION YEARS AGO
Isolated sequence of ten bases (see p.27) in common ancestor's DNA.

25 MILLION YEARS AGO
Descendent lineages have diverged. Each has had a single base mutation, so the lineages differ by two bases.

PRESENT DAY
Descendent lineages have diverged again by another base mutation. The lineages now differ by four bases.

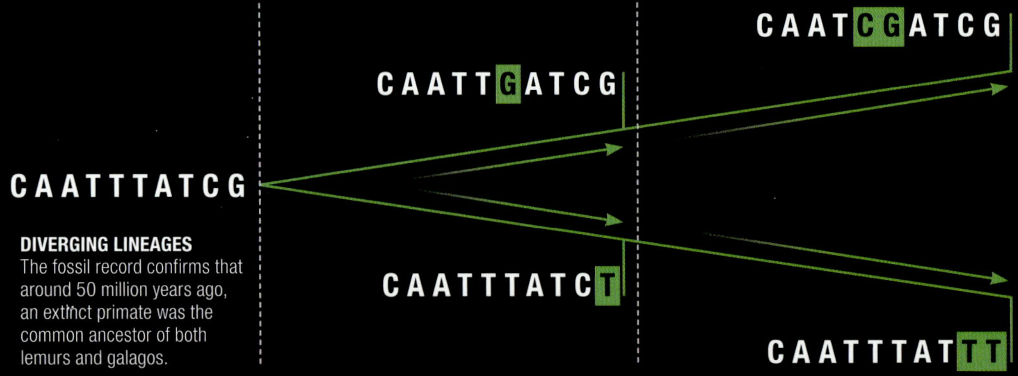

DIVERGING LINEAGES
The fossil record confirms that around 50 million years ago, an extinct primate was the common ancestor of both lemurs and galagos.

LEMUR AND GALAGO
These related tree-dwelling primates are now geographically and genetically distinct, due to numerous genetic divergences.

Classification of life

Taxonomic classification is still hotly debated in the scientific community, and no one definitive method is yet recognized globally. Problems with Linnaean classification have arisen during the last 100 years, due to the knowledge that many organisms have similar anatomical characteristics as a result of parallel or convergent evolution. By contrast, phylogenetic classification or cladistics, developed by the German entomologist Willi Hennig in the 1950s and 1960s, identifies related groups of organisms descended from a common ancestor via the identification of new characteristics in their most likely evolutionary order. For the purposes of this book, all living things are grouped into three domains: bacteria, archaea, and eukaryotes. Protists and invertebrates are shown below within dotted lines because although they are useful informal groupings they are not true taxa.

Bacteria

DOMAIN Bacteria
KINGDOMS 10
SPECIES Several million

This major group of single-celled micro-organisms lack a defined cell nucleus. They have a great abundance and diversity of form, and have colonized virtually every habitat, from deep in Earth's crust to ocean-floor sediments and the living tissues of most organisms. Bacteria are essential to animal digestive systems, but they can also cause fatal diseases. There are millions of bacterial cells within a single pinch of soil, where they play a vital role in recycling nutrients.

Archaea

DOMAIN Archaea
KINGDOMS 3
SPECIES Several million

The microscopic and single-celled archaea have no defined cell nucleus or specialized structures. Their characterization as a group has been based on molecular data, and they have been intensely studied over recent years because of their remarkable tolerance to extreme environments and their various means of obtaining energy. The latter range from photosynthesis (converting sunlight into energy sources), through various modes of chemosynthesis: transforming metal ions, carbon, and hydrogen into a "food" source.

Eukaryotes

DOMAIN Eukarya
KINGDOMS At least 15
SPECIES 2 million

Most familiar living organisms are eukaryotes, ranging from the microscopic, single-celled amoeba to the giant redwood tree and the blue whale. This vast group of organisms all have common features in each body cell. The eukaryote cell is unique in having a complex structure that includes a distinct nucleus surrounded by a membrane; it also possesses mitochondria: structures that conduct the biochemical processes of respiration and energy production. Reproduction in eukaryotes involves the separation of duplicated chromosomes.

Protists

KINGDOMS At least 10
SPECIES More than 100,000

Single-celled protists are a diverse group of micro-organisms that were viewed traditionally as a major group. However, details of their biology show that they have very little in common; some are plant-like, others animal-like, and yet others more like fungal slime moulds, indicating that they have evolved from separate origins.

Red seaweeds

KINGDOM Rhodophyta
CLASSES 1 or more
SPECIES 5,500

The evolutionary relationships of red, brown, and green algae have long been problematic. Modern studies place them in very different groups. Red algae (rhodophytes) are one of the largest, most ancient eukaryote groups, with a fossil record extending back over 1.2 billion years.

Brown seaweeds

KINGDOM Phaeophyta
CLASSES 1
SPECIES 2,000

Mainly large, multicelled, cool-water organisms. Some grow up to 60m (197ft) and play an important role in the ecology of coastal waters; others are free-floating and form the unique habitats of tropical sargassum. The group is distinguished by genetic data and their chloroplasts: cell structures that conduct photosynthesis.

Plants

KINGDOM Plantae
DIVISIONS 6
SPECIES 283,000

The remarkable abundance and diversity of plants includes all the familiar forms, from green seaweed, mosses, and grasses to giant redwoods. Most plants obtain energy from sunlight through photosynthesis, but may not have a single evolutionary origin because of fundamental differences in their modes of reproduction.

Fungi

KINGDOM Fungi
PHYLA 4
SPECIES 600,000

This diverse, abundant group has a fossil record extending back to Devonian times and includes microscopic single-celled and macroscopic multicelled organisms such as mushrooms, some with fruiting bodies many metres in size. Typically, they use spores for reproduction and have cell walls of chitin rather than cellulose.

Animals

KINGDOM Animalia
PHYLA About 30
SPECIES Over 1.5 million

The group name "animals" is deeply embedded in our language and common culture, with roots going back to classical times. The word's Latin root means "having breath", and has been used traditionally to distinguish plants from animals. However, since scientists have recognized that primitive life is much more complex than was previously thought, the dividing line between plants and animals has become blurred. As a result, modern usage of the name "animal" tends to restrict it to multicelled eukaryotes that must consume other organisms as food.

Invertebrates

PHYLA About 30
SPECIES Over 1 million

This group name is based on a negative attribute that differentiates all animals that are not vertebrates: that is, they do not have a backbone. As such, it has no place in modern phylogenetic classification, but is still a commonly used "catch-all" word for about 30 phyla, ranging from sponges and flatworms to arthropods and echinoderms.

Chordates

PHYLUM Chordata
SUBPHYLA 3
SPECIES 51,550

The chordates all have a distinct body plan – at least in their embryonic stages, if not always retained in adults. A stiffening rod, called a notochord, extends dorsally from front to back. Its fibrous tissue makes it flexible, and the paired muscle blocks on either side bend it into the sinuous, sideways curves that are associated with swimming. Above the notochord lies a dorsal hollow nerve cord, and below that lies the intestine. Most chordates also have gill slits and a tail at some developmental stage.

Lancelets

SUBPHYLUM Cephalochordata
CLASSES 1
SPECIES 50

Flattened, leaf-shaped lancelets grow to around 8cm (3in) long. There are no defined fins or head, and the front is marked by a cavity perforated by slits through which sea water is pumped for feeding and breathing. These filter-feeders burrow tail-first into the sea bed. Their poor fossil record extends to Cambrian times.

Tunicates (sea squirts and salps)

SUBPHYLUM Urochordata
CLASSES 4
SPECIES 2,000

Small, marine filter-feeding chordates that grow up to 15cm (6in) long, generally characterized by a notochord and tail in their larval, free-swimming stage. Typically, adults show a radical change in body form: the ascideans attach to the sea bed, the salps lose their tails and notochord but remain free-swimming, and the appendicularians remain free-swimming and retain both tail and notochord.

Vertebrates

SUBPHYLUM Vertebrata
CLASSES 7
SPECIES 49,500

In these animals, the notochord is surrounded by segmented skeletal elements to form a strengthened, articulated vertebral column. They are also characterized by nervous regulation of the heart, muscles for eye movement, a sensory system of lateral lines, and two semicircular canals in the inner ear. They are divided into two major groups: the jawed gnathostomes and the jawless petromyzontiforms.

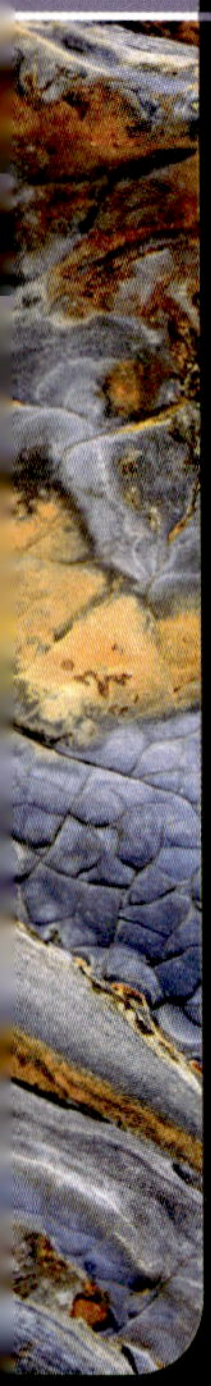

Mass Extinctions

In Charles Darwin's day there was plenty of evidence to suggest that life had changed radically over time, but little to indicate that evolution had been seriously disrupted by major extinction events that wiped out a huge percentage of all life on Earth. Yet, as we now know, the growth in the overall diversity of life has suffered several major setbacks, especially at the end of the Permian Period 251 million years ago, after which recovery took the best part of 100 million years.

Causes

There have been at least five mass extinctions in the last 540 million years. Much effort has been expended on searching for their causes, because there is no reason why one should not happen again. The most famous extinction event, at the end of the Cretaceous Period, was marked by a meteorite impact and volcanic eruptions. Although many other impacts have been found, none can readily be linked to extinction events. The most common factors in mass extinctions are large-scale eruptions of lava and volcanic gases, and the absence of oxygen in the oceans. Volcanic gases, especially carbon dioxide, are connected with short-term regional warming, medium-term acid rain, ozone depletion, and long-term global warming. This atmosphere–ocean link is the most likely reason for mass extinctions, including those in which impacts played a role and where glaciation was a factor.

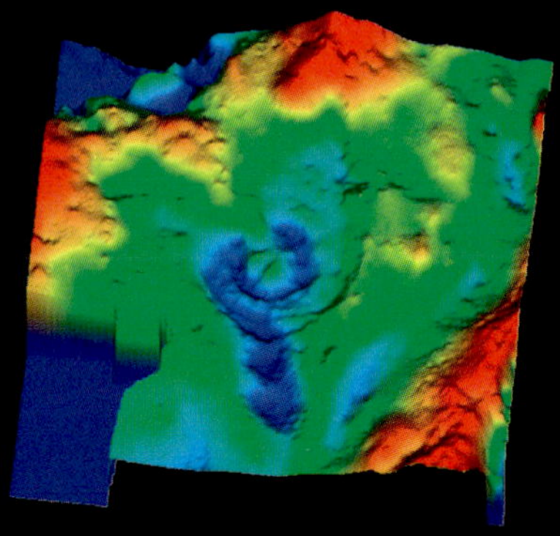

CHICXULUB CRATER, YUCATÁN PENINSULA
When an asteroid ploughed into the Gulf of Mexico 65 million years ago, it was around 10km (6 miles) in diameter – large enough to create a crater 240km (150 miles) wide and set off a mass-extinction event.

PERMIAN–TRIASSIC BOUNDARY CLAYS
Scientists take samples from a cliff in Italy. The dark brown notch halfway up is evidence of vegetation destroyed suddenly at the end of the Permian Period.

K–T BOUNDARY
Found worldwide, this thin, grey band is enriched in iridium, an asteroidal element. It is evidence of the impact that led to the extinction at the end of the Cretaceous Period.

GOSSES BLUFF
The hills of Gosses Bluff in Australia are the remnants of an impact crater that was originally 22km (14 miles) wide, created 142 million years ago. The rocks show a type of fracture associated with catastrophic impacts, yet there is no link to any extinction event. The Chicxulub crater (above right), by comparison, is ten times larger.

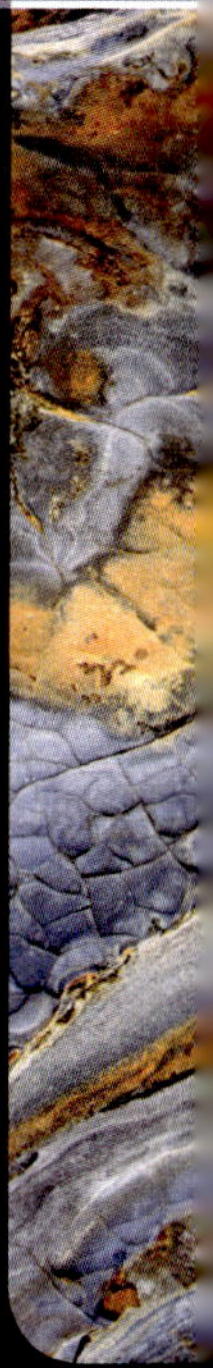

Major mass extinctions

Since the end of the nineteenth century, it has been shown that the history of life has included evolutionary "booms" as well as "busts" of catastrophic extinctions alongside a more gradual background of species origination and extinction. The most famous "bust" episode occurred at the end of the Cretaceous Period. It resulted in the extinction of dinosaurs, various marine-reptile groups, flying pterosaurs, and various invertebrates, and has been linked to a major impact event and volcanic activity. However, an even greater extinction event, at the end of the Permian Period, saw the extinction of up to 96 per cent of marine species, 70 per cent of terrestrial vertebrate species, and 83 per cent of insect species. An earlier mass extinction, at the end of the Ordovician, coincides with a rapid fall and rise in sea level related to ice-cap growth and melting.

LAST OF THE DINOSAURS
Fossil records show that *Corythosaurus*, a 10m- (33ft-) long plant-eater, was one of the last dinosaurs to walk the Earth.

IRON PYRITES
Seabed rocks in Greenland contain iron pyrites formed in the absence of oxygen. They are a record of the oxygen-poor conditions that led to an extinction at the end of Permian times.

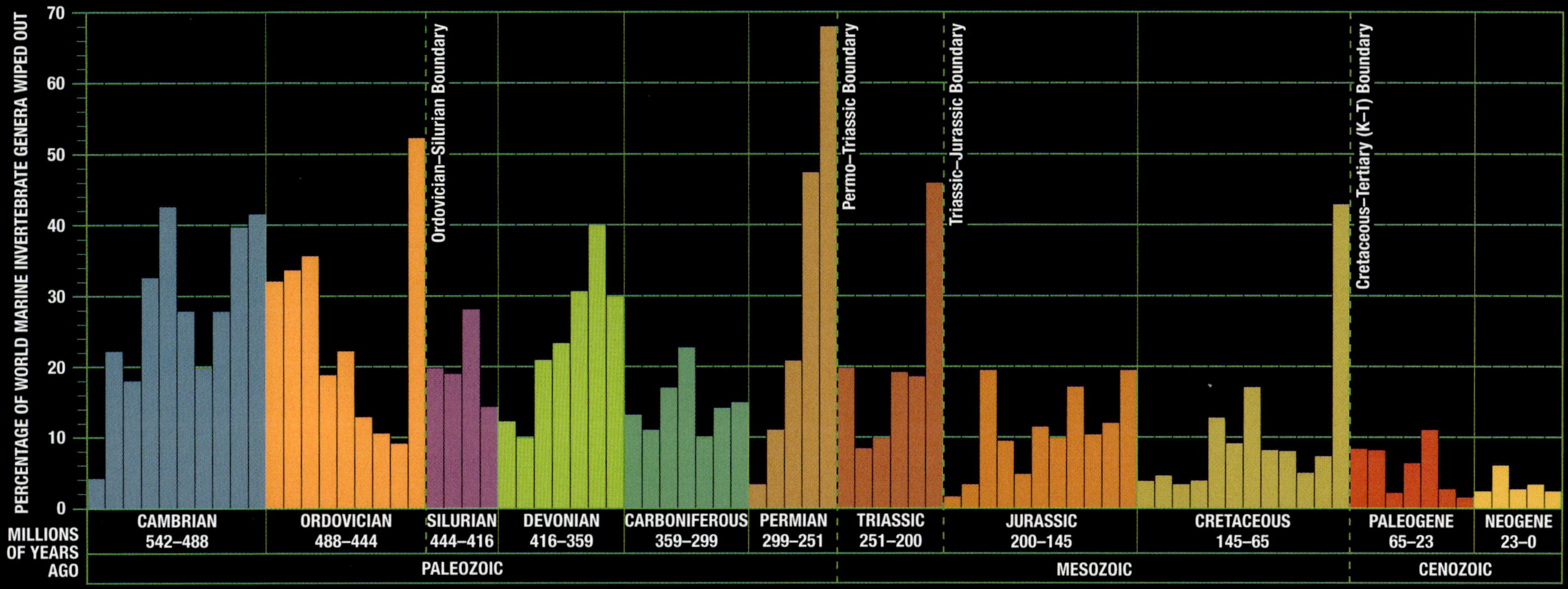

RECORD OF EXTINCTIONS
The state of fossil preservation makes analysis of data associated with extinction very difficult. The most reliable, continuous records come from the remains of sea creatures such as ammonites.

Recovery from mass extinctions

The fossil record provides evidence of mass extinctions but also of the subsequent recovery of life. For example, coral reefs were particularly affected by the end-Permian extinction. Their collapse had a major impact on oceanic food chains, but new skeleton-forming corals evolved, slowly re-establishing reef structures. Similarly, the collapse of terrestrial ecosystems at the end of the Cretaceous Period is marked by the extinction of the dinosaurs, yet the recovery of vegetation is evident from fern spores: plants adapted to survive the environmental disaster.

PERMIAN TO TRIASSIC
Land animals were greatly affected by the Permian extinction. Even successful plant-eaters such as these two *Diictodons* (left), found huddled in their burrow, died out, but their evolutionary descendant, *Lystrosaurus*, (above) survived.

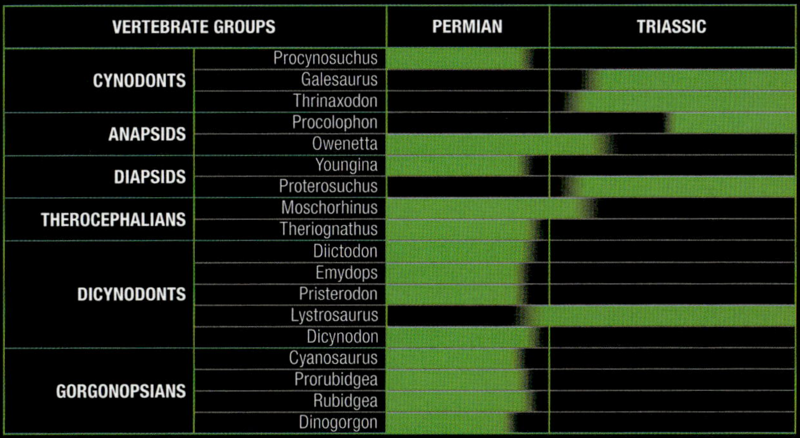

FOSSIL RANGES OF VERTEBRATE GROUPS IN THE KAROO
The scale of extinction at the end of the Permian Period is reflected in the abundant fossils of land animals preserved in the Karoo region of South Africa. A few groups were completely wiped out, but enough survived for life to continue.

Recent extinctions

Since the last glacial maximum, most of Earth's larger land animals have disappeared, except those that remain in Africa. It was thought that a changing climate led to the demise of cold-adapted animals such as the woolly mammoth, woolly rhinoceros, and giant deer. However, recent dating shows that their extinction coincided with the arrival of modern humans in various regions. Similarly, the New Zealand giant moa did not become extinct until one or two centuries after the arrival of modern humans, around 1250. However, climate change may also have played a role in some areas.

GOLDEN TOADS
Today, amphibians are particularly vulnerable to extinction. Discovered in Costa Rica's mountainous tropical forest in 1966, the golden toad has not been seen since 1989.

WOOLLY RHINO
Coelodonta, the woolly rhinoceros, lived in ice age North America and Eurasia. It became extinct 10,000 years ago – after humans arrived.

Types of Fossils

Fossil remains are preserved by such a diversity of geological processes that their form and appearance can be very deceptive. In fact, it has taken over 200 years for fossil experts to understand exactly how fossils are formed and how to interpret them accurately.

What is a fossil?

A fossil is the remains of a once-living organism that has been buried and preserved in rock. The word "fossil" comes from the Latin *fossa,* meaning "dug up". Originally, fossils were thought to consist of both organic and inorganic matter, but by the end of the seventeenth century scientists generally agreed they were of purely organic origin. There are three fossil types: body, trace, and chemical fossils. Body fossils include hard parts such as shells, bones, teeth, and some plant material, while trace fossils are marks preserved in hardened sediment that record evidence of past life, such as footprints. Finally, chemical fossils consist of degraded organic substances from the molecules of living things, such as fossil DNA recovered from bone or teeth. No matter how they form, many fossils can be identified at species level.

OTZI THE ICEMAN
The frozen cadaver of a 5,300-year-old Neolithic hunter, found in glacial ice high in the Tyrolean Alps, was so well preserved that it was initially recovered by police. But, just like the meat in a domestic freezer, this body – despite its great age – is not a true fossil because it has not undergone fossilization processes.

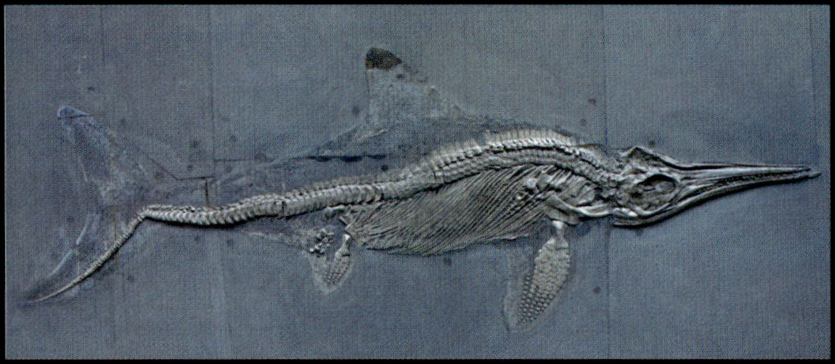

WELL PRESERVED
This Jurassic ichthyosaur, *Stenopterygius megacephalus,* is a true fossil. It might look almost perfectly preserved, but what appears to be skin has been replaced by a microbial film.

Body fossils and trace fossils

The most easily visible, familiar fossils are body fossils made from the shells and skeletons of animals that lived in shallow seas millions of years ago. Micro-organisms also leave body fossils. These can only be seen with a microscope but include huge numbers of plant and algae components such as pollen and spores. Trace fossils, such as the burrows of animals and spaces left by plant root systems, are also common. However, these are rarely preserved with the organism that made them, so an entirely different naming system had to be invented to deal with them. Yet trace fossils can be very useful providers of biological and ecological information – especially about extinct animals such as dinosaurs. For example, the study of dinosaur trackways is helping to resolve arguments about how fast these creatures could run (see p.38).

TRACE FOSSILS
Trace fossils, such as these 400-million-year-old tetrapod footprints on Valentia Island, Ireland (top), or fossil faeces, known as coprolites, of *Tyrannosaurus rex* (above), provide tantalizing clues about the day-to-day lives of extinct creatures.

BODY FOSSILS
The body fossil of *Homeosaurus* (right), a 20cm- (8in-) long Jurassic reptile, reveals its lizard-like skeleton. A tooth (below) is all that is left of *Carcharodon auriculatus*, an extinct ancestor of the modern great white shark. Sharks' cartilaginous skeletons are not normally fossilized, but one tooth is often enough to identify a shark species.

How fossils form

Organic remains are fossilized in many ways. There is almost always a loss of information as the remains decay and are buried, compressed, and sometimes heated. At best, soft tissues such as feathers can be fossilized, but at worst, all that is left is a trace, such as a footprint. The chances of any remains being fossilized are actually very slim, especially if organisms have no easily preservable "hard parts" like shell or bone. Organic remains are commonly preserved in places where they will have little or no disturbance, such as in still or slow-moving water. Even delicate tissues can be preserved when decay is slowed or even halted by low oxygen levels, low temperatures, or a natural preservative such as tar or resin.

FOSSILS REVEALED
Wind and rain can expose fossils by removing the surrounding soil. Here, the petrified trunks of a Triassic conifer have been brought to light in Arizona.

UNALTERED PRESERVATION

Occasionally, the unaltered remains of organisms are completely preserved, particularly in cases where the hard parts are made of minerals that do not degrade over time. For example, shells and skeletons made of calcite, found in echinoderms such as starfish, are more stable than those composed of aragonite, which is found in molluscs. Silica is another stable mineral that is found in some sponges and microscopic algae known as diatoms. Organic material is essentially unstable, so shells lose not only soft tissue but also their colour, except under special circumstances, such as when they are covered with resin. Even then, however, microbes can still degrade a fossil, although it may look perfect on the outside.

FOSSILIZED DIATOMS

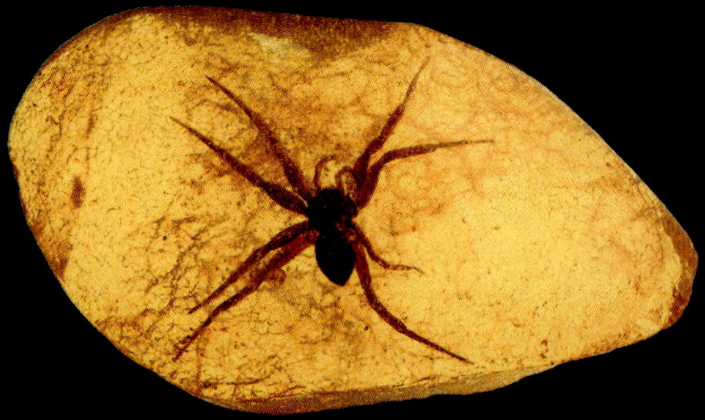

SPIDER PRESERVED IN AMBER

TRITON SHELL

TRITON FOSSIL

RECRYSTALLIZATION

Many fossils are preserved as a result of recrystallization (the formation of new crystals), especially when a skeleton or shell is made of an unstable crystalline mineral with a closely related but more stable form, such as aragonite. Aragonite is a very common constituent of mollusc shells and coral skeletons, but it usually recrystallizes into the more stable mineral calcite, which has a slightly different atomic structure. This process does not alter the surface details of the fossil; in fact, even some internal details may be preserved, which means the fossil can be identified and its structure and growth can be studied.

THECOSMILIA CORAL COLONY

EUOMPHALUS (TOP VIEW)

EUOMPHALUS (CROSS-SECTION)

CARBONIZATION

The complex organic compounds that make up plant tissues are prone to a chemical change called carbonization, which results in a loss of their watery components yet leaves a carbon residue that is much more stable over time. When a plant is covered by sediment, the burial compresses and may even heat its remains. Pressure squeezes out any liquid and gas, and flattens the plant tissues; heat may turn these into charcoal, leaving only carbon behind in the form of a compression fossil. Over time, even this carbon residue may be lost, leaving in its place only an impression and outline of the plant tissue. In fine-grained sediment, this is often enough to preserve detailed information so that the species can be identified, and this process is particularly effective in preserving leaves and delicate animal forms.

NEUROPTERIS SCHEUCHZERI LEAF

BASSWOOD STEMS AND BERRIES

REPLACEMENT

The complete replacement of an organism's structure with another material is called replacement. Replacement is common during the long, complex process of fossilization or petrification (see pp.36–37). Burial of fossil remains and transformation of the surrounding sediment into rock involves physical and chemical changes, the latter resulting from the groundwater that saturates the sediment. Groundwater containing dissolved chemicals commonly substitutes them, atom by atom, for original fossil minerals so that fine details are often preserved. Silica often replaces and fills in cell spaces of woody fossil plant stems, while the iron-sulphide mineral pyrite replaces the aragonite of shells of molluscs such as ammonites.

PETRIFIED LOG CROSS-SECTION

AMMOLITE (PRESERVED MOTHER OF PEARL)

PYRITIZED AMMONITE

IMPRESSIONS, MOULDS, AND CASTS

When an organism is pressed into soft sediment, it creates a negative impression, and many of its external details are recorded in negative relief. Because this impression effectively forms an external mould of the fossil, any sediment that later fills it in, or fills in the spaces within the fossil itself, such as the hollow inside a seashell or skull, forms an internal mould with details preserved in positive relief. If fossil material degrades and disappears altogether after the surrounding sediment hardens, this mould still preserves a record of the fossil's internal and external structures. Scientists often use natural moulds to make artificial casts of such "missing" fossils in order to study their structure in greater detail.

IMPRESSION OF AN EOCENE POPLAR LEAF

EXTERNAL MOULD OF CYCLOSPHAEROMA

INTERNAL MOULD OF CYCLOSPHAEROMA

SHELLS IN SANDSTONE

SCIENCE

VIRTUAL FOSSILS

Traditionally, palaeontologists prepare fossils by laboriously removing sediment stuck to their surfaces. This is difficult and sometimes impossible, especially around delicate structures. Much effort has gone into finding less destructive means of revealing small, fragile fossils. Chemical cleaning can be effective, but even this is not always feasible. New techniques employing non-destructive computed tomography (CT scans) and digital processing generate three-dimensional images from a series of two-dimensional X-ray scans. But where X-rays cannot differentiate between the fossil and its surrounding rock, the physical grinding of slices through the fossil produces a stack of images that can be electronically transformed into a virtual fossil.

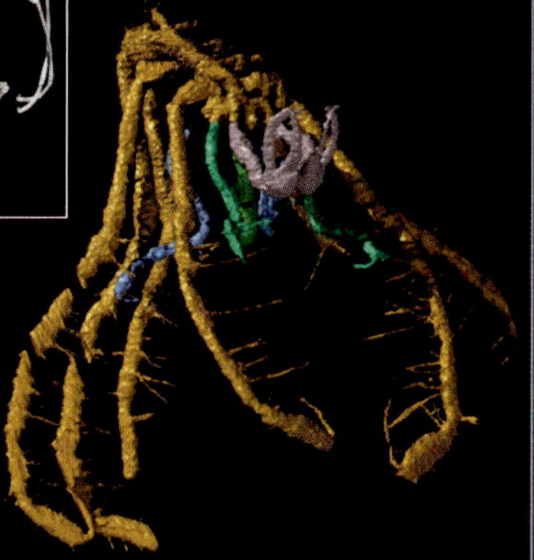

REAL AND RECONSTRUCTED
The modern sea spider, *Nymphon gracile* (above), is remarkably like its Silurian ancestor, *Haliestes dasos* (right), whose image has been digitally reconstructed from slices produced by grinding.

The making of a fossil

As well as supporting living organisms, bodies of sea water and fresh water can also preserve their remains, buried within sediments at the bottom. Much depends upon conditions at the boundary between the sediment and the water, but wherever fine-grained, muddy sediments are deposited, preservation can be remarkably good. For example, the covering of tough, bony scales on the body and fins of a type of marine fish called a coelacanth may be preserved more or less intact, along with the bones of the head, if the carcass comes to rest in water without enough oxygen to support scavengers. The different stages in the fossilization process are shown here. However, in well-oxygenated water, scavengers may consume the body altogether, leaving just some bones and scales preserved in their fossilized faeces.

LIVING COELACANTH

FOSSILIZED COELACANTH

FROM FISH TO FOSSIL
Coelacanths are a primitive kind of bony fish, with fleshy fins and thick, heavy scales. The mineralization of the fish's body after death could take as little as several decades, while the formation of the rock around the fossil may take millions of years.

DEATH

The death of an organism, such as this coelacanth, may be caused by predation or disease or a natural disaster. Initially, a dead coelacanth will float to the surface, buoyed up by gas released by internal decay. Upon rupture of the body cavity, the body sinks until it lies on top of the sediment. From the moment of the fish's death, scavengers may begin to break down or remove the soft tissues. The scales may fall away, exposing the skeleton. Only the toughest body parts, such as bones, teeth, and scales, are likely to survive long enough to be fossilized, except in exceptional circumstances. As well as attack by living organisms, the remains may be subjected to physical damage – for example, caused by the action of waves and currents.

BURIAL

Where and how, and how quickly organic remains are buried makes a great difference to their preservation as fossils. If the coelacanth settles in the fine-grained muds of a quiet lagoon, most of its scale-clad body may be preserved. However, more typically, the body will be disturbed by water action and it will be broken apart and worn down before it is finally buried. The illustration, right, shows a crosssection of a sediment-covered scale.

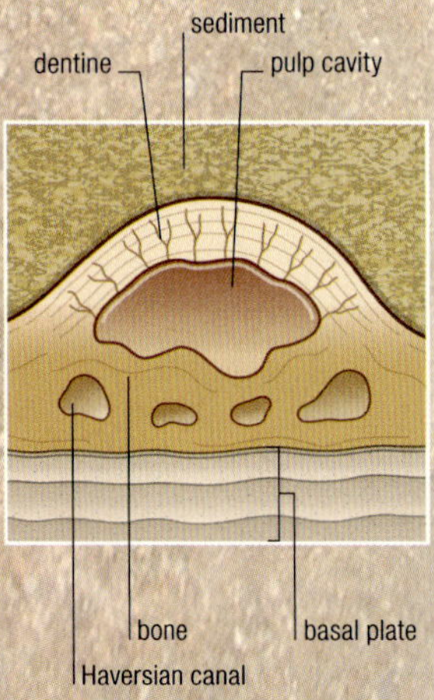

Where and **how**, and **how quickly** organic remains are **buried** makes a **great difference** to their **preservation** as **fossils**.

DECAY

Decay is a continuing process from the moment of death to final fossilization as a mineral residue within the surrounding rock. Depending upon the rate of burial, most decay and loss of soft tissue occurs before burial. However, many biodegrading organisms such as bacteria live in sediment, and these continue the process of decay and change after the body has been entombed, often removing all trace of soft tissue. But, as the sediment's biochemical environment is often different from that of the surface, other chemical changes can occur and preserve details of any remaining soft tissues, such as gills, muscles, and guts. As the body decays, sediment continues to settle, forming separate layers.

REPLACEMENT

The mineral components of tissues may be replaced by different minerals, such as calcium carbonate or phosphate, supplied by water in the sediment (see right). The pressure of successive layers of sediment will compress the coelacanth's body. If the water enters the body before compression, its original form will be preserved. If it enters after compression, just a flattened outline of the body will be preserved.

ROCK FORMATION

Finally, hard, stratified rock is formed. The sediment and fossil remains are cemented together by minerals, which fill spaces and replace original materials. The tough coelacanth scales fossilize better than the thin scales of the many other species of more advanced bony fish, such as members of the salmon family.

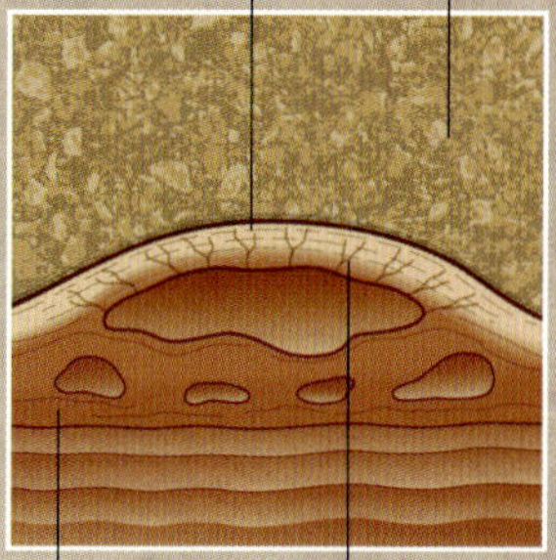

Information in the Fossil Record

Fossils reveal extraordinary and wonderful details about ancient life. They tell us about what living organisms looked like, how they behaved, and the environments they lived in. They can also be used to estimate the ages of the rocks in which they are found.

Life in the past

While only a small portion of life on Earth has been preserved in the fossil record, fossils still provide most of the data we have about prehistoric life and its development within the planet's changing environments. The fossil record favours creatures with hard, preservable parts, yet it has revealed the outlines of the history and evolution of major groups of organisms, including extinct ones such as the trilobites, ammonites, dinosaurs, seed ferns, and our own earliest ancestors. We now have a record of evolution over the last 600 million years, showing how life developed in the seas, emerged onto land, and how ecological relationships were established between the first plants and animals. Despite catastrophic extinction events, life has evolved and adapted to fill Earth's available and habitable niches, from ocean depths to mountaintops.

FISH-EATING FISH
This predatory fish, *Mioplosus labracoides*, from the Green River Formation in Wyoming, bit off more than it could chew when it tried to swallow its prey 50 million years ago. The scarcity of other *Mioplosus* fossils indicates that this species was a solitary predator.

FOSSIL FOOTPRINTS
Much can be learned about ancient animals' behaviour by studying the preserved marks they left in the ground, such as these dinosaur footprints. Found in Middle Jurassic-age strata at Ardley in Oxfordshire, England, they form trackways up to 180m (600ft) long. Trace fossils like these help scientists to calculate how fast extinct animals such as dinosaurs walked (see below).

DINOSAUR LOCOMOTION
One of the trackways at Ardley provided the first evidence of the changing gait and speed of a large predatory theropod – the dinosaur group that includes giant carnivores such as *Tyrannosaurus rex*. The footprints record a bipedal theropod (one walking on two legs) breaking into a run from a walk. Calculations show that it increased its speed from 6.8 to 29.2kph (4.2 to 18.1mph).

WALKING

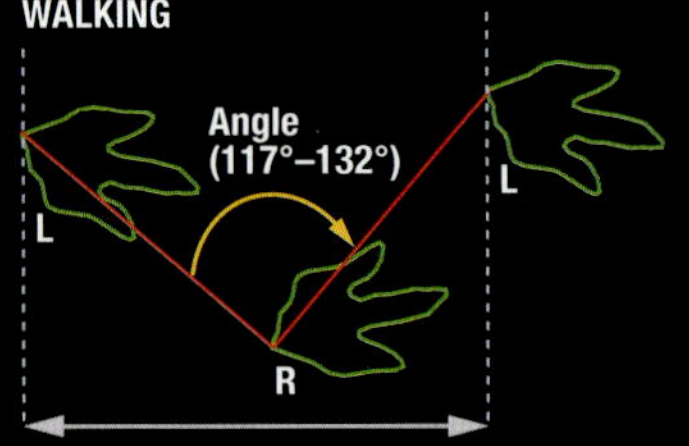

RUNNING

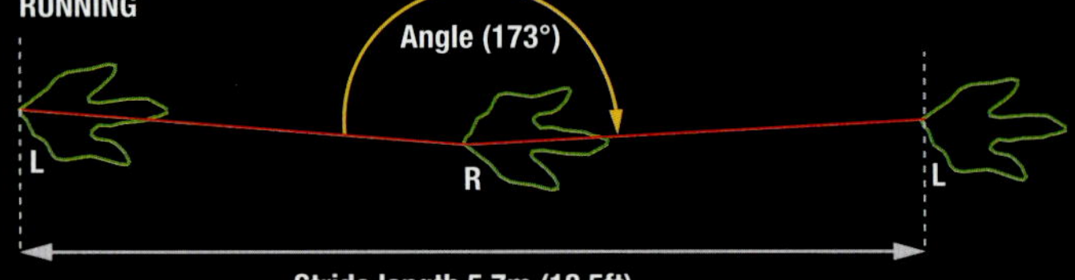

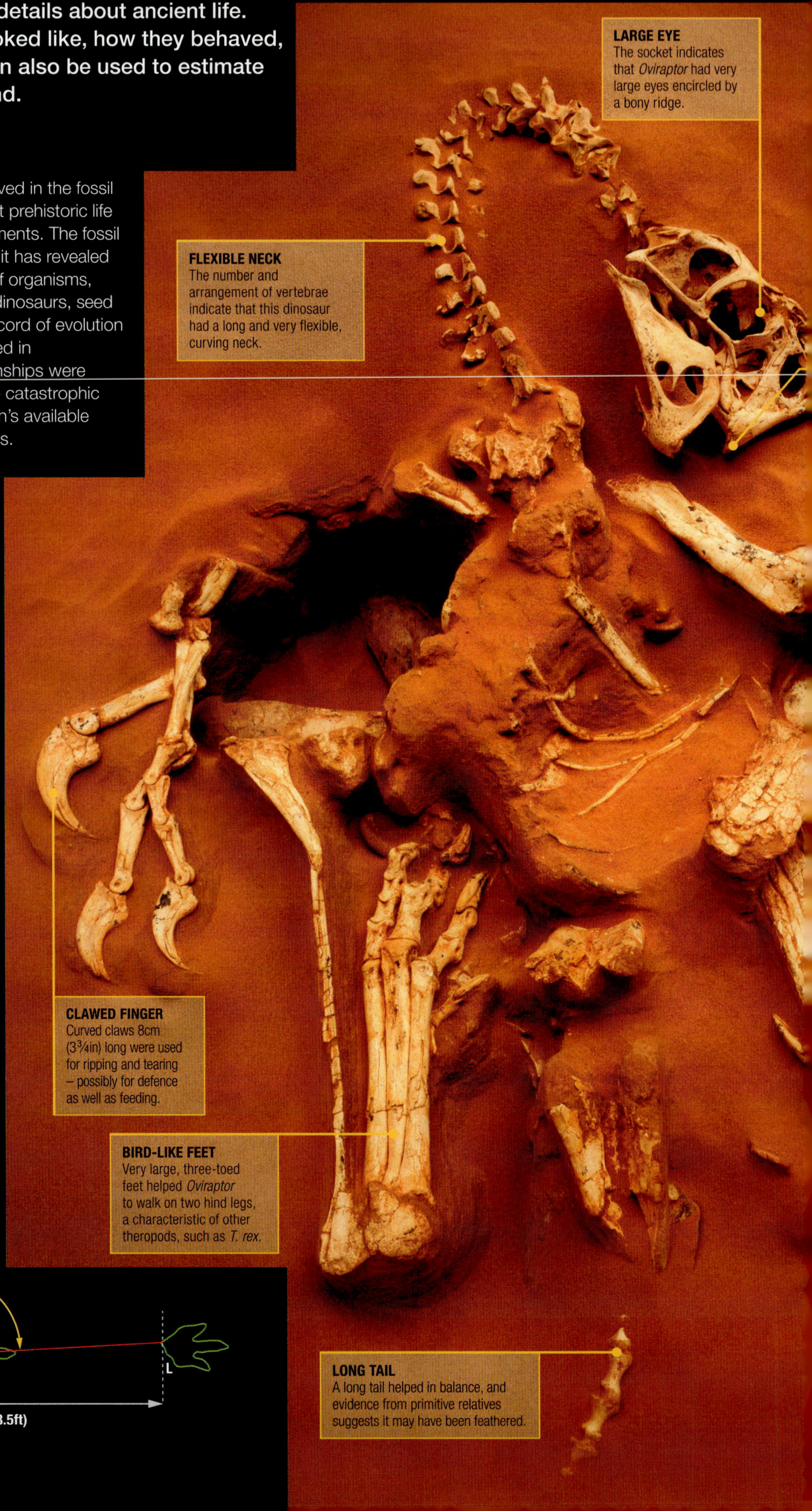

LARGE EYE
The socket indicates that *Oviraptor* had very large eyes encircled by a bony ridge.

FLEXIBLE NECK
The number and arrangement of vertebrae indicate that this dinosaur had a long and very flexible, curving neck.

CLAWED FINGER
Curved claws 8cm ($3\frac{3}{4}$in) long were used for ripping and tearing – possibly for defence as well as feeding.

BIRD-LIKE FEET
Very large, three-toed feet helped *Oviraptor* to walk on two hind legs, a characteristic of other theropods, such as *T. rex*.

LONG TAIL
A long tail helped in balance, and evidence from primitive relatives suggests it may have been feathered.

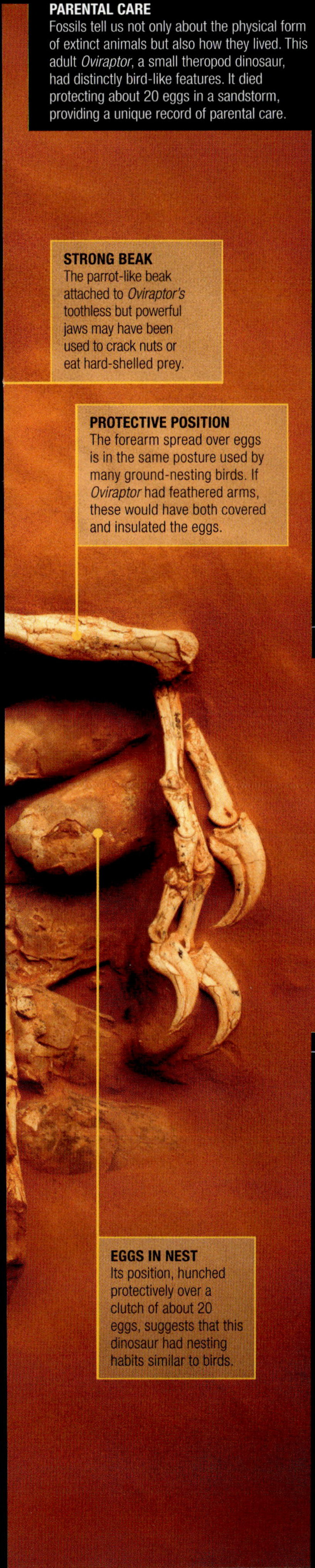

PARENTAL CARE
Fossils tell us not only about the physical form of extinct animals but also how they lived. This adult *Oviraptor*, a small theropod dinosaur, had distinctly bird-like features. It died protecting about 20 eggs in a sandstorm, providing a unique record of parental care.

STRONG BEAK
The parrot-like beak attached to *Oviraptor's* toothless but powerful jaws may have been used to crack nuts or eat hard-shelled prey.

PROTECTIVE POSITION
The forearm spread over eggs is in the same posture used by many ground-nesting birds. If *Oviraptor* had feathered arms, these would have both covered and insulated the eggs.

EGGS IN NEST
Its position, hunched protectively over a clutch of about 20 eggs, suggests that this dinosaur had nesting habits similar to birds.

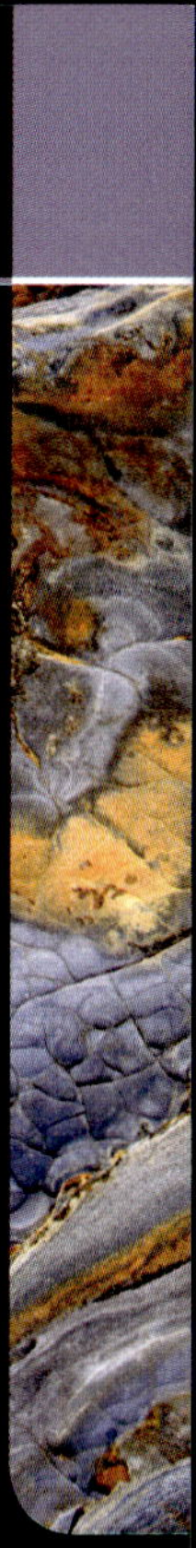

Dating

Even before Darwin and Wallace proposed their theory of evolution, it was recognized that fossils change over time, with different ones occurring in different rock layers. By the early nineteenth century, fossils were being used to characterize successive intervals and periods of geological time; for example, certain species of sea urchin, ammonite, and bivalve were known to be characteristic of the Cretaceous Period while the Cambrian was noted for particular trilobites and brachiopods. Relative fossil dating has since been highly refined by using "index species" as reference points. Ideally, index species are common, widespread, easily identifiable, and rapidly evolving, such as graptolites and ammonites. Scientists use them to distinguish well-defined intervals of time in rock layers or strata. However, finding an accurate age for these intervals required radioactive dating of suitable rocks in the stratigraphic rock record. This has now been achieved with considerable accuracy for a record of the last 542 million years.

INDEX FOSSILS
The succession of fossils in stratigraphic sequences changes with environments and the passage of time. Fossils that characterize particular environments and time intervals are known as index fossils, and they can be used to match strata of the same type and age.

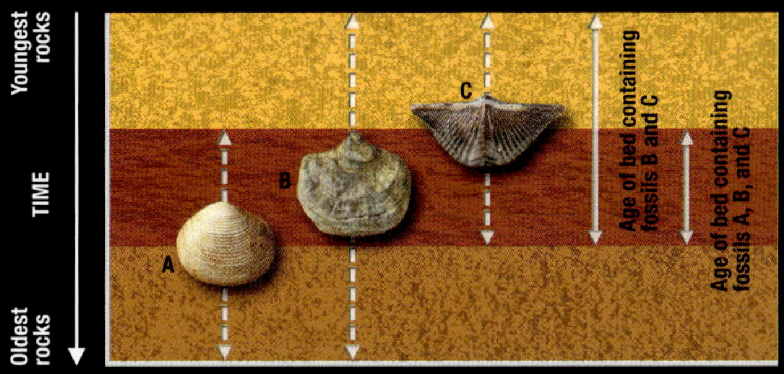

FOSSIL ASSEMBLAGES
In a sequence of rocks without index fossils, groups of other fossils may sometimes indicate age. Such fossils are particularly useful if their age ranges overlap; and the more fossils there are, the more precisely the rock can be dated.

MAIDENHAIR TREE
The maidenhair tree, *Ginkgo biloba* (right), is a living fossil whose ancestors (above) served as dinosaur food. Ginkgos prefer warm climates and were abundant in Jurassic Britain. Their subsequent disappearance points to climate change.

Reconstructing past environments

Interpreting the fossil record is vital for reconstructing environments and their development over time. Such interpretations are based on the knowledge that all organisms have certain inherent constraints, such as the amount of light, heat, or water needed to survive and reproduce. For example, strata containing tropical marine fossils – reef corals, sea lilies, and calcareous algae – were deposited in shallow tropical seas. Thus, any limestone in Britain and Ireland that contains such fossils was formed between Silurian and early Jurassic times, when these areas lay in low-latitude subtropical waters. By contrast, the bones of cold-adapted mammoths and giant deer found in much younger deposits (less than a million years old) are indicative of cold, high-latitude areas. Analysis of pollen and beetle remains from fossil deposits shows evidence of species whose habitats are restricted to high, tundra-type environments, indicating climate change associated with Quaternary ice ages.

Bias in the fossil record

Fossils do not preserve a fair sample of the life of the past because fossilization favours organisms with preservable parts such as shells, bones, teeth, and certain tough plant tissues. Furthermore, the rock record is full of gaps that reflect past changes in sea level and plate movements. It also shows a bias towards the deposits of shallow-water continental-shelf seas and inland lake basins. Despite the greater size of oceans compared to land and shallow seas, very few ocean-floor rocks or fossils are preserved, because they have been subducted by plate movements (see p.18). Overall, the fossil record tends to under-represent the past life of ocean floors, terrestrial uplands, and any organisms with soft bodies (such as jellyfish), or delicate tissues (such as flowers). However, now that this is understood, palaeontologists have redressed the balance by seeking examples of under-represented environments.

SCAVENGED BONES
Other factors prevent the formation of fossils. In the natural world, almost any food item will be scavenged by some specialist organism. Today, these range from microbes to hyenas (above), whose powerful jaws destroy bones for their marrow.

DINOSAUR QUARRY
Granted National Monument status in 1915, Dinosaur Quarry in Utah, USA, is one of the world's richest sources of dinosaur bones. Paleontologists have uncovered hundreds of bones from several species of Jurassic dinosaurs in this former river channel, including those of the giant, long-necked, quadrupedal sauropods.

Key Fossil Sites

Centuries of scientific investigation have uncovered a wealth of fossil evidence at sites around the world. Some of these places have provided particularly revealing glimpses into the past, greatly advancing our understanding of evolution. A selection of these sites is featured here.

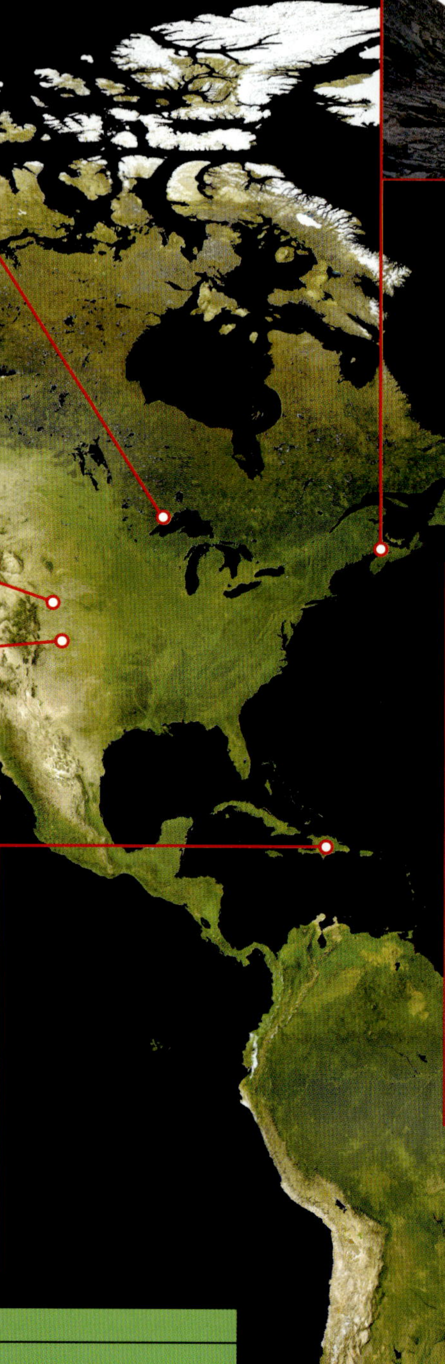

BURGESS SHALE, CANADA
This UNESCO World Heritage Site, discovered in 1909 by American palaeontologist Charles Walcott, has yielded thousands of Middle Cambrian marine fossils from around 505 million years ago, many with soft parts such as gills preserved.

GUNFLINT CHERT, CANADA
The discovery in the 1950s of 2-billion-year-old marine microfossils was the first convincing evidence of life in Precambrian times.

JOGGINS, CANADA
This UNESCO World Heritage Site is home to a 313-million-year-old Carboniferous coal forest that was once prone to wildfire. Fossils of some of the first egg-laying, lizard-like animals have been found in the hollows of trees preserved here.

RHYNIE CHERT, UK
Chert from 408-million-year-old deposits has revealed fossils of the oldest bog community, including primitive plants and early arthropods.

GREEN RIVER, USA
The 54-million-year-old lake deposits of the Green River formation preserve a wide range of Eocene fossils, from turtles, snakes, and fish to plants and insects.

JURASSIC COAST, UK
Fossils of Jurassic marine and flying reptiles were first found here in the early nineteenth century by the Anning family of professional collectors. A UNESCO World Heritage Site, it preserves Mesozoic strata and fossils in coastal cliffs.

MORRISON FORMATION, USA
Late Jurassic strata (148–155 million years old) have yielded a succession of terrestrial floodplain fossils, including the bones and footprints of giant sauropods such as *Diplodocus* and *Apatosaurus*, as well as the plants they ate.

LA BREA TAR PITS, USA
Pleistocene-age tar trapped thousands of animals, from sabre-tooth cats to condors, perfectly preserving their bones, often with entire skeletons.

DOMINICAN AMBER
Paleogene-age plant resin captured a huge number of tropical forest insects, sometimes leaving their original colouring intact. These range from ants and bees to weevils and cockroaches – all alive around 30–40 million years ago.

KEY STUDY

ANTARCTIC FOSSIL SITES

Antarctica was once part of the Gondwana supercontinent. In Permian times, the remains of *Glossopteris* plants that had flourished in the region were transformed into the coal that is found today in the Antarctic Mountains. Such deposits continued into the Triassic Period; the remains of *Lystrosaurus*, a mammal-like reptile, have been found alongside cycad and ginkgo plant fossils. By Cretaceous times, when Gondwana began to break apart, Antarctica moved near its present polar position. Warm climates worldwide allowed the forests to grow even in these high latitudes. They were populated with plant-eating dinosaurs despite the long, dark winter months.

SANTANA, BRAZIL
The Cretaceous Santana and older Crato formations between them have yielded fossils from a diverse group of fish, insects, and plants, including examples of preserved soft tissue. The land is quarried in places for its fossils; many are sold commercially.

MESSEL PIT, GERMANY,
The 52-million-year-old fossils of this UNESCO World Heritage Site record Eocene life in and around a volcanic lake, including primitive horses and other extinct mammals.

SOLNHOFEN, GERMANY
Solnhofen's limestone is a UNESCO World Heritage Site, protected due to its abundant, beautifully preserved Jurassic fossils of life in and around a shallow lagoon. These range from the primitive bird *Archaeopteryx* to small dinosaurs and fish.

BIOGRAPHY

XU XING

The Chinese palaeontologist Xu Xing (b.1969) has been one of the most active researchers of the remarkable new dinosaurs found in China during the past few decades. With Chinese colleagues and Western specialist collaborators, Xu Xing has described over 30 new genera of dinosaurs and related fossils. These include feathered dinosaurs such as *Dilong*, a 1.6m- (5¼ft-) long relative of *Tyrannosaurus rex*; *Microraptor*, an 80cm- (31in-) long dromaeosaur with wing-like feathers on its arms and legs; and *Mei*, the 50cm- (19½in-) long troodontid found in a bird-like sleeping position.

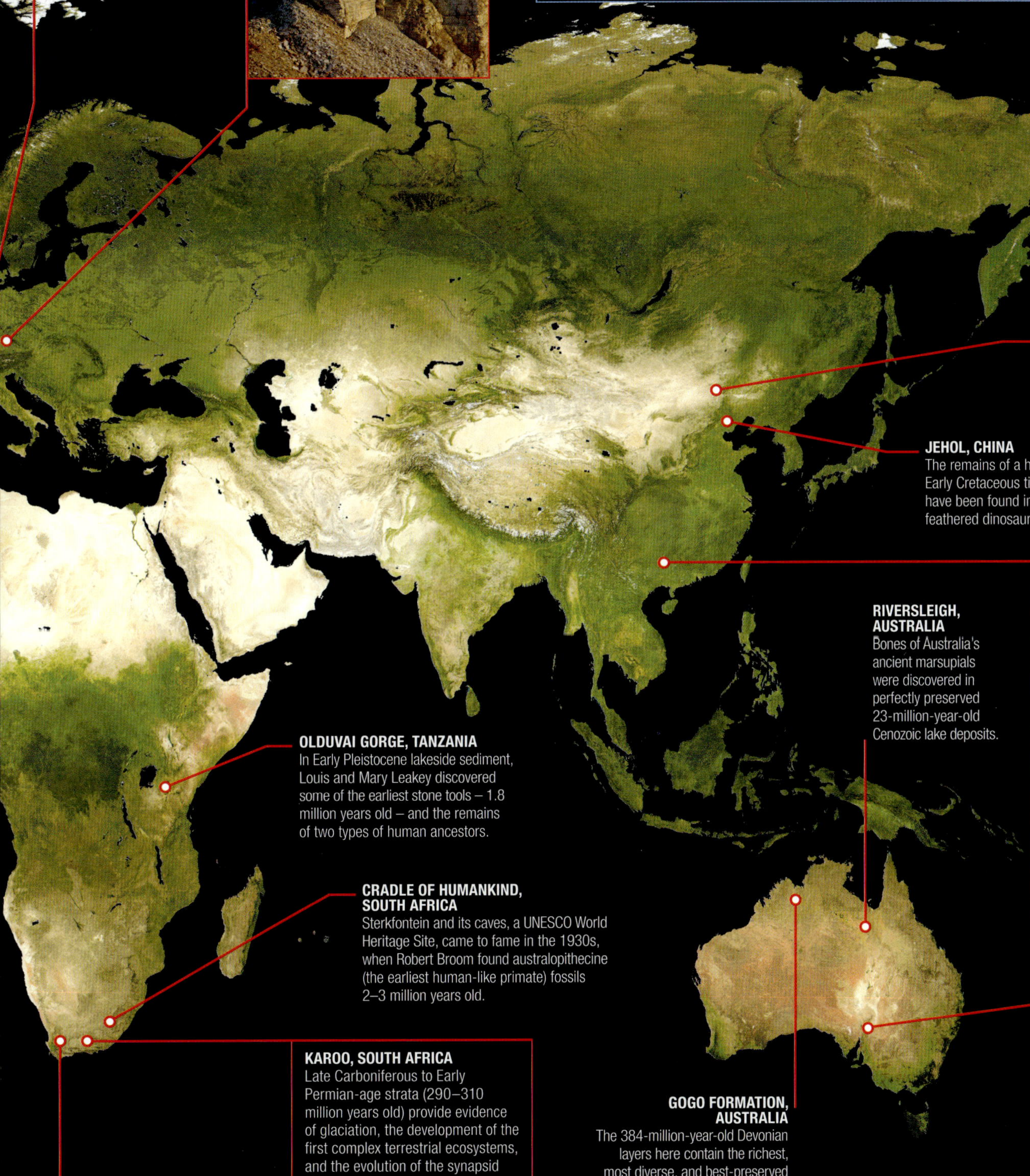

KOLYMA RIVER, RUSSIA
The entire body of a baby mammoth, 40,000 years old and still covered in hair, was found frozen in the permafrost here by a Siberian gold miner in 1977.

GOBI DESERT, MONGOLIA
Late Cretaceous deposits discovered by Roy Chapman Andrews in the 1920s have since yielded thousands of exquisite fossils, including entire dinosaur skeletons and tiny mammals.

JEHOL, CHINA
The remains of a huge range of animals from Early Cretaceous times (128 million years ago) have been found in Liaoning Province, including feathered dinosaurs, birds, and early mammals.

DOUSHANTUO, CHINA
Late Proterozoic acritarch spores (below) are among the 565–590-million-year-old microfossils to have been found here, as well as seaweeds, sponges, and cnidarians (a group that includes jellyfish, anemones, and corals).

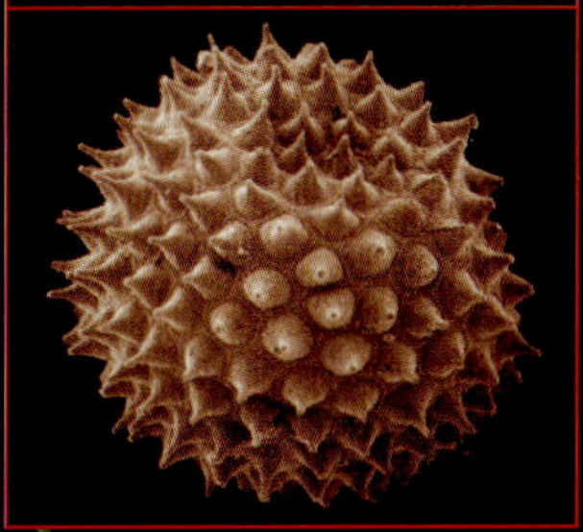

RIVERSLEIGH, AUSTRALIA
Bones of Australia's ancient marsupials were discovered in perfectly preserved 23-million-year-old Cenozoic lake deposits.

OLDUVAI GORGE, TANZANIA
In Early Pleistocene lakeside sediment, Louis and Mary Leakey discovered some of the earliest stone tools – 1.8 million years old – and the remains of two types of human ancestors.

CRADLE OF HUMANKIND, SOUTH AFRICA
Sterkfontein and its caves, a UNESCO World Heritage Site, came to fame in the 1930s, when Robert Broom found australopithecine (the earliest human-like primate) fossils 2–3 million years old.

EDIACARA HILLS, AUSTRALIA
Australian geologist Reg Sprigg found 542–575-million-year-old fossil jellyfish *Mawsonites spriggi* (below) here in the 1940s – the first example of this Late Proterozoic creature. Other specimens have since been discovered worldwide.

KAROO, SOUTH AFRICA
Late Carboniferous to Early Permian-age strata (290–310 million years old) provide evidence of glaciation, the development of the first complex terrestrial ecosystems, and the evolution of the synapsid ancestors of today's mammals.

GOGO FORMATION, AUSTRALIA
The 384-million-year-old Devonian layers here contain the richest, most diverse, and best-preserved fish fossils of this age in the world. They record the early evolution of jawed fish, including the lobe-finned ancestors of tetrapods.

SOOM SHALE, SOUTH AFRICA
Here 450-million-year-old cold-water deposits contain fossils of some unique Ordovician animals, including the giant conodont *Promissum*: a 40cm- (16in-) long, eel-shaped, primitive jawless vertebrate.

The Geological Timescale

The geological timescale is a scheme for dividing the history of Earth into named units. It is a hierarchical system, in which the largest divisions are called eons, followed in order of size by eras, periods, epochs, and ages. A standard timescale is an essential tool for geologists – working without one would be like studying history without a calendar. The refinement of the timescale using modern dating methods has allowed geologists to construct an accurate chronology of the evolution of life and other events in Earth's history.

The history of the timescale

Early ideas about geological time were driven by curiosity and the practicalities of mining. In the late eighteenth century, German mineralogist Abraham Werner proposed a succession of rocks deposited by a universal flood: first, primitive igneous rocks were laid down, followed by secondary sedimentary layers (strata), then tertiary surface deposits. However, in the nineteenth century, scientists realized that strata could be characterized by their fossils. They used this knowledge to draw the first geological maps, which included vertical sections showing the relative timescale for deposits of successive rock types in each country – such as coal and chalk, which are widespread in Europe. Although attempts had been made to establish Earth's geological age, there was still no means of dating the rocks themselves. The knowledge that fossil layers could be matched up across national borders led to the development of internationally recognized divisions of geological time. Fossil strata were divided into a series of periods from the Cambrian to the Quaternary, with older rocks classified as Precambrian. Divisions of smaller units (epochs and ages) and larger ones (eras and eons) were added later.

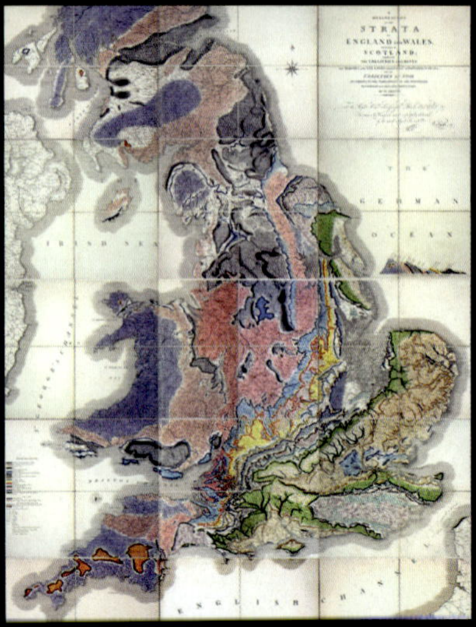

THE FIRST GEOLOGICAL MAP
The first modern geological map of an entire country was compiled by the English geologist and surveyor William Smith in 1815. Smith pioneered the use of fossils to establish the chronology of strata.

The modern timescale

The modern geological timescale provides an internationally recognized chronology of Earth's history for over 3.8 billion years. The timescale allows a geologist to go anywhere in the world, examine the strata, identify the fossils within them, and give them an approximate age within the global scheme. It also lets geologists communicate with the confidence that they are referring to the same events, strata, and time periods. The modern timescale combines a number of schemes: lithostratigraphic (based on changing sedimentary rock types and sequences); biostratigraphic (evolving fossils); chronostratigraphic (radiometric dating); magneto-stratigraphic (the changing polarity of Earth's magnetic field); and the history of global changes in ocean and atmospheric chemistry preserved in the sedimentary rock record. On some timescales, the time before the Archean eon is recognized as an informal unit known as the Hadean. The terms "Upper", "Middle", and "Lower" may also be replaced by "Early", "Middle", and "Late".

PRECAMBRIAN

EON	ERA	PERIOD	MILLIONS OF YEARS AGO
ARCHEAN	EOARCHEAN		4,000
ARCHEAN	PALEOARCHEAN		3,600
ARCHEAN	MESOARCHEAN		3,200
ARCHEAN	NEOARCHEAN		2,800
PROTEROZOIC	PALEOPROTEROZOIC	Siderian	2,500
PROTEROZOIC	PALEOPROTEROZOIC	Rhyacian	2,300
PROTEROZOIC	PALEOPROTEROZOIC	Orosirian	2,050
PROTEROZOIC	PALEOPROTEROZOIC	Statherian	1,800
PROTEROZOIC	MESOPROTEROZOIC	Calymmian	1,600
PROTEROZOIC	MESOPROTEROZOIC	Ectasian	1,400
PROTEROZOIC	MESOPROTEROZOIC	Stenian	1,200
PROTEROZOIC	NEOPROTEROZOIC	Tonian	1,000
PROTEROZOIC	NEOPROTEROZOIC	Cryogenian	850
PROTEROZOIC	NEOPROTEROZOIC	Ediacaran	635
			542

EON	ERA	PERIOD	EPOCH	AGE	MILLIONS OF YEARS AGO
PHANEROZOIC	PALEOZOIC	Carboniferous	Mississippian – Lower	Tournaisian	359.2
PHANEROZOIC	PALEOZOIC	Carboniferous	Mississippian – Middle	Visean	345.3
PHANEROZOIC	PALEOZOIC	Carboniferous	Mississippian – Upper	Serpukhovian	328.3
PHANEROZOIC	PALEOZOIC	Carboniferous	Pennsylvanian – Lower	Bashkirian	318.1
PHANEROZOIC	PALEOZOIC	Carboniferous	Pennsylvanian – Middle	Moscovian	311.7
PHANEROZOIC	PALEOZOIC	Carboniferous	Pennsylvanian – Upper	Kasimovian	307.2
PHANEROZOIC	PALEOZOIC	Carboniferous	Pennsylvanian – Upper	Gzhelian	303.4
PHANEROZOIC	PALEOZOIC	Permian	Cisuralian	Asselian	299.0
PHANEROZOIC	PALEOZOIC	Permian	Cisuralian	Sakmarian	294.6
PHANEROZOIC	PALEOZOIC	Permian	Cisuralian	Artinskian	284.4
PHANEROZOIC	PALEOZOIC	Permian	Cisuralian	Kungurian	275.6
PHANEROZOIC	PALEOZOIC	Permian	Guadalupian	Roadian	270.6
PHANEROZOIC	PALEOZOIC	Permian	Guadalupian	Wordian	268.0
PHANEROZOIC	PALEOZOIC	Permian	Guadalupian	Capitanian	265.8
PHANEROZOIC	PALEOZOIC	Permian	Lopingian	Wuchiapingian	260.4
PHANEROZOIC	PALEOZOIC	Permian	Lopingian	Changhsingian	253.8
PHANEROZOIC	MESOZOIC	Triassic	Lower	Induan	251.0
PHANEROZOIC	MESOZOIC	Triassic	Lower	Olenikian	249.5
PHANEROZOIC	MESOZOIC	Triassic	Middle	Anisian	245.9
PHANEROZOIC	MESOZOIC	Triassic	Middle	Ladinian	237.0
PHANEROZOIC	MESOZOIC	Triassic	Upper	Carnian	228.7
PHANEROZOIC	MESOZOIC	Triassic	Upper	Norian	216.5
PHANEROZOIC	MESOZOIC	Triassic	Upper	Rhaetian	203.6
PHANEROZOIC	MESOZOIC	Jurassic	Lower	Hettangian	199.6
PHANEROZOIC	MESOZOIC	Jurassic	Lower	Sinemurian	196.5
PHANEROZOIC	MESOZOIC	Jurassic	Lower	Pliensbachian	189.6
PHANEROZOIC	MESOZOIC	Jurassic	Lower	Toarcian	183.0
PHANEROZOIC	MESOZOIC	Jurassic	Middle	Aalenian	175.6
PHANEROZOIC	MESOZOIC	Jurassic	Middle	Bajocian	171.6
PHANEROZOIC	MESOZOIC	Jurassic	Middle	Bathonian	167.7
PHANEROZOIC	MESOZOIC	Jurassic	Middle	Callovian	164.7
PHANEROZOIC	MESOZOIC	Jurassic	Upper	Oxfordian	161.2
PHANEROZOIC	MESOZOIC	Jurassic	Upper	Kimmeridgian	155.6

Dating the timescale

By the mid-nineteenth century, geologists realized that the rock layers that form the stratigraphic record of Earth's history must have taken hundreds of millions of years to accumulate, given that they are many tens of kilometres thick. Yet there was still no reliable means of dating them; for example, British physicist William Thomson's empirical methods showed that Earth was probably 20–100 million years old. The discovery of radioactivity demonstrated why Thomson's measures were far too short. Radiometric dating made it possible to calculate dates for the crystallization of certain minerals. In 1904, Ernest Rutherford obtained the first mineral-based radiometric age of 500 million years; by 1911 the first radiometric timescale had been compiled. Today's timescale, based on thousands of measurements, has been linked to the stratigraphic rock record to produce an International Stratigraphic Chart, with radiometric-based dates applied to the recognized stratigraphic boundaries.

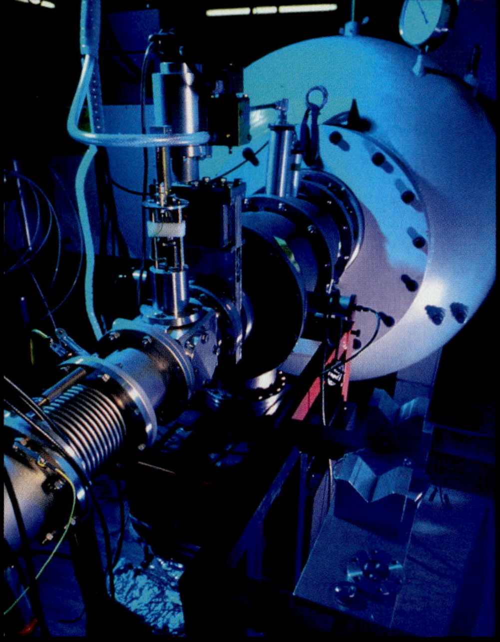

LINEAR ACCELERATOR
This equipment can determine the number of atoms of radiocarbon isotope 14C, also known as carbon-14, in carbon-containing materials up to 50,000 years old, which allows them to be dated.

SCIENCE

MAGNETOSTRATIGRAPHY

In recent decades, a new kind of chronology has been developed, based on periodic reversals in the polarity of Earth's magnetic field. When some igneous and sedimentary rocks form, iron-rich minerals in the rocks align themselves to the prevailing magnetic field. By combining studies of the magnetic alignment of rocks on land with radiometric dating, geologists have produced a timescale of magnetic reversals. This timescale has been used to date rocks from the ocean floor. In particular, the dating of rocks of similar magnetism either side of spreading ridges (see below) has provided key evidence to support the theory of plate tectonics.

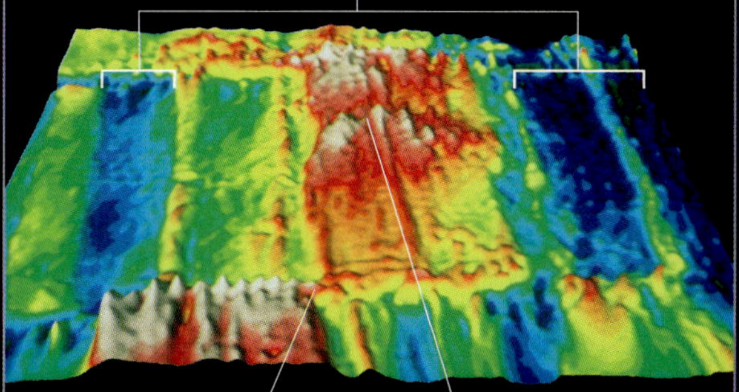

EON	ERA	PERIOD	EPOCH	AGE	MILLIONS OF YEARS AGO
PHANEROZOIC	PALEOZOIC	Cambrian	Lower Series	Fortunian	542
				Stage 2	528
			Series 2	Stage 3	521
				Stage 4	515
			Series 3	Stage 5	510
				Drumian	506.5
				Guzhangian	503
			Furongian	Paibian	499
				Stage 9	496
				Stage 10	492
		Ordovician	Lower	Tremadocian	488.3
				Floian	478.6
			Middle	Dapingian	471.8
				Darriwilian	468.1
			Upper	Sandbian	460.9
				Katian	455.8
				Hirnantian	445.6
		Silurian	Llandovery	Rhuddanian	443.7
				Aeronian	439.0
				Telychian	436.0
			Wenlock	Sheinwoodian	428.2
				Homerian	426.2
			Ludlow	Gorstian	422.9
				Ludfordian	421.3
			Pridoli		418.7
		Devonian	Lower	Lochkovian	416.0
				Pragian	411.2
				Emsian	407.0
			Middle	Eifelian	397.5
				Givetian	391.8
			Upper	Frasnian	385.3
				Famennian	374.5
					359.2

EON	ERA	PERIOD	EPOCH	AGE	MILLIONS OF YEARS AGO
PHANEROZOIC	MESOZOIC	Cretaceous	Lower	Berriasian	145.5
				Valanginian	140.2
				Hauterivian	133.9
				Barremian	130.0
				Aptian	125.0
				Albian	112.0
			Upper	Cenomanian	99.6
				Turonian	93.6
				Coniacian	88.6
				Santonian	85.8
				Campanian	83.5
				Maastrichtian	70.6
	CENOZOIC	Paleogene	Paleocene	Danian	65.5
				Selandian	61.1
				Thanetian	58.7
			Eocene	Ypresian	55.8
				Lutetian	48.6
				Bartonian	40.4
				Priabonian	37.2
			Oligocene	Rupelian	33.9
				Chattian	28.4
		Neogene	Miocene	Aquitanian	23.03
				Burdigalian	20.43
				Langhian	15.97
				Serravallian	13.82
				Tortonian	11.608
				Messinian	7.246
			Pliocene	Zanclean	5.332
				Piacenzian	3.600
		Quaternary		Gelasian	2.588
			Pleistocene	Calabrian	1.806
				Ionian	0.781
				Upper	0.126
			Holocene		0.0117

LAYERS OF TIME
Over what could be as little as just a few million years, the waters of the Colorado River, USA, carved out the Grand Canyon. Cutting down about 1,800m (5,900ft) through sedimentary rock layers of the Colorado Plateau, it has revealed a glimpse of Earth history that extends back about 1.8 billion years into the depths of Precambrian time.

FE

ON EARTH

ARCHEAN

Archean

The Archean Eon represents a period in Earth's history from 4 billion to 2.5 billion years ago. While these are arbitrarily chosen figures, the beginning of the Eon coincides with the end of the so-called "Great Cometary Bombardment" of the Earth, while the end coincides with the "Great Oxygenation Event", when the fundamental nature of the Earth's atmosphere changed forever.

KOMATI RIVER VALLEY
Rocks from some of Earth's oldest volcanic crust are found here in the Barberton Greenstone Belt of southern Africa, renowned for its gold deposits.

Oceans and continents

The beginning of the Archean was marked by the end of a bombardment of Earth by comets or asteroids. Determining the positions of continents and oceans at this distance in time is an extremely difficult business. Very few rocks of this age remain, and those that do have, in many cases, been profoundly altered. Indirect evidence for the earliest rocks comes from crystals of the mineral zircon, eroded from granites and redeposited in younger sedimentary rocks. These formed during the early Archean, or even in the preceding Hadean period. The oldest of these is a single crystal from Western Australia that has been dated as more than 4 billion years old, which supports the argument that at this extremely early period in the Earth's history rocks were crystallizing, slabs of continental crust were forming and, just as significantly, there was water on the planet. However, it may not have been until about 3 billion years ago that a land mass of any appreciable size was formed. Before this, fledgling continents probably came together and were then rapidly eroded or subducted, leaving only a few traces in the rock record. The first continent from which rocks still exist has been called "Ur". It comprises the ancient cratons, also known as shields, of central-southern Africa, the Pilbara region of Western Australia, India, and small parts of Antarctica. About half a billion years later, at the end of the Archean, a second continent called "Arctica" formed, consisting of the cratons of present-day Canada, Greenland, and Siberia. Reconstructing where these early continents were situated on the globe is very difficult. During the late Archean, Ur may have been situated in high latitudes, whereas Arctica was in low latitudes. The Archean oceans are likely to have been more extensive than they are today because the continents were smaller. They also differed from modern oceans in that they were about 1.5 to 2 times more saline. Moreover, they were devoid of dissolved oxygen. The only life that oceans could support was in the form of single-celled organisms, probably anaerobic bacteria and archaea, which do not require oxygen. Evidence for their existence comes from fossils of filamentous bacteria and structures called stromatolites found in rocks as old as 3.49 billion years in present-day Australia and South Africa (see opposite page).

PILBARA REGION, AUSTRALIA
Some of the world's oldest sedimentary and volcanic rocks, dating back almost 3.5 billion years, are found in the mineral-rich Pilbara Region in northwest Australia.

FOSSILIZED MUD POOL
Gas escape structures in a fossilized mud pool in the Barberton Greenstone Belt show that heat-loving bacteria colonized early oceanic crust, as early as 3.5 billion years ago.

Climate

For much of the Archean, the Earth's atmosphere is thought to have been devoid of oxygen. Instead, it was composed primarily of nitrogen, methane, and carbon dioxide. The Sun was much weaker then, emitting less radiation, so had it not been for these greenhouse gases, the world would have been a ball of frozen rock. In fact, research indicates that an atmosphere rich in greenhouse gases helped to create extremely high global temperatures, especially early in this period. These could have been within the range 55–80°C (131–176°F). Ocean temperatures would have been correspondingly high, and the only forms of life capable of living in the hot, salty seas were thermophilic (heat-loving) anaerobic bacteria and archaea. Although the climate was very hot overall, about 2.9 billion years ago temperatures appear to have fallen drastically for a relatively short time and the Earth may even have experienced its first glaciation. Geological evidence suggests that during the last few hundred million years of the Archean temperatures returned to their previously high levels.

VOLCANIC GASES
Volcanoes, many of them submarine, erupted hot gases that formed the early atmosphere, but there was little free atmospheric oxygen until the end of the Archean.

STROMATOLITES
Living stromatolites can be found forming today in Shark Bay, Western Australia, where they are commonly regarded as the work of cyanobacterial communities.

Life in the Archean

Any life forms that may have lived in the Archean were likely to have been very tough, and very small, like modern eubacteria and archaebacteria. Recent work on the DNA of these modern forms has shown that they are probably the most primitive of life forms found today. They are known to use the energy stored within the rocks and minerals of the Earth as sources from which to create the basic building blocks of life. Fossils from this period are very rare, but of supreme importance for our understanding of how the Earth became a habitable planet. Some of the earliest evidence for cells in rocks of the Archean comes from South Africa and Western Australia. These ancient rocks, dated to 3.49 to 3.43 billion years ago, show that simple bacteria were probably living in settings lacking in oxygen, but high in silica and metallic sulphides, especially volcanic environments. All microfossils of this age are rather controversial, but simple filamentous forms have been found in cherts by the Komati River, southern Africa, and around Strelley Pool in the Pilbara region of Australia. Recently, tiny tubular microborings, which may well have been produced by bacteria, have been discovered in basaltic pillow lavas. In Sulphur Springs, Australia, slightly younger clusters of thread-like fossils preserved in iron sulphide (pyrite) have been found.

BLACK SMOKER
Columns of black smoke rich in iron sulphides pour out of a vent on the sea floor where temperatures can reach over 400°C (752°F). Life may have emerged and thrived around such vents, which were common in the Archean.

PILLOW LAVA
These rocks on the edge of the Kruger National Park in South Africa are about 3.49 billion years old. The pillow lava has intervening black chert that contains tiny structures resembling bacterial microfossils.

These deposits are about 3.2 billion years old, and comprise sulphide-rich cherts that formed around submarine hot springs, much like those of modern black smokers. The fossils themselves comprise tiny cylindrical threads about one micron wide, trapped between layers of iron sulphide that may have formed part of a black smoker chimney. A second line of evidence is from stromatolites. Microbial stromatolites (see above) often form distinctly crinkly layers, and stromatolites of this kind have only been found in the rock record after about 3 billion years ago. Some of the most ancient examples are known from the Strelley Pool Chert, dated to about 3.43 billion years ago, although these do not show the crinkly layers consistent with a microbial origin. It is only after about 2.6 billion years ago that cyanobacterial fossils appeared, leading to the build-up of atmospheric oxygen by means of photosynthesis. Before this, life may therefore have been widespread, but it seems to have consisted largely of forms living on the enormous supplies of internal energy from the young Earth.

SIMPLE FILAMENT
This microfossil from the Pilbara region of Australia shows a folded filament with cell-like structures made of carbon, comparable to modern iron bacteria.

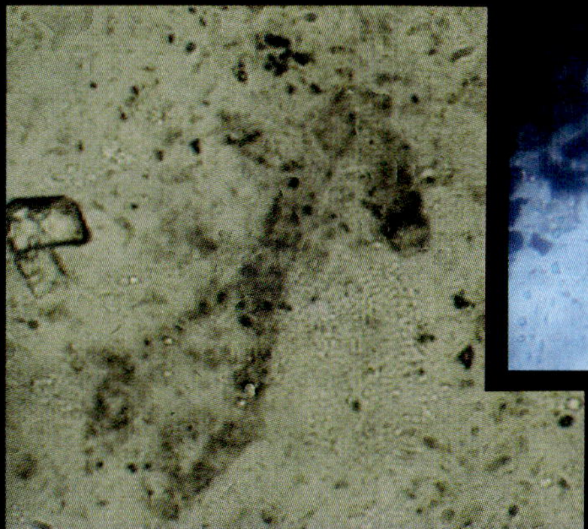

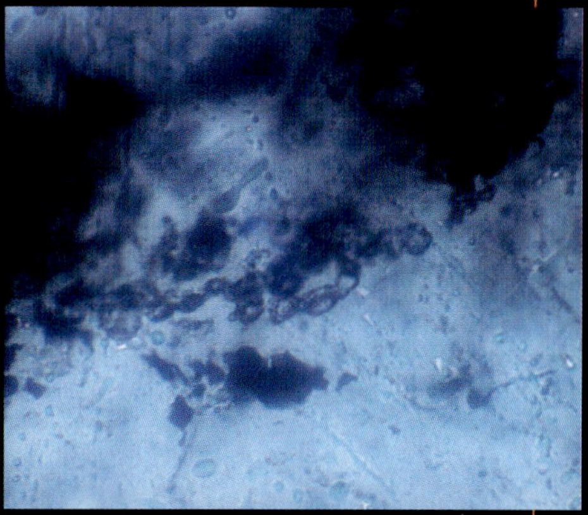

BARREL-SHAPED CELLS
These microfossils from the Strelley Pool Chert in the Pilbara region show a chain of cell-like structures similar to modern purple bacteria.

PROTEROZOIC

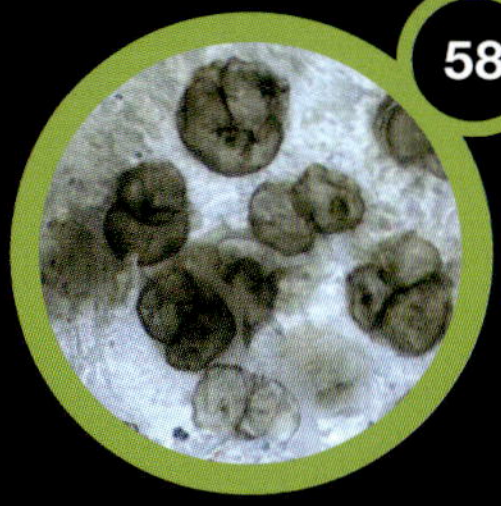

58 Microscopic life

60 Invertebrates

Proterozoic

The Proterozoic Eon covers an unimaginable stretch of time from 2.5 billion to 542 million years ago. The activity of photosynthetic microbes, begun in the Archean, transformed Earth into a planet with an oxygenated atmosphere and oceans. The scene was set for the subsequent evolution of life in the seas, on land, and in the air. Bacteria and archaea were joined by the first simple animals and plants.

Oceans and continents

Over the two billion years of the Proterozoic, continents evolved and fragmented, but left evidence in the rocks that has enabled experts to reconstruct how the globe might have looked. A world map can be pieced together by dating early mountain chains, thrown up by continental collisions and long since eroded. The continents were on the move. Ur continued to grow until 1.5 billion years ago, absorbing today's Zimbabwe, northern India, and the Yilgarn block of western Australia. About 1.6 to 1.3 billion years ago, Arctica accreted more continental blocks to form a larger continent called Nena. A third continent, Atlantica, was formed about 2 billion years ago, adding the Tanzanian block about 1.3 billion years ago. Ur, Nena, and Atlantica may have come together to form the continent of Columbia, before separating and reuniting about 1 billion years ago to form the supercontinent of Rodinia. This in turn split 300 million years later, opening the Panthalassic Ocean. By the end of the Proterozoic, these isolated continental blocks had rejoined in the supercontinent of Pannotia. As continents were continually reshaped, the erosion of new mountain chains deposited sediments rich in minerals in the oceans. Micro-organisms in the shallower seas gradually changed the composition of the atmosphere, and by the end of the Proterozoic soft-bodied animals began to leave their imprints on the sea floor.

STROMATOLITES
In Shark Bay, Australia, bacterial colonies called stromatolites release oxygen as a by-product when they photosynthesize. Stromatolites flourished in shallow waters surrounding Proterozoic continents.

GRANITE DOMES
Pink granite outcrops at Enchanted Rock in Texas, USA, date back to the Proterozoic. This ancient batholith (body of intrusive igneous rock) is an estimated one billion years old.

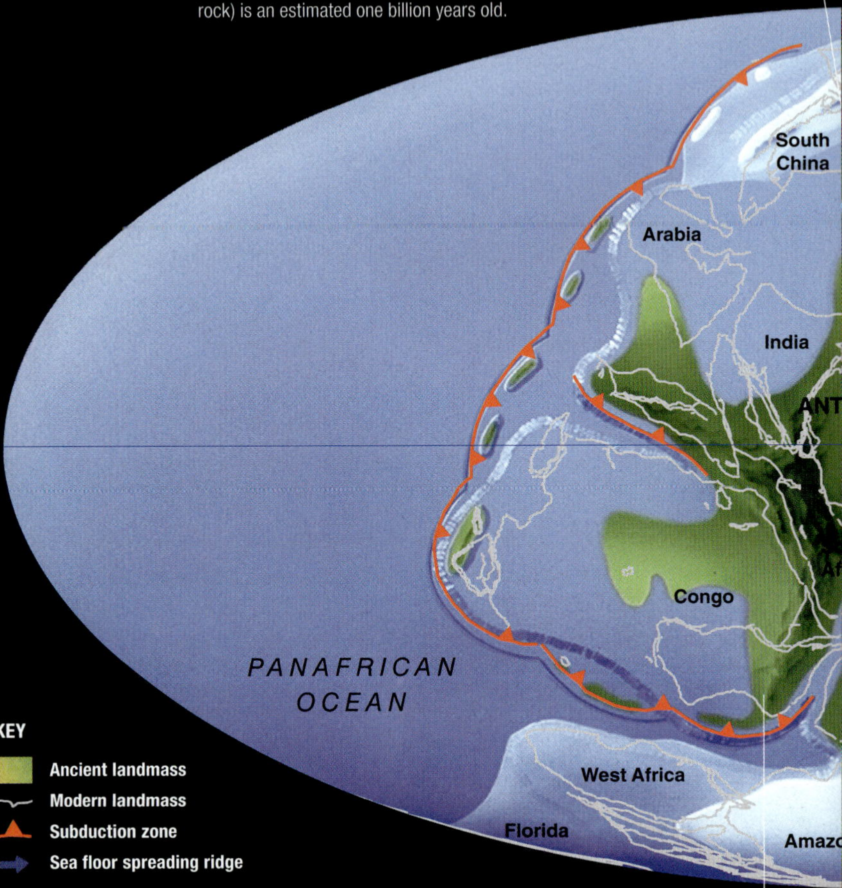

PROTEROZOIC WORLD MAP
Continents formed, joined, and moved apart in the Proterozoic period. When Rodinia divided, its North American section moved south to the South Pole.

PALEOPROTEROZOIC

MYA 2500 2000

MICROSCOPIC LIFE

● **2400** First cyanobacteria (photosynthesizing single-celled micro-organisms without nucleii – prokaryotes)

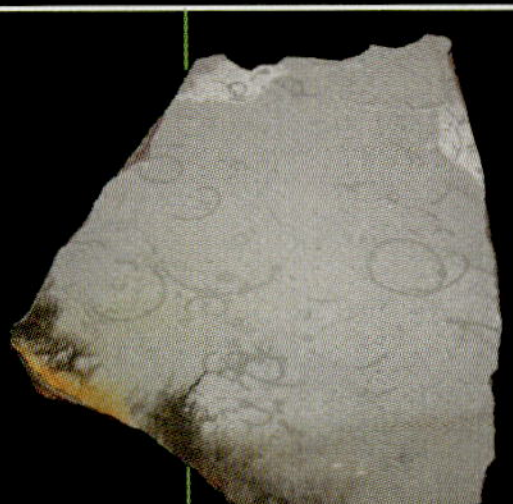

GRYPANIA

● **1850** First eukaryote (organism with cell or cells containing nucleus and other organelles)

2700–2300 Oxygenation event

RED ALGAE
Red algae on the surface of Lake Magadi, Kenya, bear a close resemblance to *Bangiomorpha*, the oldest known multicellular genus in the fossil record, from a locality in arctic Canada.

Climate

The early Proterozoic was probably a little cooler than the Archean, but global temperatures may still have been around 40°C (104°F). They appear to have declined rapidly, resulting in a major glaciation event between 2.2 and 2.4 billion years ago. This dramatic cooling may have been triggered by the evolution of photosynthetic cyanobacteria. After a protracted warm phase from 2.2 billion to 950 million years ago, Earth was plunged into another series of major glaciations, many of which may also have been global. Evidence for this has been found on all continents today. Glacial evidence on some continents once located in low latitudes has led to the scenario of a "Snowball Earth", frozen from pole to pole. Yet, during the last few hundred million years of the Proterozoic, conditions were often tropical, especially at low latitudes. Intense glacial periods, perhaps for a million years, were followed by "greenhouse" periods when global temperatures soared. At the end of the Proterozoic, temperatures had returned to pre-glaciation levels.

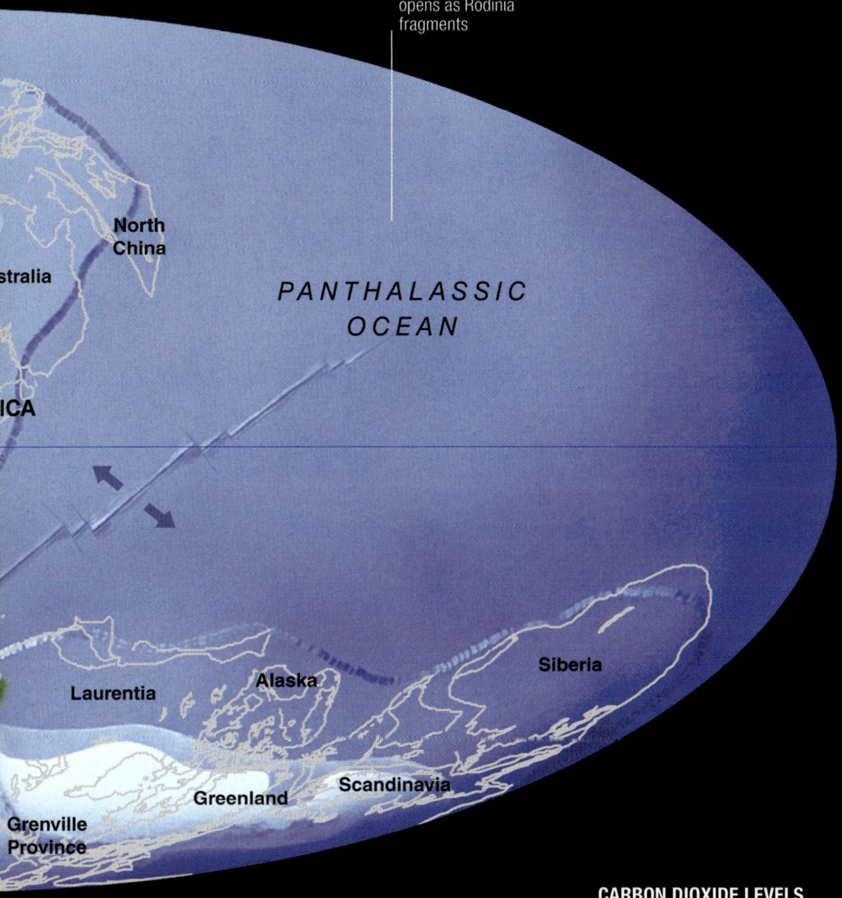

BRACHINA GORGE
A sequence 4 km (2.5 miles) thick of tillites (glacial sediments) in the Brachina Gorge, Flinders Ranges, South Australia, provides evidence of extensive glaciation during the late Proterozoic.

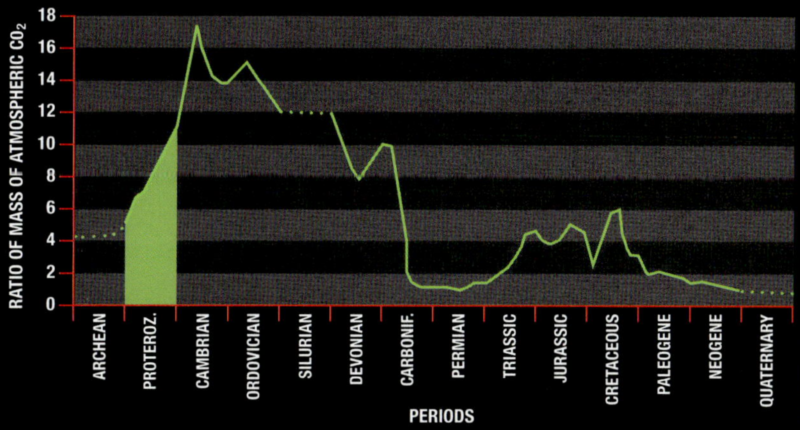

CARBON DIOXIDE LEVELS
After a long warm phase, global temperatures declined. During the course of the Proterozoic, carbon dioxide levels increased.

MESOPROTEROZOIC | **NEOPROTEROZOIC**

00 | 1000 | 500 | MYA

- **1500** First structurally complex eukaryote, possible fungus
- **1400** Great increase in diversity of stromatolites

STROMATOLITES

- **1200** Microscopic colonization of land; first multicellular red algae (below)

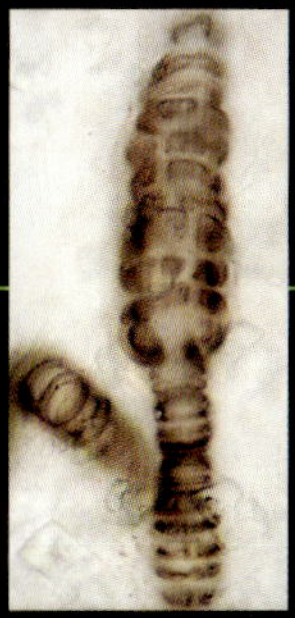

BANGIOMORPHA

- **1100** Earliest dinoflagellate (single-celled eukaryotic micro-organism with flagellae)
- **1000** First vaucherian algae (*Palaeovaucheria*)
- **750** First protozoans (single-celled non-photosynthesizing eukaryotes) e.g. *Melanocyrillium*
- **750–700** First calcareous algae
- **713–635** First indirect evidence for metazoans (multi-celled animals): chemical biomarkers suggest demosponges
- **560** First fungi

MICROSCOPIC LIFE

DICKINSONIA

- **565–540** First Ediacaran biota (e.g. *Dickinsonia*)
- **555** First anthozoans
- **550** First evidence for ctenophores (comb jellies), sponges; first anthozoans (group containing corals, sea anemones)

INVERTEBRATES

PROTEROZOIC MICROSCOPIC LIFE

During the Proterozoic, evolution was almost wholly concerned with making large improvements to the cell. Such changes included the development of a nucleus, oxygenic photosynthesis, and the ability to reproduce sexually. The resulting specialized cells became the ancestors of fungi, plants, and animals.

Early in the Proterozoic, life was entirely microscopic and single-celled. Microfossils of this age are common in rocks called stromatolites, which are fossilized, cabbage-like growths of bacteria. They can be compared with bacteria found living today in salty lagoons and around hot springs. The cells have no nucleus or evidence for sexual reproduction and are known as prokaryotes (meaning "without a nucleus"). Some Proterozoic bacteria did not need oxygen, which suggests conditions on Earth's surface were very toxic. Others resembled oxygen-producing "algae" called cyanobacteria. The earliest examples of these appeared around the same time as a rapid build-up of atmospheric oxygen, which made Earth far more habitable.

GUNFLINT CHERT
Early microfossils are preserved in cherts. These rocks contain some of the first recorded fossils. This slice of Gunflint chert shows stromatolite columns that were formed by bacteria.

Cells gradually became larger, and more diverse and specialized. These complex cells are called eukaryotes (meaning "with a nucleus"). The transformation of simple cells into these more advanced cells was probably the longest and hardest step in evolution. By the end of the Proterozoic, large, soft-bodied organisms were colonizing the sea bed.

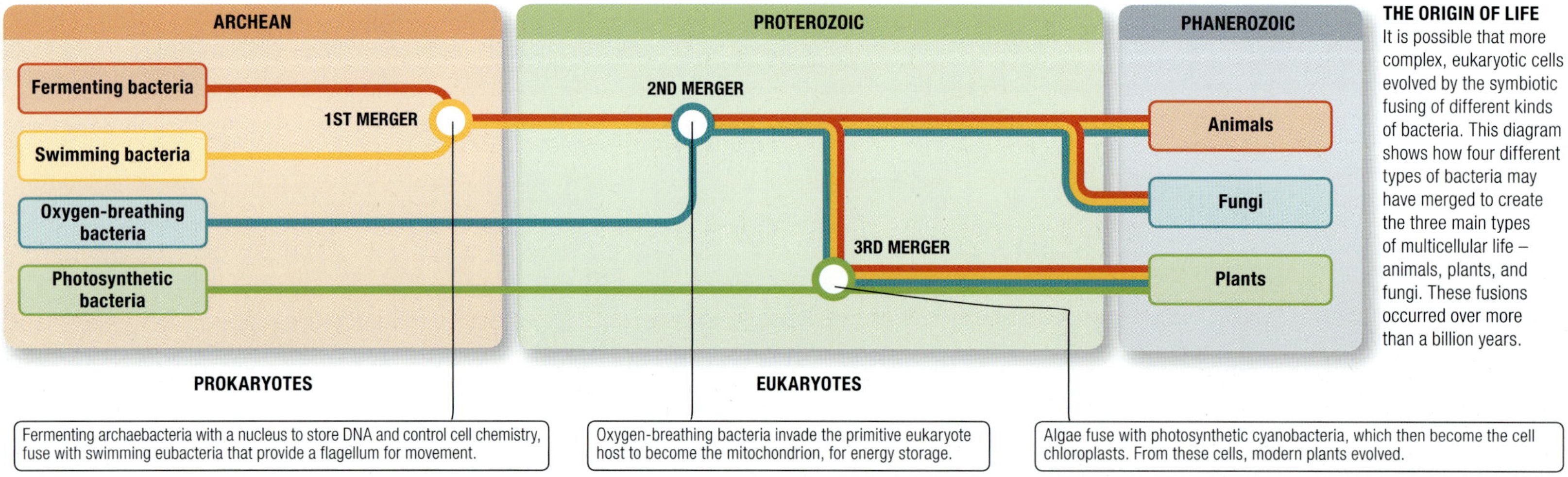

THE ORIGIN OF LIFE
It is possible that more complex, eukaryotic cells evolved by the symbiotic fusing of different kinds of bacteria. This diagram shows how four different types of bacteria may have merged to create the three main types of multicellular life – animals, plants, and fungi. These fusions occurred over more than a billion years.

GROUPS OVERVIEW

The fossil record for the Proterozoic era reveals a world dominated by microscopic bacteria. Complex eukaryotes only emerged after 1900 mya, once there was enough oxygen to support them. Later revolutions involved feeding by ingestion at around 750 mya, and the evolution of a through-gut at around 545 mya.

CYANOBACTERIA
Most early fossils are photosynthetic bacterial cells. They were less than 2 microns across, and gathered together to form microbial mats that helped to build stromatolites. They made Earth into the habitable planet we know today by ensuring that life had a constant supply of food from sunlight, and by producing oxygen.

ALGAE
Larger, more complex photosynthetic algal cells appeared after 1400 mya. These were over 20 microns wide, and evolved to form far larger colonies. They expanded the ecological zone occupied by cyanobacteria by developing leaves and root-like systems. By 450 mya, some had moved onto land, eventually evolving into the modern plant groups.

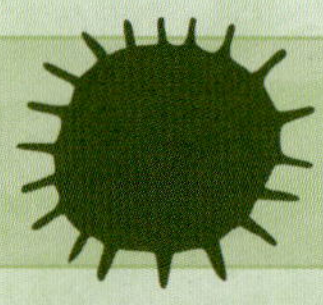

PROTOZOA
Protozoans were single cells that could ingest other creatures. They can be recognized by their tiny, flask-like shells, and did not appear until after 750 mya. The evolution of ingestion enabled more rapid recycling of the organic matter grown by late Proterozoic algae. Some protozoans merged into colonies that specialized into early fungi and animals.

SPONGES
Colonies of cells that were differentiated into specific tissues to form simple animals did not appear until the end of the Proterozoic. The earliest recognizable fossils are sponge spicules, which appeared by about 543 mya. Sponges are filter-feeders, and helped "clean-up" the water column and the sea floor at the end of the Precambrian.

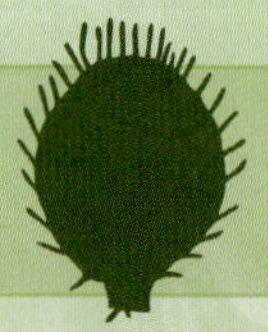

Gunflintia

GROUP Cyanobacteria
DATE Proterozoic to present
SIZE 5 microns wide
LOCATION Canada, Australia

Gunflintia is the first fossil to appear abundantly anywhere in the rock record. It was among the first oxygen-producing cyanobacteria. Photosynthetic organisms like this helped to raise the oxygen in the atmosphere and make the Earth a more habitable place for later oxygen-users such as protozoans, plants, and animals.

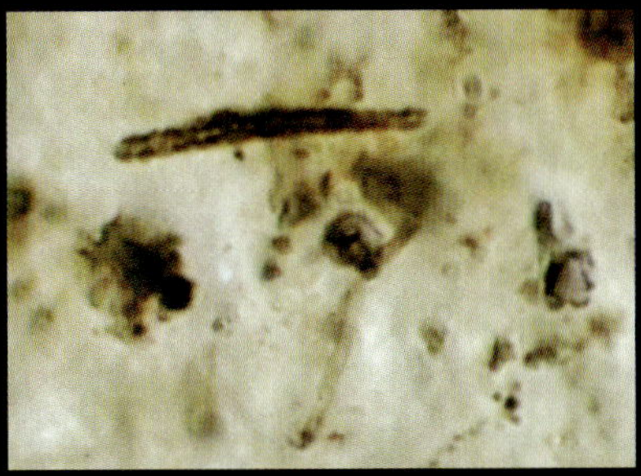

COMMONLY FOUND
Filaments of *Gunflintia* are seen here with rounded spheres of *Huroniospora*. *Gunflintia* is one of the most common microfossils in Gunflint Chert.

Eoentophysalis

GROUP Cyanobacteria
DATE Proterozoic to present
SIZE 5 microns wide
LOCATION Worldwide

Fossilized *Eoentophysalis* occurs widely in shallow-water cherty rocks from the Proterozoic. It is very similar to a modern-day coccoid cyanobacterium called *Entophysalis*, which can still be found in shallow, salty lagoons. Although bacteria have simple structures, they are extremely cosmopolitan in the way they live.

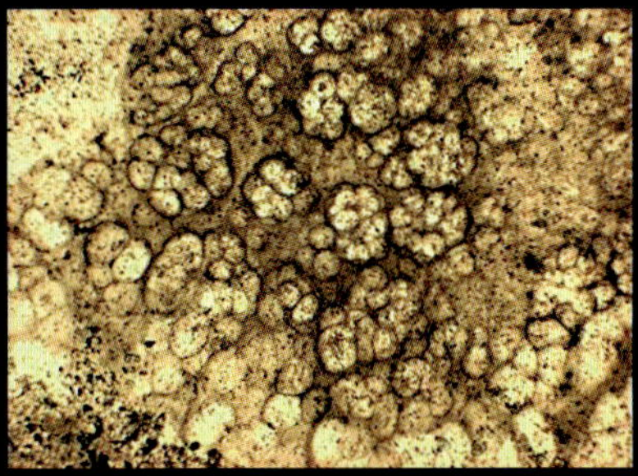

BUNCHES OF GRAPES
Not all cyanobacterial microfossils were filamentous like *Gunflintia*. Some, like *Eoentophysalis*, were arranged in packets, like bunches of grapes.

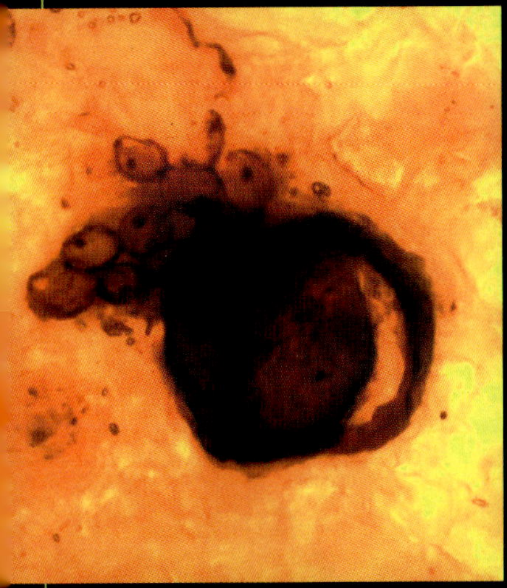

Torridonophycus

GROUP Chlorophyceae
DATE Proterozoic to present
SIZE 20 microns wide
LOCATION Scotland

This algal microfossil is seen here escaping from a round, bag-like structure called an acritarch, which helped to protect the enclosed organism from cold, drying out, or lack of oxygen. Their presence suggests the surface of the land was starting to turn green during springtime as far back as a billion years ago.

GREEN ALGA
It is thought the dark blob inside each cell could be the remains of specialized subunits (organelles), suggesting this is a eukaryotic green alga.

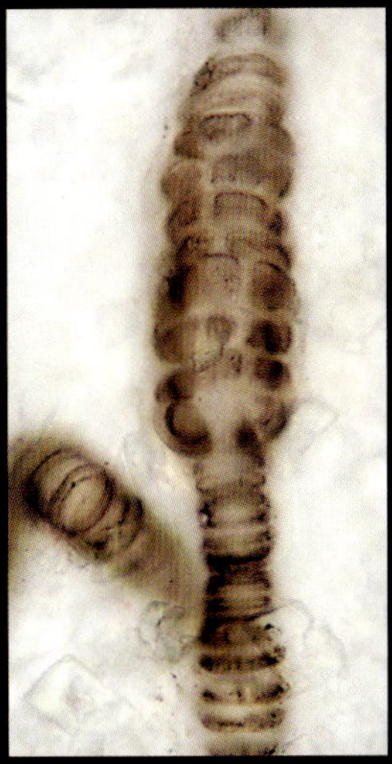

Bangiomorpha

GROUP Rhodophyceae
DATE Proterozoic to pesent
SIZE 20 microns wide
LOCATION Canada

Multicellular filaments, similar to those of modern red algae, first appeared around 1,200 million years ago. *Bangiomorpha* had specialized reproductive structures, and a primitive holdfast that attached it to the sea floor and allowed it to rise upwards towards the sunlight. The evolution of sexual reproduction and cellular differentiation can be taken to mark the long march towards the more complex body plans of plants and animals.

STACKED CELLS
Large cells, similar to those typically found in modern day red algae, are stacked like a pile of plates in this fossilized *Bangiomorpha*.

Melanocyrillium sp.

GROUP Amoebozoa
DATE Late Proterozoic to present
SIZE 60 microns long
LOCATION America, Northern Europe

Melanocyrillium can be compared to a modern group of single-celled organisms called testate amoebae. Like the amoeba, they had a nucleus, mitochondria, an elastic body wall, and pseudopods for feeding on other organisms. The addition of an outer shell provided protection against dessication and attack.

ORGANIC BAGS
These vase-shaped microfossils typically consist of flask-like bags of organic material. Some have additional mud particles or secreted silica scales.

Megasphaera

GROUP Possible Animalia
DATE Late Proterozoic
SIZE 500 microns wide
LOCATION China

Originally interpreted as green algae, these are now thought to be the eggs and embryos of early animals, or possibly the cells of sulphur-oxidizing bacteria. Despite their exquisite preservation, their affinities are still debated. They may be the embryos of sponges of the Ediacara biota, which appears just above them in the rock record.

ANIMAL EMBRYO?
Inside the sculptured wall of *Megasphaera* can be seen a cluster of cells that looks like the early stages of division in an animal embryo.

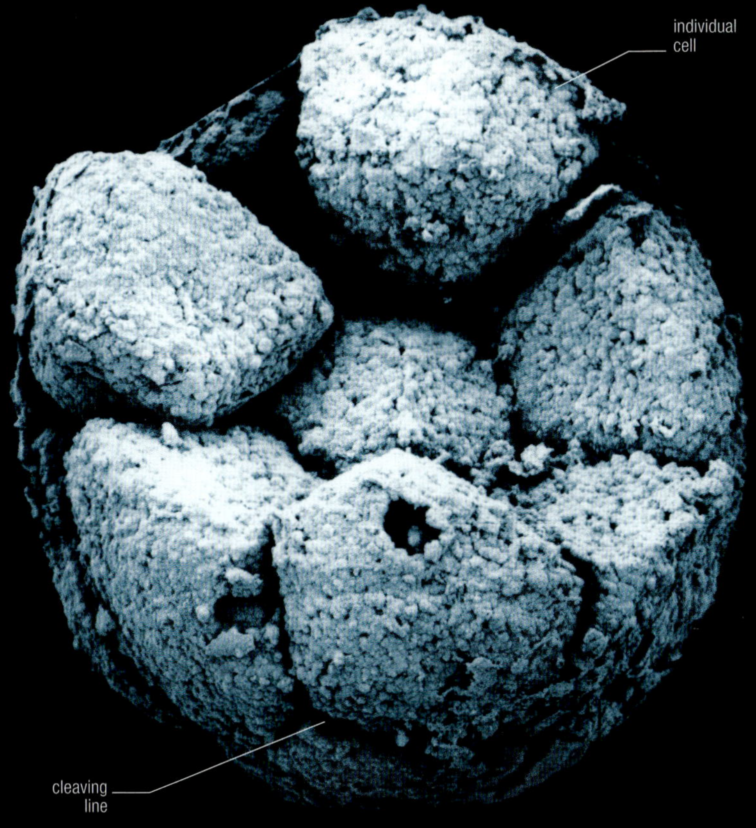

Protospongia

GROUP Porifera
DATE Late Proterozoic to Cambrian
SIZE 1mm (1/16in) long
LOCATION Worldwide

Modern sponges are widely agreed to be descendents of the earliest cell colonies that evolved into modern animals. The earliest fossil evidence for sponges are tiny sponge spicules, which first appeared very close to the end of the Proterozoic era. They probably evolved as protection, as well as support, during the great arms race at the start of the Cambrian explosion (see p.68).

DISTINCTIVE SHAPES
Some sponge spicules are very distinctive, particularly these cross-shaped forms that first appeared at the very end of the Ediacaran Period.

PROTEROZOIC INVERTEBRATES

Until 1946 the only known Proterozoic fossils were layered deposits called stromatolites, which were formed by the growth of blue-green algae. However, in 1946, the Ediacaran fauna was discovered in Australia, and an entirely new perspective on the nature of the Late Precambrian opened up.

Ediacaran fauna

The Late Precambrian was generally a time of simple ecosystems with short food chains, consisting of bacteria, algae, and protists, which were not fossilized. It was also a time of successive glaciations, some of which were so severe that they covered the Earth with ice, and gave rise to the "Snowball Earth" hypothesis. But about 580 million years ago the Ediacaran fauna (see p.63) began to flourish. This consisted of a diverse range of organisms in many shapes and sizes, most of which had a "quilted" surface and lacked a gut. Over 30 genera have been identified, but opinion is still divided as to whether they are remote ancestors of modern animals, such as jellyfish, sea-pens, or worms, or if they are the relics of a failed evolutionary experiment, unrelated to modern organisms – almost as though they were life from another planet. If this is the case then these Vendozoa, as they are known, may have had photosynthetic algae within their tissues, or fed by absorbing nutrients across their surfaces.

PUZZLING DISC
It is not known if this flat, disc-like fossil, *Cyclomedusa*, was a jellyfish or a now-extinct Vendozoa.

EDIACARA HILLS
North of Adelaide, South Australia, are the Ediacara Hills, where a range of Late Precambrian fossils, the Ediacaran fauna, were described in 1947.

Fine preservation

Apart from the Ediacara Hills in Australia, these Late Precambrian fossils have been found in many other parts of the world, including Canada, Russia, and England. The Ediacara Hills have yielded many thousands of specimens that are preserved in very fine sand – an important factor in the survival of the fossils. They appear at the interface of the thin silt and hard sandstone, which suggests that the living animals were stranded on tidal mudflats or in shallow tidal pools and preserved when the sand covered them. All Ediacaran organisms lived in shallow, well-lit waters where they could photosynthesize if necessary.

SEA-FLOOR DWELLER
The common Ediacaran fossil *Charniodiscus* had a "quilted" surface and was anchored to the sea floor, rising vertically.

Charnia

GROUP Unclassified
DATE Late Precambrian
SIZE 0.15–2m (6in–6½ft) tall
LOCATION England, Australia, Canada, Russia

In 1957, a schoolboy, Roger Mason, found a *Charnia* fossil in Late Precambrian rocks in Leicestershire, England. It was described a year later as a probable alga, but since then its similarity to modern sea pens, which are a type of octocoral, has generally been accepted. Another interpretation is that *Charnia* was a type of quilted organism that evolved in the Late Precambrian, to which the general name Vendobionta is given. *Charnia* has a bilaterally symmetrical, feather-shaped frond with a series of side branches that remain in contact with each other along their entire length. These branches are arranged at an angle of about 45 degrees to the vertical axis. They have regular subdivisions that, if the organism were similar to a sea pen, would have housed the polyps. Some specimens have been found with a stem at the base, and in the related fossil *Charniodiscus*, the stem is connected to a basal disc. This example has led to the theory that the disc-shaped fossils, *Aspidella* (see p.62) and *Medusinites*, are the anchoring discs of *Charnia*-like fronds. Although the fronds typically broke up when the organism died, the discs, which were already buried in the sediment, were fossilized. Recently, the suggestion that *Charnia* was similar to sea pens has been questioned, as the mode of budding at the tip of the colony appears to differ from that of modern sea pens.

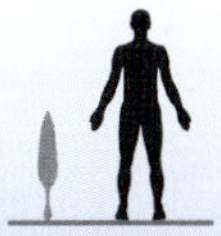

CHARNIA SP.
Like most Precambrian fossils, all we know about *Chamia* comes from impressions made by the outside of its body in soft sediments. No internal structures have been preserved.

CHARNIA IN LIFE
It has been suggested that *Chamia*'s fronds were shaped to catch sunlight for photosynthetic alga or bacteria living inside – and were, therefore, green.

Cyclomedusa

GROUP Possible Cnidarian
DATE Late Precambrian
SIZE 2.5–30cm (1–12in) across
LOCATION Australia, Russia, China, Mexico, Canada, British Isles, Norway

Cyclomedusa was found in 1946 at Ediacara in South Australia's Flinders Ranges. It was one of the first Ediacaran fossils to be described (see panel, opposite). At first misidentified as a Cambrian jellyfish, it was not dated as Precambrian until the late 1950s. Today, little is known about *Cyclomedusa*, despite it being a common fossil. It is usually circular with concentric ridges and radial markings. Odd-shaped specimens have been found with two or more centres, suggesting that the growth of one individual affected that of another nearby.

CYCLOMEDUSA RADIATA
The central boss often looks broken, which lends weight to the theory that *Cyclomedusa* was merely the anchoring disc of a larger organism.

SCIENCE
AN EARLY JELLYFISH?

Cyclomedusa's impressions may look like a round jellyfish, but they lack many of their features. True jellyfish have four-fold radial symmetry, but *Cyclomedusa* does not. *Cyclomedusa* also has no tentacles, which might be because its tentacles were so delicate. Their fragility is one reason why true jellyfish fossils are rare – so why are so many *Cyclomedusa* fossils known?

Aspidella

GROUP Possible Cnidarian
DATE Late Precambrian
SIZE 0.1–5cm (1/16–2in) across
LOCATION Canada, British Isles

When *Aspidella* was discovered in Canada in 1872, it was the first Precambrian fossil to be described. Later, geologists decided it was really a pseudofossil – a fossil-like impression produced by sedimentary processes. It was only after the Ediacaran faunas of Australia were described in the 1960s that *Aspidella* was re-examined, and geologists concluded that it was indeed a fossil. Initially classified as a jellyfish, *Aspidella* is now thought to have been the anchoring disc of a frond-like organism, such as *Charnia* (see p.61). The problem with this theory is that many more discs have been found than fronds. Some locations yield hundreds of *Aspidella* discs but no fossilized fronds at all.

ASPIDELLA SP.
The fossil is a small disc, often with a raised central boss, and radial and concentric ridges.

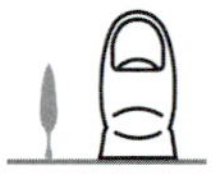

Yelovichnus

GROUP Unclassified
DATE Late Precambrian
SIZE 1–15cm (3/8–6in) long
LOCATION Canada, Russia, British Isles, China, Australia

YELOVICHNUS SP.
The raised ridges of *Yelovichnus* have to be photographed under low-angle lighting to be visible.

These enigmatic fossils were first thought to be feeding trails. Their meandering nature appeared to indicate a sophisticated method of feeding, whereby an animal, often an annelid worm or a mollusc, grazes methodically over the sea bed and removes surface food. However, better-preserved examples of *Yelovichnus* imply a different story. Unlike true feeding trails, there is no evidence of a turning point in the meander. Instead, the fossils appear to be collapsed, segmented tubes. It has more recently been suggested that they represent spirally organized organisms, possibly algae.

Dickinsonia

GROUP Possible Placozoan
DATE Late Precambrian
SIZE 1–100cm (3/8in–3ft 3in) long
LOCATION Australia, Russia

Dickinsonia is one of the most perplexing fossils found in the Ediacaran fauna. At first sight it appears to be segmented, with a distinct head and tail end. This led to the belief that it was a segmented marine worm. Hundreds of examples showing all growth stages and in various types of preservation have been collected – the largest specimens known are about one metre (3ft 3in) in length. However, no convincing gut or other internal structures have ever been found. This has led to the conclusion that *Dickinsonia* fed by absorbing food through its entire undersurface. More recently, it has been suggested that this strange animal may be a placozoan – a group of animals that is represented by only one living species. Placozoans have only four types of cell, which exist in two layers. It is possible that they may be a halfway stage between sponges and Eumetozoa, animals with true tissues and specialized cells.

No one knows which end of *Dickinsonia* was its head – or whether it had one at all.

DICKINSONIA COSTATA
Numerous segments radiated from a central groove. The larger segments at one end may have been more mature than those at the opposite end.

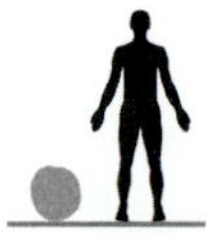

Parvancorina

GROUP Unclassified
DATE Late Precambrian
SIZE 1–2.5cm (3/8–1in) long
LOCATION Australia, Russia

Parvancorina is an Ediacaran fossil with a shield-shaped front end, which appears to be a distinct head. The body has a central axial ridge along its length and shows some evidence of segmentation. Up to 10 pairs of possible appendages have been identified on some *Parvancorina* fossils. Many show a distinct set of growth stages. A large number of specimens were found with their head shields facing in the direction of the water current, implying a type of feeding strategy. *Parvancorina* has a distinct similarity in shape to the Cambrian *Marrella* (see pp.74–75).

PARVANCORINA MINCHAMI
The shield-shaped head region has led to suggestions that *Parvancorina* was an ancestor of trilobites.

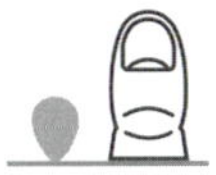

Tribrachidium

GROUP Unclassified
DATE Late Precambrian
SIZE 2–5cm (3/4–2in) across
LOCATION Australia, Russia, Canada

Tribrachidium is a mysterious Ediacaran fossil. It is a disc-shaped organism with three raised "arms" on its surface and a raised border. It also has a suggestion of bristle-like structures around the edge. *Tribrachidium* has tri-radiate symmetry – it is symmetrical along three lines. No tri-radiate animals existed throughout the Phanerozoic, but it is possible that an animal with tri-radiate symmetry may have preceded the echinoderms, which have five-fold radial symmetry. However, *Tribrachidium* lacked the calcareous plates that are a characteristic of echinoderms.

TRIBRACHIDIUM HERALDICUM
One theory is that *Tribrachidium* was a holdfast to a *Charnia*-like specimen (see p.61).

Spriggina

GROUP Unclassified
DATE Late Precambrian
SIZE 3cm (1 1/4in) long
LOCATION Australia, Russia

Spriggina was a segmented animal with what appears to be a clear head and tail end. The head is horseshoe shaped and the body consists of around 40 segments, with a prominent mid-line along its back and a small tail. Fossils have been found with various degrees of curvature, which shows that the body was flexible. Debate continues about what *Spriggina* really was. Some have likened it to modern marine annelid worms, while others have suggested that it could be a proto-trilobite, as its head has spines that are similar to the genal (cheek) spines of trilobites. For now, *Spriggina* is probably best classed within the Bilateralia, a group that includes all animals that are bilaterally symmetrical at some point in their life cycle.

KEY FIND
EDIACARAN FAUNA

In 1946, geologist Reginald Sprigg was examining ore prospects in the Flinders Ranges of South Australia when he discovered a group of fossils in the Ediacaran Hills. He described these soft-bodied animals as jellyfish from the Early Cambrian. Their actual Precambrian age was soon established, and this diverse fauna, including *Spriggina*, became known as the Ediacaran fauna.

Is it a **worm**, a **proto-trilobite**, or completely **unrelated** to either? The debate over ***Spriggina*** continues.

SPRIGGINA FLOUNDERSI
One of the big problems with *Spriggina* being a primitive trilobite or other arthropod is that its body segments are not arranged in opposite pairs, but are slightly offset against each other down the mid-line.

CAMBRIAN

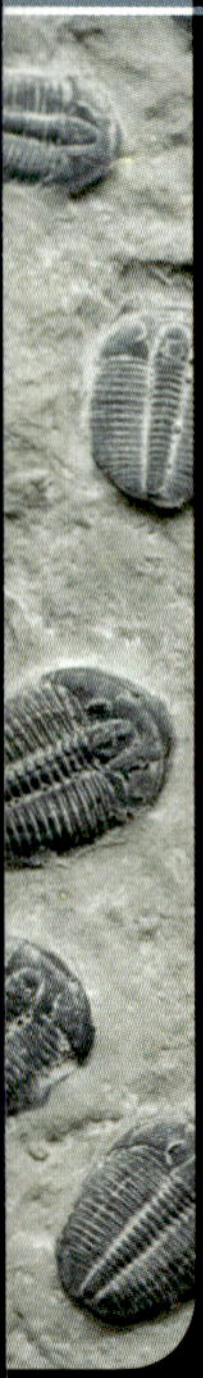

Cambrian

The Cambrian Period, between 542 and 488 million years ago, was one of rapid plate movement. Most importantly, there was an extraordinary explosion in animal diversity. Many of these new forms evolved the ability to biomineralize; this probably occurred in response to the evolution of predators. Soft-bodied animals from this period are preserved in localities such as Canada's Burgess Shale and Chenjiang in China.

KAZAKHSTAN TODAY
Kazakhstan, now a large, dry, landlocked country in Central Asia, was a group of microcontinents in Cambrian times. These arid hills lie east of the shrinking Aral Sea.

Oceans and continents

By the beginning of the Cambrian, Pannotia had begun to break up, while another supercontinent, Gondwana, was being assembled. This comprised much of present-day South America, Africa, Madagascar, India, Australia, and Antarctica. At this time the continents appear to have been moving at a relatively fast rate of about 30cm (12in) per year. As Pannotia fragmented, the Iapetus Ocean opened and widened between the continents of Laurentia (present-day North America), Baltica (northern Europe), and Siberia. To the north of Laurentia lay the Panthalassic Ocean. During the Early and Middle Cambrian, Laurentia drifted rapidly from polar to tropical latitudes so that it straddled the Equator. To the east of the other continents, Gondwana, the largest land mass, stretched from the South Pole to the Equator. This rotated through 90 degrees, driven by movement of the Earth's lithospheric plates. Other smaller continental blocks existed, such as Kazakhstan and China, and part of present-day Southeast Asia. Most were of relatively low relief and were surrounded by extensive shallow seas. During periods of sea level rise, continents were inundated, apart from Gondwana, eastern Siberia, and Kazakhstan, which were mountainous. Life proliferated in warm, shallow-water marine environments. Complex animals with mineralized skeletons evolved, such as sponges, brachiopods, corals, molluscs, echinoderms, and arthropods. Changes in ocean chemistry, greater diversity of plankton species at the base of the food chain, and pressure from predators may all have contributed to the Cambrian Explosion (see p.70).

APPALACHIAN MOUNTAINS
Sediments that consolidated into rocks uplifted in the Appalachians, USA, were deposited near the start of the Cambrian Period (542 million years ago) on the shores of the new Iapetus Ocean.

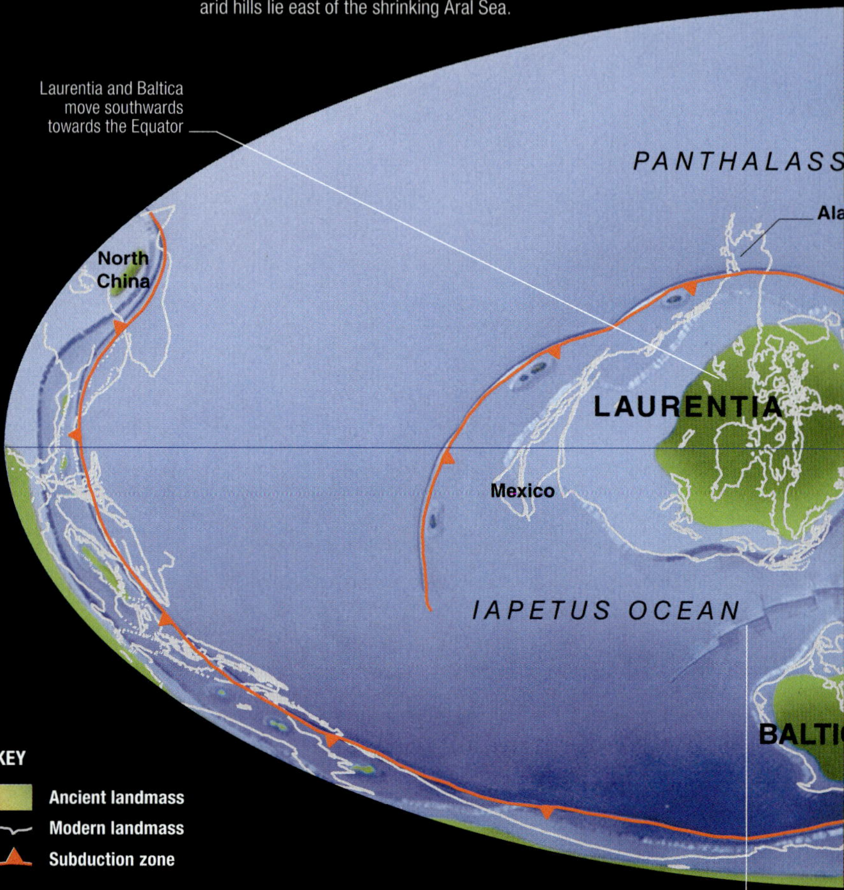

CAMBRIAN WORLD MAP
By the end of the Cambrian, much of the north was covered by the Panthalassic Ocean. The Iapetus Ocean had opened, and Gondwana had rotated.

	TERRENEUVIAN			SERIES 2
MYA	540	530		520
MICROSCOPIC LIFE				
INVERTEBRATES	**540** First small shelly fossils: tommotiids, halkieriids, wiwaxids; **535** First trilobites (e.g. below); other arthropods: bradoriids, brachiopods (orthids), crustaceans; molluscs (monoplacophorans, bivalves), echinoderms, cephalochordates, conularids, lobopodians, priapulids, nematodes, chelicerates, rostroconchs, onychophorans (marine), foraminifers, radiolarians	**530** First rhombiferan and first edrioasteroids (extinct echinoderms); first arthropod trace fossils on the sea floor	**525** First graptolites (e.g. *Dictyonema*); Chengjiang biota	**520** First pentamerid brachiopods (lamp shells)
VERTEBRATES	**535** First chordates (agnathan fish)			

SMALL SHELLY FOSSILS

OLENUS GIBBOSUS

DICTYONEMA

CROSS-BEDDED DUNES
Cross-bedded sandstone formations, such as these in Utah, USA, provide evidence of arid landscapes in the past, where dry winds swept across desert dunes.

EVAPORITES FORMED
When sea water evaporates, it deposits minerals such as salt and gypsum. Such thick evaporite layers indicate high temperatures.

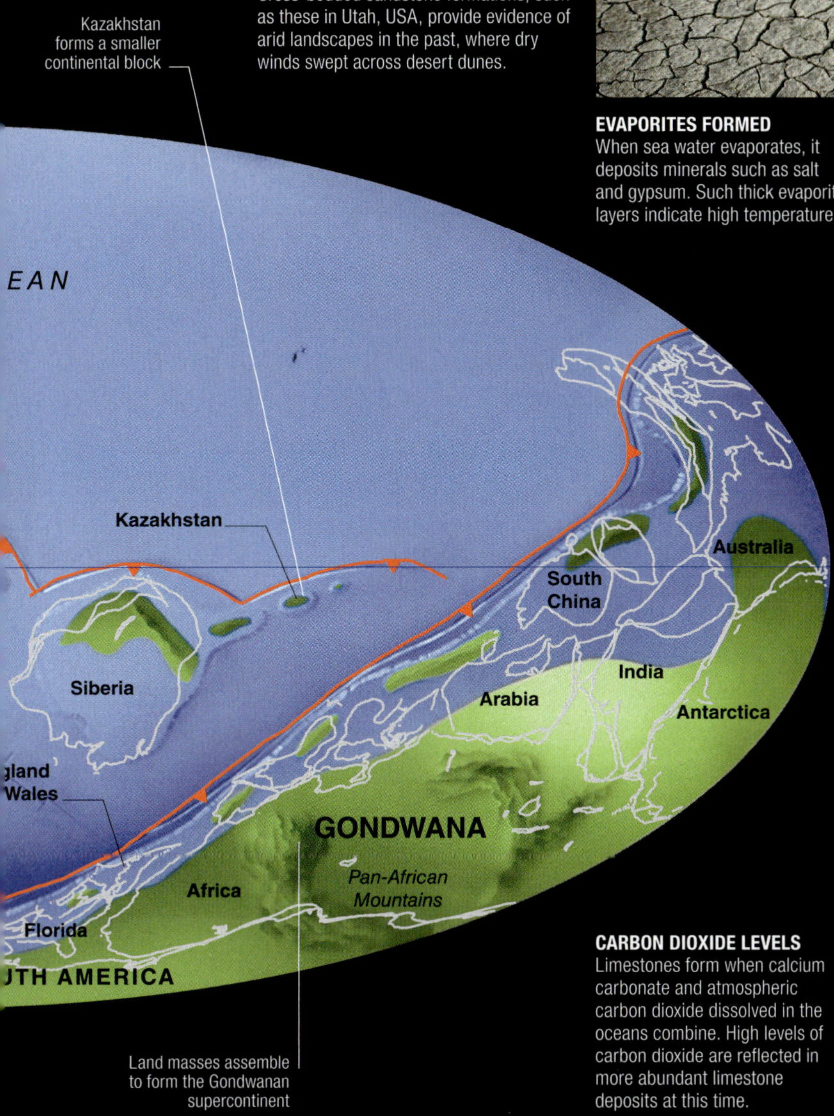

Climate

Detailed knowledge about Earth's climate during the Cambrian Period is rather limited. Carbon dioxide levels were high, up to 15 times present-day levels, and there were huge accumulations of evaporites. Both factors provide evidence for high global temperatures. One estimate suggests that by the Late Cambrian these averaged 50–60°C (122–140°F). Many low-lying continents experienced arid conditions produced by wind blowing from the land and low rainfall. Few indicators for hot, humid climates have been found in the Cambrian geological record, although there are examples in Laurentia of laterite and bauxite, known to form in the tropics in these conditions. Instead, particularly along the western margin of the Asian blocks, appropriate rainfall, topographic, and wind conditions existed for the development of arid climates. Dry and humid regions alike were located in low latitudes near the Equator. With no continents located directly over the poles, and oceans covering a larger part of the globe (about 85 per cent, compared with about 70 per cent today), the climate is likely to have been relatively equitable overall. There would have been less variation than today in temperature from the poles to the tropics; conditions were mainly tropical and subtropical, and glaciations did not occur. Evidence for this view is provided by the presence of limestone deposits up to a latitude of 50 degrees.

BURGESS SHALE
The Burgess Shale area, in the Canadian Rocky Mountains, was part of ancient Laurentia. The spectacular Cambrian fauna preserved here would have lived in warm seas just south of the Equator.

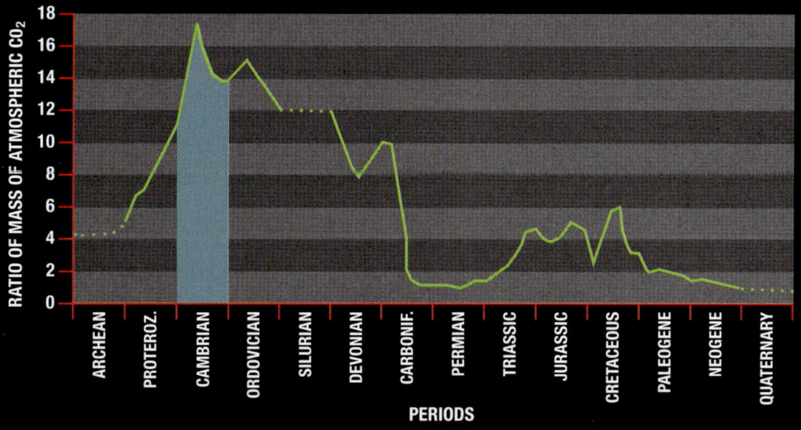

CARBON DIOXIDE LEVELS
Limestones form when calcium carbonate and atmospheric carbon dioxide dissolved in the oceans combine. High levels of carbon dioxide are reflected in more abundant limestone deposits at this time.

SERIES 3	FURONGIAN

510 | 500 | 490 | MYA

MICROSCOPIC LIFE

INVERTEBRATES

- **510** First cephalopods (nautiloids) and chitons (polyplacophorans)
- **505** Fossilization of Burgess Shale (e.g. *Marrella* and *Wiwaxia*)

WIWAXIA

MARRELLA

- **500** First conodonts, archaeogastropods (sea snails); first bellerophontine gastropods, and ostracods (tiny crustaceans); first arthropod trace fossils on land

VERTEBRATES

CAMBRIAN MICROSCOPIC LIFE

The beginning of the Cambrian witnessed an incredible event in evolution – in just a few million years animal groups, from worms to fishes, appeared. Microscopic remains of these creatures have been remarkably well preserved. These small, shelly fossils are evidence of this avalanche of evolution.

Before the Cambrian, most animals lacked either jaws or a through-gut, which meant they had no anus. The evolution of chewing, and then of predation, started an arms race that rapidly transformed ecosystems around the world.

Jaws and protection

Among the first tooth-like elements to appear were those of *Protohertzina*, belonging to a creature similar to a modern arrow worm. External skeletons quickly evolved for protection. The tiny, cap-shaped shells of *Maikhanella*, for example, were made from clusters of spines called sclerites. Later, other animals, such as brachiopods, merged their sclerites to form a single, solid shell. Fossils of soft tissues, including the cyanobacteria and other algae, can be found in Cambrian phosphates, but after the Cambrian, such remarkable preservation became rare, possibly because more scavenging was taking place on the sea floor.

The Cambrian cascade

A multitude of new, small organisms, collectively known as plankton, began to drift freely through the Cambrian seas and oceans, transforming the food chain. The great oceans provided a valuable means of dispersal, as well as a degree of protection from grazers and predators. Radiolarian zooplankton, with their delicate silica skeletons, appeared where they could feed on other small, drifting organisms. They began to form the biological silica deposits called radiolarian cherts. Diverse types of phytoplankton also thrived, using sunlight to make their own food by means of photosynthesis. Many secreted a protective armour called resting cysts, and fell to the sea floor, where they were readily preserved. All of these creatures provided an abundant supply of food for the new types of animal that appeared during the Cambrian Explosion.

CAMBRIAN OUTCROP
This red limestone outcrop in Siberia contains a multitude of microfossil remains from the huge number of new species that came about during the Cambrian Explosion.

GROUPS OVERVIEW

The microfossil record from this period shows a world that was increasingly influenced by multicellular animals with a through-gut system. These organisms are called bilaterians. Their earliest remains can often be found as microfossils when the Cambrian limestones in which they are preserved are digested in weak acids in a laboratory.

PROTOCONODONTS
The tiny jaws of *Protohertzina* are similar to those of some modern, predatory arrow worms. They appeared all over the world at the very start of the Cambrian. It is possible that these fearsome jaws caused other creatures to build protective shells or burrow into the sea floor, starting a kind of "arms race" that turned into the Cambrian Explosion.

SMALL SHELLY FOSSILS
The earliest protective skeletons were often tubular, like tiny drinking straws. It can be difficult to tell what species these remains come from, but it seems likely they were built by creatures that were related to modern coral polyps, marine fan worms, or even deep sea vent worms. They flourished on the sea floor until the trilobites and other scavengers evolved.

MICROMOLLUSCS
The sclerites of enigmatic multi-element skeletons and tiny cap-shaped molluscs became abundant soon after the start of the Cambrian period. They created a huge variety of shell shapes, and while they can look like clams or garden snails, evidence suggests that these tiny molluscs lacked many of the features seen in molluscs of the later Paleozoic.

MICROARTHROPODS
Arthopods diversified explosively, leaving a trail of phosphatized remains, sometimes including legs, gills, and soft parts. Scanning electron microscope images have revealed the rapid evolution of a variety of limbs in Cambrian arthropods. Later forms evolved specialized appendages, like those of modern spider crabs and lobsters.

Protohertzina

GROUP Chaetognathans
DATE Early Cambrian
SIZE 2mm (1/16in) long
LOCATION Worldwide

This microfossil appears near the base of Cambrian rocks. Each tooth-like "protoconodont" is made of calcium phosphate, and has a hollow basal portion for attachment. It is shield-shaped in cross section and has a long, curved spine. Occasionally, they are found in clusters that closely resemble the portcullis-like jaws of modern arrow worms. This suggests that this was an early predator.

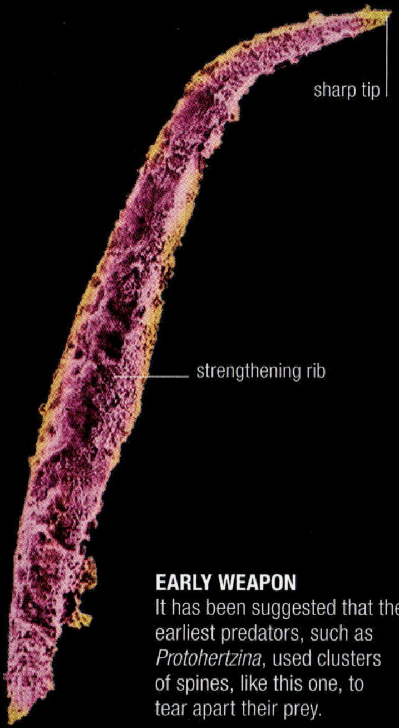

EARLY WEAPON
It has been suggested that the earliest predators, such as *Protohertzina*, used clusters of spines, like this one, to tear apart their prey.

Maikhanella

GROUP Molluscs
DATE Early Cambrian
SIZE 2mm (1/16in) long
LOCATION Worldwide

This cap-shaped fossil had a weakly mineralized shell that was phosphatized after its death. The pineapple-like shell was made from tightly packed spines called sclerites. At first, as in this example, these sclerites were not strongly fused together. Sometimes they are found separately, when they are called *Siphogonuchites* or *Halkieria*. After several million years of evolution, the spines became more strongly fused into a single, rigid shell that was a similar shape to that of a modern limpet. In Greenland, examples have been found where the slug-like owner had been carrying two *Maikhanella*-like caps, surrounded by a skirt of *Siphogonuchites*-like sclerites. This complex arrangement is a reminder that much experimentation was still taking place at this time. *Maikhanella* probably grazed on algae that grew on the Cambrian sea floor.

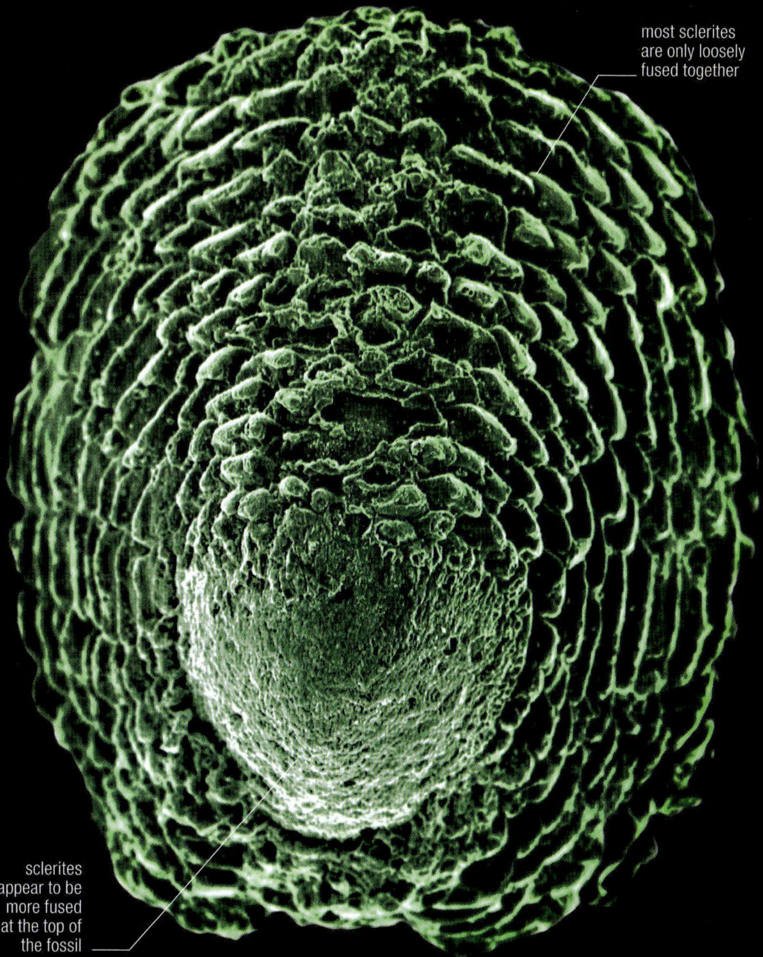

SNAIL ANCESTOR
Maikhanella is one of the earliest molluscs. It is thought that modern snails and cuttlefish evolved from forms like the one shown here, which is just 1mm (1/16in) across.

Aldanella

GROUP Molluscs
DATE Early Cambrian
SIZE 3mm (1/8in) wide
LOCATION Worldwide

Modern-looking mollusc shells evolved with suprising rapidity during the course of the early Cambrian period. This resulted in coiled forms, such as the one shown here, which appeared across a wide area from Canada to China. It occurred alongside other molluscs that crudely resemble much younger bivalves and even coiled cephalopods. Although little *Aldanella* looks similar to a modern garden snail, there is no evidence that it was closely related to Gastropoda. The owner of this shell probably grazed on detritus near the surface of the Cambrian sea floor.

OXFORD FIND
This *Aldanella* shell was fossilized around 530 million years ago. It was discovered in a borehole in Oxford, England while its discoverers were searching for coal.

Anabarites

GROUP Possible Cnidarian
DATE Early Cambrian
SIZE 5mm (3/16in) long
LOCATION Worldwide

This tubular fossil had a calcareous shell made of aragonite. In cross section it is a distinctive, trilobed, cloverleaf shape. Like many small, shelly fossils from the early Cambrian, its biological affinities remain uncertain, but it could have been close to the cnidarian groups of jellyfish and corals. It may have lived in colonies, embedded in the sediment, and feeding on organic matter in the water column.

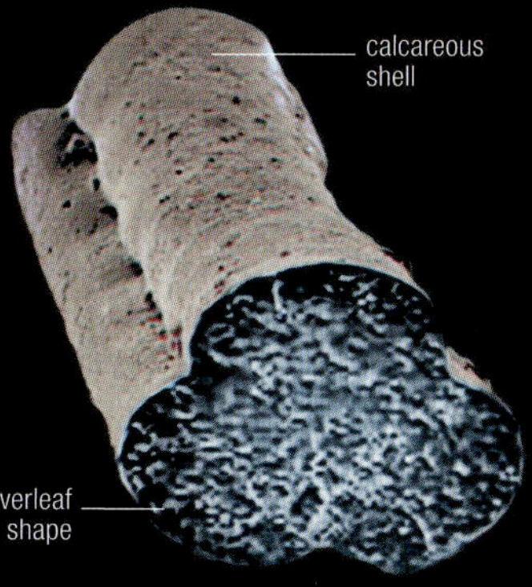

UNUSUAL SHAPE
Shells of *Anabarites* have a curious trilobed cross section, which is quite unlike anything alive today. Calcium-phosphate-filled microfossils like this are found at the very start of the Cambrian.

Orsten arthropod

GROUP Arthropoda
DATE Late Cambrian
SIZE 3mm (1/8in) long
LOCATION Sweden, China

Although the Cambrian is known as the age of trilobites, many other arthropod groups also existed. Some have been remarkably well preserved within phosphatic nodules. When these nodules are broken up and immersed in weak acids, remarkable fossils, seldom found in younger rocks, are released. Examination under a scanning electron microscope can reveal the details of their tiny legs and feather-like gills. This little ostracod-like fossil probably fed on the soup-like layer that had settled on the dark, cold sea floor.

PERFECT DETAIL
Numerous pairs of jointed limbs, with tiny hairs for trapping food particles, can be clearly seen in this scanning electron microscope image, which has been taken from below.

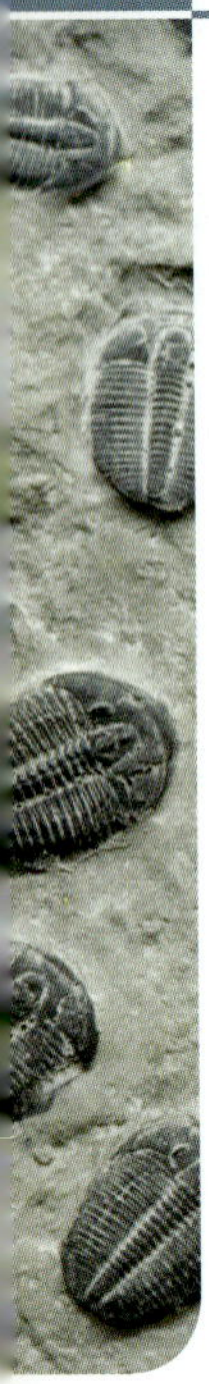

CAMBRIAN INVERTEBRATES

The Cambrian was a critical time in evolutionary history as diverse marine invertebrates began to appear. Some had eyes and strong jaws that enabled them to live as the first active predators. Other Cambrian evolutionary fauna, such as trilobites, flourished, but declined from the Ordovician onwards.

The "explosion" of life

The Cambrian Explosion was the first truly significant evolutionary event in the history of life, heralding the origin and proliferation of hard-shelled marine invertebrates. First, following the extinction of the Ediacaran fauna (see p.63) about 543 million years ago, came the "small, shelly fossils", which are represented by phosphatic horns and coils, plates, and tubes. These persisted until around 525 million years ago, when the first trilobites, inarticulate brachiopods, and other hard-shelled fossils started to appear. But these represent a mere fraction of the vast diversity of Cambrian life. The Middle Cambrian Burgess Shale fauna, discovered in British Columbia, Canada, in 1909 (see p.74), and the Lower Cambrian Chengjiang fauna, found in Yunnan, China, in 1984, testify to a plethora of body plans, especially in arthropods and "lobopods", such as the terrestrial velvet worm *Peripatus*, which are rarely preserved.

BURGESS SHALE ARTHROPOD
This small arthropod, *Marrella splendens*, is from the Burgess Shale. It is just one of a large number of different arthropods that have been found there.

SHELLY REMAINS
These small, shelly fossils are hyolithids, which formed an integral part of the Cambrian evolutionary fauna, and are often found in great numbers.

Diversity in the Cambrian seas

The Burgess Shale and the Chengjiang faunas were dominated by a diverse range of marine invertebrates with soft bodies or unmineralized shells. They lived alongside the more familiar trilobites, which were common at the time. The Cambrian seas would have abounded with these soft-bodied animals, but their chances of preservation would normally be slim. By the Late Cambrian crustaceans and other faunas had also started to appear.

MIDDLE CAMBRIAN TRILOBITE
This trilobite, *Ellipsocephalus hoffi*, was found in the Czech Republic. Large numbers of this species are often found grouped together.

GROUPS OVERVIEW

Many marine invertebrate groups first appeared in the Cambrian. The diverse fauna included early forms of sponges that did not survive to the end of the period and brachiopod forms that are, in contrast, still living today. The Cambrian saw the earliest forms of trilobite evolve at the beginning of the period, while echinoderms were also abundant.

ARCHAEOCYATHIDS
These extinct fossil sponges were only in existence during the Early and Middle Cambrian. They were regularly conical in form, with two perforated cones (one inside the other) that were connected by partitions and originally lined with feeding cells. They often formed small reefs.

INARTICULATE BRACHIOPODS
These brachiopods, some forms of which are still living today, have a two-valved shell and a muscular stalk, or pedicle. They are suspension-feeders, filtering out particles of food from the water surrounding them. Some genera, such as the "living fossil" *Lingula*, have hardly changed in 500 million years.

ECHINODERMS
Cambrian echinoderms do not appear to have closely resembled the echinoids, or crinoids, that are still in existence today, yet the structure of their calcitic plates was identical. Echinoderms were an important component of the Cambrian evolutionary fauna.

TRILOBITES
These now extinct creatures first arose in the Early Cambrian and persisted through to the Permian. They were arthropods with a typical three-lobed calcitic shell and compound eyes; the legs and gills were very delicate and were preserved in just a small number of species.

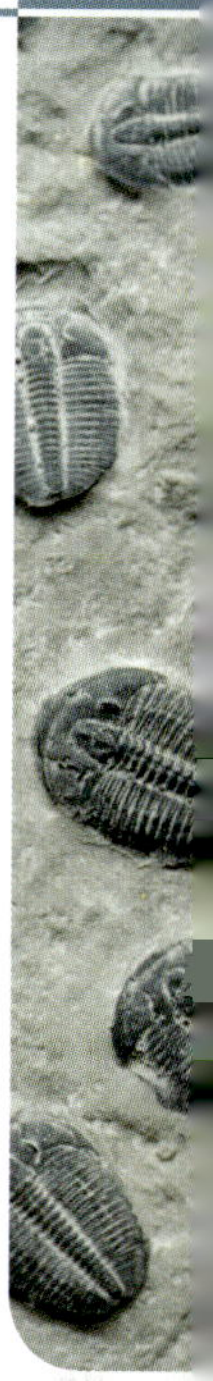

Archaeocyathids

GROUP Archaeocyathids
DATE Early to Middle Cambrian
SIZE 5–30cm (2–12in) long
LOCATION Antarctica, Australia, North America, S. Europe, Russia

Archaeocyathids were the first reef-building organisms. They were confined to tropical areas and were only in existence for a relatively short period, but they diversified rapidly during this time. Some forms were solitary, while others were colonial. They were made up of two calcite cones, arranged one inside the other with a space between them. The cones were held apart by calcitic septa, or partitions, that crossed the space. The outer cone attached to the sea bed via a root-like structure at its base.

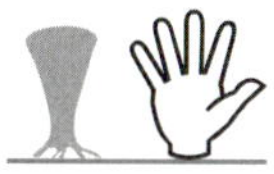

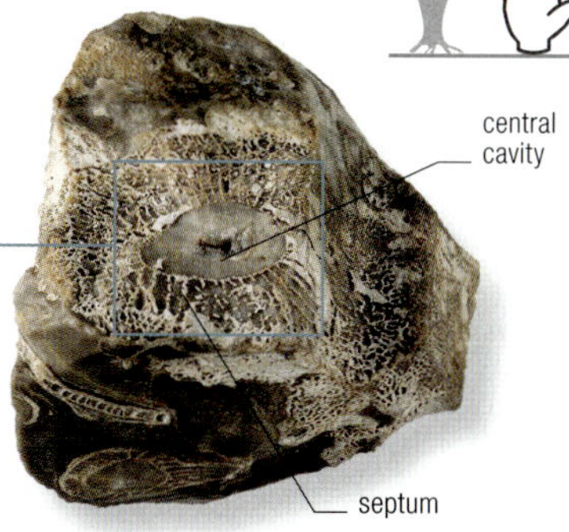

ARCHAEOCYATHID
These sponge-like organisms comprised two downward-facing cones that were both perforated. They relied on the flow of water to transport food and oxygen to them, in contrast to some sponges that can generate their own water currents.

Ottoia

GROUP Worms
DATE Middle Cambrian to present
SIZE 4–8cm (1½–3¼in) long
LOCATION Canada

Ottoia is one of the fossils found in Canada's Burgess Shale that can be associated with a living animal group – the marine priapulid worms. *Ottoia* is the most common Burgess Shale priapulid worm, with around 1,500 known specimens. It gathered food with a tube-like organ called a proboscis, which was equipped with spiny hooks and could be turned inside out. Many fossilized specimens have a strongly curved body, which has led to suggestions that, like its modern counterparts, *Ottoia* lived in a U-shaped burrow and extended its proboscis to catch prey. Several specimens have been found with their gut contents intact, which includes small, shelly hyolithids and even members of their own species, suggesting that Ottoia was cannibalistic.

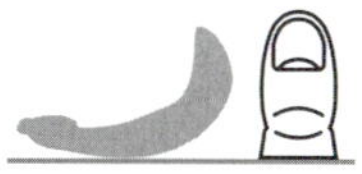

MUDDY TOMBS
Ottoia is thought to have lived passively in its burrow. Mudslides may have carried them into deep water where large numbers were ultimately entombed.

Echmatocrinus

GROUP Anthozoans
DATE Middle Cambrian
SIZE 2.5cm (1in) wide, below the tentacles
LOCATION Canada

Echmatocrinus is an unusual fossil, found only in the Burgess Shale formation in Canada. The surface of its long, conical body was covered with thin, polygonal plates or scales that were arranged irregularly. There were seven to nine plated arms or tentacles that were attached to the top of the body. Long, thin, non mineralized branches occurred on alternate sides of these arms. *Echmatocrinus* has been difficult to classify. When it was first described it was thought to be a crinoid (sea lily), but the absence of distinctive echinoderm features, such as five-fold symmetry, has led to other interpretations. Some palaeonotologists suggest it could actually be an octocoral (a subclass of corals).

Echmatocrinus has **been difficult to classify**. When it was **first described**, it was **thought to be a crinoid**, but it could actually be an **octocoral**.

BRANCHING OUT
Echmatocrinus had a long, conical body with a number of tentacles that were covered with thin branches. Its body tapered towards the bottom, where it would have attached itself to the substrate.

Small Shelly Fossils

GROUP Small Shelly Fossils
DATE Early Cambrian
SIZE Up to 1cm (⅜in) wide
LOCATION Worldwide

Small Shelly Fossils is the name given to a wide variety of different fossils, including closely packed groups of small shells that occur in some of the earliest Cambrian rocks in many parts of the world. Sometimes they form beds around 2cm (¾in) thick, where the shells had been swept together by sea bed currents. While some of these fossils are whole organisms, many are small parts of larger organisms, such as plates or spines belonging to molluscs or several enigmatic groups of extinct marine invertebrates.

TINY TUBES
This specimen shows a number of small, flattened, conical tubes, known as hyolithids. The dark coloration towards the top has been caused by an abundance of the mineral phosphate.

Lingulella

GROUP Brachiopods
DATE Early Cambrian to Middle Ordovician
SIZE 1–2.5cm (3/8–1in) long
LOCATION Worldwide

Lingulella was an inarticulate brachiopod, which means that it used muscles to hold its valves (shells) in place, rather than teeth or sockets. It was attached to the substrate by a fleshy stalk that emerged from an opening in its pedicle (lower) valve. *Lingulella*'s shell was elongated and the beak area was pointed. The surface of the shell had fine growth lines and fossils show delicate radial striation on the inner layers.

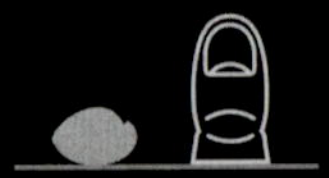

VERTICAL BURROW
Like its modern-day relatives, *Lingulella* probably lived in vertical burrows. This specimen became fossilized in situ.

LIVING RELATIVE
LINGULA

Lingula is a modern brachiopod that can be traced back to the Silurian. Unusually for a brachiopod, it lives buried within sediment on the sea bed. It is anchored by a long, fleshy pedicle that emerges between its two valves; it has no special opening like *Lingulella*. *Lingula* is adaptable and can re-anchor itself if disturbed by current activity.

Bohemiella

GROUP Brachiopods
DATE Middle Cambrian
SIZE 1–2cm (3/8–3/4in) long
LOCATION Czech Republic

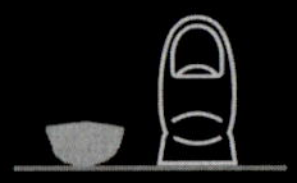

Bohemiella was an orthid brachiopod, a group of articulate brachiopods that lived throughout the Paleozoic era. Its valves (shells) were elongated in a crossways direction. The pedicle (lower) valve was less concave than the brachial (upper) valve – or sometimes almost flat. There was a small pair of teeth on the pedicle valve and corresponding sockets in the brachial valve. These had long projections on their inside edges to hold the teeth in position.

INNER VALVE
This inside mould of *Bohemiella*'s pedicle valve shows scars where the shell's opening muscles were attached in life.

Wiwaxia

GROUP Molluscs or worms
DATE Middle Cambrian
SIZE 3–5cm (1¼–2in) long
LOCATION Canada

Wiwaxia was a striking animal with a body that was symmetrical about its length, elliptical viewed from the top, and squarish in cross section. Its upper surface was covered with overlapping rows of protective, armour-like plates, called sclerites, and two rows of long spines. Its lower surface lacked any form of protection. *Wiwaxia*'s mouth contained a food-gathering apparatus with two or three rows of rear-facing, conical teeth. These were probably used for scraping algae or bacteria from the sea bed, or for collecting food particles from the surrounding water. Because of the similarities between this apparatus and the radula of molluscs, it is thought that *Wiwaxia* may be related to molluscs. An alternative hypothesis suggests that it is related to worms.

STRIKING SPINES
Although only up to 5cm (2in) long, *Wiwaxia* was a striking animal with an armour-plated body and upwardly projecting spines. Its mouth was located on the underside of its body.

MYSTERIOUS FOSSIL
Wiwaxia is a mysterious fossil. Found in the Burgess Shale fossil bed, it dates to about 500 million years ago. Its plates and spines were not mineralized and appear to have been fibrous in texture.

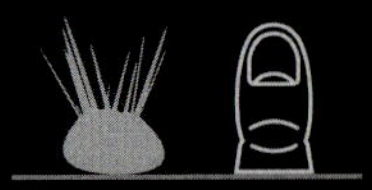

Pojetaia

GROUP Bivalves
DATE Early Cambrian
SIZE 1–2mm (1/16in) long
LOCATION Australia

Early Cambrian bivalves are not only very rare – only two genera are known – but also very small. *Pojetaia* is currently the oldest bivalve ever discovered. Its shell was almost circular with well-defined beaks on each valve (shell) and a straight hinge line. The outside of the shell had growth increments and some faint radial ribs. Internally, the shell had two adductor muscles that opened and closed the valves. The hinge had five or six teeth, and sockets in each valve helped the valves to align.

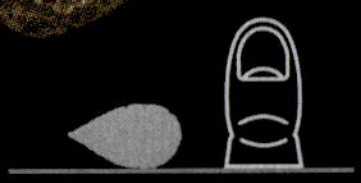

SHALE FOSSIL
Helicoplacus fossils have been found most commonly in shales.

Helicoplacus

GROUP Echinoderms
DATE Lower Cambrian
SIZE 2.5–4cm (1–1½in) long
LOCATION USA

Helicoplacus was a bizarre echinoderm that represents a very early, and ultimately unsuccessful, body plan. Unlike other echinoderms, it lacked radial symmetry of any kind. Tiny plates were arranged spirally around its shell, which in its resting state was pear-shaped. In fossils, the shell plates are usually found separated, suggesting that they were not fused together. Instead, the body had the capacity to expand and contract, and during expansion the plates separated.

ANATOMY
EXPANDING BODY

As *Helicoplacus*'s plates were not rigidly fused, they may have allowed its body to expand. Recent ideas suggest that the sharper end of its body was partly buried in the sea bed, so expansion of the body may have assisted in feeding or respiration.

Hallucigenia

GROUP Arthropods
DATE Middle Cambrian
SIZE Up to 2.5cm (1in) long
LOCATION Canada, China

When *Hallucigenia* was first studied in the 1970s, reconstructions showed the animal walking on rigid, stilt-like, paired "limbs", with a single row of fleshy projections along the top of its back. Nothing of its kind had ever been found before. However, later discoveries showed that earlier reconstructions had, in fact, turned the organism upside down. *Hallucigenia* had an elongated body, with a rounded "head" at one end, and a long, fleshy "tail". It had seven pairs of stiff, pointed spines along the upper side of its body, along with a cluster of small projections, situated close to its rear end.

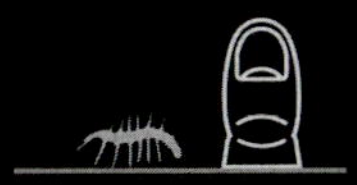

VELVET WORM
Reconstructed with its fleshy limbs on its underside, *Hallucigenia* can be recognized as a velvet worm, of which there are living relatives in today's oceans.

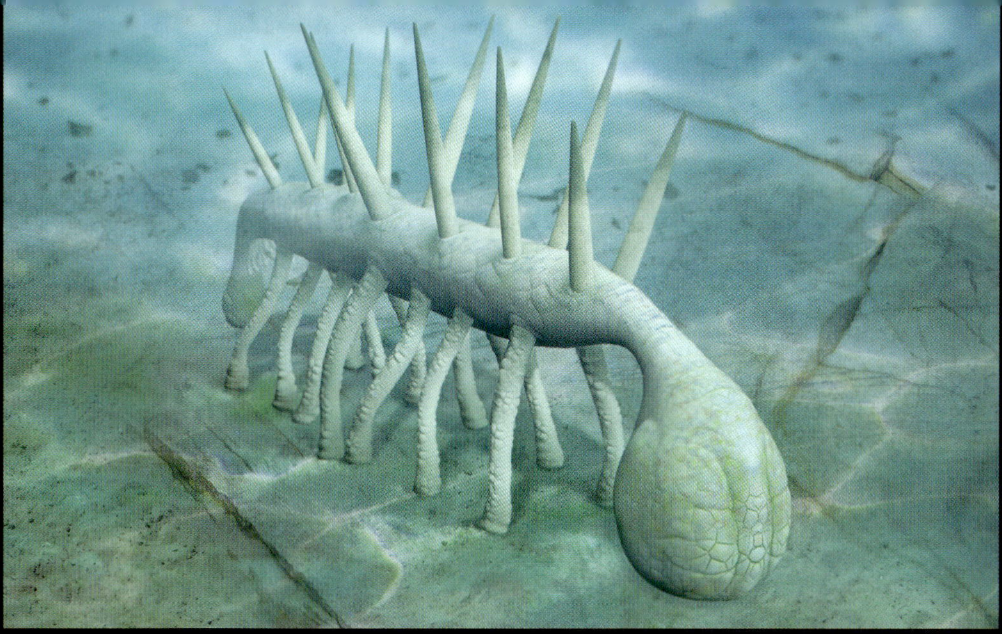

Opabinia

GROUP Arthropods
DATE Middle Cambrian
SIZE About 6.5cm (2½in) long
LOCATION Canada

Opabinia is one of the strangest animals to have been found in Canada's Burgess Shale fossil bed – it is quite unlike any other. Its head had five prominent eyes – two sets of pairs, and one eye that was central. Extending from the front of its head was a long, flexible, trunk-like feature, or proboscis, which ended in a pod-shaped organ bearing small spines that were probably used to grasp prey. *Opabinia* would have used its proboscis to pass its prey up to its mouth. Its elongated body comprised 16 segments, each of which possessed a flap-like lateral lobe, with gills on the underside. Its tail had three flaps that projected from either side.

BIZARRE APPEARANCE
Opabinia's form was so bizarre that when a reconstruction was presented to the Palaeontological Association at its meeting in Oxford, England, in December 1972, its appearance provoked laughter.

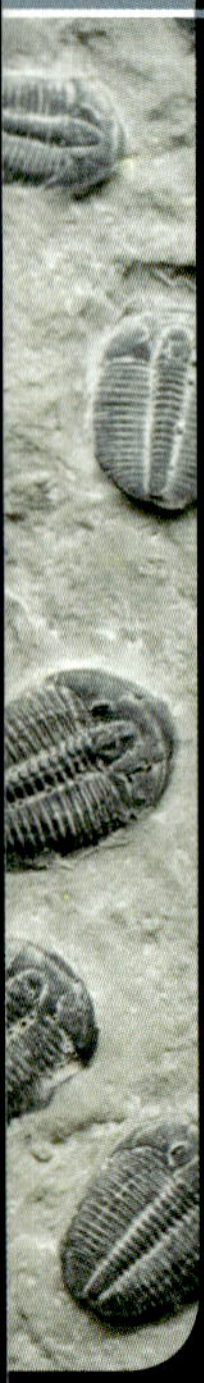

Marrella

GROUP Arthropods
DATE Middle Cambrian
SIZE Up to 2cm (¾in) long
LOCATION Canada

Marrella is the most abundant Burgess Shale fossil (see below), with over 15,000 known specimens. Curiously, this Canadian site is the only place it has been found in the world. *Marrella* was given the informal name "lace crab" by the American palaeontologist Charles Walcott (see p.77) due to its "feathery" appearance. It had two distinctive pairs of large, backward-facing spines, one pair running alongside the body, the other, above it. Two pairs of antennae arose from the front of its body, one very long, the other shorter and stouter. The body was composed of 20 segments; each segment had pairs of identical limbs, and a gill branch. The identical limbs suggest that it was a primitive arthropod.

MARRELLA SP.
The headshield and two sets of backward-facing spines are clearly visible on the above specimen. The discovery of *Marrella* is particularly significant as it may be a descendent of a common ancestor of the crustaceans, trilobites, and chelicerates.

SEA BED SCAVENGER
Marrella is thought to have swum along the sea bed or just above it, feeding on tiny particles of organic material. Its jointed legs contained feathery gill branches that formed part of its respiratory system. The teeth-like projections on the inner pair of spines are evident in this illustration.

KEY SITE
BURGESS SHALE

The 510 million-year-old Burgess Shale in British Columbia, Canada, is one of the world's most significant sources of fossils, as it contains well-preserved soft-bodied organisms – a very rare occurrence. It provides a unique window into the full diversity of Cambrian life, with a wealth of different types of fossils. The Burgess fauna provides important evidence for the theory of the "Cambrian Explosion", which was a rapid diversification of marine life (see p.70).

Olenellus

GROUP Arthropods
DATE Middle Cambrian
SIZE Up to 6cm (2¼in) long
LOCATION North America, Greenland, Scotland

Olenellus was a trilobite with a semicircular headshield and a tapering glabella (central head bulge) with four pairs of backwards-pointing furrows. Its eyes were crescent shaped, and their front ends merged with the glabella's frontal lobe. Its thorax had 18 segments, and the third was wider and longer than the others, ending in spine. One of the earliest trilobites in the geological record, *Olenellus* was a key fossil used by Canadian geologist Tuzo Wilson to show that there was once a "proto-Atlantic Ocean" that separated much of continental North America, western Newfoundland, and the northern British Isles from Eastern Newfoundland and the southern British Isles.

OLENELLUS THOMSONI
Although its head was well-developed and its eyes large and crescent shaped, *Olenellus* possessed only a small, poorly formed tail.

Olenus

GROUP Arthropods
DATE Late Cambrian
SIZE Up to 4cm (1½in) long
LOCATION British Isles, Norway, Sweden, Denmark, Newfoundland, Texas, South Korea, Australia

Olenus is a trilobite fossil commonly found in dark mudstones, which were deposited on the sea floor in environments with low oxygen levels. It had up to 15 thoracic segments, with very wide pleural or side lobes. These are thought to have supported extended gills, which would have helped the animal absorb the maximum amount of oxygen possible in such an environment. Evidence also suggests that *Olenus* and its relatives may have developed a symbiotic relationship with sulphur bacteria, either by feeding on them directly, or by absorbing nutrients directly from them.

OLENUS GIBBOSUS
Olenus had a rounded headshield with an oblong glabella (central part), and small, crescent-shaped eyes. Its tail was much smaller than its headshield.

Ellipsocephalus

GROUP Arthropods
DATE Middle Cambrian
SIZE Up to 4cm (1½in) long
LOCATION Sweden, Czech Republic, Morocco, Canada

This trilobite fossil is commonly found in large clusters in Cambrian rocks in the Prague district of the Czech Republic. Many specimens are complete, but lack free cheeks, which suggests that they were moults, and that the animals congregated to shed their old exoskeletons in order to grow in size. *Ellipsocephalus*'s headshield had a prominent, smooth glabella with slightly concave sides. Like *Olenus* (above), its eyes were small and crescent shaped. The border of the headshield was narrow, defined by a rather narrow, shallow furrow.

E. HOFFI
Ellipsocephalus had a thorax divided into 12 well-defined segments.

Paradoxides

GROUP Arthropods
DATE Middle Cambrian
SIZE Up to 45cm (17½in) long
LOCATION Europe, North America, British Isles, Morocco, Turkey, Siberia

Paradoxides was one of the largest known genera of trilobites, and it was probably a predator, high in the Cambrian food chain. Underneath the widest part of the glabella lay the hypostome, a large, plate-like structure supporting the stomach, whose size and shape suggests predatory activity.

PARADOXIDES BOHEMICUS
Paradoxides's thorax consisted of 18 to 21 segments, all of which ended in long spines, with those at the tail end longer than the others.

Elrathia

GROUP Arthropods
DATE Middle Cambrian
SIZE Up to 4.5cm (1¾in) long
LOCATION W. USA

ELRATHIA KINGII
Elrathia's exoskeleton is usually flattened due to its preservation in shale.

Elrathia is one of the best-known trilobites in North America. Its headshield was semicircular, with a short, conical glabella and two pairs of short, shallow furrows. The crescent-shaped eyes were set some distance from the glabella, near the front. There were 13 segments in its thorax, and its tailshield was much smaller than its head. The Pahvant Ute Indians of Utah, western USA, used *Elrathia kingii* fossils in amulets to ward off evil.

Tomagnostus

GROUP Arthropods
DATE Middle Cambrian
SIZE Up to 2cm (¾in) long
LOCATION Sweden, Denmark, British Isles, Czech Republic, Newfoundland, E. USA, Siberia, Australia

Tomagnostus is one of many agnostoid trilobites with an almost global distribution, and is thus valuable in correlating Cambrian rocks across wide areas. It probably lived in the open ocean, but it occurs as a fossil in association with more "local" trilobites in different regions. Agnostoids were highly specialized and distinctive members of the trilobite group.

TOMAGNOSTUS FISSUS
The head- and tailshields of *Tomagnostus* were almost equal in size and shape.

Anomalocaris

GROUP Arthropods
DATE Middle Cambrian
SIZE Up to 1m (3ft 3in) long
LOCATION Canada, S. China

Anomalocaris was the largest animal in the Cambrian ecosystem that became fossilized in what is now the Burgess Shale in British Columbia, Canada. The animal's head had two eyes, in front of which was a pair of segmented, downwards-curving appendages. Each of the segments of these appendages had a pair of spiny projections on its underside. *Anomalocaris*'s mouth was situated on the underside of the head, and consisted of a circle of elongated plates. Behind each eye were three small flaps. The body was divided into eight segments, each of which had side flaps, and the small tail featured an upturned fan of flaps. Fragments of *Anomalocaris* were originally thought to belong to separate animals. The front appendages were interpreted as the segmented abdomen of a shrimp-like crustacean, while the circular mouthparts were thought to be a jellyfish.

ANOMALOCARIS SP.
Viewing this specimen with its segmented body, it is easy to see why scientists thought *Anomalocaris* was a type of shrimp or crustacean when it was discovered in the Canadian Rockies. In fact, its genus name means "anomalous shrimp", so-called because its appendages were unsegmented, unlike in true shrimps.

1m The length of the largest specimens of *Anomalocaris*. Its size and formidable mouthparts made it a top predator of the Cambrian seas.

TOP PREDATOR
Although it was probably not the swiftest of swimmers, many scientists believe that *Anomalocaris* was at the top of the food chain in the Cambrian seas. Its large size, of up to 1m (3ft 3in) or more in length, and formidable circular mouthparts, allowed it to capture smaller prey, such as trilobites, with ease.

BIOGRAPHY
CHARLES WALCOTT

American invertebrate palaeontologist Charles Doolittle Walcott (1850–1927, centre) was a keen fossil collector as a boy in New York State. With little formal education, he forged a highly successful career as secretary of the Smithsonian Institution and director of the US Geological Survey. His most famous discovery, in 1909, was of superbly preserved fossils of Cambrian animals, including *Anomalocaris*, in the Burgess Shale in the Canadian Rockies.

CAMBRIAN VERTEBRATES

The Early Cambrian was a crucial period in the history of life on Earth – witnessing an explosion in animal diversity. About 540 million years ago the seas were home to the first wave of new animal groups, some of which would eventually die out, but others would lead to the evolution of the first vertebrates.

The start of the Cambrian is marked by the appearance of a vast array of animal forms in marine sedimentary rocks that are unknown in older rocks. This explosion in diversity saw the arrival of most of the major groups of animals or animal body plans. They included the first chordates – animals that at some point in their life cycle possess a notochord, the precursor of a backbone – and the first vertebrates.

The first vertebrates

Vertebrates are a group of animals that possess a backbone (vertebral column), which surrounds the notochord and provides support for the body. They also have a skull (cranium) that encloses the brain, eyes, olfactory organs, and internal ears. Another important characteristic of vertebrates is that parts of the head, gill arches, and nerves form from neural crest cells. These cells migrate from the nerve cord early in the development of the embryo and travel to different parts of the animal to form these structures. Other vertebrate features include the nervous control of the heart, a set of muscles to control eye movement (extrinsic muscles), at least two semicircular canals in the inner ear, and a lateral line system running along the head and body that has sensory organs called neuromasts. All extinct jawless fish are vertebrates. Of the two living groups, lampreys are considered to be vertebrates, but many authorities exclude hagfish because of their poorly developed vertebral column. However, hagfish do have a cranium.

Until quite recently it was thought that vertebrates arose in the Ordovician. In 1984, a remarkable Early Cambrian fossil site was discovered in Chengjiang, Yunnan Province, China. Most of the 180-plus species found there are invertebrates, but in 1999, two 530-million-year-old vertebrates were announced to the world (see below and opposite).

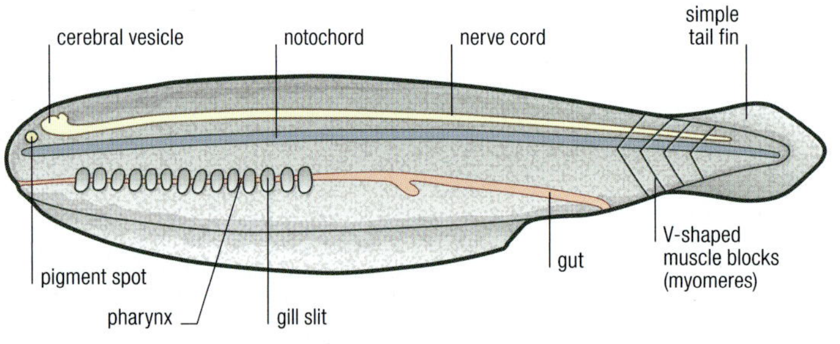

EARLY NON-VERTEBRATE CHORDATE

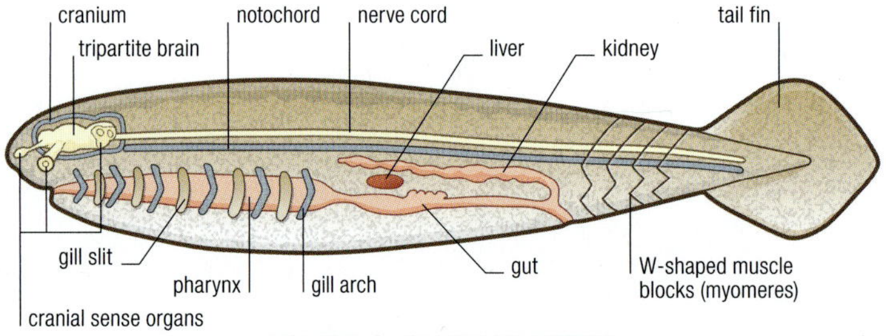

HYPOTHETICAL PRIMITIVE VERTEBRATE

NON-VERTEBRATE AND VERTEBRATE
These two schematic diagrams compare the basic body plans of a non-vertebrate chordate and a hypothetical early vertebrate. Note the presence of a brain protected by a cranium, sense organs, and gill arches in the latter.

GROUPS OVERVIEW

The Cambrian explosion produced the first cephalochordates and vertebrates, both members of the Phylum Chordata. All chordates have a distinct body plan, at least in their embryonic stages. A stiffening rod, called a notochord, extends dorsally from front to back. Above the notochord lies a hollow nerve cord, and below it lies the gut.

CEPHALOCHORDATES
Cephalochordates retain a notochord even as adults, but they lack a vertebral column and neural crest cells. The first cephalochordate, *Pikaia*, swam in the sea about 530 million years ago. It is very similar to the only living cephalochordates, the lancelets (or *Branchiostoma*; see right), in having gill slits and body muscles arranged in a V-shaped pattern.

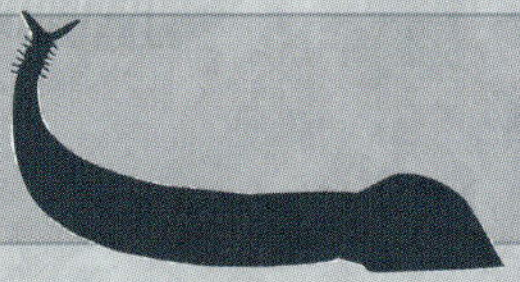

MYLLOKUNMINGIDS
The first vertebrates, *Myllokunmingia* and *Haikouichthys*, are often grouped in the same taxon. They are known from over 500 specimens found in the Chengjiang shales. They possess small vertebral elements that represent the first vertebral column (backbone). Other vertebrate-like features include a clearly defined head and structures that supported the gill slits.

LANCELETS
Resembling long leaves, lancelets usually lie half-buried in sediment with their heads sticking out as they filter feed. The characteristic V-shaped muscle blocks that run along both sides of the body can be seen through their transparent skin.

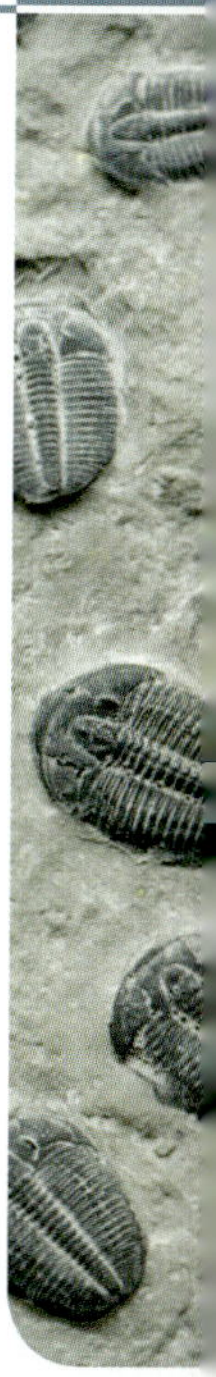

Pikaia

GROUP Cephalochordates
DATE Early Cambrian
SIZE 5cm (2in) long
LOCATION Canada

A marine animal that lived about 530 million years ago, *Pikaia* is one of the earliest-known chordates, and belongs to the subphylum cephalochordata (see opposite). In appearance, *Pikaia* is similar to the modern lancelet, differing only in having a pair of antenna-like structures on its head. *Pikaia* was small and delicate, with a dorsal nerve cord running from front to back under which lay the supporting notochord. V-shaped muscles ran along the sides of its flattened body. A narrow fin membrane extended down the rear two-thirds of the body, broadening into a single tail fin that tapered to a point.

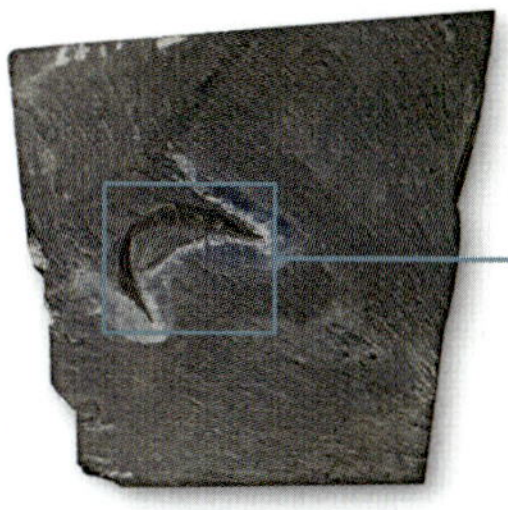

CANADIAN FOSSIL
Pikaia was discovered in the Burgess Shale deposits in British Columbia (see p.74), where it lived alongside a variety of invertebrates.

UNLIKELY ANCESTOR
Pikaia may not look very exciting but it is of great interest to scientists. It clearly shows traces of a notochord, dorsal nerve cord, and muscle blocks – key features in the evolution of all vertebrates.

Myllokunmingia

GROUP Myllokunmingids
DATE Early Cambrian
SIZE 2–3cm (¾–1¼in) long
LOCATION China

The oldest vertebrates by about 30 million years, *Myllokunmingia* and *Haikouichthys* (see below) were tiny, jawless, marine fish. *Myllokunmingia* was similar in form to *Haikouichthys* but less slender. It had a distinct head, and its body had backward-facing, V-shaped muscle blocks (myomeres). However, it differed from *Haikouichthys* in having pouch-like structures associated with its five or six gills – *Haikouichthys* had gill bars (and possibly more gills). *Myllokunmingia* fossils suggest that it had a cartilaginous skull and some primitive vertebral elements. Some parts of the digestive system are preserved but not the mouth, and no tail parts have been found. *Myllokunmingia* had a triangular dorsal fin that inclined gently upwards a short distance behind the head. Further back, on the underside of the body were fin folds (precursors of fins).

Haikouichthys

GROUP Myllokunmingids
DATE Early Cambrian
SIZE 2.5cm (1in) long
LOCATION China

Haikouichthys is thought to be one of the most primitive fish without jaws (agnathan). Preserved in the 530-million-year-old marine sediments of Chengjiang (see opposite), *Haikouichthys* is unlike any other agnathan. It had a rounded extension to the head which bore sensory organs – the eyes, and possibly nasal sacs and otic capsules associated with hearing. On the sides of its body, it had at least six, and possibly up to nine, gill slits supported by gill bars. V-shaped muscle blocks are also visible in places. *Haikouichthys* had a notochord but there is also evidence of vertebra-like elements, similar to those seen in modern lampreys. A number of circular structures on the lower part of the body may have been slime organs as these are also seen in hagfish.

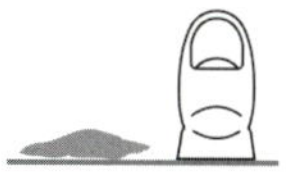

CONTINUOUS FIN
Haikouichthys had a prominent dorsal fin and there is some evidence that it was supported by radials. It was continuous with the tail fin, which in turn was joined to the narrower ventral fin fold.

530 The age in **millions of years** of Haikouichthys and Myllokunmingia. They had small **vertebral elements** that represent the **first backbone**.

ORDOVICIAN

Ordovician

During this period, the nature of marine faunas changed dramatically. Cambrian faunas were replaced by a more diverse assemblage in the mid-Ordovician, and animals, in the form of arthropods, first colonized the land. At the end of the Ordovician, the first of five major mass extinctions of flora and fauna occurred. Trilobites, echinoderms, brachiopods, graptolites, bryozoans, and reef-builders were particularly badly hit.

ARTHROPOD TRACKS
These trackways were made by an early arthropod on wet, rippled sand about 450 million years ago. They are preserved in the Tumblagooda Sandstone, Western Australia.

Oceans and continents

The rapid plate movement and volcanism of the Cambrian Period continued, resulting in a profound reorganization of the positions of the continents and ocean basins. Extensive shallow epicontinental seas with almost flat sea beds surrounded the main landmasses. Into these seas were shed vast amounts of sediment eroded from barren mountain belts still largely devoid of plant life. Rising sea levels and tectonic activity reduced smaller landmasses to a series of archipelagos. Four major continents existed, Gondwana, Laurentia, Baltica, and Siberia. By the Middle Ordovician, Siberia had moved from its southern hemisphere position to the northern hemisphere, nearer Laurentia (which lay across the Equator). The vast Panthalassic Ocean remained to the north of these landmasses. Gondwana still extended from north of the Equator to the South Pole. During the Ordovician, it rotated anticlockwise carrying the areas that form today's Australia and part of Antarctica into the northern hemisphere. The small islands of the South China block lay off Gondwana's western margin. In the Paleo-Tethys Ocean, between Gondwana and Laurentia, the smaller landmass of Baltica shifted into lower latitudes. Meanwhile, the Iapetus Ocean continued to widen. The Ordovician landscape was one of barren continents, with frequent volcanic eruptions and earthquakes, and coastlines that were constantly reshaped. The wide, shallow seas were inhabited by extensive coral reefs and a great diversity of marine invertebrates, until many species were wiped out in a mass extinction at the end of the Ordovician.

MOUNT ST. HELENS ERUPTION
Towering columns of ash were thrown up by Mount St. Helens, Washington State, USA, when it erupted in 1980. Ordovician volcanism was on an immense scale, producing the largest ashfalls known.

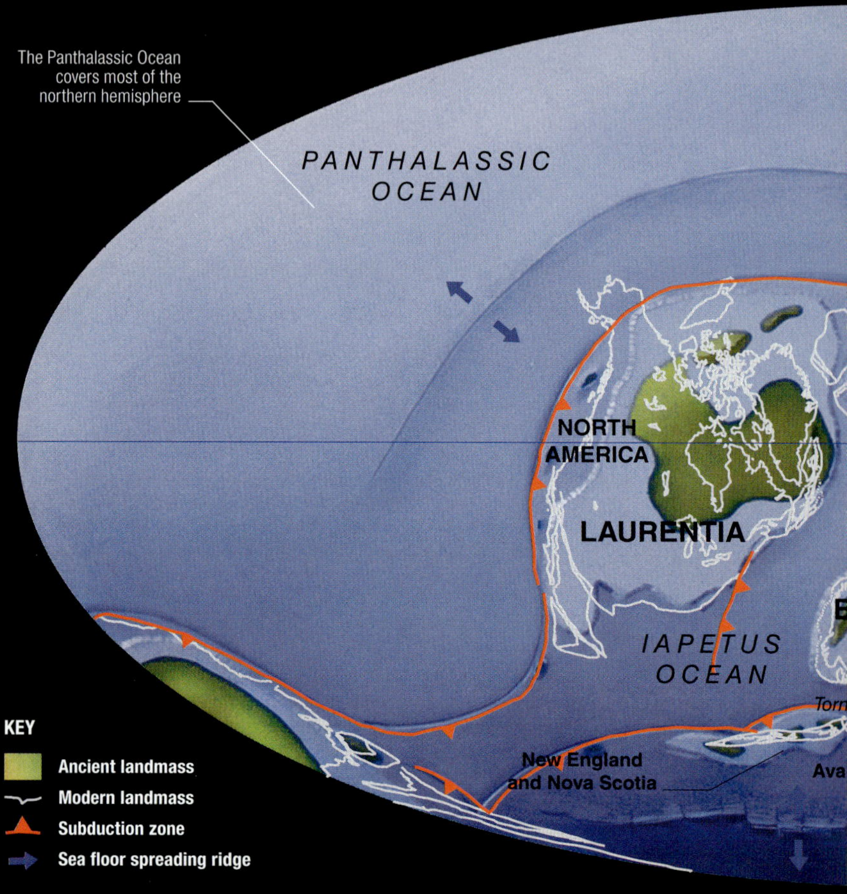

ORDOVICIAN WORLD MAP
Continents moved by intense tectonic activity. Gondwana rotated, Baltica shifted southwards, and Siberia moved closer to Laurentia.

EARLY | **MIDDLE**

MYA 490 | 480 | 470

PLANTS

- 475 First non-vascular plant spores

INVERTEBRATES

- 480 First asteroids (starfish, left); first ophiuroids (brittlestars); first crinoids; first chitinozoans; diversification of trilobites and nonarticulate brachiopods; first strophomenid brachiopods

STENASTER

ORTHOGRAPTUS CALCARATUS

- 475 First stenolaemate bryozoans; first graptolite (above) and nautiloid radiations; first eurypterids (sea scorpions); first craniid brachiopods
- 470 First rhynchonellid brachiopods

PLATYSTROPHIA

VERTEBRATES

- 485 First vertebrates with bone – the jawless fishes

RED SEA TEMPERATURE
The waters of the Red Sea and the Persian Gulf today can reach a temperature of 42°C (108°F) at their upper limit. Similar temperatures were found in oceans at the beginning of the Ordovician, before they began to cool.

CORAL REEF
Reef-building animals secreted calcium carbonate, building up structures that formed the basis for new ecosystems in Ordovician seas.

Climate

Like the Cambrian Period that preceded it, the Ordovician was initially very warm, with ocean temperatures of about 42°C (108°F). However, a prolonged period of cooling soon began, with temperatures dropping to 23°C (73°F) by the end of the Early Ordovician. Greenhouse conditions gave way to those of modern equatorial oceans. Temperatures remained reasonably constant for about 25 million years during the Middle to Late Ordovician. This period of relative stability was followed a period of rapid temperature decline, due to extensive glaciation close to the end of the Ordovician. Evidence for this is widespread, in Gondwanan rocks in Arabia, the Sahara, western Africa, Canada, and South America. Laurentia, situated at a much lower latitudinal position, was not affected by the glaciation. Glacial deposits indicate that continental ice sheets existed in Africa and Brazil, and Alpine-type glaciers formed in the Andes region. The glaciation was at a high latitude, centred in the South Pole, where southern Gondwana was located at the end of the Ordovician. It began and was most intense in the area of the supercontinent that was to become North Africa. Conditions during the glacial period were arid, becoming more humid during the Ordovician's final phase. Life prospered during most of the Ordovician, until Earth re-entered this icehouse phase. By the Early Silurian, modern equatorial temperatures had returned.

FOSSILS ON EVEREST
Fossil fragments show that some limestone strata on Mount Everest's summit were laid down in warm Ordovician seas. Uplifted 60 million years ago, rocks on this peak are scoured by glaciers today.

Gondwana rotates, taking parts of Australia and Antarctica north of the Equator

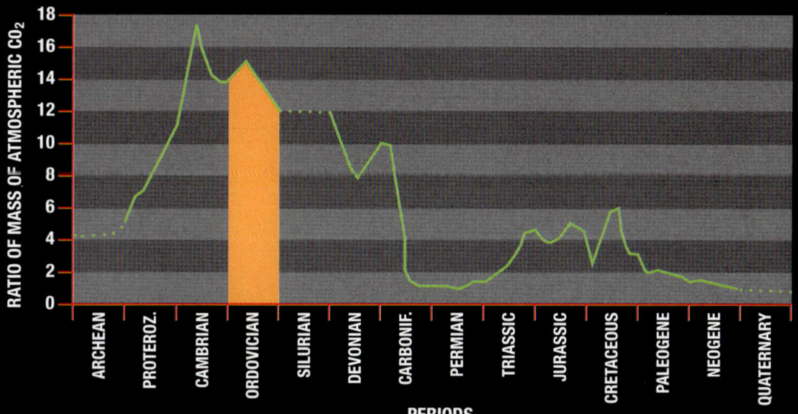

CARBON DIOXIDE LEVELS
While temperatures were high, carbon dioxide levels would have been considerably higher than today, but declined when the Ordovician world entered a period of intense cold.

LATE

460 | 450 | 440 | MYA

PLANTS

INVERTEBRATES

VERTEBRATES

AMBONYCHIA RADIATA

465 Gastropod radiation in Laurentia; bivalves (above) spread in Gondwana; radiation of conodonts, bryozoans, bivalves, acritarchs, first articulated brachiopods; first atrypid brachiopods

460 Major echinoderm (below), ostracod, and coral radiations; metazoan reefs develop

MALOCYSTITES MURCHISONI

455 First spiriferid brachiopods

450 First land arthropod burrows – millipedes; first complete known conodont animal; first echinoids

445 Mass extinction – 50% of species go, mainly trilobite (below) and echinoderm families, and genera and species of these. Also corals, nautiloids, brachiopods, graptolites, conodonts, and acritarchs

TRIARTHUS

ORDOVICIAN INVERTEBRATES

An important evolutionary phase during the Ordovician records the most significant increase in marine life during Earth's history. However, in contrast with the Cambrian "explosion", relatively few new invertebrate body plans were introduced during this period.

The Great Ordovician Biodiversity Event, during which marine biodiversity tripled, was one of the two most important evolutionary events in the history of life, second only to the Cambrian "explosion" (see p.70).

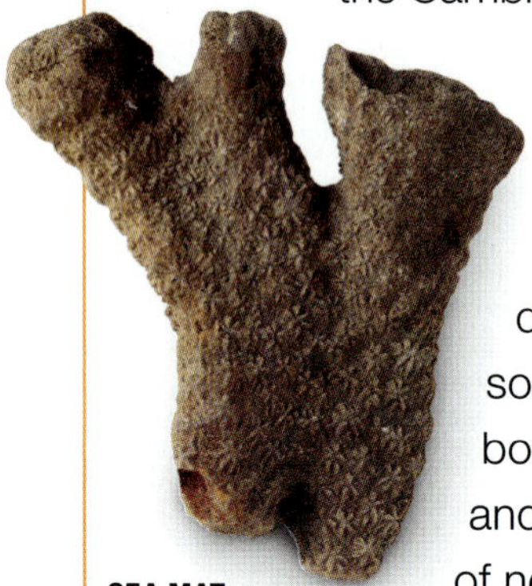

SEA MAT
This "stony" bryozoan had a strong, calcitic skeleton. Bryozoans first arose in the early Ordovician.

Great biodiversification

During the Early and Middle Ordovician there was a phenomenal proliferation and diversification of marine life, phased over some 25 million years. This resulted from both geological and biological processes, and the development of new ecosystems. In particular, there were more continents than at any other time in the Paleozoic, each with surrounding continental shelves. These continents provided a great number of new habitats. Sea levels were high, the continental shelves were flooded, and warm climates prevailed until the end of the Ordovician. Moreover there were profound evolutionary changes, both in the plankton (possibly fuelled by a great input of mineral-rich volcanic material), and in the animals that lived in or near the sea bed, especially the suspension feeders. During this time, the first large reefs, constructed by algae, were established.

FILTER FEEDER
This strophomenid brachiopod, typical of the Early Paleozoic, lay flat on the sea floor and had two calcitic valves, which let water currents in and out.

Mass extinction

Towards the end of the Ordovician, an intense though short-lived glaciation event took place, with ice extending to fairly low latitudes. Sea levels were drastically lowered, much living space was lost on the continental shelves, and tropical faunas suffered particularly through decreased temperatures. The cool-water forms, however, were less affected since they tracked the rims of the ice.

ARTHROPOD ADAPTATION
Trilobites diversified in the Ordovician. This specimen has thin lines crossing the head that enabled the exoskeleton to break easily when the animal moulted.

GROUPS OVERVIEW

The Ordovician saw tremendous radiation of species, but the end of this period was a time of mass extinction due to glaciation. The open ocean was never re-colonized by trilobites thereafter. The subsequent Silurian invertebrate faunas were, in some ways, impoverished versions of those of the Ordovician.

BRYOZOANS
Bryozoans made their first appearance in the Ordovician. Colonial organisms, they consist of small, soft-bodied animals residing in cavities within a calcite skeleton. While the Ordovician forms are extinct, modern bryozoans are an important component of modern-day ocean fauna.

ORTHID BRACHIOPODS
Orthids are the simplest of the Ordovician articulated brachiopods. They were anchored by a pedicle to the sea floor, with their two calcite valves held upright and open during feeding. They were affected by the extinction that occurred at the end of the Ordovician, but survived until the end of the Permian.

BIVALVES
Ordovician bivalves were quite diverse, mainly inhabiting nearshore environments where the otherwise dominant brachiopods were less common. Most of them were filter feeders or fed on bacteria extracted from the sediment, in the same way as the modern *Nucula*.

GRAPTOLITES
These form the main preservable component of the plankton from the Early Ordovician to about the middle of the Devonian. They are colonial animals and are excellent zonal fossils. They are usually flattened, but rare three-dimensional preservation shows them to be of surprisingly complex form.

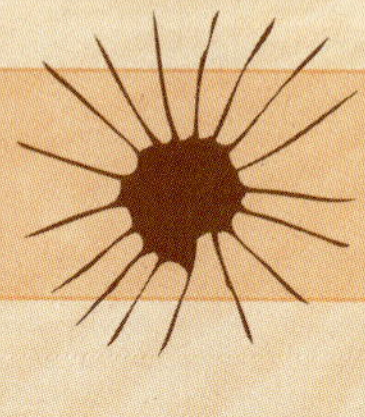

Sponge anchorage spines

GROUP Sponges
DATE Cambrian to Devonian
SIZE Up to 4.5cm (1¾in) long
LOCATION Worldwide

During the Ordovician period, large areas of the sea bed were covered with loose mud that was unsuitable for many of the bottom-dwelling animals of the time. However, one group was able to colonize the sea floor – the glass sponges, so called because they possessed skeletons consisting of silica. These simple animals evolved extremely long spicules, or spines, to anchor themselves in the soft mud so that they could exploit the multitude of tiny food particles in the surrounding waters; these spicules also enabled them to maintain their comparatively small sponge bodies above the sediment without being swept away.

SPONGE ANCHORAGE SPICULES
As the spicules of these glass sponges were often very long and fairly thick, they are far more conspicuous fossils than those of the small sponge bodies they supported.

CATENIPORA SP.
The corallites have thick walls with horizontal partitions, which look like chains when seen in cross section.

The corallite structure looks like an array of tiny panpipes.

Catenipora

GROUP Anthozoans
DATE Late Ordovician to Late Silurian
SIZE Corallite 1–1.5mm (1/16in) across
LOCATION Worldwide

Catenipora was a colonial tabulate coral (see p.100) composed of elliptical corallites arranged in a single series. A soft-bodied polyp (tiny animal with tentacles and a mouth) would have lived in each corallite. The rows of corallites join and divide, forming a network with spaces (lacunae) in between, often filled with sediment. From the side, the corallite structure looks like an array of tiny panpipes. *Catenipora* lived in warm shallow seas, possibly with only the top of the corallites poking above the surface of the sediment.

Constellaria

GROUP Bryozoans
DATE Middle Ordovician to Early Silurian
SIZE Branch 1–1.5cm (⅜–½in) across
LOCATION Worldwide

Constellaria was a branching bryozoan colony. The branches were erect and fairly thick but were sometimes compressed in one direction. The surface of the colony was covered in distinctive star-shaped mounds called maculae. These structures have been interpreted as "chimneys" to expel water away from the surface of the colony, after it had been filtered to obtain food for the organism.

CONSTELLARIA SP.
Every branch surface was covered in tiny, star-shaped maculae.

Diplotrypa

GROUP Bryozoans
DATE Ordovician to Silurian
SIZE Colonies up to 10cm (4in) across
LOCATION Worldwide

Diplotrypa was a trepostome bryozoan that formed large dome-shaped colonies. The fossilized colony surface is covered with many circular holes. These were the openings to long, tubular structures called autozooecia, which were divided into parts by thin diaphragms. In life, soft-bodied feeding zooids (autozooids) lived within these autozooecia. Each autozooid had a lophophore, which consisted of a ring of tentacles that surrounded the mouth. Food particles were driven towards the mouth by small, hair-like cilia on the tentacles, which created a flow of water by rapidly beating up and down. When the zooids were not feeding, the tentacles were withdrawn into the autozooecia for protection. Between the autozooecia there are smaller tubes that would have housed the non-feeding zooids.

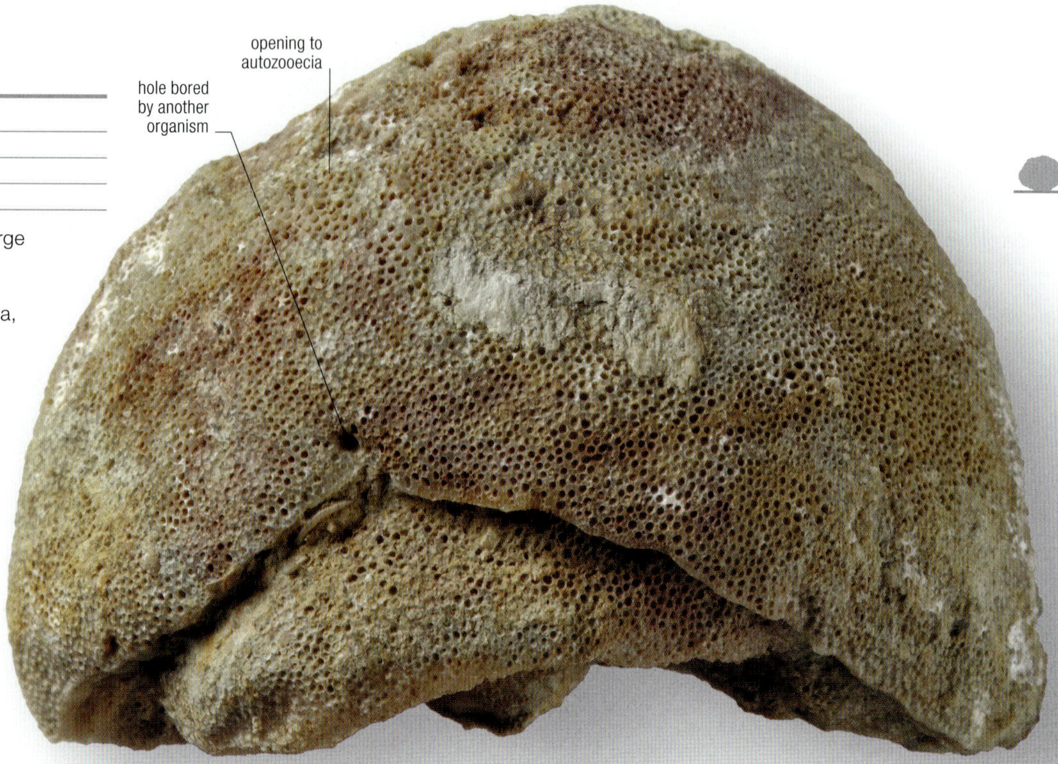

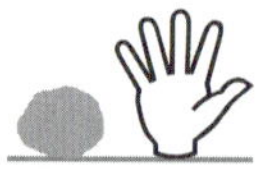

DIPLOTRYPA SP.
Diplotrypa often featured large holes in the colony, which were produced by another organism boring into it, possibly a polychaete worm.

Dinorthis

GROUP Brachiopods
DATE Middle Ordovician
SIZE Up to 3.5cm (1½in) long
LOCATION Europe, Russia, North America

Dinorthis was an orthid brachiopod, which was a group that ranged through the whole of the Paleozoic. The most notable feature of *Dinorthis* are the sharp radial ribs, which are equally prominent on both valves. Another characteristic is that the smaller (brachial) valve is more strongly convex than the larger (pedicle) valve, and in some species, the latter may even be concave. The margin of the shell is rounded, so that in outline, the shell has the shape of a letter D.

DINORTHIS ANTICOSTIENSIS
The ribs are very well defined on both valves of *Dinorthis*. The beaks of both valves project slightly above the straight hinge-line.

Rafinesquina

GROUP Brachiopods
DATE Middle to Late Ordovician
SIZE Up to 4cm (1½in) long
LOCATION Europe, Asia, North America

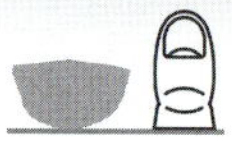

Rafinesquina was a common brachiopod, ornamented with strong concentric ribs. The larger half of the shell (the pedicle valve), through which its stalk (pedicle) passed, is convex in shape, while the smaller half (the brachial valve), is concave and fits neatly inside it. This would have meant that the animal living inside was very thin. The beaks project slightly above the centre of the straight hinge-line.

R. ALTERNATA
The hinge-line on this valve is clearly visible in this fossil brachiopod.

Platystrophia

GROUP Brachiopods
DATE Middle Ordovician to Late Silurian
SIZE Up to 4cm (1½in) long
LOCATION Worldwide

Platystrophia was closely related to *Dinorthis*, but unlike *Dinorthis* (see above), both halves of its shell (valves) were strongly convex, which makes it appear quite thick when seen side-on. Both valves had very strong radial ribs, and there was a strong outward fold in the smaller (brachial) valve, which was matched by a corresponding inward fold in the larger (pedicle) valve. Some species developed fine pimples on the surface of both the brachial and pedicle valves, although these are not visible on the specimen below.

PLATYSTROPHIA PONDEROSA
The flexible stalk that anchored *Platystrophia* to the sea bed was attached to the shell at the V-shaped point in the centre of the hinge-line.

Endoceras

GROUP Cephalopods
DATE Middle to Late Ordovician
SIZE Up to 9m (30ft) long
LOCATION North America, N. Europe, Russia, E. Asia

Endoceras belonged to the same group as the modern-day *Nautilus*. It had a long, conical shell that was divided internally into many chambers, which were joined by a long tube (siphuncle) that emptied water from the chambers. The largest chamber at the wider end was undivided and housed the soft, muscular body of the animal. *Endoceras* would have swum horizontally, moving by expelling water rapidly out of a funnel, in the same way as a modern *Nautilus*. It would have been a formidable predator, and at up to 9m (30ft) long, was one of the largest animals of the Ordovician seas.

ENDOCERAS PROETFFORME
This cross section shows where the siphuncle would have been.

DIVIDING LINES
This section of shell shows the partitions (septa) that would have divided the internal chambers.

Orthonyboceras

GROUP Cephalopods
DATE Middle to Late Ordovician
SIZE 25cm (10in) long
LOCATION North America, Asia

During the Ordovician, there were many types of nautiloid cephalopods and this group evolved rapidly. *Orthonyboceras* was one of the more modest-sized members of this group, and is similar in some ways to *Endoceras* (see opposite), as both shells are fairly straight and contain many individual chambers. *Orthonyboceras* would have had a problem with excess buoyancy, which it counteracted by laying down calcium carbonate in its internal chambers to make itself heavier.

ORTHONYBOCERAS COVINGTONENSE
The wall (septa) of each chamber had a hole through which a long strand of soft tissue (siphuncle) passed. In this cross section, the siphuncle looks like a row of beads.

Maclurites

GROUP Gastropods
DATE Ordovician
SIZE Up to 7cm (2¾in) across
LOCATION North America, Europe, N.E. Asia

Maclurites had a large, heavy shell and a distinctive shape. One side of the shell was almost flat, while the opposite side was made concave by a wide umbilicus. Unlike most modern gastropods, it is thought *Maclurites* led a sedentary lifestyle. Rather than grazing, it removed tiny particles of food from the water.

MACLURITES SP.
Unlike most gastropods, *Maclurites*' shallow umbilicus faced upwards.

Ambonychia

GROUP Bivalves
DATE Late Ordovician
SIZE Up to 5cm (2in) long
LOCATION USA, Europe

Like modern-day mussels, *Ambonychia* spent its adult life fixed to a rock or similarly hard surface by means of a mass of strong, sticky filaments called a byssus. *Ambonychia* was particularly common in the shallow seas that existed over eastern North America and parts of the Baltic region during the late Ordovician, as the firm shelves provided ideal conditions for byssal attachment. The surface of both of its valves was covered with quite strong radial ribs.

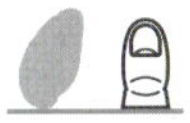

AMBONYCHIA RADIATA
Some of the radial ribs are still visible on this specimen.

Praenucula

GROUP Bivalves
DATE Middle to Late Ordovician
SIZE Up to 2.5cm (1in) long
LOCATION North America, Europe

Praenucula was a primitive bivalve mollusc, distantly related to the modern genus *Nucula*, which includes small shellfish known as nut clams. Unlike more advanced bivalves, *Praenucula* was a deposit-feeder, using special appendages to sweep food particles from the sea bed towards its mouth. Its gills were only involved with respiration, in contrast to more advanced bivalves that have modified gill margins that act like sieves to filter tiny particles of food from the water around them. In *Praenucula* fossils, both halves of the shell (valves) are convex and the beaks are almost centrally placed along the hinge-line.

PRAENUCULA SP.
This internal mould shows the pointed beak, which is positioned almost at the centre of the hinge-line.

Isorophusella

GROUP Echinoderms
DATE Middle Ordovician
SIZE 2cm (¾in) across
LOCATION Canada

Isorophusella belonged to an extinct group of echinoderms called edrioasteroids. Like its modern-day relatives, such as starfish, *Isorophusella*'s body had five-fold symmetry. On the surface were the five radiating ambulacra that carried the tube feet responsible for both feeding and respiration, though the food grooves and the mouth are hidden under the plates of the ambulacra. Around the periphery was a raised rim of scaly plates.

ISOROPHUSELLA INCONDITA
This species would have been attached by its underside to either the sea bed or a larger shell.

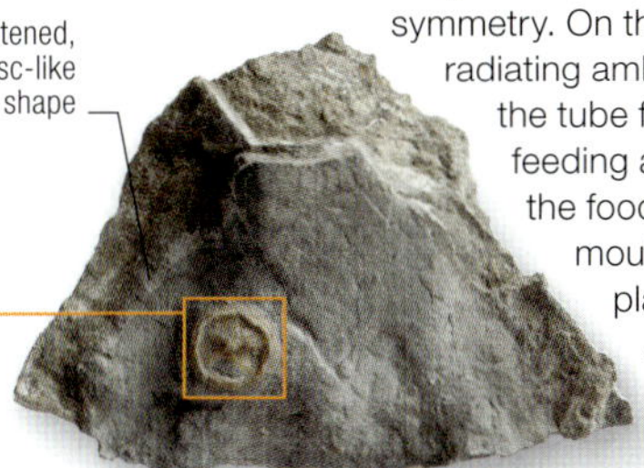

Malocystites

GROUP Echinoderms
DATE Middle Ordovician
SIZE 2.5cm (1in) high
LOCATION North America

This globular-shaped echinoderm belonged to a rare, short-lived class called the Paracrinoidea. They differed from crinoids (sea lilies), which still exist today, in that they possessed fixed arms, rather than free-moving ones. They also had fixed, rigid plating over the upper surface of the cup. Unusually, the anal opening (rather than the mouth) was in the central position on the upper surface of the cup. One arm originated from each end of the mouth and they branched irregularly over the entire surface of the cup.

MALOCYSTITES MURCHISONI
The cup was made up of around 30 irregular, polygonal plates. The anal opening was in the centre, and the slit-like mouth was to one side.

Eodalmanitina

GROUP Arthropods
DATE Middle Ordovician
SIZE Up to 4cm ($1\frac{1}{2}$in)
LOCATION France, Portugal, Spain

The headshield of this striking trilobite was dominated by its prominent crescent-shaped eyes, which had large, individual lenses. The central region of the head (glabella) was weakly inflated, with marked lateral indentations. The thorax was composed of 11 segments. After the sixth segment, they became gradually narrower. On the eighth and ninth segments of the specimen shown, there is some damage that reveals the surface of the joint beneath the preceding segment. The tailshield was triangular and ended in a short, narrow spine.

The eyes of *Eodalmanitina* were composed in a series of large lenses, arranged in vertical files.

EODALMANITINA MACROPHTALMA
The triangular cheeks of *Eodalmanitina* terminate in short, backward-facing spines. Part of the right hand one can be seen on this specimen.

Cyclopyge

GROUP Arthropods
DATE Early to Late Ordovician
SIZE Up to 3cm (1¼in) long
LOCATION Wales, England, France, Belgium, Czech Republic, Kazakhstan, S. China

The most characteristic feature of this trilobite was its enormous eyes, which occupied the whole of each side of the headshield. They were composed of hundreds of tiny lenses, and the surfaces were curved in such a way as to give almost 360 degree vision. This is in contrast to most trilobite eyes, where the field of vision was confined to a much narrower band.

SHORT BODY
The thorax is composed of five narrow segments, with a wide axis and narrow pleurae.

TRIARTHRUS EATONI
Unusual conditions of preservation in one fossil bed meant that the antennae and limbs have been preserved.

Triarthrus

GROUP Arthropods
DATE Late Ordovician
SIZE Up to 3cm (1¼in) long
LOCATION N.E. USA, Norway, Sweden, S.W. China

Triarthrus is one of only several trilobites that have been found with limbs preserved. It had a pair of slender antennae on the headshield, and behind these were three pairs of two-branched (biramous) limbs. Each limb consisted of a walking leg and a gill branch, with which it breathed. Similar limbs were present on each body (thoracic) segment, and on the tailshield. The limb segment at the base of each limb (coxa) had an array of spines on its inner side. Acting together, the coxae were probably used to pass food forwards to the mouth, which lay underneath the headshield.

Trinucleus

GROUP Arthropods
DATE Middle Ordovician
SIZE Up to 3cm (1¼in) long
LOCATION Wales, England

A trinucleid trilobite was among the first trilobites ever to be illustrated, in 1698. This group was one of the most widespread among trilobites in the Ordovician. The distinctive headshield is made up of three smooth lobes, around which is a fringe. The function of the fringe is a mystery. It may have been a sensory device, or was possibly involved in filtering food particles out of the water. The lack of eyes is not a primitive feature. Many different lineages of trilobites lost their visual organs.

TRINUCLEUS FIMBRIATUS
The style of the fringe varies significantly between different genera of trinucleid trilobites.

Orthograptus

GROUP Graptolites
DATE Late Ordovician to Early Silurian
SIZE Up to 6cm (2¼in) long
LOCATION Worldwide

Orthograptus is a biserial graptolite, which means that its single branch (stipe) had two rows of cups (thecae), arranged back-to-back, on either side of it. It is commonly found flattened and compressed, but in life it was oval or rectangular in cross section. The thecae have either an angular opening or, in some cases, a pair of short, slender spines protruding from the lip. Some species have three short spines at one end. *Orthograptus* specimens are most commonly preserved flattened in shale.

ORTHOGRAPTUS CALCARATUS
This graptolite has serrated edges, caused by the flattened openings of the tubular thecae.

D. MURCHISONI
The openings of the thecae are only on the inner edges of the stipes.

Didymograptus

GROUP Graptolites
DATE Middle Ordovician
SIZE Up to 5cm (2in) long
LOCATION Worldwide

Didymograptus includes a group of graptolites with two branches (stipes). Examples like this are commonly known as "tuning fork" graptolites, because of their shape. The key to identifying graptolites lies in the way in which the first few cups (thecae) in the colony are arranged, and this requires well-preserved specimens. The sicula – the theca that accommodates the first soft-bodied individual (zooid) in the colony to develop from the embryo – lies at the origin of the two branches in *Didymograptus*. The stipes are almost vertical and up to 40 or more thecae may be present.

Rhabdinopora

GROUP Graptolites
DATE Early Ordovician
SIZE Up to 12cm (4½in) long
LOCATION Worldwide

Rhabdinopora appear fossilized as triangular networks of branches that radiate from an apex. In life, the colony was conical, but it is invariably preserved flattened on its side, except in the case of small, young colonies that can be preserved in plan view, giving them a star-like pattern. *Rhabdinopora* belongs to the dendroid group of graptolites. Most of these lived on the sea bed, but *Rhabdinopora* evolved a pelagic lifestyle, inhabiting the upper layers of the open sea. It was widespread in the early Ordovician oceans, and is an important fossil for recognizing Lower Ordovician rocks across the world.

RHABDINOPORA SOCIALIS
At the apex is the sicula, a tiny conical cup that was occupied by the first zooid (soft bodied part) to develop.

Selenopeltis

GROUP Arthropods

DATE Early to Late Ordovician

SIZE Up to 12cm (4 ½in) long

LOCATION British Isles, France, Iberia, Morocco, Czech Republic, Turkey

Selenopeltis was an unusual spiny trilobite. The headshield was almost rectangular and the central region of the head (glabella) tapered slightly towards the front. The eyes were small and situated about halfway between the glabella and the side of the headshield. The cheeks were drawn out into long, slender, genal spines that extended back beyond the tip of the tailshield. The thorax was made up of nine segments. The central lobe of each segment had a large, convex lobe at the side. The tip of each segment (pleurae) had distinctive, arched ridges that reached beyond the end of the segments into long, slender spines. Spines at the front extended as far, or even farther, back than the tip of the tailshield. Each segment also had shorter, curved spines underneath the longer spines. The tailshield was short and much smaller than the head. It had a short axis with two or three rings, the first giving rise to a curved, raised rib that extended to form a backward-pointing spine.

SPINY TRILOBITES
This slab of rock contains examples of three different trilobite genera: *Selenopeltis*, which has long spines on both flanks of the body (see large specimen, far left); *Dalmanitina*, which is smaller with one long spine extending back from the tip of the tail (directly below *Selenopeltis*); and *Calymenella*, which is large and elongated with an even, round outline (bottom, right).

Selenopeltis **may have favoured cooler waters along the margins of Gondwana, vanishing from the more northerly latitudes after the Mid Ordovician, when plate tectonic movements took them further south.**

EARLY
MIDDLE
LATE
Tremadocian
Floian
Dapingian
Darriwilian
Sandbian
Katian
Hirnantian

ORDOVICIAN VERTEBRATES

Many organisms increased in size, strength, and speed during the Ordovician, although most marine life remained quite small. The seas saw the arrival of jawless fish that sported a major development in vertebrate evolution: bony plates and scales.

Microvertebrate faunas contain animals that are known from bits of bone, scales, or teeth, generally obtained by breaking down large quantities of rock (often by acid preparation), then patiently sorting through what remains. Ordovician rocks from around the world have proved to be a rich source of microvertebrate fauna, which have revealed new lines of vertebrate evolution.

The origin of bones and scales

Bony plates, scales, and possibly teeth appeared for the first time in vertebrates in the Ordovician, but the reasons for the evolution of these characteristic vertebrate features are still poorly known. Bone is composed of calcium phosphate and some scientists have suggested that it evolved as a way for the animal to remove excess phosphorus from its body, and perhaps to then use the bony skeleton as a phosphorous reservoir, available to be re-absorbed when needed (phosphorus is an important component in several physiological processes). Bony plates and scales could also have evolved as a means of providing protection against predators or parasites, or as insulation for electroreceptors (sensory organs sensitive to electrical signals).

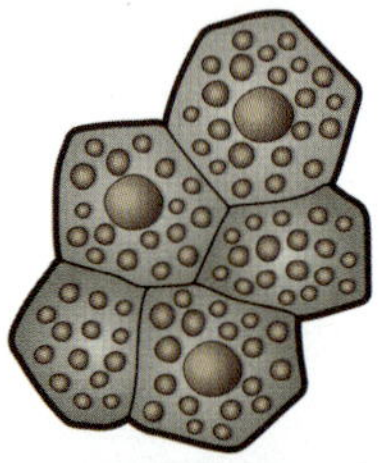
ASTRASPIS

PORASPIS

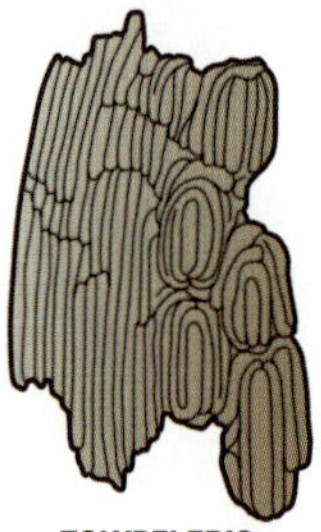
TOLYPELEPIS

DEVELOPMENTS IN FISH DERMAL PLATE ORNAMENTATION
The headshield of *Astraspis* had polygonal plates with a central tubercle around which successive zones of tubercles were produced as the fish grew. In the Silurian fish *Poraspis*, the plates had longitudinal dentine ridges and only appeared when it was fully grown. *Tolypelepis* (Late Silurian) had both types of plate.

The conodont question

Conodont elements are microfossils that are widespread in marine sedimentary rocks from the Late Cambrian to the Late Triassic. They were first described in 1856 as tiny tooth-like elements, and became a useful tool in the dating of rocks. However, their origin remained a mystery until 1982, when a fortuitous discovery in a collection finally associated conodont elements with an actual animal. This "conodont animal", from Carboniferous rocks in Scotland, possessed V-shaped muscle blocks and what appeared to be eyes at the front of the body. Another conodont animal, from Ordovician rocks in South Africa, appears to have had muscles associated with the eyes. Held within the body, in the approximate region of the gill arches, were the conodont elements. Their assembled arrangement and the wear patterns on these elements were suggestive of a feeding mechanism. Conodont elements are said to be made of bone and enamel, and all of these features are characteristic of vertebrates. Despite this, the identification of conodonts as true vertebrates remains controversial.

GROUPS OVERVIEW

The jawless fish comprise eight groups of fossil fish and two living groups, the hagfish and lampreys (see p.78). They arose in the Ordovician with the arandaspids and astraspids, followed by the anaspids, thelodonts, galeaspids, heterostracans, osteostracans, and pituriaspids. None of these fossil agnathans survived beyond the Devonian (see p.30).

ARANDASPIDS
Arandaspids are the earliest-known jawless fish, appearing about 470 million years ago. Until the discovery of *Sacabambaspis*, *Arandaspis* was the only known Ordovician vertebrate from the southern hemisphere. Their headshields consisted of bony dorsal and ventral plates separated by a row of smaller plates. Like the astraspids, they died out in the Late Ordovician.

ASTRASPIDS
Astraspids were first described by the renowned American palaeontologist Charles Walcott in 1892, after he found fragments of bony plates in the Harding Sandstone in Colorado, USA. *Astraspis*, whose name means "star shield", is distinguished by the ornamentation on the plate's surface (see above). Astraspids appeared about 450 million years ago.

1 mm The average size of the tooth-like **conodont elements** that have been found. The **actual animals** were mostly 2–5.5cm long.

Arandaspis

GROUP Agnathans
DATE Early Ordovician
SIZE 20cm (8in) long
LOCATION Australia

Arandaspis is named after the Aranda Aboriginal people, who live in the Alice Springs area of Australia where it was discovered. It was preserved as impressions in 470-million-year-old shallow-water sandstone deposits, making *Arandaspis* one of the earliest known jawless fish. Shaped like a flattened oval, its bony headshield was separated into upper and lower halves by a series of about 14 small, square, horizontally running plates that covered and protected its gill openings. The eyes were small and faced forwards, and some palaeontologists think that nostrils opened between them. *Arandaspis* had no teeth in its ventrally situated mouth and it probably fed by filtering organic debris and micro-organisms from the sea floor. Its body was covered with long scales. To date, no tail has been found.

Astraspis

GROUP Agnathans
DATE Late Ordovician
SIZE 13–15cm (5–6in) long
LOCATION Central North America

The oldest-known American vertebrate, *Astraspis* had a distinctive headshield and tapering body. The upper half of the shield had five conspicuous longitudinal ridges and numerous polygonal plates (see opposite). The plate ornamentation and closely overlapping scales may have enhanced the fish's streamlined shape, which suggests that *Astraspis* lived in a marine habitat with strong currents or a tidal influence.

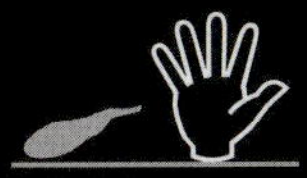

UNPROTECTED GILLS
Unlike *Sacabambaspis* and *Arandaspis*, the gill openings of *Astraspis* were not covered and the eyes were situated on the sides rather than at the front of the head.

Sacabambaspis

GROUP Agnathans
DATE Early Ordovician
SIZE 30cm (12in) long
LOCATION Bolivia

Discovered in 1986, *Sacabambaspis* was a jawless fish that inhabited coastal areas of a shallow sea that once extended across parts of North America. The head region was broad, with front-facing, close-set eyes. The upper and lower parts of the headshield were clearly demarcated by about 20 small plates, in between which the gill openings were hidden. The tapering body ended in a unique tail fin: in addition to a dorsal and ventral fin web, a long, rod-like extension of the notochord had a small fin web at its tip.

SALTWATER FISH
Sacabambaspis was a bottom-feeder that lived in salty coastal waters. The positioning and accumulation of the fossils found suggest the fish were killed by a sudden influx of fresh water that lowered the salinity beyond their tolerance level.

Because it lacked paired fins, *Sacabambaspis* was probably a poor swimmer.

SILURIAN

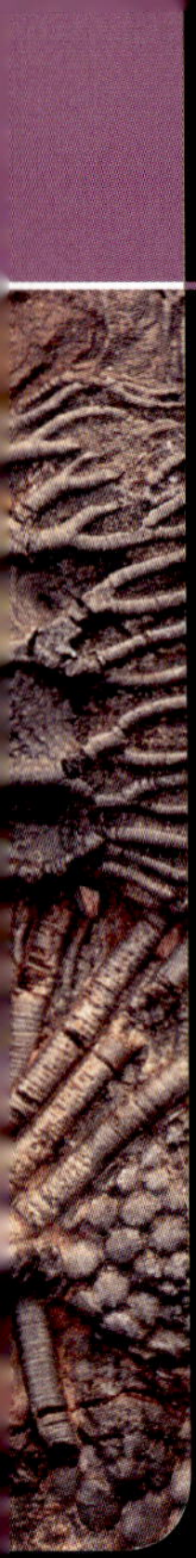

Silurian

The Siluran period was a time of consolidation for the many forms of marine invertebrates that had evolved, and of recovery from the Ordovician extinction event. Huge coral reefs, comparable in extent to Australia's Great Barrier Reef, flourished in sunlit tropical seas. New genera of fishes appeared in the oceans and fresh water, and the first small vascular plants began to colonize coastal areas of the continents.

NIAGARA DOLOMITE
Water rushes over hard dolomites at Niagara Falls, part of a long escarpment around the Michigan Basin. These rocks formed in tropical Silurian seas.

Oceans and continents

During the Silurian, the Iapetus Ocean began to close, as Baltica and another smaller land mass, Avalonia, moved north to collide obliquely on the southern and eastern margins of Laurentia. Avalonia essentially comprised present-day southern Britain and Ireland. Many pre-existing island arcs were lost during this collision, dragged down beneath plate margins by the powerful forces of subduction. The Rheic Ocean opened and widened through the Silurian to the south of this new, enlarged landmass and to the north of Gondwana. One of the results of the northerly movement of Baltica into Laurentia was a continued northward movement of Siberia into higher latitudes in the shrinking Panthalassic Ocean. The North and South China blocks started to shift away from the northern margin of Gondwana during the Silurian, and moved north across the Paleo-Tethys Ocean. Gondwana rotated into an even more southerly, polewards orientation, carrying present-day Australia southwards to lie across the Equator and moving all of Antarctica into the southern hemisphere. The closing of ocean basins and the rapid melting of ice sheets brought about a signicant rise in sea levels, helping to expand shallow sea environments for corals and fishes. Graptolites restablished themselves after the Ordovician extinction, and many species of these colonial animals were pelagic, drifting on global ocean currents. Their rapid evolution, short species lifespan, and widespread distribution make them invaluable index fossils for correlating rocks and past marine environments around the world.

DINGLE PENINSULA
Silurian clastic rocks in these spectacular cliffs on Ireland's Dingle Peninsula were formed when sediments were eroded, accumulated, and consolidated in ancient Avalonia.

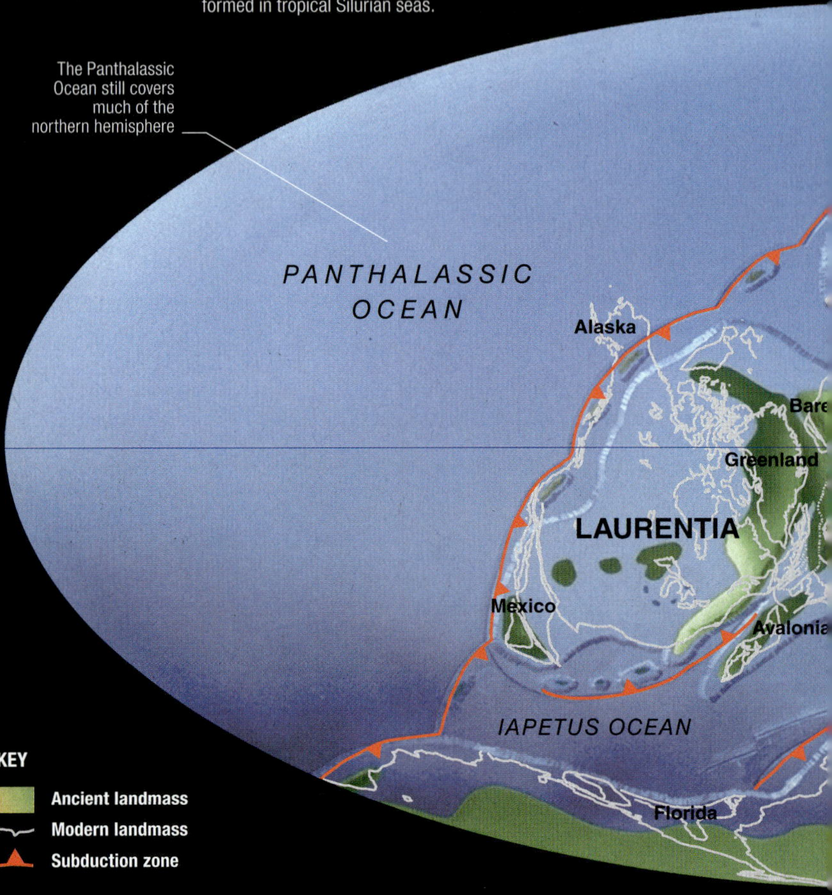

KEY
- Ancient landmass
- Modern landmass
- Subduction zone

SILURIAN WORLD MAP
Gondwana occupied the south, while Baltica and Avalonia continued northwards to collide with Laurentia, closing the north of the Iapetus Ocean.

LLANDOVERY

MYA	440	430
PLANTS	**440** Earliest spores of vascular (higher) plants	
INVERTEBRATES	**440–435** Progressive decrease in graptolite diversity	**430** First lucinid bivalves, e.g. *Cardiola*; **430–425** Major decrease in graptolite diversity
VERTEBRATES	**440** First agnathan fishes: Heterostraci, Galeaspida, and Pituriaspida; **435** First thelodont, agnathan fish with tooth-like scales	**430** Earliest "protosharks", earliest jawed fishes (acanthodians) and placoderm fishes; first osteostrachan agnathan fish; first anaspid agnathan fish

CARDIOLA

GLACIAL MELT
Glaciers advanced and retreated several times during the early Silurian. As the climate became warmer, the large-scale melting of glacial ice and ice caps brought about a rise in sea levels.

Climate

Estimating palaeotemperatures is extremely difficult. One study of fossil brachiopods puts global temperatures during the Silurian at between 34 and 64°C (93–147°F), while other studies suggest 21–45°C (70–113°F). However, the general consensus is that conditions were very warm. Recent research has also shown that the climate may have been more variable than was once thought. Glaciation that began in the Late Ordovician continued, with four major episodes of advancing ice during the first 15 million years of the Silurian. When the ice was at its maximum extent, it was cold at high latitudes and cool and humid nearer the Equator, whereas in the interglacial periods high latitudes were cool, and lower latitudes were warm and arid. During the later Silurian, limestones extended from their earlier limit of up to 35 degrees latitude to reach 50 degrees by the Carboniferous. Their wide distribution, particularly when deposited in low reef environments, indicates that warm oceans, and tropical and subtropical conditions expanded. Reefs were also widespread, extending northwards as far as 50 degrees.

LIMESTONE FORMATIONS
Fossil-rich limestones (above) on the Swedish island of Gotland date back to the late Silurian. Monumental eroded sea stacks (left), 8–10m (26 –33ft) high, stand in ranks on the beach of the tiny island of Faro, off Gotland's northern tip.

A new ocean, the Paleo-Tethys Ocean, begins to open

Gondwana rotates towards the South Pole, moving Australia across the Equator and Antarctica into the southern hemisphere

CARBON DIOXIDE LEVELS
Limestone deposits and widespread reefs to high latitudes suggest a return to greenhouse conditions. Earth entered a warm period, with higher levels of carbon dioxide than today.

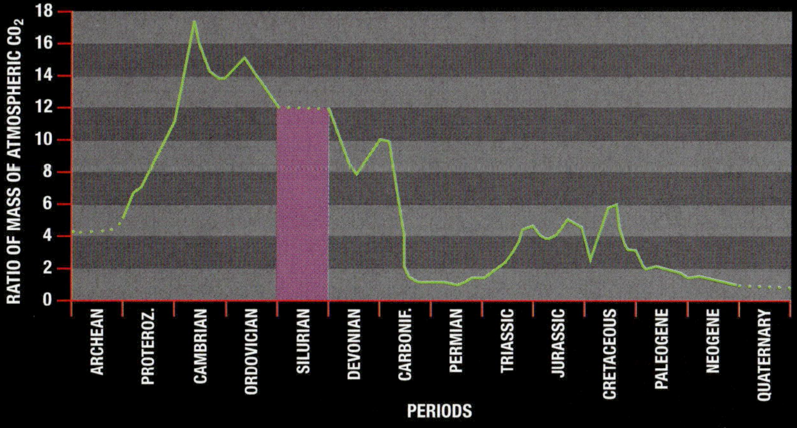

CK | LUDLOW | PRÍDOLÍ

420 | 410 | MYA

PLANTS

25 Oldest land plant
COOKSONIA

INVERTEBRATES

420 First trigonotarbid arachnid; first land scorpion

DIMEROCRINITES

VERTEBRATES

420 First ray-finned fish

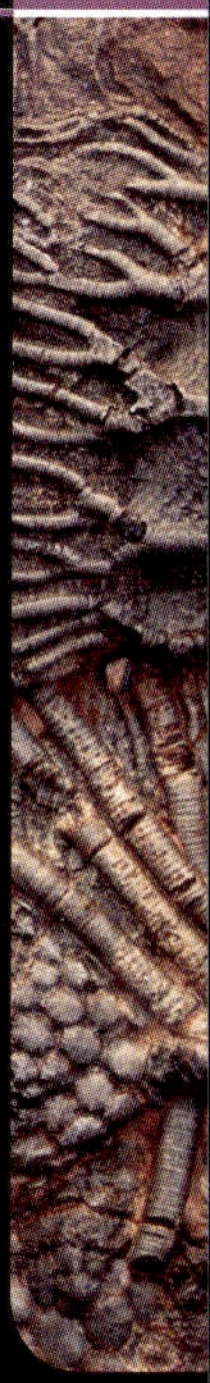

SILURIAN PLANTS

The first evidence of plants on land is from the Ordovician, but fossils of land plants become steadily more common through the Silurian. By the end of the Silurian, the diversification of plants on land and the creation of new kinds of land ecosystems was well underway.

There is clear evidence for a massive explosion of animal life in the sea around the beginning of the Cambrian, but the first evidence of well-developed ecosystems on land, with complex animals and plants, came much later, in the Silurian.

First steps onto land

The earliest indication that plants were beginning to colonize the land is provided by microscopic spores isolated from Ordovician rocks. These spores have a tough and resistant wall. This would have helped the living cell survive desiccation and dispersal through the air. The resistant wall also enhances the preservation of spores in the fossil record. From the Ordovician and Silurian onwards, spores, as well as pollen grains, which are their equivalent in seed plants, are a very important part of the plant fossil record.

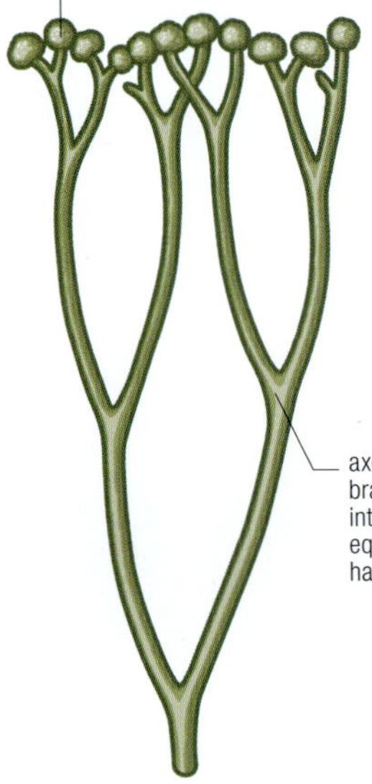

COOKSONIA
Several different kinds of early land plants have been included in the genus *Cooksonia*, but they all have small, slender, equally branched stems that bear sporangia at their tips

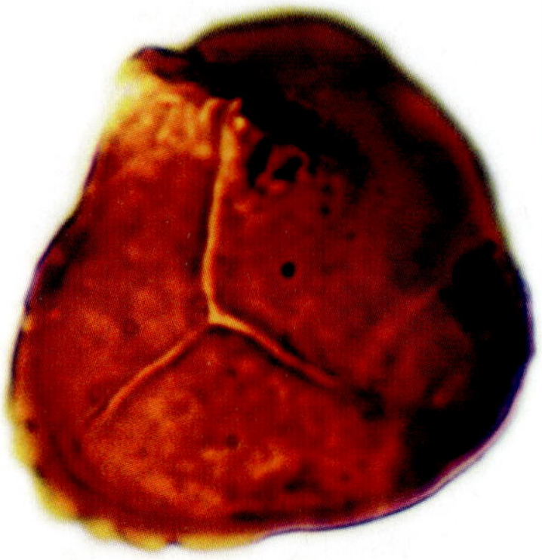

LAND PLANT SPORE
The spores of living and ancient land plants are produced in groups of four (tetrads). Each spore has a three-armed mark on one surface that is formed by contact with the other three spores in the tetrad.

Spores of early land plants become more common during the Silurian and are joined eventually by plant fossils assigned to the genus *Cooksonia*. Although these fossils are usually tiny, their branching axes and terminal sporangia (see left) are visible to the naked eye. Some *Cooksonia* fossils show the first evidence of specialized, water-conducting cells in the centre of their axes. These cells have special internal thickenings of the cell wall that stopped them collapsing as water was carried through them to the aerial parts of the plant.

Freshwater origins

Several lines of evidence suggest that the land surface was colonized from fresh water rather than directly from the oceans. The earliest land plants may have been a special kind of green alga that was adapted to living in temporary freshwater pools. The resistant-walled spores may have been an early adaptation that allowed these plants to disperse from pool to pool.

GROUPS OVERVIEW

Fossil plants from the Silurian are relatively rare, and usually rather small and poorly understood, but there is no doubt that the Silurian was a critical time in the early diversification of land plants. Some of the pioneering land plants of the Silurian may have been closely related to living mosses, liverworts, and hornworts, but few of the early fossils from this time are known well enough to assess this in detail.

RHYNIOPHYTES
Rhyniophytes are a very varied group of extinct primitive land plants from the Silurian and Devonian. Different rhyniophytes may be related to different modern groups and include the early genus *Cooksonia*, as well as better-preserved fossil plants from the Devonian, such as *Rhynia*.

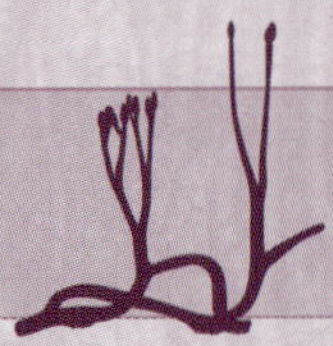

LYCOPHYTES
Lycophytes were the first group of land plants to diverge from rhyniophyte-like ancestors. They include living clubmosses as well as an important group of extinct plants known as the zosterophylls. The early lycophyte *Baragwanathia* is known from the Late Silurian of Australia.

The colonization of the land by freshwater plants appears to have only happened once in the history of plant life.

Psilophyton

GROUP Invertebrates
DATE Silurian
SIZE 20cm (8in) long
LOCATION Sweden

Fossils of early plants are sometimes difficult to distinguish from other objects, such as mineral growths or animal remains. *Psilophyton hedei* is a classic example of the confusion this can create. When it was found, it was identified as the world's earliest plant, but it is now thought to be a colony of animals related to graptolites. This is only one of several fossils classified as *Psilophyton*. Others, such as *P. burnotense* (see p.119), are known to be plants.

PSILOPHYTON HEDEI
Found on the Swedish island of Gotland, this plant-like fossil is probably a colony of marine invertebrates, borne on a branching "stalk".

Macroalgae

GROUP Algae
DATE Precambrian to present
SIZE Up to 80m (260ft) across
LOCATION Worldwide

Algae make up a large, heterogenous group of organisms that include the closest relatives of green plants. They evolved in water, where most still live today. The earliest forms were microscopic and single-celled, and ancestral to larger algae, including seaweeds. Red algae often form hard deposits of calcium carbonate that fossilize well, as do some green algae. Brown algae leave fewer fossils because they are uncalcified and less robust.

RED MACROALGA
Red seaweeds are the most common fossilized macroalgae. They typically grow beneath the water's surface, where they are more protected.

Cooksonia

GROUP Rhyniophytes
DATE Late Silurian to Devonian
SIZE 1–5cm (3/8–2in) high
LOCATION Worldwide

Growing not much taller than a pin, *Cooksonia* is one of the world's earliest-known land plants. It had slender, leafless branches with Y-shaped forks, topped by capsules that released microscopic spores. Some fossils have a dark stripe in their stems. This may be the remains of vascular tissue – the internal plumbing that most living plants have for moving water. *Cooksonia* grew on estuarine mud and other damp, low-lying habitats, forming dense mats. The first fossil specimen was found in Wales in 1934. Since then, a number of species have been discovered around the world.

COOKSONIA HEMISPHAERICA
Originally found in siltstone and mudstone in Wales, this species of *Cooksonia* is named after the shape of its spore capsules.

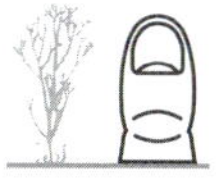

Baragwanathia

GROUP Lycophytes
DATE Late Silurian to Early Devonian
SIZE 25cm (10in) high
LOCATION Worldwide

For the Silurian, *Baragwanathia* was an unusually large and complex plant, with upright branches that grew from ground-hugging stems. Its surface was covered with simple, spirally arranged leaves – a feature that has persisted in lycophytes until the present day. Like all early land plants, *Baragwanathia* reproduced by growing spores. These formed in capsules tucked into its axils – the points where the leaves meet the stems.

BARAGWANATHIA LONGIFOLIA
Baragwanathia's simple leaves were up to 4cm (1½in) long. Its stems were several centimetres across, making this plant much larger than most others in Silurian times.

The entire surface of the plant was covered with simple, spirally arranged leaves.

SILURIAN INVERTEBRATES

The Silurian was a short period of time lasting only about 28 million years, but during this time the land was invaded by both plants and animals. It was a "greenhouse" period, with warm tropical seas bearing rich and diverse faunas, but it was also punctuated by a number of short-lived glaciations.

Plants colonized the land during the Silurian, the earliest, such as *Cooksonia*, being just a few centimetres high, with photosynthetic stems and bearing spore capsules. These were confined to damp places. Where there were plants, animals followed, and Silurian millipedes are known both from fossils and the tracks they made when walking. Marine Silurian scorpions also eventually made the transition onto land.

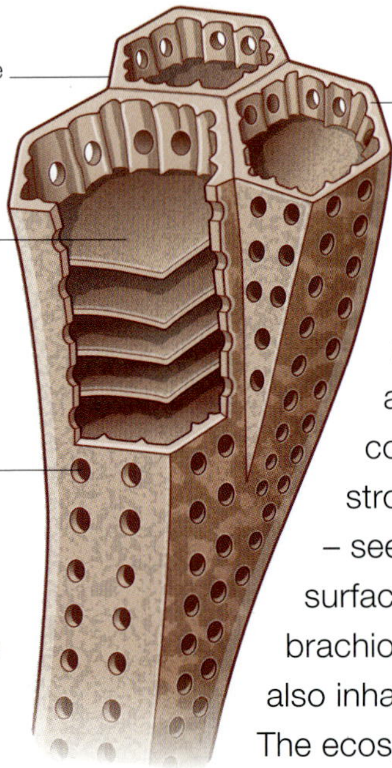

TABULATE CORAL
In this type of coral, the polyps were set in calcareous corallites; each corallite comprised a tube with a flat platform (tabula) below the polyp. Living tissue ran through the pores, connecting the individual polyps.

Extensive reef development

During the Silurian, extensive reef systems developed, following those of the Ordovician (see p.84). Silurian reefs are particularly well known from North America, the Swedish island of Gotland, and the Welsh Borderland. They grew as elongated barrier reefs or as smaller patch reefs – the latter can often be seen in vertical quarry faces, one stacked above the other, and consisting of pale, largely unbedded, fine, calcareous sediment, within a normal bedded sequence. These are not strictly coral reefs, for like most Paleozoic build-ups – as they are properly known – they were constructed by algae. Corals and stromatoporids (calcareous layered structures – see *Stromatopora*, opposite) grew on the surface of these build-ups; likewise crinoids, brachiopods, trilobites, snails, and bivalves, which also inhabited the sea floor between the build-ups. The ecosystem was greatly diverse, with patrolling cephalopods as the top predators. These Silurian environments have been the source of some of the world's best-preserved fossils.

REEF FORMATION
Algal reefs, such as the one above, are formed as one layer of carbonate after another is deposited. Paleozoic build-ups probably only rose a few metres above the level of the surrounding sea floor.

RETIOLITES GEINITZIANUS
This was a planktonic graptolite, whose skeleton consisted of a network of thin collagen strings, presumably surrounding the soft-bodied zooids. This may have aided buoyancy.

GROUPS OVERVIEW

Some invertebrate groups began to make the transition to land during the Silurian but it was also a time of great adaptation for marine invertebrates. Colonies of tabulate corals emerged, creating important habitats for many other invertebrate species. Nautiloid, graptolite, and some arthropod groups adapted successfully to the changing environments during this period.

TABULATE CORALS
One of the two major Paleozoic coral groups (the other being rugose corals), tabulate corals became extinct in the Permian. They are always colonial, never solitary, and were an important component of Silurian shallow-water fauna. Septa, the vertical partitions which divide the corallites, are usually small or absent.

NAUTILOIDS
Most of the Silurian nautiloid cephalopods had straight or curving shells; a few were fully coiled. These were the top predators of the time. They are distinguished by their straight, simple sutures (lines where the internal partitions join the inside of the shell), which are visible on internal moulds.

GRAPTOLITES
These continued to flourish throughout the Silurian and into the Devonian. Some Ordovician types – with thecae (which housed the zooids) on both sides of the stem – lasted to the Early Silurian; most Silurian graptolites, however, had thecae on one side only (monograptids). Some became large and elaborate.

ARTHROPODS
Trilobites continued through the Silurian – often being quite modified in form – and adapted to various environments. There were also eurypterids (water scorpions) living in shallow and often brackish or fresh water, and pod-shrimps, crustaceans like the living *Nebalia*.

Stromatopora

GROUP Sponges
DATE Silurian to Devonian
SIZE Colonies 5cm–2m (2in–6½ft) wide
LOCATION Worldwide

Stromatoporoids were sponge-like marine creatures that ranged from small to massive, and examples are commonly found in both Silurian and Devonian limestones. Made up of vertical tubes intersected by transverse structures, *Stromatopora* had closely spaced calcitic plates with strong vertical pillars and star-shaped elements called astrorhizae (these cannot be seen in this example). Each astrorhiza had a central, circular opening from which furrows radiated in an irregular series of markings. For years palaeontologists were unsure of how to group stromatoporoids. A modern group of sponges, the sclerosponges, have exhalant canals similar to astrorhizae, but they also have an aragonite skeleton, while that of stromatoporoids is calcitic. Therefore stromatoporoids are best regarded as a separate class of sponge.

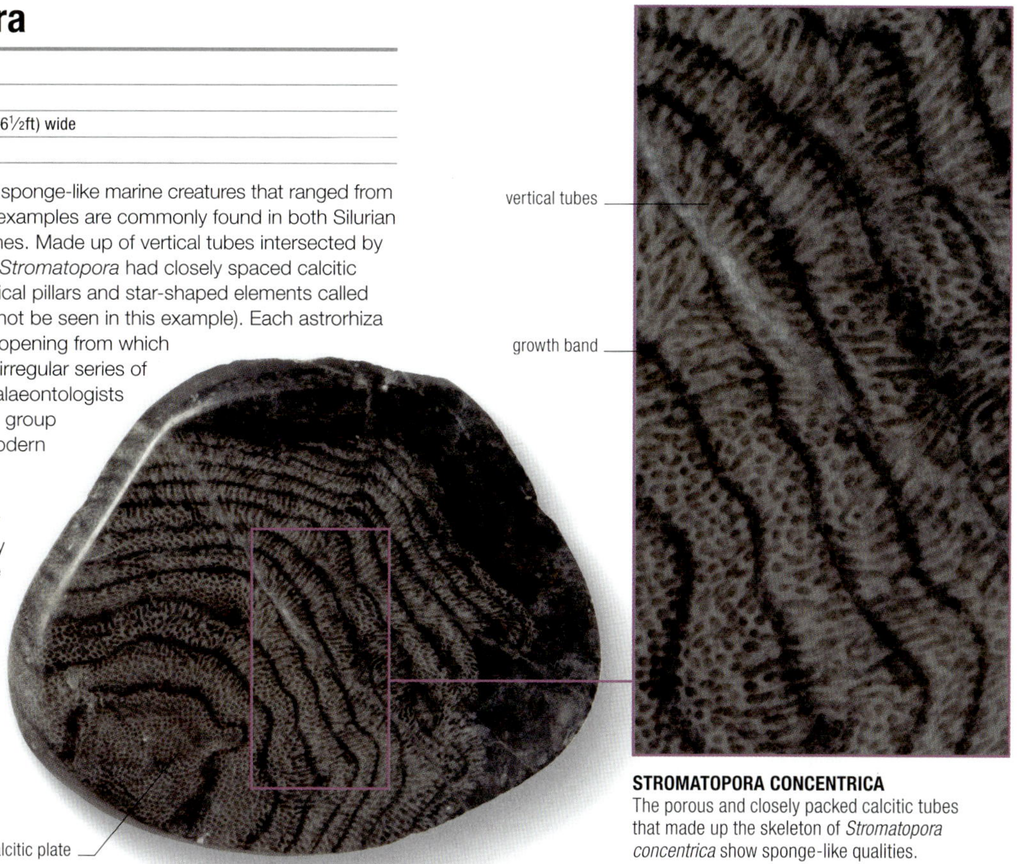

STROMATOPORA CONCENTRICA
The porous and closely packed calcitic tubes that made up the skeleton of *Stromatopora concentrica* show sponge-like qualities.

LIVING RELATIVE
SCLEROSPONGES

Sclerosponges are a widespread group of tropical marine sponges found at depths of from just below the ocean surface to 200m (656ft). They can be over 1m (3ft 3in) across and over 1,000 years old. Their banded skeletons can be used to determine past levels of carbon dioxide in sea water because, unlike corals, their skeletons reflect exactly the isotopes in the surrounding water. Recent evidence from genetic sequencing shows that sclerosponges are not a unified group, but belong to at least two different groups of "demosponges", which, independently, evolved skeletons composed of calcium carbonate.

Favosites

GROUP Anthozoans
DATE Late Ordovician to Middle Devonian
SIZE Corallite 1–2mm (1/16in) in diameter
LOCATION Worldwide

Favosites was a tabulate coral that lived in warm, shallow seas and formed colonies in a variety of shapes; this example is a flattened hemispherical form. The individual skeletons or corallites that made up the colony were many-sided in cross section. Their walls were thin and closely packed, giving the colony a honeycomb appearance. The corallite walls also had up to four longitudinal rows of pores. The septa, the vertical partitions that divided the inside of the corallites, were short and commonly in the form of rows of spines. Horizontal partitions (tabulae) were also present. It is unusual to find fossilized remains of the soft body parts, although calcified polyps have been found in examples of Silurian *Favosites*.

FAVOSITES SP.
The corallites or skeletons gave the colony a honeycomb appearance.

Heliolites

GROUP Anthozoans
DATE Middle Ordovician to Middle Devonian
SIZE Corallite 1–2mm (1/16in) in diameter
LOCATION Worldwide

Heliolites was a tabulate coral that lived in warm, shallow seas. It existed in different forms, which could be branching or in massive colonies. The soft-bodied, coral polyps would have rested on top of cylindrical skeletons called corallites. In cross section, corallites were smooth and circular, or had a scalloped edge made up of 12 segments. Twelve small, longitudinal partitions (septa) may have been present, extending a short distance from the corallite wall. There were horizontal partitions (tabulae) in the central regions of the corallites. Between the corallites, the skeletal tissue was made up of small, polygonal tubes divided with transverse diaphragms.

HELIOLITES SP.
Coral polyps once rested on the cylindrical structures known as corallites, which made up the framework of *Heliolites*.

KEY SITE
WENLOCK EDGE

Wenlock Edge is a British limestone ridge of mid-Silurian age that runs for 29km (18 miles) southwest to northeast along the Anglo-Welsh borders. Wenlock limestone has been studied for centuries. It was deposited when the area was covered in warm, shallow, tropical seas containing reef-like structures (bioherms) made up of stromatoporoids, corals, and bryozoans. These are found today as well-preserved fossils, along with those of crinoids, brachiopods, trilobites, and molluscs. The limestone has been extensively quarried, which has left large rock faces exposed for geological study.

Goniophyllum

GROUP Anthozoans
DATE Early to Middle Silurian
SIZE Calice 1.5cm (½in) in diameter
LOCATION Europe, North America

Goniophyllum was a solitary rugose coral, with a distinctive four-sided top section. The soft-bodied coral polyp lived in a concave cavity, called a calice. A lid-like structure (operculum), made of four thick, triangular plates, covered the calice, but this is not usually preserved in fossils. Contracting muscles shut the lid so that the polyp was protected when not feeding. The calice was deep, with short, thick partitions (septa), and its horizontal partitions (tabulae) and small, curved plates were thickened in zones.

GONIOPHYLLUM PYRAMIDALE
Four plates, not shown here, on the top of the coral could close so that the polyp was protected.

Ptilodictya

GROUP Bryozoans
DATE Late Ordovician to Early Devonian
SIZE Branch 2–15mm (1/16–½in) wide
LOCATION Worldwide

Ptilodictya was an upright bryozoan colony, made up of one straight or slightly curved branch. At the base was a conical socket that allowed the colony to articulate. Rectangular holes on the surface opened to the autozooecia, where the individual, soft-bodied animals known as autozooids lived. These feeding autozooids had a lophophore – a ring of tentacles surrounding their mouths, which could be withdrawn into the autozooecia when the zooid was not feeding.

PTILODICTYA LANCEOLATA
Small openings on the colony surface led to chambers where the autozooids lived.

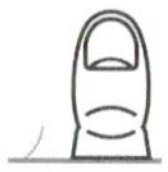

Leptaena

GROUP Brachiopods
DATE Middle Ordovician to Devonian
SIZE 1.5–4cm (½–1½in) long
LOCATION Worldwide

Leptaena was a genus of brachiopods that had an outward-curving pedicle valve or shell with an inward-curving brachial valve just inside it. The hinge-line was straight, and the beaks of both shells projected above it. The pedicle opening, where the fleshy stalk attaching the animal to a surface protruded, was located just below the pedicle valve beak. The shell's surface had strong concentric ribs crossed with fine radial ribs. Toward the front edge of the shell was a sharp bend in both valves.

LEPTAENA RHOMBOIDALIS
Leptaena's smaller brachial valve sat closely inside the pedicle valve, leaving room for only a very compressed body.

Pentamerus

GROUP Brachiopods
DATE Silurian
SIZE 2.5–6cm (1–2¼in) long
LOCATION North America, British Isles, Northern Europe, Russia, China

Pentamerus was large, with two outward-curving valves or shells; its length was usually greater than its width. The pedicle valve beak was prominent; the opening where the animal's fleshy stalk protruded lay just beneath it. Its shell surface was almost smooth, but featured fine radial ribs and fine growth lines. Inside, a thin dividing wall (septum) was located nearest the pedicle beak, and supported a large, spoon-like structure that acted as a muscle-attachment area. This feature is characteristic of pentamerid brachiopods. *Pentamerus* often lived in large groups.

Atrypa

GROUP Brachiopods
DATE Early Silurian to Late Devonian
SIZE 2–3cm (¾–1¼in) long
LOCATION Worldwide

Atrypa was a brachiopod or lamp shell with a flat or slightly outwardly curved pedicle or larger shell (valve) and a strongly outwardly curved smaller or brachial valve. The beaks of both shells projected slightly above the hinge-line. The shell surface was characterized by strong, concentric ribs crossed by radial ribs. There was a slight outward fold in the brachial valve and a corresponding fold or sinus in the pedicle valve. The supports for the animal's feeding organ, the lophophore, were located inside the brachial shell (see panel, right), and were arranged in spirals directed towards the centre of the shell.

ATRYPA SP.
Layers of concentric ribs characterize *Atrypa*, and these are crossed by radiating ribs – a trait found in modern bivalves.

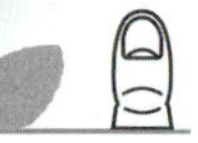

ANATOMY
LOPHOPHORE

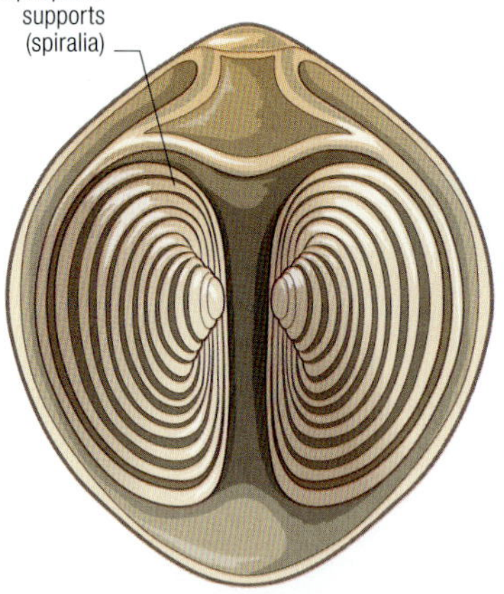

Like all other brachiopods, *Atrypa* fed by means of an organ called the lophophore, which carried large numbers of minute, tentacle-like cilia that trapped food particles and swept them into the mouth. The greater the lophophore's length, the better the animal's chance of collecting more food. Arranging the lophophore in a spiral provided an efficient way of getting a long lophophore into a small area, but it needed some support to keep turns of that spiral separated. Thus, in *Atrypa,* the lophophore supports took the form of two delicate, calcite spirals, the spiralia, located in the interior of the brachial valve.

Gomphoceras

GROUP Cephalopods
DATE Middle Silurian
SIZE 7.5–15cm (3–6in)
LOCATION Europe

Gomphoceras was a marine mollusc known as a cephalopod, the same class to which a chambered nautilus belongs. The thin walls inside the shell were closely spaced, and it had a large body chamber that narrowed towards the shell opening at maturity. The aperture is missing from this example, but from other specimens we know that this final opening was small, with little space for tentacles. This suggests that, at maturity, the animal could no longer feed, and died after mating, as do many modern cephalopods.

BODY SHAPE
Its awkward shape suggests *Gomphoceras* was not an agile carnivore; the animal may have fed on carrion.

Cardiola

GROUP Bivalves
DATE Silurian to Middle Devonian
SIZE 1.2–3cm ($\frac{3}{8}$– $1\frac{1}{4}$in) long
LOCATION Africa, Europe, North America

Cardiola was a strongly ribbed bivalve with well-marked growth rings. Both shells (valves) were of equal size, with prominent beaks. A smooth, triangular area lay beneath each beak, with fine growth-parallel ribs across it. Internally, two adductor muscles controlled closure of the valves, but there were no hinge teeth. A ligament would have opened the valves when the adductor muscles relaxed. *Cardiola* was probably attached to the sea bed by threads known as the byssus, which it retained from its initial settling as a larva.

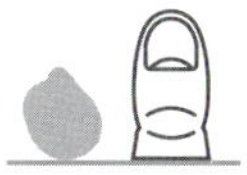

CARDIOLA INTERRUPTA
Cardiola fossils such as these are rarely found with other fossils, suggesting a specialized environment.

Gissocrinus

GROUP Echinoderms
DATE Middle Silurian to Early Devonian
SIZE Cup with complete arms 7cm ($2\frac{3}{4}$in) high
LOCATION Europe, USA

Echinoderms known as crinoids, which is what *Gissocrinus* was, were marine animals related to sea urchins and starfish. *Gissocrinus* had a small cup made of three circles of plates. In the upper circle, the five radial plates had wide, crescent-shaped upper surfaces; these served as articulation points for the animal's arms, which extended upwards from this point. A sixth plate in this circle, at the back of the cup, marked the base of a tubular structure, which ended with the anal opening at its top. The arms branched several times, and the branches were always equal in length. In this example, however, only the lower branches have been preserved.

G. INVOLUTUS
Although only its lower parts are shown here, *Gissocrinus* had up to 32 branches.

Pseudocrinites

GROUP Echinoderms
DATE Late Silurian to Early Devonian
SIZE Cup 1.5–3cm ($\frac{1}{2}$–$1\frac{1}{4}$in) across
LOCATION Europe, North America

Pseudocrinites was shaped like a round or oval inflated disc, with a flat rim formed by two, thin-plated zones called ambulacra. It was anchored to the sea bed by a jointed stem. The mouth was set opposite the stem attachment in the centre of the oral surface. The ambulacra began at each side of the mouth, and in some specimens reached down as far as the stem. In well-preserved examples, a number of short, articulated appendages called brachioles can be seen. The sockets where they were attached show that they were arranged alternately on each side of the ambulacra. The anus, an opening into the water vascular system, and a gonopore (for release of eggs or sperm) were also located on the oral surface. The hard, external shell features rhomboid-shaped areas that are situated across plate junctions, and pairs of slit-like openings that penetrate the shell; the function of these was probably respiratory. Respiratory structures scattered over the cup surface are characteristic features of the class Cystoidea, to which *Pseudocrinites* belongs.

PSEUDOCRINITES BIFASCIATUS
This fossil shows rhomboid-shaped respiratory structures that characterize *Pseudocrinites*, but some cystoids had many pairs of pores over the cup surface.

Pseudocrinites used its **jointed stem** to **anchor** itself to the sea bed, where it lived as a **suspension-feeder.**

Lapworthura

GROUP Echinoderms
DATE Late Ordovician to Middle Silurian
SIZE 10–12cm (4–4½in) across
LOCATION Europe

Lapworthura was a brittlestar with a large, central disc area. All of its body openings were located on the animal's lower (oral) surface, and its mouth was situated centrally. *Lapworthura's* arms were relatively stout for a brittlestar, and its internal skeleton of calcareous plates, known as ossicles, were arranged in pairs opposite each other. Fossilization of brittlestars is usually dependent on the animal being buried alive by the sudden influx of sediment: for example, during a storm. If this does not happen, the soft, outer skin of the organism decays, and the ossicles separate and disperse. Brittlestars, starfish, and sea urchins are all echinoderms; like modern brittlestars, *Lapworthura* was probably a carnivore.

LAPWORTHURA MILTONI
Unlike modern brittlestar species such as *Ophiothrix fragilis*, *Lapworthura miltoni*'s arms were relatively short and stout.

CENTRAL STAR
Lapworthura miltoni's large, star-like disc, was where the animal's mouth was located.

LIVING RELATIVE
GREEN BRITTLESTAR

Brittlestars date back to the Ordovician and are represented today by 2,000 species found from polar to equatorial regions. Their tube feet are primarily sensory. They occur mainly below 500m (1,640ft), ranging to abyssal depths. There are, however, some shallow-water species, one of which, the green brittlestar (*Ophiarachna incrassata*), is sometimes kept in aquariums – although, as owners have discovered to their cost, it has a penchant for small fish and crustaceans.

Dalmanites

GROUP Arthropods
DATE Early to Late Silurian
SIZE Up to 10cm (4in) long
LOCATION Worldwide

The trilobite *Dalmanites* had a large headshield with a prominent, forward-widening glabella in the centre that featured two pairs of narrow furrows as well as deep, oblique ones. Its eyes were large and set to the back of the head, with prominent lenses. Its cheeks were gently convex, and extended into robust, broad-based spines. Under the headshield, and attached to it beneath the glabella, was a large hypostome: a calcified structure thought to have enclosed the stomach. The thorax was segmented, and the broadly triangular tailshield ended in a spine.

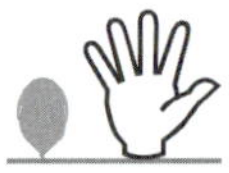

DALMANITES CAUDATUS
The thorax of *Dalmanites* was made up of 11 separate segments, clearly visible here.

ANATOMY
ENROLLMENT

Like woodlice, most trilobites had the ability to roll themselves into a ball in order to protect their soft undersides. This image depicts *Dalmanites* rolled up and seen from above, with its tail spine pointing forwards from underneath the head. This spine may have aided moulting: all trilobites periodically shed their hard exoskeleton to accommodate growth. In fact, most trilobite fossils are moults rather than dead carcasses.

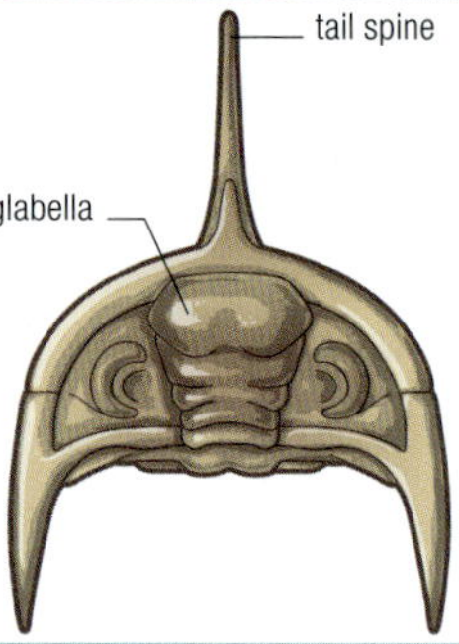

Encrinurus

GROUP Arthropods
DATE Silurian
SIZE Up to 5cm (2in) long
LOCATION Worldwide

Encrinurus was a small trilobite that was widespread in the world's seas throughout the Silurian. As a fossil, it is immediately recognizable from the many prominent pustules on its headshield, which have given it a nickname – the "strawberry-headed trilobite". The central part of the headshield, the glabella, was inflated and pear-shaped, widening towards the front. It had several furrows, but these are not easy to see on this specimen. The eyes (only the bases are preserved here) sat on the ends of short stalks, suggesting this trilobite may have spent much of its time half-buried in soft sediment on the sea bed, with only its eyes protruding above the surface.

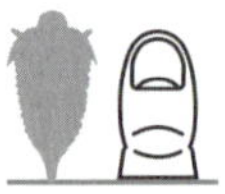

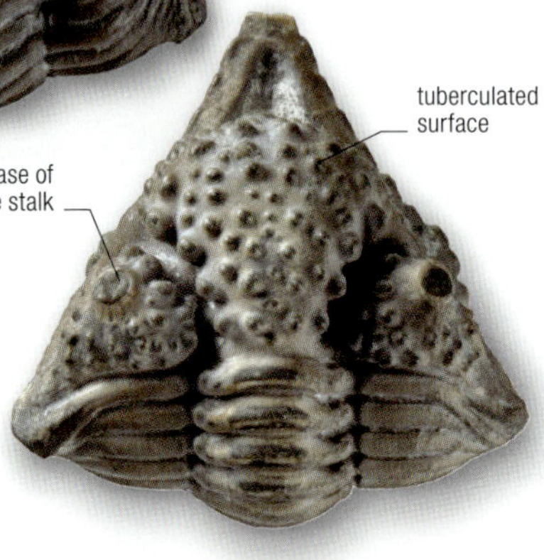

ENCRINURUS TABULATUS
This specimen of *Encrinurus* is partially enrolled and shows the "strawberry" headshield in front of the animal's thorax. The tailshield (above) is triangular with a large number of narrow rings.

EXALLASPIS SP.
Although only the head and a small part of the thorax are visible here, the backwards-curved spines extending from the ends of the thorax can clearly be seen.

Exallaspis

GROUP Arthropods
DATE Middle to Late Silurian
SIZE Up to 2.5cm (1in)
LOCATION Worldwide

This trilobite's headshield had a prominent central raised area (glabella) that narrowed towards the front, with three distinct lobes; the hindmost was the largest. A curved, raised ridge ran from the front of the glabella to the rather small eye, which was located on the back part of the headshield, about halfway between the glabella and the lateral edge. The thorax was composed of 10 segments, which laterally tapered into long, curving spines on both sides. The tailshield was small, and had two large, backwards-pointing spines with a row of four small ones between them. There were also two short spines to the front of the large ones. On the specimen shown above, however, most of the thorax and the tailshield are missing.

Calymene

GROUP Arthropods
DATE Silurian
SIZE Up to about 6cm (2¼in) long
LOCATION Worldwide

One of the most familiar trilobites, *Calymene* had a roughly semicircular headshield dominated by a prominent glabella (the raised, central part) that narrowed towards the front and was broadly bell-shaped. The glabella had three lobes running along each of its two outer edges, and the hindmost of these was the largest. Its eyes were small and located roughly opposite the mid-point of the glabella. This was such an abundant fossil in the limestone quarries of Dudley, England, that it became a local emblem.

***Calymene* is often referred to as the "Dudley Locust" or "Dudley Bug".**

CALYMENE BLUMENBACHII
Calymene's thorax was made up of 13 segments, and its tailshield was smaller than its rounded head.

PROTECTIVE ROLL
Many *Calymene* fossils, like the one above, have been found rolled up. Like modern woodlice, most trilobites did this for protection.

Monograptus

GROUP Graptolites
DATE Early Silurian
SIZE Up to 5cm (2in) long
LOCATION Worldwide

Monograptus is characterized by having arm-like structures (thecae) on only one side of its stipe (branch). The thecae housed the individual soft-bodied zooids of the colony. This genus appeared in the geological record at the beginning of the Silurian Period, about 443 million years ago, and a large number of species of *Monograptus* and related genera evolved rapidly over the following 30 million years. Many of these are of worldwide distribution. Graptolites were colonial hemichordates, a small group that is allied to vertebrates. Living hemichordates are comparatively rare, but the abundance of graptolite fossils in many rocks shows that they were once widespread in the planet's oceans.

SCIENCE

ZONE FOSSILS

Most graptolites were planktonic, and were therefore widely distributed in Ordovician and Silurian oceans. They evolved rapidly and individual species existed for a relatively short time, usually about two million years. These attributes make graptolites ideal zone fossils. Individual species or groups of species can be used to calibrate rock successions, and correlate them across the world. In Silurian rocks, around 40 graptolite zones have been recognized, each with an average length of about 0.7 million years.

MONOGRAPTUS TRIANGULATUS
In this species the thecae are long, tubular, and widely spaced on the colony's outer curve.

MONOGRAPTUS CONVOLUTUS
Here, the colony forms a spiral, and its long thecae, on the outside of the curve, give it the appearance of a watch spring.

Paracarcinosoma

GROUP Arthropods
DATE Late Silurian
SIZE Most a few centimetres long, but giants of over 1m (3ft 3in) long are known
LOCATION British Isles

Eurypterids like *Paracarcinosoma* were scorpion-like animals, and the majority inhabited fresh or brackish water. In some, the first pair of limbs ended in pincers armed with sharp, pointed teeth on their inner edges. The animal's small headshield is not well preserved on this specimen, but was roughly rectangular, and its small eyes were located close to the front. The mouth lay underneath, and the first pair of appendages, the small chelicerae, lay in front of it. *Paracarcinosoma* had four pairs of spiny walking legs, and the sixth, hindmost pair were swimming legs with paddle-like ends. Its abdomen was divided into two parts: a broad, oval pre-abdomen, made up of seven segments; and a narrower, nearly cylindrical postabdomen, with five segments. Behind the abdomen was a pointed terminal segment called a telson.

PARACARCINOSOMA OBESA
Paracarcinosoma fossils are sometimes referred to as "water scorpions" – and the similarity of shape is apparent here.

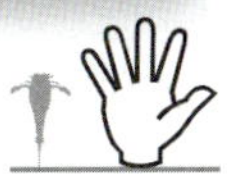

SILURIAN VERTEBRATES

Jawless fish (agnathans) were the most common vertebrates of the Silurian. However, new vertebrate groups appeared and diversified during the period. The oceans were home to a key development – jaws – and by the end of the Silurian, all the major groups of vertebrates had evolved.

The period of glaciation that ended the Ordovician extended into the Silurian, but gradually the ice melted, sea levels rose, and the oceans became warmer. While plants and invertebrates sought to establish themselves on land, vertebrates continued to take advantage of the hospitable environment offered by the oceans. And it was the Silurian seas that saw what is probably the most important advance in vertebrate history: the evolution of jaws.

The advantages of jaws

The possession of jaws enabled the development of new types of behaviour. With jaws, an object can be grasped firmly, and if those jaws have teeth, food can be sliced into pieces small enough to swallow, while hard items can be ground down. Eating plants was now an option, and many jawed vertebrates became larger than their jawless contemporaries. With or without teeth, a strong pair of jaws can be a powerful weapon – for killing or defence – but jaws can also be used to hold on tight to a partner during courtship or mating, and to carry young. Objects can be manipulated with jaws – holes can be dug, stones can be moved, and pieces of plants can be used to build nests. The possibilities for jawed vertebrates were many and varied.

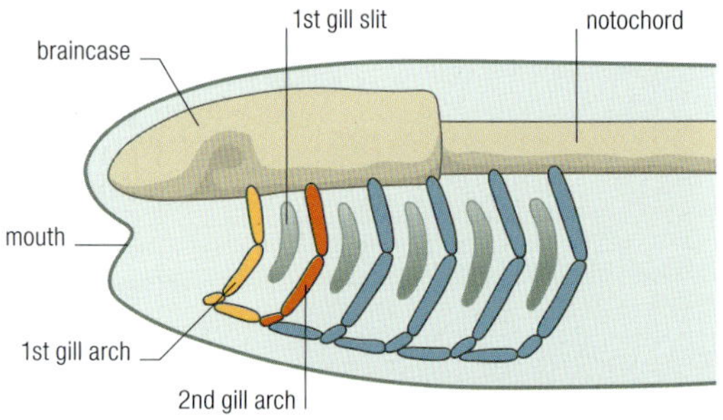

AGNATHOUS (JAWLESS) VERTEBRATE

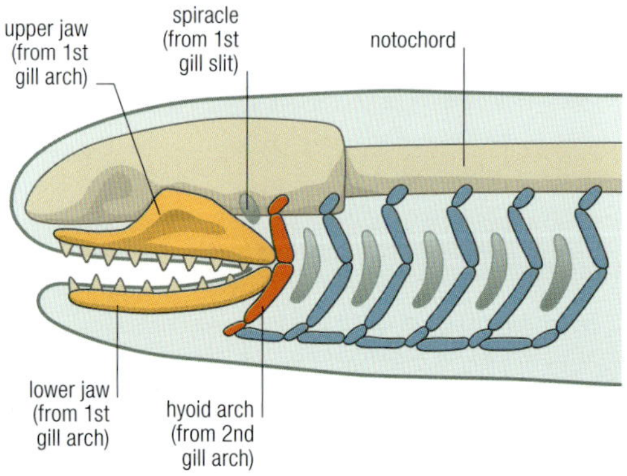

GNATHOSTOME (JAWED) VERTEBRATE

EVOLUTION OF THE VERTEBRATE JAW
In agnathans, the first and second gill arches support the first gill slit. In jawed vertebrates, the first gill arch has become a pair of jaws and the first gill slit a spiracle, through which water can be pumped over gills. The jaws can also bear teeth.

Fossil finds in China

The sarcopterygians, or lobe-finned fish, are important jawed vertebrates because they are the group from which tetrapods (four-legged animals) evolved (see p.128). They were thought to have first appeared in the Early Devonian, but in 1997 parts of the skull and lower jaw of a new fish, *Psarolepis*, were found in China's Yunnan Province, and these proved that sarcopterygians first evolved in the Late Silurian. However, the main radiation of sarcopterygians, including several other new and primitive forms found more recently in China, occurred in the Early Devonian.

GROUPS OVERVIEW

New jawless fish groups in the Silurian included the anaspids, thelodonts, osteostracans, and galeaspids. Osteostracans had pectoral fins supported by a shoulder girdle, a feature shared with jawed vertebrates but absent in all other jawless vertebrates. This suggests that osteostracans are the sister taxon to the jawed vertebrates. Of the jawed fish, acanthodians, placoderms, and chondrichthyans were present in the Early Silurian, while the actinopterygians (ray-finned fish) arose in the Late Silurian.

ANASPIDS
These jawless fish are found primarily in Silurian deposits in the northern hemisphere. Some anaspids, such as *Birkenia*, possess large dorsal scales but lack paired fins. Others, such as *Pharyngolepis*, have paired fin-like structures extending along the ventral surface of the body. *Jamoytius* and *Euphanerops* are unusual in having more than 30 gill arches.

OSTEOSTRACANS
Osteostracans first appeared in the Middle Silurian and, like the anaspids, were restricted to the northern hemisphere. They are characterized by a large, bony headshield, with dorsal eyes. The headshield was drawn out laterally into spines, forming a horseshoe, although these spines were absent in primitive members of the group, such as *Ateleaspis*.

THELODONTS
Thelodonts arose in the Early Silurian and survived until the Late Devonian, and these jawless fish had a worldwide distribution. They are well-known from isolated scales (microvertebrates), but more complete specimens have also been found, such as *Loganelia*, showing the head, tail, gill arches, and evidence of paired flaps that may have been fins.

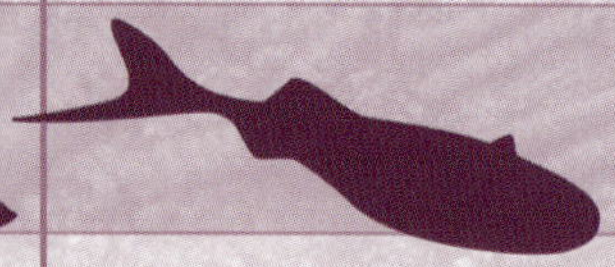

Birkenia

GROUP Agnathans
DATE Late Silurian to Early Devonian
SIZE 15cm (6in) long
LOCATION Europe

A small jawless fish, *Birkenia* had a laterally compressed, "fusiform" body (wider in the middle than at both ends). It was covered in elongated scales arranged in distinct rows. Unusually, those on the hind dorsal flanks sloped down and back instead of down and forwards. Large scales ran along the top of the body; some pointed forwards, others backwards, but in the middle was a double-headed scale pointing in both directions. *Birkenia* had a well-developed anal fin, and small eyes with a single nasal opening between them. It lived in fresh water and was probably an active swimmer that fed on detritus (the remains of dead plants and animals).

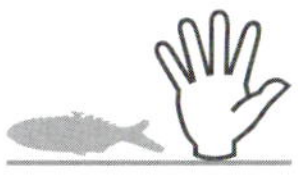

UNIQUE SCALES
Birkenia's distinctive arrangement of dorsal scales is unique among anaspids, the group to which it belongs.

UNUSUAL TAIL
Birkenia's hypocercal tail, in which the larger lower lobe was supported by the downwardly directed notochord, initially confused an early palaeontologist, who reconstructed the fish upside down.

Ateleaspis

GROUP Agnathans
DATE Early Silurian to Early Devonian
SIZE 15–20cm (6–8in) long
LOCATION Scotland, Norway, Russia

Ateleaspis was a primitive osteostracan and is thought to have lived in sheltered seas or river mouths. It is the earliest-known fish with paired appendages – its pectoral fins. It had two dorsal fins: the front one was covered in scales; the hind fin was larger, with webbed spines. *Ateleaspis*'s head was protected by a bony shield, and bony scales covered the body. The mouth was on the underside of the head, which suggests *Ateleaspis* was a bottom-feeder.

10 The number of **gill pairs** that were positioned **beneath** the body of *Ateleaspis*.

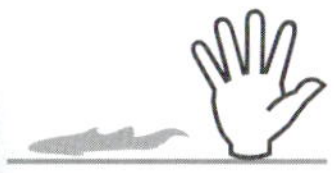

FLATTENED BOTTOM-FEEDER
Ateleaspis's body was flattened, and the forward portion was flatter than its hind section, suggesting it was a bottom-feeder.

Loganellia

GROUP Agnathans
DATE Late Silurian
SIZE 10–20cm (4–8in) long
LOCATION Europe

Loganellia was a thelodont, a flattish fish, with a body entirely covered in scales, and a hypocercal tail. Its eyes were small and far apart. The position of its mouth on the underside of its head suggests that it fed on the sea bed by sifting through the mud. Paired flaps on the side of the head may have functioned as fins.

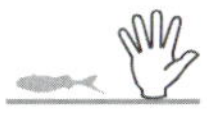

FORKED TAIL
Loganellia had a wide, flat head and a long, forked tail, in which the lower lobe was much longer than the upper one (hypocercal).

Climatius

GROUP Acanthodians
DATE Late Silurian to Early Devonian
SIZE 12cm ($4^1/_2$in) long
LOCATION British Isles

Climatius was a short fish with many short spines. Both pectoral fins had a spine, as did the front and rear dorsal fins and the anal fin, and there were five pairs of spines underneath the body. Bony armour covered its "shoulders". Its tail fin was shark-like, with a larger upper lobe; its eyes were large, and its nasal openings small. *Climatius* had small grinding teeth, which suggest it fed on tiny creatures.

Andreolepis

GROUP Actinopterygians
DATE Late Silurian
SIZE 9cm ($3^1/_2$in) long
LOCATION Europe, Asia

The earliest known ray-finned fish, *Andreolepis* had sharp, pointed teeth that were interspersed with smaller tooth-like projections and arranged in several rows. Old teeth were not shed; instead, new ones were added along the jaw's inner edge. Its scales were diamond-shaped and ridged, and carried a thin layer of a substance called ganoine, which is similar to the enamel found on a tooth.

DEVONIAN

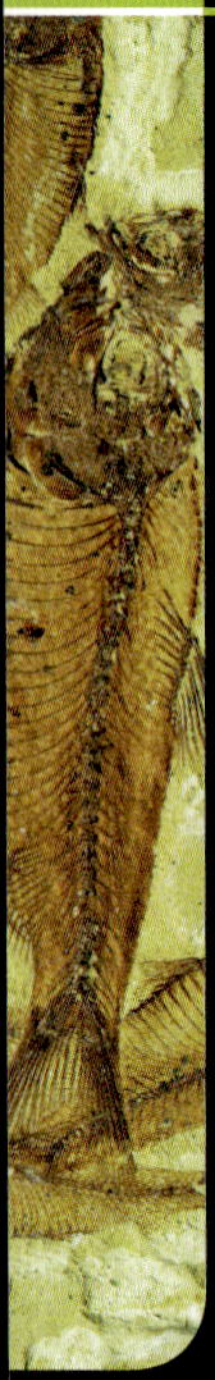

Devonian

The Devonian Period, often called the "Age of Fish" in Earth's history, is characterized by an evolutionary radiation in some fish groups, including strange jawless fish and armoured placoderms. Equally signficant was the great increase in the diversity of plants, which spread from the coastlines of the continents across the land and created new terrestrial ecosystems – the world's first forests.

CALEDONIDES MOUNTAINS
The once massive Caledonides are heavily eroded, but still extend from Scandinavia to the Appalachians in North America.

Oceans and continents

By the middle of the Devonian, about 400 million years ago, the Iapetus Ocean had completely disappeared, as Baltica and Laurentia collided. This collision unleashed massive tectonic forces that continued to reshape the continental landscape, forcing up the Calédonide mountains in today's Scandinavia, northern Britain, and Greenland, and the beginning of the Northern Appalachian chain in eastern North America. Gondwana rotated in a clockwise direction around an axis centred on Australia, bringing the western margin of the continent closer to the Equator and to Laurentia. Land masses gradually became greener as ferns and tree-like plants formed forests and swamps, adding organic material to soils and creating new habitats for invertebrates. The Rhynie Chert in Scotland provides an exceptional insight into the flora of a Mid-Devonian peat bog. During the Late Devonian, there were major changes in oceanic geochemistry, with periods when the oceans became depleted of oxygen. Rapid plant diversification may have enriched rivers with high levels of nutrients that were then carried to the seas. As reefs prefer low-nutrient conditions, this may have contributed to a decline in their diversity during the Devonian mass exinction.

OLD RED SANDSTONE
Sediments eroded from the Caledonide Mountains, and deposited in the basins between, created thick beds of Old Red Sandstone that typify the Devonian.

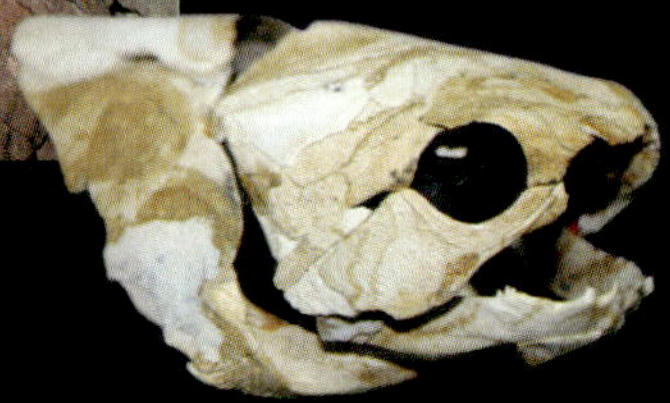

DEVONIAN MOTHER FISH
This early placoderm fish, preserved in rocks in the Gogo formation in Western Australia, lived on a Devonian barrier reef.

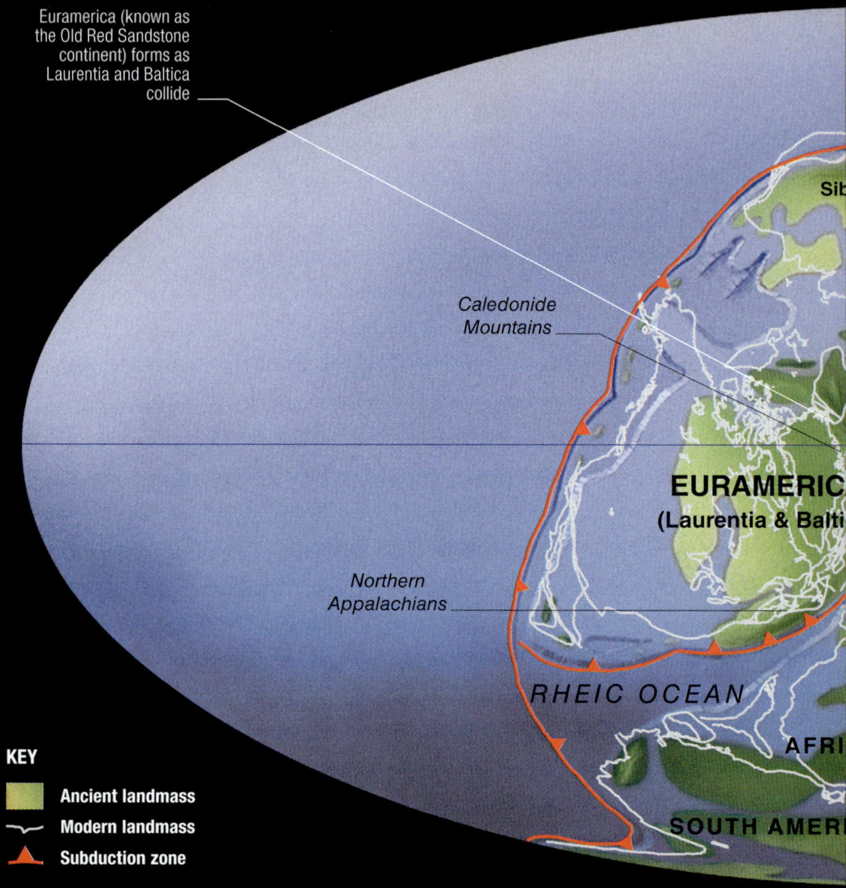

DEVONIAN WORLD MAP
Baltica and Laurentia combined to form Euramerica and close the Iapetus Ocean. This led to the creation of the Caledonide mountain chain.

	EARLY		MIDDLE	
MYA	**420**	**410**	**400**	**390**
PLANTS	415 First zosterophylls (primitive plants) ZOSTEROPHYLLUM	410 First lycophytes and trimerophytes BARAGWANATHIA	400 First sphenopsids (horsetails); development of tree-like habit in some plants 395 First lichen, first charophytes (stoneworts)	ARCHAEOPTERIS
INVERTEBRATES	415 First terebratulid brachiopods	410 First nautilid nautiloids	395 First harvestman arachnid, first mite, first hexapods (springtail); first ammonoids (left) SOLICLYMENIA	
VERTEBRATES		410 First true fish teeth	400 Major diversification of placoderm fishes; first true sharks; evolution of air gulping in lungfish	

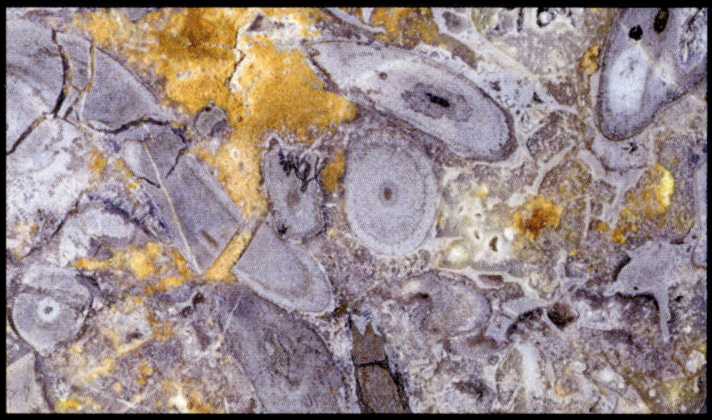

RHYNIE CHERT FOSSILS
Silicified fossils of Early Devonian age are found beautifully preserved in Scotland's Rhynie Chert. This thin section shows stems of one of the first land plants, *Rhynia.*

Climate

The global climate was generally relatively warm, arid, and dry, and the overall Equator to pole temperature gradient was smaller than today. In the Early Devonian, much of Laurentia and the northeastern part of Gondwana were hot and arid. Large parts of what are now Australia, North America, and Siberia were covered by warm, shallow seas. Analysis of fossil brachiopods using oxygen isotopes (see p.23) suggests that in the Early Devonian, ocean temperatures were about 26°C (79°F), comparable with modern day oceans. Later in the Devonian they could have risen to about 30°C (86°F). In higher southern latitudes across Gondwana the climate was warm–temperate. The polar regions were cool, but without any ice cover. As Laurentia and Gondwana moved closer together, a wider tropical belt became established north of the Equator. Here, the first thick coal deposits reveal that rainforests stretched across what is now the Canadian Arctic and southern China. The Siberian continent to the north, and the northern parts of the converging continents of Laurentia and Gondwana remained arid, but the more southerly parts of Gondwana became more temperate and more humid, with some of the higher, mountainous regions close to the Devonian South Pole, in what is now the Amazon Basin, becoming cold enough for glaciers to develop.

Gondwana rotates in a clockwise direction, with Australia at its axis

WINDJANA GORGE, THE KIMBERLEY
Spectacular Windjana Gorge in the Kimberley region, in Australia's northwest corner exposes the remains of a large barrier reef from Devonian times, when the area was a shallow sea, fringed by reefs.

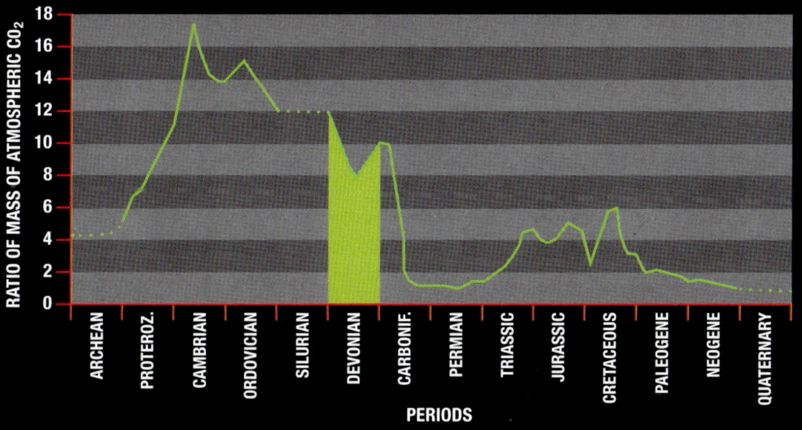

CARBON DIOXIDE LEVELS
The diversification of land plants during the Devonian Period may have contributed to the drop in carbon dioxide levels. This was accompanied by cooler temperatures.

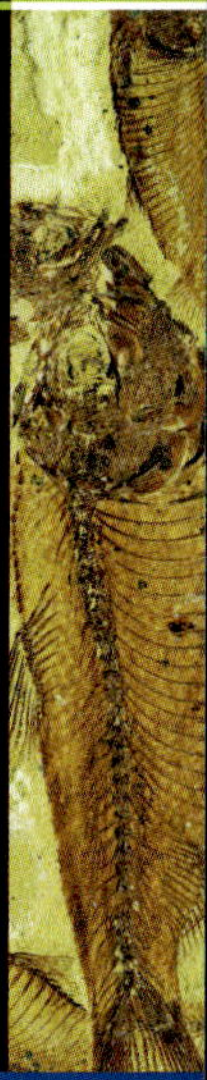

LATE

	380	370	360	350 MYA
85 First rogymnosperms ree-like plants, left); rst development of najor forests			360 Floras dominated by seed ferns; first ferns	**PLANTS**
	380 First spiders (*Attercopus*), terrestrial millipedes, centipedes, pseudoscorpions; first goniatites (earliest ammonoids); first nuculid bivalves; 375 First mytilid bivalves	365 Mass extinction: of up to 70% of species, 50% of genera; 3 of 5 trilobite orders: corynexochids, odontopleurids, harpetids; nearly 90% of brachiopod genera, including atrypids (left), orthids, pentamerids; major loss of reefs	360 First crabs	**INVERTEBRATES**
85 First lampreys	380 First advanced amphibian-like, lobe-finned fishes, such as *Tiktaalik*; 375 First signs of viviparity (embryo develops inside mother) in fishes; first amphibian; first coelocanth	365 Placoderms (armoured jawed fishes) become extinct		**VERTEBRATES**

ATRYPA

DEVONIAN PLANTS

The development of land plants initiated during the Ordovician and Silurian continued rapidly during the Devonian, with an explosion of diversity and the appearance of many new and different kinds of plants. By the end of the Devonian, the forerunners of many groups of living plants had already evolved.

Land plants evolved from ancestors that lived in fresh water and key adaptations that help control water loss are present in various forms in early land plants. Early animals faced the same difficulty as they followed plants into the new kinds of ecosystems that were forming on land. Several different groups of animals moved on to land independently, but the colonization of land appears to have occurred only once in the evolutionary history of plants.

Adapting to life on land

The most important fossil occurrence for understanding the structure of early land plants, as well as for studying many aspects of early land ecosystems, was discovered in 1914 near the village of Rhynie, Scotland. The Rhynie Chert is an ancient peat bog that is preserved in silica. Nearby volcanic activity appears to have been the cause of the remarkable preservation. In the Rhynie Chert, early land plants, along with fungi, algae, and a few early land arthropods, are preserved in their position of growth, exactly where they once lived almost 400 million years ago. The fossil plants in the Rhynie Chert are exquisitely preserved. They include several different kinds of rhyniophytes, such as *Aglaophyton* and *Rhynia*, as well as the early lycopod *Asteroxylon*. Rhyniophytes have simple branching axes, similar to those of *Cooksonia*, but with more variation in the form of the branching. Spores are produced in elongated sporangia at the branch tips. Axes of *Asteroxylon* are covered with small, flattened, flap-like leaves, occasionally with small, kidney-shaped sporangia scattered among them. *Asteroxylon* is very similar to the modern clubmoss *Huperzia*. None of these

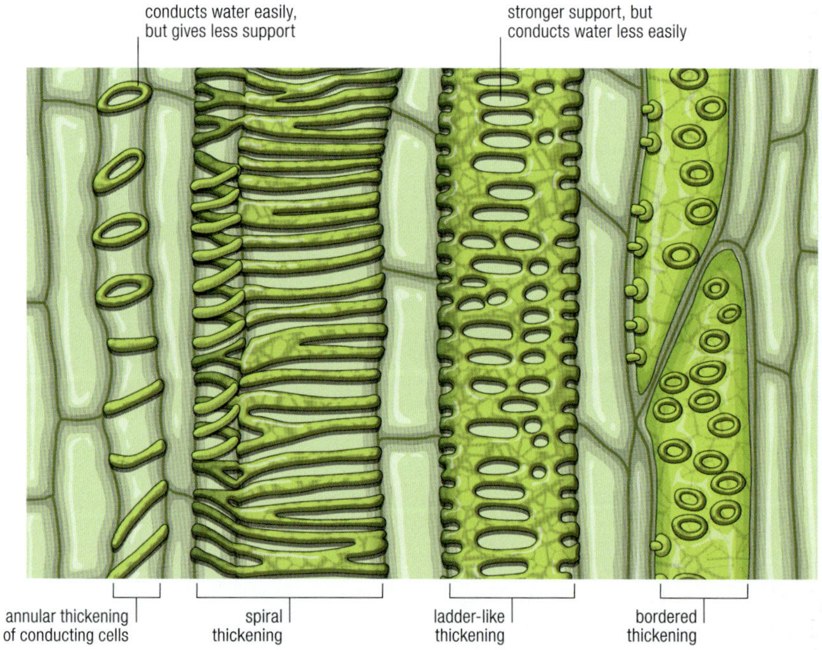

WATER-CONDUCTING CELLS
The specialized cells that originally evolved for water conduction later became co-opted to provide support, resulting in several different kinds of cells specialized for different functions.

GROUPS OVERVIEW

The Devonian saw the rapid appearance of many new and different kinds of plants. All were probably derived from a single common ancestor, exploiting the new ecological opportunities made possible by the ability to maintain a sufficient supply of water. The Devonian diversification of land plants is the botanical equivalent of the Cambrian Explosion (see p.70).

GREEN ALGAE
Many different lines of evidence indicate that various groups of freshwater green algae, especially the stonecrops and similar charophytes, are the closest living relatives of land plants. Beautifully preserved fossil algae very similar to living stonecrops are known from the Early Devonian Rhynie Chert.

RHYNIOPHYTES
Several different kinds of simple, early land plants are grouped together as rhyniophytes. The earliest forms were intermediate between mosses and more complex land plants. Branching of the sporophyte or spore-producing phase, which is seen in all rhyniophytes, resulted in the production of many sporangia rather than just one, which increased spore output.

ZOSTEROPHYLLS
Zosterophylls are a very prominent group of early land plants during the Early and Middle Devonian. Their kidney-shaped sporangia and the development of their water-conducting tissues indicate that they are closely related to living clubmosses. They differ from clubmosses in their flattened stems and the absence of leaves.

CLUBMOSSES
The earliest clubmosses are from the Late Silurian but the group diversifies rapidly during the Devonian, quickly differentiating into the three main subgroups that are recognized today. *Asteroxylon* is one of the best-known early clubmosses. Clubmosses have a continuous fossil record from the Devonian up to the present day.

On land, water is scarce. Key problems faced by all land plants is how to get water, how to keep it, and how to use it efficiently.

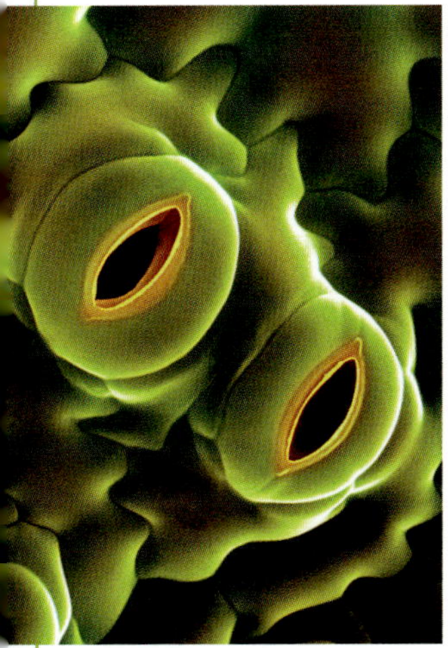

STOMATA
Land plants need to take in carbon dioxide for photosynthesis, but must also not lose too much water. Stomata are adjustable valves that open and close to control the exchange of gases between air spaces inside the plant and the atmosphere.

early land plants are very large, but all have elongated cells in the centre of their axes that helped conduct water from the lower parts of the plants, which were rooted in the soil, to the upper parts of the plant growing in the air. In some rhyniophytes, such as *Aglaophyton*, these water-conducting cells are similar to those of living mosses, which lack internal thickenings of their cell walls. In other early land plants, such as *Rhynia* and *Asteroxylon*, these internal thickenings are present, making these cells more similar to the specialized water-conducting cells of living ferns and lycopods. The outer surfaces of early land plants from the Rhynie Chert and other fossil localities are covered by a waxy cuticle, which helped reduce water loss from the plant into the air.

The struggle for light

During the Middle and Late Devonian, rhyniophytes were eventually displaced by larger plants with more complex branching, such as *Psilophyton*. Larger plants may have been more successful as the struggle for light intensified in early land ecosystems. *Psilophyton* and similar plants also show a new kind of growth form, with a single main axis and smaller lateral branches. These lateral branches subsequently became modified to form the different kinds of leaves seen in living ferns, horsetails, and seed plants. As the size of land plants continued to increase during the Devonian, more conducting cells were needed to keep the upper parts of the plants supplied with water. Increasingly, these cells also provided the internal support needed to keep the plants upright. In the Middle and Late Devonian, a new innovation was the ability to produce large quantities of water-conducting cells throughout the life of the plant. This gave some plants the capacity to produce significant amounts of woody tissue. The earliest true trees with woody trunks appear for the first time in the Late Devonian.

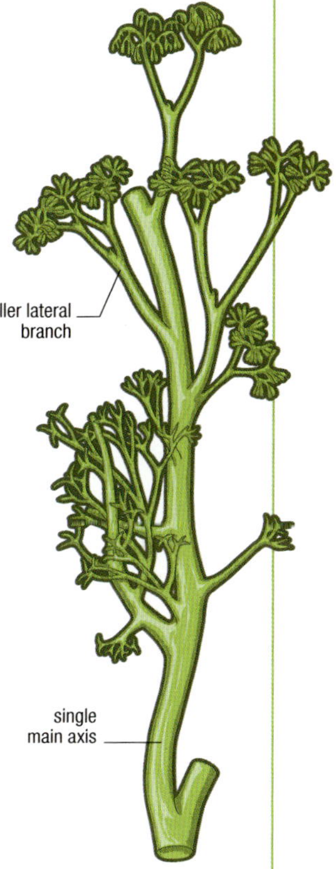

BRANCHING SYSTEMS
Psilophyton, from the Early and Middle Devonian, was larger than most rhyniophytes. It also had more complex branching. This developed into a new growth pattern in which one branch dominated while others became lateral branch systems with limited growth capacity.

PRIMITIVE EUPHYLLOPHYTES
During the Devonian, the relatively simple rhyniophytes were replaced by larger forms – primitive euphyllophytes – with much more complex patterns of branching. It was these more complex forms that ultimately evolved into living horsetails, ferns, and seed plants.

HORSETAILS
Towards the end of the Devonian, there were several different kinds of primitive euphyllophytes with complex branching that were the forerunners of modern-day horsetails. By the end of the Devonian, *Archaeocalamites* showed many clear similarities to members of the living genus *Equisetum*.

PROGYMNOSPERMS
A key innovation seen among several groups of fossil plants from the Late Devonian was the capacity to produce additional water-conducting tissues throughout the life of the plant. These tissues comprise the wood of modern trees. One of these groups of Late Devonian plants was the progymnosperms.

PRIMITIVE SEED PLANTS
The oldest seed plants are known from the Late Devonian and probably evolved from progymnosperm ancestors. All the early seed plants are rather similar, making it likely that seeds evolved only once in the history of land plants. The origin of seeds was an important step towards new kinds of reproduction that were less dependent on water.

Prototaxites

GROUP Fungi
DATE Late Silurian to Devonian
SIZE Up to 8m (26ft) high
LOCATION Worldwide

Prototaxites is one of the world's most enigmatic macrofossils. It was first described in the late 1850s by Canadian geologist John William Dawson, who thought it was fossilized rotting wood. He named it *Prototaxites*, "first yew". *Prototaxites* fossils are sometimes found intact, or broken into short, cylindrical sections that typically have concentric markings resembling tree rings. However, under a microscope, their structure is clearly unlike wood; instead of plant cells with clearly defined walls, they have microscopic tubes that run vertically inside the "trunk". Individually, these tubes are thinner than a human hair, but they form a dense mass that can be over 1m (3ft 3in) across. Dawson interpreted these as fungal threads feeding on dead wood. However, later researchers dismissed his suggestion that *Prototaxites* was a tree, or even a plant. At various times, it has been interpreted as a kelp-like alga, a lichen, and most recently, as the giant fruiting body of a fungus.

P. LOGANI CROSS SECTION
It is easy to see why *Prototaxites* was taken for fossilized wood. Its surface texture strongly resembles tree bark, just as its internal structure mimics a tree's concentric growth rings. Only under the microscope do the differences become clear.

PROTOTAXITES LOGANI
At up to 8m (26ft) high, *Prototaxites* must have been quite imposing. Whatever its true classification, there can be no doubt that it dominated the Devonian landscape.

SCIENCE
MYSTERIOUS IDENTITY

In 1872, botanist William Carruthers identified *Prototaxites* as the stalk of a giant, kelp-like alga. In the 20th century, this theory was largely accepted, but in 2001, Francis Hueber of the Smithsonian Institution published a fossil analysis with clear evidence of three different fungal threads. We now know *Prototaxites* lived by absorbing nutrients: a fungal habit.

Parka

GROUP Algae
DATE Late Silurian to Early Devonian
SIZE Up to 7cm (2¾in) across
LOCATION Worldwide

With a rounded outline and net-like surface, *Parka* is an intriguing fossil. In the past, it was mistaken for many objects, including the eggs of prehistoric arthropods or fish. However, its microscopic structure reveals that it was an alga, with a hard outer coat covering a flat body only a few dozen cells thick. The fossil's surface discs are tiny compartments, each filled with thousands of spores. Anatomical and chemical features link *Parka* to a group of algae called the coleochaetes. These are probably the closest living relatives of green plants.

Parka was made up of many **disc-like structures**, each containing **thousands of microscopic spores.**

PARKA DECIPIENS
Each individual disc contained thousands of microscopic spores. These would have been dispersed and then each would have germinated to produce a new plant.

Aglaophyton

GROUP Rhyniophytes
DATE Early Devonian
SIZE 18cm (7in) high
LOCATION Scotland

One of the earliest terrestrial plants, *Aglaophyton* flourished near hot springs 396 million years ago. Anchored by microscopic hairs, its creeping rhizomes produced upright branches that repeatedly divided into two. *Aglaophyton* had a waterproof outer covering to prevent dehydration, and microscopic pores (stomata) that could be closed by their special guard cells. Spore-producing, egg-shaped capsules grew on its tips.

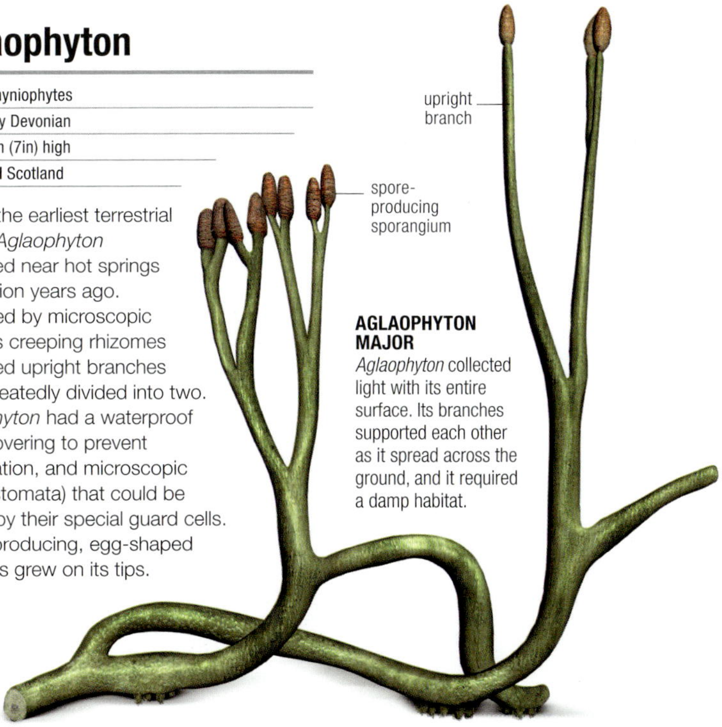

AGLAOPHYTON MAJOR
Aglaophyton collected light with its entire surface. Its branches supported each other as it spread across the ground, and it required a damp habitat.

Rhynia

GROUP Rhyniophytes
DATE Early Devonian
SIZE 18cm (7in) high
LOCATION Scotland

Rhynia is one of the best-known early plants, with horizontal and upright branches but no true roots or leaves. The horizontal branches, called rhizomes, spread out as they colonized the ground and its upright branches divided repeatedly, creating a low-growing tangle that intercepted a maximum amount of light. It was one of the first vascular plants, with specialized tissues for carrying water and dissolved substances. It also had a waterproof outer covering and stomata: microscopic pores that could be opened or closed to regulate water loss and gas exchange. The tiny nodules scattered over its stems are less easily explained: these have been identified as signs of damage, dormant branches, and secretory structures.

KEY SITE

THE RHYNIE CHERT

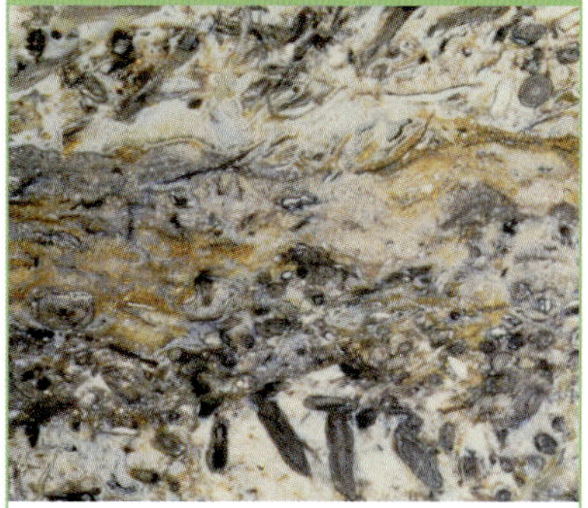

In 1914, Dr William Mackie noticed some fossils in a garden wall in the Scottish village of Rhynie. The fossils were in fine-grained chert; when examined, they revealed an ecosystem containing some of the world's earliest-known terrestrial plants, as well as fossilized fungi and many extinct arthropods.

RHYNIE FLORA
Rhynia and other primitive plants grew close to silica-rich hot springs. The springwater would periodically engulf them – preserving them as it cooled and its silica crystallized.

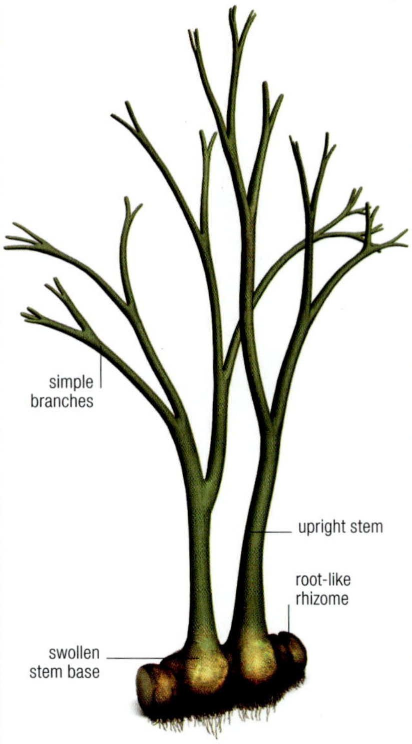

Horneophyton

GROUP Rhyniophytes
DATE Early Devonian
SIZE 20cm (8in) high
LOCATION Scotland

With its slender branches, *Horneophyton* resembled several other early plants found in Scotland's Rhynie Chert. However, its fossils show two unique features: swollen stem bases and lobed spore-producing organs (sporangia). Each of the lobes was cylindrical with a central internal column. This structure is found in living mosses, but the spore- bearing part of the *Horneophyton* life cycle is free-living – an evolutionary breakthrough lacking in mosses. This set of features has made it difficult to classify. However, there is no doubt that it was successful in the Early Devonian, when it grew in miniature thickets on damp ground.

HORNEOPHYTON LIGNIERI
Growing at or just beneath ground level, *Horneophyton*'s root-like rhizome may have helped to keep the plant in place. They probably also facilitated growth in dense thickets.

Renalia

GROUP Rhyniophytes
DATE Early Devonian
SIZE 30cm (12in) tall
LOCATION Canada

Many early land plants had dichotomous growth, meaning that their stems divided into equal halves. However, *Renalia* exhibited a different kind of growth, with unequal branching creating more complex shapes. Each of its stems had a leading growing point, with small branches splitting off on either side. These side branches bore the plant's spore-forming organs (sporangia). Fossils of *Renalia* come from the Gaspé Peninsula in Quebec, where they were first found in the 1970s. Exactly where it fits in plant evolution is uncertain, although its spore capsules may be a link with simple plants called zosterophylls.

RENALIA HUEBERI
Renalia's kidney-shaped sporangia were several millimetres across – a substantial size for an Early Devonian plant. When the spores inside were ripe, the capsules split open to release them into the air.

Zosterophyllum

GROUP Zosterophylls
DATE Late Silurian to Middle Devonian
SIZE 25cm (10in) high
LOCATION Worldwide

Preserved in fine-grained rock, fossils of *Zosterophyllum* have been found in many parts of the world. They first appeared in the Silurian, and by the Devonian many different species had evolved. Known collectively as zosterophylls, they covered large areas of damp, low-lying ground.

Fossil zosterophylls were first thought to be related to *Zostera*, or eel-grass, which is one of the few flowering plants that grows in the sea. However, we now know they are not closely related. Zosterophylls may have been ancestors of lycopods, the plant group that included the giant clubmosses of coal-forming forests.

ZOSTEROPHYLLUM RHENANUM
Zosterophylls reproduced by releasing spores. The spore-forming organs, or sporangia, were located along the sides of stems, not at the tips. Some were grouped together in clusters or primitive cones.

ZOSTEROPHYLLUM RHENANUM
Zosterophyllum had a smooth stem that branched repeatedly from near the base. It also had an internal vascular system like the ones used by most of today's plants to carry water.

Discalis

GROUP Zosterophylls
DATE Early Devonian
SIZE 30cm (12in) high
LOCATION China

Pressed flat by ancient silt, fossils of *Discalis* show a plant that flourished nearly 400 million years ago. This zosterophyll was discovered in China in the late 1980s. Like other zosterophylls, *Discalis* did not have true roots or leaves and was made up of heavily branched stems. It grew by uncurling across the ground. The stems often formed H-shaped or K-shaped branches, as well as simply dividing in two. The branches may have helped to bind the plant into a sturdy tuft. *Discalis* had spore capsules (sporangia) arranged in open spirals along the sides of its stems. Each sporangium was the size of a pea.

***Discalis's* spiky stems are a mystery. Perhaps they were an early example of a plant protecting itself from animals.**

DISCALIS LONGISTIPA
The stems of *Discalis* were covered with minute spines, known as enations. These enations had button-shaped tips.

Sawdonia

GROUP Zosterophylls
DATE Devonian
SIZE 30cm (12in) high
LOCATION Northern hemisphere

During the nineteenth century, *Sawdonia* was the subject of a case of mistaken identity, when two very different fossil plants were combined. Today, it is recognized as a typical zosterophyll, with creeping, root-like rhizomes and upright stems. These grew by uncoiling, and formed paired branches. *Sawdonia*'s water-conducting system may have given it support – a double role that is a key feature of modern vascular plants. *Sawdonia*'s kidney-shaped spore capsules split into two halves to release their spores.

SAWDONIA ORNATA
Like many zosterophylls, *Sawdonia's* stems are covered with tiny spines. Even with the naked eye, the spines often show up well on fossils, giving the stems a distinctive saw-toothed edge.

Sciadophyton

GROUP Rhyniophytes
DATE Early Devonian
SIZE 5cm (2in) high
LOCATION Worldwide

The spore-producing form, or sporophyte, was just one phase in the life cycle of a primitive plant. The sporophyte dispersed spores, which grew into gametophytes, a second phase concerned with sexual reproduction. Gametophytes are seldom preserved as fossils, but *Sciadophyton* is one rare example. It had stalks ending in cups where the male and female sex cells (gametes) were produced. After rain, mature male cells swam through the water to fertilize the female ones – and sprout into new sporophytes. Gametophytes and sporophytes generally look very different. Several different types of primitive land plants seem to have produced *Sciadophyton*-like gametophytes, possibly including some zosterophylls.

SCIADOPHYTON
At least a dozen stalks grew from *Sciadophyton's* central point, and the upper surface of each stalk's cup held the gametangia: the organs that produced the eggs or the sperm.

Protobarinophyton

GROUP Zosterophylls
DATE Devonian
SIZE 30cm (12in) high
LOCATION Worldwide

This low-growing plant does not fit neatly into any group of Devonian plants. Some botanists classify it as a zosterophyll, but others think it belongs to a sister group of its own. Its stems resemble those of zosterophylls – they lack true roots or leaves – but *Protobarinophyton*'s spore-producing capsules (sporangia) are arranged in clusters not seen in other zosterophylls. Seen under a powerful microscope, *Protobarinophyton* has another peculiarity. It was heterosporous, meaning that it produced two different sizes of spores. Heterosporous plants normally use separate sporangia to make different spores. *Protobarinophyton* is unusual in that its single spore capsules made both sizes.

PROTOBARINOPHYTON LINDLARENSIS
The cone-like clusters of spore capsules are called strobili, from the Greek word for "cone". Fossils show that the strobili's capsules were arranged in pairs.

Minarodendron

GROUP Lycophytes
DATE Middle Devonian
SIZE 25cm (10in) high
LOCATION China

Formerly (and incorrectly) known as *Protolepidodendron*, this lycopod is one of many early plants discovered in China. As its original name indicates, this plant was the forerunner of greater things to come. With its relatives, *Minarodendron* belonged to the line that eventually produced *Lepidodendron* (see p.145) and other giant clubmosses of Carboniferous times. *Minarodendron*'s ground-hugging stems occasionally grew upwards towards the light. These stems carried the spore capsules cradled in the uppermost leaves. When the spores were ripe, the capsules split open, shedding them into the wind.

MINARODENDRON CATHAYSIENSE
This was a low-growing plant with rope-like stems that would have trailed across the ground. Like other lycopods, *Minarodendron* had tiny leaves called microphylls. The microphylls were arranged in spirals on the stems.

Asteroxylon

GROUP Lycophytes
DATE Early to Middle Devonian
SIZE 50cm (19½in) high
LOCATION Europe

With its covering of small, scale-like leaves, *Asteroxylon mackiei* is the most complex plant that has been found in the Rhynie Chert in Scotland; another *Asteroxylon* species has also since been found across northern Europe. Its name, meaning "star wood", is a reference to the shape of its water-carrying vascular system, which was star-like in cross section. The root-like rhizomes under the ground were thinner than the stems above it – an arrangement that echoes the proportions of living plants. *Asteroxylon*'s stems were covered in leaf-like scales. Like leaves, their purpose was to collect light, but the scales were constructed differently from true leaves. Instead, they were probably an early form of microphyll – the leaf-like flaps that are a feature of modern lycophytes.

LIVING RELATIVES
LYCOPHYTES

Living lycophytes follow the pattern pioneered by *Asteroxylon* and have simple leaves crowded around their stems. The 1,000 living lycophyte species include clubmosses, as shown here, spikemosses, (which are not mosses at all), and a group of pond-dwelling plants called quillworts.

ASTEROXYLON MACKIEI
Asteroxylon was one of the tallest plants of the Early Devonian. Unlike most plants from this time, its stems had an obvious main shoot, with smaller branches at intervals along its sides.

Psilophyton

GROUP Psilophytes
DATE Early to Late Devonian
SIZE 60cm (23½in) high
LOCATION Worldwide

First identified in 1859, *Psilophyton* was one of the earliest of the euphyllophytes, a group that includes the great majority of living plants. Many different species have been found across the world. Some were smooth-stemmed, others were covered in spines. However, one species, *P. hedei*, described from Silurian rocks, turned out to be a colony of marine invertebrates, rather than a plant (see p.99).

PSILOPHYTON BURNOTENSE
Like most early plants, *Psilophyton* was leafless. It grew in a twisted mesh of stems. Spores were produced from spore capsules borne on the fine tips of the side branches.

Cladoxylon

GROUP Cladoxylopsids
DATE Middle Devonian to Early Carboniferous
SIZE 30cm (12in) high
LOCATION Worldwide

First described in 1856, *Cladoxylon* belonged to one of the earliest groups of plants to develop into trees. *Cladoxylon* itself was probably a low-growing plant, with stems less than 10cm (4in) thick, but some of its relatives had stout trunks and grew much higher. *Cladoxylon*'s miniature "trunk" was reinforced by thick strands of water-conducting tissue, which repeatedly divided and rejoined as they travelled the length of the stem. This arrangement gave the plant extra strength.

CLADOXYLON SCOPARIUM
Leafless *Cladoxylon* collected light with bare branches. In fossils, these branches look brown, but in life, they would have been bright green.

Hyenia

GROUP Cladoxylopsids
DATE Middle Devonian
SIZE 50cm (19½in) high
LOCATION Worldwide

This Devonian plant was once thought to be a primitive horsetail but is now believed to be related to the ancestors of today's ferns. Fossils of *Hyenia* show a plant with a short upright stem not unlike those of other Devonian plants. However, some of its ground-hugging, root-like rhizomes were 5cm (2in) thick – huge compared to those of other primitive plants. One rhizome, found in the 1970s, was almost 2m (6½ft) long, showing that *Hyenia* could colonize large areas of ground. *Hyenia* also spread by shedding spores. Its spore capsules were at the branch tips. The spores escaped through a lengthways split in the capsule when they were ripe.

HYENIA ELEGANS
At first glance, the flattened structures that surround the stem of *Hyenia* appear to be leaves or fronds. However, they are actually fine branches, which are subdivided many times from the plant's central stem.

Calamophyton

GROUP Cladoxylopsids
DATE Middle Devonian
SIZE 3m (9¾ft) high
LOCATION Worldwide

Trees have evolved independently in many groups of plants, and *Calamophyton* and its relatives were among the earliest. Trees were tall for two reasons: it gave them an edge in the competition for light, and it also helped them to scatter their spores over a wide area. However, it took more time and energy for them to grow. *Calamophyton*'s relatives include *Wattieza* – often described as the world's first tree. It reached a height of 12m (39ft). A grove of *Wattieza* stumps was unearthed in New York State in 1870.

CALAMOPHYTON PRIMAEVUM
Calamophyton had a stout trunk and a network of stems that branched in all directions. However, it is best known from compression fossils, where the stems and branches are pressed flat.

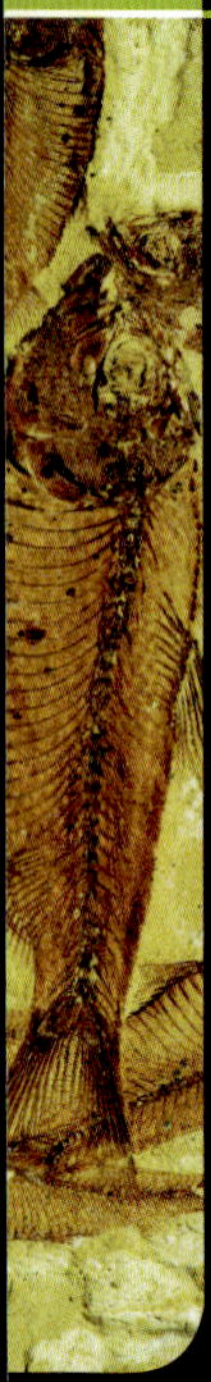

Archaeopteris

GROUP Progymnosperms
DATE Late Devonian
SIZE 8m (26ft) high
LOCATION Worldwide

With its tall trunk and spreading branches, *Archaeopteris* is the first tree to have formed forests on a truly global scale. It is also one of the first plants known to have had dense wood and true leaves. Study of *Archaeopteris* began in the second half of the nineteenth century, when its fossilized leaves were thought to belong to a low-growing fern. They were named *Archaeopteris*, or "ancient wing", because of their feathery appearance – a name similar to the fossil bird *Archaeopteryx* (see p.264). Nearly a hundred years passed before it was linked with fossilized trunks that showed it to be a tree. *Archaeopteris* had a similar outline and type of wood to many modern conifers. However, it belonged to an earlier group known as the progymnosperms.

The **trunk** of *Archaeopteris* could be over **8m high** and **1.5m thick**.

FROND STRUCTURE
Archaeopteris's sweeping fronds bore thousands of flat leaflets. Its spores developed in cone-shaped clusters of capsules, usually at the base of each frond.

Aneurophyton

GROUP Progymnosperms
DATE Middle to Late Devonian
SIZE 3m (9¾ft) high
LOCATION Northern hemisphere

This bushy plant was related to *Archaeopteris*, but it represents an earlier stage in progymnosperm evolution, before the development of true leaves. One sign of this is that its stems branched in many different planes; this feature helped leafless plants like *Aneurophyton* to trap more light. Unlike many other progymnosperms, *Aneurophyton* had only a small amount of wood, which suggests that it was a low-growing or straggling plant, rather than a substantial tree. Elongated, complex organs (sporangia) that were borne in clusters produced spores. Divided into two groups, these organs, which were attached to shared stalks, looked like stubby fingers in a pair of hands.

ANEUROPHYTON SP.
Like other progymnosperms, *Aneurophyton* had a growth layer, or cambium, beneath the surface of its stems. This made the stems thicken as they lengthened – a characteristic shown by nearly all living trees.

Rellimia

GROUP Progymnosperms
DATE Middle Devonian
SIZE 1–2m (3¼–6½ft) high
LOCATION Northern hemisphere

Fossils of *Rellimia* are widespread in sedimentary rocks during the Middle Devonian. During the nineteenth century, it was thought to be a seaweed, before being reclassified as a fern. Today, it is recognized as an early progymnosperm, making it one of the earliest plants with woody stems. *Rellimia* was a shrub or small tree, and had twig-like branches up to 2.5cm (1in) across. Like *Aneurophyton* (above), it did not have true leaves. Most fossils show the branches only, but a few include the spore-producing structures that grew when the plant was mature. Each cluster of spore capsules was about the size of a golf ball, with tightly in-rolled branches like the stems of a fern.

RELLIMIA SP.
The photosynthetic appendages of *Rellimia* were like thin, much-branched twigs rather than flat, bladed leaves. These fossils have the appearance of a confused mass of tiny branches.

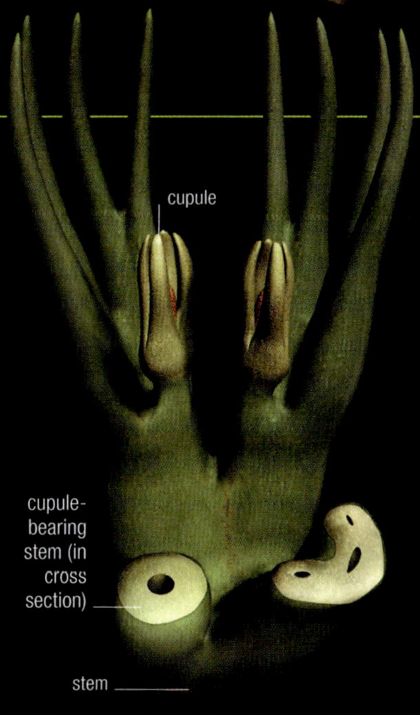

Elkinsia

GROUP Primitive seed plant
DATE Late Devonian
SIZE 1m (3ft 3in) high
LOCATION USA

Elkinsia marks a great turning point in evolution, because it is one of the earliest plants known to have produced seeds. It had straggling stems with two different kinds of fronds: some spread out to catch the light, while others divided into fine branches, which ended in structures called cupules. The cupules contained seeds arranged in groups of four; each of these seeds were about 7mm (¼in) long. Unlike simple spores, seeds are complex packages of living cells, containing food reserves and an embryo plant.

Chamaedendron

GROUP Lycophytes
DATE Late Devonian
SIZE 1.5m (5ft) high
LOCATION China

Meaning "dwarf tree", *Chamaedendron* belonged to a group of lycophytes that had forked leaflets arranged spirally around their stems. Initially, they were low-growing, but their descendants included the great trees that dominated Carboniferous swamps (see pp.145–47). As these plants grew taller, they evolved features to give them support, including a woody stem.

CHAMAEDENDRON MULTISPORANGIUM
The simple leaves or microphylls that covered *Chamaedendron*'s stems are a feature that still identify lycophytes today.

DEVONIAN INVERTEBRATES

The Devonian was the time when the land turned green. From the small land plants of the Early Devonian came the tall forests of later periods, and with them came animal inhabitants, including invertebrates. But this period was a time of change in the marine realm, culminating in the Late Devonian mass extinction.

The plants of the Early Devonian were small, vascular, and generally leafless, though some had "scale leaves" like those of clubmosses living today. At Rhynie, in Scotland, an area of preserved hot springs (see p.115), many early plants have been fossilized in chert, a type of sedimentary rock.

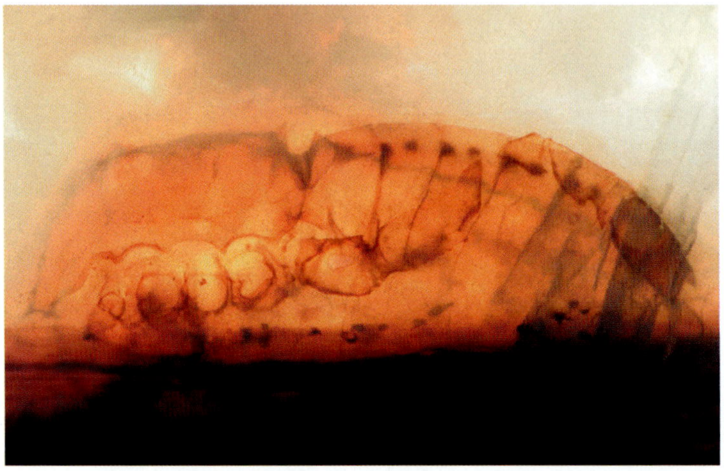

TRIGONOTARBID
This spider-like *Palaeocharinus* was preserved within the Rhynie chert. The large fangs and stout legs suggest that it was a ground predator. Some specimens have been found within the fossilized stems of plants, which would have provided excellent cover for ambushing prey.

Spread of terrestrial organisms

The simple plants of the Devonian were confined to damp environments, though by the Late Devonian a greater range of habitats had evolved and a number of animals appeared, including spider-like, arachnid forms (trigonotarbids), flightless insects, myriapods, and mites. Many invertebrates colonized the new, tall-tree environments that were the precursors of the great Carboniferous forests.

DEVONIAN SPIRIFERID
The two valves in this brachiopod opened along a straight hinge, so that sea water could be sucked in for feeding and respiration.

Late Devonian mass extinction

Throughout the Devonian the globe cooled slowly. Representatives of several invertebrate groups, such as trilobites, gradually disappeared in stages through time, due to a combination of environmental changes. Before the end of the Devonian there was a dramatic catastrophe, the second of the great five Phanerozoic mass extinctions. The causes of this are still much debated. Sea levels had risen, but there is also some evidence for changes in ocean currents that may have disturbed stagnant bottom water and poisoned the upper waters of the sea. There is also increasing evidence for multiple meteorite impacts around the time of the extinction event.

GROUPS OVERVIEW

Some invertebrate groups, such as sponges, become much more abundant in the Devonian. Corals continued to evolve throughout the period, with tabulate and rugose forms becoming widespread. Spiriferid brachiopods flourished, and some species grew quite large during this time of global cooling. However, other groups, such as the trilobites, did not fare so well.

SPONGES
These originated in the Cambrian, and were widespread by the Devonian. Sponges suck water in through pores in their bag-like body, and strain off food particles using specialized cells. Tiny, spike-like structures called spicules supported the body, and these are often found in sediment.

ANTHOZOANS
Both rugose and tabulate corals flourished in the Devonian. They are unrelated to modern corals, but fed in the same way, catching small animals with their stinging cells, which lined their tentacles. Rugose corals were solitary or compound; tabulates were always compound.

SPIRIFERID BRACHIOPODS
These were typical of the Devonian and Carboniferous and superficially resembled bivalves. Inside the valve were two conical spires, pointing opposite ways, through which sea water was filtered and food particles strained off. A stout pedicle attached the brachiopod to the sea floor.

EURYPTERIDS
These large "water scorpions", often measuring at least 2m ($6^1/_2$ft) in length, inhabited fresh and brackish waters, as well as marginal marine environments. Most of them were ferocious predators. It is possible that some may have been able to crawl on land for short distances.

Heliophyllum

GROUP Anthozoans
DATE Early to Middle Devonian
SIZE Average diameter of cup 3cm (1¼in)
LOCATION Europe, North Africa, USA, Australia

Heliophyllum was a rugose coral that lived in the oceans more than 385 million years ago. As a fossil, it is normally found singly. Its skeleton, the corallum, is conical, although slightly irregular in shape. There are fine growth ridges on the outer walls. The shape of *Heliophyllum*'s corallum indicates its growth history. Decreases in its diameter are the result of environmental stress and bends reflect changes in growth orientation. The soft-bodied coral polyp would have lived in a shallow concave cavity at the top, known as a calice.

HELIOPHYLLUM SP.
The shallow cup (calice) at the top of the fossil is where the coral animal sat in life.

ANATOMY
RUGOSE CORAL

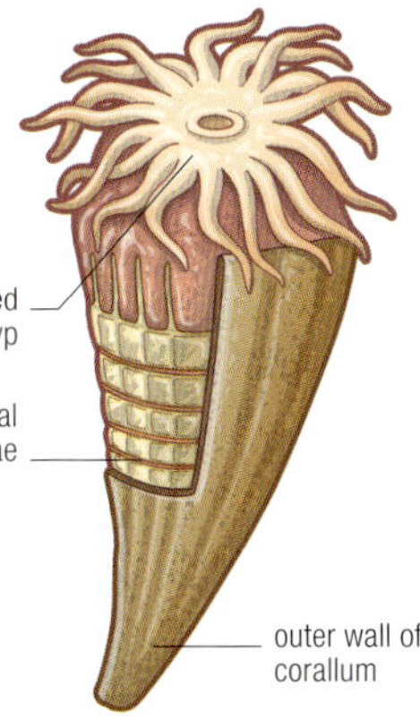

Rugose corals existed in solitary or colonial forms. Solitary ones were often horn-shaped and are informally known as horn corals. The soft-bodied polyp rested in the concave calice at the top of the corallum. The inside of the coral was subdivided with radial plates (septa). The septa were typically of two kinds, major and minor, which alternated with each other. As the coral grew, horizontal partitions (tabulae) and small arched plates (dissepiments) were secreted.

Calceola

GROUP Anthozoans
DATE Early to Middle Devonian
SIZE Up to 5cm (2in) long
LOCATION Europe, Africa, Asia, Australia

Calceola was a small, solitary rugose coral. Its fossils are roughly semicircular in cross section and the corallum or skeleton is often slightly curved. In life, the coral would have rested on the curved side of the skeleton, with a soft-bodied polyp (individual member of a coral colony) sitting in the calice at the top. A lid-like structure, the operculum, covered the concave calice. This could be drawn closed by muscles contracting, so that the polyp was totally enclosed and protected when it was not feeding.

CALCEOLA SP.
In this specimen the hinged, lid-like operculum is preserved. When the coral was feeding, the lid would have been open, but drawn shut if the polyp sensed danger.

Hydnoceras

GROUP Sponges
DATE Late Devonian
SIZE Up to 25cm (10in) long
LOCATION Europe, USA

Hydnoceras was a vase-shaped glass sponge that lived in the oceans between 385 and 360 million years ago. Seen in cross section, it had twin walls, one inside the other. The inner wall surrounded a large central chamber. Like living glass sponges (see panel, right), the body of *Hydnoceras* had a skeleton made up of numerous similarly shaped structures known as spicules, made from silica. *Hydnoceras* spicules had six rays, each ray forming an angle of 90 degrees with the next, so that every spicule looked like a cross with a vertical bar running through its centre. Some large outer spicules are visible on *Hydnoceras*, forming a surface pattern of squares. Inside each square are smaller squares.

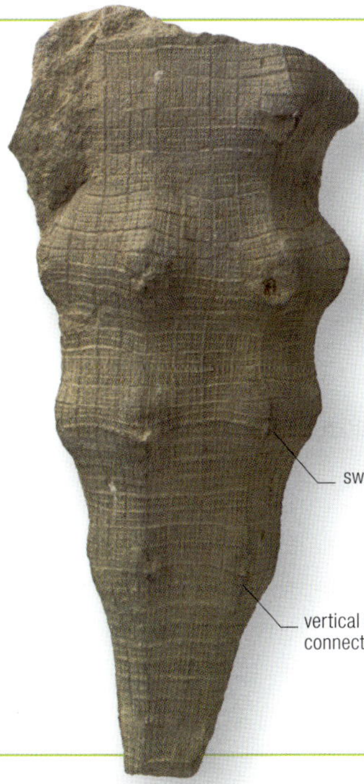

HYDNOCERAS TUBEROSUM
Hydnoceras fossils, such as this, have a distinctive shape, with several waist-like constrictions around the body. Between these "waists" are expanded areas, each carrying about eight swollen nodes.

LIVING RELATIVE
GLASS SPONGES

Hexactinellid sponges have a skeleton composed of typically six-rayed spicules of clear opaline silica fused together. The clear spicules give these animals their name of glass sponges – perhaps the most famous of the living species being Venus's flower basket, *Euplectella aspergillum*. In some parts of the world glass sponges form large reefs, such as those off the coast of British Columbia. They live at a range of depths but are mainly found in waters of about 200m (660ft).

PLEURODICTYUM PROBLEMATICUM
Many *Pleurodictyum* colonies, such as this one, had live-in lodgers, *Hicetes* worms, which left behind remains of the tubes they inhabited.

It is thought that worms lived within *Pleurodictyum* colonies for protection from predators, but did not harm the coral animals around them.

Pleurodictyum

GROUP Anthozoans
DATE Late Silurian to Middle Devonian
SIZE About 3cm (1¼in) across
LOCATION Worldwide

Pleurodictyum was a tabulate coral that formed small dome- or disc-shaped colonies. The base of a *Pleurodictyum* colony was flat or slightly concave, and this is seen in its fossils. The individual coral animals, or polyps, in a colony were relatively large and their skeletons (corallites) are clearly visible. The corallites had thick walls, which grew in contact with one another; they were perforated by many tunnels. Within the corallite, there were numerous dividing ridges, rows of spines, or spiny ridges. The corallites themselves were large and short.

Phillipsastrea

GROUP Anthozoans
DATE Devonian
SIZE Average distance between centres of corallites 1.2cm (½in)
LOCATION Worldwide

Phillipsastrea was a colonial rugose coral. Its individual corallites did not have separate walls, and vertical partitions (septa) from neighbouring corallites were shared. The septa varied in length; many were long and curving, but others were very short. Complete and incomplete horizontal partitions (tabulae) were present at the centre of the corallites, where the soft-bodied polyps were located. Small, horseshoe-shaped plates were found in zones surrounding the centre.

PHILLIPSASTREA SP. Individual corallites shared septa, giving *Phillipsastrea* its distinctive structure. The soft-bodied polyps lived in the centre of each corallite.

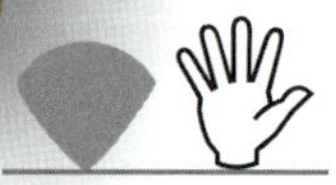

Rensselaerina

GROUP Brachiopods
DATE Early Devonian
SIZE Up to 5cm (2in) long
LOCATION North America

Rensselaerina was an elongated brachiopod with a prominent beak and a smooth shell. Inside the smaller (brachial) valve, there was a long calcitic loop, which extended across two thirds of the valve's length. In life, this loop supported the animal's feeding apparatus, the lophophore. The shell surface is punctuated by numerous minute perforations through which bristles would have protruded when the animal was alive.

R. HARAYANANA *Rensselaerina* has a smooth shell. Faint radial ribs are visible at the edges of this specimen.

Ptychomaletoechia

GROUP Brachiopods
DATE Late Devonian
SIZE Up to 2.5cm (1in) high
LOCATION Europe, Asia, North America

Ptychomaletoechia was a small- to medium-sized rhynchonellid brachiopod. Rhynchonellid brachiopods can be identified by the nut-like shape of their shells, and usually their strong ornamentation. Both valves are convex, although the smaller (brachial) valve is more inflated than the larger (pedicle) valve. The pedicle valve has a prominent beak and a small pedicle opening. In life, this opening housed the pedicle, a fleshy stalk that anchored the animal to the sea bed. The entire shell was ornamented with strong, radial ribs.

P. OMALIUSI *Ptychomaletoechia* had distinct ribs running the length of both valves. Near the centre of the smaller valve's edge there is a fold, with a matching infold on the larger valve.

Mucrospirifer

GROUP Brachiopods
DATE Middle Devonian
SIZE Up to 8cm (3¼in) across
LOCATION Worldwide

Mucrospirifer was a spiriferid brachiopod with an unusually long hinge-line. The widest point of the fossil shell, the hinge-line, runs between the two valves, coming to a point at each end. In the centre of the smaller (brachial) valve is a clear, V-shaped fold; there is a corresponding infold on the shell's larger (pedicle) valve. On either side of the brachial valve's central fold, there are strong radial ribs. The ribs themselves are crossed with growth lines. The growth lines are far stronger near the lower edge of the specimen shown here, suggesting that there was a pause in growth late on in this brachiopod's life.

MUCROSPIRIFER MUCRONATA Its distinctive winged shape gave *Mucrospirifer* its modern nickname of "butterfly shell".

ANATOMY

COILED LOPHOPHORES

Spiriferid brachiopods such as *Mucrospirifer* were elongated along their hinge-line. Shell models have shown that if open shells are situated facing the current, then internal spiral eddies (whirlpools) are produced. This meant that *Mucrospirifer's* coiled, calcite lophophore supports were able to maximize the food-gathering capacity of the animal's feeding structure, called the lophophore. The lophophore filtered food particles, such as organic matter or minute organisms, from the surrounding sea water.

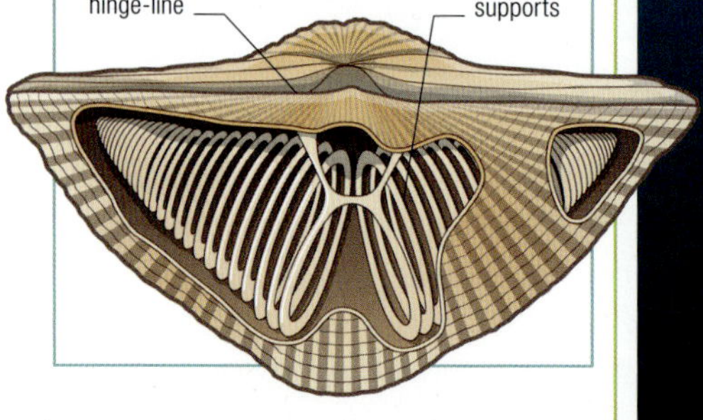

5cm The maximum **shell diameter** of the tiny, **squid-like** animal known as ***Soliclymenia.***

SOLICLYMENIA PARADOXA
This specimen clearly shows the distinctive triangular coiling. Not all *Soliclymenia* species exhibit this – some have the more usual circular coiling.

Soliclymenia

GROUP Cephalopods

DATE Late Devonian

SIZE Up to 5cm (2in) across

LOCATION Europe, N. Africa

The strange-looking *Soliclymenia* was one of very few ammonoids with triangular coiling. The whorls in its concave shell barely overlap, resulting in a wide and shallow umbilicus (navel-like indent) at the centre of the coil. The whorls are ornamented by densely spaced ribs. *Soliclymenia* was a member of a group of early ammonoids that evolved in Europe or North Africa. Fossils from this group are distributed only in these areas and also only in the Late Devonian period. Unusually for ammonoids, their siphuncle – the internal tube that connected the living shell to the earlier chambers – migrated from the ventral (underside) of the shell to the dorsal (upper side) during early growth.

MINIATURE SQUID
With its soft, muscular body, the living *Soliclymenia* resembled a tiny squid. The animal used its long tentacles for capturing prey.

Murchisonia

GROUP Gastropods
DATE Ordovician to Triassic
SIZE Up to 5cm (2in) long
LOCATION Europe, North America, Asia, Australia

Murchisonia existed on Earth longer than almost any other gastropod genus: a period of more than 200 million years. It was a sea creature that fed on algae and similar marine growth. *Murchisonia*'s shell was the only body part that fossilized, and this is notable for the spiral ridge, known as the slit band, that runs around the centre of the whorl. This ridge can be traced forward to a slit at the edge of the mouth of the shell, a feature shown by many primitive gastropods, including the Jurassic *Pleurotomaria* (see p.239). In the specimen shown here, the whorls are ornamented with long nodes, although not all *Murchisonia* fossils exhibit these. Indeed, *Murchisonia* generally showed very little in the way of shell ornamentation. In the Middle Devonian, however, there was a sudden burst of evolution that produced many more ornamented forms.

MURCHISONIA BELINEATA
The slit band, which runs around the middle of each whorl, is clearly visible on this specimen. A slit in the shell's mouth would have enabled the animal to direct the exhalant current from the body cavity away from the head region.

Cheiloceras

GROUP Cephalopods
DATE Late Devonian
SIZE Up to 4.5cm (1¾in) long
LOCATION Europe, Africa, Australia

Cheiloceras was a small goniatite. Its shell was tightly coiled, with a small, navel-like hollow, known as the umbilicus, at its centre. The coiling of its shell was involute, meaning that the inner coils were hidden by the outermost one. In cross section *Cheiloceras* was somewhat flattened. Around the whorl were irregularly spaced constrictions, almost as if someone had pulled a string tight around the shell and left an impression. The significance of these constrictions is unknown, although it has been suggested that they may represent short periods of restricted growth. The shell also had faint ribs running around it, much less visible in *Cheiloceras* fossils than in those of many other ammonites. The suture line was gently curved.

CHEILOCERAS VERNEUILI
Cheiloceras was so tightly coiled that it seemed to be made up of only one large outer coil, as seen here.

Cupressocrinites

GROUP Echinoderms
DATE Middle to Late Devonian
SIZE Cup height up to 5cm (2in)
LOCATION Europe, Morocco, USA, Australia

Cupressocrinites was an unusual crinoid, exhibiting perfect five-sided symmetry in its cup. It had three circles of plates in the cup at the end of its stalk, a feature that helps to identify its fossils. Its arms were flexible, and used to remove tiny particles of food from the water.

In fossils, the edges of the arms have articulation points for the small pinnules, the filament-like extensions that gathered food particles, although the pinnules themselves are never visible. The arms were massive and could fold over the cup, presumably to protect it either from predation or from fine sediment flowing onto the grooves within it, where food from the arms collected. The stem was roughly square in cross section, with a central canal and four peripheral canals along its length. These are visible in fossils where the stem has been broken.

CUPRESSOCRINITES CRASSUS
In this specimen of *Cupressocrinites*, the stem is clearly visible.

Phacops

GROUP Arthropods
DATE Middle to Late Devonian
SIZE Up to about 6cm (2¼in) long
LOCATION Worldwide

Phacops was one of the most characteristic trilobites of the Devonian period. Its headshield was dominated by a large, bun-shaped, central raised area known as a glabella, which in many species had a bumpy surface. The eyes were located close to the back of the glabella, and were large and crescent shaped. They carried large, individual lenses arranged in a hexagonal pattern across the surface, and can easily be seen with the naked eye. Outside the eye, the cheek fell away at a steep angle. The hypostome, a broadly oval structure, lay beneath the front part of the glabella, and was rigidly attached to the rest of the headshield at its front end. The thorax had 11 segments, and the tailshield was smaller than the head, with eight or nine well-defined segments on its axis, and five to six ribs on the flanks, defined by deep furrows.

PHACOPS AFRICANUS
Like modern woodlice, most trilobites could roll up to protect their softer undersides, and *Phacops* was no exception.

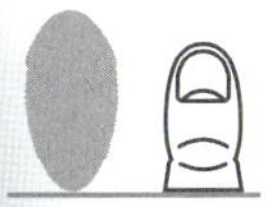

SEGMENTED BODY
The thorax was divided into 11 segments. The pleurae (the side parts of each segment) had rounded ends. The tailshield was smoothly curved.

LARGE GLABELLA
One of *Phacops*'s most prominent features was its bun-shaped glabella, the raised area between its eyes, which in this specimen had a tuberculated surface.

ANATOMY
SOPHISTICATED VISION

Phacops was characterized by eyes that had individual lenses arranged in a hexagonal pattern, with each separated by thin walls of cuticle. Research in the early 1970s revealed that, in life, these lenses were composed of calcite crystals, which were orientated so that they behaved like glass. Each lens had two components: a bowl-like lower unit and an upper one that was convex on both sides. Experimental models have demonstrated that the slight difference in refraction of the two units produced a sharp focus. Thus, these trilobites had highly sophisticated vision, and it is possible that it was used in a predatory feeding mode.

Paralejurus

GROUP Arthropods
DATE Early Devonian
SIZE Up to about 15cm (6in) long
LOCATION C. Europe, C. Asia, Morocco

Paralejurus was a trilobite with a crescent-shaped headshield, and a glabella that widened dramatically at the front. Its eyes were located towards the back of its head, and its cheeks were broad, without a defined border, and ending in a sharp angle on the sides. The thorax consisted of ten broad segments, and its fan-shaped tailshield was markedly longer than its headshield. *Paralejurus* belongs to a group of trilobites with distinctive fan-shaped tails, which might have been used to propel the animals through water.

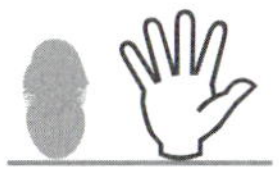

PARALEJURUS SP.
The ten segments of the thorax are easily seen on this specimen, as is its distinctive, sharp-angled cheek.

Lioharpes

GROUP Arthropods
DATE Middle Devonian
SIZE Up to about 7.5cm (3in) long
LOCATION C. Europe, C. Asia

Lioharpes was a trilobite with a large, horseshoe-shaped headshield. Its conical glabella occupied a little over half its length. Its tiny, circular eyes lay opposite the front of the glabella, on the inner part of its smooth cheeks. The outer cheeks sloped steeply to the edge of a broad, flat fringe. This had a large number of tiny pits, and extended backwards into a long, broad, dagger-shaped spine (not shown on this specimen). Its thorax had at least 28 narrow segments, and its tail was small. The headshield of *Lioharpes* is of quite high relief, and this may have been related to its function as a filter chamber for feeding on small particles of food.

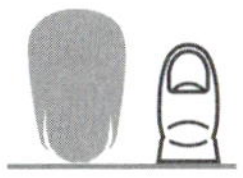

LIOHARPES VENULOSA
With its horseshoe-shaped headshield and distinctive fringe, *Lioharpes* is easy to identify.

DEVONIAN VERTEBRATES

Often called the "Age of Fish", the Devonian saw a major radiation of fish groups both in marine and freshwater environments. By the Late Devonian, the stage was set for an invasion of the land, with the arrival of the first four-legged vertebrates, the tetrapods.

Devonian seas teemed with life, and a wide variety of fish evolved and spread around the globe, including the first true sharks. Jawless fish, such as the osteostracans, were the first to explore estuaries and lagoons and were followed by jawed predators, including placoderms and the sarcopterygians (bony fish with lobed fins). It was the sarcopterygians that gave rise to the first vertebrates to venture onto land.

The origin of tetrapods

Tetrapods are four-legged animals, and include amphibians, reptiles, birds, and mammals. The earliest-known tetrapods appeared in the Late Devonian about 365 million years ago, in the form of *Acanthostega* and *Ichthyostega*. They are assigned to the Class Sarcopterygii and evolved from the lobe-finned fish. Recent research has established that the fish most closely related to the tetrapods is *Tiktaalik roseae*, whose fossils were found in 375-million-year-old rocks on Ellesmere Island in the Canadian Arctic in 2004. *Tiktaalik* and another closely related sarcopterygian fish, *Panderichthys rhombolepis*, which lived about 380 million years ago, have flattened skulls and dorsally placed eyes, similar to *Acanthostega*. One feature that links *Tiktaalik* more closely to *Acanthostega* is that its skull had become separated from the bony shoulder girdle so that the head could be moved more freely; *Panderichthys* retained the fish-like connection to the shoulder girdle.

The discovery of the fossils of *Acanthostega* and *Ichthyostega* overturned some entrenched misconceptions about the evolution of the first tetrapods. Originally, it was thought that tetrapods had five fingers and toes ("digits") when hands and feet evolved from fish fins. However, *Acanthostega* has eight digits, while *Ichthyostega* has seven (see right), indicating that the number of digits was variable and only became fixed at five later in tetrapod evolution. In addition, the idea that tetrapods first evolved when they lumbered onto the land also proved to be incorrect. The fossil specimens of *Acanthostega* and *Ichthyostega* were found in rocks in Greenland that were clearly deposited

BRACED FOR WALKING
Frogfish are actinopterygians that live in coral reefs. Instead of swimming, they use their pelvic and pectoral fins to "walk". As they move about, their left and right pelvic fins touch the coral independently, in a similar way to the walk of a tetrapod.

GROUPS OVERVIEW

The Age of Fish saw the beginning – and the end – of many groups of fish. Placoderms were a hugely successful and diverse group, but they only survived for about 50 million years, whereas sharks became important fish and are still represented by about 400 species today. But the main event in terms of vertebrate evolution was the arrival of the tetrapods.

OSTEOSTRACANS
By the Early Devonian, osteostracans had diversified and were widely distributed in North America, Europe, and Asia. These jawless fish typically had bony, horseshoe-shaped headshields. They lived in marine or estuarine environments and were mostly bottom-dwellers. Osteostracans died out in the Late Devonian, about 370 million years ago.

HETEROSTRACANS
These heavily armoured jawless fish arose during the Early Silurian. The most common subgroup in the Devonian was the pteraspids, characterized by large dorsal and ventral bony shields, but they were only found in the northern hemisphere. By the Middle Devonian, most heterostracans were extinct, although the psammosteids persisted until the end of the period.

ARTHRODIRES
Arthrodires are one of the largest groups of placoderms. These early jawed fish had articulated armoured plates on their head and neck and were found worldwide in the Devonian. One of the most famous is *Dunkleosteus*, which at 6m (20ft) long, was one of the largest Devonian vertebrates. This fearsome predator had one of the most powerful bites ever known.

ANTIARCHS
Another group of placoderms, antiarchs had unusual front fins that were covered in bony plates; their exact function is still a matter of speculation. Antiarch jaws were small and weak compared to those of arthrodires and they are thought to have scooped up mud containing tiny invertebrates. Antiarchs, along with all the rest of the placoderms, became extinct at the end of the Devonian.

The earliest tetrapods evolved in estuaries in the Late Devonian. They retained tail fins and internal gills – basically, they were fish with legs.

ICHTHYOSTEGA FOSSIL LIMB

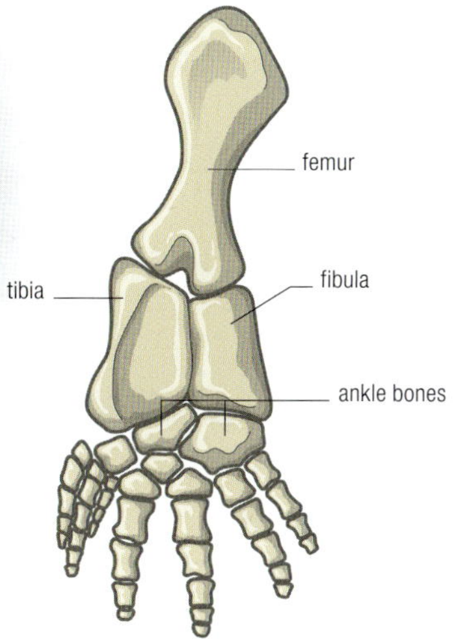

EARLY TETRAPOD LIMB
Ichthyostega had seven digits on its hind limbs (no *Ichthyostega* fossil hands have been found). The structure of this paddle-like limb indicates that it was probably of more use for swimming than walking, rather like a seal's flipper.

in a river environment, indicating that these early tetrapods lived in streams or small rivers. The fossils are well preserved and nearly complete rather than being disarticulated, meaning that these animals died and were buried very close to where they lived, rather than being transported any great distance. Also, their paddle-like limbs were more suited to swimming than walking, and *Acanthostega* had a fish-like tail. Although they both had lungs and could breathe air, they retained internal gills. It seems that tetrapods in the Devonian were gradually adapting to the final transition from water to land; their conquest of the land would not be completed until the Carboniferous.

The origins of internal fertilization

The Gogo Formation in Western Australia is the remains of an extensive Devonian reef system (see p.43). Many placoderms have been preserved here in limestone nodules. These nodules can be dissolved with acetic acid, revealing the fossilized fish within. The preservation is so fine that placoderm muscle tissue has recently been discovered, as well as evidence of embryos within the body of the mother. In *Materpiscis attenboroughi*, the embryo is connected to its mother via an umbilical cord. These embryos provide some of the earliest evidence of internal fertilization in the fossil record. By contrast, external fertilization is where sperm from the male mixes with eggs in the water and the embryos develop outside of the female.

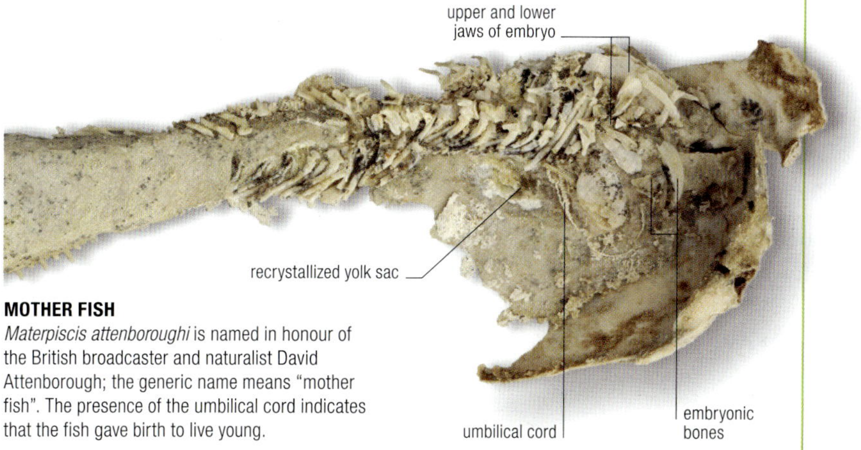

MOTHER FISH
Materpiscis attenboroughi is named in honour of the British broadcaster and naturalist David Attenborough; the generic name means "mother fish". The presence of the umbilical cord indicates that the fish gave birth to live young.

ACANTHODIANS
Acanthodians are characterized by the large, bony spines in front of their fins (they are also known as "spiny sharks", but they are not true sharks). Early acanthodians were exclusively marine, but by the Middle Devonian these jawed fish had colonized freshwater systems. Their distinctive diamond-shaped scales are common in the fossil record but skull and jaw bones are rare.

ACTINOPTERYGIANS
Actinopterygians, or ray-finned fish, first appeared in the Late Silurian; today, this highly diverse group has more than 28,000 species. As their name suggests, these bony fish have ray fins built from a fan of narrow bone or cartilage rods. *Cheirolepis*, from the Middle Devonian, is a primitive actinopterygian because its teeth lack the mineralized cap found in other actinopterygians.

ACTINISTIANS
These sarcopterygians arose in the Devonian and were highly successful. Coelacanths were thought to have died out with the last-known fossil genus, *Macropoma*, from the Cretaceous, until the spectacular discovery of a living coelacanth, *Latimera chalumnae*, in a fisherman's catch off eastern South Africa in 1938. In 1997, a second species was found in Indonesia.

DIPNOANS
This group of sarcopterygians are known as lungfish. Their "lung" is a modified swimbladder, which as well as aiding buoyancy (its usual function in fish) also absorbs oxygen and removes wastes. Dipnoans had a worldwide distribution in the Devonian. Today, there are three genera of living lungfish found in Australia (*Neoceratodus*), Africa (*Protopterus*), and South America (*Lepidosiren*).

TETRAPODS
Tetrapods include all four-legged vertebrates. When they evolved in the Late Devonian, they were still reliant on water for survival. Current research shows that today's lungfish are more closely related to tetrapods than any of the coelacanths. However, numerous fossil sarcopterygians are even closer relatives, the most recent discovery being *Tiktaalik roseae*.

Stethacanthus

GROUP Chondricthyans
DATE Late Devonian to Early Carboniferous
SIZE 1.5m (5ft) long
LOCATION North America, Scotland

Although the outline of *Stethacanthus* was similar to that of modern sharks, there are striking differences. The tail fin is almost symmetrical – in modern sharks, the upper lobe is usually much larger than the lower lobe. The pectoral fins had a long, distinctive rod, called the "whip", which trailed behind the fin. Another distinctive feature is the large structure projecting from its back. Sometimes called the "ironing board" because of its shape, its upper surface was studded with a dense patch of denticles – tiny, tooth-like granules embedded in the skin. The only other place denticles appeared was in a patch on the head. These features, probably present only in males, may have been important for gripping females during mating.

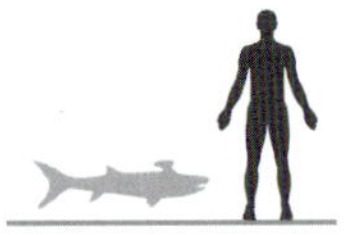

Weigeltaspis

GROUP Agnathans
DATE Devonian
SIZE Approximately 10cm (4in) long
LOCATION Europe, North America

A primitive heterostracan (see p.128), *Weigeltaspis* lacked both jaws and paired appendages. The body scales, which were largest behind the single pair of branchial openings (gills) and smallest on the tail, bore a complex surface ornament. The substantial tail fin suggests that *Weigeltaspis* was likely to have been a relatively strong swimmer – the fin web was interleaved with at least five horizontal, scaly, finger-like projections, which would haved aided propulsion. *Weigeltaspis* was a relatively small fish that lacked the spines seen in later genera. It was probably a plankton feeder.

EFFICIENT SWIMMER
This *Weigeltaspis* fossil from the Ukraine clearly shows the streamlined shape and large tail of an active swimmer.

Drepanaspis

GROUP Agnathans
DATE Early Devonian
SIZE 35cm (14in) long
LOCATION Europe

One of the earliest known psammosteid fish, *Drepanaspis* was small compared to its Late Devonian descendents, some of which reached 2m (6½ft) in length. Its flattened headshield featured a mosaic of small, scale-like plates that separated the midline plates from the paired ones on either side. These small plates appeared as the fish matured and were entirely absent in juveniles. Its flattened shape suggests it was a bottom-dwelling fish that foraged on the ocean floor. However, it is unclear how it fed because its jawless mouth pointed upwards. *Drepanaspis* had no fins other than the ones on its tail.

PADDLE-SHAPED FISH
Distinctly paddle-shaped when viewed from above, *Drepanaspis* was heavily armoured and had a flattened headshield. This fossil of *Drepanaspis gemuendenaspis* is from Germany.

Protopteraspis

GROUP Agnathans
DATE Late Silurian to Early Devonian
SIZE Unknown
LOCATION North America, Europe, Australia

The jawless pteraspids can be divided into five families, including Protopteraspididae, to which *Protopteraspis* belongs. This somewhat unspecialized fish had a narrow, rounded snout that was less elongated than those of many related genera. It also lacked the large, rear-facing projections, called cornua, that extended back from the sides of the bony headshields of other pteraspids. Instead, it had small points on each side, just behind the single gill opening. Like its relatives, however, *Protopteraspis* did possess a medium-sized dorsal spine. The plates covering the headshield had concentric, serrated ridges of dentine (calcified tissue). The headshield formed at an early age and grew by addition to the plate edges. In adults, the plates became fused together. *Protopteraspis*'s body scales were small and diamond shaped.

BOTTOM-DWELLING FISH
The fish on the right is *Protopteraspis gosseleti*, but the identity of its companion in death is not known. Because it was slightly flattened from top to bottom, *Protopteraspis* is thought to have been a bottom-dweller. It lived in fresh water.

Zenaspis

GROUP Agnathans
DATE Early Devonian
SIZE 25cm (10in) long
LOCATION Europe

Zenaspis was a large osteostracan (see p.128) with a horseshoe-shaped head, which is also seen in the more primitive *Ateleaspis* (see p.107). *Zenaspis* differed from *Ateleaspis* in having long, narrow cornua (backward-pointing projections) on the head, and the lateral sensory fields were also somewhat narrower. The rear edge of the bony headshield was drawn up into a central crest that continued as a row of dorsal-ridge scales, then formed a single dorsal fin towards the back of the body. The pectoral fins were smaller than those of *Ateleaspis*, and had narrow, rather than broad, bases. The body scales were also much larger. The mouth (on the underside of the fish) lacked teeth, and instead was lined with elongated oral plates. These merged into the scales covering the chamber that housed paired gill openings. Like *Ateleaspis*, *Zenaspis* was a bottom-dweller that lived in shallow seas or river-mouths.

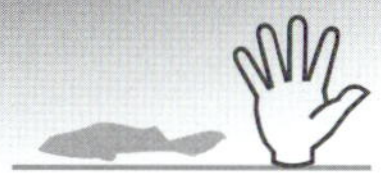

DISTINCTIVE HEADSHIELD
The headshield of *Zenaspis* can be clearly distinguished from its body, which is covered in bony scales. The eyes were close together and situated on top of the head where, as a bottom-dweller, *Zenaspis* could keep a look out for predators.

Osteostracans **first appeared** in the Middle Silurian. They had **diversified** into **many forms** by the Early Devonian, but became extinct during the Late Devonian.

Lunaspis

GROUP Placoderms
DATE Early Devonian
SIZE 10–30cm (4–12in) long
LOCATION Germany

The heavily armoured *Lunaspis* lived in the shallows of a sea that covered parts of Europe 400 million years ago. This distinctive flattened fish had plates that were heavily ornamented with concentric rings, and separate head- and trunk shields linked at a rudimentary hinge joint. The median dorsal plate of the headshield was very long, and extended forwards as far as the backs of the eye sockets, which were on the top of its head. Because the jaws have not been preserved, little can be said of its diet, but its flattened shape suggests that *Lunaspis* lived on the sea bed. Rigid pectoral spines projected at about 45 degrees from the body and extended to the back edge of the trunk shield. Instead of a dorsal fin, *Lunaspis* had three large ridge scales behind the median dorsal plate. The rest of the body was covered with bony scales that decreased in size towards the fish's tail, which appears to have been divided into two equal parts.

Gemuendina

GROUP Placoderms
DATE Early Devonian
SIZE 25–30cm (10–12in) long
LOCATION Germany

A flattened fish with large, wing-like pectoral fins, *Gemuendina* was a bottom-dweller with a shape similar to modern rays. Its eyes were on the top of its head, with nostril openings between them, and it had a wide mouth. Two large plates, the submarginals, covered the gill region on its sides. The only other identifiable plates on the head were the suborbitals to the front and outsides of the eye sockets and the paranuchals, which carried sensory canals. The bony covering of the cranium had no distinguishable plates. All of these areas were surrounded by a mosaic of small plates called tesserae. The trunk shield was very short and did not extend beyond the opening for the narrow-based pectoral fin. The pelvic fins comprised small, semicircular flaps, while the dorsal fin was reduced to a single dorsal spine. The anal fin was absent, and the tail, like the body, was long, tapering, and covered in scales. Larger tooth-like scales were also present, randomly scattered over the fins.

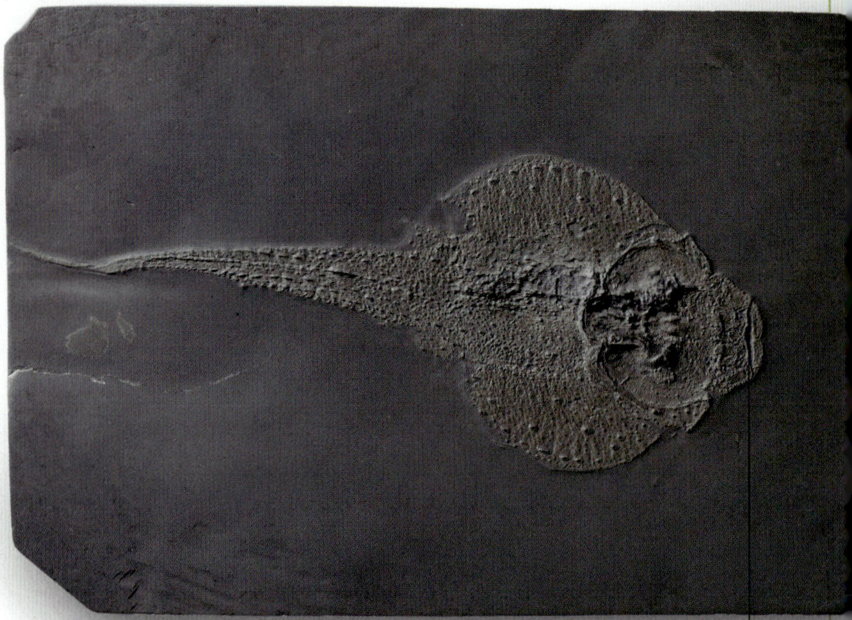

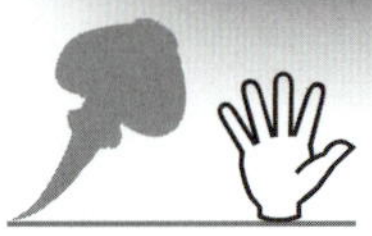

ADAPTATED FOR LIFE ON THE SEA BED
On this well-preserved specimen of *Gemuendina stuertzi*, the dorsally situated sense organs (eyes and nostrils) and mouth are clearly evident. Their positions are adaptations for living on the sea floor.

Rhamphodopsis

GROUP Placoderms
DATE Middle Devonian
SIZE 12cm (4½in long)
LOCATION Scotland

Unlike many placoderms, *Rhamphodopsis* was not heavily armoured. Its body tapered into a whiplike tail. This small freshwater fish has the distinction of being the earliest example of a fish in which the sexes can be distinguished by the form of their pelvic fins. Like sharks, the males had claspers – rod-like structures that aided internal fertilization. In females, the pelvic fins were covered in large scales. *Rhamphodopsis* had strong, crushing tooth plates, which initially led scientists to think that ptyctodontids, such as *Rhamphodopsis*, were closely related to sharks. However, anatomically they share many more features with placoderms, so ptyctodontids are assigned to that class.

Dicksonosteus

GROUP Placoderms
DATE Early Devonian
SIZE 10cm (4in) long
LOCATION Norway

This primitive arthrodire (see p.128) had a long trunk shield and long, curving, spinal plates that formed the front of an opening (the pectoral fenestra) for the pectoral fin. In *Dicksonosteus*, the median dorsal plate of the trunk shield was long and narrow and ended in a rounded lobe, rather than the spine seen in *Coccosteus* (see opposite). The headshield had a small nuchal plate centred at the back, and a pair of larger central plates. The armour as a whole was flattened dorsoventrally and adorned with concentric rows of tubercles. The two pairs of upper tooth plates were covered in robust, tooth-like denticles, corresponding to those at the front of the lower jaw. The body tapered, but the tail is unknown.

Rolfosteus

GROUP Placoderms
DATE Late Devonian
SIZE 30cm (12in) long
LOCATION Australia

Known from a single complete specimen and a few bone fragments, *Rolfosteus* was a bizarre, long-snouted fish that lived among the reefs of Western Australia. The rostral plate was greatly elongated, forming a tapering tube that comprised half the length of the headshield. The function of this tubular snout is unknown. It may have been used to expose buried prey on the sandy sea bed, or to provide better streamlining so the fish could hunt shrimps living near the surface. *Rolfosteus* had flattened, crushing tooth plates at the back of the mouth, suggesting that small crustaceans and shellfish formed at least part of its diet. The snout could also have been a sexual characteristic of males that was used for display to attract females during the breeding season.

Coccosteus

GROUP Placoderms
DATE Middle to Late Devonian
SIZE 40cm ($15\frac{1}{2}$in) long
LOCATION North America, Europe

Probably the best-known of all arthrodires, *Coccosteus* was first described in 1841, after which 49 species were identified, although many were later reclassified. This armoured fish had large pectoral, dorsal, and tail fins and a powerful tail, suggesting that it was an able swimmer. The two pairs of tooth plates in the upper jaw initially had teeth, but these were gradually worn down by contact with the lower jaw to form sharp, blade-like cutting edges. Stomach contents reveal that *Coccosteus* was a predator of both acanthodians (see p.129) and juvenile arthrodires. It may have lain in wait on the sea bed and ambushed prey, using its tail to power the attack. *Coccosteus* had a long, tapering spine on the median dorsal plate and many other smaller spines on the underside.

SKULL PLATES
The different plates on the roof of the skull are clearly evident in this specimen, as are some of the sensory lines that may be seen as grooves running across the surface of some plates.

DOUBLE JOINT
The head- and trunk shields of arthrodires are linked by a simple articulation on each side that allows the head to move only in a vertical plane (up and down). The "socket" is on the headshield and the "hinge" is on the trunk shield.

Dunkleosteus

GROUP Placoderms
DATE Late Devonian
SIZE 6m (20ft) long
LOCATION USA, Europe, Morocco

Dunkleosteus was a heavily armoured fish. Its body armour did not enclose its pectoral fins at the rear. This development allowed the fin base to increase in length and aided mobility, suggesting that *Dunkleosteus* was an active hunter in the shallow seas in which it lived. The upper and lower parts of its trunk shield were connected only by small, non-projecting spinal plates. Its lower jaw and front upper tooth plate eventually wore down, leaving fang-like points. *Dunkleosteus* lacked ornamentation in most species, but its plates often bear bite marks and puncture wounds, indicating that, despite its size, it was itself hunted.

DISTINCTIVE PLATES
Dunkleosteus was covered in bony plates that were up to 5cm (2in) thick. The median dorsal plate (top) contrasts with *Coccosteus* (see p.133) in that it has a rounded posterior margin and lacks a posteriorly directed spine. The headshield (below) is equally distinctive, with its inwardly curving rear.

FORMIDABLE PREDATOR
This huge fish had a massive blade-like lower jaw and upper tooth plates with sharp shearing edges. *Dunkleosteus* could open widely when taking hold of prey, its gape aided by the unarmoured space between the head- and trunk shields.

Bothriolepis

GROUP Placoderms
DATE Early Devonian to Early Carboniferous
SIZE 30cm (12in) long; a few reached 1m (3ft 3in)
LOCATION Australia, North America, Europe, China, Greenland, Antarctica

This early antiarch (see p.128) had a somewhat clumsy appearance. It had a box-like trunk shield with almost vertical sides and a flat bottom. The "roof" may have varied among species. The pectoral appendages were not fin-like, but were gently curving spine-like structures, covered with small bony plates.

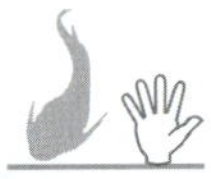

LONG APPENDAGES
The long pectoral appendages of *Bothriolepis* often extended beyond the rear of the trunk.

Pterichthyodes

GROUP Placoderms
DATE Middle Devonian
SIZE 20–30cm (8–12in) long
LOCATION Scotland

Pterichthyodes was a small, heavily armoured fish with distinctive pectoral appendages, not unlike *Bothriolepis* (left) in appearance. However, the head-to-trunk proportions were different in that *Pterichthyodes* had a longer body and relatively smaller head, and the pectoral appendages were shorter in length. Having its eyes on the top of the head (and a ventrally flattened trunk shield) suggests that *Pterichthyodes* was a bottom-dweller. The body behind the armoured parts tapered and was covered with overlapping cycloid scales. The single dorsal fin was triangular and was supported anteriorly by large arrow-shaped scales.

LAKE DWELLER
It has been suggested that Pterichyodes may have used its pectoral appendages to crawl along the bottom of the ancient Scottish lakes in which it lived.

Cheiracanthus

GROUP Acanthodians
DATE Middle Devonian
SIZE 30cm (12in) long
LOCATION Scotland, Antarctica

Cheiracanthus was a freshwater fish. An active swimmer, it patrolled the middle depths of lakes and rivers, feeding on small prey items caught in its gaping jaws. It had no teeth, so presumably filtered water through its long gill rakers to extract food. *Cheiracanthus* had large eyes, a large mouth, a deep body, and an upturned (heterocercal) tail. One of its defining characteristics is the lack of intermediate spines on the belly. Each of its fins was protected by a spine on the anterior edge. Its scales, which were small, non-overlapping, and ribbed, have been found at many sites. Fin spines, which were loosely attached to the body, have also been discovered. Complete specimens are preserved only in the Old Red Sandstone deposits of Scotland.

Ischnacanthus

GROUP Acanthodians
DATE Early Devonian
SIZE Unknown
LOCATION Scotland, Canada

Ischnacanthus had stout teeth that were larger towards the front. There was a whorl of smaller teeth where the two jawbones met. This freshwater-lake fish was slender and streamlined and had well-developed fins. The spines, which were long and narrow, are found only in association with fins; there were no intermediate spines. The armouring of the head and shoulder area is reduced compared to some earlier acanthodians, making *Ischnacanthus* capable of manoeuvring at speed while hunting or fleeing capture. The body was covered with small, polygonal scales, and the tail was shark-like.

TRIANGULAR TEETH
Ischnacanthus had triangular-shaped teeth in its upper and lower jaws. Teeth were added at the front of the jaws and so were larger as they had not been worn done.

Cheirolepis

GROUP Actinopterygians
DATE Middle Devonian
SIZE Up to 50cm (19½in) long
LOCATION Scotland, Canada

A ray-finned fish, *Cheirolepis* is the most primitive member of the actinopterygian group known from complete specimens. It lived in shallow lakes that existed about 380 million years ago and had a long body covered in tiny interlocking diamond-shaped scales containing ganoine, a substance similar to tooth enamel. Its pectoral fins were fleshy and more like those of a modern lungfish. *Cheirolepis* was probably capable of swimming at high speeds to catch its prey, which it trapped in the many sharp teeth that lined its long jaws. The length of the jaws suggests that it could open its mouth wide and take in sizeable food items, possibly up to two-thirds of its own size. Stomach contents have revealed a diet of fish – including its own kind.

RAY-FINNED FISH
Cheirolepis had only one dorsal fin and a shark-like tail with a notochord (skeletal rod) extending into the longer upper lobe. Most of the fin web hung downwards.

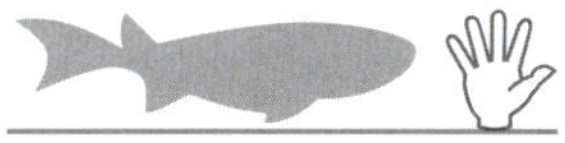

Dipterus

GROUP Sarcopterygians
DATE Middle Devonian
SIZE 35cm (14in) long
LOCATION Scotland, North America

Lungfish were widespread and diverse in Devonian times. Early forms were found in coastal saltwater sediments, but by the Late Devonian they had moved almost exclusively to fresh water. *Dipterus* had a well-armoured head, the top of which was covered with a mosaic of small bones. Some bones carried lines of pits that may have had a sensory or nutritional function. There was a large pair of tooth plates in the upper and lower jaws, across which ran rows of robust teeth that were probably used for crushing shellfish. A large plate covered the gill chamber, suggesting that gill breathing in lungfish was more important in Devonian times than breathing air.

COVERED IN SCALES
Rounded scales, coated with a thin, shiny later of cosmine (a type of dentine) covered *Dipterus*. It had two dorsal fins far down its back, one anal fin, and long, thin pelvic and pectoral fins. The tail was shark-like in structure.

Fleurantia

GROUP Sarcopterygians
DATE Late Devonian
SIZE 25cm (10in) long
LOCATION Canada

Named after Point Fleurant in Miguasha National Park, Canada (see p.139), *Fleurantia* was an early lungfish. Its snout was longer and more slender than those of modern forms, and it also lacked the robust tooth plates that are typical of many lungfish species. Instead, it had rows of conical teeth and small denticles (tooth-like projections lining its palate). The body scales were large and round and lacked cosmine (see *Osteolepis*, p.139). *Fleurantia*'s largest fins were on its back (see below). Like *Dipterus* (see p.135), *Fleurantia* had an anal fin, long, thin pectoral and pelvic fins, and a heterocercal tail (the upper lobe is larger then the lower one). The body was deep and laterally compressed, which suggests that *Fleurantia* was a fast swimmer that hunted prey rather than foraging along the bottom for food. *Fleurantia* lived in fresh or brackish water along with another lungfish, *Scaumenacia*, with which it was initially confused.

DORSAL FINS
Fleurantia had two dorsal fins. The front fin was small and lobe-shaped, while the second was considerably elongated, running along about 25 per cent of the fish's total body length.

Holoptychius

GROUP Sarcopterygians
DATE Late Devonian
SIZE 2m (6½ft) long
LOCATION North America, Greenland, Latvia, Lithuania, Estonia, Russia

Holoptychius was a large fish that was a dominant and clearly very successful predator. Its remains have been found throughout the Late Devonian in several parts of the world. The main type of fossils that are found are sheets of large, bony scales, preserved as the body of the fish disintegrated – a single *Holoptychius* scale can be as big as a saucer. *Holoptychius* had long, paired fins, rimmed by a fringe of fin rays, and its skull had a joint across the top of its head that allowed the snout to be raised even further when the mouth was opened. *Holoptychius* lived alongside fish such as placoderms, on which it may have fed, and it also ventured into some of the same habitats as the tetrapods, the earliest creatures with four legs (see pp.128–29).

MASS GRAVE
Here, many *Holoptychius* have died together, along with other fish. The cause of death is unknown. *Holoptychius* belongs to the porolepifomes, whose nearest relatives are the modern lungfishes, although there is little more than a superficial resemblance between *Holoptychius* and any living species.

Tristichopterus

GROUP Sarcopterygians
DATE Middle Devonian
SIZE 30cm (12in) long
LOCATION Scotland

Tristichopterus was a tetrapodomorph fish, and the most primitive member of the Tristichopteridae, a family with about ten genera, including *Eusthenopteron* (see p.138) and a huge predator called *Hyneria*. At 30cm (12in) long, *Tristichopterus* was a relatively small member of the family. Its primitive features included a very slightly asymmetrical tail, although the tail did have the three-lobed (trifid) pattern characteristic of the tristichopterids. Other features – aside from its age and location – that differentiate *Tristichopterus* from the better-known *Eusthenopteron* include the proportions of its head, as the front part of the skull is relatively short compared to that of *Eusthenopteron*.

Although they shared some features with primitive tetrapods, tristichopterids are not thought to be their direct ancestors.

Panderichthys

GROUP Sarcopterygians
DATE Middle to Late Devonian
SIZE 1.5m (5ft) long
LOCATION Latvia, Lithuania, Estonia, Russia

Panderichthys is one of three known Devonian fish that show many transitional, tetrapod-like features. Compared to other tetrapodomorph fish of the time, such as *Eusthenopteron* (see p.138), *Panderichthys* was more flattened top to bottom. And compared to most other fish, it had a long snout and large eyes placed close together on the top of its head. Similarly, the rear part of the skull that covered the ear region and brain was comparatively short relative to the snout. The hinged joint across the skull that was present in many primitive tetrapodomorph fish was absent in *Panderichthys*. It is likely that it breathed air using a lung-like air bladder to a great extent, supplementing the use of its gills for respiration. *Panderichthys* retained fin rays on its paired fins, and scales all over its body, but fins on the back and tail were absent or reduced in size.

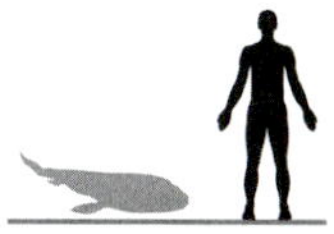

LOBED FINS
As a lobe-finned fish, the fins of *Tristichopterus* contained robust bony supports rather than flimsy rays of bone or cartilage as in the fins of most modern fish. The pectoral fins, towards the front of the body, appear to be the most rounded, while the fins further back are more pointed and streamlined.

Eusthenopteron

GROUP Sarcopterygians
DATE Late Devonian
SIZE 1.5m (5ft) long
LOCATION North America, Greenland, Scotland, Latvia, Lithuania, Estonia

The skeleton of *Eusthenopteron* exhibits clear evidence of the link between fish and land vertebrates. It was a late member of the Tristichopteridae, an extinct family of lobe-finned fishes. The small bones at the base of the pectoral and pelvic fins of *Eusthenopteron* can be traced to the bones – by now greatly enlarged – of the front- and hind limbs of early tetrapods, such as *Icthyostega* (see opposite). The structure of its vertebrae and its skull – especially the nostrils – also link this fish to the tetrapods. Long before modern scanning techniques and computer analysis were available, the details of *Eusthenopteron*'s skull were obtained using a more painstaking method. A fossil skull was ground down, millimetre by millimetre, and several photographs were taken at each stage. The sequence of images was used to make replicas of every thin section of skull out of slices of wax. The fossil was destroyed in the process, but the complete wax model produced allowed scientists to study the fish's skull, inside and out.

FOSSIL IMPRESSION
The soft internal parts of *Eusthenopteron* have not been preserved along with the hard skeleton. However, this specimen, which is an impression of the outside of the body, reveals that the fish was covered in large scales.

Tiktaalik

GROUP Sarcopterygians
DATE Late Devonian
SIZE 3m (9¾ft) long
LOCATION Canada

In many respects, *Tiktaalik* is a fish that shows even more tetrapod-like features than *Panderichthys* (see p.137). For example, *Panderichthys* has a series of bones that joins the skull to the shoulder girdle. There is also a series of bones that covers the gill region. These bones are absent in tetrapods, and most of them are also absent in *Tiktaalik* – at least, they have never been found. And many of the tetrapod features that *Tiktaalik* does share with *Panderichthys* appear to be at a more advanced stage in the transition to land vertebrates. For example, it has an even longer snout and shorter rear portion of the skull than *Panderichthys* and a wider round notch (the spiracular notch) at the back of the skull.

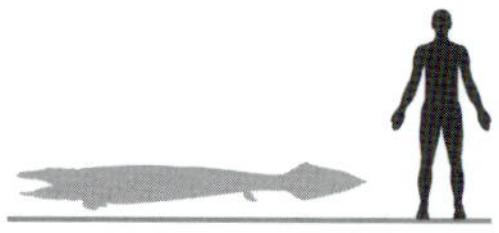

Acanthostega

GROUP Early tetrapods
DATE Late Devonian
SIZE 1m (3ft 3in) long
LOCATION Greenland

Acanthostega is a famous Devonian tetrapod. It had four limbs that ended in eight fingers or toes, but it also had some primitive features, such as fin rays along both the top and bottom of its tail. *Acanthostega* had a well-developed gill skeleton, so it probably used its gills for respiration as well as having had an air bladder for breathing air. The ear was similar to that of other tetrapods and included a stapes (see p.182); in fish, the equivalent bone is involved with operating the gills.

BEST AS PADDLES
The limbs do not look like they were capable of bearing *Acanthostega*'s weight and were probably used as paddles. *Acanthostega* was probably an ambush predator that lived in shallow rivers.

The name *Eusthenopteron* means "good strong fin".

AMBUSH PREDATOR
Eusthenopteron is believed to have been an ambush predator, hiding in aquatic vegetation to attack passing prey, as pike do today. Like *Osteolepis* and *Holoptychius*, it had a hinge across the top of its skull to to aid in prey capture.

CASE STUDY

MIGUASHA NATIONAL PARK

Miguasha, in Quebec, Canada, is one of the richest sites for Late Devonian fish fossils. The fish lived in coastal to brackish marginal marine environments, and representatives of all the major fish groups of the time, except chondrichthyans, are preserved in the rocks. Tetrapodomorphs such as *Eusthenopteron* have been discovered there, including a particularly important one called *Elpistostege*, a fish that seems to have been very like *Tiktaalik*. As well as fish, many invertebrates, plants, algae, and other micro-organisms combine to give a very detailed picture of Devonian life of the time.

Osteolepis

GROUP Sarcopterygians
DATE Middle Devonian
SIZE 50cm (19½in) long
LOCATION Scotland, Latvia, Lithuania, Estonia

Osteolepis is one of the best-known early members of the tetrapodomorphs, the group that gave rise to the four-legged tetrapods. It was one of the first Devonian fish to be discovered and as a result lends its name to the Osteolepididae, an extinct family of lobe-finned fish that are a sister group to the tristichopterids, such as *Eusthenopteron* (opposite). Its fossils are common in the Middle Devonian rocks of Scotland, where it lived in a large, shallow lake known as the Orcadie Basin. Its bony scales and skull bones were covered with a kind of glossy, enamel-like substance called cosmine. Cosmine was full of pores, which are thought to have been the openings to a sensory system for detecting water currents around the fish. *Osteolepis*'s skull had a mobile joint across the top of the head, like that of *Holoptychius* (see p.136), which allowed the fish to open its mouth very wide. *Osteolepis* lived alongside many other kinds of fish, including the lungfish *Dipterus* and several placoderms.

HETEROCERCAL TAIL
The tail of *Osteolepis* had a distinctive heterocercal form, in which the upper lobe was longer than the lower one.

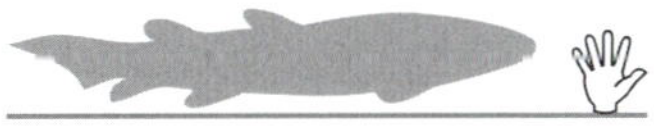

SQUARE SCALES
The name *Osteolepis* means "bone scale", and the fish had roughly square scales that were arranged in rows running from top to bottom.

Ichthyostega

GROUP Early tetrapods
DATE Late Devonian
SIZE 1.5m (5ft) long
LOCATION Greenland

Ichthyostega was the first Devonian tetrapod to be discovered. Its fossils come from rocks in East Greenland, and it lived at the same time as *Acanthostega*. *Ichthyostega* has many unique features that have puzzled palaeontologists for decades, but new research is clarifying some aspects of its structure. Its ear region, for example, appears to have been adapted for hearing in water rather than in air. *Ichthyostega* certainly spent some of its time in water, but its vertebral column seems to have been adapted to a form of terrestrial locomotion. Its body was encased in broad, overlapping ribs, which together with its robust shoulders and forearms, suggest that strong muscles were used to pull the animal along on land. Its paddle-like hind limbs were better-adapted for swimming.

***Ichthyostega*'s large teeth point to it being a predator, but did it hunt on land, in water – or both?**

SEVEN-TOED FOOT
Ichthyostega had three small toes at the leading edge of its foot followed by four stout ones. The foot was held so that the leading edge would have cut into the sand or mud on which it walked, to give it some purchase.

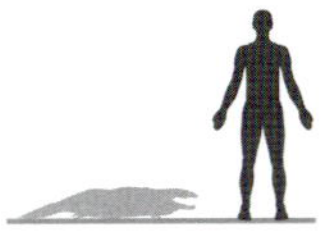

CARBONIFEROUS

Carboniferous

Carboniferous, meaning "carbon-bearing", is an appropriate name for a period when lush swamps laid down huge coal deposits across the globe. Yet this was largely an icehouse world, with extensive ice sheets that persisted for tens of millions of years. Earth's crust continued to shift, as two massive continents, Eurasia and Gondwana, slowly came together to form an even larger continental land mass, Pangaea.

EARLY CARBONIFEROUS LIMESTONE
The huge limestone amphitheatre towering above Malham Cove, North Yorkshire, UK, consists of fossilized creatures deposited on the beds of Carboniferous seas.

Oceans and continents

During the Early Carboniferous the continent of Euramerica moved even closer towards Gondwana, reducing the Rheic Ocean to little more than a narrow seaway between the western end of Gondwana and the southwestern tip of Euramerica. The mountain-building that resulted culminated in the formation of the Appalachian and Variscan chains. The Paleo-Tethys Ocean was encircled by Euramerica and Gondwana to its west, and the smaller islands of North China and South China to the east. Gondwana extended from low latitudes in the southern hemisphere to the South Pole, where a major ice cap began to develop in the Early Carboniferous. By the Late Carboniferous, Euramerica and Gondwana had merged completely to form the supercontinent of Pangaea, extending from high in the northern hemisphere across the Equator to the South Pole. The ice caps around the South Pole spread to cover much of the former continent of Gondwana. At this time the Panthalassic Ocean lay to the west of Pangaea, and the Paleo-Tethys Ocean to the east. On the continents, forests continued to expand and, with them, land animals diversified, including invertebrates, amphibians, and the first vertebrates capable of laying eggs on land. Large-scale glaciation affected sea levels, which rose and fell with the advances and retreats of ice, affecting coastal and offshore habitats. There were periods when sea water flooded into coastal swamps, and others when shallow bays, deltas, and inlets dried out. These fluctuations are reflected in Carboniferous sedimentary deposits, including alternating bands of shale and coal.

URAL MOUNTAINS
The Ural Mountains, now extending 2,500 km (1,500 miles) across west–central Russia were created by the collison between Kazakhstania, Siberia, and Baltica 250 to 350 million years ago.

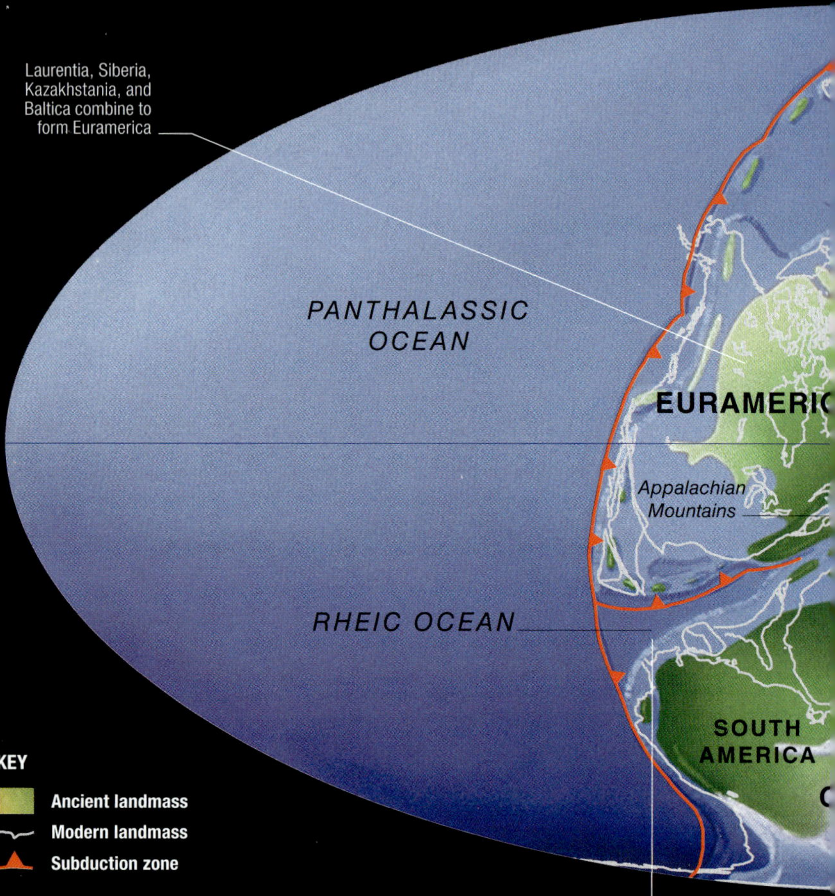

CARBONIFEROUS WORLD MAP
The Applachian and Variscan mountains formed as Gondwana and Euramerica converged. In the Late Carboniferous they merged into Pangaea.

MISSISSIPPIAN

MYA 350 | 340 | 330

PLANTS

350–340 Lycophytes and seed ferns dominate coastal swamps; calamites trees (horsetails) along streams; sphenopsids (right), seed ferns, and ferns diversify

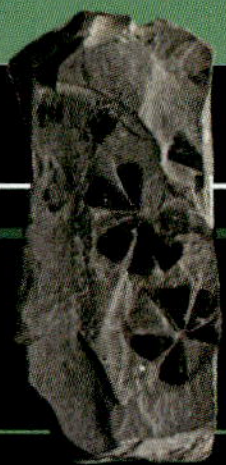

SPHENOPHYLLUM EMARGINATUM

INVERTEBRATES

350 First pulmonate gastropods; trilobites decline to one order – the proetids (right); first limid bivalves

EOCYPHINIUM

350–340 Insects evolved wings

VERTEBRATES

350 First large sharks (right), first ratfishes and chimaerids, first hagfish

PETALODUS TOOTH

340 Diversification of amphibians (right)

MEGALOCEPHALUS

330 *Paleothyris*, first amniote (egg-bearing vertebrate)

ICE SHEET
Massive ice sheets advanced and melted at intervals during the Carboniferous. The maximum extent of glacial ice in this icehouse world was between 305 and 315 million years ago.

Climate

During the Early Carboniferous overall global temperatures continued to increase. A broad tropical belt extended from Euramerica across the Proto-Tethys to eastern Gondwana. This was separated by a narrow arid belt from the cooler and more temperate conditions that extended into higher latitudes. Although there is some evidence for localized glaciations in southern Gondwana beginning around 350 million years ago, these were of limited extent. The formation of Pangaea had a major impact on air circulation patterns and therefore on climate, because such a vast land mass would have experienced climatic extremes. A major glaciation began about 330 million years ago, centred on the southern Gondwanan part of the supercontinent of Pangaea. Through the Carboniferous the size of the glaciated regions grew, stretching from South America and Australia into other regions of the Gondwanan part of Pangaea, namely present-day India, Africa, and Antarctica, and reaching as far as 35° latitude. As the ice sheets expanded, the tropical belt in low latitudes reduced in width. An arid belt separated this from the cool temperate region, the warm temperate belt having disappeared. The end of this great ice age is not known with any certainty, but it was at some time between 255 and 270 million years ago, in the Permian Period. Despite the glaciation, in lower latitudes close to the Equator and during the many interglacial episodes, conditions were warmer and more humid, resulting in the formation of substantial coal beds. Widely deposited evaporites in some areas of lower latitude also demonstrate the existence of these warmer regions.

WESTPHALIAN COAL BED
As cold and warm periods alternated, coastal swamps were flooded, forming peat bogs that became coal seams, like this one in Germany.

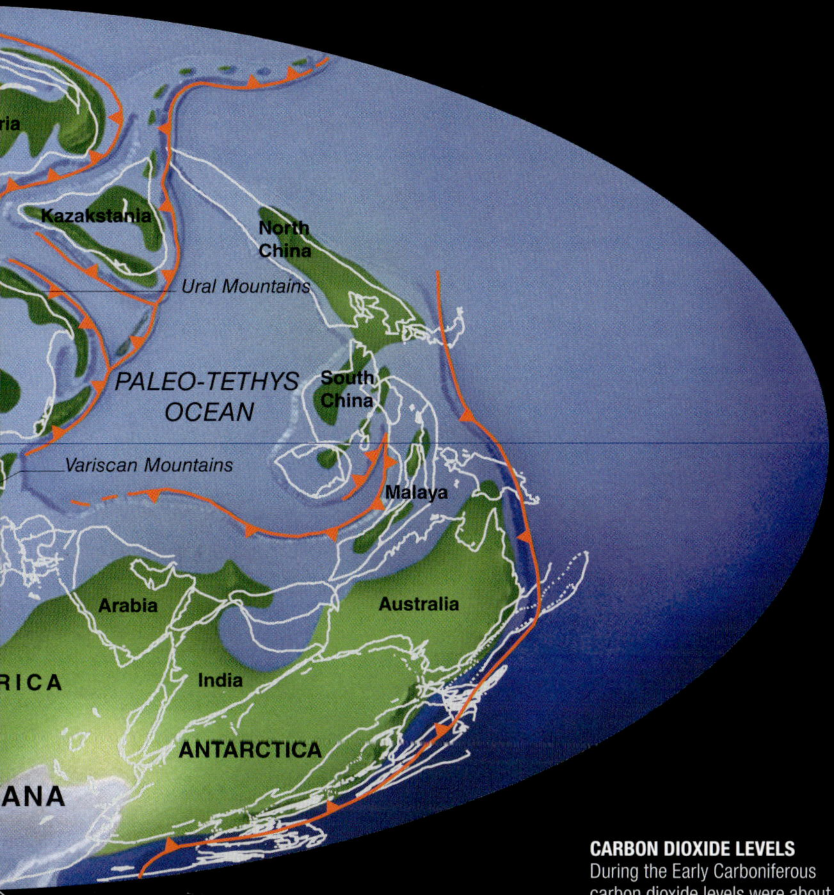

Large ice cap on the South Pole

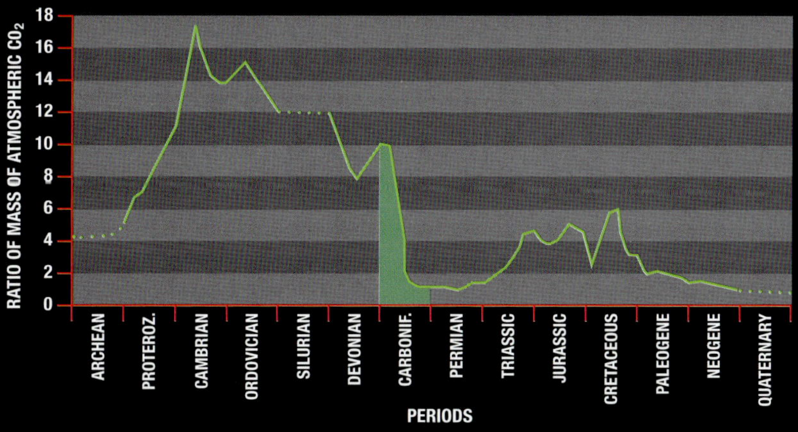

CARBON DIOXIDE LEVELS
During the Early Carboniferous carbon dioxide levels were about 1,500 parts per million. They then dropped rapidly, and by the Middle Carboniferous they had declined to 350 ppm.

PENNSYLVANIAN

320 | 310 | 300 | MYA

PLANTS

- **320** First conifers and glossopterids; lianas and epiphytes develop in forests dominated by lycopsids, cordaites, and seed plants; small ferns (right) dominate ground cover

ZEILLERIA FRENZII

INVERTEBRATES

- **325** First dragonflies; diversification of winged insects and development of plant-eating by insects
- **320** First giant arthropleurids (large millipede-like arthropods); millipedes, scorpions, spiders diversify; blattoids (roaches, e.g. *Archimylacris*) dominant insects
- **310** Winged insects become more diverse; giant dragonflies (with 63cm/25in wingspan); last edrioasteroids

ARCHIMYLACRIS

VERTEBRATES

- **310** First herbivorous tetrapods (diadectid cotylosaurs); dominant amphibians are temnospondyls; first synapsids
- **305** First diapsids (a type of reptile, e.g. *Petrolacosaurus*)

CARBONIFEROUS PLANTS

By the Carboniferous, many different kinds of large and complex plants had evolved. Giant clubmoss trees dominated the coal swamps. Tree ferns and giant horsetails, as well as several groups of primitive seed plants, grew on riverbanks and in drier areas.

Carboniferous plants reached levels of structural diversity and biological sophistication well beyond that of their Devonian precursors. Tree-like forms developed in several groups. There were also herbaceous plants, as well as scramblers and climbers. Carboniferous plants formed new and more complex ecosystems.

COAL BALLS
Much of our knowledge of Carboniferous plants comes from the study of coal balls – calcified pieces of peat that contain the preserved remains of coal swamp plants. This specimen shows a cross section through the fossil stem *Medullosa*.

Seed plants

Several different groups of seed plants are recognized for the first time during the Carboniferous. Lyginopterids had fern-like leaves and generally produced relatively small seeds, similar to those of the earliest seed plants from the Devonian. As in these very early seed plants, the pollen grains of lyginopterids were indistinguishable from the spores of primitive euphyllophytes (see p.113). Medullosans were generally larger plants, often with large, fern-like leaves. They had unusually large pollen grains and some also produced very large seeds. Cordaites, which may be closely related to living conifers, had strap-shaped leaves with parallel veins. They produced small seeds and, like conifers, their pollen grains had either one or two attached air sacs, which would have given them additional buoyancy both in air and in fluids. They may have had a pollination system similar to that of living conifers. Some cordaites were trees, while others were scrambling plants. None of these Carboniferous seed plants appears to have produced seeds that could remain dormant for long periods.

Coal swamps

Giant lycopods, such as *Lepidodendron* and *Lepidophloios*, dominated the tropical coal swamps of eastern North America and Europe during the Carboniferous. These remarkable plants grew in a very strange way, quite unlike that of any living tree. They also reproduced in several different ways. The coal swamps, and the strange plants that inhabited them, declined in importance towards the end of the Carboniferous as the wet tropical conditions in which they flourished gradually became increasingly restricted.

LEPIDODENDRON
Impressions of the bark of *Lepidodendron* and other clubmoss trees are among the most common and characteristic Carboniferous fossils. The diamond-shaped marks were formed by the cushions on which the leaves were borne.

GROUPS OVERVIEW

In the Carboniferous, further diversification of land plants resulted in several groups that can be clearly related to living plants. However, all the Carboniferous seed plants appear to represent a relatively primitive level of organization that is less specialized than what is seen in any seed plant living today.

CLUBMOSSES
The diversification of clubmosses continued through the Carboniferous. In particular, ancient relatives of modern quillworts (*Isoetes*) evolved into very large and tree-like forms. Unlike most modern trees, these lycopods had relatively little wood and were instead supported by bark-like tissue on the outside of the stem.

HORSETAILS
By the Carboniferous, horsetails had evolved into tree-like forms, although these were much smaller than the trees of the giant clubmosses. The stems of these plants are usually assigned to the genus *Calamites*, while the leaves are assigned to *Annularia*. These ancient horsetails grew in similar habitats to their modern relative *Equisetum*.

FERNS
Ferns probably arose from among the euphyllophytes of the Middle and Late Devonian, but by the Carboniferous they had evolved into many different forms. Especially important were a group of tree ferns with stems assigned to the genus *Psaronius* and leaves assigned to the genus *Pecopteris*.

MEDULLOSANS
Medullosans were one of several groups of primitive seed plants that flourished during the Carboniferous. They were shrubs or small trees with large, fern-like leaves. The name of the group comes from the name given to their stems: *Medullosa*. The leaves are usually assigned to the genera *Neuropteris* and *Alethopteris*.

Oncolites

GROUP Algae
DATE Precambrian to present
SIZE Up to 15cm (6in) across
LOCATION Worldwide

Oncolites are fossilized structures created by cyanobacteria or blue-green algae. These micro-organisms need light to grow, and they typically live in warm shallow sea water, where they encrust hard objects such as pieces of shell. They secrete mucus and calcium carbonate. As this sediment accumulates around them, they migrate to the outside to stay in the light. Over time, this creates a layered structure with a rounded outline.

ONCOLITES SP.
With their rounded shape, *Oncolites* often resemble shells. There are numerous *Oncolites* spheres in this rock.

Thallites

GROUP Bryophytes
DATE Devonian to present
SIZE Spread up to 15cm (6in)
LOCATION Worldwide

Only a few cells thick when alive, *Thallites* is often little more than a film of carbon when fossilized. Precise identification is often uncertain: some *Thallites* specimens are probably algae, while others are liverworts. Liverworts are among the simplest of land plants and do not have roots, stems, or leaves. Their plant body – a flat green ribbon known as a thallus – clings to solid surfaces by tiny hairs or rhizoids, and spreads by branching into two. Plants this simple do not have an internal system for moving water, and need to live in damp habitats.

THALLITES HALLEI
The delicate pattern seen in this Permian species shows how the plant repeatedly divided and spread across the surface of the rock to which it clings.

Lepidodendron

GROUP Lycophytes
DATE Carboniferous
SIZE 40m (130ft) tall
LOCATION Worldwide

With its scaly bark and towering trunk, *Lepidodendron* is one of the most famous of all prehistoric plants. A giant clubmoss, its fossils show many features linking it with its smaller relatives from Devonian times. Its narrow leaves were spirally arranged and it reproduced by forming spores inside cones. It spent half of its life as a branchless pole that emerged from the forest floor. As it reached maturity, its upward growth slowed, and branches appeared. Finally, cones formed at the branch tips, and upward growth stopped, as the plant put all its energy into making and shedding spores. In some species, the tree died once this task was complete.

TOWERING TRUNKS
In the lush conditions of the Carboniferous, *Lepidodendron aculeatum* could reach about 40m (130ft) in just 20 years.

LEPIDODENDRON ACULEATUM
Lepidodendron is, strictly, the name given to the bark, which has a pattern of embossed diamond-shaped scars.

ULODENDRON MAJUS
The name *Ulodendron* is used for stems that show scars. Unlike most living trees, giant clubmosses did not heal scars by growing new bark.

LEPIDO-STROBUS OLYRI
Lepidostrobus is the generic name used for the cones of *Lepidodendron*, which grew at the branch tips.

STIGMARIA FICOIDES
Stigmaria are fossilized stumps and rooting systems of giant clubmoss trees, buried under seams of coal.

LEPIDOSTROBO-PHYLLUM
This scale, with a sporangium at its base, is one of the smallest visible parts of a *Lepidodendron*.

Sigillaria

GROUP Lycophytes
DATE Carboniferous to Permian
SIZE 25m (82ft) tall
LOCATION Worldwide

Like *Lepidodendron* (see p.145), *Sigillaria* is known first and foremost for its fossilized bark. In one of the commonest forms, this looks like an embossed replica of a honeycomb, with six-sided cells arranged with almost mathematical precision. Each cell – or scar – marks the position of a leaf cushion, each of which bore a single grass-like leaf. As the tree grew, the oldest leaves were shed, to be replaced by new ones growing higher up the trunk. As a young plant, *Sigillaria* echoed *Lepidodendron*'s growth pattern. But as they reached canopy height, the two trees matured in quite different ways. *Lepidodendron* formed a high crown, with many leafy branches, but *Sigillaria* forked only once, or occasionally twice, producing limbs that looked like huge bottlebrushes. Like all giant clubmosses, it reproduced by forming spores inside cones.

SIGILLARIA MAMMILLARIS
The fossilized bark of *Sigillaria* can often be found among coal-bearing rocks. In the fossil above, the regular pattern formed by the scars where the tree's leaves were once attached is clearly visible.

SIGILLARIA ALVEOLARIS
In this species the leaf bases sit in furrows, separated by vertical ribs.

FOREST GIANT
Giant clubmosses such as *Sigillaria* dominated the swampy forests of the Carboniferous, but had dwindled dramatically by the end of the period.

SELAGINELLA SP.
Easily mistaken for a branch tip, *Selaginella* had a flat, creeping form with small leaves.

Selaginella

GROUP Lycophytes
DATE Carboniferous to present
SIZE 15cm (6in) tall
LOCATION Worldwide

Tree-sized clubmosses were not the only successful lycophytes during the Carboniferous period. Fossils show that creeping forms – such as *Selaginella* – also flourished. With its small, spirally arranged leaves of different sizes, and dichotomous, branching growth, *Selaginella* had a strong resemblance to the low-growing clubmosses with the same name that have survived to the present day. *Selaginella* is a relatively rare fossil, and the earliest specimens have been dated to the Early Carboniferous. In Coal Age rocks, shoots of *Selaginella* should not be confused with the tips of branches from giant clubmosses.

Lepidophloios

GROUP Lycophytes
DATE Carboniferous
SIZE 25m (82ft) tall
LOCATION Worldwide

Bearing a striking resemblance to a snake's skin, this fossilized bark comes from a giant clubmoss that lived at the same time as *Lepidodendron* (see p.145). Although not as tall as its more famous relative, *Lepidophloios* was just as widespread. It grew in peat-rich forests rather than muddy estuarine swamps. *Lepidophloios* produced the largest spores of any primitive vascular plant, and these were probably fertilized aquatically following dispersal.

LEPIDOPHLOIOS SCOTICUS
The shape of the scars shows that *Lepidophloios*'s leaves were relatively flat and grew in a tight spiral.

Bowmanites

GROUP Sphenophytes
DATE Carboniferous to Permian
SIZE Cone up to 8cm (3¼in) long
LOCATION Worldwide

This name is used for the fossilized cones of *Sphenophyllum* (see right). These soft cones formed on short, fertile shoots, with up to 20 closely set whorls of bracts, or sporangium-bearing leaves. The spores were microscopic, and a single cone could produce many thousands. Those few that were able to germinate produced the gamete-forming body, which, in turn, produced male and female cells. Then, once fertilization occurred, a new spore-forming plant started to grow.

BOWMANITES SP.
In most *Bowmanites* cones, the bracts are sharply upturned at their tips. This fossil shows a mature cone, which has opened out to release its spores.

Sphenophyllum

GROUP Sphenophytes
DATE Carboniferous to Permian
SIZE 1m (3ft 3in) tall
LOCATION Worldwide

During the Carboniferous period, not all horsetails evolved into tree-like forms. Some – including *Sphenophyllum* – were low-growing or climbing plants that lived in damp places and on forest floors. Despite being soft-stemmed, they were ideally placed to form fossils, and their preserved remains are common in Coal Age rocks all over the world. *Sphenophyllum*'s leaves were arranged in whorls, but they had a wide variety of shapes. Some resemble the "pressed flower" appearance of *Annularia* leaves (see p.148), while others end in small hooks that would have helped them to climb.

SPHENOPHYLLUM EMARGINATUM
Sphenophyllum literally means "wedge-shaped leaves" – an indication that this plant looked quite different to the horsetails alive today.

Calamites

GROUP Sphenophytes
DATE Carboniferous to Permian
SIZE Up to 20m (66ft) tall
LOCATION Worldwide

Once the name given to its stem fossils, but now commonly used for the whole plant, *Calamites* was a conspicuous, bushy tree of Carboniferous swamps. *Calamites* had the typical horsetail outline, with a ridged stem divided into jointed segments, carrying branches and leaves arranged in circular whorls. Mature specimens could reach heights of 20m (66ft) with stems over 60cm (23½in) across.

CALAMITES CARINATUS
Moulds of *Calamites*'s stems, such as the one shown here, formed when the tree died and sediment infiltrated its core.

ASTEROPHYLLITES EQUISETIFORMIS
Literally meaning "star leaves", *Asterophyllites* is fossilized foliage left by *Calamites* and its relatives. The needle-shaped, upswept leaves are arranged in whorls of 4 to 40.

ANNULARIA SINENSIS
Another species of fossilized foliage from *Calamites*, this well-known fossil has elegant, evenly spaced whorls of leaves that can look like printed designs.

Psaronius

GROUP Ferns
DATE Late Carboniferous to Early Permian
SIZE Up to 10m (33ft) tall
LOCATION Worldwide

With its 3m- (9¾ft-) long fronds and elegantly flaring trunk, *Psaronius* would have been one of the most graceful ferns of Carboniferous times. Like today's tree ferns, it had a single, unbranched trunk, topped by an arching crown of fronds that uncoiled as they grew. The trunk grew differently to those of most other trees. Its core was made of a mass of woody segments, surrounded by thick, fibrous roots that ran down from higher on the trunk. This jacket of roots worked like wood, thickening as the tree gained height. The name *Psaronius* was originally applied to these fossilized stems, but has since been used for the entire plant. *Psaronius* seems to have grown in a wide variety of habitats, from wet, low-lying ground to much drier environments.

PSARONIUS INFARCTUS
The unusual structure of *Psaronius*'s trunk shows a central mass of woody segments surrounded by a thick, outer covering of roots.

PECOPTERIS SP.
Named after the Greek word for a comb, *Pecopteris* are the fossilized fronds of *Psaronius*. Each frond had rows of symmetrical leaflets, with parallel edges and a central vein. The fronds unfurled like a shepherd's crook from a central growing point.

PECOPTERIS MAZONIANUM
This form of foliage is named after Mazon Creek, an exceptionally rich fossil site near Morris, Illinois, USA. During the Carboniferous, this region was part of a tropical flood plain that built up deep layers of fine sediment. Its shale and ironstone have yielded over 100 genera of fossil plants.

Sphenopteris

GROUP Ferns / Seed plants
DATE Late Devonian to Cretaceous
SIZE Frond up to 50cm (19½in) long
LOCATION Worldwide

Sphenopteris foliage grew on two distinct kinds of plants. Some grew on ferns, where the fronds produced spores, but most belonged to seed plants with fern-like leaves. One of the best-known of these primitive seed plants is *Lyginopteris oldhamia*, which comes from Carboniferous rocks in Britain. It was covered with tiny club-shaped glands, distinguishing it from all other prehistoric plants. *Lyginopteris* has an important place in botanical history. Its discovery, made in the early 1900s, proved the existence of previously unrecognized groups of extinct seed plants.

SPHENOPTERIS ADIANTOIDES
Whether it was produced by a fern or a seed plant, *Sphenopteris* foliage has a feathery outline, with alternately arranged leaf segments that are winged at their base. It can be found in rocks throughout the Carboniferous period.

The foliage of *Sphenopteris* grew on two distinct kinds of plants: ferns and seed plants.

Zeilleria

GROUP Ferns
DATE Late Carboniferous
SIZE 50cm (19½in) tall
LOCATION Worldwide

During the Carboniferous period, many plants evolved finely divided fronds, and *Zeilleria's* were some of the most delicate. *Zeilleria* grew in shady, swampy ground. It produced spores on the underside of its leaves, but it may also have spread by shedding specialized leaflets, which would then have taken root. This way of growing is used today by some ferns with highly divided fronds, particularly in wet habitats.

ZEILLERIA SP.
As is clearly shown in this beautiful fossil, *Zeilleria's* numerous fronds created dense patterns that look like crystalline growths between layers of mudstone or shale.

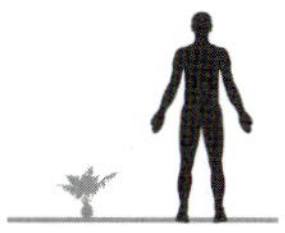

Medullosa

GROUP Medullosans
DATE Carboniferous to Permian
SIZE 3–5m (10–17ft) tall
LOCATION Europe, North America

With seeds as big as eggs, *Medullosa* is one of the most intriguing plants of the Carboniferous. Unlike most other seed plants, it combined fern-like foliage with a seed-forming lifestyle, though it did not grow flowers. It grew by producing new fronds at the tip of its stem, while the oldest fell away. To reproduce, it formed large seeds and huge numbers of unusually large pollen grains. The grains would have been too heavy to drift far in the wind. Instead it has been suggested that insects may have carried them between plants, in an early example of the partnership that was to become so important to flowering plants.

MEDULLOSA LEUCKARTII
Medullosa grew by producing new fronds at the tip of its stem. Old fronds fell away leaving a spiral of leaf bases filled with resin, as seen in this stem cross section, which together formed the trunk.

TRIGONOCARPUS SP.
Medullosa produced large seeds with three prominent ribs, which hung beneath its fronds.

IMPOSING FERN
Medullosa was one of the tallest seed plants with fern-like leaves. It had a strong, stable trunk and finely divided fronds that could grow to more than 1m (3ft 3in) long.

Potoniea

GROUP Medullosans
DATE Late Carboniferous
SIZE Up to 5m (16½ft) tall
LOCATION Worldwide

Medullosans produced their pollen in specialized organs that hung from their leaves. Many of these, including *Potoniea*, are known from fossils. They grew in a wide variety of shapes, from simple branching structures to objects that look like figs or showerheads several centimetres wide. All had the same function: to disperse pollen efficiently so that seeds could be formed. The pollen-producing organs are often found in isolation, which makes it hard to link them to particular parent plants.

POTONIEA SP.
Potoniea fossils are often found in the same rocks as *Macroneuropteris* fronds (see right). The fine, finger-like, pollen-producing structures are just visible.

Macroneuropteris

GROUP Medullosans
DATE Carboniferous
SIZE Frond up to 2.5m (8ft) long
LOCATION Worldwide

Medullosans were not the largest plants in the Carboniferous period, but their leaves are some of the commonest fossils from that time. *Macroneuropteris* is the name given for one particular form of their foliage, which includes some of the earliest-known leaves. In *Macroneuropteris* foliage, the leaflets have heart-shaped bases and are only attached by their stalks, rather than their blades. They also have a fine tracery of spreading veins.

TEXTURED SURFACE
The finely-veined surface, typical of *Macroneuropteris* foliage, and the prominent midrib are clearly visible on this fossilized leaflet.

MACRONEUROPTERIS SCHEUCHZERI
This fossilized leaflet has tiny bite marks along its edges, perhaps caused by a small, leaf-eating arthropod.

Cyclopteris

GROUP Medullosans
DATE Carboniferous
SIZE Leaflet up to 10cm (4in) across
LOCATION Worldwide

With their rounded outlines, these isolated leaflets look like tiny shells, but they have veins that radiate out like the ribs of a fan. Several kinds of medullosans seem to have produced *Cyclopteris* leaflets. One theory is that these leaflets grew on young plants, in the dim light of the forest floor. Once the plant had grown taller it produced more typical, fern-like fronds. Another theory is that the rounded leaflets protected the rest of the frond as it uncoiled.

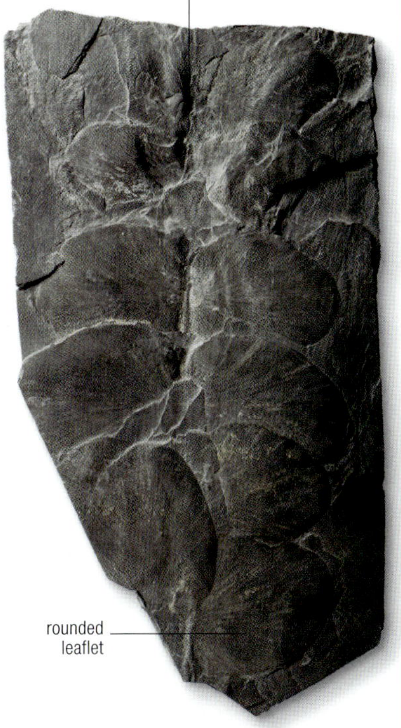

CYCLOPTERIS ORBICULARIS
Cyclopteris fossils look quite different to the leaflets of typical seed ferns. This fossil shows a small group fixed to a common stem, but they are also found as isolated leaflets in Carboniferous rocks.

Alethopteris

GROUP Medullosans
DATE Carboniferous to Early Permian
SIZE Frond up to 7.5m (25ft) long
LOCATION Worldwide

Compared to ferns, medullosans were robust plants, with strong stems and a crown of large leaves. *Alethopteris* leaves were the largest of all. One specimen, found in a coal mine in northern France, was nearly 7.5m (25ft) long and 2m (6½ft) wide, making it one of the largest leaves ever discovered from Paleozoic times. *Alethopteris* leaflets could be up to 5cm (2in) long, and they typically flared out near the stem. Leaves of this type were produced by *Medullosa* and its relatives. The tough leaf bases merged to form strong, woody stems. At certain times of the year, the leaves would have been weighed down by the developing seeds.

ALETHOPTERIS SULLIVANTII
The leaflets in this fossil are not separate, but joined together at their bases to create a continuous flap of tissue around the entire leaf segment.

ALETHOPTERIS SERLII
The thick, strongly veined leaflets typical of *Alethopteris* are clearly seen in this fossilized leaf fragment. The individual leaflets are all connected at their base.

Callipteridium

GROUP Medullosans
DATE Late Carboniferous to Early Permian
SIZE Frond up to 3m (9¾ft) long
LOCATION Worldwide

Meaning "beautiful fern", *Callipteridium* foliage was strikingly convex when seen from above. Individually, *Callipteridium* fossils are often less than 15cm (6in) long, but in life, the leaves could be 20 times that length, subdividing three or four times. Their leaflets were attached by broad bases, and had a deeply sunken midvein. Like most fossil leaves, *Callipteridium* are almost always found detached from their parent plant, and unlike the fronds of true ferns, none of them shows structures for making spores. Instead, at least two species have been found with ovules, showing that they were grown by seed-ferns or pteridosperms. Compared to fern fronds, the leaves of seed plant fronds had a tough outer cuticle or "skin". This made them more resistant to dry conditions, and increased their chances of forming fossils.

Mariopteris

GROUP Lyginopterids
DATE Late Carboniferous
SIZE Frond up to 50cm (19½in) long
LOCATION Worldwide

Usually found as isolated fragments, *Mariopteris* foliage belonged to a collection of climbing forms of seed plant, which grew up other plants in their quest for light. These plants had narrow, pliable stems, and many of them also had grappling hooks on the ends of their leaflets, in some cases up to 4cm (1½in) long. *Mariopteris* fronds also had long, bare stalks between the leaflets – the ideal shape for locking the plant in place once the frond had fully formed. Fossils show that the remainder of the leaf was divided into four roughly equal parts – an unusual feature that helps to identify this group of plants.

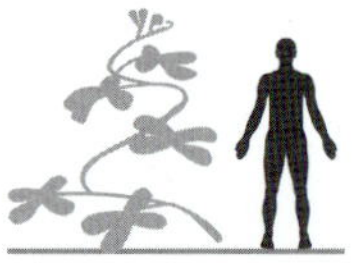

MARIOPTERIS MURICATA
The leaflets on *Mariopteris* fronds varied in size, shape, and number.

Odontopteris

GROUP Medullosans
DATE Late Carboniferous to Early Permian
SIZE Frond up to 1m (3ft 3in) long
LOCATION Europe, North America, China

Over a thousand different seed plant leaves have been identified from fossil foliage, the majority coming from Coal Age rocks. *Odontopteris* is one of the most common types. Most of these fronds were less than a metre long, and were attached to plants with *Medullosa* stems (see p.150). However, these plants were much smaller in size than the *Medullosa* trees, and would have scrambled through low-growing vegetation, or used the larger trees for support. They were common in both swamps and on drier ground, especially towards the end of the Carboniferous.

ODONTOPTERIS SUBCRENULATA
Often found in shale, this specimen is from Shanxi Province in China.

Plagiozamites

GROUP Noeggerathialeans
DATE Late Carboniferous to Permian
SIZE 1m (3ft 3in) tall
LOCATION Worldwide

With their two matching ranks of leaflets, *Plagiozamites* could easily be mistaken for the leaf of a cycad. However, this kind of foliage was also produced by noeggerathialeans – a group of extinct plants that have proved difficult to classify. Noeggerathialeans have been linked to horsetails and to ferns, but recent research suggests that they were more closely allied to progymnosperms.

PLAGIOZAMITES OBLONGIFOLIUS
Plagiozamites leaves have been found close to fossilized cones, making it likely that both were grown by the same plant. So far, no stems have been found.

Noeggerathialeans **reproduced** by forming **spores**, which developed on the scales of long, **cylindrical cones.**

ODONTOPTERIS STRADONICENSIS
Odontopteris leaflets are attached to the leaf's midrib by a broad base, and have a spreading network of veins.

Utrechtia

GROUP Conifers
DATE Late Carboniferous to Early Permian
SIZE 10–25m (33–82ft) tall
LOCATION Worldwide

Over a dozen species of *Utrechtia* have been identified from fossilized twigs. A tree with a slender trunk, its branches were arranged in widely spaced whorls and were covered with a spiral of small needles. These were flanked by short shoots that created a spreading, feathery outline. Like other early conifers, *Utrechtia* lived in dry ground rather than the low-lying swamps of Carboniferous times.

UTRECHTIA PINIFORMIS
Utrechtia's side branches, such as the one shown here, housed cones on their tips. The male cones were pendulous and the female ones were borne upright.

Walchia

GROUP Conifers
DATE Late Carboniferous to Early Permian
SIZE 10–25m (33–82ft) tall
LOCATION Worldwide

The name *Walchia* is used for foliage of this type in which preservation of detail is insufficient to enable accurate assignment to an extinct family. Walchian conifers were common forest trees in the late Carboniferous period. They had a straight, upright trunk, with leafy branches growing out in horizontal tiers. Each branch was covered with needle-like leaves, just a few millimetres long. Conifers became more widespread towards the end of the Coal Age, as the global climate dried.

WALCHIA ANHARDTII
On its recent growth, *Walchia* had opposite ranks of dwarf shoots, giving the tips of the branches a feathery outline.

Noeggerathia

GROUP Noeggerathialeans
DATE Late Carboniferous to Early Permian
SIZE 1m (3ft 3in) tall
LOCATION Worldwide

Named after the German geologist Johann Jacob Noeggerath (1788–1877), this plant was first identified in the 1820s from its distinctive, compound leaves. Each one has two rows of leaflets, arranged alternately or in opposite pairs, creating a frond up to 30cm (12in) long. The leaves are thought to have grown on a short trunk, and may have been tough and leathery despite their fragile appearance.

NOEGGERATHIA FOLIOSA
Unlike other species of the genus, *Noeggerathia foliosa* has distinctive, rounded leaflets.

Cordaites

GROUP Gymnosperms
DATE Carboniferous to Permian
SIZE Up to 45m (148ft) tall
LOCATION Europe, North America, China

The cordaites were a major group of plants that shared many features with conifers. They had woody stems and produced seeds in cones, but instead of scales or needles, they grew strap-shaped leaves up to 1m (3ft 3in) long. Their male and female cones grew on separate branches, and they often produced heart-shaped (cordate) seeds. As a group, cordaites grew in a wide range of habitats. Many were low-growing shrubs, but they also included towering, tropical trees that could reach heights of 45m (148ft).

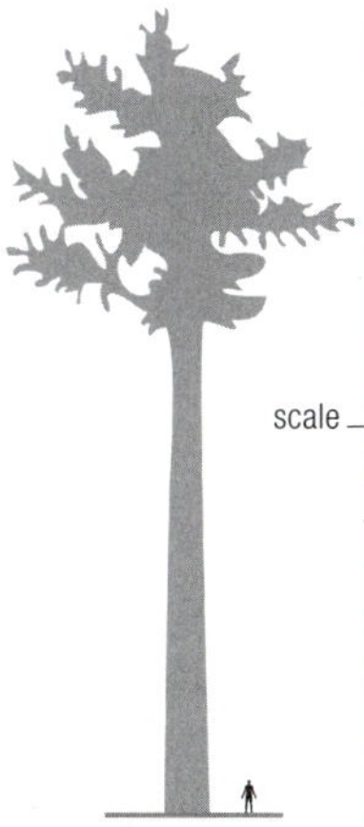

CORDAITANTHUS
This fossil genus name is used for the cone-bearing structures of cordaites.

CORDAITANTHUS CONES
Seen close to life size, *Cordaitanthus* cones reveal a spiral of short, overlapping scales. In these female cones, the furthest scales would each have formed a single seed.

CARBONIFEROUS INVERTEBRATES

The Carboniferous was a time of great change. Global cooling resulted in a series of glaciations across the immense Gondwanan continent in the southern hemisphere, a cycle that continued into the Permian. However, marine life continued to evolve vigorously, and massive forests emerged.

In the Carboniferous, great forests spread over huge areas, with giant clubmosses, horsetails, conifer relatives, ferns, and seed ferns flourishing. They culminated in the coal-swamp forests of the Late Carboniferous, which provided havens for amphibians and early reptiles, and also for invertebrates, including eurypterids, dragonflies, and millipedes. Some arthropods grew very large, since the oxygen levels in the atmosphere reached 28 per cent or more.

FERNY FOREST
Invertebrates, especially arthopods, flourished in new Carboniferous ecosystems, among which were coal-swamp forests that were dominated by tree ferns.

Forest hazards

The hazards of forest fires increased greatly and so plants adapted to rapid regeneration once the flames had died down. In the Lower Carboniferous of central Scotland, a complete ecosystem was preserved, including scorpions up to 90cm (35in) long, mites, and harvestmen, along with intact amphibian and early reptile skeletons. These, together with many plants, were deposited in a stagnant lake fed by hot springs, into which the animals fled during a forest fire.

TABULATE CORAL
This Carboniferous coral shows many calcite-walled corallites in which the polyps originally resided.

Marine ecosystems

The seas of the Carboniferous were dominated by brachiopods, rugose and tabulate corals, crinoids, bryozoans, and fish. Such mature ecosystems began in the Devonian and continued into the Permian. Goniatite fossils are common in algal reefs, and in dark shale (a type of sedimentary rock) where they are often found with bivalves. Successive glaciations of the Gondwanan continent, then centred on the South Pole, are recorded in the northern hemisphere as rhythmic sedimentary cycles, a result of fluctuating sea levels as ice sheets expanded and shrunk.

COCKROACH
This fossilized early cockroach, *Archimylacris eggintoni*, flourished in moist, forest habitats.

GROUPS OVERVIEW

Some invertebrate groups flourished in the changeable conditions of the Carboniferous. On land, lush conditions allowed some insect species to reach vast sizes. Of the marine groups, rugose corals and blastoids were abundant and spread over wide areas. Goniatites thrived in extensive reef habitats and new forms of echinoderms appeared.

RUGOSE CORALS

Rugose corals were very abundant in the Carboniferous, often forming extensive beds and thickets. Unlike modern corals, the main septa (radial partitions) were laid down on four lines only, so that the coral was biradially symmetrical.

GONIATITES

Goniatites were cephalopods whose spiral shells had gas-filled internal chambers and a body chamber in which the main part of the animal resided. The suture lines where the internal partitions met the inside of the shell had a zig-zag form.

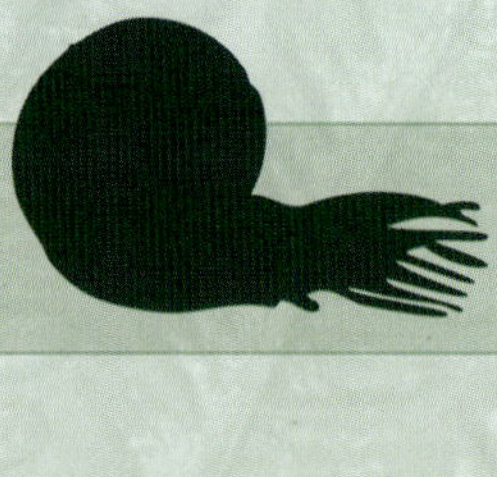

BLASTOIDS

These were relatives of crinoids but generally smaller. They had short stalks and a bud-shaped, tentacled crown within which was a remarkably complex respiratory system. Blastoid fossils are often found abundantly in thin beds, traceable over wide areas.

CRINOIDS

These "sea-lilies", some forms of which still exist today, may look like plants, but they are actually calcite-plated echinoderms. Crinoids have long stalks of stacked calcite discs, ending in a flower-like crown with extended arms for feeding. Some limestones consist almost entirely of crinoid debris.

The Sephardis had participated fully in the cultural life of Amsterdam for centuries and thought of themselves as Dutch. The Antwerp Jews, on the other hand, had been there only a short while. They were Ashkenazis, fleeing persecution in Eastern Europe, and had come to Antwerp as a port of embarkation for the New World, arriving with nothing. Barred from trades and professions in their places of origin, commodity trading was all they knew. Those who stayed on in Antwerp found a place for themselves as diamond dealers and brokers, not least of all because a man could carry his entire inventory in his pockets if forced to decamp. Orthodox in their faith, conscious of their identity, their survival instinct sharpened by adversity, they fled Belgium ahead of the Nazi invasion, while the Amsterdam Jews, believing that as Dutch they were safe, were more easily rounded up by the Nazis. Few survived to return to Amsterdam after the war, and there was little for them to return to. Both the manufacturing and dealing in Amsterdam had been largely in Jewish hands before the war, whereas in Antwerp there was a tradition of Flemish manufacturing, and there was therefore a manufacturing base for the Jewish dealers to come back to.

Sus Dielens, now in his nineties, was one of the first to take up the trade when diamond cutting originally came to the Kempen, and he still polishes the occasional stone.

Even before the war ended the refugee diamond dealers began flocking back to Antwerp. By the end of March 1945 over three thousand had returned, and by the end of the year more than eleven thousand were again active in the city. Men who are among Antwerp's leading diamantaires today, former president of the Hoge Raad voor Diamant Alfred Lachowsky, president of the Federatie der Belgische Diamantbeurzen Sylvain Zucker, and the specialist macle cleavers Charles and Isi Trau, were given a fresh start as members of two groups of Belgian ex-servicemen awarded sights by the CSO.

"My uncle Romi Goldmuntz, a good friend of Sir Ernest Oppenheimer, was instrumental in getting these sights," recounts retired dealer in rough diamonds Sylvain Goldmuntz. "He felt that we who had served in the war, while others had continued in the business, deserved a chance. He put together two firms of ex-servicemen and introduced them to the Syndicate. Symbolically the first sights after the war were received at the Town Hall, and Sir Ernest came over for the occasion."

During the early 1960s the small manufacturers in the Kempen began experiencing difficulty in competing for rough with emergent Indian manufacturers, who were prepared to pay higher prices for smaller goods on the free market in order to get a foothold. Although the CSO took steps to give the Kempen cutters access to more CSO rough, the cutters also had to contend on the sales side with the problems of manufacturing small-sized goods, which have a high labour content, in competition with the burgeoning Indian manufacturing industry, which is not burdened with European wage bills and strict social security regulations. In the event, manufacturing in the Kempen declined steadily. Since 1968 the number of active Belgian diamond cutters has fallen from almost twenty thousand to about five thousand in 1989.

The Kempen remained marginally competitive for a time because the largely family enterprises could dodge paying the high legal wage and social security contributions and could get away with working much longer hours than the law allowed. But when these regulations were enforced more strictly, manufacturing in the Kempen became too expensive to be competitive in the goods which traditionally they had manufactured. The problem was exacerbated by their having to pay wages in strengthening Belgian francs out of income from sales in weakening US dollars, the currency of the trade. Recently, however, the strengthening of the dollar has done something to restore their competitive edge.

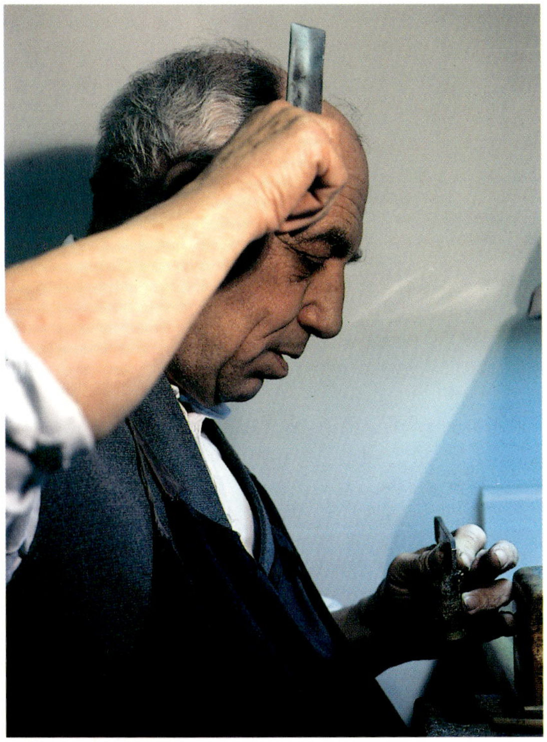

Diels and his family, who work with him, have so far survived the decline of the Kempen by specializing in larger goods and the flat, usually triangular, twinned diamonds called macles. "The demand for larger, quality goods is more constant than for inferior goods, which are more sensitive to market fluctuations, because of the greater volume produced," he says. "We've made money polishing macles, which sell cheaply in the rough because they are difficult to work. Quite a few of the older people have kept going in this way, but how much longer the younger people continue working with diamonds in the Kempen is uncertain. When opportunities arise for employment in other industries they take them."

Renowned as a cleaver of big stones, Isi Beck (left) is regarded as having the perfect "feel". His blade slotted in the kerf he has cut with another stone in a 34.35-carat West African diamond, he deftly strikes a single blow that cleanly splits it in two. Macles like this one (below left), which are difficult to manufacture because of the conflicting grains of the twinned crystals, are a speciality of Antwerp.

In Antwerp itself, where manufacturers have traditionally specialized in larger and more difficult stones, the industry has remained strong, because labour costs are less critical; the skill of the cutter in obtaining the maximum yield from an expensive stone outweighs the higher cost of his labour. Antwerp is able to hold its own against other cutting centres by maximizing its skill and specializing in macles and larger stones of good quality.

As a diamond trading centre, Antwerp is still the largest. "We have the tradition, the infrastructure and the facilities that others lack," says Lachowsky. "Rough and polished diamonds come from all over the world to be traded on our market. Buyers can get a complete range of polished: Indian, Israeli and Russian, as well as the characteristic Antwerp production."

Antwerp's yearly trading turnover of gem quality diamonds in the late 1980s exceeded $10 billion. Of official rough imports in the region of 70 million carats, almost half come by way of London. In terms of value, the Soviet Union and India each provide about 20 per cent of the polished diamonds imported, while the United States and Israel supply 10 and 14 per cent respectively. By volume, however, India supplies the Antwerp market with some 1.4 million carats of polished diamonds, compared with about 500 000 carats from the Soviet Union, 460 000 carats from Israel and 325 000 carats from the United States. The value of polished exports amounts to about $3 billion a year, of which some 35 per cent goes to the United States, 20 per cent to Japan and 11 per cent to Hong Kong.

When large quantities of Russian polished diamonds first appeared on the Antwerp market in the sixties local manufacturers felt threatened. Whereas their own profits depended on how much of the original stone they could conserve in cutting, the state-run Soviet industry, not having to compete for its rough, could be more extravagant and produce quantities of uniform size and cut. "Fortunately they learnt that if you sell cheap there is no end to it," says Abraham Fischler, a leading manufacturer and chairman of the Beurs voor Diamanthandel. "Now their prices are more market related; anyone wanting uniform Russian polished today generally has to pay a premium to get it."

Diamond manufacturing in India, in addition to its effect on manufacturing in the

There is hard bargaining (right) in a corner of the Beurs voor Diamanthandel, but the mood on the trading floor is convivial amid eddying cigar smoke and the hubbub of friendly discussion. The informal sector of the diamond trade does business in Antwerp cafés (below right), in much the same atmosphere in which deals were done before the bourses were established. Eduard Poot doubtless knows a thing or two about what is going on.

ROSY BLUE
ROSY BLUE

Dilip Mehta, a leading Indian diamantaire in Antwerp, who in 1973 founded Rosy Blue diamond trading company, one of an international group of diamond companies owned by his family. Here he discusses the prices of parcels of polished with foreign buyers.

Kempen, has also altered diamond trading in Antwerp. Where formerly the maharajahs came to buy polished diamonds in the city, now the Indian manufacturers come chiefly to buy rough for their industry, and most of it returns as polished for sale in Antwerp to international buyers. A number of Indians have established companies in Antwerp; together they now handle almost a third of the diamond trade there.

First was Gembel, established in 1950 by Kirtilal Mehta, whose family had supplied diamonds and jewellery to maharajahs in India before Independence. Headed today by his son Rashmi Mehta, it is chiefly a wholesale business, with manufacturing establishments in Antwerp, Tel Aviv and Bombay, and selling offices run by members of his family in New York and Hong Kong. Jayam, headed by Mafatlal Mehta and his son Madhu, has become one of the top diamond companies in the world.

Rosy Blue, a major diamond trading company in Antwerp, was formed in 1973 by Dilip Mehta to buy rough for his family's manufacturing business in Bombay and sell its manufactured product in Antwerp. Today the company deals also in Russian and other polished, and has sales offices in New York, Tokyo and Hong Kong. "We buy and sell rough, we manufacture what we can't sell quickly if we see profit in it, and we buy and sell polished," says Dilip Mehta. "We do business; we have no priorities."

Being a small country that depends on international trade, Belgium has created a trading climate that is as free as possible. Rough and polished diamonds may be imported and exported with a minimum of formality, free of duties and value-added taxes, and there are no restrictions on the movement of currency.

The whole business in Antwerp is conveniently concentrated in three adjoining streets, having moved from Pelikaanstraat into Schupstraat, Hoveniersstraat and Rijstraat. Here the merchants, manufacturers, brokers and international diamond companies have their premises, together with the four diamond exchanges and the two major banks that specialize in financing diamond deals and arranging the insurance and shipping of diamonds. Here too are the premises of the Hoge Raad voor Diamant (Diamond High Council), the central representative body for the entire industry.

"The High Council was formed in 1973 as a single mouthpiece for all branches of diamond dealing and manufacturing," says Alfred Lachowsky. "The trade unions and bodies that represent these branches elect delegates to the Council, which also looks after street security, provides public relations and an internationally recognized diamond certification service, operates an institute of gemmology and technical research centre for diamonds, and administers the Diamond Office, through which, by law, all diamond imports and exports must pass."

Originally established by the government towards the end of World War II at the request of the British government to control all diamond business and guard against diamonds entering liberated parts of Belgium passing into enemy hands, the Diamond Office was afterwards retained to gather statistical data on diamond imports and exports, and was placed under the administration of the High Council in 1974. It handles some twenty thousand import and fifty thousand export shipments a year for more than three thousand diamond firms.

At the request of dealers, manufacturers and purchasers, the certification department reports on upward of sixty thousand polished diamonds a year. "On the Antwerp market nowadays it's difficult to sell a stone of good quality above one carat without a certificate," says the department's manager, Peter De Jong. "Our certificate confirms that a stone is a diamond and describes its main quality characteristics – colour, clarity, carats and cut – according to the International Diamond Council's rules for grading polished diamonds, which are based on similar definitions to those of the Gemological Institute of America. The scientific aspect is supervised by Antwerp University."

Financed half by the government and half by the constituent bodies of the High Council, the Diamond Scientific and Technical Research Centre studies techniques for improving methods of processing diamonds and develops prototypes of new machines. These machines are made and sold to the industry by a separate company under the aegis of the High Council, and manufacturers who buy them receive government assistance, in the form of an interest-free extended payment system. The Institute of Gemmology, in collaboration with Antwerp University, offers training in the identification, analysis and evaluation of diamonds.

The four exchanges, or bourses, were formerly the principal places where diamond business was transacted, but today more business is done in the private offices of the manufacturers and dealers. Broker of rough diamonds Dany Kiek, who is president of the brokers' organization within the Syndikaat der Belgische Diamant Nijverheid, attributes the change to the disappearance of many cleavers and small manufacturers in the slump of 1980. "They would buy a small parcel on the bourse, cleave or manufacture it, and then put it back on the market," he says, "but they no longer have the money it takes to get into the business." According to Sylvain Zucker, it has become difficult to work without an office because clients are now more exacting about their requirements. "Customers wanting half-carat stones no longer accept 48-pointers, so each stone must be weighed individually."

Educating the next generation of diamantaires is considered essential to the continuity of the industry in Antwerp, where there are six diamond training schools. At one of them, the Stedelijk Instituut voor Technisch Onderwijs, lecturer R. Cleerbout talks to bearded Hans Wins, public affairs manager of European Gem Services, about a parcel of rough diamonds for the students to practise on.

"Using telefax and telex, people can work far faster from their offices," says Fischler. "The whole procedure of obtaining a parcel in a bourse is too time consuming. If someone places an order with us we can prepare the goods, and he needs to come for only a day to examine the goods and reach an agreement, whereas it would take him at least a week to find what he needs in a bourse. The role of the bourses today is more as a meeting place, not so much for doing business, but where people can talk and keep in touch with what's happening in the market through personal contact and gossip. Information is vital in the diamond business."

Although loose rough, polished or industrial diamonds of any size or quality may be traded in any of the exchanges, they have tended historically to specialize. "The Vrije Diamanthandel is the meeting place of Kempen manufacturers who come in on Fridays to sell what they have polished during the week and buy more rough. It no longer has its own trading floor and shares the facilities of the Diamantclub van Antwerpen, the most prestigious but least active of the bourses, which deals mainly in better quality gem rough. Principally polished diamonds are traded in the Beurs voor Diamanthandel, while the Antwerpsche Diamantkring, the most active bourse, specializes in lower qualities of polishable rough and industrial diamonds. The bourses are united in the Federatie der Belgische Diamantbeurzen and in the International Federation of Diamond Bourses, which links twenty diamond bourses world-wide. "The main function of the federations has been to ensure that we have the same view of the business and subscribe to the same rules of ethics," says Zucker, who has been president of the Belgian Federation for over a decade and of the Diamond Club for twice as long.

Membership criteria are strict, because, says Fischler, "To be a member of a bourse is like having a card stating that we have faith in you and can recommend you to others." The bourses can maintain a high ethical level among members because the penalty for dishonour is loss of good name, without which a diamantaire might as well seek another line of business. Dishonesty merits expulsion, while the price of not living up to one's word is suspension. Disciplinary action taken in one bourse is automatically followed by all; the offender's name is posted on the noticeboards in bourses around the world, and nobody will have anything to do with him. "Our business is often done without sales documents," says Zucker, "but we have our word and must defend it to the utmost."

Bourses provide arbitration in settling commercial differences between members and, through the Federation, between their members and the members of other bourses. Members who through no fault of their own get into financial difficulties are not penalized; the bourses have commissions to help negotiate agreements with creditors,

Essexella

GROUP Scyphozoans
DATE Late Carboniferous
SIZE 8–12cm (3¼–4½in) across
LOCATION USA

Essexella is one of the rarest types of fossils – a jellyfish. Many other supposed jellyfish in the fossil record are of dubious affinity, but *Essexella* is beyond doubt. It is one of the commonest fossils in the Mazon Creek fauna of Illinois, USA, and some specimens have been obtained with a preserved rim of tentacles around the disc. Like other fossils in this fauna, they owe their exceptional preservation to rapid burial and then replacement with siderite (ferrous carbonate) following initial bacterial action. This mineralization left the fossils within siderite concretions, and protected them from further decomposition and decay. *Essexella* would have hunted and fed much like modern jellyfish – by using the stinging cells along the length of its tentacles to gather prey before transferring them to the mouth.

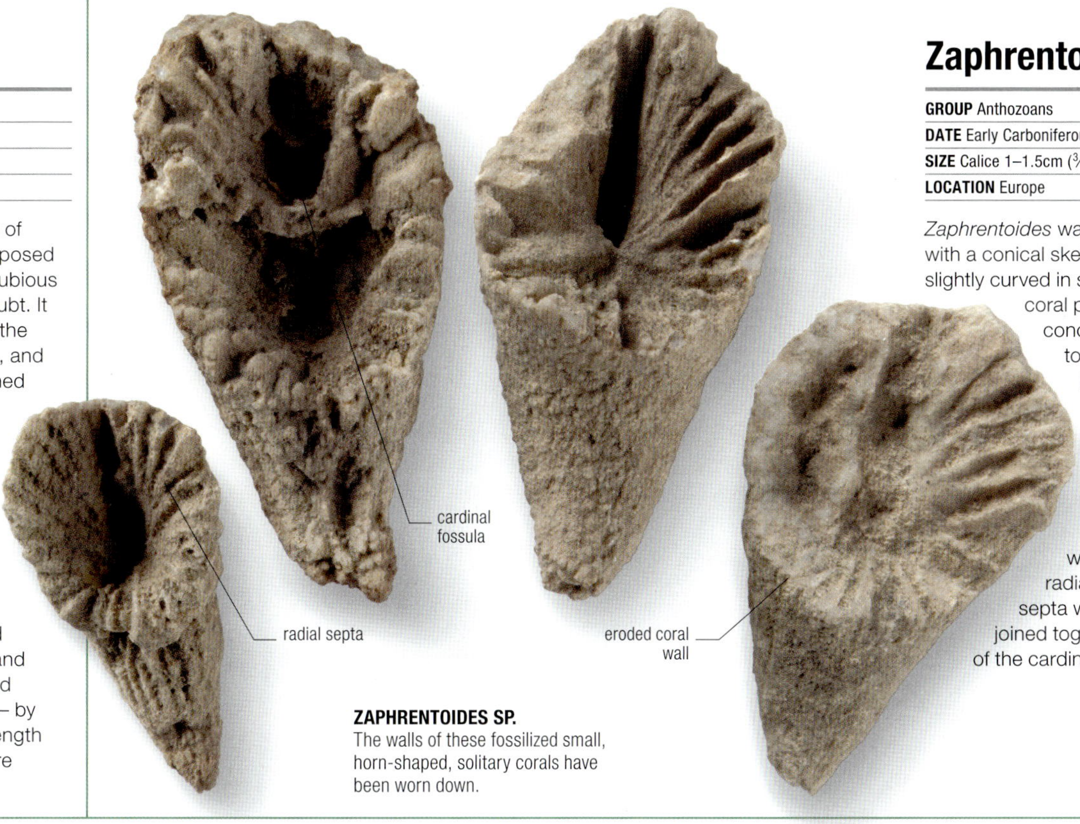

ZAPHRENTOIDES SP.
The walls of these fossilized small, horn-shaped, solitary corals have been worn down.

Zaphrentoides

GROUP Anthozoans
DATE Early Carboniferous
SIZE Calice 1–1.5cm (3/8–1/2in) across
LOCATION Europe

Zaphrentoides was a small, solitary coral, with a conical skeleton (corallum) that was slightly curved in shape. The soft-bodied coral polyp rested in the concave dip (calice) at the top of the corallum in the living animal. There was a distinctive large, deep space (cardinal fossula) on the convex side of the corallum. This was located between the vertical plates, or septa, which were arranged in a radial pattern. The major septa were elongated and joined together around the cavity of the cardinal fossula.

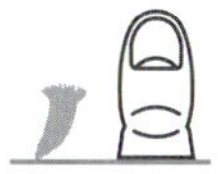

Siphonophyllia

GROUP Anthozoans
DATE Carboniferous
SIZE Up to 1m (3ft 3in) long
LOCATION Europe, North Africa, Asia

Siphonophyllia was a very large, solitary rugose coral. It was curved and conical in early stages of growth, often becoming straighter and cylindrical later in life. The soft-bodied coral polyp was situated in the concave calice at the top of the skeleton (corallum). Within the corallum there were many vertical plates (septa) arranged in a radial pattern. Steeply inclined, curved plates (dissepiments) were found at the edge of the corallum. On the Sligo coast, in Ireland, *Siphonophyllia*'s abundance and its snake-like appearance has earned one rock the nickname Serpent Rock.

The snake-like appearance of *Siphonophyllia* has led to one rock being named Serpent Rock.

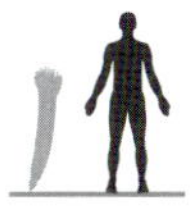

SIPHONOPHYLLIA SP.
This fossilized coral is found in shallow water or in calcareous shales and limestones. The long, slender septa can be seen running through the corallum, towards the outer edge in this specimen.

PARTITIONS
This fossil shows the closely spaced horizontal partitions (tabulae) in the centre of the corallum.

Syringopora

GROUP Anthozoans
DATE Late Ordovician to Carboniferous
SIZE Corallites 1–2mm (1/16in) across
LOCATION Worldwide

Syringopora was a tabulate coral (see p.100), consisting of cylindrical tubes (corallites) that were roughly parallel to each other. The tubes were not in direct contact with each other, but were connected by smaller horizontal tubes (tubuli). The corallites, which housed the soft-bodied, tentacled polyps at the top, contained spiny partitions (septa) arranged in longitudinal rows. Along the length of the corallite were funnel-shaped plates (tabulae). As the colony matured it would have formed new corallites and expanded sideways.

SYRINGOPORA RETICULATA
This coral was made up of long, cylindrical corallites that were connected to each other by smaller horizontal tubes.

As the colony **matured,** it would have **formed new corallites** and expanded sideways.

Archimedes

GROUP Bryozoans
DATE Early Carboniferous to Early Permian
SIZE Colony height over 20cm (8in)
LOCATION Worldwide

Archimedes was a fenestrate bryozoan colony (see p.179) that would have stood upright on the sea floor. It is thought to have lived in shallow waters on carbonate muds. Its structure is immediately recognizable due to the distinctive screw-shaped axis. Thin, net-like structures called fronds were attached to the axis, forming a continuous spiral. The fronds were made up of branches and on one side, the autozooecia – chambers that housed the soft-bodied individuals (zooids) of the colony – were found in two rows. The branches were connected by partitions, which formed rectangular open spaces between them called fenestrules, hence fenestrate bryozoan.

ARCHIMEDES SP.
The fronds are very delicate structures, which means that usually only the screw-shaped axis of the fossil is preserved.

WHAT'S IN A NAME?
Archimedes is named after the Greek mathematician because the screw-shaped axis resembles the water pump he invented.

Productus

GROUP Brachiopods
DATE Carboniferous
SIZE Up to 7.5cm (3in) long
LOCATION Europe, Asia

Productus was a slightly unusual brachiopod. As an adult it had no anchoring pedicle, but instead relied on the weight of its pedicle valve (the half of its shell from which a pedicle would emerge in most brachiopods) to keep it stable on the sea bed. Some species of *Productus* had spines on this valve that helped to hold them firmly in position.

PRODUCTUS SP.
On the outside of the pedicle valve are obvious ribs, which are crossed by shallow folds.

Pugnax

GROUP Brachiopods
DATE Devonian to Carboniferous
SIZE Up to 4cm (1½in) long
LOCATION Europe

Like most brachiopods, *Pugnax* was attached to the sea bed via a flexible stem or pedicle emerging through the pedicle valve. The outside of most of the shell is essentially smooth and devoid of all ornamentation except for some very fine growth lines and short ribs around the margin of each of its valves.

PUGNAX ACUMINATUS
This view from beneath shows the sinus (infold) of the pedicle valve and the outward fold of the brachial valve.

Fossundecima

GROUP Worms
DATE Late Carboniferous
SIZE Up to 6cm (2¼in) long
LOCATION North America

Fossundecima was a bristle worm (or polychaete) that lived around 300 million years ago in the seas over what is now North America. One distinctive feature was a flexible feeding organ (proboscis) that it could push out from its mouth. This was equipped with powerful jaws, which strongly suggests that it was a predatory carnivore. Behind the head were 15–20 abdominal segments, each carrying bristles, and a final, smaller tail segment. *Fossundecima* moved by use of the paired lobes (parapodia) on each of its abdominal segments; the parapodia were also thought to be used in respiration.

Vestinautilus

GROUP Cephalopods
DATE Early Carboniferous
SIZE Up to 12.5cm (4¾in) across
LOCATION Europe, North America

Vestinautilus was a coiled, nautilus-like cephalopod. The whorls of its shell overlapped only slightly, resulting in a wide umbilicus (naval-like structure); this had a central perforation as the initial coil was not tight. *Vestinautilus* fossils show a prominent spiral ridge running around the outside of each whorl and distinctive, V-shaped suture lines. The ventral region (outside margin) of the whorl is extremely broad, and in mature specimens, the whorls sometimes lose contact with each other as the spiral opens out.

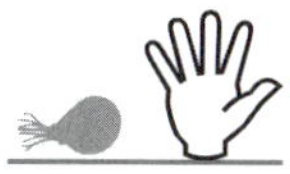

Vestinautilus fossils show a **prominent spiral ridge** running around the outside of **each whorl** and distinctive, **V-shaped** suture lines.

VESTINAUTILUS CARINIFEROUS
This internal mould has been preserved in limestone. It clearly shows the suture lines and the ridge, but the body chamber has broken off.

Goniatites

GROUP Cephalopods
DATE Early Carboniferous
SIZE Up to 3.5cm (1½in) across
LOCATION Europe, Asia, N. Africa, North America

Goniatites was an inflated, rounded ammonoid with very tight coiling, so that each whorl almost entirely overlapped the preceding one. This type of coiling is said to be involute and results in a small, deep depression (umbilicus) at the centre of the coil. The shell is missing from the specimen shown here; it is an internal mould, so the septal sutures are readily visible. These mark the line where the septa, the walls between the internal chambers, joined the inside of the shell. The septal sutures are very angular – a characteristic *Goniatites* feature. The goniatites were a group of ammonoids that lived from the middle of the Devonian to the end of the Permian. Details of suture lines can be used to distinguish between different goniatites. The earliest ones, from the Middle Devonian, have very simple sutures, whereas some of the latest ones from the Permian have sutures that are very complex.

GONIATITES CRENISTRIA
This internal mould clearly shows *Goniatites'* angular sutures and small umbilicus.

BUOYANCY AID
Goniatites used its many internal chambers as buoyancy aids, but the shape of its shell suggests that it was not a swift swimmer.

The pattern of the suture lines identifies different goniatites.

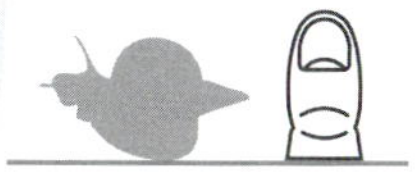

STRAPAROLLUS SP.
Like some modern snails, the whorls of *Straparollus* are relatively loose with an open spiral and completely lack ornamentation.

Straparollus

GROUP Gastropods
DATE Carboniferous
SIZE Up to 6cm (2¼in) across
LOCATION North America, Europe, Australia

Straparollus was a conical gastropod mollusc with whorls that are rounded in cross section, but have a slight "shoulder". The shell surface lacks ornamentation, apart from fine growth lines. In some Carboniferous limestones, mollusc fossils are rare, as the calcium carbonate shells of most are made of the mineral aragonite, which later dissolved. *Straparollus*, however, had an outer layer of calcite, the other mineral form of calcium carbonate.

Bellerophon

GROUP Gastropods
DATE Silurian to Early Triassic
SIZE Up to 5cm (2in) across
LOCATION Worldwide

Bellerophon is unusual for a gastropod in that its shell was coiled in a plane spiral. The whorls were very rounded, and later whorls enveloped the preceding ones. The shell was essentially smooth. Scars in some fossils show paired muscles (again, a departure from the gastropod norm), and this has led to suggestions that *Bellerophon* belongs to a more primitive mollusc group, the tergomyans. Yet some *Bellerophon* species show an aperture slit and slit band, which suggest it may have been a gastropod. The debate is still open.

BELLEROPHON SP.
Bellerophon's smooth, unadorned shell shows a plane spiral symmetry that is rare in other species of gastropods.

Conocardium

GROUP Rostroconchs
DATE Devonian to Permian
SIZE Up to 15cm (6in) long
LOCATION Worldwide

Conocardium looks like a bivalve, but is in fact a rostroconch. Rostroconchs differ from bivalves in that they developed from a single-valved protoconch (initial shell), whereas bivalves developed from a bivalved protoconch. Also, rostroconch shells were fused along the front mid-line – the hinge region of bivalves – so the two valves gaped open permanently. Like many bivalves, rostroconchs could use a muscular foot to burrow or move. They probably fed on plankton and other particles of organic matter from the surrounding sea water, but exactly how they did this is unknown.

CONOCARDIUM SP.
The shell surface of *Conocardium* was ornamented by fine ribs, and the shell itself was thick.

DUNBARELLA SP.
With its rounded outline and fine, radiating ribs, *Dunbarella* is similar in some ways to scallops of the present day.

Dunbarella

GROUP Bivalves
DATE Carboniferous
SIZE Up to 4cm (1½in) long
LOCATION Europe, North America

Dunbarella's shell was rounded but had a straight dorsal line produced by ear-like shell extensions. Many fine ribs radiated from the centre of the line, crossed by growth lines parallel to the outer valve edges. All *Dunbarella* were attached to the sea bed with a byssus (like a mussel's "beard") when young, but some broke free at maturity, resting on the sea floor with their right valve lowermost. The ribs of the right valve split in two as the shell grew, while new ribs arose on the left valve in between previously formed ones.

Carbonicola

GROUP Bivalves
DATE Late Carboniferous
SIZE Up to 4cm (1½in) long
LOCATION W. Europe, Russia

A freshwater bivalve that lived in lakes and swamps, *Carbonicola*'s shell usually had beaks of differing heights. The shell tapered towards one end, and inside each valve were two adductor muscle scars: one circular and deep, the other slightly larger but shallower. Each valve usually had one tooth and one socket, but some forms had two and others none at all. The ligament that held the shell open lay behind the beaks.

CARBONICOLA PSEUDOROBUSTA
Carbonicola's shell tapered at one end. Its exterior was ornamented only with fine growth lines, which are just visible on this specimen.

Woodocrinus

GROUP Echinoderms
DATE Carboniferous
SIZE Cup and arms 6–10cm (2¼–4in) high
LOCATION Europe

Woodocrinus was a crinoid with a small cup consisting of three circlets of plates. Each arm normally branched twice, giving 20 arms in total. The ossicles or segments that made up the arms were stout, but short, so that each arm was composed of numerous short, circular plates stacked on top of each other. Different arms branched at different heights above the cup, and as a whole they appeared blunt and thick. A plated anal tube rose from the base of the cup. The column that supported the cup was circular in cross section and featured regular annual thickening of its component ossicles, which gave the stem a ribbed appearance.

Woodocrinus grew in **extensive "crinoid forests"** in shallow seas.

WOODOCRINUS SP.
The currents that supplied crinoids with food were also responsible for breaking up their bodies after death. This fossil shows the arm and stem ossicles of *Woodocrinus* beginning to separate and drift apart.

Actinocrinites

GROUP Echinoderms
DATE Carboniferous
SIZE Cup up to 3cm (1¼in) across
LOCATION Europe

This picture shows the upper surface of the crinoid *Actinocrinites*. The junctions between the plates have been inked in to show them clearly. In the centre is the slightly domed upper surface of the cup; in most crinoids this is leathery, with some calcite plates, but in *Actinocrinites* it was made up entirely of fused calcite plates. The side of the cup is enlarged by the lower arm (brachial) plates and extra plates (interbrachial) between them, and the bases of four arms can be seen. *Actinocrinites* would have lived on reefs, anchored by a stalk-like structure.

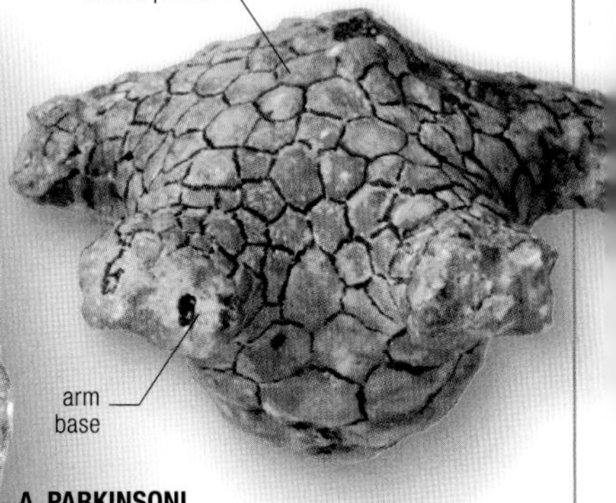

A. PARKINSONI
From the five arm bases (one is hidden here) of *Actinocrinites*, about 30 arms branched out in total.

Pentremites

GROUP Echinoderms
DATE Early Carboniferous
SIZE Cup up to 2.5cm (1in) high
LOCATION North America, South America, Europe

Pentremites was an echinoderm known as a blastoid, which lived fixed to the sea bed by a root at the base of a jointed stem. Its cup, shown here, consisted of three circlets of plates and five petal-shaped ambulacra. Along the edges of the ambulacra were numerous long, slender tube feet that conveyed food particles to a central food groove. From here, food was moved to the star-shaped mouth, shown at the top of this fossil. The tube feet in *Pentremites* were used entirely for feeding.

PENTREMITES PYRIFORMIS
The five openings around *Pentremites*'s mouth are known as spiracles. These served as the exits from the animal's respiratory chambers.

Xyloiulus

GROUP Arthropods
DATE Late Carboniferous
SIZE Up to 6cm (2¼in) long
LOCATION North America, Europe

This millipede is one of several kinds known from Carboniferous rocks. Its head is not well preserved in this specimen, but the long, tubular, segmented body is clearly visible. Each segment is similar, and has two pairs of slender appendages; parts of many of these can be seen. Fossil millipedes from rocks of mid-Silurian age in Scotland are the oldest land animals known. Some, from the Carboniferous coal swamps, reached a huge size: up to 30cm (12in) long.

XYLOIULUS SP.
This fossil *Xyloiulus* is preserved in a clay-ironstone nodule.

Archegonus

GROUP Arthropods
DATE Early to Middle Carboniferous
SIZE Up to 4cm (1½in) long
LOCATION England, C. Europe, Spain, Portugal, N.W. Africa, Asia, S. China

Archegonus had a rectangular or conical glabella (the central part of the headshield) with shallow, narrow furrows. Its eyes were crescent-shaped, rather small, and located close to the glabella. Its cheeks had short spines on their posterior corners. *Archegonus* had nine thoracic segments, and its tailshield was a similar size to its head. Its broad sides had at least six pairs of rather weak ribs. This trilobite is found in sediments that were deposited in fairly deep water; its small eyes also suggest that it probably inhabited this type of environment, with low light levels.

A. NEHDENENSIS
Archegonus species have been used to zone some Carboniferous rocks.

Hesslerides

GROUP Arthropods
DATE Middle Carboniferous
SIZE Up to 3–4cm (1¼–1½in)
LOCATION North America

Hesslerides was a trilobite with a large, granulated glabella (central raised part) and crescent-shaped eyes, set close to the rear of its headshield. On the inner part of its cheeks, a narrow ridge ran around the lower margin of the eye. The cheeks sloped down towards a flattish border. Its thorax consisted of nine segments and its tailshield was about as long as its head; the middle lobe had 12 narrow segments, and the side lobes had 10 pairs of ribs, each with a deep furrow. "Three lobes" – two side or pleural ones, and a central or axial lobe – are what the word "trilobite" means.

HESSLERIDES BUFO
The eyes of this species were set to the back of its headshield, close to the large glabella.

Archimylacris

GROUP Arthropods
DATE Late Carboniferous
SIZE 2–5cm (¾–2in) long
LOCATION Europe, North America

This fossil of *Archimylacris*, a cockroach-like insect, shows two pairs of wings and parts of its body. The front part of the rather crushed thorax can also be seen between the wings, although the head itself is missing. The forewings are the smooth areas on either side of the fossil, and these formed a protective, toughened covering for the body when the animal was not flying. Between them, the thinner, more delicate hind wings show a distinctive vein pattern. This is important in identification.

ARCHIMYLACRIS EGGINTONI
The wings of this specimen of *Archimylacris* are fully developed, indicating that this was an adult; immature stages did not have full wings.

Pleophrynus

GROUP Arthropods
DATE Early Devonian to Late Carboniferous
SIZE 1.5–4cm (½–1½in) long
LOCATION North America

Pleophrynus was an extinct, spider-like arachnid called a trigonotarbid. Unlike true spiders, it lacked poison and silk glands. Its hard, outer shell (or carapace) was pear shaped, and tapered to a point at the front, close to which two small eye clusters were situated. Remains of some of its eight limbs can be seen here originating underneath the head (prosoma). Its abdomen (opisthosoma) was covered in rigid plates, which afforded protection for the animal's underlying soft tissues, and each of its six segments was ornamented with rows of bumps or tubercles. The back margin of the carapace had four short, spur-like projections.

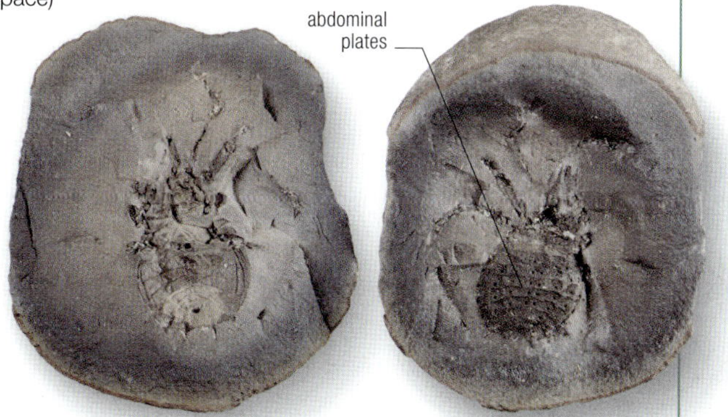

PLEOPHRYNUS SP.
Pleophrynus protected its soft body parts with rigid plates that covered its abdomen, as well as with four short, spiky projections.

Tealliocaris

GROUP Arthropods
DATE Early Carboniferous
SIZE Up to 5cm (2in) long
LOCATION Scotland

Tealliocaris was a shrimp-like crustacean that inhabited very shallow, probably brackish, water on tidal flats. The hard upper shell (or carapace) occupied the left half of this specimen, which is flattened from the side. The pair of first antennae had two short branches, and lay just above the very long, slender, and delicate second antennae (one is preserved here). Some of the thoracic legs can be seen projecting from underneath the carapace; each had two branches. The curved, flexible abdomen was about the same length as the carapace, and was composed of five segments, under which were five pairs of swimming limbs or "swimmerets".

TEALLIOCARIS WOODWARDI
Although flattened, this fossil is well preserved, showing outlines of some of the five lobes of the distinctive tail fan.

Euproops

GROUP Arthropods
DATE Carboniferous
SIZE Up to 5cm (2in) long
LOCATION Europe, C. USA

Euproops was an early relative of the horseshoe crab. Its headshield was crescent shaped, with broad, pointed spines at the cusps. Two side ridges on its head bowed outwards at the front ends, where its tiny eyes were situated. These ridges tapered into long, pointed spines.

EUPROOPS ROTUNDATUS
The oval abdomen had a narrow, segmented central ridge, and these segments ran onto the animal's flanks. Each rib ended in a short spine.

CARBONIFEROUS VERTEBRATES

From their origins in the Devonian, tetrapods diversified within the coal swamps of the Carboniferous, adapting to different lifestyles, including those of aquatic opportunists and terrestrial predators. In the oceans, highly armoured fish were replaced by more active swimmers with much thinner scales.

AMPHIBAMUS
This little temnospondyl had advanced features seen today in frogs and salamanders, including a very large tympanic ear, large eyes, and specialized, hinged teeth. And, just like frogs and salamanders, *Amphibamus* had only four fingers.

The Carboniferous period could be called the "Age of Coal", after the thick swamps of coal-forming plants that grew on land. The end of the Devonian saw the extinction of placoderms, heterostracans, and osteostracans (see pp.128–29). In the Carboniferous, advanced jawed fish, including chondrichthyans, spiny sharks (acanthodians), and ray-finned fish in particular, dominated the seas, lakes, and rivers. On land, tetrapods flourished, perhaps aided by the arthropod colonization. The Carboniferous opens with a 20-million-year period of poor fossilization known as Romer's Gap, after the American palaeontologist Alfred Romer. The fossil record then reveals a wide diversity of fauna, so there must have been a major radiation of species during those Gap years.

The first terrestrial fauna

The earliest tetrapods evolved in estuaries during the Late Devonian, and their anatomy suggests they had to live in water. They retained tail fins and internal gills – basically, they were fish with legs. During and shortly after Romer's Gap, the tetrapods changed; whereas they were still adapted to live in the water, many, including the embolomeres and *Crassigyrinus* (see p.165), retained their tail fins, but seem to have lost their internal gills, which suggests they were more capable of coming onto land. The famous fossil site of East Kirkton in Scotland (see p.168) provides ample evidence of the world's oldest-known vertebrate terrestrial community. The fossil fauna recovered includes a highly diverse collection of tetrapods, with many animals more specialized to live on land. Temnospondyls first appear here, including a species that lacks many of the hallmarks of life in the water. Lepospondyls are also present, represented by the snake-like aïstopod *Ophiderpeton*, which would have been as comfortable on land as in the water. But most striking are the reptile-like animals, including *Westlothiana* (see p.168), which was hailed as the first amniote (see opposite), though this interpretation is disputed. If correct, it would have represented the first time reproduction was freed from needing water – a necessary step towards fully occupying the land. However, in the Late Carboniferous, undisputed amniotes became prominent members of a truly terrestrial fauna.

GROUPS OVERVIEW

The first major radiation of tetrapods took place in the Early Carboniferous, and another followed 40 million years later in the Late Carboniferous. This two-step diversification permitted the full occupation of the land, and would lead to the emergence of the modern herbivore-dominated fauna in the Permian.

BAPHETIDS
Appearing in the Early Carboniferous, baphetids (also known as loxommatids) are immediately recognizable by their keyhole-shaped eye sockets. *Megalocephalus* and other baphetids retain the sensory canals on the skulls that indicate an aquatic way of life. Skeletal fossils currently being studied will provide more information as this group is mostly known from skulls.

WHATCHEERIIDS
Another of the Early Carboniferous tetrapods, Whatcheeriids were perhaps the first accomplished terrestrial tetrapods. Animals such as *Pederpes* had tall eye sockets and lacked the sensory structures on the skull that are common in aquatic animals. These tetrapods have only been discovered recently, and more remains to be learned about them.

CRASSYGYRINIDS
A large, Early Carboniferous aquatic predator, this is one of the more bizarre tetrapods. With a gigantic head, elongate trunk, a tail fin, and tiny forelimbs, *Crassigyrinus* probably lurked in the murky shallows waiting for fish to swim by. However, it lacks the structures associated with internal gills, and so is more advanced than the first tetrapods.

TEMNOSPONDYLS
One of the most successful groups of early tetrapods, temnospondyls lived in most Carboniferous environments. Some were specialized aquatic bottom-dwellers, others lurked at the water's edge (like crocodiles), and still others were terrestrial. They ranged in length from 12cm (5in) to 1.5m (5ft). One type, represented by *Amphibamus*, gave rise to frogs and salamanders.

The development of the amniotic egg was a critical step in the expansion of tetrapods across all parts of the land.

Evolution of the shelled egg

The first tetrapods inherited simple eggs from their fish ancestors that were fertilized outside the body. Water was critical for reproduction, because the thin membrane surrounding the egg would not protect it against drying out on land. Water was also necessary to carry sperm to the eggs. This mode of reproduction is called "amphibious", and is still seen in frogs and salamanders. In order to become fully terrestrial, it was necessary to protect the egg against drying. This was accomplished by the amniotes (reptiles, birds, and mammals) with the addition of three new membranes to the egg: the chorion, amnion, and allantois. These membranes, together with the yolk sac, also help support the growing embryo by providing nutrients and collecting waste. Around the outermost membrane lies the shell, which may be leathery and flexible (as in many turtles and lizards) or mineralized and hard (best-known in birds, but also found in crocodiles and many tortoises). Amniotes further became independent from water by their mode of reproduction. Instead of releasing sperm into water over eggs, the male places sperm within the reproductive system of the female, where it travels to and fertilizes the egg. After fertilization, the extra membranes and shell are formed, and the egg is laid on dry land. Among amniotes, mammals have gone one step further, doing away with the shell and retaining the egg inside the mother (monotremes are the exception, as they still lay eggs; see p.357). The extra membranes are modified to interact with tissues of the mother in order to directly nourish the developing foetus. This modification is called the placenta.

NO WATER REQUIRED
The hard shells and extra internal membranes allow this tortoise's eggs to be laid in the ground without drying out. This was a critical step in the tetrapod radiation onto all parts of the land.

OPHIACODON
One of the better-known early amniotes, this crocodile-like predator is related to the sail-backed *Dimetrodon*, which lived during the Permian. *Ophiacodon* is a synapsid, which is the evolutionary lineage that leads to mammals.

LEPOSPONDYLS
Another diverse group of early tetrapods, lepospondyls were usually small, but they assumed many body forms. One of the earliest-known tetrapods was, ironically, a limbless aïstopod, and the first microsaurs were mistaken for early amniotes (thus their name). The aquatic nectrideans were newt-like, while the elongate lysorophians would curl into burrows and aestivate during dry spells.

EMBOLOMERES
These large, aquatic tetrapods, such as *Proterogyrinus*, were ubiquitous during the Carboniferous. Some early forms retained tail fins, and their moderately elongate skulls would help for catching fish. Features in their limb bones show more adaptations towards terrestrial life, and many scientists have considered these tetrapods to be early members of the lineage leading to amniotes.

AMNIOTES
Highly adapted for life on land, the first amniotes appeared in the Late Carboniferous, making their homes in hollow tree stumps. Amniotes are comprised of three groups: reptiles, synapsids (see right), and a collection of early species from before the reptile–synapsid split. Some of the first amniotes were also the first tetrapod plant-eaters (herbivores).

DIAPSIDS
The largest group of reptiles, diapsids are named after two openings in the back of their skulls. Diapsids alive today include lizards, snakes, crocodilians, and birds. Turtles, which arose in the Middle Triassic and lack the skull openings, are also probably diapsids. The first diapsids are found in Late Carboniferous rocks from the central United States.

SYNAPSIDS
The lineage leading to mammals, synapsids are named after a single opening in the back of their skull. The first fossils are isolated fragments found in the tree stumps with the first amniotes (see left), meaning that amniotes rapidly diversified soon after they evolved. Carboniferous synapsids include active predators and some of the first herbivores.

Falcatus

GROUP Chondrichthyans
DATE Carboniferous
SIZE 30cm (12in) long
LOCATION USA

This genus, which is closely related to *Stethacanthus* (see p.130), is known from numerous small specimens preserved in limestone sea beds. *Falcatus* is named after the large, forward-pointing, sickle-shaped spine that is armed with pointed denticles located above the head. This spine is only known in specimens above a certain size, indicating that it was a sexual feature typically found in males and was most likely used for display. The almost symmetrical tail was used for cruising efficiently through the water.

GROUP SWIMMERS
Many specimens of *Falcatus* have been found together, which suggests these sharks travelled in shoals.

ANATOMY
SENSE ORGANS

Falcatus had well-adapted senses for life in the murky depths of the sea. It had large eyes ringed by rectangular calcified plates, which probably helped to support the eye and acted as areas for muscle attachment. *Falcatus* also possessed a soft snout, or rostrum, which probably housed special sense organs that detected prey by picking up electrical impulses caused by movement of the prey's muscles.

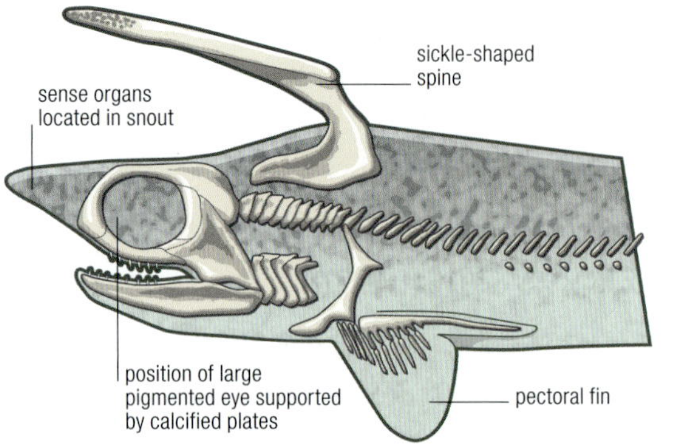

Echinochimaera

GROUP Chondrichthyans
DATE Early Carboniferous
SIZE 30cm (12in) long
LOCATION USA

Echinochimaera is known from two species that can be distinguished by differences in their upper dentition. Both have been found in marine shallow-water sediments. It was a small fish, built for manoeuvrability rather than speed, with a tapering body, relatively large pectoral and pelvic fins, and a paddle-shaped tail. It had two dorsal fins, the front one with a mobile spine on its leading edge. The body was completely covered with scales. Pairs of enlarged scales, called scutes, extended from the rear dorsal fin to the base of the tail. There were also pairs of backward-projecting scutes above the eyes. Males were much larger than females and had different features, including antler-like eye scutes.

Scute translates as "shield" and explains the protective quality of these special scales.

SHOWING OFF
The larger males had some elaborate features that could have been used for recognition or displaying during the breeding season.

SLIM FISH
The unusual shape of *Belantsea* meant it was well-adapted for moving slowly through the water while grazing.

Belantsea

GROUP Chondrichthyans
DATE Late Carboniferous
SIZE 70cm (28in) long
LOCATION USA

Belantsea belongs to a rather unusual group of fish with cartilaginous skeletons – the petalodonts. These relatives of sharks had a distinctive set of teeth. Rather than having lots of teeth arranged in numerous rows, *Belantsea* had only seven teeth in the top and bottom jaws. Furthermore, the teeth were not shed while feeding, but were kept throughout life. Each tooth had specially reinforced, serrated cutting edges, ideal for grazing on hard materials such as sponges. The leaf-like body shape of *Belantsea* is very distinctive. Its small tail suggests slow motion, but the large fins and the deep body indicate good manoeuvrability.

Discoserra

GROUP Actinopterygians
DATE Early Carboniferous
SIZE 60cm (23½in) long
LOCATION USA

Discoserra was a tropical fish that inhabited a shallow bay now known as Bear Gulch in Montana, USA. It had a deep, disc-shaped body that was highly laterally compressed – it was not built for speed or endurance, but for precise manoeuvring. The head was small, with large eyes and a small mouth. *Discoserra* had many large scales (scutes) on the front half of its body. Each dorsal scute had a forward-facing hook. The scales were held together by peg-and-socket joints.

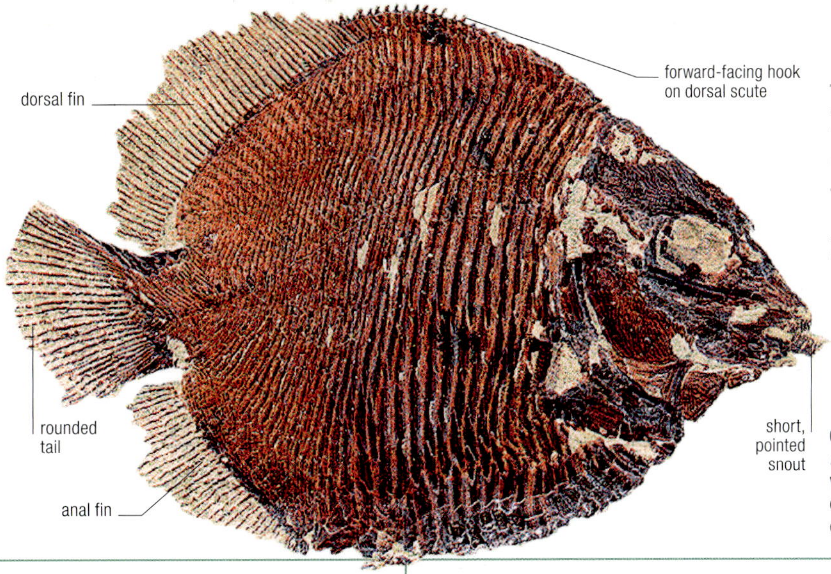

***Discoserra* was not built for speed or endurance, but for precise manoeuvring.**

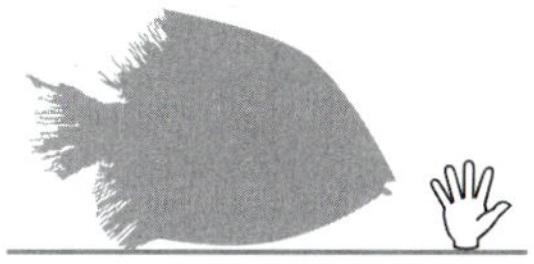

CORAL FEEDER
Discoserra had many features that indicated it was adapted for feeding among sponges and corals, including a short snout with long, comb-like teeth for sucking or grazing.

Megalocephalus

GROUP Early tetrapods
DATE Late Carboniferous
SIZE 1.5m (5ft) long
LOCATION British Isles

Megalocephalus, like most baphetids, is known only from its skull. The baphetids were a family of presumably aquatic predators from the Carboniferous that combined some widespread primitive features with several unique ones.

The skull of *Megalocephalus* (the name translates as "big head") is about 30cm (12in) long and has large, pointed teeth. One unique feature of baphetids, including *Megalocephalus*, is the strange elongated extension to their eye sockets (orbits), the true reason for which is unknown.

UNUSUAL SKULL
Suggested functions for the extended eye socket include housing for a salt gland or electric organ, or to allow for extra movement of jaw-closing muscles.

Pederpes

GROUP Early tetrapods
DATE Early Carboniferous
SIZE 1m (3ft 3in) long
LOCATION Scotland

Pederpes is known from a single specimen discovered in the west of Scotland near Dumbarton and is the only tetrapod known from an articulated skeleton dating from a time known as Romer's Gap (see p.162). It shows a mix of primitive and advanced features. It probably had the usual five toes on its hind limb, and a foot that appears adapted for terrestrial walking. By contrast, its hands may have had more than five fingers, though only two tiny ones have been found. Its ear region was like that of *Acanthostega* (see p.138).

FEET FORWARD
The joints in *Pederpes'* feet pointed forwards, which suggests that it could walk on land.

Whatcheeria

GROUP Early tetrapods
DATE Early Carboniferous
SIZE 1m (3ft 3in) long
LOCATION USA

Whatcheeria is a close relative of *Pederpes* but it lived at a later time. It is known from a large amount of fossil material including several almost complete skeletons and many isolated bones. Many other tetrapods and fish have been found at the same site in Iowa, USA. *Whatcheeria* seems to have been more aquatic than *Pederpes*, with a powerful tail that provided propulsion. Much of its anatomy still remains to be described, including its limbs.

Crassigyrinus

GROUP Early tetrapods
DATE Middle Carboniferous
SIZE 3–4m (9¾–13ft) long
LOCATION Scotland, possibly USA

Crassigyrinus was one of the largest predatory animals in the swamps of the Middle Carboniferous. Its enormous gaping mouth housed rows of very large teeth, positioned mostly near the front so that prey would have been firmly impaled when it closed its jaws in a fast, snap-trap movement. Its large head, elongated body, and deep tail suggest that *Crassigyrinus* was fully aquatic, as do its tiny forelimbs and relatively small hind limbs. In some respects, the skull of *Crassigyrinus* is quite primitive, and its evolutionary relationships are still debated.

LYING IN WAIT
Crassigyrinus may have been an ambush predator, hiding in vegetation or among rocks while waiting for its prey to swim past.

Balanerpeton

GROUP Temnospondyls
DATE Early Carboniferous
SIZE 50cm (19½in) long
LOCATION Scotland

Balanerpeton was the earliest member of a large group of fossil amphibians called the temnospondyls. It would have looked like a large salamander, but its middle ear and tympanic membrane more closely resembled those of a frog. *Balanerpeton* was the first known land animal to have had an ear capable of hearing airborne sound waves, which means it could have heard its insect prey moving nearby. Its name means "the creeping animal from the hot springs", as its remains were found in what is believed to have been a volcanic crater lake, surrounded by hot springs.

FLATTENED SKULL
The indentations for the tympanic membranes can be seen on either side of the back of the skull.

Cochleosaurus

GROUP Temnospondyls
DATE Late Carboniferous
SIZE 1.5m (5ft) long
LOCATION Czech Republic, Canada

Many *Cochleosaurus* fossils have been found at the Nýrany coal mine in the Czech Republic, which contains the remains of a Carboniferous swamp. *Cochleosaurus* was one of the first crocodile-like temnospondyls. It lurked at the edge of the swamp, and ambushed prey such as fish and smaller amphibians. It was the most common large animal living in the swamp, making it the top predator in this type of habitat. Similar temnospondyls filled this niche until the end of the Triassic, when they were replaced by true crocodylomorphs (see p.244). The name *Cochleosaurus* means "spoon-reptile" and refers to two spoon- like bony extensions to the back of the skull, which were hidden under the skin.

Branchiosaurus

GROUP Temnospondyls
DATE Late Carboniferous
SIZE 15cm (6in) long
LOCATION Czech Republic, France

The *Branchiosaurus* specimens found at the Nýrany coal mine had gills – the name means "gilled reptile" – and it is not clear whether these individuals would have remained aquatic throughout their lives or would have become terrestrial as adults. In some populations of *Apateon* (a later relative of *Branchiosaurus*) the animals remained aquatic as they grew, while in other populations they transformed into terrestrial forms. These alternative life cycles can be seen in some species of newt today, in which metamorphosis into an air-breathing terrestrial form only occurs when conditions are right. For many years, branchiosaurs were all thought to be the larvae of much larger temnospondyls, such as *Eyrops* (see p.184), but they are now known to be a distinct group.

15 cm The average length of *Branchiosaurus,* making it one of the **smallest** of the temnospondyls.

Phlegethontia

GROUP Lepospondyls
DATE Late Carboniferous
SIZE 70cm (28in)
LOCATION USA, Czech Republic

Phlegethontia is the best-known aïstopod, a group of legless amphibians that resembled snakes. It had rows of small spiked teeth, similar to those found in some non-venomous snakes today. Its backbone had more than 200 vertebrae, two-thirds of which were in the tail alone.

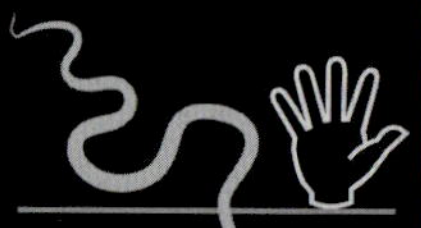

SNAKE-LIKE AMPHIBIAN
Phlegethontia was probably terrestrial, living at the edges of ponds and lakes and venturing out onto aquatic plants in search of prey.

Amphibamus

GROUP Temnospondyls
DATE Late Carboniferous
SIZE 12cm (4½in) long
LOCATION USA

Amphibamus was one of the first Carboniferous amphibians discovered. It was the first temnospondyl to have the type of teeth that are seen in frogs and salamanders today, and is thought to be closely related to the ancestor of these two groups. Its tall, thin teeth each had two tiny spikes at the tip and were hinged partway down their length. *Amphibamus* means "equal legs", owing to its long front and hind limbs being the same size, like those of salamanders. But it was small by temnospondyl standards.

At the site where *Amphibamus* was discovered, palaeontologists have also found several tiny amphibian fossils that look like *Branchiosaurus* (see opposite), but share features with *Amphibamus*. These may be *Amphibamus* larvae, which suggests some degree of metamorphosis was already happening in this early amphibian. For example, these fossils have a long tail with a fin, like a tadpole, while the adult *Amphibamus* had a short stumpy tail that was more in common with the missing tail in modern frogs and toads.

This little temnospondyl had advanced features seen today in frogs and salamanders.

DELTA DWELLER
Amphibamus is from a site called Mazon Creek in Illinois, USA, which represents a Carboniferous river delta. It is likely that *Amphibamus* lived in the creeks and along the banks of the delta.

Microbrachis

GROUP Lepospondyls
DATE Late Carboniferous
SIZE 30cm (12in) long
LOCATION Czech Republic

More than 100 specimens of *Microbrachis*, of all ages and sizes, have been found at Nýrany. All of them had internal gills and a lateral-line sensory system. Clearly this microsaur was aquatic and spent its entire life in the waters of the swamp, unlike most microsaurs, which were terrestrial.

AT HOME IN THE WATER
Microbrachis, meaning "small arms", had tiny limbs, a flattened tail to aid swimming, fish-like gills, and lateral-line canals along its flanks. All of these are adaptations for a fully aquatic lifestyle.

Proterogyrinus

GROUP Reptiliomorphs
DATE Early Carboniferous
SIZE 1.5m (5ft) long
LOCATION USA, Scotland

Proterogyrinus was one of the earliest anthracosaurs ("coal lizards"), a group of amphibians that were top predators for much of the Carboniferous. Although it ate fish, *Proterogyrinus* was not fully aquatic. It was one of the first vertebrates to have the long curved ribs and muscles that are used to pump air in and out of the lungs. This evolutionary adaptation developed after vertebrates had moved onto the land.

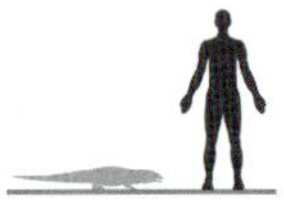

EARLY WRIGGLER
Proterogyrinus had a long, undulating body and tail – hence its name, which means "early wriggler". Its robust limbs indicate that *Proterogyrinus* could walk on land and so was not restricted to a wholly aquatic lifestyle.

LIZZIE THE LIZARD
At first glance, *Westlothiana* resembles a lizard, hence its nickname of "Lizzie the lizard", but its ankles are characteristic of an early tetrapod.

Westlothiana

GROUP Early tetrapods
DATE Early Carboniferous
SIZE 25cm (10in) long
LOCATION Scotland

Westlothiana was a small terrestrial animal whose remains have been found at East Kirkton in the West Lothian area of Scotland. It had an elongated body with very short but slender limbs, strongly built vertebrae, and curving ribs that would have enclosed most of the body. Only about five specimens are known, and the skull is badly preserved in all of them. When first discovered, *Westlothiana* was hailed as "the earliest reptile", but key parts of the skull, such as the palate and ear bones, are needed to confirm its evolutionary position. However, *Westlothiana* does appear to be one of the earliest relatives of the modern amniotes (see p.163).

KEY SITES

EAST KIRKTON

The Early Carboniferous rocks from East Kirkton have given us a window into an assemblage of early terrestrial tetrapods. Volcanic activity at the time meant ash falls, fires, warm mineral-rich springs, and poisonous gases killed and preserved many plants, arthropods, and tetrapods. Among the tetrapods were the earliest members of lineages leading to modern amphibians and reptiles, including *Westlothiana* and *Balanerpeton*.

Ophiacodon

GROUP Synapsids
DATE Latest Carboniferous to Early Permian
SIZE 3m (9¾ft) long
LOCATION USA

Ophiacodon was one of the largest early synapsids (see p.163). It resembled a crocodile, with long jaws and numerous pointed teeth that were ideal for catching fish. Flattened finger and toe bones suggest that it was partly aquatic and

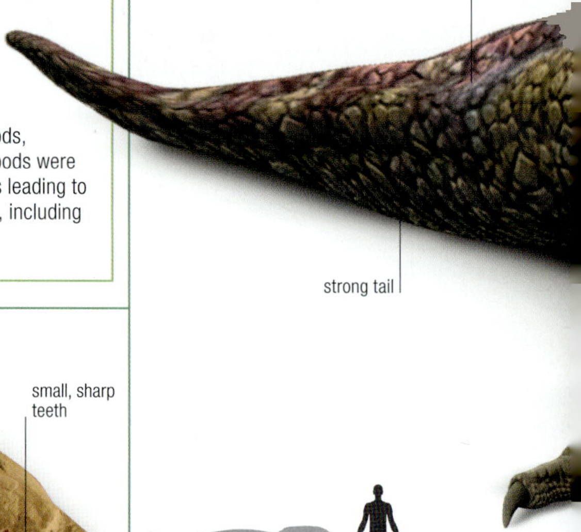

BIGGER THAN ITS BITE
With its large size and crocodile-like jaws, *Ophiacodon* looks like a fearsome predator, but it seems to have mostly caught fish and small vertebrates rather than large prey.

Paleothyris

GROUP Eureptiles
DATE Late Carboniferous
SIZE 25cm (10in) long
LOCATION Canada

A primitive reptile, *Paleothyris* was a small, lizard-like insectivore that lived in forested regions. Its skeletons have been found in the hollow trunks of lycopsids (giant early relatives of clubmosses), buried in sediment from a flash flood. *Paleothyris* is similar to another early reptile, *Hylonomus*, in being anapsid (having no holes in the back of the skull), but it is five million years younger.

AGILE HUNTER
The limbs, body, and teeth of *Paleothyris* indicate that it would have been an agile hunter, easily able to track down and capture its insect prey.

Spinoaequalis

GROUP Diapsids
DATE Late Carboniferous
SIZE 25cm (10in) long
LOCATION USA

Spinoaequalis is known from a single skeleton found in a quarry at Hamilton, Kansas, USA. It was an early diapsid (see p.163), and had long legs and toes and a distinct neck. More unusually, instead of narrowing to a point, its tail remained deep along its length and was vertically flattened. Long, bony spines above and below the vertebrae provided extra muscle-attachment points, making the tail a powerful tool for swimming.

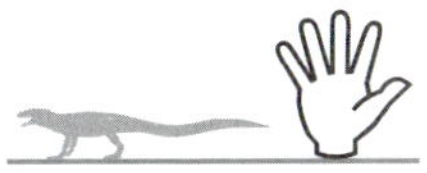

BACK IN THE WATER
Spinoaequalis was the first reptile to return to the water and live alongside amphibians, though it was not fully aquatic and would have returned to dry land to breed.

could use its limbs as paddles. Unlike later pelycosaurs, such as *Dimetrodon* (see pp.188–89), *Ophiacodon* was not well-adapted for killing large animals. It lived like a crocodile, lurking at the water's edge to ambush smaller terrestrial vertebrates as well as catch fish. However, being a synapsid, *Ophiacodon* was much more closely related to mammals than to modern crocodiles. This is an example of convergent evolution, where adaptations to a particular way of life evolve separately in different groups of animals.

166 The number of sharp teeth in *Ophiacodon*'s long snout.

KEY SITES
JOGGINS FOSSIL CLIFFS

These famous cliffs in Nova Scotia, Canada, preserve the oldest fossil amniotes, namely *Protoclepsydrops*, a close relative of *Ophiacodon*, and *Hylonomus*, the earliest true reptile. On this site, 300 million years ago, small animals used hollowed-out tree stumps as nesting sites. Today, their remains give us our first clear insight into the record of small terrestrial vertebrates.

PERMIAN

Permian

The last period of the Paleozoic era was dominated by the giant supercontinent of Pangaea, into which Laurasia and Gondwana finally merged. The Permian was named after the Russian region of Perm, because its limits were defined by strata found in the Ural Mountains. It ended with the largest mass extinction of the last half a billion years, when at least 90 per cent of land and oceanic organisms disappeared.

URAL MOUNTAINS
The forest-covered western slopes of the Urals, which stretch from north to south in west–central Asia, are especially noted for their extensive Permian strata.

Oceans and continents

The sheer size of the Pangaean supercontinent had significant implications for the world's climate and the spread of fauna and flora. At low latitudes, the Central Pangaean Mountains extended in an east–west orientation, which greatly influenced atmospheric currents and hence weather patterns. During the Late Carboniferous, this mountain belt had been located across the tropics and was the source of the coal deposits formed from the lush rainforests that grew in the wet, tropical climate. By Permian times, it had moved further north into more arid regions. The mountains blocked the humid, equatorial winds, with the result that deserts formed in the northern part of Pangaea (now the central part of North America and northern Europe). Sandstone strata reveal the existence of huge expanses of dunes. As the climate became drier, large mammal-like reptiles appeared, better adapted to coping with the new conditions. In less arid regions, plants continued to diversify. Upland forests of conifers, ferns, and seed ferns established themselves and large expanses of the plant *Glossopteris* fringed cool polar regions. Most of Pangaea extended south of the Equator, and included the former continent of Gondwana. Its vast land mass gave some faunas the opportunity to spread over a wide area. On the northern section of Pangaea, in Siberia, one of Earth's largest eruptions of magma covered a huge area with flood basalts and was accompanied by outpourings of ash and gases, especially sulphur dioxide and water vapour. This coincided with the Permian mass extinction, when the oceans became starved of oxygen and most marine species died out.

TROPICAL RAINFOREST
Lush tropical forests, similar to those that flourished during the Carboniferous, still existed around the Equator during the Permian, in areas that are now part of China.

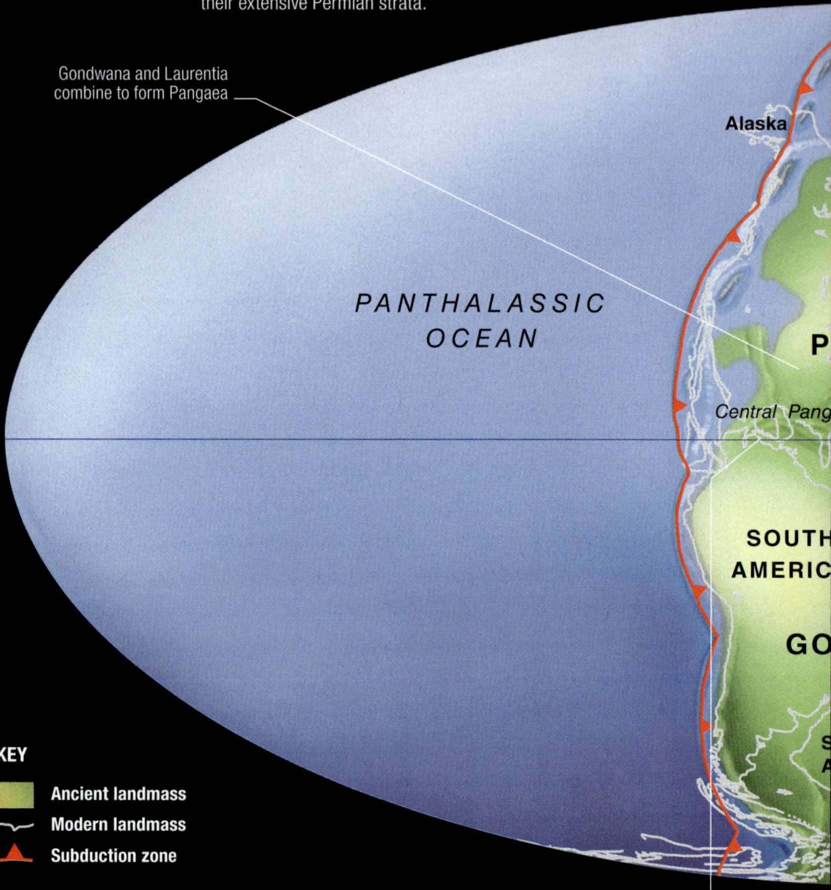

PERMIAN WORLD MAP
Pangaea became a vast supercontinent that extended from pole to pole. The once tropical Central Pangaean Mountains moved northwards.

CISURALIAN

MYA 300 | 290 | 280

PLANTS

290 Conifers and seed plants diversify

280 Seed plants and conifers increase in diversity; lepidodendrids and sphenopsids decrease in diversity

DWYKA TILLITE
Tillites are conglomerates composed of boulders, cobbles, and sand scraped up by glaciers. The Dwyka Tillite in the Karoo Basin in South Africa dates from the Permian period.

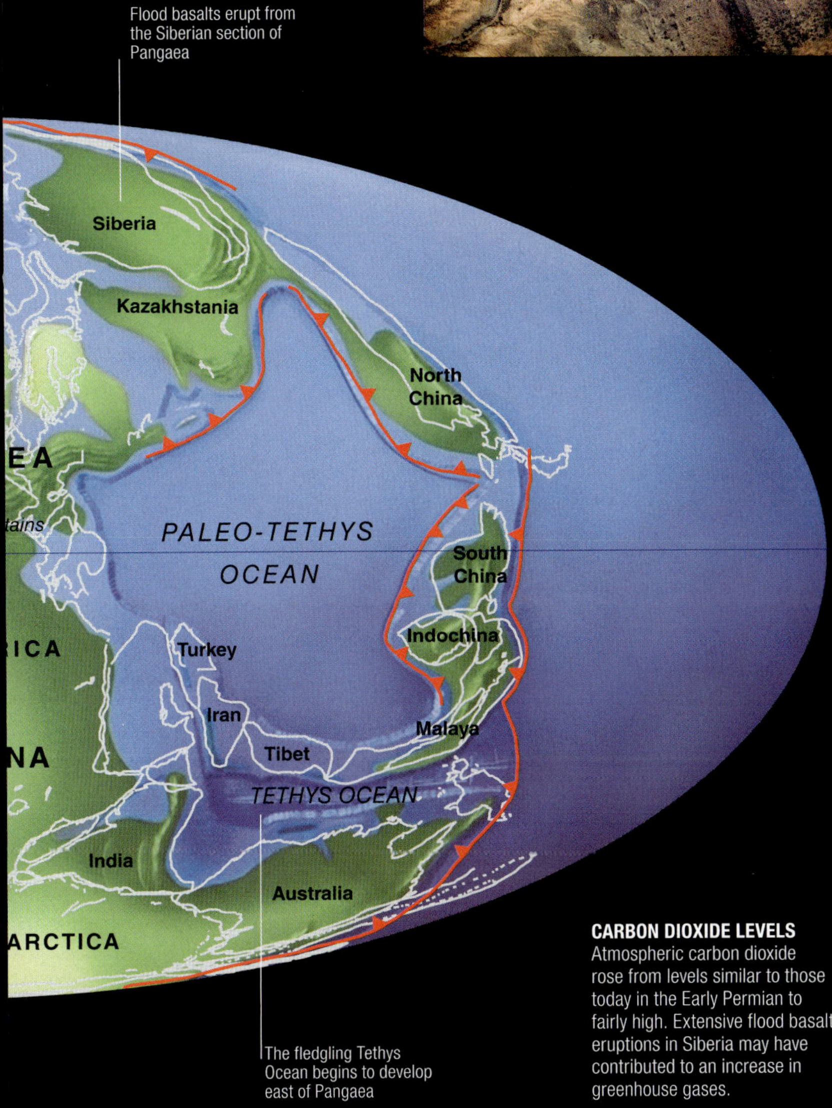

Climate

The glaciation that so influenced the Late Carboniferous world continued, coming to an end in the early part of the Late Permian. Ice sheets covered all the continents that once formed part of Gondwana. Many glacial centres were involved, with repeated ice advances and retreats. Evidence for this comes from glacial deposits in South Africa, India, and Tasmania. The major Permian glacial episode spanned about 2 million years, coming to an end in a period of just a few thousand. From the Late Carboniferous to the Permian, the major centres of ice dispersal migrated with the movement of Pangaea over the South Pole. The main centres of glaciation in the Late Carboniferous were in South America, India, the southern part of Africa, and western Antarctica, but they now shifted towards Australia. While in the Early Permian much of the southern hemisphere was covered by ice, the climate in mid-latitudes in southern and northern Pangaea was relatively cool temperate to arid, with only a narrow tropical belt in equatorial regions. During warmer interglacial periods, coal-forming rainforests were quite widespread in lower latitudes. By the Late Permian, the ice sheets that had covered the southern hemisphere landmass had all but disappeared, although small ice caps had developed at the North Pole. Much of Pangaea was very dry, with vast deserts in areas that are now South America and Africa in the south, and North America and northern Europe in the north.

SIBERIAN TRAPS
Vast outpourings of basaltic magma continued for thousands of years to form what is now the Siberian Traps. The environmental stress this caused may have contributed to the Permian mass extinction.

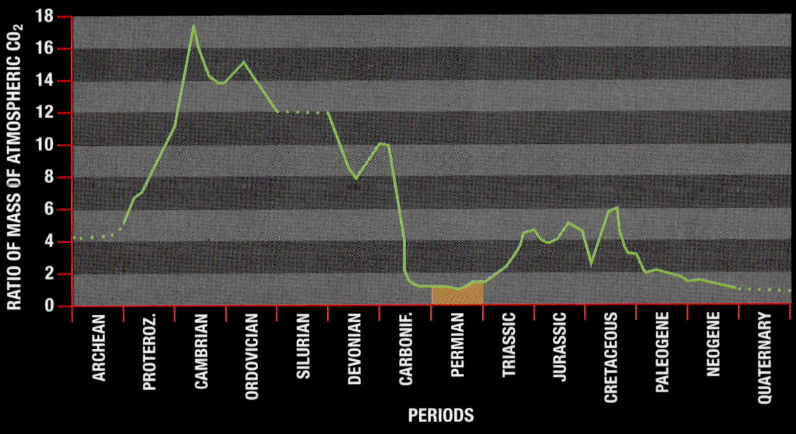

CARBON DIOXIDE LEVELS
Atmospheric carbon dioxide rose from levels similar to those today in the Early Permian to fairly high. Extensive flood basalt eruptions in Siberia may have contributed to an increase in greenhouse gases.

GUADALUPIAN | **LOPINGIAN**

270 | 260 | 250 | MYA

PLANTS

260 Decrease in *Glossopteris* flora

GLOSSOPTERIS

250 Mass extinction: 50% decline in plant species diversity, mostly affecting woodlands

INVERTEBRATES

275 Ammonoid diversity begins to reduce; first ceratitid ammonoids

XENODISCUS

255 First pectinid and carditid bivalves

250 Mass extinction: 96% of all species; all fusulinid foraminifera, rugose and tabulate corals, trilobites, pareiasaurs, eurypterids, receptaculitids; goniatites; strophomenid brachiopods; 5 orders of insects. Almost extinct: echinoids, crinoids, stenolaemate bryozoans, articulated brachiopods

VERTEBRATES

275 First pareiasaur reptiles (below); first therapsid (mammal-like) reptiles; extinction of pelycosaur reptiles

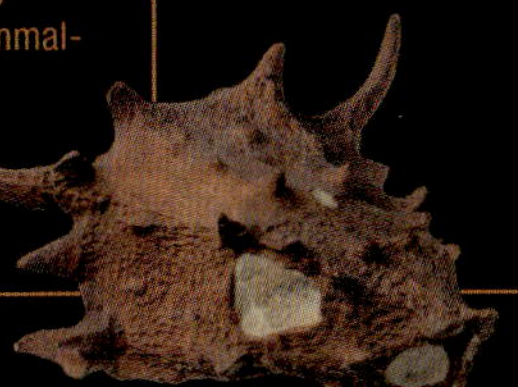

ELGINIA

DICYNODONT

260 Decrease in diversity of terrestrial and aquatic amphibians; increase in therapsid diversity, especially herbivores like dicynodonts

250 Two-thirds of amphibians, reptiles, and therapsids become extinct

PERMIAN PLANTS

The drying climates of the Permian finally destroyed the ancient coal swamps that were characteristic of the Carboniferous. This led to the extinction of many characteristic Carboniferous plants. However, new kinds of habitats provided fresh opportunities for the expansion of other plant lineages.

During the Permian, the giant tree lycopods and the habitats in which they thrived declined and eventually disappeared. Medullosans, cordaites, and many other groups of archaic plants also became extinct during the Permian. However, conifers and other new kinds of plants diversified and set the stage for further evolutionary innovation through the Mesozoic.

New kinds of seed plants

Several new groups of seed plants appeared during the Permian. Most are quite poorly understood, but glossopterids are one of the better known of these groups. They are characteristic of areas that were once at high latitudes in the southern hemisphere. Recognition of their distinctive leaves in Australia, South Africa, South America, and Antarctica was important early evidence for continental drift (see p.21). Cycads also made their first appearance in the Permian and subsequently became important in the Mesozoic. Several groups of poorly understood Permian seed plants may be early relatives of these true cycads. In China, an enigmatic group called gigantopterids are especially interesting. Gigantopterids had large leaves with specialized net venation very similar to that seen in angiosperms. Some of these gigantopterids were probably climbers in Permian equatorial forests.

SEED-BEARING LEAF OF CYCAS
Living *Cycas*, one of about ten genera of living cycads, has seed-bearing leaves that are very similar to those of fossil cycads from the Permian.

FOSSIL CONIFER
The shoots and leaves of the earliest conifers were very similar to those of some living conifers in the monkey-puzzle family, such as the Norfolk Island Pine.

The rise of conifers

The earliest conifers are from the Late Carboniferous. They had normal, conifer-like leafy shoots but their cones were much less compact than those of any modern conifer. The cones of early conifers have been compared to the cones of cordaites (see p.153). The early conifers of the Carboniferous occurred in drier environments outside the lowland coal swamps, and as the climate dried in the Permian, conifers and other groups from these areas became more important. Conifers may also have been the first seed plants in which the seed could remain dormant prior to germination.

GROUPS OVERVIEW

The Permian was a time of transition in the evolution of land plants. The ancient plant lineages of earlier in the Paleozoic declined and were replaced by other groups that were apparently better adapted to the drier conditions of the Permian. Among seed plants, the lyginopterids, medullosans, and cordaites had all disappeared by the end of this period.

FERNS
The marattialean tree ferns of the Carboniferous persisted into the Permian. They eventually became much less prominent, but nevertheless persisted through the Mesozoic and Cenozoic to the present. In the drier climates of the Permian, new kinds of ferns appeared for the first time, including forms that are similar to the modern royal ferns.

CONIFERS
The earliest conifers are from the Carboniferous, although they were rare at this time. Conifers flourished and diversified in the drier climates of the Permian. The earliest conifers had normal-looking conifer foliage but their cones, and perhaps other aspects of their reproduction, were very different from any living conifer.

CYCADS
Fossil cycads that are clearly related to the living group appeared for the first time during the Permian. There is much about these earliest cycads that is still not well understood, but their seed-bearing leaves were very similar to those of the living genus *Cycas*.

GLOSSOPTERIDS
This was one of several new seed-plant groups that appeared in the Permian. Glossopterids' characteristic leaves are common in fossil floras from high latitudes in the southern hemisphere. This group was important in the vegetation of cooler climates during the major Permian glaciation of the southern hemisphere.

Oligocarpia

GROUP Ferns
DATE Carboniferous to Permian
SIZE Up to 50cm (19½ in) tall
LOCATION Worldwide

This ground-hugging fern is similar to a family of living ferns that are mainly found in the tropics. Known as forking ferns, they have fronds that grow in pairs, and creeping rhizomes or underground stems. Like all ferns, *Oligocarpia* spread from place to place by spores, which developed on the undersides of its fronds. However, recent fossil finds show that its rhizomes may have been just as important, allowing it to invade new areas of ground in the way that bracken does today. Unusually for a prehistoric species, these fossils also show the same plant at many different stages of development. These include adult ferns, and tiny plantlets with a minute rosette of leaves.

Oligocarpia spread from place to place by **spores dispersed** through the **air**.

OLIGOCARPIA GOTHANII
This beautifully preserved specimen was collected in China in the 1920s. Fossils in the same strata suggest that it lived in a wetland habitat, such as a low-lying river plain. Here, periodic flooding may have created the ideal conditions for fossilization – almost all of the plant has been preserved.

Protoblechnum

GROUP Peltasperms
DATE Permian
SIZE Frond 2–5m (6½–16½ft) long
LOCATION Worldwide

Protoblechnum fronds have broad, webbed leaflets – a feature found in some living hard ferns or deer ferns of the genus *Blechnum*, which gives this type of foliage its scientific name. However, instead of belonging to ferns, *Protoblechnum* fronds belonged to a group of plants called peltasperms. Most of these plants were shrubs or small trees. They grew in a variety of habitats, but their foliage would have made them particularly well adapted to areas with an annual dry season, such as subtropical river plains. Their seeds were usually a few millimetres long.

PROBLECHNUM SP.
This fossilized *Protoblechnum* species shows the typical webbed leaves. Peltasperms, such as this, bore their seeds in small, umbrella-shaped organs that were often arranged in open columns or spikes.

Tubicaulis

GROUP Ferns
DATE Late Carboniferous to Permian
SIZE Up to 2m (6½ft) high
LOCATION Worldwide

Tubicaulis fossils are the preserved trunks of an ancient group of ferns, dating back over 300 million years. In cross section, the fossils show a dense mass of stems embedded in a matrix of roots. Each stem is enclosed by an oval sheath and contains a C-shaped stalk. *Tubicaulis* ferns were upright or scrambling, anchoring themselves with roots that grew from the stems near the base of their leaves. Most lived on shady swamp floors, but some species may have grown high up on trees to collect light.

TUBICAULIS SOLENITES
This cross section reveals that, unlike in living ferns, *Tubicaulis* stems were constructed "back to front" – the "C" faced outwards rather than inwards.

Glossopteris

GROUP Glossopterids
DATE Permian to Triassic
SIZE 4–8m (13–26ft)
LOCATION Southern hemisphere

Since its discovery in 1828, by French palaeobotanist Adolphe Brongniart, *Glossopteris* has become one of the most studied prehistoric plants. The first *Glossopteris* fossils were found in India, but others have been unearthed in all the land masses of the southern hemisphere, providing striking biological evidence for continental drift. *Glossopteris* means "tongue fern", which describes its leaf shape. Dozens of species have been identified, most being shrubs or small trees. Despite its name, *Glossopteris* reproduced by seed, not spores, which developed in stalked organs attached to the leaves.

LOW-LYING FORESTS
With its height and foliage, *Glossopteris* was the dominant plant across much of the southern hemisphere. It formed immense forests in low-lying ground. Some species were subtropical, but many grew at high latitudes, where winters were long and cold.

AUSTRALOXYLON SP.
Australoxylon is fossilized wood of glossopterids. It has been found in many parts of the southern hemisphere. In structure, it was similar to the wood found in modern conifers.

VERTEBRARIA INDICA
With their cylindrical shape and conspicuous joints, *Vertebraria* fossils resemble animal vertebrae, rather than the fossilized roots of *Glossopteris* trees. They are found from India and Africa, to Australia, South America, and Antarctica.

TONGUE-SHAPED LEAVES
Glossopteris is identified mainly by its leaves, which were up to 30cm (12in) long, with a strong midrib and diverging veins. They were arranged in spirals around the ends of the branches.

Callipteris

GROUP Peltasperms
DATE Late Carboniferous to Permian
SIZE Leaf 80cm (31in) long
LOCATION Worldwide

Belonging to the same group of seed ferns as *Protoblechnum* (see p.175), callipterids are well known from fossils of their leaves and seed-producing organs. Their leaves were up to 1m (3ft 3in) long. As well as having their leaf segments arranged in opposite pairs, they often had small leaflets attached to the central midrib. *Callipteris* bore its seeds on the underside of fan-shaped scales, which were arranged in a spiral around a vertical stalk.

CALLIPTERIS CONFERTA
Callipteris conferta was one of the most common seed plants of the Permian. It is easy to identify due to the regular sequences of paired leaf segments attached to a large, central rib.

Gigantopteris

GROUP Gigantopterids
DATE Permian to Triassic
SIZE Leaf up to 50cm (19½in) long
LOCATION Southeast Asia, North America

The gigantopterids are an enigmatic group of plants, known mainly from fossils found in Asia and North America. They had woody stems, sometimes with spines, and many had leaves like those of flowering plants. They died out in the Early Triassic, and little is known about how they reproduced, which makes it hard to judge exactly where they fit in plant evolution.

GIGANTOPTERIS NICOTIANAEFOLIA
The fossil above is a species called *Gigantopteris nicotianaefolia*, which was so named because its leaf segments resemble those of tobacco plants.

Tingia

GROUP Noeggerathialeans
DATE Early Permian
SIZE Cone up to 15cm (6in) long
LOCATION China

This foliage belonged to fern-like plants that reproduced by growing spores. In *Tingia elegans*, the leaves bore leaflets in four rows – the two rows on the upper surface were largest, masking the smaller ones underneath. The plant's cones were attached at the base of the leaves. As fossils, noeggerathialeans are relatively rare, although cones and leaves have been found together, showing that they came from the same plant. They appeared in the Late Carboniferous, and became extinct during the Triassic. Although they show links with progymnosperms, their place in plant evolution is not clear.

TINGIA ELEGANS
Tingia is known only from the fossil remains of its leaves, like those shown here, and of its cones. *Tingia* reproduced by means of spores, rather than through the production of seeds.

Psygmophyllum

GROUP Ginkgos
DATE Permian
SIZE Leaf up to 10cm (4in) long
LOCATION Worldwide

From the Triassic onwards, many fossils show how ginkgos evolved variations of fan-shaped leaves. Earlier fossils are far rarer. *Psygmophyllum* may have been an ancient forerunner of living ginkgo, as its leaves superficially have the classic ginkgo outline. However, this resemblance is probably misleading: no fossils have been found of its fruits or seeds, which makes it likely that the leaves come from a more primitive plant.

PSYGMOPHYLLUM MULTIPARTITUM
Although the leaves of *P. multipartitum* have the usual fan-shaped outline of living ginkgo, it is unclear whether this species and ginkgo were closely related.

Ullmannia

GROUP Conifers
DATE Late Permian
SIZE 10m (33ft) high
LOCATION Worldwide

Conifers first appeared in the Carboniferous period, but they became more widespread during the Permian, with its cooler, drier global climate. *Ullmannia* was one of these Permian conifers, with narrow, widely spreading leaves. Its male cones produced pollen grains that had a microscopic, wing-like sac to help them drift on the wind. Like most conifers, its seed-bearing cones were woody, measuring up to 6cm (2¼in) long. The seeds developed on the upper surface of rounded scales.

ULLMANNIA BRONNI
Ullmannia had narrow, widely spreading leaves, not dissimilar to those of the modern monkey puzzle tree. The leaves were arranged in a spiral around a thick, central stem.

PERMIAN INVERTEBRATES

Most Permian sediments encountered are red beds of desert or river origin, with marine sediments less common. This is because of a drastic lowering of sea level, especially during later Permian times, and the emergence of the continental shelves. The consequences for invertebrate marine life were disastrous.

Sea-level lowering

The Carboniferous–Permian glaciation, which covered so much of the land in the southern hemisphere, was largely over by mid-Permian times, and marine faunas proliferated. Yet by the late Permian the Earth witnessed the most acute biological crisis of all time. The mainly tropical crinoid-brachiopod-, and bryozoan-dominated ecosystems of the later Permian were destroyed completely; a process that took place over some 10 million years. All the continents were assembled into the giant super continent of Pangaea, which must have affected climate and oceanic circulation, and it was no doubt an important factor in generating extinction. But most importantly, sinking, inactive mid-oceanic ridges caused the global sea level to fall by up to 280m (920ft), with great loss of habitat, and little sedimentation: warm, shallow seas became reduced to certain parts of the world only. Massive volcanic eruptions contributed to environmental instability and thus to ecosystem collapse. It was several million years before normal marine sedimentation was restored.

PERMIAN REEF
This gigantic reef in Texas, USA, was built up by calcareous algae, but carried its own fauna of marine invertebrates and fish.

FENESTELLA PLEBEIA
This common Carboniferous–Permian bryozoan fed by filtering food from currents flowing through colonies that grew upwards on the near vertical reef-front.

Giant algal reefs

In northern England, Texas, and elsewhere, vast algal reefs – rich in unusual brachiopods and bryozoans – bordered inland seas, connected to the main ocean by narrow channels. As the sea level fell, the reefs died and were stranded high and dry. The inland seas dried out, with great quantities of salt deposited as evaporites, sometimes to be flooded again, at least for a time. This caused a biological catastrophe on an unparalleled scale, and the worst crisis through which life on Earth ever passed.

SCHIZODUS SP.
Permian bivalves, such as this mould of *Schizodus*, are quite common, and were not greatly affected by the Permian mass extinction event.

GROUPS OVERVIEW

Throughout this period most mollusc groups suffered comparatively little; bivalves, gastropods, and cephalopods continued to do well. Bryozoans, however, did not fare so well during the Permian extinction. Brachiopods were abundant in the early Permian and ammonoids continued to evolve, becoming increasingly complex and diverse.

BRYOZOANS
Permian bryozoans were normally marine, colonial invertebrates, but there were some freshwater genera. They all usually have calcite skeletons, which house large numbers of tiny zooids; most of these feed with a ring of tentacles; some are specialized for other functions. Several groups were badly affected by the Permian extinction.

BRACHIOPODS
These marine invertebrates were sea-floor suspension feeders, and although they also had two valves, they had quite a different body plan to the bivalves (see right). They were a dominant component of the sea-floor ecosystem until the mass extinction at the end of the Permian; some species are still in existence to this day.

AMMONOIDS
Several groups of these highly successful cephalopods flourished in the Permian, with "goniatitic" (strongly angled) suture lines (see p.202), and they gave rise to the ceratitid ammonoids that dominated the Triassic.

BIVALVES
Sea-floor dwelling or burrowing bivalve clams, which fed by sucking in currents of water laden with food particles, proliferated during this period. With the decline of the brachiopods at the end of the Permian these became the dominant sea-floor suspension feeders.

Rectifenestella

GROUP Bryozoans
DATE Devonian to Permian
SIZE 5cm (2in) average height
LOCATION Worldwide

Rectifenestella was an upright, fenestrate bryozoan colony. It was conical or fan-shaped, with a net-like appearance, and was composed of thin branches, each of which divided into two. These branches were linked by small, usually horizontal partitions, which produced rectangular holes in the colony. Circular openings on one side of the colony were arranged in two rows. These were the openings to the autozooecia, where the colony's individual soft-bodied animals (autozooids) lived. Each autozooid had a ring of tentacles (lophophore) around its mouth. Small, hair-like extensions (cilia) on these tentacles would beat to create a flow of water that drove food particles towards the autozooid's mouth. The filtered water then flowed out of the rectangular holes to the reverse side of the colony. *Rectifenestella* was supported by root-like spines, which occurred at the base of the colony, on the opposite side to the feeding autozooids, but these cannot be seen on the specimen below.

DIRECTION OF FLOW
Filtered water flowed through *Rectifenestella's* fenestrules and out of the back of the colony.

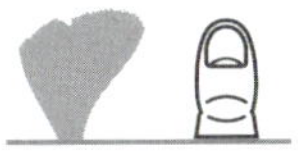

RECTIFENESTELLA RETIFORMIS
This net-like colony can be seen next to *Acanthocladia anceps*. Both species were wiped out in the end-Permian extinction.

ANATOMY
FENESTRATE COLONIES

Bryozoan colonies form a variety of different shapes. Some grow as massive lumps that encrust the sea floor, while others are branching or globe shaped. The extinct fenestrate bryozoans were named after the way they grew – in distinctive net-like patterns. Their thin branches were connected by partitions called dissepiments, which formed open spaces called fenestrules, from the Latin word for "window". The upright branches often divided in two to create a fan. Small openings, in two or more rows, were located on the front surface of these branches. These were the entrances to the chambers (autozooecia) that housed the colony's tiny, soft-bodied animals.

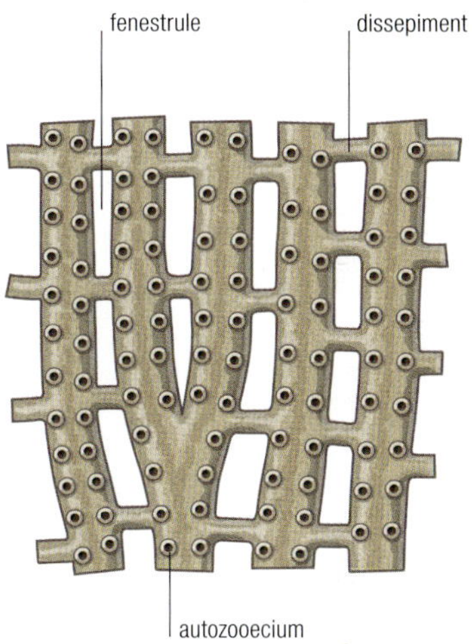

Acanthocladia

GROUP Bryozoans
DATE Carboniferous to Permian
SIZE 2cm (¾in) average height
LOCATION Europe, North America, Asia

Acanthocladia was an upright, often bush-like, fenestrate bryozoan colony. The branches grew from a central axis and split in two. Sometimes, smaller branches fused together and irregular-shaped holes (fenestrules) would develop between them. There were rows of circular openings on one side of the colony that housed soft-bodied autozooids. When the autozooids were not feeding, their tentacles could be withdrawn. The reverse of the colony was smooth, or had fine striations.

ACANTHOCLADIA ANCEPS
The colony had a bush-like appearance with branches growing from a central axis.

Thamnopora

GROUP Anthozoans
DATE Devonian to Permian
SIZE Corallite 1–2mm (1⁄16in) wide
LOCATION Worldwide

Thamnopora was a branching tabulate coral. The corallites, which in life housed the soft-bodied polyps, were circular in cross section and curved outwards from the centre of the branch. The polyp would have rested on the thin horizontal partitions (tabulae). The walls of the corallites were perforated with pores and increased in thickness towards the edge of the colony.

THAMNOPORA WILKINSONI
The walls between the corallites of this tabulate coral were very thin.

HORRIDONIA HORRIDUS
This brachiopod was widest between the auricles at each end of the hinge-line.

Horridonia

GROUP Brachiopods
DATE Permian
SIZE 2.5–8cm (1–3¼in) long
LOCATION Europe, Asia, Arctic, Australia

Horridonia was a strophomenid brachiopod, related to *Productus* from the Carboniferous (see p.157). The slightly concave brachial valve, which was sharply curved at its anterior margin, sat inside the larger, convex pedicle valve. Only the immature *Horridonia* had a stalk-like appendage (pedicle) to attach it to the sea bed. This withered as the animal matured, so the adult relied largely on the weight of its pedicle valve to stay put on the sea bed.

Coledium

GROUP Brachiopods
DATE Middle Devonian to Middle Permian
SIZE 0.8–1.5cm (5⁄16–½in) wide
LOCATION USA, Indonesia

Coledium was a small rhynchonellid brachiopod with an opening beneath the beak of its larger (pedicle) valve. There was an outward fold on the forward edge of the smaller (brachial) valve, which corresponded with an infold (sinus) on the forward edge of the pedicle valve. The inside of the shell had a thin wall that rose above the floor of the pedicle valve. In life, this would have supported a small, spoon-shaped muscle-attachment site in the middle of the valve.

COLEDIUM HUMBLETONENSIS
This species had weak, radial ribs on each valve. Here, the brachial valve is in the foreground, with the beak of the pedicle valve visible at the top.

Xenodiscus

GROUP Cephalopods
DATE Late Permian
SIZE 6–10cm (2¼–4in) across
LOCATION Pakistan, Indonesia

Xenodiscus was an ammonoid with a compressed, disc-shaped (discoidal) shell and flattened sides. The whorls on the shell overlapped slightly and it had a wide, shallow centre (umbilicus). It had weak radial ribs that were developed on the early whorls, but were more faint on the mature body chamber. The point where the internal partitions (septa) were joined to the shell is called the suture line, and it is the form of this line that separated *Xenodiscus* from the goniatites (a group of ammonoids existing at the same time). The type of suture line shown below is "ceratitic" and *Xenodiscus* is a ceratitid ammonoid – a group that evolved in the Middle Permian. Unlike the goniatites, it survived the end-Permian extinctions, but the ceratitids eventually died out at the end of the Triassic.

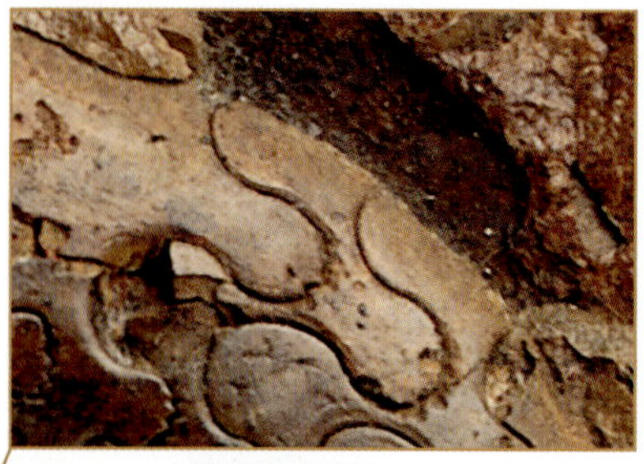

SUTURE LINES
The suture-line folds that face the aperture are rounded, while those facing backwards are finely serrated.

XENODISCUS SP.
The configuration of the whorls and the broad umbilicus mean that this genus is referred to as having an evolute shell. The body-chamber has natural cracks infilled by minerals.

Juresanites

GROUP Cephalopods
DATE Early Permian
SIZE 7–10cm (2¾–4in) across
LOCATION Russia, Australia

Juresanites was a stout, round goniatite. Each whorl partly overlapped the one before it and it had a broad and fairly deep centre. This type of shell is known as moderately involute. The shell surface was smooth and there were no visible patterns. Although many Permian goniatites had similar sutures to *Juresanites*, others had much more complex sutures, closer to some of the Mesozoic ammonoids (see p.298).

JURESANITES JACKSONI
The suture lines on this specimen are visible and are folded as in many of the goniatites. The outside edge of the whorls (venter) is rounded.

Permophorus

GROUP Bivalves
DATE Early Carboniferous to Early Permian
SIZE 2–4cm (¾–1½in) long
LOCATION Worldwide

PERMOPHORUS ALBEQUUS
This internal mould shows the impression where the anterior of the two muscles that close the shell was anchored.

Permophorus was an almost rectangular bivalve that has rounded anterior and posterior extremities. The beaks on both of the valves, or shells, were low and positioned towards the anterior end of the shell. The shell surface was smooth, but there were occasionally concentric ribs and faint radial patterns at the posterior end. The shell also had an angular shoulder that ran back at a slant towards the hinge. Internal features that cannot be seen on the specimen above included two well-developed teeth beneath the beaks on each valve and two more teeth parallel to them on the dorsal margin. In life, these would have helped hold the two valves together.

Deltoblastus

GROUP Echinoderms
DATE Permian
SIZE 1.5–2.5cm (½–1in) long
LOCATION Indonesia, Sicily, Oman

Deltoblastus was one of the last blastoid echinoderms to have existed before the group became extinct. In life, *Deltoblastus* would have been attached to the ocean floor by a thin stalk. Compared with the similar-looking *Pentremites* (see p.160), which existed in the Carboniferous, there are some notable differences, such as the more developed upper circlet of plates of *Deltoblastus*. Although it occurred much later than *Pentremites*, *Deltoblastus* is a survivor of a more primitive group of blastoids.

DELTOBLASTUS PERMICUS
The five sunken ambulacra carried numerous long, slender extensions used in feeding.

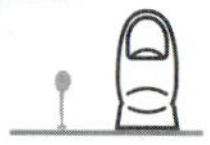

Ditomopyge

GROUP Arthropods
DATE Late Carboniferous to Late Permian
SIZE 2.5–3cm (1–1¼in) long
LOCATION North America, Europe, Asia, W. Australia

Ditomopyge had a distinctively prominent central region of the head or glabella that widened towards the front. Deep furrows in the rear of the glabella divided it into three lobes, a feature that distinguished it from the Carboniferous trilobite *Hesslerides* (see p.161), which was otherwise quite similar. The thorax had nine segments, with a raised axis and side parts (pleurae) that turned downwards. The tail shield was about as long as the head and had 14 well-marked rings. The last trilobites, including *Ditomopyge*, became extinct at the end of the Permian, following a long decline. They formed only a minor part of marine Permian faunas, although they have been found locally (for example in the Crimea) in fairly large numbers.

REINFORCING THE HEAD
The long, broad-based projection running from the side of the head is the genal spine, which bolstered the structural strength of the head. The eyes are the kidney-shaped mounds located towards the side of the head.

PERMIAN VERTEBRATES

Opening with a murmur and ending with a bang, the Permian saw the rise of advanced synapsids. In the oceans, ray-finned fish continued to diversify, while terrestrial vertebrates evolved a new type of predator, the cynodont. Meanwhile, herbivores were perfecting sophisticated chewing mechanisms.

The rise of advanced synapsids (mammal-like reptiles) was the most important evolutionary event of the Permian, because it was critical to the emergence of mammals and, ultimately, modern humans. As the Permian progressed, the terrestrial fauna took on a more familiar set-up, with relatively few predators feeding on the more numerous herbivores.

Cynodonts and mammalness

The top predators of the Late Carboniferous were sprawling in posture, demonstrating a less active mode of life, similar to many modern reptiles. By the Late Permian, synapsid predators had evolved features that enabled active hunting: their limbs had elongated and were placed more directly under the body; their teeth had differentiated into incisors, canines, and cheek teeth; and their eyes had moved further forward, improving stereoscopic vision. These early cynodonts had nearly all the characteristics of mammals. Nowhere is the evolution of "mammalness" more clear than in the middle ear. Mammals have three middle ear bones: the malleus, incus, and stapes. All other tetrapods only have the stapes. The fossil record of synapsids shows how these two extra bones, once part of the jaw articulation, were freed from their jaw-bracing function to become integrated into the middle ear region, where they form part of the sound-detecting apparatus used in hearing.

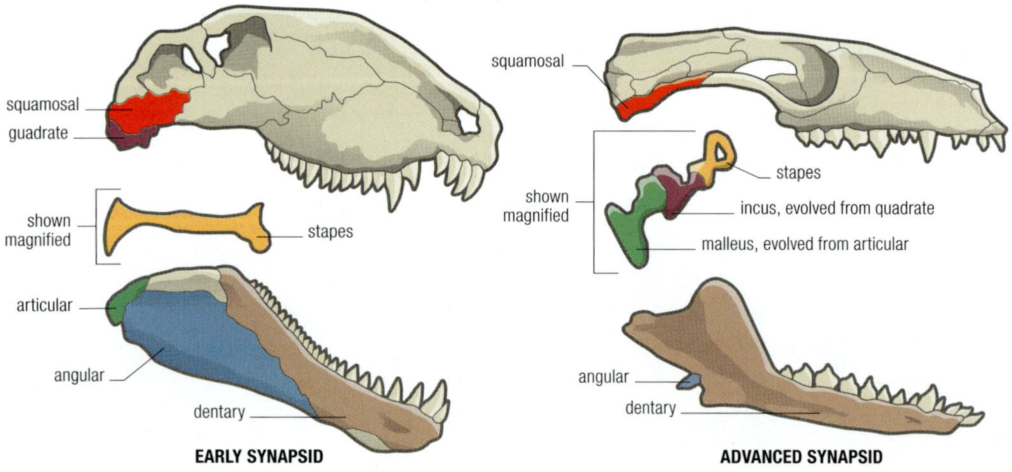

The great dying

The Permian ended with the greatest mass extinction of all, with an estimated 95 per cent of all marine and 70 per cent of terrestrial species dying out after volcanic eruptions poisoned the oceans and atmosphere (see p.32).

THE EVOLUTION OF EAR BONES
Most tetrapods (far left) only have a single middle ear bone, the stapes, which articulates with the tympanum. The jaw hinges between the quadrate and articular. In advanced synapsids (including mammals, left), the jaw hinges between the dentary and squamosal, freeing the quadrate and articular to become incorporated into the middle ear.

GROUPS OVERVIEW

The Permian started with a fauna almost indistinguishable from that of the Late Carboniferous, and ended with a synapsid-dominated fauna with lots of herbivores and fewer predators. The groups shown here, just four of many, highlight important innovations in vertebrate evolution during the Permian's 48 million years.

TEMNOSPONDYLS
During the Permian, the amphibious temnospondyls either became specialized for life on land or perfected life in water. Some, like *Eryops*, became quite alligator-like semi-aquatic animals, whereas others, such as *Archegosaurus*, kept to a life in lakes. Fully terrestrial species include *Cacops*, whose back was protected by a double row of armour plates.

LEPOSPONDYLS
This group of small tetrapod amphibians occupied many lifestyles. The salamander-like microsaurs wandered the land or became specialized burrowers. The snake-like lysorophians and adelospondylids had reduced limbs and an aquatic lifestyle. Nectrideans, like the boomerang-headed *Diplocaulus*, were also aquatic and were the largest of the group.

EDAPHOSAURIDS
These early synapsids, including the sail-backed *Edaphosaurus*, were herbivorous. Their teeth became short, closely-packed, and blunt, for cropping and grinding. And their bellies became extended to host the bacteria in their guts necessary to digest plants. Eventually, herbivores became the largest part of the terrestrial vertebrate fauna.

CYNODONTS
These advanced synapsids arose in the Late Permian. Early cynodonts, such as *Procynosuchus*, show a number of innovations towards a mammal-like condition, including hair, skull construction, and limb placement. These and other skeletal changes indicate that the higher metabolic rate of mammals was in the process of being acquired.

SPINY SHARK
Acanthodians are also known as "spiny sharks", although they are not true sharks. *Acanthodes* had fewer spines than many of its relatives, but they would still have served to protect it.

Acanthodes

GROUP Acanthodians
DATE Early Carboniferous to Middle Permian
SIZE Usually 20cm (8in) long; a few 2m (6½ft) long
LOCATION Europe

Although a small, slender-bodied fish, *Acanthodes* was well protected from predators, with paired spines on the front edges of its pectoral and pelvic fins as well as projecting from its dorsal and anal fins. More specific characteristics included a single dorsal fin set far down its back near the base of its shark-like tail, and closely packed, roughly rectangular scales. It had no teeth, however. *Acanthodes* inhabited fresh water, where it lived as a fast-swimming filter-feeder.

Xenacanthus

GROUP Chondrichthyans
DATE Late Carboniferous to Early Permian
SIZE 70cm (28in) long
LOCATION Europe, USA

Xenacanth sharks have been described mostly on the basis of isolated teeth found in marine and freshwater rocks from all over the world. As in all other shark groups, two pairs of fins projected from the belly region: the pectoral fins closest to the head, and the pelvic fins closest to the tail. These acted as hydrofoils, lifting the shark as it moved through the water. The male had a pair of rod-like claspers projecting backwards from the pelvic fins, used to deposit sperm directly into the female's body during mating. The final fin on the underside of *Xenacanthus* was the anal fin, which was divided into two parts. Its tail was completely in line with the rest of the body and the upper and lower lobes were about the same size.

HEAD SPINE
Xenacanthus's head spine is unusual in that it is attached to the back of the head instead of supporting the dorsal fin. It was probably a defensive structure, and may well have had poison glands associated with it.

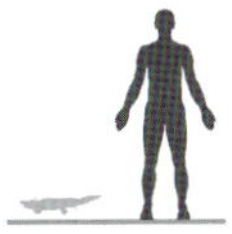

NARROW JAWS
With its narrow jawline and crocodile-like proportions, *Archegosaurus* was probably primarily a fish-eater, feeding on species such as *Acanthodes* in the waters of deep lakes.

Archegosaurus

GROUP Temnospondyls
DATE Early Permian
SIZE 1.5m (5ft) long
LOCATION Germany

Archegosaurus was discovered in 1847 and was one of the first Permian amphibians to be described by scientists. Its name means "founder of the reptile race", and it resembled a small, narrow-snouted crocodile. Almost all *Archegosaurus* fossils come from deposits that formed on the bed of a deep lake. Modern narrow-snouted crocodiles such as gharials are fish-eaters, and *Archegosaurus* probably hunted in the same way. Some archegosaur specimens have been found with *Acanthodes* bones inside their bodies. When *Archegosaurus* was first found, it was the earliest animal known to have legs (and toes); however, it shared many features with fish, which led some scientists to suggest it was a transition species. Today, *Archegosaurus* is regarded as an early amphibian.

Eryops

GROUP Temnospondyls
DATE Late Carboniferous to Early Permian
SIZE 2m (6½ft) long
LOCATION North America

Eryops is one of the best-known of the temnospondyl amphibians. A large number of specimens have been collected, and because the animal was large and robust these specimens are remarkably well preserved. *Eryops* was also unusual among early amphibians in that the entire skeleton was ossified (made from bone), which is far more readily fossilized. Its contemporaries had more cartilage in their skeletons, and because these were not mineralized, their complete anatomy has not survived. *Eryops* first appeared in the Late Carboniferous of North America, but most specimens that have been found come from the Early Permian. *Eryops* was one of the top predators in its ecosystem. It resembled a short, fat alligator and almost certainly fed on fish and small amphibians. The long rectangular muzzle is the basis of its name, which means "drawn-out face". *Eryops* probably underwent a change of diet during the Permian. Early specimens have a muzzle shaped like a rounded shovel with small, almost marginal teeth suitable for seizing small fish. However, later *Eryops*, including the one shown here, had a longer, more rectangular snout with clumps of larger teeth at intervals, much like those found in modern crocodiles, which feed on larger animals caught at waterholes and in rivers.

KEY SITE

TEXAS RED BEDS

Since the 1870s, the major source of Early Permian amphibian and reptile fossils has been a series of formations known as the Texas Red Beds. They extend across northeast Texas into Oklahoma and represent flood deposits, pools, lake-beds, and river accumulations on a tropical, coastal flood plain. The fauna they contain is rich and includes a range of aquatic, amphibious, and terrestrial animals, many of which have been found as complete skeletons. Fossils that have been found include the shark *Xenacanthus*; amphibians *Diplocaulus*, *Seymouria*, and *Eryops*; and the reptile *Captorhinus*.

MISSING RIBCAGE
Like modern salamanders, and unlike reptiles and mammals, the ribs of *Eryops* were short, straight rods that did not wrap around the chest to form a ribcage. The ribs were mainly for body support and not used for filling the lungs with air. Like salamanders, *Eryops* would have filled its huge mouth with air and pushed it back into the lungs.

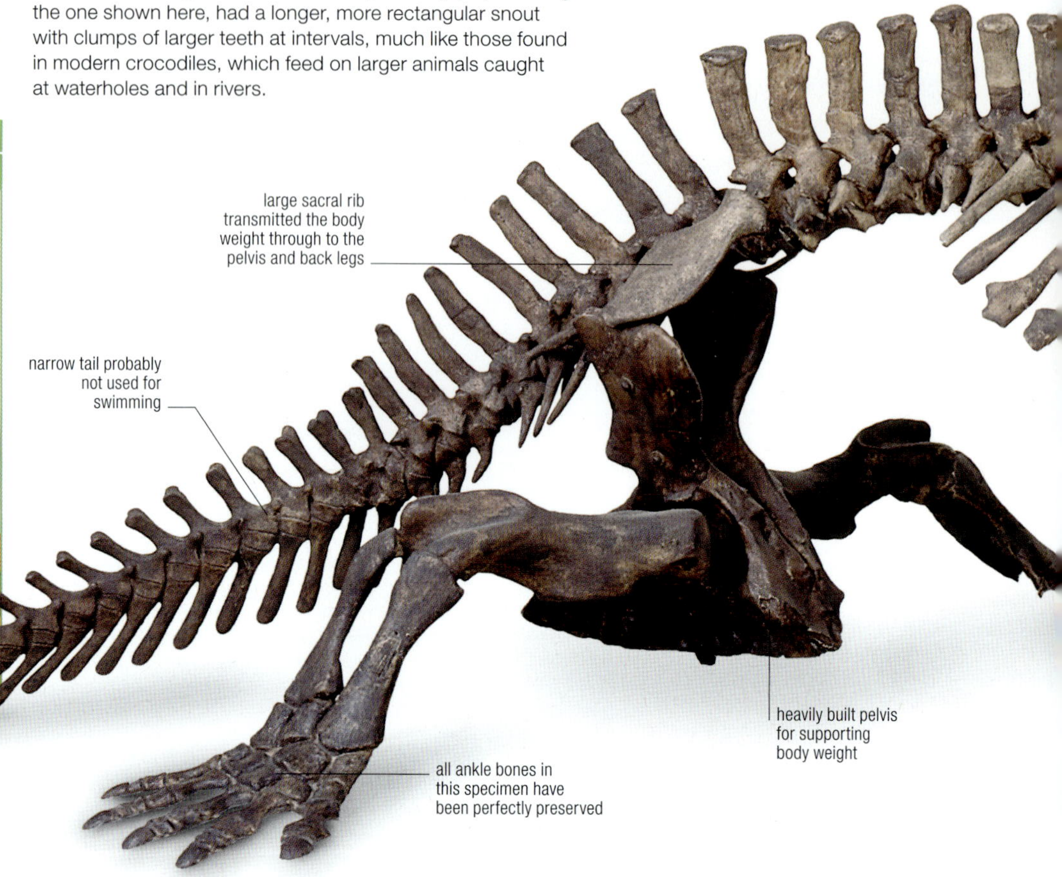

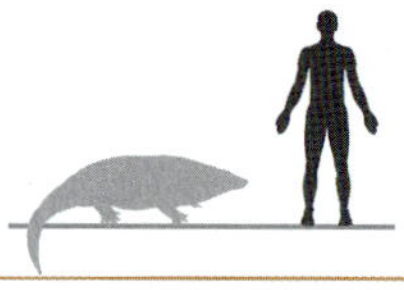

Sclerocephalus

GROUP Temnospondyls
DATE Early Permian
SIZE 1.5m (5ft) long
LOCATION Germany, Czech Republic

Sclerocephalus was a smaller relative of *Eryops* (above). It lived in large, shallow, tropical alpine lakes that developed between rows of mountains across what is now Central Europe. In some cases the water became stagnant, and the large adults and small larvae of *Sclerocephalus* died together in large numbers. This fossilized specimen is a part-grown juvenile.

WELL PRESERVED
In this fossil, the outline of the body is seen as a black shape on the rock. When the animal died it was buried quickly, along with the bacteria that were decomposing it. They were reduced to a black coal-like film, which solidified.

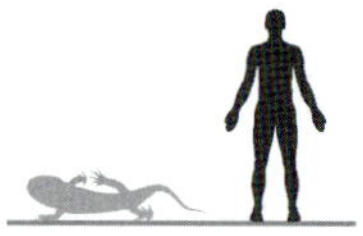

Seymouria

GROUP Tetrapods
DATE Early Permian
SIZE 80cm (31in) long
LOCATION USA, Germany

Seymouria was a small carnivore that, for many years, was considered to be close to the amphibian–reptile transition. There was much debate as to whether it was a reptile or not until it was noticed that its smaller relative, *Discosauriscus* (right), had external gills and lateral-line canals in its juvenile stage. This pointed to *Seymouria* being pre-reptile, and it is now considered to be a primitive amphibian tetrapod that paralleled reptiles in some of its adaptations to terrestrial life.

For many years, **scientists** thought *Seymouria* was a **primitive reptile.**

THICK HEAD
Male *Seymouria* had thick skulls that may have been used to butt rivals during mating contests. Although they probably spent much of their time on land, the female would almost certainly have returned to the water to lay her eggs.

Unusually among amphibians, the entire skeleton of *Eryops* was made of bone, which means that complete specimens have been preserved.

Diplocaulus

GROUP Lepospondyls
DATE Early Permian
SIZE 1m (3ft 3in) long
LOCATION USA, Morocco

Diplocaulus is one of the most common fossils found in the Texas Red Beds (left), but it has no living relatives. From the neck backwards, it resembles a fat salamander, but its skull is a strange boomerang shape, with long, flattened "horns" on each side. One idea is that these horns acted like hydrofoils – by using them to manipulate the water current *Diplocaulus* could gain lift or downward thrust, which enabled it to rise and sink with little effort.

GROWING HORNS
A range of skulls have been found that show how the "horns" developed as *Diplocaulus* grew. In young animals they are short and point backwards, but they suddenly grew outwards when the animal was half-grown.

Discosauriscus

GROUP Tetrapods
DATE Early Permian
SIZE 40cm ($15^1/_2$in) long
LOCATION Czech Republic, Germany, France

Discosauriscus was a smaller relative of *Seymouria* (left). Unlike its larger cousin, *Discosauriscus* seems to have been aquatic throughout its life, having external gills when it was young and lateral-line canals throughout its life. Most specimens of *Discosauriscus* have been found in the Czech Republic, in an area where the surface rocks represent a shallow Early Permian lake that became stagnant, killing all the fish and amphibians in it. Thousands of *Discosauriscus* have been collected from this area, making it one of the most common fossil amphibians in existence.

LITTLE-DISC LIZARD
Each vertebra has a small disc-like structure called a pleurocentrum. These gave *Discosauriscus* its name, which means "little disc-lizard".

Orobates

GROUP Reptiliomorphs
DATE Early Permian
SIZE 1m (3ft 3in) long
LOCATION Germany

Orobates was one of the first, most primitive vertebrate herbivores. It is known from a few complete skeletons found near the village of Bromacker in Germany. The Bromacker finds represent one of the first truly terrestrial reptile and amphibian communities found in the fossil record. The name *Orobates* means "mountain walker" and this animal is believed to have lived in rugged, terrestrial habitats. *Orobates* is the most primitive member of the family Diadectidae. Diadectids were all herbivores and large for their time. They had broad back teeth for side-to-side chewing of vegetation, and longer front teeth for stripping foliage off branches.

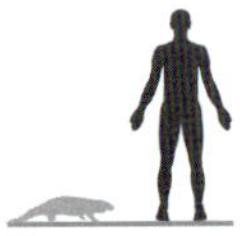

MOUNTAIN WALKER
With its broad feet and stocky limbs, *Orobates* is thought to have been able to cope with rugged terrain as it foraged for the vegetation that made up its diet.

Captorhinus

GROUP Eureptiles
DATE Early Permian
SIZE 50cm (19½in) long
LOCATION USA

Captorhinus was a primitive reptile with batteries of crushing teeth that were continuously replaced. It could have been an omnivore, feeding on smaller animals but also capable of grinding leaves or stripping twigs. Later Permian captorhinids were almost certainly plant-eaters.

LIZARD-LIKE SKULL
The skull shape of *Captorhinus* was not unlike that of a modern lizard, which it also resembled in body type.

MID-SIZED HUNTER
Although not large compared with many dinosaur predators, *Varanops* was nonetheless roughly the size of some modern monitor lizards, with similar predatory behaviour.

Mesosaurus

GROUP Parareptiles
DATE Early Permian
SIZE 1m (3ft 3in) long
LOCATION South Africa, South America

Mesosaurus was the first aquatic reptile. Vertebrates adapted for life on land (amniotes) evolved in the Late Carboniferous (see p.163). About 20 million years later, one group of amniotes, including *Mesosaurus,* re-adapted to life in the water. *Mesosaurus* was a reptile, so had waterproof, probably scaly, skin and long limbs. It had to come to the surface to breathe, as it had lungs instead of gills. Its tail was long and flat like a crocodile's, and its feet were large and could be used in swimming. Almost 200 long, needle-shaped teeth trapped small fish and crustaceans as *Mesosaurus* swept its long, narrow snout rapidly sideways through the water to capture prey.

BIOGRAPHY
ALFRED WEGENER

German meteorologist Alfred Wegener (1880–1930) first published the theory of continental drift in 1915. He observed that the coastline of West Africa fitted neatly into South America's east coast, proposing that Earth's land masses were once joined in one vast supercontinent, Pangaea, before separating about 300 million years ago. *Mesosaurus* fossils helped support his theories (see p.21).

STREAMLINED BODY
With its long, streamlined body and powerful tail, *Mesosaurus* would have been a strong swimmer, yet its relatively small teeth meant that it probably survived on equally small prey.

Scutosaurus

GROUP Parareptiles
DATE Late Permian
SIZE 2m (6½ft) long
LOCATION Russia

Scutosaurus was a large herbivore belonging to the pareiasaur group. All pareiasaurs had long, thick ribs forming a barrel-shaped chest. The large torso contained an extensive digestive tract to break down tough plant material; blunt teeth were capable of cropping vegetation, not chewing it. Its skull was broad and flat, and short, down-pointing tusks hung from its lower jaw and the sides of its face.

PILLAR-LIKE LEGS
Although its name means "scute lizard", due to the small, bony plates embedded in its skin, *Scutosaurus*'s legs were not lizard-like, but nearly vertical, like pillars.

Varanops

GROUP Synapsids
DATE Early Permian
SIZE 1.2m (4ft) long
LOCATION USA, Russia, southern hemisphere

Compared to other pelycosaurs, *Varanops* and its relatives were active and agile, with long limbs and sharp, backwards-curved teeth. They were voracious predators and also the most widely distributed of the primitive synapsids. Their remains have been found in the USA, Russia, and across the countries of the southern hemisphere. They outlasted even the primitive synapsids known as caseids, and *Elliotsmithia*, a relative of *Varanops*, with a primitive, lizard-like body plan, would have been an unusual sight among the Late Permian, mammal-like therapsids.

In 2006, a skeleton of *Varanops* was reported to have bite marks, and scientists theorized that its remains had been chewed on by a scavenger before being buried and fossilized. The shape of the bite marks shows the scavenger to have been a large temnospondyl amphibian. This is the oldest evidence of scavenging among land vertebrates.

220 **The number of sharp, curved teeth belonging to *Varanosaurus*, a relative of *Varanops*.**

FEARSOME FANGS
This fossil *Dimetrodon* skull shows the features of a deadly predator. The hole behind the eye socket for jaw muscle attachment is a feature shared by mammals.

This **successful predator** is one of the **commonest fossils of its age.**

DANGEROUS PREDATOR
Dimetrodon had a large sail on its back, which was supported by rod-like extensions of the vertebrae. This evolved separately from the sail of *Edaphosaurus* (see p.191) and lacked side branches.

Dimetrodon

GROUP Synapsids
DATE Early Permian
SIZE 3.2m (10½ft) long
LOCATION Germany, USA

Dimetrodon is often included in dinosaur books, and thus commonly mistaken for a dinosaur. It is an early synapsid, and in fact more closely related to mammals, with which it shares some anatomical features, such as a single hole in the skull behind the eye. It lived in the Early Permian, at least 40 million years before the first dinosaurs and lacks dinosaur features, such as upright legs. *Dimetrodon* was among the largest carnivorous pelycosaurs (see panel, right) and was the most fearsome predator of its time. Its high skull and relatively short snout gave it a powerful bite. *Dimetrodon* means "two-sized tooth" and refers to the two sizes of teeth in the jaw. The two canine-like fangs in the upper jaw were roughly located in the position of canine teeth of modern mammals, such as dogs and humans, and their roots extended through the bone all the way up the side of the face. Although they are not as striking as the sabre-teeth of big cats, they were powerful and deadly. In all *Dimetrodon* had 80 large, pointed teeth, which show that it was specialized for killing other large land vertebrates.

MAKING TRACKS
These five-toed trackways may have been made by *Dimetrodon* – which was one of the commonest animals in its ecosystem. However, it is difficult to assign tracks to their makers as body fossils are almost never found with the footprints.

ANATOMY
SAIL-BACKED REPTILE

Pelycosaur, which means "sail reptile", is an informal name for early synapsids, although not all pelycosaurs had sails. *Dimetrodon*'s sail helped to increase its body's surface area so that it could warm up more quickly in the sun. The cold-blooded *Dimetrodon* would have been sluggish in the morning before basking, and the sail may have helped it to warm up before its prey did.

CISURALIAN: Asselian, Sakmarian, Artinskian, Kungurian
GUADLUPIAN: Roadian, Wordian, Capitanian
LOPINGIAN: Whuchipingian, Chandhsingian

Moschops

GROUP Synapsids
DATE Late Permian
SIZE 2.5m (8¼ft) long
LOCATION South Africa

Moschops was a thick-skulled herbivore with stout legs and thick ribs that formed a large, barrel-shaped chest. It had a very large gut for digesting tough plant matter. Overall, *Moschops* was a heavily built animal and full-grown adults probably had little to fear from the predators of the day. The dinocephalians, which included *Moschops*, were one of the earliest therapsid groups to evolve and were very common in the first part of the Late Permian. They diversified to exploit many of the ecological roles formerly carried out by pelycosaurs (see p.189). Carnivorous and herbivorous dinocephalians are known, but all of them have peculiar front teeth. Unlike those of other animals, the upper incisors do not overlap the lower incisors, but interlock with them, allowing effective nipping.

ANATOMY

THICK SKULL

The skull of *Moschops* was astonishingly thick. In a 30cm- (12in-) long skull, the bone of the top surface is more than 10cm (4in) thick. This was an adaptation to allow ritualized head-butting, possibly in contests between males competing for mates, as in modern-day bighorn sheep. The neck joint and vertebrae were adapted for transmitting the force of the blow without resulting in whiplash. One old theory was that thickening of the skull occurred due to overactive pituitary glands. This was supposed to lead to blindness as the space for the eye was reduced. Today, skull thickening is recognized as an adaptation to social behaviour.

EARLY HERDS
Moschops may have formed small social groups or herds. Indeed, discoveries have been made of several individuals fossilized together; the thick skull is probably an adaptation for social behaviour, such as ritualized combat.

Eothyris

GROUP Synapsids
DATE Early Permian
SIZE Skull 6cm (2¼in) long
LOCATION USA

The mysterious *Eothyris* is only known from a single broad, flat skull described in 1937. It is distinctive due to two large, fang-like teeth on each side of the upper jaw. The reason for this peculiar arrangement is not known. The remaining teeth were also sharply pointed, so *Eothyris* was a carnivore, although the prey must have been small because *Eothyris* was shorter than 30cm (12in). It is possible that it hunted insects and other vertebrates smaller than itself. Surprisingly, *Eothyris* is related to the giant herbivore *Cotylorhynchus* (see below).

The short, broad skull of *Eothyris* gave it a rapid, snapping bite.

PROMINENT FANGS
Eothyris was a small, lizard-like, early synapsid with peculiar double-fangs in the upper jaw.

Cotylorhynchus

GROUP Synapsids
DATE Early Permian
SIZE 4m (13ft) long
LOCATION USA

Cotylorhynchus was the largest of the caseids, a specialized group of herbivorous pelycosaurs. It had the smallest skull relative to its body size of any synapsid. Although *Cotylorhynchus* had blunt, expanded teeth for crushing tough plants, most of the digestive work was done by the enormous gut, held within a barrel-shaped ribcage. Caseids were highly successful and survived longer than most other pelycosaurs. They were still living in the Late Permian when therapsids had become dominant.

Edaphosaurus

GROUP Synapsids
DATE Late Carboniferous to Early Permian
SIZE 3.3m (10¾ft) long
LOCATION Czech and Slovak Republics, Germany, USA

Edaphosaurus was a herbivorous pelycosaur with numerous blunt teeth that formed crushing plates. The teeth were not just around the jaws but also on the roof of the mouth and sides of the lower jawbone. Although it was herbivorous, *Edaphosaurus* was not related to *Cotylorhynchus* (see left). One of the most distinctive features of *Edaphosaurus* is the "sail" crest on its back. Unlike the sail of *Dimetrodon* (see p.189), small prongs branched off from the main rods that supported the sail, giving them the appearance of twigs. Despite this difference, the sails of *Dimetrodon* and *Edaphosaurus* probably evolved for the same purpose – trapping heat while basking. This meant that *Edaphosaurus* could warm its body quickly in the morning sun and become active before many predators.

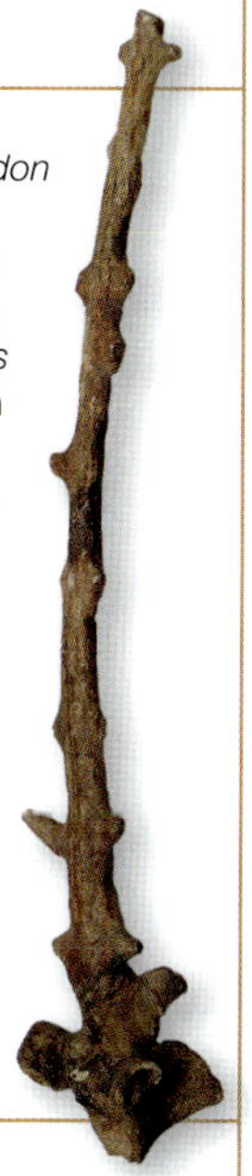

BACK SAIL
The neural spines of *Edaphosaurus's* backbone extended upwards. The spines were covered in skin, producing the characteristic "sail" on its back. The spines also had smaller prongs branching off.

Robertia

GROUP Synapsids
DATE Late Permian
SIZE 42cm (16½in) long
LOCATION South Africa

NEARLY TOOTHLESS
Apart from the canines, *Robertia* lacked teeth. Instead it had a horny beak, like that of a turtle. The skeleton was adapted to dig for food.

The dicynodonts were an extraordinarily diverse group of specialized herbivores that survived for longer than all other therapsid groups except the cynodonts. Dicynodonts continued into the Late Triassic, while cynodonts evolved into today's mammals. The earliest dicynodont known from good fossils is *Robertia*. It was relatively small, about the size of a house cat. Despite its early appearance it shows recognizable dicynodont features. For example, it has two large canine tusks, although the other teeth are absent. It is likely that the canine tusks were used for digging.

Procynosuchus

GROUP Synapsids
DATE Late Permian
SIZE 50cm (19½in) long
LOCATION South Africa, Zambia

Procynosuchus is one of the earliest cynodonts, a group of therapsids that gave rise to mammals. It shows the first signs of mammal-like cheek teeth, differentiated into premolars and molars. Other therapsids, and most reptiles, have only one kind of tooth in the side of the jaw. Despite its advanced teeth, the body was still primitive.

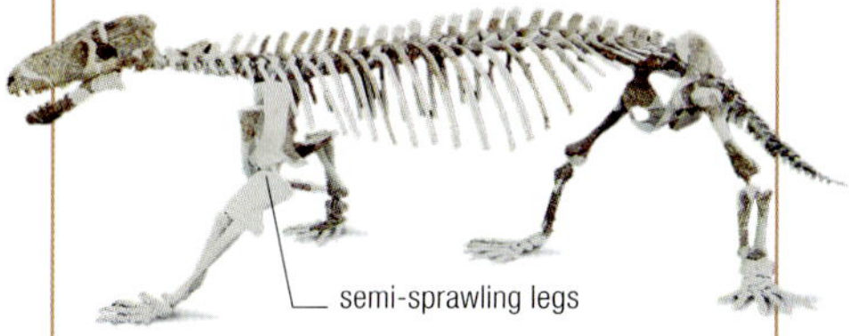

PRIMITIVE BODY
Procynosuchus had a long tail and a more reptilian, semi-sprawling stance.

Pelanomodon

GROUP Synapsids
DATE Late Permian
SIZE 1m (3ft 3in) long
LOCATION South Africa

Dicynodonts such as *Pelanomodon* were stocky, pig-like animals with shorter tails than most other early synapsids. Their jaw mechanism was unusual, with a sliding jaw joint, and jaw-closing muscles located behind the jaw and attached around the rim of the skull opening behind the eye (temporal fenestra). As well as chewing food by moving the jaw up and down, like humans and many other animals, dicynodonts could also move the jaw backwards with each chewing stroke. This produced a powerful grinding motion that pulverized the tough plant material on which they fed. This adaptation must have been effective as dicynodonts were one of the most successful groups of herbivores in Earth's history. At about 40cm (15½in) long, *Diictodon* (see far right) was a small dicynodont. It made corkscrew-shaped burrows over 50cm (19½in) deep that have also been found fossilized. These become wider as they coil downwards, and end in an expanded chamber. They are remarkably similar to those made 220 million years later by the primitive beaver *Palaeocastor* (see p.408). *Pelanomodon* was a more typical, non-burrowing dicynodont, but may have dug in the ground for nutritious roots and tubers to eat. The remains of some dicynodonts are so common that they are used by geologists to indicate the age of the rocks around them. Dating rocks based on the fossils they contain is known as biostratigraphy (see pp.39 and 44). Both *Pelanomodon* and *Diictodon* occur in the *Cistecephalus* Assemblage Zone, a layer of rock named after a particularly abundant dicynodont. Because *Cistecephalus*, *Diictodon*, and *Pelanomodon* were only around for a few million years, the rock strata in which they are discovered is easily assigned. Slightly older rocks belong in the *Tropidostoma* Assemblage Zone and younger rocks are in the *Dicynodon* Assemblage Zone.

Dicynodont fossils are so abundant that they are used to date the rock in which they are found.

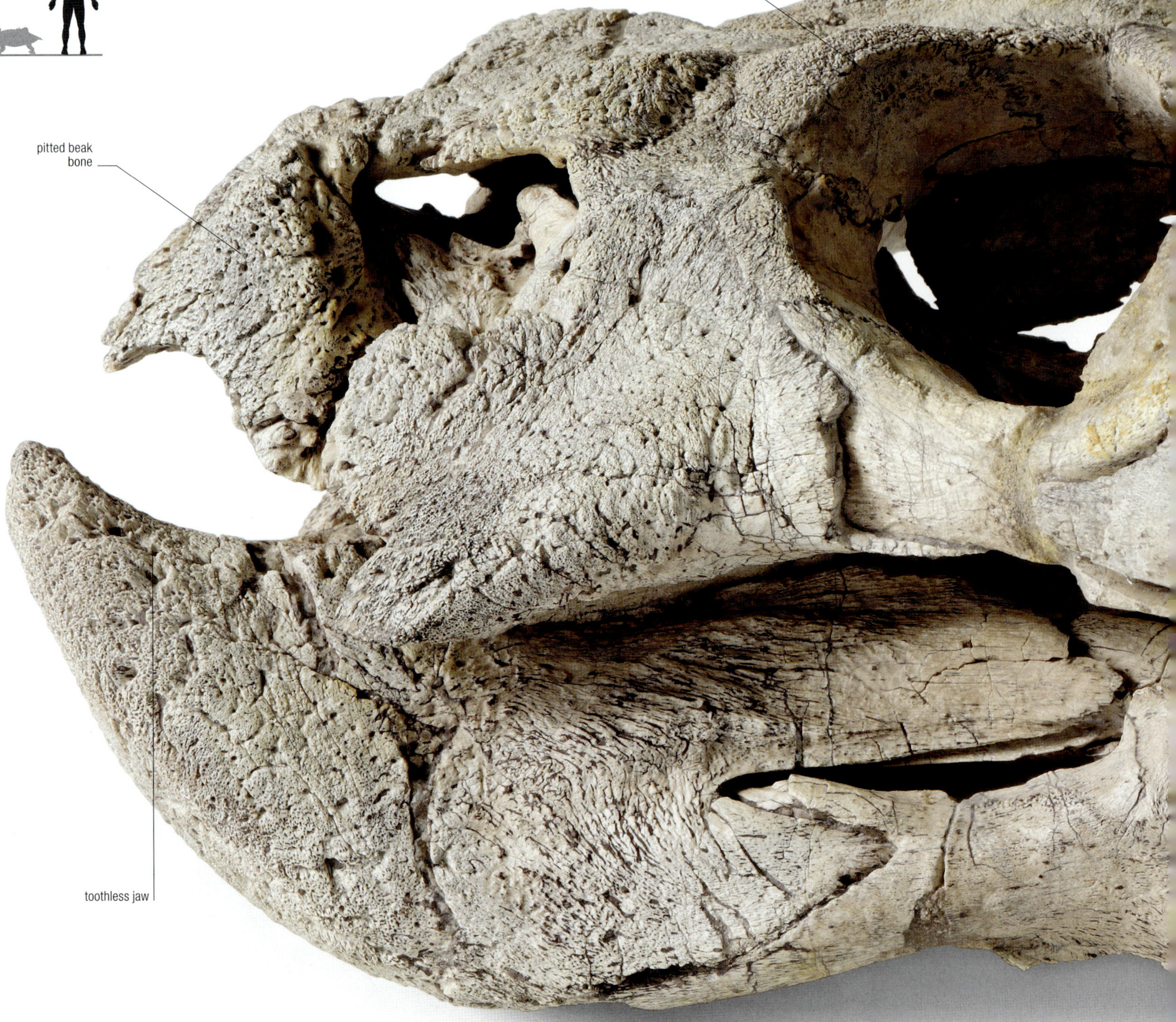

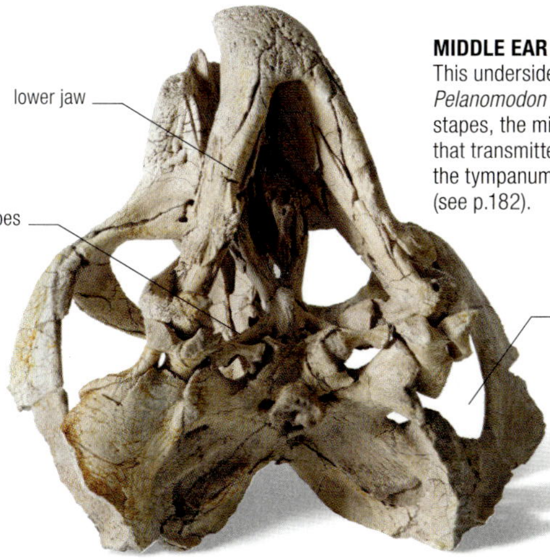

MIDDLE EAR BONE
This underside view of a *Pelanomodon* skull reveals the stapes, the middle ear bone that transmitted vibrations from the tympanum to the inner ear (see p.182).

TUSKED DICYNODONT
This skull belongs to a tusked dicynodont called *Diictodon.* Although it died over 250 million years ago, the skull is in such good condition it looks as if it died recently.

TURTLE-LIKE BEAK
Pelanomodon was a tuskless dicynodont that had a turtle-like beak for cropping plants. This fossil is remarkably well-preserved and shows the pitted texture of the beak bone, which is due to the blood vessels that supplied nutrients to it.

TRIASSIC

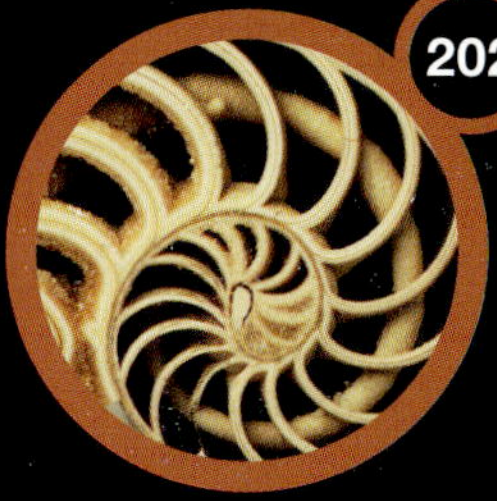

Triassic

The Triassic Period was the time of recovery after the devastation of the end-Permian faunal and floral crisis. This was also a period of very high global temperatures. Into the empty ecosystems on land, dinosaurs and mammals evolved, and were able to spread rapidly. In the sea, modern corals appeared, while sea urchins began to diversify after having almost become extinct at the end of the Permian.

TRIASSIC SEDIMENTARY ROCKS
Newark Group sedimentary rocks form outcrops along the North American east coast stretching from South Carolina to Nova Scotia. Red beds laid down in streams alternate with thinner black deposits from lake floors.

Oceans and continents

The world's land masses were still combined in the supercontinent of Pangaea, which extended from pole to pole. This reached its maximum extent around the time of the Middle to Late Triassic boundary when, due to lower sea levels, Earth's land area was greater. Throughout the Triassic, Pangaea was moving steadily northwards and, as a consequence, Siberia moved to the North Pole. A narrow sea which during the Permian had separated Europe and Kazakhstania, closed causing further mountain-building in the Urals during the Early Triassic. As Pangaea moved north, it also rotated anticlockwise, taking North and South China in a northerly direction. Meanwhile, Cimmeria moved north, crossing the Equator, widening the rift between it and the southeastern part of Pangaea, and expanding the Tethys Ocean. The Paleo-Tethys Ocean began to contract as Cimmeria spread north at a faster pace than the eastern extension of Pangaea that is now North China. Pangaea, comprising all the component parts of Gondwana, still extended close to the South Pole. By the end of the Triassic, the world map was changing. Pangaea began to break up, a process that was to continue through the rest of the Mesozoic into the Cenozoic. Triassic life on land and in the sea is preserved in some spectacular fossil localities. Early Triassic land vertebrates are found in the Urals in Russia and the Karoo Basin, South Africa, while the San Giorgio beds in Switzerland and Italy contain beautifully preserved aquatic reptiles and ammonites from the Middle Triassic. The agile dinosaur *Coelophysis* would have lived on river floodplains in the Late Triassic.

NEWARK GROUP RED BEDS
These deep beds of conglomerate rock made up of non-marine sediments formed as Pangaea began to break up. They contain ripple marks, mud cracks, and even the occasional dinosaur footprint.

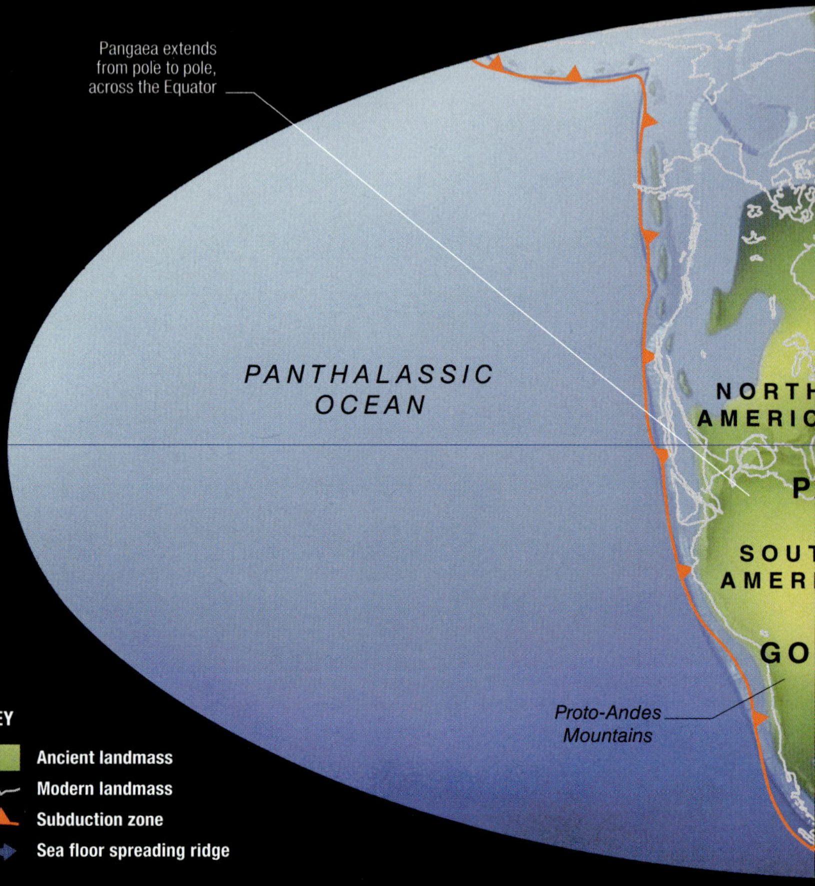

TRIASSIC WORLD MAP
Pangaea reached its maximum extent around the Middle to Late Triassic boundary. By the end of the Triassic, it was beginning to break up.

Gondwana begins to fragment, separating the southern tip of South America from southern Africa and western Antarctica

EARLY		MIDDLE
MYA 250	240	230

MICROSCOPIC LIFE

PLANTS
250–220 Sphenopsid and lycophyte plants decrease in diversity; *Dicroidium* flora diversifies

DICROIDIUM

PACHYPTERIS

INVERTEBRATES
- 245 First phylloceratine ammonoids; first isocrinids (type of sea lily)
- 240 First scleractinian corals; last goniatites; increase in diversity of ophiuroids; first pteriid (left) and cuspidariid bivalves
- 235 First ostreid bivalves

OXYTOMA

VERTEBRATES
250–220 mammal-like reptiles (left) increase in body size
- 240 Increase in diversity of gomphodont cynodonts and rhynchosaurs

CYNOGNATHUS

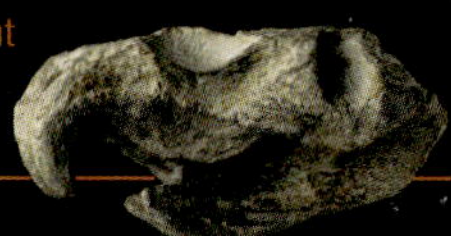
HYPERODAPEDON

SCALE-LEAF CONIFER
The vegetation in the Triassic was adapated to dry conditions. The flora included scale-leaved conifers, similar to this Lawson cypress, and cuticle-thickened seed ferns.

Europe and Kazakhstania join, as Siberia moves to the North Pole

North and South China move northwards as Pangaea rotates anticlockwise

Rift between Cimmeria and Pangaea widens, enlarging Tethys Ocean

Climate

After the end of the Permian Period, rapid global warming in the Early Triassic produced one of the hottest periods in Earth's history. Some estimates put subtropical marine surface temperatures as high as 38°C (100°F). During the Triassic, the large size of the supercontinent Pangaea gave rise to extreme climatic zonation, with some areas becoming very dry, whereas others were influenced by monsoonal conditions. One consequence of this zonation was the separation of plants into distinct northern and southern realms. However, this endemism (belonging to a particular place) was reduced through the Triassic. Climates were very warm, even in high latitudes, and both poles were ice free and were warm even in winter. The presence of coal deposits in high northern and high southern latitudes suggests that these regions were wetter than lower latitudes, probably being warm temperate. However, much of the continent was dry, especially in lower latitudes, and the arid-adapted nature of the vegetation reflects this. In the Early Triassic, arid areas extended to relatively high latitudes, perhaps as high as 50° both north and south of the Equator. During the Triassic the equatorial tropical belt grew and extended from Cimmeria northwest across the European and western Siberian parts of Pangaea. As it did so the arid belts narrowed, the northern and southern warm temperate belts expanding into lower latitudes, especially in the southern hemisphere.

MONTE SAN GIORGIO FAUNA
Fossils found preserved in this pyramid-shaped mountain, on Lake Lugano in Canton Ticino, Switzerland, are the best-known record of life in a lagoon during the Middle Triassic.

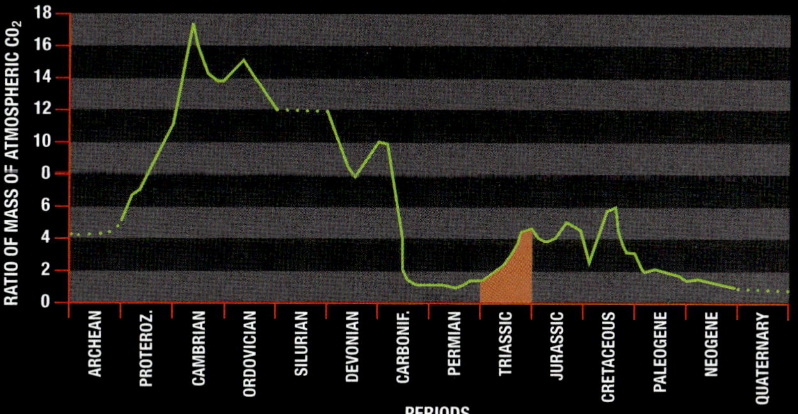

CARBON DIOXIDE LEVELS
The graph of carbon dioxide levels rises steeply during the Triassic, as Earth entered a greenhouse phase and global temperatures increased dramatically.

LATE

220 | 210 | 200 | MYA

MICROSCOPIC LIFE
225 First coccolithophores (single-celled marine planktonic algae)

PLANTS
225 Cycads, bennettitaleans, and conifers increase in diversity
215 Some plant extinctions

INVERTEBRATES
225 First cardiid bivalves

CARDINIA

220–200 Increase in diversity of echinoids
220 First flies
215 Last trepostome bryozoans; last orthocerid nautiloids (right); first pentacrinitids

CENOCERAS

210 First thraciid bivalves
200 Last ceratitid ammonoids

CERATITES

VERTEBRATES
225 First dinosaurs (prosauropods); first teleost (ray-finned bony) fish
220 First turtles (*Odontochelys*)
220–200 Ornithischian dinosaur diversity increases; diversity of mammal-like reptiles decreases; gastric processing herbivorous reptiles diversify
215 First mammals, e.g. *Eozostrodon*; some vertebrate extinctions
200 Major extinction of terrestrial vertebrates and many large amphibians

TRIASSIC PLANTS

The transition from the Permian to Triassic saw the demise of glossopterids, as well as other more ancient groups of plants from the Paleozoic. However, the Triassic brought further modernization of the flora, as well as the appearance and diversification of new groups of ferns and seed plants.

Relatively little is known about the plants of the Early Triassic, but rich floras of the Late Triassic, for example from Greenland and South Africa, contain new kinds of ferns and many new groups of seed plants. Much research needs to be done to better understand these ancient seed plants, including clarifying how they relate to each other, and their possible living relatives.

PETRIFIED FOREST
Massive fossil logs in the Petrified Forest, Arizona, USA, which have been almost entirely replaced by silica, clearly show the great size attained by some of the ancient conifers of the Triassic.

New seed plants

Many new kinds of seed plant come into the fossil record for the first time during the Triassic. Studies of the Molteno flora of South Africa show a range of different reproductive structures for these plants. Many of them are difficult to understand in terms of the living plant groups to which they may be closely related. Especially prominent among the Triassic seed plants are corystosperms, which are particularly diverse in the southern hemisphere. Bennettitaleans are another important group of seed plants to appear for the first time in the Triassic. Their leaves are superficially similar to those of living cycads, but the reproductive structures of the two groups are quite different. Those of cycads are simple cones, while those of bennettitaleans are often flower-like. Some bennettitalean "flowers" produced only pollen or seeds, and a few were bisexual.

Modern seed plants and ferns

Cycads and ferns continued to diversify through the Triassic. This period also sees the first reliable evidence of the ginkgo lineage. Conifer-like leafy shoots are common in Triassic floras. Some of these plants are clearly related to living conifers, but the relationships of others are much less certain. Various groups of plants during the Triassic may have had conifer-like foliage but very different kinds of reproductive structures.

NORFOLK ISLAND PINE
The living conifer commonly known as the Norfolk Island pine (*Araucaria heterophylla*) has leafy shoots that are similar to those of early conifers.

GROUPS OVERVIEW

Triassic plants looked quite different from those of the Permian. Conifers were still very prominent during the Triassic, but they were joined by different kinds of seed plants that often had more or less pinnate leaves. During the Triassic, ferns also began to take on a more modern aspect.

FERNS
The Triassic saw a new wave of fern evolution. Earlier forms from the Paleozoic, such as the royal ferns and marattialeans, were joined for the first time by families of so-called primitive leptosporangiate ferns, such as Dipteridaceae.

CONIFERS
The more ancient conifers of the Permian were replaced by more diverse conifers during the Triassic. These voltzialean conifers show the next steps in the evolution of modern conifer cones from the more lax reproductive structures that were seen in the late Paleozoic.

CYCADS
Cycads survived the extinction that occurred at the end of the Permian, and by the Late Triassic they were diverse and also geographically widespread. They are important in Jurassic vegetation from all over the world.

OTHER SEED PLANTS
The seed-plant group that is most characteristic of the Triassic is the corystosperms. They are especially important in the southern hemisphere. Other seed-plant groups that appeared for the first time during the Triassic include ginkgos, bennettitaleans, and caytonialeans.

Hexagonocaulon

GROUP Bryophytes
DATE Middle Triassic
SIZE 10cm (4in) across
LOCATION Southern hemisphere

This simple plant is known from several locations in the southern hemisphere, including Triassic rocks in the Antarctic Peninsula. *Hexagonocaulon* was a flat-bodied, or thallose, liverwort, similar to many liverwort species alive today. It lacked leaves and roots, and anchored itself with fine hairs, or rhizoids, which formed tufts along its underside. Like modern liverworts *Hexagonocaulon* did not have any internal vessels for conducting water. As a result, the plant was restricted to damp habitats. Most plant fossils are formed by the sporophyte, or spore-producing phase in the life cycle. The intervening phase, known as the gametophyte, is typically much smaller, and is rarely found as a fossil. In mosses and liverworts this situation is reversed. The fossil shown here represents *Hexagonocaulon*'s dominant gametophyte form. The smaller sporophyte grew on the female gametophyte, where it released spores. The spores grew into new gametophyte individuals.

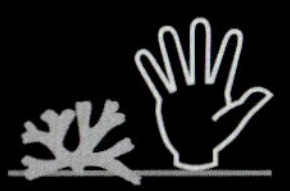

HEXAGONOCAULON MINUTUM
The entire plant was only a few cells thick and formed a flat thallus covering the ground. A thallus is a simple plant body without an obvious stem or tip.

Pleuromeia

GROUP Lycophytes
DATE Triassic
SIZE 2m (6½ft) high
LOCATION Worldwide

At the beginning of the Triassic, Earth's vegetation was emerging from the greatest of all mass extinctions. Many plants died out, but *Pleuromeia* benefited from the catastrophic change. Fossils show that it grew right across the world, appearing in a wide variety of different habitats left vacant by the disappearance of competing plants. *Pleuromeia* belonged to the lycophytes, a group that included the giant clubmosses that formed the coal swamps of the Carboniferous period (see p.144). Although *Pleuromeia* was a tree, it was built on a smaller scale. It had a single, unbranched trunk, topped by a tuft of grass-like leaves. Its root system consisted of four bulb-shaped lobes, connected to rootlets that fanned out through the soil. *Pleuromeia* reproduced with spores from cones. Some species produced several cones, but many had just one on top of the stem.

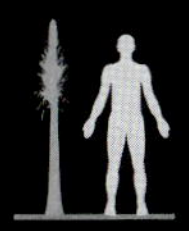

Fossils show that *Pleuromeia* grew right across the world in the Triassic.

PLEUROMEIA
As it grew, *Pleuromeia*'s oldest leaves were discarded, leaving a ruff of broken bases. As the tree grew, the ruff became smooth leaf scars covering the trunk.

Langcamia

GROUP Ferns
DATE Triassic
SIZE Leaf up to 1m (3ft 3in) long
LOCATION Europe, USA, North Vietnam

This fern is similar to *Cladophlebis* from the Jurassic (see p.228). It is not known how many times its leaf (frond) divides in total, but the specimen shown here is almost certainly only a portion of an entire frond. The segments of the frond, called pinnules, are about 6–10mm (1/4–3/8in) long and 4mm (3/16in) wide, with rounded tips. These pinnules have a primary vein that runs to their tips, as well as veins orientated sideways, that form a distinct network. This type of net venation is also characteristic of the Carboniferous and Permian primitive seed plant *Lonchopteris*.

LANGCAMIA SP.
Like most ferns, the fronds of *Langcamia* are divided into many parallel segments (pinnae), each of which have rounded, alternately attached pinnules.

Thaumatopteris

GROUP Ferns
DATE Late Triassic to Early Jurassic
SIZE 1m (3ft 3in) high
LOCATION Worldwide

Also known as *Dictyophyllum*, this common fossil plant is often used as a geological marker, helping to identify layers of rock that were formed at a given point in time. Some *Thaumatopteris* species straddle the Triassic–Jurassic boundary, but many fossils appear only at the point where the Jurassic period began. *Thaumatopteris* can be identified by the complex network of veins in its long fronds. The veins divide and rejoin, forming a mesh on the leaf surface.

THAUMATOPTERIS SCHENKII
The mesh of veins seen on *Thaumatopteris*'s leaves is unusual for ferns, and is more like the anatomy of flowering plants.

Dicroidium

GROUP Corystosperms
DATE Triassic
SIZE 4–30m (13–98ft) high
LOCATION Southern hemisphere

Dicroidium belonged to a group of plants called the corystosperms, which were seed plants of uncertain relationship found mainly in the southern hemisphere. They had fern-like leaves but produced seeds instead of spores. *Dicroidium* and other corystosperms evolved in Gondwana, which was the southern landmass of the supercontinent of Pangaea during the Triassic (see p.196). A fully grown *Dicroidium* would have been beyond the reach of ground-based herbivores, but all corystosperms would have been browsed by animals at some stage in their lives. These browsers included the Triassic dicynodont *Lystrosaurus* (see p.220), whose fossils have been found in the same rocks as corystosperm remains in Antarctica.

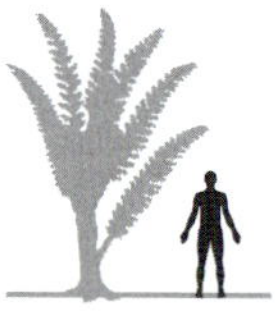

DICROIDIUM SP.
Like all corystosperms, *Dicroidium's* foliage was made up of pairs of leaflets growing either side of a central stalk. The stalks joined at Y-shaped forks, and a network of veins ran through the leaves.

Dicroidium's **woody stems** were thought to grow to just a few metres, but **recent finds** show that these plants could reach a height of **30m.**

Dioonitocarpidium

GROUP Cycads
DATE Late Triassic
SIZE Cone scale 5cm (2in) long
LOCATION Worldwide

The earliest fossils of cycads date back to the Permian period, but they became more common at the dawning of the Triassic. Typically shaped like squat palm trees, cycads such as *Dioonitocarpidium* reproduce by growing cones. Individual plants are either male or female. The cones of male cycads release pollen grains, which fertilize the female cones. The seeds then develop inside the female cone.

DIOONITOCARPIDIUM LILIENSTERNII
Dioonitocarpidium fossils are the scales from female cones. They have finely divided tips, and often show traces of seeds.

Baiera

GROUP Ginkgos
DATE Triassic to Jurassic
SIZE Leaf up to 15cm (6in) long
LOCATION Worldwide

Ginkgos are not only living fossils – they form part of an ancient group of flowerless seed plants called the gymnosperms, which also include conifers and cycads. They have a fossil history stretching back more than 200 million years. During this time, evolution has created countless variations on the theme of fan-shaped leaves. With its highly divided leaves, *Baiera* looks very different to its living relative, the well-known *Ginkgo* or maidenhair tree (see p.39).

BAIERA MUNSTERIANA
In living ginkgos, the fan-like leaf is almost complete, apart from a central notch, but in *Baiera* it was split into separate ribs.

Stachyotaxus

GROUP Conifers
DATE Late Triassic to Early Jurassic
SIZE 10m (33ft) high
LOCATION Northern hemisphere

A shrub or a small tree, *Stachyotaxus* was one of the most abundant conifers during the Late Triassic, but died out along with its relatives in the Jurassic. It was evergreen and had narrow leaves arranged in opposing ranks. Their narrow width made the leaves resistant to drought, in much the same way as the needles of today's conifers. *Stachyotaxus* produced seeds using cones. Pollen from male cones was spread by the wind and fertilized the female seed-bearing cones. The pollen grains were spherical and lacked the sacs or wing-like scales that pine pollen uses to drift in the wind.

STACHYOTAXUS SEPTENTRIONALIS
The female cones of *Stachyotaxus* were about 10cm (4in) long. Nothing is known about the structure of the male cone.

Voltzia

GROUP Conifers
DATE Triassic
SIZE 5m (16½ft) high
LOCATION Worldwide

The Triassic was a time of great change for conifers. In this period, most of the earlier conifers of the Paleozoic were dying out, and new ones emerged and diversified into the conifers that survive until the present day. The small tree or shrub *Voltzia* belonged to one of the early, now-extinct groups, but the structure of its cones shows that it had links to today's forms. *Voltzia*'s male cone had pollen cases arranged in a spiral around the central core. The seed-producing scales of the female cones were fused together. This was a continuation of the evolutionary trend in which cones developed originally from dwarf shoots. The *Voltzia* conifer is named after the Voltzia sandstone of northeast France, where it is a common find.

VOLTZIA COBURGENSIS
Voltzia coburgensis had short scales that hugged the stems, although some other species also had sprays of needles several centimetres long.

TRIASSIC INVERTEBRATES

The great Late Permian mass extinction cleared the field for many new kinds of invertebrate marine organisms – extinctions are both destructive and constructive. But the Triassic was itself terminated by yet another mass extinction, the fourth during the Phanerozoic.

Recovery from Permian extinctions

It took several million years for marine life to recover fully from the Permian catastrophe. Among the first post-extinction life forms were stromatolites – it was like a return to the Precambrian. Although Paleozoic corals, trilobites, and most brachiopod groups were gone for ever, some brachiopods, along with ammonoids, bivalves, gastropods, and other marine invertebrates had survived in refuges around the world. New kinds of invertebrates would soon join them.

TRIASSIC OYSTER
This is a fossilized Triassic oyster cemented to a rock surface. These specialized bivalves arose during the Triassic and similar species of living oyster remain very important today.

Inception of modern faunas

By the early Triassic, most elements of the Cambrian fauna had vanished, and the Paleozoic fauna had been severely damaged, especially with the loss of so many groups of brachiopods, whose role as dominant sea-floor filter feeders was taken over by bivalves. It was during the Triassic that the modern evolutionary fauna, which had begun in a small way during the Ordovician, became dominant. Paleozoic echinoids were replaced by new versions, and scleractinian corals, derived independently from soft-bodied sea anemones, appeared. Ceratitid ammonoids flourished, but these, like many other marine animals, died out during the important mass extinctions towards the end of the Triassic. It seems that there were several extinction events during the later Triassic, though their causes are not well understood. Possible causes are climatic change and falling sea levels, leading to habitat loss.

OXYTOMA FALLAX
This fossilized group of small Triassic bivalves consists of a single species. Such monospecific groups are typical of stressed environments.

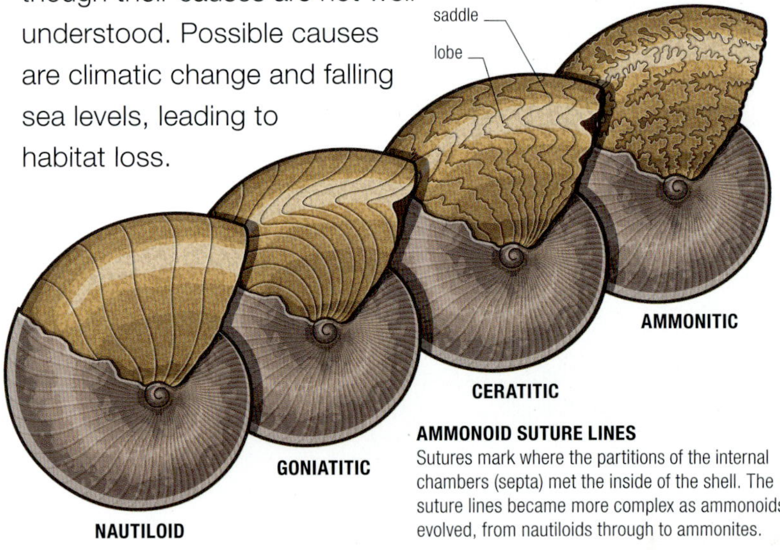

AMMONOID SUTURE LINES
Sutures mark where the partitions of the internal chambers (septa) met the inside of the shell. The suture lines became more complex as ammonoids evolved, from nautiloids through to ammonites.

GROUPS OVERVIEW

Groups of invertebrates that persisted through the Permian mass extinction into the Triassic continued to do well, with some groups of brachiopods and gastropods surviving to the present day. Bivalves flourished in the turbulent Triassic environments as they were able to adapt relatively quickly to the difficult, fluctuating conditions.

BRACHIOPODS
Following the Permian catastrophe only a few brachiopod groups, chiefly terebratulides and rhynchonellides, survived to the present day. Strophomenides died out at the end of the Triassic.

GASTROPODS
The earliest gastropods are Cambrian in age, and they could be considered as part of the Cambrian evolutionary fauna. They did well after the Permian mass extinction, persisted through the Triassic, and are an important component of the marine fauna today.

CERATITIDS
These are typical Triassic ammonoids. The suture line has forwardly directed semicircular "saddles", and backwardly-pointing frilled "lobes" (see above). Ceratitids died out at the end of the Triassic, and only a few ammonoid groups continued into the Jurassic.

BIVALVES
These became the important sea-floor shelled filter-feeders after the Permian, taking over from brachiopods like ships that pass in the night. Some Triassic bivalves are tolerant of changing salinity, evidently a factor in their survival in stressed conditions.

Arcticopora

GROUP Bryozoans
DATE Triassic
SIZE Colony 2mm (1/16in) across
LOCATION Europe, Asia, North America

This tiny bryozoan colony was made up of tubular structures called autozooecia, which housed many soft-bodied individuals (autozooids). Each autozooid fed using a lophophore – a ring of tentacles around the mouth. After feeding, the lophophores were withdrawn into the autozooecia for protection. In the outer parts of the colony, the autozooecia had thickened walls, and were partitioned with diaphragms. The colony walls contained calcite rods, which projected like spines above the surface of the colony.

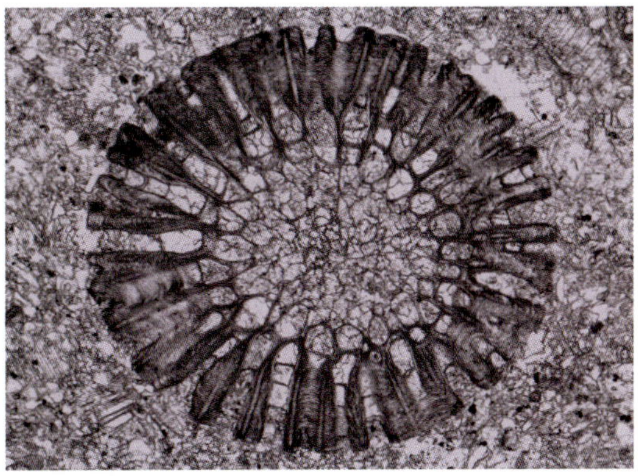

ARCTICOPORA CHRISTIEI
This specimen is a cross-section of an upright colony, showing the tubular autozooecia, which housed the feeding zooids.

Monophyllites

GROUP Cephalopods
DATE Middle to Late Triassic
SIZE 7–11cm (2¾–4½in) across
LOCATION Worldwide

Monophyllites was one of the earliest members of the phylloceratid ammonoids. This group evolved in the Early Triassic from *Meekoceras* (see below right) and survived to the end of the Cretaceous, changing little during this time. Phylloceratids are characterized by the leaf-shaped terminations of the forward-facing folds (saddles) of the suture lines. The suture lines mark where the walls (septa) of the internal chambers join the inside of the shell. Like most phylloceratids, *Monophyllites* has a disc-shaped form, with slightly overlapping whorls. Distribution patterns show that this group preferred open seas to more shallow water.

MONOPHYLLITES SPHAEROPHYLLUS
Like most other phylloceratids, *Monophyllites* has no pattern on the outside of its shell, except for strong growth lines, which are S-shaped.

Coenothyris

GROUP Brachiopods
DATE Triassic
SIZE 1–2cm (⅜–¾in) long
LOCATION Europe, Middle East

This small brachiopod had perforations in its upper shell that in life housed small bristles. The function of the bristles may have been sensory or to aid respiration. Inside the smaller (brachial) valve, a loop of calcite would have supported the lophophore, which was the animal's feeding organ. There is a small opening in the larger (pedicle) valve, through which a muscular stalk, the pedicle, would have extended to attach the animal to the sea bed. A thickened ring extends around the inside of the pedicle opening. The outer edge of the smaller valve is gently folded outwards; the larger valve has a corresponding infold. The main pattern on the shell is a series of growth lines, which are crossed by fine radial ribs. Some specimens of this genus have been found with bands of colour running across the valves – a feature rarely seen in fossils.

COENOTHYRIS VULGARIS
Both valves are convex, but the larger valve curves outwards slightly more than the smaller valve.

Preflorianites

GROUP Cephalopods
DATE Early Triassic
SIZE 2–5cm (¾–2in) across
LOCATION USA, Albania, Indonesia

Preflorianites was a ceratitid ammonoid from the Early Triassic. It had a simple ceratitic suture, meaning that the folded joints between the internal walls (septa) and the shell have a serrated pattern where they face away from the shell's living chamber or aperture. The whorls (coils) that make up the spiral shell only overlap each other slightly. Shells like this, where most of each whorl is visible, are described as evolute. The whorls are rounded in cross-section and converge into a rounded venter, which is the outer curve of the last whorl.

PREFLORIANITES TOULAI
Strong radial ribs cross all of the shell's whorls, but are most visible on the inner whorls. The ribs do not reach all the way round to the venter.

Meekoceras

GROUP Cephalopods
DATE Early Triassic
SIZE 3–10cm (1¼–4in) across
LOCATION USA, Indonesia

Meekoceras was a smooth ceratitid ammonite. The genus was discovered in the western USA in 1879 by C.A. White. It is usually fairly tightly coiled and has a small umbilicus – the navel-like structure at the centre of the shell. Such coiling is said to be involute. As the animal matures, the coiling becomes less tight and the umbilicus widens. The sides of the shell are slightly convex, and the ventral (outer) side of the shell is usually flat.

MEEKOCERAS GRACILITATUS
The shell is ornamented by very fine growth lines; as is usual in molluscs, each one of these lines probably represents a day's growth.

Ceratites

GROUP Cephalopods
DATE Middle Triassic
SIZE 7–15cm (2¾–6in) across
LOCATION France, Germany, Spain, Italy, Romania

Ceratites gives its name to the Ceratitina – a major suborder of ammonoids that originated during the Middle Permian. The great majority of Triassic ammonoids belonged to this suborder, and they became a very successful group that flourished until their extinction at the end of the Triassic. The suture lines, which were strongly folded on *Ceratites,* marked the point where the internal walls (septa) dividing the shell met the inside of the shell. The folds that faced towards the shell opening (saddles) were simple, smooth curves, while those facing backwards (lobes) were finely serrated. A suture line with this combination is known as "ceratitic". The last two suture lines on the specimen shown below are closer together than the others, which shows that this ammonoid was fully grown. This specimen's large size suggests that it was probably female.

CERATITES NODOSUS
This fossil is an internal mould. It formed from sediment infilling the shell and shows the detailed features of the interior wall.

ANATOMY
SUTURE LINES

As a shelled cephalopod grows, it produces a new septum behind its body that is cemented to the inside of the shell. The shell is strengthened by the folding of the junction of each septum with the shell, so that the length of the cementation (the suture line) increases. Suture lines (which are only visible on internal moulds) become more elaborate from the Late Paleozoic goniatites, through the Permian and Triassic ceratites, to the Jurassic and Cretaceous ammonites.

Arcomya

GROUP Bivalves
DATE Middle Triassic to Late Cretaceous
SIZE 1–3cm (⅜–1¼in) across
LOCATION Worldwide

The two valves, or shells, of *Arcomya* were elongated and the two beaks lay very close together. The hinge lacked interlocking teeth, and there would have been a strong external ligament to open and close the valves. *Arcomya* lived buried deeply in the sea bed, with two siphons that extended up to the sea floor emerging from its permanently gaping posterior end. One siphon sucked in oxygenated water that also carried food particles, while the other expelled de-oxygenated water and waste.

ARCOMYA SP.
Both of the valves show a series of fine growth lines. The hinge, which enabled the two valves to open and close, is also visible.

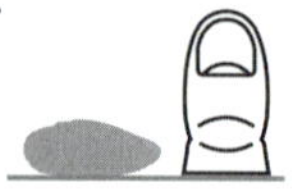

Pustulifer

GROUP Gastropods
DATE Middle to Late Triassic
SIZE 3.5–24cm (1½–9½in) long
LOCATION Peru

This gastropod had a very tall and narrow shell that is referred to as "turreted". The patterns on the outside of the shell consisted of two rows of bumps, or tubercles, which were arranged in two spirals. One row ran around the upper margin of each whorl and the other around the lower margin. Irregular growth lines ran across the whorls, in some cases connecting the tubercles on either side of the whorl margins to the whorl sutures, the line where the whorls meet. This genus is coiled dextrally, meaning that the shell opening is on the right when the point (spire) of the shell is facing upwards. However, the side of the shell with the aperture is damaged in this specimen so the opposite side is shown. There would have also been a short inhalant canal, for water intake, at the mouth of the shell.

This gastropod has a very tall and narrow shell that is referred to as "turreted".

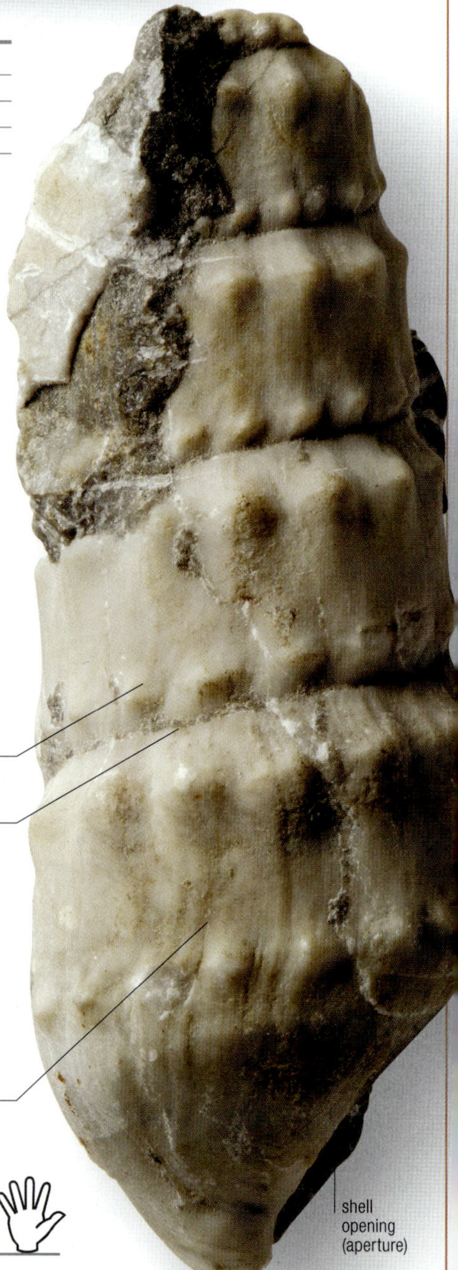

PUSTULIFER SP.
This narrow and acutely angled specimen shows very little overlap of one whorl by the succeeding one.

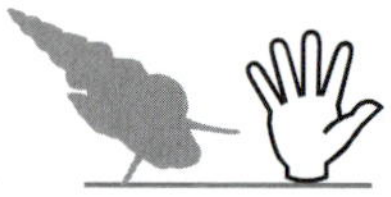

FEEDING FANS
Encrinus extended each of its ten arms to form a feeding fan to trap plankton that drifted close enough. If threatened by a predator, it could close its arms tightly together.

Encrinus

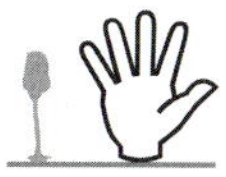

GROUP Echinoderms
DATE Middle Triassic
SIZE Cup 4–6cm (1½–2¼in) long
LOCATION Europe

Encrinus was a crinoid with a large cup consisting of three circlets of plates. The cup displays five-fold symmetry and the base is slightly concave. There are ten arms that branch from the top of the cup, which would have borne many smaller branches called pinnules. When the arms and pinnules were splayed, they formed a feeding fan. In life, the mouth was hidden beneath a leathery dome, called the tegmen, that formed the roof of the cup. Particles of food were passed along the arms via food grooves that led to further food grooves below the tegmen; these were lined with tiny, hair-like structures called cilia that led to the mouth. The stem is circular in cross-section and features larger, regularly spaced bone-like structures (ossicles) between the more frequent, smaller ones. *Encrinus* is the last survivor of a very large group of crinoids that can be traced back to the Early Ordovician. Most of the members of this group died out in the end-Permian extinction.

ENCRINUS LILIIFORMIS
The long, cylindrical stem was attached to the sea bed. The evenly spaced enlarged ossicles would have provided structural support.

TRIASSIC VERTEBRATES

The Triassic was a critical time in vertebrate evolution. It is regarded as the birth of modern ecosystems, as successful living groups such as mammals, turtles, and archosaurs arose in the aftermath of the devastating Permo–Triassic extinction, which killed off many primitive reptile and amphibian groups.

The list of vertebrates that arose during the Triassic is staggering: dinosaurs, crocodylomorphs, pterosaurs, mammals, turtles, ichthyosaurs, and frogs. The Triassic, which stretched from 250–200 million years ago, was a time of transition. Archaic groups that had dominated Permian ecosystems, such as many "mammal-like reptiles", had faded into extinction, and the arid landscape of Triassic Pangaea was the perfect breeding ground for entirely new groups of vertebrates, many of which remain successful today.

THE GREAT SURVIVOR
The stocky, pig-like *Lystrosaurus* looks fairly ordinary, but this early mammal cousin thrived across the globe in the Early Triassic. It is a classic example of a "disaster taxon".

Extinction recovery

The Permo–Triassic extinction was the greatest catastrophe in Earth's history, as up to 95 per cent of all species became extinct after volcanic eruptions poisoned the atmosphere. Among the departed were several groups of land-living vertebrates, most notably many synapsids ("mammal-like reptiles"), insect-eating reptiles, and fish-eating amphibians. In the aftermath, many of the world's ecosystems were barren. The evolutionary clock had been reset and new groups had the opportunity to evolve, expand, and dominate. Within the first few million years of the extinction, ecosystems remained in disarray and a select few vertebrates spread across the globe. These so-called "disaster taxa", such as *Lystrosaurus* (see left), were well-equipped to withstand a toxic world as they were ecological generalists that could withstand a variety of climates. However, as the atmosphere returned to normal, ecosystems stabilized and major groups like the dinosaurs appeared.

The ruling reptiles

The archosaurs, or "ruling reptiles", include living birds and crocodilians, along with a range of extinct groups restricted to the Mesozoic, such as the dinosaurs. Archosaurs originated approximately 245 million years ago and quickly spread across

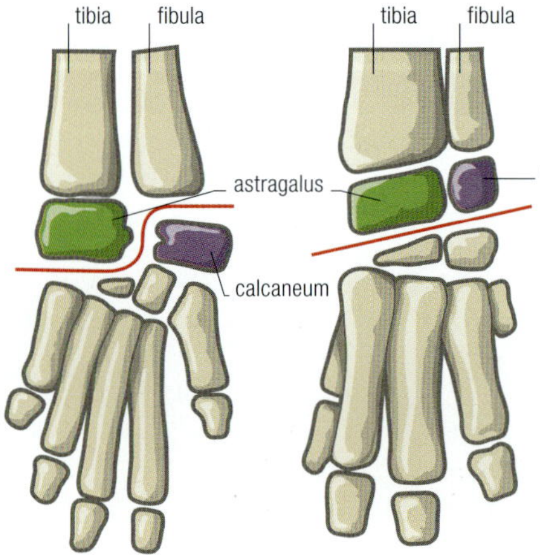

PRIMITIVE ANKLES
On the left is the crurotarsal ankle of crocodiles, in which there is rotation between the astragalus and calcaneum bones of the ankle. On the right is the mesotarsal ankle of dinosaurs, pterosaurs, and birds, in which the ankle bones together form a hinge that rotates against the foot.

GROUPS OVERVIEW

Many important groups of vertebrates originated in the Triassic, such as dinosaurs, crocodylomorphs, and other reptile groups. Other groups, such as the cynodonts and temnospondyl amphibians, became amazingly diverse after holding on and enduring the Permo–Triassic extinction.

TEMNOSPONDYLS
One of the largest and most important groups of early amphibians, temnospondyls originated during the Carboniferous but flourished during the Triassic. They lived across the globe, ranged in size from tiny to enormous, and prospered both on land and in the water. The largest temnospondyls, such as the German *Mastodonsaurus*, were up to 4m (13ft) in length.

RHYNCHOSAURS
Rhynchosaurs were some of the most unusual reptiles of the Triassic. These barrel-gutted herbivores only thrived for a short period of time, becoming extinct about 220 million years ago, but they were exceptionally abundant and were the primary large herbivores in many ecosystems. They sheared plants with their beaks and several rows of teeth on the roof of the mouth.

PHYTOSAURS
One of the most familiar of the crocodile-line archosaurs, phytosaurs prospered during the Late Triassic. These long-snouted and semi-aquatic predators resemble living crocodiles and probably had a similar lifestyle, hunting for fish and small reptiles around the water's edge. However, this resemblance is only superficial, and is a prime case of convergent evolution.

AETOSAURS
Like the phytosaurs, aetosaurs were a subgroup of crocodile-line archosaurs that were exceptionally common during the last 30 million years of the Late Triassic but died out at the Triassic–Jurassic extinction. Aetosaurs were heavily armoured and resembled giant military tanks. Most aetosaurs were herbivorous, and fed on low-growing plants, but some may have eaten meat.

Birds actually evolved from the "lizard-hipped" theropods rather than the "bird-hipped" dinosaurs.

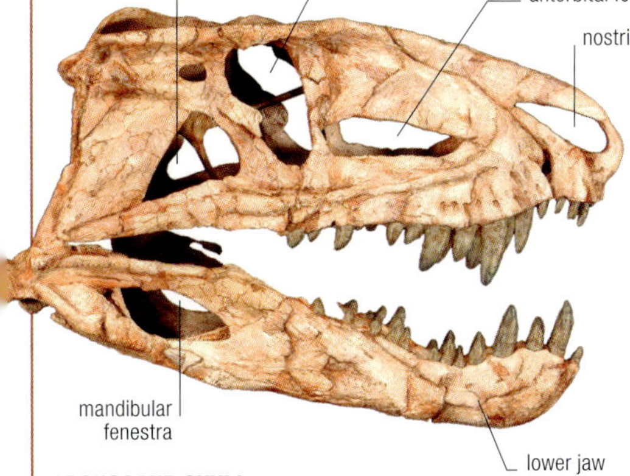

ARCHOSAUR SKULL
Postosuchus was a large predator. Its skull has a number of archosaur features, such as the antorbital fenestra and mandibular fenestra (skull openings).

the globe, diversifying into an array of different species, which generally can be divided into two groups based on the anatomy of the ankle. The bird-like archosaurs, which include dinosaurs, have a straight hinge joint between the ankle bones and the foot (a mesotarsal ankle), which allows for rapid locomotion (see left). However, the crocodile-line archosaurs have a ball-and-socket joint between the astragalus and calcaneum bones of the ankle (a crurotarsal ankle). This joint allows the two bones to rotate against each other, which offered a degree of manoeuvrability but prevented most crocodile-line archosaurs from attaining high speed.

Dinosaurs

Dinosaurs are the most familiar of the extinct archosaur groups and are closely related to the flying pterosaurs. The earliest-known dinosaurs appeared at the beginning of the Late Triassic, approximately 230 million years ago. Dinosaurs quickly diversified into an array of different body forms, but it took them much longer to multiply into a large number of species and to become exceptionally abundant in individual ecosystems.

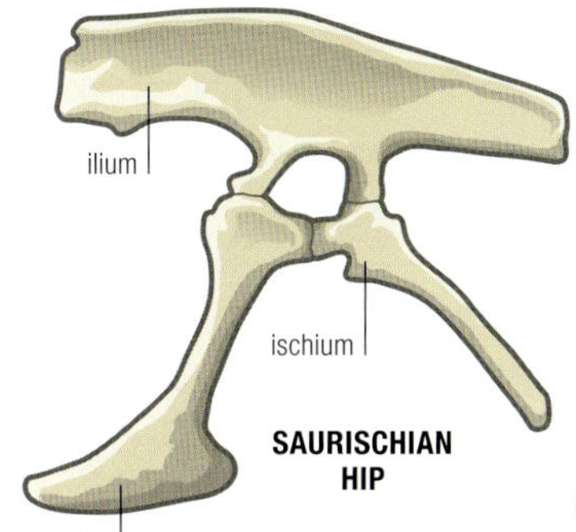

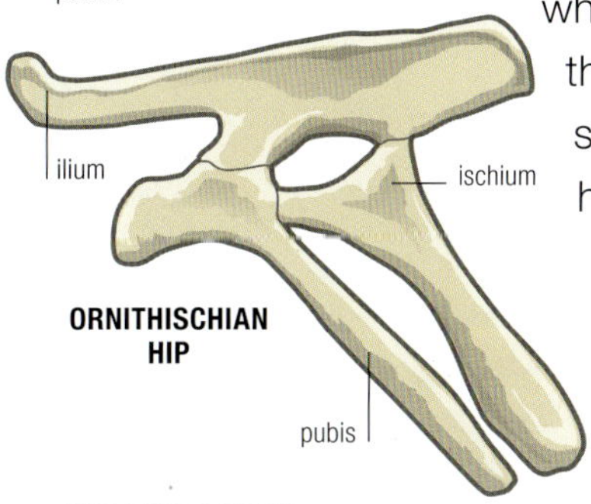

TWO TYPES OF HIP
In the saurischian hip of theropods and sauropodomorphs (top), the pubis points forwards. In the ornithischian hip, the pubis points backwards and lies against the ischium. These hips define the two major groups of dinosaurs.

The most primitive dinosaurs, such as *Herrerasaurus* and *Eoraptor*, were sleek, bipedal predators boasting an arsenal of serrated teeth and sharp claws. From these small, carnivorous ancestors dinosaurs proceeded to divide into two major groups: the saurischians and ornithischians. The saurischians, which include the predatory theropods and the long-necked sauropodomorphs, possess a "lizard hip", in which the pubis bone points forwards. The ornithischians, or "bird-hipped" dinosaurs, have a modified pelvis in which the pubis points backwards, as in living birds. However, in one of the great ironies of paleontology, birds actually evolved from the "lizard-hipped" theropods.

RAUISUCHIANS
The rauisuchians, yet another group of crocodile-line archosaurs, were the keystone predators of most Triassic terrestrial ecosystems. Over 25 species have been found, and include giant quadrupedal hunters, sleek bipedal omnivores, and lumbering beasts with deep sails on their backs. They probably occupied the large-predator niche later filled by theropods.

SPHENOSUCHIANS
Although they barely look like their later relatives, the sphenosuchians were some of the earliest and most primitive members of the crocodylomorph group. Sphenosuchians, which flourished during the Late Triassic and Early Jurassic, were sleek predators that walked upright (some could walk on two legs), could run fast, and resembled greyhounds in their overall proportions.

THEROPODS
Perhaps the most recognizable group of dinosaurs, the theropods include predators such as *Tyrannosaurus*, *Allosaurus*, and *Velociraptor*. Theropods first evolved in the Triassic but did not evolve into giant sizes until later, during the Jurassic. Most Triassic theropods, such as *Coelophysis*, were only a metre or two long and hunted small prey in the shadow of the giant rauisuchians.

CYNODONTS
Cynodonts are a large group that includes true mammals. The first cynodonts evolved in the Permian, but many groups prospered during the Triassic. Their characteristic mammalian features included hair, a large brain, and upright posture. Many were small, but some were enormous and filled the large-herbivore niche before the evolution of sauropodomorph dinosaurs.

Saurichthys

GROUP Actinopterygians
DATE Early to Late Triassic
SIZE 1m (3ft 3in) long
LOCATION All continents except Antarctica

Saurichthys, meaning "lizard fish", was a fast, open-water hunter, similar to present-day barracuda. This primitive ray-finned fish was a top predator. Its streamlined shape, posteriorly positioned pelvic fins, and forked tail can be clearly seen in this fossil. Whether it hunted in groups, like barracuda, or laid in wait for its prey is not known. One of its closest modern relatives is the sturgeon.

LONG SNOUT
Its long snout made up over 50 per cent of the total head length and its narrow jaws were lined with sharp, conical teeth.

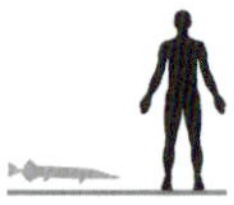

Mastodonsaurus

GROUP Temnospondyls
DATE Middle to Late Triassic
SIZE 6m (20ft) long
LOCATION Europe, Russia

Mastodonsaurus was the first early amphibian to be recognized and described, in 1828. It is far larger than any other fossil amphibian ever found. Although the skull is crocodile-like in shape, and very large, it is more like a frog's in its construction. It was assumed at first that *Mastodonsaurus* had a squat, rounded body like a frog. It was only when bones from the body were found that it became apparent that the creature resembled a thickset crocodile with a gigantic head and no neck.

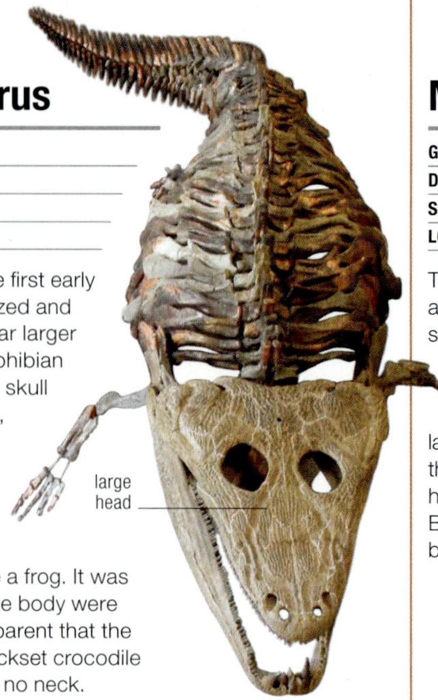

LARGEST AMPHIBIAN
At 6m (20ft) in length, *Mastodonsaurus* is the largest amphibian to have ever lived. Fossil finds in coastal lagoons suggest it was saltwater tolerant.

Metoposaurus

GROUP Temnospondyls
DATE Late Triassic
SIZE 2m (6½ft) long
LOCATION Europe, India, North America

The metoposaurs were a family of amphibians that were widespread for a short time in the Late Triassic. This was when the supercontinent Pangea was just beginning to split up and was criss-crossed by rift valleys filled with lakes and rivers. *Metoposaurus* spread along these waterways, which explains why they have been found in many parts of the world. By the end of the Triassic, metoposaurs had been replaced by early crocodiles.

PADDLING SWIMMERS
It has been suggested that *Metoposaurus* swam by using its limbs as paddles, rather than undulating its whole body like a crocodile.

Odontochelys

GROUP Turtles
DATE Late Triassic
SIZE 40cm (15½in) long
LOCATION China

Turtles have a classic body plan that is not seen in any other group. This includes a short skull, a tiny tail, and a thick shell that surrounds the body. The mystery of how the turtle shell evolved has been clarified by *Odontochelys*, the oldest and most primitive turtle ever discovered. It had a fully formed plastron (the lower part of the shell), but lacked a carapace (the upper part). This shows that the shell began as armour on the belly, with the carapace forming much later as the ribs became thicker and broader and fused with bony plates (osteoderms) inside the skin.

OLDEST TURTLE
Odontochelys is the oldest-known turtle. It lived in nearshore waters, suggesting that turtles evolved in the sea.

Diphydontosaurus

GROUP Lepidosaurs
DATE Late Triassic
SIZE 10cm (4in) long
LOCATION British Isles, Italy

A nearly complete Late Triassic ecosystem has been reconstructed from the rich fossil deposits of the Bristol Channel region that lies between England and Wales. About 205 million years ago, this area was home to a labyrinth of small caves that formed after the thick limestone bedrock was eroded by acidic rainwater. The remains of a menagerie of small animals have been found in these dark caverns, having been drowned and washed inside by monsoon floods. They ranged in size from small sauropodomorph dinosaurs, which were up to 3m (9¾ft) long, to minute sphenodontian reptiles at just 10cm (4in) long. The most common sphenodontian was *Diphydontosaurus*. Once a diverse group, the tuatara of New Zealand is now the only living member. *Diphydontosaurus* was far smaller than the tuatara, which may have been an adaptation for hunting down its insect prey in crevices.

AGILE HUNTER
The sharp claws on its feet would have helped *Diphydontosaurus* scratch around for insects to eat.

Lariosaurus

GROUP Sauropterygians
DATE Middle to Late Triassic
SIZE 50–70cm (19½–28in) long
LOCATION Italy

Nothosaurs filled the same general niche that seals do today. Like seals, these reptiles primarily lived in the water, but they also came ashore to rest. *Lariosaurus* was one of the smallest known nothosaurs. Its front feet were webbed flippers and, uniquely, its hind feet retained five individual toes. This suggests that *Lariosaurus* was more active on land than other nothosaurs, and it may have hunted small shoreline animals as well as fish.

WEBBED FEET
Like all nothosaurs, *Lariosaurus* had webbed feet and an elongated neck. In the water it used its long neck to lunge after fish, catching them with its needle-sharp teeth.

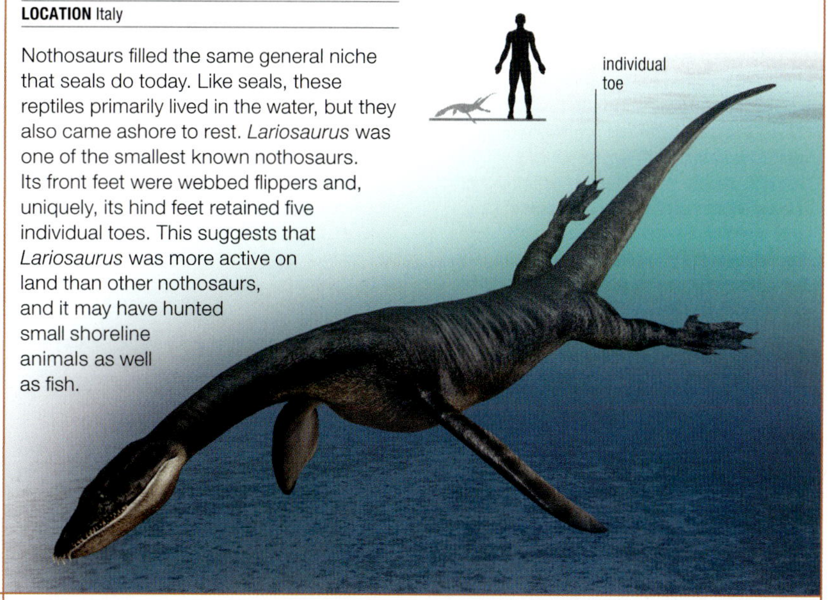

The only living sphenodontian reptile is the tuatara from New Zealand.

LIVING RELATIVE
TUATARA

Sphenodontian reptiles such as *Diphydontosaurus* were once common, but today these scaly predators are restricted to two species of tuatara, found only in New Zealand. The tuatara, known scientifically as *Sphenodon*, is highly endangered. It is often considered a "living fossil", as it is one of the most primitive living reptiles and greatly resembles its fossil cousins that lived over 70 million years ago.

Placodus

GROUP Sauropterygians
DATE Middle Triassic
SIZE 2–3m (6½–9¾ft) long
LOCATION Germany

Living alongside nothosaurs in the warm, shallow seas that covered Europe in the Middle Triassic were some peculiar reptiles called placodonts. The most common, and best-studied, placodont is *Placodus*. It had all the characteristic features of the placodonts: a large, barrel-chested body; webbed hands and feet that formed paddles; and a long, deep tail. The hands, feet, and tail were specially adapted for swimming, and indicate that *Placodus* was probably an accomplished swimmer despite its bulky frame. It is also likely that *Placodus*, like many other marine reptiles, spent time out of the water feeding and mating, although it was probably a clumsy creature on land. Placodont skulls, including that of *Placodus*, look quite unusual. They feature distinct protruding teeth at the front of the jaw, and a series of smooth, peg-like teeth on the roof of the mouth. This strange collection of teeth is unlike anything found in any other reptiles. It is thought the front teeth were used for spearing fish and the ones in the roof of the mouth were for crushing the hard shells of molluscs.

PERFECT TEETH
The bizarre Triassic reptile *Placodus* foraged in the shallow waters along nearshore reefs looking for molluscs and fish to eat. Its unusual teeth were perfectly designed for dealing with its prey.

Pachypleurosaurus

GROUP Sauropterygians
DATE Middle Triassic
SIZE 30–40cm (12–15½in) long
LOCATION Italy, Switzerland

There are two subgroups of nothosaur: nothosaurids and pachypleurosaurids. The most characteristic pachypleurosaurid is *Pachypleurosaurus*. Its body was long and thin with an elongated neck, and its limbs acted as robust paddles and for steering and stability in the water. This agile reptile swam by laterally undulating its slim yet muscular torso.

FAST MOVER
The small and gracile *Pachypleurosaurus* used its muscular body to dart quickly through the water while chasing after small prey.

Mixosaurus

GROUP Ichthyosaurs
DATE Early Triassic
SIZE Up to 1m (3ft 3in) long
LOCATION North America, Europe, Asia

One of the oldest-known members of the ichthyosaur group, *Mixosaurus* lived right across the globe during the Early Triassic. It had the characteristic ichthyosaur body plan (the word ichthyosaur means "fish lizard"), which resembles the dolphins of today. Most notably, the snout was long and slender, the torso was streamlined, the limbs were modified into paddles, the tail was deep, and there was a dorsal fin for stability in the water.

Mixosaurus was undoubtedly a fast swimmer and used its speed to surge through shoals of fish, which made up the bulk of its diet. Compared to most other ichthyosaurs, *Mixosaurus* was small, with most individuals falling short of 1m (3ft 3in) in length; however, some later cousins reached more than four times that size. Not only was *Mixosaurus* one of the smallest ichthyosaurs, but it was also among the most primitive. In fact, *Mixosaurus* was so named because George Baur, the German paleontologist who first studied it, thought it showed a "mixture" of primitive and advanced characteristics.

Procolophon

GROUP Primitive amniotes
DATE Early Triassic
SIZE 30–35cm (12–14in) long
LOCATION South Africa, Antarctica

The small frame of *Procolophon* and other procolophonid reptiles belies their ability to survive global catastrophe. Although unspectacular at first glance, these sluggish reptiles were one of the few groups to endure the devastating Permo–Triassic extinction. After surviving the most destructive global die-off in Earth's history, procolophonids evolved into a new array of species. One of these was *Procolophon*, the namesake of the group.

Procolophon was no larger than a small iguana. It had a wide body and small tail, and its broad, triangular skull is similar to that of many burrowing reptiles living today. Perhaps procolophonids survived the extinction by hiding away in burrows, sequestering themselves from intense bouts of heat and cold and showers of acid rain.

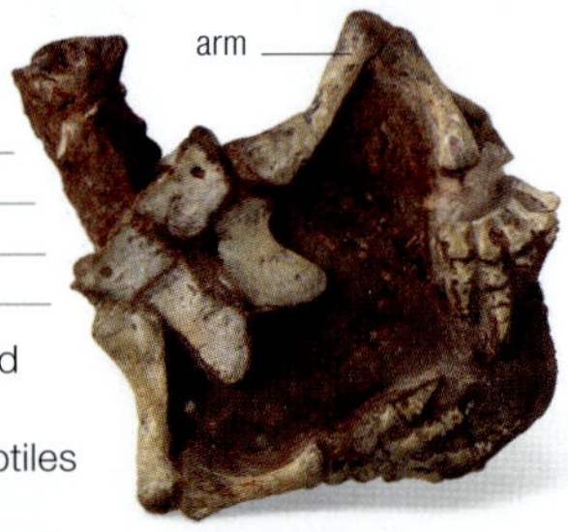

FORELIMBS
The forelimbs of *Procolophon* were small and sleek, but they may have been strong enough to dig burrows.

BURROW DWELLER
The wide, triangular skull and squat body of *Procolophon* is similar to that of many modern lizards, such as the gila monster of North America – another reptile that often seeks shelter in burrows.

Hyperodapedon

GROUP Rhynchosaurs
DATE Late Triassic
SIZE 1.2–1.5m (4–5ft) long
LOCATION Scotland

Rhynchosaurs were just one of many groups of herbivorous reptiles that lived during the Late Triassic. They were also among the main land-dwelling herbivores that existed around the globe immediately prior to the origination of dinosaurs, as well as during the first few million years of dinosaur evolution. One of the best-studied rhynchosaurs is *Hyperodapedon*, a species represented by over 35 skeletons from 230-million-year-old rocks discovered in Elgin, Scotland.

The skull of *Hyperodapedon* was broad and triangular when viewed from above. It was fronted by a sharp "beak", comprised of a pair of tusks, which had razor-sharp edges that formed a scissor-like mechanism designed to shear through vegetation. Further back in the jaw, rows of teeth at the sides of the snout and on the roof of the mouth ground down food before it was swallowed. The body of *Hyperodapedon* was short and stocky, and its limbs may have been used to dig small holes, which would have enabled this barrel-chested herbivore to access nutritious tubers and roots. Rhynchosaurs became extinct during the Late Triassic. It is thought their disappearance allowed dinosaurs to replace them as the dominant land-based herbivores.

BEAKED LIZARD
A large nasal cavity means *Hyperodapedon* probably had a good sense of smell, and large eye sockets indicate good vision. This squat, rather heavy-set lizard also had claws on its back legs that were suitable for "scratch digging", possibly for food.

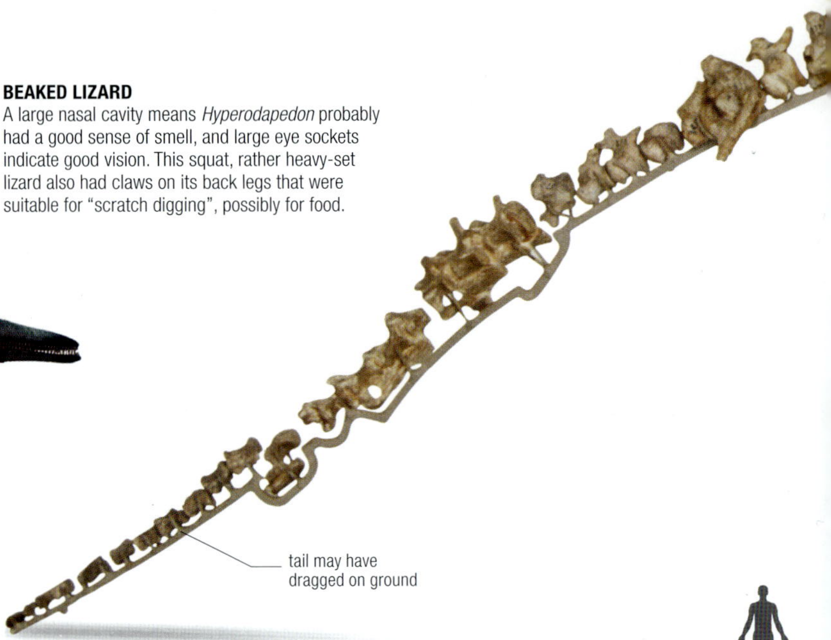

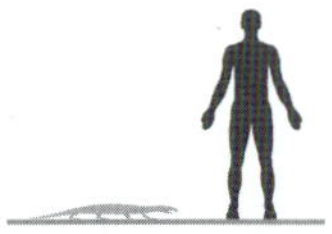

Tanystropheus

GROUP Tanystrophids
DATE Middle Triassic
SIZE 5.5–6.5m (18–21½ft) long
LOCATION Europe and the Middle East

Tanystropheus must rank among the most bizarre reptiles of all time, instantly recognizable by its outsized, noodle-like neck that was longer than its body and tail combined. This extraordinary neck was more than 3m (9¾ft) long – about twice the height of an average man – yet *Tanystropheus* possessed only ten greatly elongated cervical vertebrae to support it. A semi-aquatic animal that fed primarily on fish, *Tanystropheus*'s neck could rapidly extend to capture fish in the murky waters of the seashore. This strategy allowed it to take fish while its body remained completely still, in some cases while it stood on dry land.

BEAKED LIZARD
Hyperodapedon had a sharp-edged "beak", which could also have helped scrape up food to its powerful tongue.

ANATOMY
PLANT-EATER

Rhynchosaurs were common, mid-sized herbivores that filled the top plant-eating niche in many ecosystems. Rhynchosaurs such as *Hyperodapedon* had an extraordinary set of anatomical adaptations that facilitated their plant-vacuuming lifestyle. The rhynchosaur skull has paired tusks (the "beak") at the front, and the roof of the mouth is covered with hundreds of teeth, which were constantly replaced throughout life in a conveyor-belt fashion. The tusks cropped and sheared plants, while the tooth fields ground and crushed the plant matter before it was swallowed.

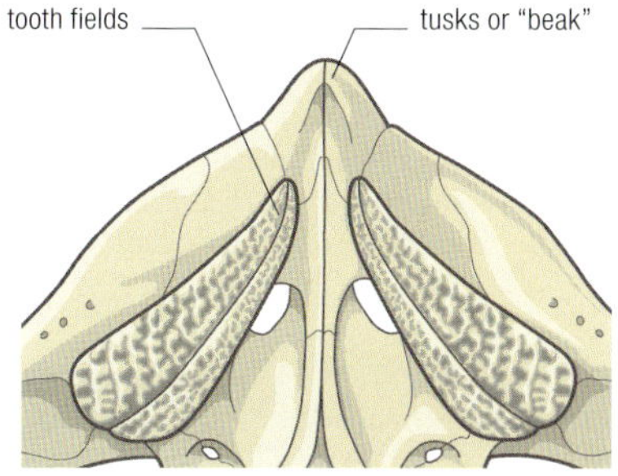

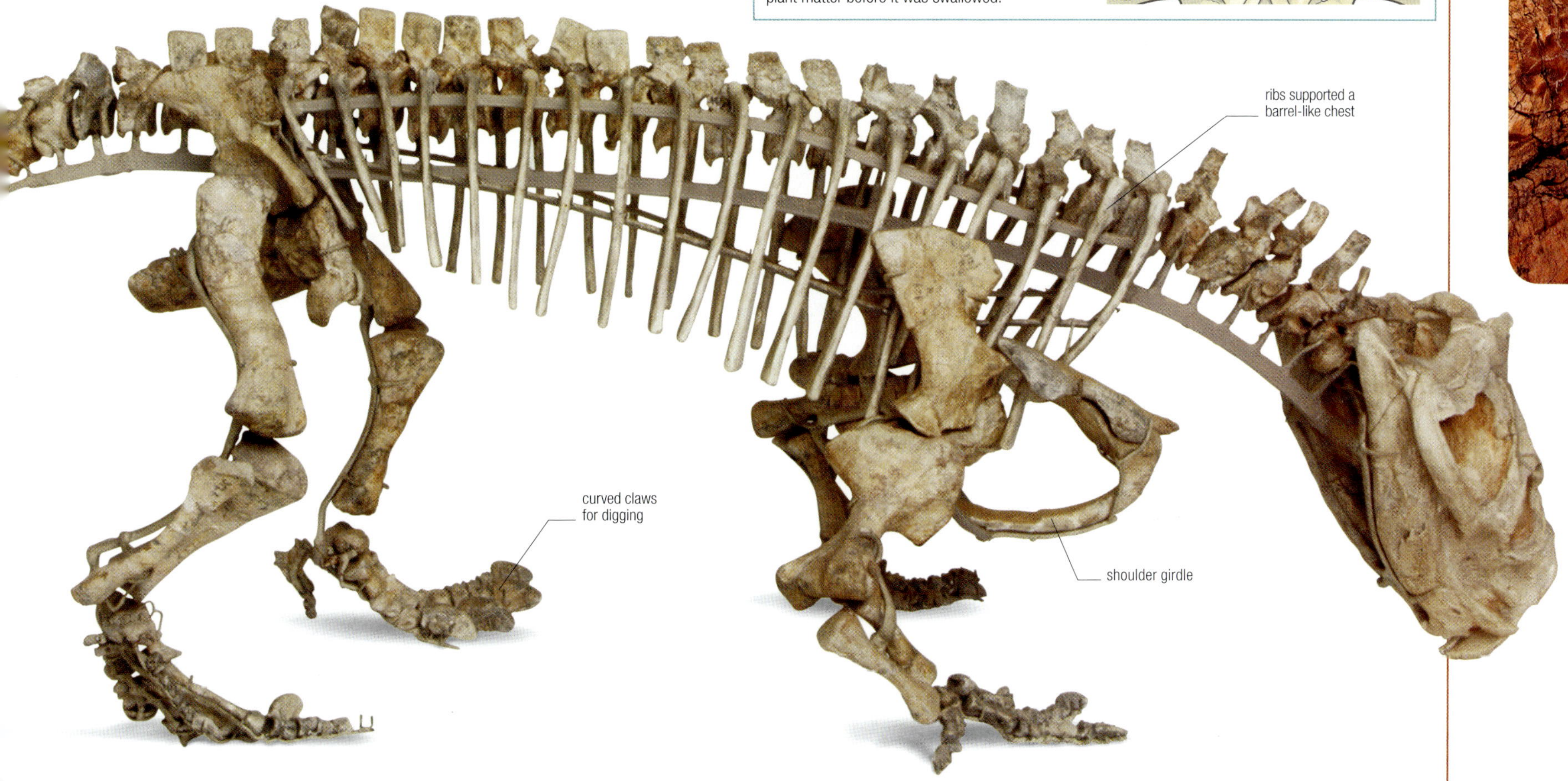

Proterosuchus

GROUP	Proterosuchids
DATE	Early Triassic
SIZE	1–2m (3¼–6½ft) long
LOCATION	South Africa

Following the Permo–Triassic mass extinction, entirely new groups of organisms emerged. One of these was *Proterosuchus*, a Komodo-dragon-sized predator that is a distant cousin of the archosaurs: the large group that contains birds, crocodiles, and dinosaurs.

The skull of *Proterosuchus* was long and narrow, with numerous small, curved teeth and enlarged eye sockets. The body was also long and thin, with a flexible, tapering tail. It is likely that *Proterosuchus* alternated between living on land and in water, giving it unparalleled access to a range of prey. Although quite small even by today's standards, *Proterosuchus* was among the largest carnivores of its day.

HOOKED MOUTH
A hook-shaped mouth and curved teeth meant that prey had little chance of escape when *Proterosuchus* struck.

Euparkeria

Group Primitive archosaurs
Date Early Triassic
Size 70cm (28in) long
Location South Africa

Euparkeria was one of many reptile species that lived during the first few million years of the Triassic in what is now South Africa. This small carnivore had a mouth full of sharp teeth and was covered in scaly skin embedded with bony plates (osteoderms) along its back and tail. A close cousin of the archosaurs, it showed many characteristics that developed more fully in dinosaurs and other archosaurs. Longer back legs allowed it to run or stand upright at times, which gave it the life-saving advantage of speed and agility, and its skull had a large opening in front of the eye that housed an enlarged space or sinus, which points to a good sense of smell.

MEAGRE DEFENCES
Despite its sharp teeth and fearsome appearance, *Euparkeria* lived in a world of many predators. Aside from speed, sharp claws on its thumbs were its only defence.

KEY SITE
KAROO SITE

Many spectacular fossils have been found in the Karoo Desert of South Africa. The footprints being cleaned below were made by a synapsid, a mammal-like reptile that preyed on *Euparkeria*. Rocks deposited during the Permian, Triassic, and Jurassic in this large, inland basin, filled with lakes and rivers, have been explored for centuries.

Gualosuchus

GROUP Proterochampsids
DATE Middle Triassic
SIZE 1–2m (3¼–6½ft) long
LOCATION Argentina

Gualosuchus lived around 235 million years ago alongside an array of dicynodonts, cynodonts, and close dinosaur cousins such as *Marasuchus* and *Lagerpeton*. The skull of *Gualosuchus* was long and very broad when viewed from the top, with large eye sockets and a long snout filled with sharp, curved teeth. *Gualosuchus* lived in and around the margins of the seas and rivers, and probably hunted fish, along with any small vertebrates that ventured too close to the shore.

Parasuchus

GROUP Phytosaurs
DATE Late Triassic
SIZE 2m (6½ft) long
LOCATION Worldwide

A crocodile-like body characterizes a group of fossil vertebrates known as phytosaurs. These reptiles, some of which grew to sizes close to that of *Tyrannosaurus rex* (see pp.322–23), lived for about 25 million years during the Late Triassic. One of the most studied is *Parasuchus*; complete skeletons have been found in India, as well as fragmentary fossils all across the world. Its name means "similar to crocodiles", which fits this sprawling, fish-eating reptile.

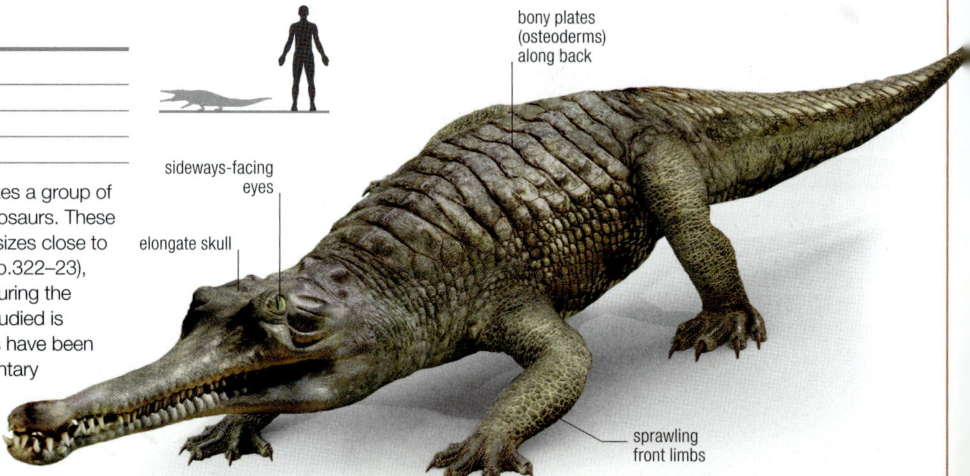

Stagonolepis

GROUP Aetosaurs
DATE Late Triassic
SIZE 3m (9¾ft) long
LOCATION Scotland and USA

Aetosaurs were exceptionally abundant from 225 to 200 million years ago. One of the most recognizable is *Stagonolepis*. Like all aetosaurs, this animal shielded its body with bony armour plates (osteoderms). The skull of *Stagonolepis* was short and deep, and ended in a shovel-like snout that was useful for uprooting tubers or slurping bugs from the mud. *Stagonolepis* is a common fossil find in the Late Triassic rocks of Elgin, Scotland.

Ornithosuchus

GROUP Primitive archosaurs
DATE Late Triassic
SIZE 1–2m (3¼–6½ft) long
LOCATION Scotland

Ornithosuchus may look like a dinosaur, but its skeleton suggests it is a crocodilian archosaur. Among the largest predators in its ecological niche 230 million years ago, when the first dinosaurs were evolving, *Ornithosuchus* spent most of its time on four legs, but had the ability to walk or run on two. Its sharp teeth show that it was a carnivore.

Postosuchus

GROUP Rauisuchians
DATE Late Triassic
SIZE 3–4.6m (9¾–15ft) long
LOCATION USA

Postosuchus is one of the best-known and best-studied members of the rauisuchians, a large group of Triassic archosaurs. Like the ornithosuchians to which *Ornithosuchus* belongs (see opposite), this group is part of the same lineage as modern crocodiles and included some of the largest predators of the day. *Postosuchus* grew to about 4.6m (15ft) in length and may have weighed as much as 680kg (1,500lb). In many ways it resembles the large theropod dinosaurs of the Jurassic and Cretaceous periods – which is why it was first described as a primitive ancestor of *Tyrannosaurus rex* (see pp.322–23). Indeed, the large skull and banana-shaped cutting teeth do resemble those of the more well-known carnivorous dinosaurs. However, its ankle anatomy and the presence of bony plates (osteoderms) down the back strongly support placing *Postosuchus* among the crocodile line of archosaurs.

Although *Postosuchus* is not a true dinosaur, it did live alongside some of the oldest theropod dinosaurs, which it probably also ate. Perhaps unexpectedly, it was much larger than these early theropods and was clearly the main predator in North America during the Late Triassic.

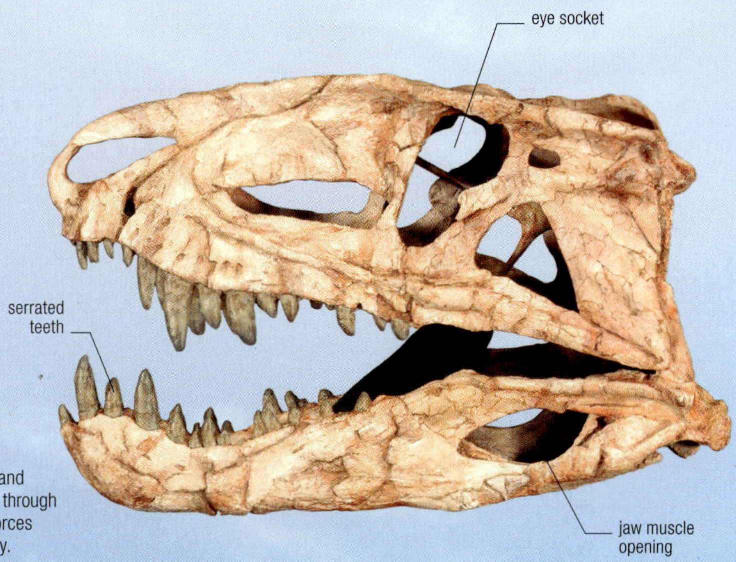

POWERFUL JAWS
The skull of *Postosuchus* is large and strong, a perfect design for tearing through flesh and withstanding the high forces associated with tackling large prey.

FORMIDABLE PREDATOR
With its sharp, serrated teeth, huge body, and powerful jaws, *Postosuchus* resembled a smaller *Tyrannosaurus*, but it was unable to move as swiftly.

Effigia

GROUP Rauisuchians
DATE Late Triassic
SIZE 2–3m (6½–9¾ ft) long
LOCATION USA

One of the weirdest-looking creatures of the Late Triassic world was *Effigia*. This archosaur was a sleek, fast-running omnivore. It lived alongside early dinosaurs such as *Coelophysis* and *Chindesaurus*, as well as large rauisuchian predators. *Effigia* greatly resembles a theropod dinosaur in its overall body plan. It was bipedal – meaning it could walk on two legs – had large attachment sites on its pelvis for strong leg muscles, possessed a long tail for balance, and had holes in its neck vertebrae for air sacs. In fact, when its close cousin *Shuvosaurus* was discovered in 1994, it was described as a Late Triassic "ostrich dinosaur", due to its bird-like, toothless skull. *Effigia* has a similar skull, capped with a beak at its front, which was perfect for cracking seeds, grinding nuts, and possibly shearing vegetation and killing small vertebrates. However, despite their resemblances to theropods, *Effigia* and *Shuvosaurus* are clearly members of the crocodile line, because they have a crocodilian ankle and many other features. This is a prime example of convergence: distantly related organisms independently evolving the same body shape in response to a similar shared lifestyle.

The word *Effigia* means "ghost", a reference to the Ghost Ranch area of New Mexico where it was discovered.

MISTAKEN IDENTITY
Although it lived at least 80 million years before true ostrich dinosaurs, *Effigia okeeffeae* was originally thought to belong to this group. In addition to referring to the area in which it was found, the "ghost" in its name is also a nod to the fact that it was effectively "invisible" to science for decades because the fossil was collected in the 1940s but wasn't examined until 2006. The species name honours American artist Georgia O'Keeffe, who lived close to Ghost Ranch.

ANATOMY

SIMILAR BODY PLAN

Effigia is a member of the crocodile-line archosaur group, but greatly resembles a theropod dinosaur. In fact, scientists once thought that a close relative, *Shuvosaurus*, was a member of the ornithomimosaur ("ostrich dinosaur") group. Like these theropods, such as *Struthiomimus* (shown here), *Effigia* was a sleek and fast animal that walked upright on two legs. It had claws on its hands and a brain specialized for intelligence and keen senses. *Effigia* also resembles these dinosaurs in that it had no teeth in its jaws, which instead were covered in a keratinous beak.

Lotosaurus

GROUP Rauisuchians
DATE Early–Middle Triassic
SIZE 1.5–2.5m (5–8¼ft) long
LOCATION China

A barrel-chested quadruped with an array of fan-like spines on its neck and back, *Lotosaurus* lived about 240 million years ago, when the world's continents were joined together. This allowed it to migrate around the globe. The *Lotosaurus* skull was lightweight, lacked teeth, and had a sharp beak. The elongated spines on its vertebrae – more than three times as long as an individual vertebra is deep – may have been used to attract mates or regulate body temperature.

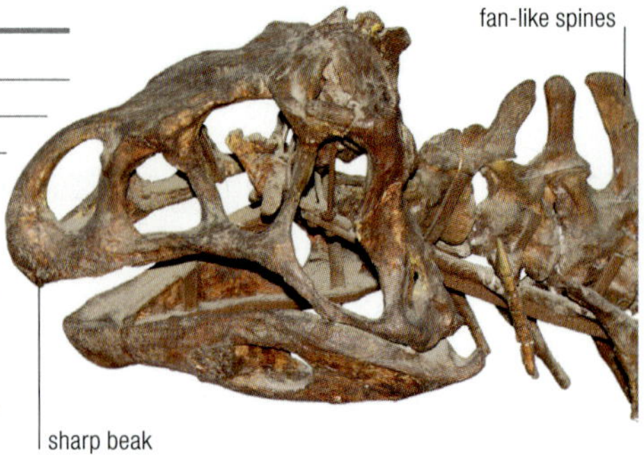

STRANGE SPINES
The long spines of the vertebrae of *Lotosaurus* are an odd feature also seen in a wide variety of unrelated reptiles.

Terrestrisuchus

GROUP Crocodylomorphs
DATE Late Triassic
SIZE 75cm–1m (30in–3ft 3in) long
LOCATION British Isles

Only an adult *Terrestrisuchus* approached 1m (3ft 3in) in length and most weighed less than 15kg (33lb). The skull was lightweight and fragile, and many limb and hand bones were pencil-thin. However, this tiny carnivore was one of the oldest relatives of the present-day crocodile, although it had a very different lifestyle to the modern reptile. *Terrestrisuchus* was a small, slender, and agile carnivore that walked with its body raised off the ground. It was able to sprint at high speeds on all four legs. In many ways, sphenosuchians resemble small theropod dinosaurs, but, like living crocodiles, they had bony plates (osteoderms) in their skin, and elongated bones in their wrists.

Herrerasaurus

GROUP Theropods
DATE Late Triassic
SIZE 3–6m (9¾–20ft) long
LOCATION Argentina

Herrerasaurus is one of the oldest, most primitive dinosaurs yet discovered. This small predator lived about 228 million years ago, and shared its ecosystem with an array of primitive reptiles that would eventually be supplanted by the dinosaurs. Even at first glance *Herrerasaurus* resembles a carnivorous dinosaur. It walked on two legs, boasted an array of sharp claws, and had jaws studded with numerous serrated teeth: it was clearly a brawny predator that could outrun and overpower its prey. It was named in honour of its discoverer Victorino Herrera, an Argentinian rancher.

Eudimorphodon

GROUP Pterosaurs
DATE Late Triassic
SIZE 1m (3ft 3in) long
LOCATION Italy, Greenland

Eudimorphodon is one of the oldest and most primitive pterosaurs, and its fossilized remains are common in the folded rocks of northern Italy's mountains. It was clearly a pterosaur, possessing all of the characteristics of this group of flying reptiles, including a greatly enlarged fourth finger that supported the wing membrane.

Eudimorphodon was a relatively small pterosaur, weighing in at about 10kg (22lb). This weight pales in comparison to the monstrous proportions of its later relatives, such as *Quetzalcoatlus* (see p.313). *Eudimorphodon* is also distinguished from most other pterosaurs by its complex teeth. Whereas most pterosaurs had simple teeth – or even lacked them altogether – *Eudimorphodon* boasted a mouthful of bizarre teeth, each with several small points called cusps. These would have been perfect for grabbing and grinding prey, and may have been an adaptation for eating large fish.

Eoraptor

GROUP Theropods
DATE Late Triassic
SIZE 1m (3ft 3in) long
LOCATION Argentina

During an expedition to Argentina in the late 1980s, a scientist named Paul Sereno and his team came across a curious fossil. No larger than a small human child, this skeleton was clearly a dinosaur, with an open hip socket, an enlarged muscle attachment on the upper arm bone (humerus), and other characteristic features. In many ways it resembled a carnivorous theropod, due to its sharp teeth, killer claws, and the ability to walk on two hind legs (bipedal gait). Sereno and his colleagues named this dinosaur *Eoraptor*, one of the most primitive dinosaurs ever found. It lived about 228 million years ago.

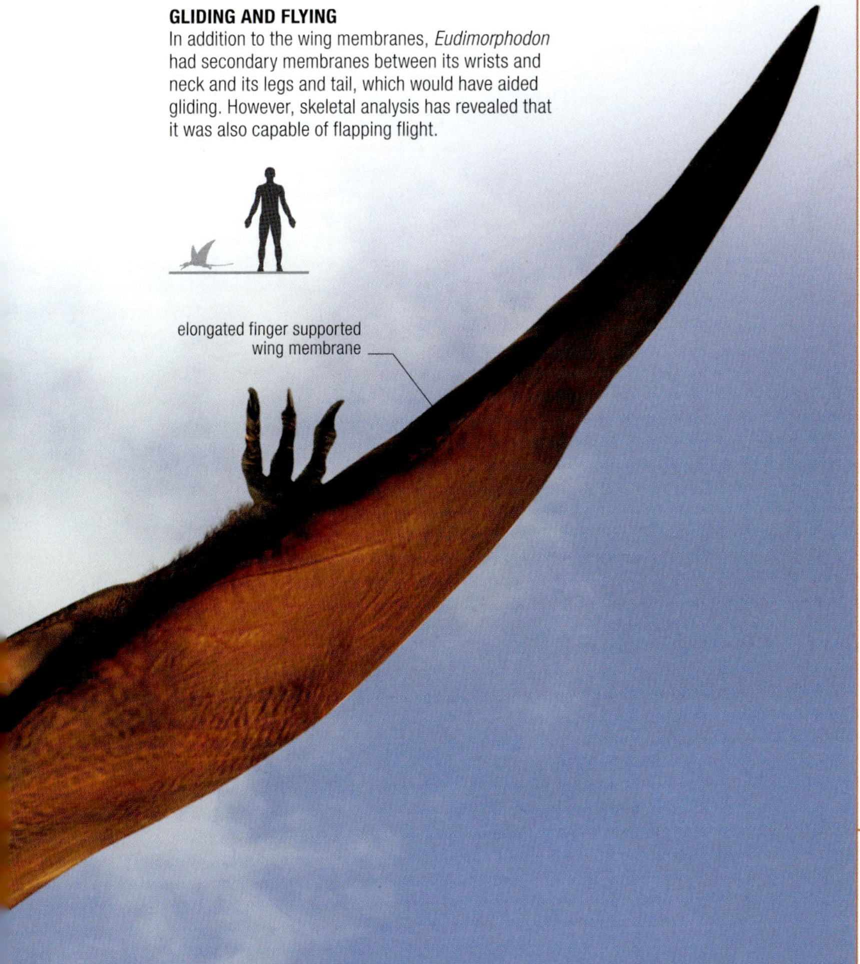

GLIDING AND FLYING
In addition to the wing membranes, *Eudimorphodon* had secondary membranes between its wrists and neck and its legs and tail, which would have aided gliding. However, skeletal analysis has revealed that it was also capable of flapping flight.

SMALL BUT DEADLY
Although small for a dinosaur, *Eoraptor* was large compared to many of the close cousins of dinosaurs that were alive at the time.

BIOGRAPHY

ROBERT BAKKER

Dr Robert Bakker (b.1945) is one of the most familiar paleontologists of the past half-century. As a student in the 1960s, he was the first scientist to argue seriously that dinosaurs, such as *Eoraptor*, were smart, agile, energetic animals – not the slow, sluggish creatures often portrayed at the time.

Chindesaurus

GROUP Theropods
DATE Late Triassic
SIZE 2–2.3m (6½–7½ft) long
LOCATION USA

Chindesaurus is a puzzling dinosaur found in 210- to 220-million-year-old rocks in the American southwest. It is thought to have been one of the foremost predators in western North America during the Late Triassic. Unfortunately, *Chindesaurus* is known only from incomplete skeletons, and no skull bones have been found. By contrast, the Argentinian species *Herrerasaurus* (see opposite) is the best- known dinosaur from the same period, with a completely preserved skeleton that has been studied in great detail. Similarities between the pelvic bones and hind limbs of *Chindesaurus* and *Herrerasaurus* show that they were probably close cousins. Because all the world's continents were joined together during the Late Triassic, it is likely that close relatives of *Herrerasaurus* lived across the globe.

Staurikosaurus

GROUP Theropods
DATE Late Triassic
SIZE 2m (6½ft) long
LOCATION Brazil

Barely larger than a human toddler, *Staurikosaurus* raced across the lowland plains of South America during the first few million years of the reign of dinosaurs. Very little is known about this nimble hunter, since only one fossil skeleton has been discovered. However, it is clear that *Staurikosaurus* was an agile and dangerous predator, and resembled a theropod dinosaur in its general body shape. It is difficult to determine whether it was a true theropod or a more primitive dinosaur that simply looked like a theropod because of shared carnivorous lifestyles. However, it is likely that *Staurikosaurus* is a close cousin of *Herrerasaurus* (see opposite). Unlike *Herrerasaurus*, however, *Staurikosaurus* was a much smaller, slighter, and lighter animal, and as a result was probably able to run much faster than its heavier relative. When it was found and described in 1970, *Staurikosaurus* was one of only a handful of dinosaurs known from South America. In fact, its name means "Lizard of the Southern Cross" – a reference to the constellation that can only be seen in the southern hemisphere. Today, hundreds of species have been named from South America.

AGILE PREDATOR
Its light, bipedal body indicates that *Staurikosaurus* would have easily been able to hunt down prey such as small lizards, insects, and primitive mammals.

Lilienstemus

GROUP Theropods
DATE Late Triassic
SIZE 5–6m (16½–20ft) long
LOCATION Germany

Liliensternus was a large, ferocious predator that lived in central Europe 210 million years ago; in fact, it was the largest land predator of its day. Its long hind legs allowed for great speed, its short arms ended in sharp-clawed hands, and its jaws were studded with serrated teeth. *Liliensternus* feasted on large prosauropod herbivores, including *Plateosaurus* (see opposite) and *Efraasia*, which inhabited wet, humid lowlands during the Late Triassic. Unfortunately, only two skeletons have been found, so it is difficult to know whether or not *Liliensternus* travelled in packs.

STALKING PREY
Liliensternus may have stalked its prosauropod prey, waiting for an opportunity to strike.

Gojirasaurus

GROUP Theropods
DATE Late Triassic
SIZE 5–7m (16½–23ft) long
LOCATION USA

Gojirasaurus prowled across the arid scrublands of North America 210 million years ago, and is named after the Japanese movie monster Godzilla (*Gojira* in Japanese). It seems strange that a 7m- (23ft-) predator is named after the colossal monster of popular culture. However, during its time in the Late Triassic, most predatory dinosaurs were small; *Gojirasaurus* stood out as the largest, strongest, and fiercest hunter of the American southwest. Unfortunately, only a few bones of *Gojirasaurus* have been found, and scientists have much to learn about this carnivore.

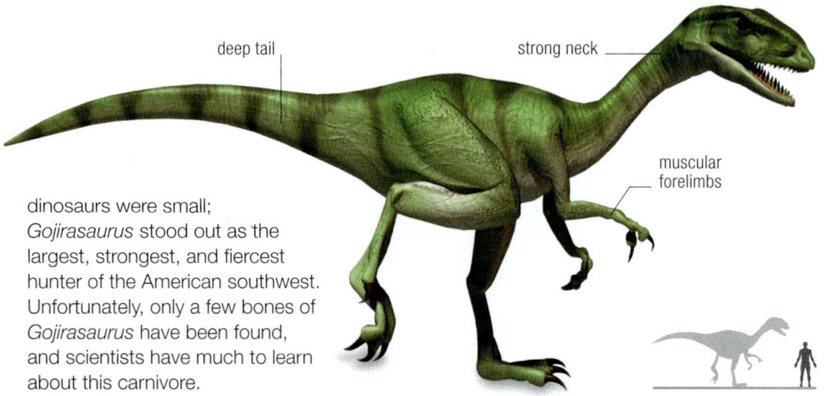

Saturnalia

GROUP Sauropodomorphs
DATE Late Triassic
SIZE Up to 2m (6½ft) long
LOCATION Brazil

This small herbivore lived 225 million years ago. Adults barely grew to 2m (6½ft) long and weighed around 12–15kg (26–33lb). A tiny, primitive cousin of later giants such as *Brachiosaurus* (see p.267), *Saturnalia* lived alongside a curious mixture of early dinosaurs and more primitive reptiles such as rhynchosaurs and dicynodonts.

SOCKET-TOOTHED LIZARD
Thecodontosaurus means "socket-toothed lizard", a name that reflects this plant-eater's blunt, leaf-shaped teeth with serrated edges.

Thecodontosaurus

GROUP Sauropodomorphs
DATE Late Triassic
SIZE 1–3m (3¼–9¾ft) long
LOCATION British Isles

About 205 million years ago, much of the Bristol Channel area of western England was filled with a network of caves. An entire ecosystem of small creatures inhabited these caverns, including some of the earliest mammals and many types of lizards. Dinosaurs also lived here, including the common, primitive sauropodomorph *Thecodontosaurus*. This skinny herbivore, which grew to no more than 3m (9¾ft) in length and weighed between 20–30kg (44–66lb), took refuge in the caves when threatened by predators. Usually this was an effective strategy, but every so often a cave collapsed, entombing *Thecodontosaurus* and other animals that were inside. Today, the fossilized remains of these animals are common discoveries around Bristol and in South Wales.

Plateosaurus

GROUP Sauropodomorphs
DATE Late Triassic
SIZE 6–10m (20–33ft) long
LOCATION Germany, Switzerland, Norway, Greenland

The prosauropod *Plateosaurus* is one of the best-understood dinosaurs. Most dinosaurs are known from only a few scraps of bone, or at most, a single skeleton, but *Plateosaurus* is represented by more than 50 complete skeletons. Most of these have come from the Late Triassic basins of Germany, which were deposited 220 million years ago by a series of large rivers. Other specimens have turned up in the glacier-covered rocks of Greenland, and even in drill cores from the bottom of the North Sea. These specimens have allowed scientists to study this dinosaur in remarkable detail. *Plateosaurus* was one of the largest prosauropods, and some individuals were 10m (33ft) long and weighed 700kg (1,540lb). Its long mouth was crammed full of leaf-shaped teeth, and these were covered in rough bumps (denticles) that were perfect for grinding vegetation. It is likely that *Plateosaurus* could walk on either two or four legs, although recent research suggests that this animal's arms and hands were not very suitable for locomotion and instead were used mainly for collecting food. Its name means "flat lizard".

Plateosaurus could **increase** or **decrease** its **growth rate**, depending on the **season**.

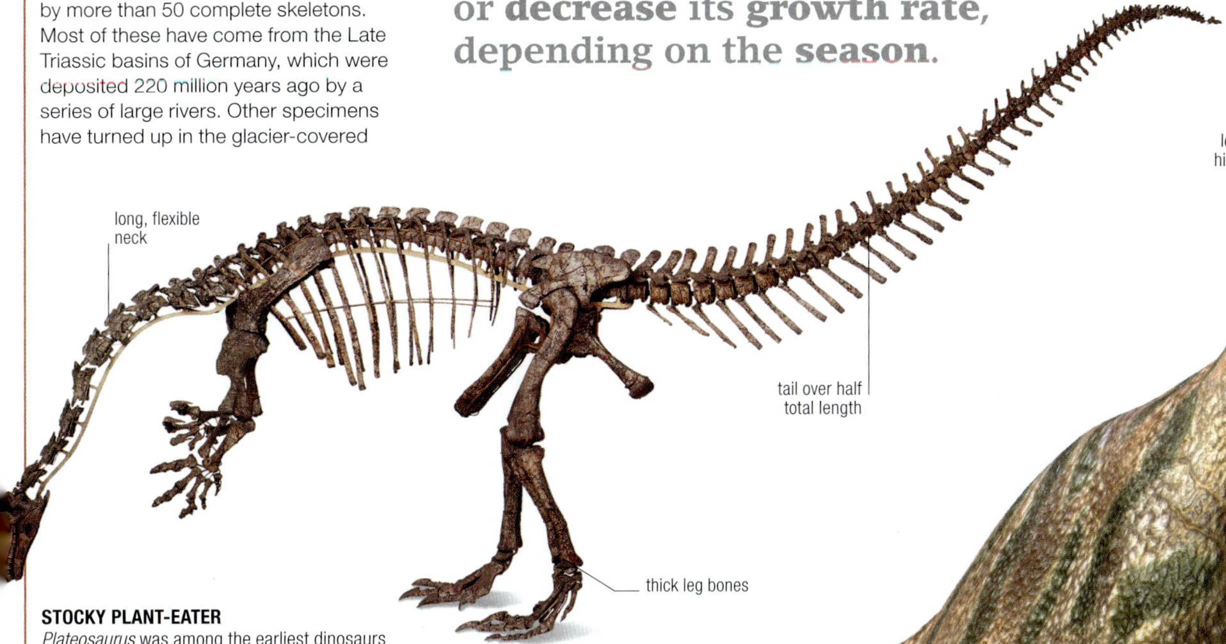

STOCKY PLANT-EATER
Plateosaurus was among the earliest dinosaurs to reach a large size – as shown by its stocky, thick leg bones.

SMALL BUT EFFECTIVE
The skull of *Plateosaurus* was relatively small and long, yet its jaws hosted numerous rough, serrated teeth that were ideally designed for grinding and tearing the plants that made up its diet.

HIGH FEEDER
Plateosaurus's sturdy, weight-bearing bone structure allowed it to stand upright, which in turn enabled it to feed off plants out of reach of other herbivorous dinosaurs and reptiles.

Eocursor

GROUP Ornithischians
DATE Late Triassic
SIZE 1m (3ft 3in) long
LOCATION South Africa

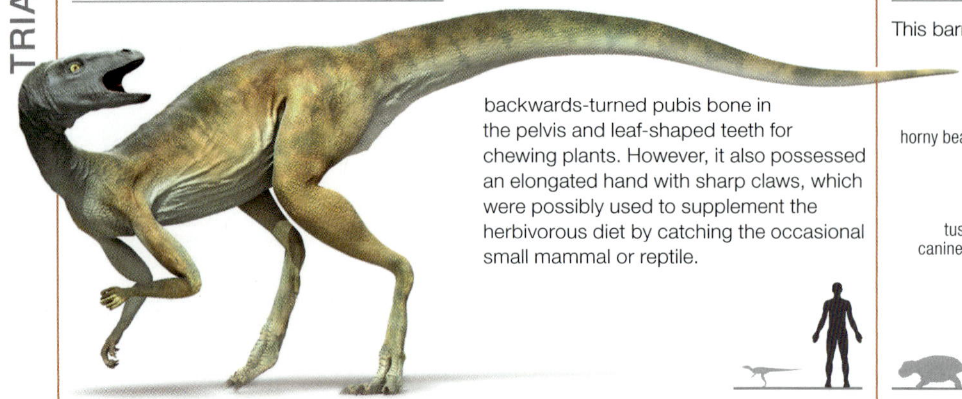

One of the few known Triassic ornithischians, *Eocursor* was a small and extremely fast herbivore from what is now South Africa. It had most of the major features of ornithischians, including a backwards-turned pubis bone in the pelvis and leaf-shaped teeth for chewing plants. However, it also possessed an elongated hand with sharp claws, which were possibly used to supplement the herbivorous diet by catching the occasional small mammal or reptile.

Lystrosaurus

GROUP Synapsids
DATE Late Permian to Early Triassic
SIZE 1m (3ft 3in) long
LOCATION Global (southern continents)

This barrel-chested, pig-like herbivore lived across half the world and survived the violent Permo-Triassic mass extinction that wiped out up to 95 per cent of all species. *Lystrosaurus* had a wide body, a slow, sprawling gait, and an enormous skull with a horny beak and two tusk-like canine teeth. The canines were probably used for display or defence, and the beak was a perfect tool for shearing through plant stems and branches. It probably weighed less than 100kg (220lb).

Placerias

GROUP Synapsids
DATE Late Triassic
SIZE 2–3.5m (6½–11½ft) long
LOCATION USA

Dicynodonts, reptile-like cousins of mammals, were some of the most successful herbivores of the Late Permian to Middle Triassic periods. One of the last surviving members of this once-great group was *Placerias*. Resembling an ancient hippopotamus, *Placerias* reached weights of as much as 2,000kg (4,400lb). It lived approximately 220 million years ago in southwestern USA and was the largest herbivore in the region. Although it lived alongside early carnivorous dinosaurs, *Placerias* was such a successful plant-eater that it was much more abundant than rival herbivores. Its skull was ornamented with two enormous tusks that were probably used for social display. Fossil discoveries suggest *Placerias* travelled in large herds.

Kannemeyeria

Group Synapsids
Date Early to Middle Triassic
Size 3m (9¾ft) long
Location South Africa, China, India, Russia

Only a few groups of therapsids survived into the Triassic Period, including the dicynodonts. The rhinoceros-sized *Kannemeyeria* was descended from one of these and had large canine tusks but otherwise lacked teeth. Instead, it possessed a beak for cropping food, and horny plates lined its mouth, just as in modern turtles. Tough plant matter was pulped by the forwards and backwards closing motion of the jaws. This body plan for herbivores was clearly successful, because it remained largely unchanged for the next 50 million years in successive dicynodonts after *Kannemeyeria*.

Thrinaxodon

GROUP Synapsids
DATE Early Triassic
SIZE 45cm (17½in) long
LOCATION South Africa, Antarctica

Thrinaxodon, a cat-sized predator, was the most common cynodont of the Early Triassic. Like mammals, it had a distinct waist in front of its hips. In addition, its broad ribs interlocked with each other, stiffening its body and making it difficult to breathe via rib movement alone. This suggests that *Thrinaxodon* used a diaphragm to breathe (a more efficient method) and points to an evolving mammal-like metabolism and similar activity levels. *Thrinaxodon's* limbs were held almost under its body, as in mammals, and this further supports the suggestion that *Thrinaxodon* was an active animal, capable of running.

MAMMAL-LIKE MOTION
Unlike in earlier synapsids, where the limbs were held out to the side in a lizard-like posture, *Thrinaxodon's* legs were placed almost underneath its body.

2,000kg The maximum weight of the hippo-like dicynodont *Placerias*.

"REPTILIAN HIPPO"
Placerias has often been called a "reptilian hippo", due to its similar body shape and weight. Over 40 *Placerias* skeletons have been unearthed in Petrified Forest National Park, Arizona, USA.

Cynognathus

Group Synapsids
Date Early to Middle Triassic
Size 1.8m (6ft) long
Location South Africa, Antarctica, Argentina

Cynognathus means "dog-jaw", and this mammal-like reptile had a large canine tooth on either side of its upper and lower jaws. A single bone, the dentary, made up most of its lower jaw, a feature more common in modern mammals than reptiles (see p.182). *Cynognathus* was probably an extreme carnivore and ate nothing but meat. It possessed blade-like teeth behind its canine teeth for slicing flesh. Its skull also had a wide area for holding large jaw muscles, suggesting that *Cynognathus* possessed a powerful bite.

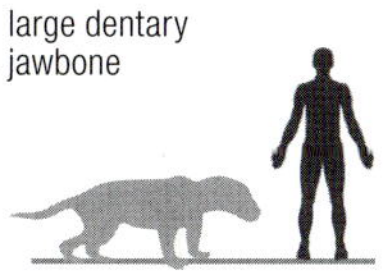

A JAW FOR HEARING
The bones that form the jaw joint in cynodonts are the same as two of the middle ear bones in mammals. The small size of these jawbones in *Cynognathus* means that they were probably already used for hearing.

Morganucodon

GROUP Synapsids
DATE Late Triassic to Early Jurassic
SIZE 9cm (3½in) long
LOCATION British Isles, China, USA

Morganucodon, one of the earliest true mammals, was described in 1949 from thousands of teeth, jaws, and fragments of bone from a South Wales quarry. A small, shrew-sized insect-eater, *Morganucodon* was probably nocturnal and had a good sense of smell. Like the most advanced cynodonts, *Morganucodon* had a double jaw joint – the new mammalian joint lying alongside the old reptilian one. The reduced size of the old jaw joint bones indicates more acute hearing in *Morganucodon*. Many early mammals retained this condition, and fossils show that the fully enclosed middle ear and single jaw joint of modern mammals evolved independently in monotremes (see p.357) on the one hand, and the marsupials and placental mammals on the other.

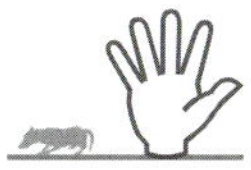

JURASSIC

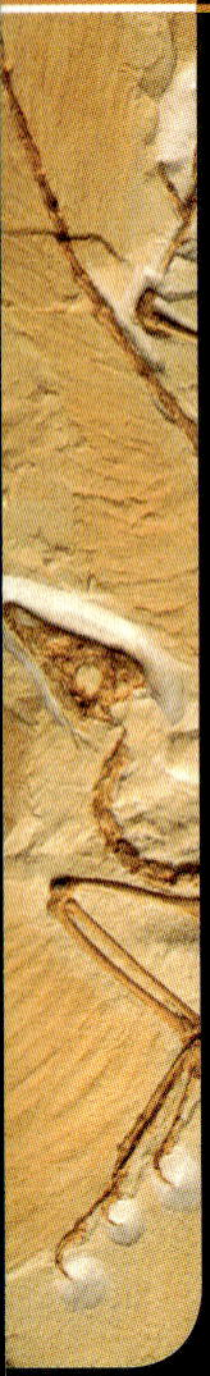

Jurassic

The word Jurassic conjures up a picture of life on a large scale, with large dinosaurs roaming the land, huge reptiles surging across the sea, and flying reptiles soaring in the air. The oceans teemed with new predators, including ammonites, belemnites, and a range of shell-crushing fish. On a much smaller scale. the plankton were evolving and changing ocean chemistry.

DURDLE DOOR
The spectacular limestone arch at Durdle Door on England's Dorset coast formed in the same tectonic event that produced the Alps.

DINOSAUR TRACKWAYS
The parallel trackways of Jurassic sauropods on an ancient lakeshore are preserved in sedimentary rocks in the Morrison Formation, in southeast Colorado, USA.

Oceans and continents

Plate movement continued to reshape continents and widen oceans during the Jurassic period. The separation of the northern and southern parts of the supercontinent Pangaea that began during the Triassic continued during the Early Jurassic, enlarging the Tethys Ocean. This ran in an east–west direction, which had a significant effect on ocean flora and fauna, and on the world's climate. Currents predominantly flowed east to west, with the result that the climate did not vary much in a north–south direction. Moreover, it allowed the east–west radiation of animals and plants, with many groups spreading mainly eastwards. Fossils found in limestones in Western Australia are therefore remarkably similar to those found on the southern coast of England. In the Middle Jurassic, another ocean, the Proto-Atlantic, began to open as America moved north-westwards. During the later Jurassic, Laurasia was separated further from Pangaea by the northwards movement of North America away from northwest Africa. This resulted in the widening of the western side of the Tethys Ocean. As Pangaea continued to be torn apart by rifting and subduction (see p.20), shelf areas around the continents expanded and shallow-water environments become globally widespread. In the oceans, new types of plankton emerged, with calcareous and siliceous skeletons, which formed limestone and silica-rich sediments. Massive limestone rocks form the Jura Mountains on the border of France and Switzerland, after which the Jurassic is named.

ZION NATIONAL PARK
Vast sand dunes accumulated in the western interior of North America throughout the Jurassic, creating the dramatic red formations exposed today in Zion National Park in Utah, USA.

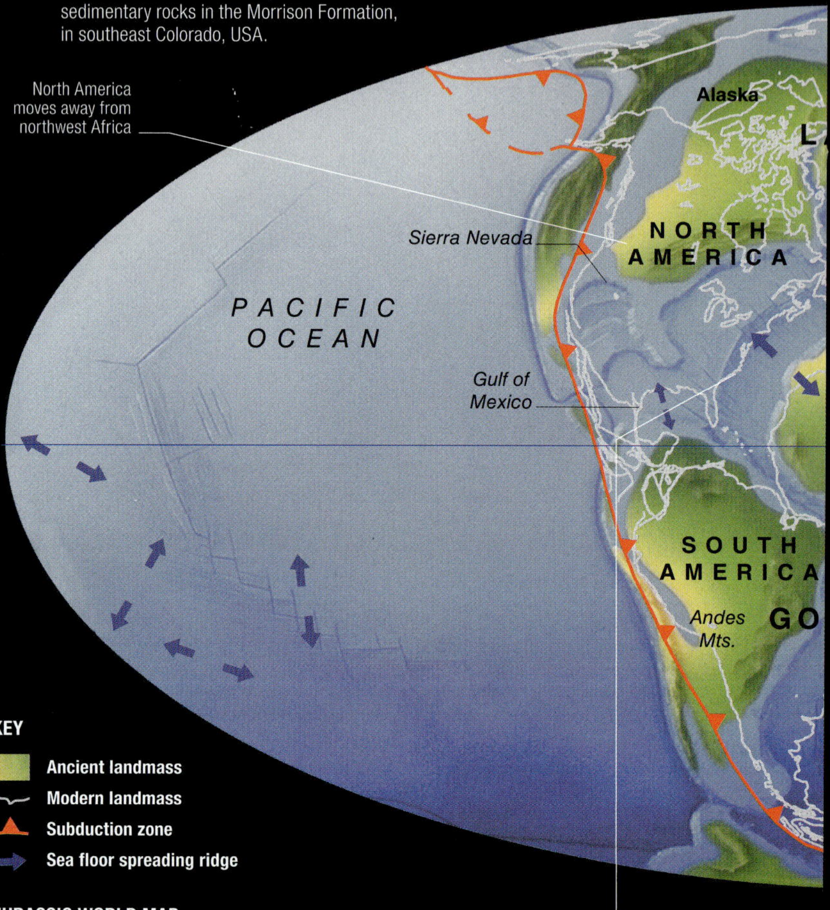

JURASSIC WORLD MAP
The northern part of Pangaea (Laurasia) and the southern part (Gondwana) continued to separate, widening the Tethys Ocean.

EARLY

MYA	200	190	180
PLANTS	 WILLIAMSONIA	195–145 Floras dominated by conifers, ginkgoales (especially northern mid- to high-latitudes), ferns; bennettitaleans, and cycads diversified (especially in lower latitudes); seed ferns declined in diversity, only caytoniaceous seed ferns dominant	
INVERTEBRATES	 PACHYTEUTHIS • 195 First lytoceratine ammonoids, ammonites, and belemnites (left)	• 190 First Lepidoptera (*Liadotaulius*), hermit crabs, modern starfish (right), and irregular echinoids; extensive development of sponge reefs; first corbulid bivalves, first tubulipore bryozoans  PENTASTERIA	• 180 First arcid bivalves • 175 Last spiriferid brachiopods; first spondylid and pholadid bivalves
VERTEBRATES	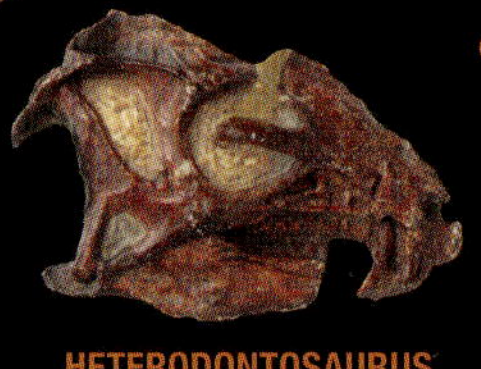 HETERODONTOSAURUS • 195 First pterosaur with specialized feeding (*Dorygnathus*); first sauropod dinosaurs; diversification of small ornithischian dinosaurs (heterodontosaurids (left), fabrosaurids, and scelidosaurids)	• 190 First pliosaurs  LIOPLEURODON	

GINKGO TREE
One species of the genus *Gingko* survives today and is popularly known as the maidenhair tree. During the Jurassic, this genus diversified and spread through Laurasia.

Climate

Global climates were warm and humid, with temperatures of up to 30°C (86°F) until the Middle Jurassic. Subtropical climates may have extended as far north as 60° latitude. Low latitudes, including present-day southwestern North America, South Africa, and South America were more arid. Middle latitudes, including present-day Western Australia, India, and the southern tip of south America, were seasonally arid. High-latitude temperatures in Australia during the Mid Jurassic probably averaged about 15–18°C (59–64°F). Higher latitudes were wetter in both the northern and southern hemispheres, with lush forests, demonstrated by the occurrence of coal deposits. In Laurasia these areas became drier during the Jurassic due to breakdown of monsoonal circulation that had been well established in the Early Jurassic. Global temperatures declined during the Late Jurassic, with relatively cool oceanic temperatures, measuring about 20°C (68°F). Pronounced seasonality, where temperatures swing from extremes of heat in summer to very cold in winter, was a feature of the Late Jurassic. At this time an arid equatorial belt spread across the Equator, up to about 45 to 50° latitude, a narrow belt of seasonally wet climates extended to 60°, and wet temperate conditions existed towards the poles. In the wetter regions, forests that consisted primarily of conifers, laid down extensive coal deposits. Mid- to high-latitude floras were dominated by *Ginkgo*-like plants and ferns, as well as conifers. Benettitaleans (cycad-like plants) and cycads were the main plants in low latitudes.

TEMPERATE RAINFOREST
Temperate forests flourished in moist conditions in both hemispheres. Conifers predominated, but there were also ferns, like these in south Australia.

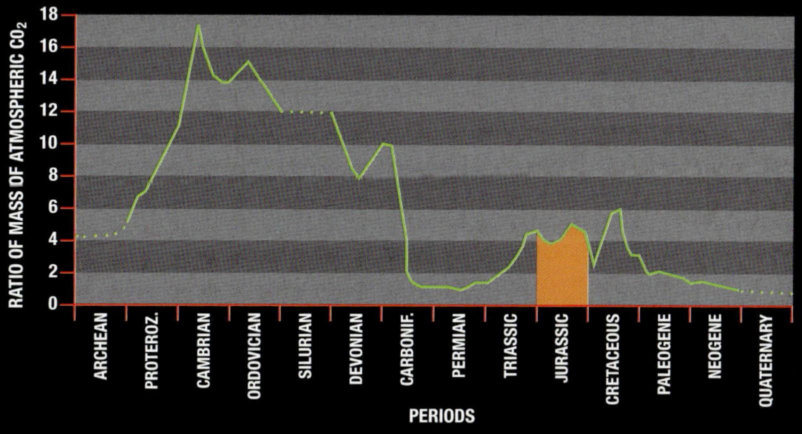

CARBON DIOXIDE LEVELS
Studies have shown that carbon dioxide levels were fairly high at the Triassic–Jurassic boundary. They dip and increase again with the warm greenhouse conditions that existed during the Jurassic.

DDLE | LATE

160 | 150 | 140 | MYA

PLANTS

SEQUOIADENDRON

INVERTEBRATES

● **165** First glycymeridid bivalves

● **155** First blood-sucking insects (ceratopogonids), rudist bivalves, and cheilosome bryozoans

VERTEBRATES

…) First salamanders and …wts; first cryptoclidid …d elasmosaurid …siosaurs; cynodonts …inct; first cladotherian …mmals; sauropod …osaurs (right) diversify

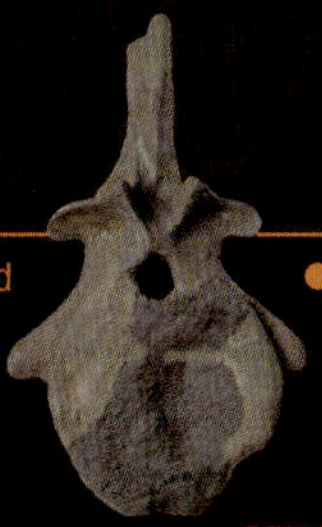

CETIOSAURUS

● **165** First rays

PTERODACTYLUS

160–145 Pterosaurs (left) diversify

● **155** First bird (*Archaeopteryx*), triconodontid and symmetrodont mammals, and true lizards; diversification of stegosaur and theropod dinosaurs

170–145 Multituberculate (extinct, rodent-like) mammals diversify

JURASSIC PLANTS

During the Jurassic, the new kinds of plants that developed in the Triassic continued to diversify. Some of the best information on Jurassic plants comes from studies of the rich and well-preserved Middle Jurassic flora found in Yorkshire, northern England.

Jurassic plants are known from many parts of the world, but fossils from several localities in northern England have told us much of what we know about Jurassic plants. This so-called Yorkshire Jurassic flora has been studied since the earliest days of palaeontology, and attempts to reconstruct extinct plants from the different dispersed parts preserved in Yorkshire flora have been especially successful.

Before flowering plants

Ferns and conifers are common in many floras from Yorkshire, but cycad-like leaves are also prominent. Some of these leaves were certainly produced by true cycads, but some are bennettitalean leaves, or may belong to other kinds of plants that are distantly related to cycads. Bennettitaleans, for example, may be closer to living Gnetales than to other groups of living seed plants. Another interesting group of extinct Jurassic seed plants are the caytonialeans. They have leaves with four leaflets, all of which have net venation similar to that of glossopterids. Caytonialeans were originally thought to be related to angiosperms, but later research showed that they lacked the key reproductive specializations characteristic of this group.

CLASSOPOLLIS POLLEN
The characteristic *Classopollis* pollen of the fossil conifer family Cheirolepidiaceae is one of the most common components of fossil pollen and spore floras in the later Jurassic and Early Cretaceous.

Plants and insects

The evolution of land plants and insects had been closely linked since the Devonian, but during the Mesozoic several living orders of insects, especially flies and beetles, diversified rapidly. Many of these insects probably fed on plants or on decaying plant matter, but some may also have been involved in pollination. By the Jurassic it is probable that insects were already involved in the transfer of pollen from one plant to another, which brought a new dimension to plant production, and a new potential for specialization. Insects play an important role in the pollination of living cycads and Gnetales. There were also Jurassic bennettitaleans that had both pollen-producing and ovule-producing parts. Such bisexual "flowers" were probably insect pollinated.

FOSSIL BEETLE
The diversification of beetles was well underway by the Jurassic. It is possible that some species may have contributed to the pollination of seed plants during this period.

GROUPS OVERVIEW

The major groups of plants that flourished during the Jurassic were very similar to those of the Late Triassic. Ferns and a variety of seed plants were the dominant groups. Conifers, cycads, and bennettitaleans were the main groups of seed plants, along with a few other smaller lineages, such as caytonialeans and czekanowskialeans.

FERNS
The dominant fern families of the Triassic, Jurassic, and Early Cretaceous were lineages that diverged very early from the main line of fern evolution. They were unusual compared to most modern ferns in that they flourished in open habitats. These ancient ferns lived in similar habitats to those occupied by grasses today.

CONIFERS
Conifers are important in most Jurassic floras and in general were more similar to the living families of conifers than to the voltzialeans from the Triassic. Living groups of conifers that can be recognized in Jurassic fossils include forms related to the swamp cypress and yew families. The monkey-puzzle family is also widely represented.

CYCADS
Cycads were diverse during the Jurassic but the best-understood fossil cycads had pollen and seed cones very similar to those of living forms. However, the fossil leaves were much smaller than those of living cycads, suggesting that the stems of these ancient forms were slender, and perhaps more highly branched.

GINKGOS
Several different kinds of *Ginkgo*-like plants are known from the Jurassic and during this interval the group reached its maximum diversity. Especially well understood are several different *Ginkgo*-like plants from the Jurassic of China. Some of these differ very little from the single living species *Ginkgo biloba*.

Selaginella

GROUP Lycophytes
DATE Carboniferous to present
SIZE 10cm (4in) high
LOCATION Worldwide

These small plants look like mosses, but they have two types of simple leaf – large and small – that are arranged in four rows along the stem. More importantly, they produce spores from spore-producing structures (sporangia), grouped into small cones, rather than from simple spore capsules, as true mosses do. The oldest-known *Selaginella* remains are over 300 million years old, making it a plant genus with one of the longest-known fossil records.

SELAGINELLA ZEILLERI
This fossilized leafy shoot of *Selaginella zeilleri* shows the rows of small and large leaves that were attached to the stem.

LIVING RELATIVE
SELAGINELLA

Modern *Selaginella* are usually either small or scrambling plants, but occasionally they grow on the branches of trees. About 700 species are alive today. Most are found in tropical and warm temperate regions, but some are found in cooler climates.

Coniopteris

GROUP Ferns
DATE Triassic to Neogene
SIZE Frond 1m (3ft 3in) long
LOCATION Worldwide

This is one of the most diverse and widespread genera of Jurassic ferns. Some fronds were vegetative, with leaflets (pinnules) that had broad lobes to collect light for photosynthesis. Other fronds, with more slender leaflets, were fertile, with small clusters of spore-producing sacs (sporangia) at the end of the leaflets. The detailed structure of the sporangia shows that these fossil ferns were related to the modern-day tree fern *Dicksonia*. At least some of the Mesozoic *Coniopteris* fronds also came from tree ferns, although others may have grown on plants that were much smaller.

CONIOPTERIS HYMENOPHYLLOIDES
Broken fragments of the dicksoniaceous fern *Coniopteris* include remains of stem and rachis, and leaflets from both fertile and sterile fronds.

Equisetites

GROUP Sphenophytes
DATE Late Carboniferous to present
SIZE 2.5m (8¼ft) tall
LOCATION Worldwide

Fossil casts of *Equisetites* stems, with clear ribs running along their length and more widely spaced "cross-bars" (nodes), are frequently found in the Jurassic. They are particularly common in rocks that formed along river-banks or the shores of lakes, where the plants probably grew at the waters edge. They look very similar to the stems of living *Equisetum*, but many are larger than any living species.

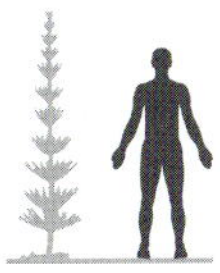

LIVING RELATIVE
HORSETAILS

Horsetails (*Equisetum*) remain a widespread and common group of plants today. They spread easily using underground stems called rhizomes. Their scaly leaves are arranged in whorls around the stem at characteristic nodes.

Phlebopteris

GROUP Ferns
DATE Late Triassic to Cretaceous
SIZE Frond 20cm (8in) wide
LOCATION Worldwide

This fossil fern is usually found as fragments of divided leaf-segments (pinnae), with slender leaflets (pinnules) attached. Occasionally complete leaves are found with several leaflets that appear to radiate from the end of a stalk. The fronds were probably produced by an underground, creeping stem (rhizome). The modern-day genus *Matonia* belongs to the same family. It has similar leaves, and also bears similar clusters of spore-producing organs (sporangia). In the Jurassic period, Matoniaceae ferns were widespread. Today, they are restricted both ecologically and geographically.

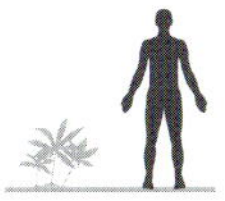

PHLEBOBTERIS SP.
Numerous small, thin pinnules attached to a slender, central axis (rachis) are visible in this fossilized pinna.

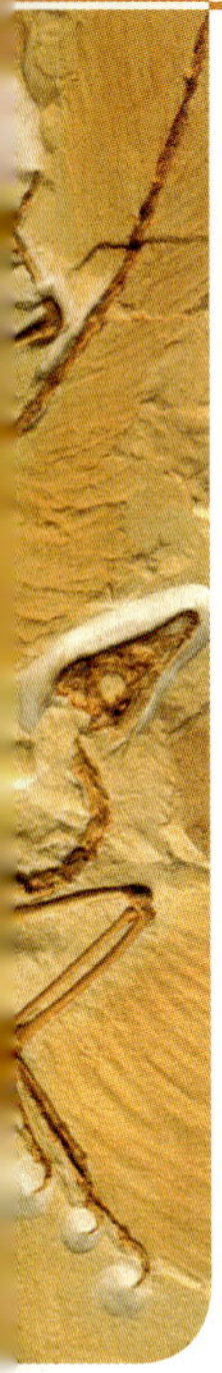

Dictyophyllum

GROUP Ferns
DATE Late Triassic to Middle Jurassic
SIZE Frond 20–30cm (8–12in) wide
LOCATION Worldwide

Just like members of the Matoniaceae, such as *Phlebopteris* (see p.227), *Dictyophyllum* had leaf-segments (pinnae) that radiated from the end of a stalk, which was attached to an underground, creeping stem (rhizome). However, this fern is distinctive in the way in which the veins formed a meshwork across the leaflets (pinnules). The veins, and the structure of the spore-producing bodies (sporangia), suggest that *Dictyophyllum* belonged to the family Dipteridaceae. Today, this family only grows in Southeast Asia. During Jurassic times *Dictyophyllum* was widespread, but it declined and eventually became extinct later in the Jurassic.

DICTYOPHYLLUM NILSSONII
This partial leaf of a Jurassic fern has characteristic meshed veins. Ferns with this type of mesh veins often favoured drier habitats.

LEAF VEINS
Each triangular leaflet has a central, thick midvein. On either side of the midvein there are lateral veins that create a polygonal mesh.

Todites

GROUP Ferns
DATE Late Triassic to Jurassic
SIZE Frond 1m (3ft 3in) long
LOCATION Europe, Central Asia, China

Todites leaves are common in Jurassic floras. These fertile leaves often have clusters of spore-bearing sacs (sporangia) on the underside of the leaflets (pinnules) that are similar to those on modern-day members of the fern family Osmundaceae. The most similar living fern is *Todea* from Australasia and South Africa. Because the sporangia are slightly different in their detailed structure, the Jurassic leaves have been given a different name, *Todites*. The vegetative leaves without sporangia are called *Cladophlebis* (see right).

TODITES SP.
The leaflets of fertile leaves, such as those seen on this fossilized leaf fragment from the Jurassic, often look quite different to the leaflets of the sterile leaves.

Cladophlebis

GROUP Ferns
DATE Triassic to Cretaceous
SIZE Frond 1m (3ft 3in) long
LOCATION Worldwide

Both extinct and modern-day members of the fern family Osmundaceae, to which *Cladophlebis* belongs, have non-fertile vegetative leaves and fertile leaves that have spore-bearing sacs (sporangia) – as seen in the living royal fern (*Osmunda regalis*). These two types of frond can look quite different and their fossils are often found separately. Palaeontologists therefore give them different genus names: *Cladophlebis* for the vegetative fronds with distinctive, triangular leaflets, and *Todites* for the fertile fronds (see left).

CLADOPHLEBIS FUKIENSIS
This fossil from China shows part of a typical sterile frond of an osmundaceous fern. Its small triangular leaflets never bore sporangia.

OSMUNDACAULIS SP.
Petrified remains of Jurassic osmundaceous fern stems allow details of their anatomy to be studied on the cut-surface. From this, it is possible to determine patterns of evolution and diversity in this group of ferns. In this example, from Queensland, Australia, several vascular strands can be seen that run through the length of the stem.

Klukia

GROUP Ferns
DATE Late Triassic to Early Cretaceous
SIZE Frond 30–50cm (12–19½in) long
LOCATION Europe, Central Asia, Japan

The distinctive, small, tongue-shaped leaflets (pinnules) of *Klukia* ferns are common in Jurassic floras. The fertile leaves are particularly distinctive. They have quite large spore-bearing sacs (sporangia) that are borne singly on the underside of the leaflets. This is quite different from most other Jurassic ferns where the sporangia occur in a series of clusters (sori) on the pinnules. Members of this genus are most closely related to the modern-day climbing ferns of the family Schizaeaceae, which grow today in tropical and subtropical areas.

***Klukia's* distinctive pinnules are often found in Jurassic floras.**

KLUKIA EXILIS
The small, circular marks on this fossil are the distinctive sporangia that are borne on the undersides of the pinnules.

Pseudoctenis

GROUP Cycads
DATE Late Permian to Cretaceous
SIZE Leaf 1m (3ft 3in) long
LOCATION Worldwide

Leaves that are divided into a series of elongate leaflets (pinnae) and attached to a central axis (rachis) are characteristic of many Mesozoic fossil floras. They are similar to the leaves of modern-day cycads and some, but not all (see p.231), are actually cycad remains. Various genera have been identified based on the shape and veining of the leaflets and how they are attached to the rachis. *Pseudoctenis* leaves were produced by true cycads. They have unbranched veins and leaflets that were attached to the sides of the rachis. Although fossilized examples are not as abundant as some other cycad leaf fossils, such as *Nilssonia*, they are very widespread. *Pseudoctenis* plants occurred in both tropical and temperate floras. Examples have even been found in Permian rocks, making them among the oldest-known cycad fossils.

PSEUDOCTENIS HERRIESII
This fossil of a Jurassic cycad leaf clearly shows a thick central rachis, with slender pinnae on either side. Each pinna has numerous fine veins running along its length.

Caytonia

GROUP Caytonialeans
DATE Late Triassic to Early Cretaceous
SIZE Cone 5cm (2in) long
LOCATION Europe, North America, Central Asia

Caytonia is the seed-bearing organ of a group of extinct plants called *Caytoniales*, which were abundant in subtropical areas during the Jurassic. Several seeds were enclosed in the protective, helmet-shaped, seed-bearing structures (cupules) that were arranged in two rows on either side of a central axis (rachis). The enclosure of the seeds within the cupules superficially resembles how today's flowering plants bear their seeds. The leaves produced by the plant also had features in common with flowering plants, particularly their mesh venation. For many years it was thought that *Caytonia* might have been a direct ancestor of the flowering plants, although this idea has now been rejected.

CAYTONIA SP.
Seed-bearing cupules are attached on either side of the main axis. Remains of several of the seeds that were borne in each cupule are just visible.

SAGENOPTERIS NILSSONIANA
The plants that produced *Caytonia* seed-bearing structures had characteristic leaves, called *Sagenopteris*, with four leaflets and mesh veins (one leaflet is missing from this example). The leaves are far more commonly found than the cones, and give a better idea of the distribution of these plants.

Androstrobus

GROUP Cycads
DATE Late Triassic to Early Cretaceous
SIZE Cone 5cm (2in) long
LOCATION Europe, Siberia

Androstrobus is the name given to pollen-producing cones that are found together with cycad leaves, such as *Pseudoctenis* (see above); they are thought to be part of the same plant. Their spirally arranged structure is thought to be made by modified leaves (sporophylls), each of which has an upturned scale at the end. The scales overlap each other to provide protection for the numerous pollen sacs (sporangia) on the underside of the sporophylls.

ANDROSTROBUS PICEOIDES
These are two pollen-producing cones of a Jurassic cycad. Each cone has spirally arranged, bract-like scales, with pollen-bearing sacs on the lower surface.

Williamsonia

GROUP Bennettitaleans
DATE Early Jurassic to Late Cretaceous
SIZE Flower 10cm (4in) long
LOCATION Worldwide

The bennettitaleans were a characteristic group of Jurassic plants. Many species had a stocky trunk, up to 2m (6½ft) high, with cycad-like leaves and distinctive flower-like reproductive structures. A particular type of seed-producing flower, known as *Williamsonia*, had a central dome-shaped structure (receptacle) to which numerous stalked seeds, separated by scales, were attached. The entire flower was surrounded by a protective layer of bracts. Often, these bracts are the only part that has been fossilized. Several aspects of the complex structure of *Williamsonia* suggest that these plants may have been related to the ancestors of flowering plants. However, the exact relationship between bennettitaleans and flowering plants is still unclear.

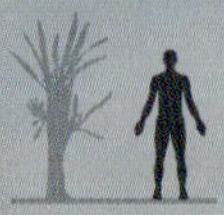

WILLIAMSONIA GIGAS
This fossil, preserved in an ironstone nodule, shows the ring of protective bracts that surrounded the bennettitalean cone – seen from the underside.

SCALY TRUNKS
The plants that bore *Williamsonia* flowers had stocky trunks with the flowers and leaves borne at the top. The scaly surface of the trunk was created by the bases of dead leaves.

Zamites

GROUP Bennettitaleans
DATE Late Triassic to Late Cretaceous
SIZE Leaf 50cm (19½in) long
LOCATION Worldwide

These commonly found plant fossils look very similar to the leaves of modern-day cycads. However, the detailed cellular structure of *Zamites*, particularly the small breathing pores (stomata) on the underside of the leaves, is quite different. It is now known that these cycad-like fossils are from plants that also bore the flower-like reproductive structures that are characteristic of all the bennettitaleans. In fact, the majority of cycad-like leaves found in Jurassic floras are likely to be a type of bennettitalean rather than true cycads. The genus *Zamites* is a good indicator of warm Jurassic climates as it only occurred in tropical or subtropical conditions.

ZAMITES GIGAS
This fossil shows typical bennettitalean leaves. It is from the Jurassic plant beds exposed on the Yorkshire coast in England.

Anomozamites

GROUP Bennettitaleans
DATE Middle Triassic to Early Cretaceous
SIZE Frond 30cm (12in) long
LOCATION Worldwide

Many bennettitaleans produced fronds that were only divided once. The different types of these plants can be distinguished by the style of the leaflets (pinnae). In contrast to *Zamites* (above), *Anomozamites* had broader pinnae that were not constricted at the base. Like *Zamites*, they also had the same type of breathing pores (stomata) on the underside that characterize all bennettitaleans. *Anomozamites* were probably more tolerant of cooler climates, as fossilized remains of the leaves have been found in Siberia.

ANOMOZAMITES MAJOR
This specimen shows a small bennettitalean leaf, with broadly attached pinnae and forking veins. This leaf form is typical of the mainly Jurassic genus *Anomozamites*.

Weltrichia

GROUP Bennettitaleans
DATE Jurassic
SIZE Cone 10cm (4in) across
LOCATION Worldwide

The bennettitalean plants that bore *Williamsonia* (see left) are also thought to have borne pollen-producing flowers that are known as *Weltrichia*. These flowers were made up of groups of long, leaf-like bracts that had pollen-sacs on their upper surface. In some species, there were also small, stalked bodies on the upper surface that may have secreted a nectar like substance to attract pollinating insects. The bracts radiated from a cup-shaped base (receptacle), giving it a flower-like appearance. *Weltrichia* has never been found directly attached to a plant and so it is hard to know exactly how the flowers were borne. They are, however, almost always found associated with *Williamsonia*, as well as with their leaves known as *Ptilophyllum*, and there is little doubt that they belonged to the same plants.

WELTRICHIA SPECTABILIS
This is an example of an almost complete male flower showing the bracts and pollen-bearing sacs. Also present is part of the leaf (known as *Ptilophyllum*) of the same plant.

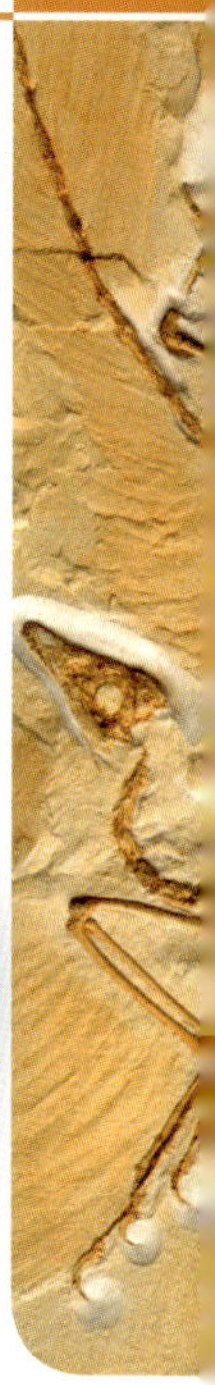

Czekanowskia

GROUP Czekanowskialeans
DATE Late Triassic to Early Cretaceous
SIZE Leaf 20cm (8in) long
LOCATION Europe, Greenland, North America, Central Asia, Siberia, China

This enigmatic genus was once thought to be related to *Ginkgo*, but it is now classified in its own extinct group: the czekanowskialeans. The leaves look superficially like clusters of pine needles, except for the fact that they fork several times. They are fused at the base to form a short shoot. The plants that bore these leaves produced seeds in loose, cone-like structures called *Leptostrobus*. Czekanowskialeans mainly seem to have favoured warm or hot, humid conditions.

LEPTOSTROBUS LUNDBLADIAE
These cones, which were produced by the plant that had *Czekanowskia* leaves, were made up of a series of small, two-valved capsules arranged around a thin axis. When mature, the capsules would open up to reveal up to five seeds.

CZEKANOWSKIA ANGUSTIFOLIA
This fossil shows a short shoot with a group of slender leaves that are fused together at the base.

Ginkgo

GROUP Ginkgos
DATE Late Triassic to present
SIZE 50m (164ft) tall
LOCATION Worldwide

Identifiable by their distinctive leaves, ginkgos were widespread and abundant in Jurassic times, but went into decline during the Late Cretaceous period. Today, there is just one living species, *Ginkgo biloba* (the maidenhair tree), which is only found growing in the wild in the mountains of China. This species is probably one of the best examples of a "living fossil" in the entire plant kingdom. Just like its fossilized cousins, the beautiful fan-shaped leaves with many fine veins are easy to recognize. Many different species of fossilized ginkgos have been identified by the shape and number of the lobes on the leaves, and the shape of the leaf as a whole. Sometimes they are found together with small seeds that are attached to the ends of stalks. These seeds and seed stalks are very similar to those produced by the living *Ginkgo*. In other species, there were more complex clusters of seeds that are known as *Karkenia*.

GINGKO SP.
This beautiful fossil from Afghanistan shows the fan-shaped leaves and radiating veins that are characteristic of most ginkgoalean plants. It is not known what sort of seeds this particular species produced so it is not clear exactly how closely related it is to today's *Ginkgo biloba*. Nevertheless, palaeontologists still usually refer to such leaves as *Ginkgo*.

SCIENCE
GINKGO BIOLOGY

Ginkgo seeds have a fleshy outer covering and a hard inner layer. They germinate better if the outer fleshy layer is removed. In prehistoric times, dinosaurs or ancient mammals may have been attracted to these succulent fruits, eaten them, and eventually dispersed the seeds in their droppings.

Podozamites

GROUP Conifers
DATE Late Triassic to Late Cretaceous
SIZE Leaf 8cm (3¼in) long
LOCATION Worldwide

The leaves of most conifers are either short and scaly, or slender needles; both types have a single vein along their length. However, some Jurassic conifers, such as *Podozamites*, had broader leaves with several veins running along their length. It was originally thought that *Podozamites* leaves belonged to cycads, but studies of the structure of their breathing pores (stomata), on the underside of leaves, suggest that they were conifers.

PODOZAMITES DISTANS
This specimen is part of a fossilized shoot of this unusual conifer. The leaves are wide and each leaf has several veins.

Elatides

GROUP Conifers
DATE Middle Jurassic to Early Cretaceous
SIZE 30m (98ft) tall
LOCATION Europe, Canada, Siberia, Central Asia, China

The remains of conifer shoots and cones are common in Jurassic floras, and can often be assigned to the same families as living conifers. *Elatides*, for example, has seed-cones that are very similar to the living "China-fir" (*Cunninghamia*), in that they both had three to five small seeds attached to each seed-producing scale. However, *Cunninghamia* has much larger and straighter leaves than the shoots that bore these Jurassic cones. *Cunninghamia* is found growing today in China and Vietnam, in conditions that are probably similar to the warm climates *Elatides* seems to have favoured in the Jurassic.

LIVING GINKGO
A few thousand years ago, *Ginkgo* almost became extinct, but cultivation in eastern Asia saved it. It was introduced to Europe and North America in the 18th century and today can be found in parks and gardens across both continents.

KARKENIA CYLINDRICA
The cone above is from Iran. It occurs with *Ginkgo*-like leaves and contains numerous seeds that are attached to a central axis.

JURASSIC INVERTEBRATES

Following the late Triassic extinctions, marine invertebrate life proliferated and diversified greatly as new ecosystems arose in the warm seas. An overall greenhouse climate prevailed, and on land there were rich forests, and also dinosaurs, flying reptiles, and the first birds.

Flourishing marine life

Typical of the Jurassic are the ammonites – these were the last group of ammonoids to evolve and they displayed a characteristic complex suture (see p.202). They spread out and diversified rapidly, genera or whole groups succeeding one another, often with a turnover rate of less than a million years. Belemnites were important in the Jurassic. Bivalves, corals, gastropods, echinoderms, bryozoans, and scleractinian corals spread out, and both terebratulid (see right) and rhynchonellid brachiopods became locally abundant. Lobster-like crustaceans became more common in some habitats. In places, large sponge reefs grew, though as yet there were no true coral reefs. The presence of large fish and marine reptiles gave a new dimension to Jurassic marine life, in contrast to that of the Carboniferous, for example. In the face of this increased predation, many groups of echinoids and bivalves adopted the strategy of burrowing into the sediment for protection.

LYME REGIS BEACH
Along this shore, remains of Lower Jurassic ammonites are abundant; the large specimen here has been eroded to show its crushed septa.

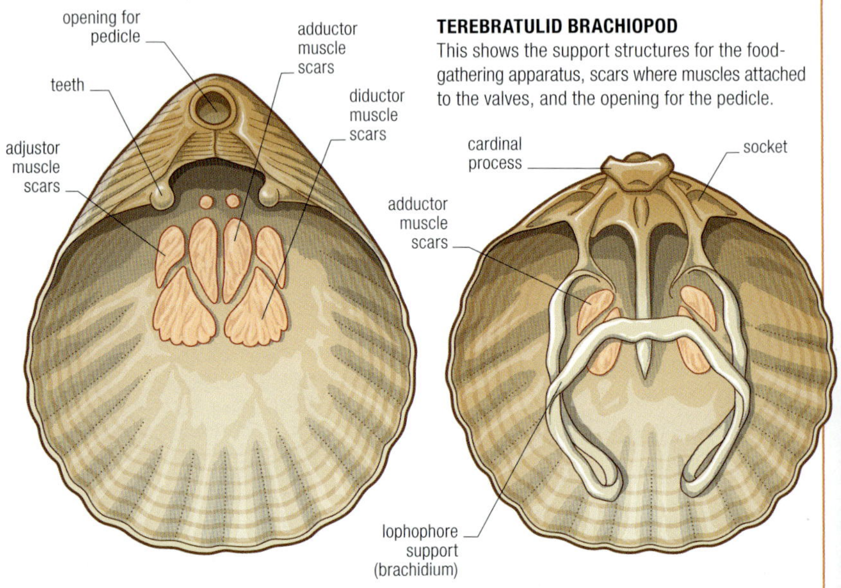

TEREBRATULID BRACHIOPOD
This shows the support structures for the food-gathering apparatus, scars where muscles attached to the valves, and the opening for the pedicle.

A biological arms race?

In all animals with voluminous and edible flesh, such as molluscs, the threat of predation has had a significant effect on their evolution. In times following a mass extinction, such as the Jurassic, the food-catching ability of predators increased dramatically. But this was matched in the prey by increasingly effective anti-predator devices. Cephalopods, for example, depend on escape or avoidance, but sessile molluscs developed other strategies, which have enabled them to survive very well.

GROUPS OVERVIEW

Many invertebrate groups flourished during the Jurassic. Ammonites appeared at the beginning of the period, and new forms of crustaceans also began to appear. Scleractinian corals continued to evolve and were very abundant in places. Brachiopods were common and fossils of the terebratulid type are often found in Jurassic limestone.

SCLERACTINIAN CORALS

These arose in the Triassic after the demise of the Paleozoic groups, and some genera survive today. The main dividing partitions (septa) in their skeletons are laid down on six lines, not four as in the rugose corals, and are radially symmetrical. They originated from a group of soft-bodied sea anemones.

TEREBRATULID BRACHIOPODS

Brachiopods were common at this time, though they had lost their Paleozoic importance. Terebratulid brachiopods (see also above) formed one of the two main groups of calcareous-shelled brachiopods in the Jurassic, and were widespread. They were attached to the sea floor by a fleshy stalk called a pedicle.

AMMONITES

The Jurassic was the heyday of these rapidly evolving coiled cephalopods, and their complex sutures are readily distinguished from those of other ammonoids. Their variegated form attests to adaptation to many environments. Ammonites are more valuable for studying rock strata than any other kind of invertebrate fossil.

CRUSTACEANS

Shrimp-like crustaceans were present in the Carboniferous, but by the Jurassic the first lobster-like forms had originated, many being predatorial. The descendants of some of these Jurassic crustaceans inhabit the deep sea to the present day.

Peronidella

GROUP Sponges
DATE Carboniferous to Cretaceous
SIZE 5–12cm (2–4$^1/_2$in) across
LOCATION Worldwide

Peronidella was a calcareous sponge, characterized by a skeleton composed of calcium carbonate spicules (in the form of calcite) that were needle-like and fused together. The rigid structure created by the spicules was cylindrical, and in most species it was also branched. In life, this sponge would have fed and breathed by drawing sea water in through pores in the outer body wall, which then passed into its central cavity. As the sea water passed through the thick body walls of the sponge, oxygen was absorbed and food particles were filtered out. With the help of many small hair-like structures called cilia that helped maintain the current, the water then passed out through an opening (osculum) along with any waste material that needed to be excreted.

In life, the sponge fed and breathed by drawing sea water in through pores in the outer body wall.

PERONIDELLA PISTILLIFORMIS
This fossil skeleton's rigid structure is formed of small spicules that are fused together. Oscula (openings) are visible, through which waste water passed out of the sponge.

Thecosmilia

GROUP Anthozoans
DATE Middle Triassic to Cretaceous
SIZE Corallites 1.5–3cm ($^1/_2$–1$^1/_4$in) across
LOCATION Worldwide

Thecosmilia was a scleractinian coral that consisted of numerous loosely packed, cylindrical individual polyp skeletons (corallites). Extra corallites were formed by a process called asexual budding, where offspring of the soft-bodied polyps grew out of the body of the parent. This process created more branches and increased the width of the colony. The outer wall (epitheca) of some of the fossilized corallites shows growth increments, but this has often worn away.

THECOSMILIA SP.
Numerous thin walls (septa) divided the inside of each individual corallite.

Thamnasteria

GROUP Anthozoans
DATE Middle Triassic to Middle Cretaceous
SIZE Colonies up to 1m (3ft 3in) across
LOCATION Worldwide

Thamnasteria was a scleractinian coral that would have lived in warm, shallow seas. It was composed of individual polygonal corallites that were closely packed together, or cerioid. Bivalves have bored into the coral skeleton shown below, leaving a number of sediment-filled cavities.

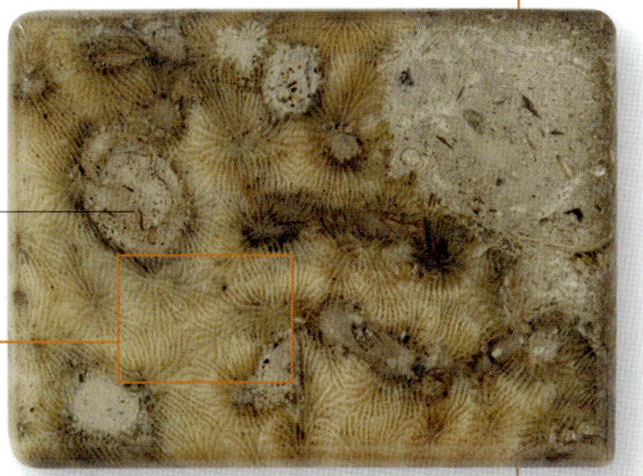

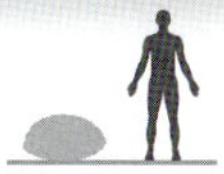

THAMNASTERIA SP.
Unlike most cerioid corals, the walls between each corallite are often missing in this genus.

Isastrea

GROUP Anthozoans
DATE Middle to Late Cretaceous
SIZE Colony up to 1m (3ft 3in) across
LOCATION Europe, Africa, North America

Isastrea was a colonial coral that belonged to a group called the hexacorals, so called because of the hexagonal shape of their polyp skeletons (corallites). *Isastrea*'s corallites were closely packed together and the walls of the adjacent corallites were fused together – an arrangement referred to as cerioid. The walls (septa) that divided up the inside of the corallites were slender and tightly spaced. *Isastrea* has several cycles of these septa arranged radially, each in multiples of six. It was a hermatypic coral, which meant that it formed reefs. Hermatypic corals need warm, clear, shallow water and live in symbiotic relationships with microscopic algae, which use carbon dioxide produced by the coral to photosynthesize and provide oxygen for the coral. *Isastrea* might have been able to withstand slightly lower temperatures, as it occurs farther north than other hermatypic corals.

Isastrea was a **hermatypic coral,** living in **large coral reefs,** often with **dozens** of **different species.**

ISASTREA SP.
This specimen clearly shows the polygonal corallites, most of which are hexagonal. The striking radial septa give it a distinctive, star-shaped appearance.

BERENICEA
Each autozooecium developed a long tube (peristome) at the entrance to the chamber.

Berenicea

GROUP Bryozoans
DATE Jurassic
SIZE Colony 1.5cm (½in) across
LOCATION Worldwide

"Berenicea" is the informal name for a type of cyclostome bryozoan, which can only be identified more precisely if brood chambers (gynozooecia), where bryozoan larvae developed, are present. "Berenicea" formed a thin, encrusting colony that was roughly circular in shape. The calcified chamber (autozooecia) that contained the feeding soft-bodied animals (autozooids) grew from the centre of the colony. These chambers were closely packed and in contact with each other.

Spiriferina

GROUP Brachiopods
DATE Carboniferous to Early Jurassic
SIZE 2.5–5.5cm (1–2¼in) long
LOCATION Europe, North America

Early Jurassic forms of *Spiriferina,* such as the specimen shown below, were the last survivors of the order Spiriferida, which first appeared in Silurian times. A long, straight hinge-line connected the valves, which were strongly patterned with four to seven, well-marked ribs that were more prominent towards the outer edge. A strong fold in the centre of the smaller (brachial) valve corresponded with an equally strong infold on the larger (pedicle) valve. Examination under a magnifying glass reveals that *Spiriferina*'s shell was perforated by minute pores (punctae).

SPIRIFERINA WALCOTTI
This *Spiriferina* shell has some well-marked growth increments and a pedicle opening, from which its attaching stalk would have emerged.

GONIOTHYRIS PHILLIPSI
This species had a well-developed circular opening beneath the beak. It had minimal patterning on its shell, except for several growth increments.

Goniothyris

GROUP Brachiopods
DATE Middle Jurassic
SIZE 2–4cm (¾–1½in) long
LOCATION Europe

Goniothyris was a terebratulid brachiopod with a well-defined triangular shape. This species is most commonly found in oolitic limestone and was clearly adapted to life attached to a firm surface, where ocean currents ensured a good supply of food particles. The shell had numerous minute perforations through which small bristles would have protruded. These may have been involved with sensory functions or respiration.

Homeorhynchia

GROUP Brachiopods
DATE Early to Middle Jurassic
SIZE 1–2.5cm (3/8–1in) wide
LOCATION Europe

This unusually shaped brachiopod belonged to the rhynchonellids. Other members of this group usually had strongly ribbed shells (valves), but *Homeorhynchia* was characterized by its smooth shell surfaces. The brachial valve and its pointed end, or beak, were more prominent than the pedicle valve, which had a small opening for the pedicle, a fleshy stalk that attached the organism to the sea bed. The centre of the brachial valve had a very strong fold that corresponded with an equally strong infold that ran down the centre of the pedicle valve. The two valves rose up in a sharp angle to the point at which they met along the junction line (commissure).

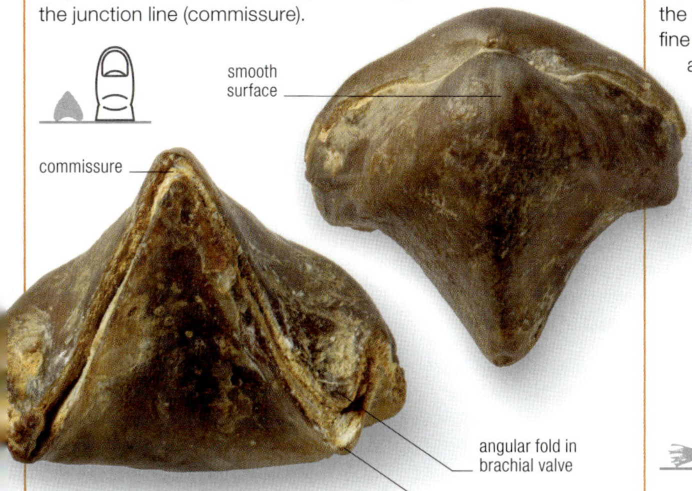

HOMEORHYNCHIA ACUTA
This species had a distinctive angular fold on the outer edge of the brachial valve, which corresponded with an infold in the pedicle valve.

Eutrephoceras

GROUP Cephalopods
DATE Middle Jurassic to Early Neogene
SIZE 12–30cm (4½–12in) wide
LOCATION Worldwide

Eutrephoceras was a tightly coiled nautiloid, whose outer whorls almost enveloped its inner whorls. It had a small and deep central part to the shell (umbilicus), and is therefore described as involute. The shell was almost spherical and the whorl sections were kidney-shaped. Inside the shell, the siphuncle, a tube of tissue that connected the inner chambers, was small in diameter and positioned roughly in the centre. The suture lines where the septa joined the shell were simple and fairly straight, but otherwise, the shell was quite plain, apart from some fine growth lines. The shell opening had a broad recess (sinus) on the inner side, through which a funnel would have protruded when *Eutrephoceras* was swimming.

EUTREPHOCERAS SP.
This internal mould clearly shows the thick, straight suture lines and the rounded outer margin. Most of the body-chamber is missing.

MICROCONCH
Smaller versions had ear-like projections called lappets. It is possible that this was a male form.

MACROCONCH
Larger forms had no projections around the shell opening. It is thought these may have been female.

Stephanoceras

GROUP Cephalopods
DATE Middle Jurassic
SIZE Microconch 2.5–7.5cm (1–3in) across, macroconch 8–30cm (3¼–12in) across
LOCATION Europe, North Africa, Middle East, Indonesia, South America, Canada, Alaska

Stephanoceras was a round-whorled ammonite. It had thick, strong ribs that radiated from the inner side of the whorl and ended in tubercles, or bumps, in the centre. *Stephanoceras* were found to occur in two sizes. Often, the smaller one of the two had ear-like projections, called lappets, on either side of the shell opening.

Dactylioceras

GROUP Cephalopods
DATE Early Jurassic
SIZE 6–8.5cm (2¼–3¼in) across
LOCATION Europe, Iran, North Africa, Arctic, Japan, Indonesia, Chile, Argentina

Dactylioceras was a heavily ribbed ammonite, with whorls that barely overlapped each other. This produced a wide and shallow central part of the shell, or umbilicus. Ammonites with this kind of whorl and umbilicus configuration are referred to as evolute. The shell had sharp, radial ribs that were dense on the inner whorls and gradually became more widely spaced on the outer whorls. On the very outer edge of the shell, many of the ribs split into two, while others remained simple and unbranched. In cross-section, these whorls appear almost circular. Ammonite fossils are abundant in some parts of northwest Europe. In medieval times people believed that these rocks were snakes that had been turned to stone and called them snakestones.

KEY SITE

THE JURASSIC COAST

The Jurassic Coast is situated on the south coast of England. It stretches from Exmouth in Devon to beyond Swanage in Dorset, and comprises 95 miles (153 km) of coastline that documents 185 million years of history, from the Triassic all the way through to the Lower Paleogene periods. The sections through the Jurassic rocks are famous for their beautifully preserved fossils. In 2001, the Jurassic Coast was made a World Heritage Site.

In **medieval** Europe, people thought **ammonites** were fossilized **snakes.**

DACTYLIOCERAS SP.
In the 19th century, a snake head was sometimes carved onto an ammonite, so that it could be sold as a fossil snake.

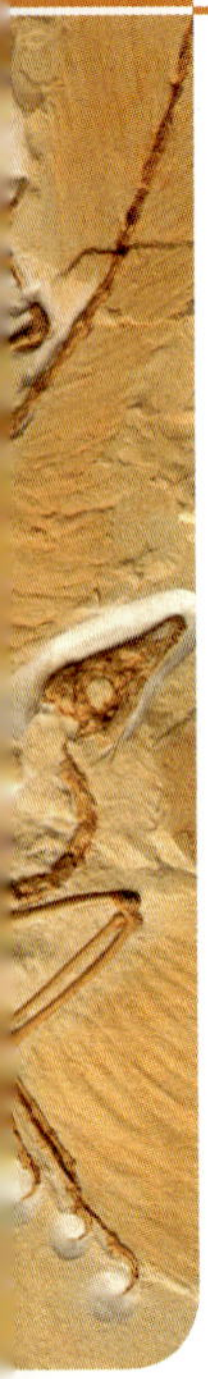

Phylloceras

GROUP Cephalopods
DATE Early Jurassic to Late Cretaceous
SIZE 7–15cm (2¾–6in) in diameter
LOCATION Worldwide

The cephalopod *Phylloceras* first appeared in the Early Jurassic and changed very little until the Early Cretaceous. The coiling in *Phylloceras* was extremely tight – its whorls almost covered each other. This created a small, deep centre or umbilicus. The outside of the animal's shell was almost smooth, but it had a large number of very fine, densely packed radial ribs that only slightly affected shell thickness. *Phylloceras* would have lived mainly in open seas, but made occasional forays into shallower waters.

PHYLLOCERAS SP.
This specimen is an internal mould, which explains why the leaf-like suture lines are visible. The outside of the shell would have been marked with very fine ribs (not shown here).

Phylloceras lived mainly in open seas, from **warm equatorial** waters to **high Arctic** latitudes.

Lytoceras

GROUP Cephalopods
DATE Early Jurassic to Late Cretaceous
SIZE 2–30cm (¾–12in) in diameter
LOCATION Worldwide

Lytoceras had round to square, barely overlapping whorls and a wide, open, navel-like umbilicus. The shell's fine, close ribs gave it a crinkled appearance. Where sharper ribs occurred, the shell was stronger and constricted inside. *Lytoceras* lived mainly in deep, open seas, only sometimes moving into shallower water. It ranged from equatorial waters to as far as Greenland and northern Alaska, and was a highly successful genus that changed very little.

L. FIMBRIATUM
A mixture of fine and thick ribs give *Lytoceras* a distinctive appearance.

Gryphaea

GROUP Bivalves
DATE Late Triassic to Late Jurassic
SIZE 2.5–15cm (1–6in) high
LOCATION Worldwide

Gryphaea was an asymmetrical member of the oyster family. Its left valve was large and outwardly curved, providing stability on the sea bed. The right valve was flat or slightly concave, and acted as a lid. A ligament controlled opening of the shell, while a single, strong adductor muscle closed the valves by contracting.

GRYPHAEA ARCUATA
Due to its odd shape, *G. arcuata* is called the "devil's toenail".

Cylindroteuthis

GROUP Cephalopods
DATE Early Jurassic to Early Cretaceous
SIZE 10–22cm (4–8½in) long
LOCATION Europe, Africa, North America, New Zealand

Cylindroteuthis was a large belemnite and is a common fossil in many Jurassic localities. The part of a belemnite most frequently preserved is its pointed guard, which was made of calcite and acted as a stiffening structure for the animal's squid-like body. Some fossils show traces of blood vessels on the outside of the guard, which indicate that it was an internal feature. A conical depression in the guard's blunt end housed the phragmocone, a kind of internal, chambered shell that helped *Cylindroteuthis* regulate its buoyancy. Well-preserved examples show some soft body parts, including the body, ten arms, and an ink sac similar to those of modern squid.

The **pointed guard** is the most commonly preserved part of a belemnite. It gave **support** to the animal's **squid-like body**.

CYLINDROTEUTHIS PUZOSIANA
This specimen shows the animal's long, tapered guard, which is attached at its blunt end (the alveolus) to the aragonite phragmocone. These hard parts were internal features.

SQUID-LIKE
With its ink sac, soft body, and ten arms, *Cylindroteuthis* was very much a squid-like creature that used its elongated guard as a buoyancy device, as well as for support.

Pleuromya

GROUP Bivalves
DATE Middle Triassic to Early Cretaceous
SIZE 2–7cm (¾–2¾in) long
LOCATION Worldwide

Pleuromya was a medium-sized bivalve with two shells of equal size. At the rear end was a permanent gape through which two siphons protruded. *Pleuromya* lived buried in sea-bed sediment, drawing in food through its inhalant siphon and expelling waste through its exhalant siphon.

PLEUROMYA JURASSI
The outside of most *Pleuromya* shells have fine growth lines, but some species have concentric ribs.

Modiolus

GROUP Bivalves
DATE Devonian to present
SIZE 1–12cm (³⁄₈–4½in) long
LOCATION Worldwide

Modiolus has adapted to a range of environments, and survives today. The beaks of its shells are to the front, and the shell surface has fine growth lines on it; in some species these become heavier ribs. It attaches to the sea bed by a byssus – horny threads often called a "beard" on mussels.

MODIOLUS BIPARTITUS
The shells of *Modiolus* get progressively wider the further back from the beak they go.

Pleurotomaria

GROUP Gastropods
DATE Early Jurassic to Early Cretaceous
SIZE 2–7.5cm (¾–3in) high
LOCATION Worldwide

PLEUROTOMARIA ANGLICA
Tubercles are clearly visible on this specimen of *P. anglica*.

Pleurotomaria was a conical gastropod. The centre of its shell was smooth, but it developed a strong ornament consisting of combinations of spiral and radial patterns. This specimen shows distinct bumps (tubercles) where the radial pattern crosses the spiral one. Around the side of the whorl is the slit band, a groove that continues around the shell to a slit in its opening that separates the waste-carrying exhalant current from the inhalant current. The animal could withdraw into its shell and close off the opening with a horny lid called an operculum.

M. CLAVELLATA
The main surface of *Myophorella*'s shell is marked with tubercles.

Myophorella

GROUP Bivalves
DATE Early Jurassic to Early Cretaceous
SIZE 4–10cm (1½–4in) long
LOCATION Worldwide

Myophorella was a wedge-shaped bivalve with a distinct shell pattern. Behind its two beaks lay the short ligament that was responsible for opening the valves. The ligament ran along a flat, lozenge-shaped section on both valves, which was smoother than the rest of the shell, and had three rows of small bumps (tubercles) on each valve. On the main surface of the shell were rows of much larger, irregular tubercles.

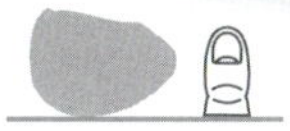

Laevitrigonia

GROUP Bivalves
DATE Early Jurassic to Late Cretaceous
SIZE 4–10cm (1½–4in) long
LOCATION Worldwide

Laevitrigonia was closely related to *Myophorella* (left); this is a fossil mould of the inside of the shell. The circular impression on the right is where the posterior adductor muscle attached – this pulled the valves to close them. The hinge teeth can be seen near the beak.

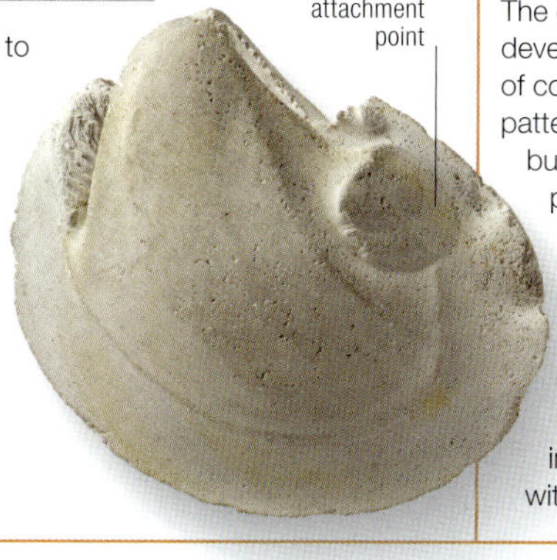

LAEVITRIGONIA GIBBOSA
The line along the curve of the shell is the pallial line, where the mantle, the membranous body covering, attached.

Pentacrinites

GROUP Echinoderms
DATE Jurassic
SIZE Arms up to 80cm (31in) across
LOCATION Europe

The crinoid *Pentacrinites* is often found in Jurassic rocks in the form of isolated pentagonal stem segments, called ossicles. The small cup is often hidden by the abundant side branches, called cirri, that grow from the stems. The cup was made of circles of regularly arranged plates and roofed by a domed membrane called the tegmen, which incorporated a number of small calcite plates. Long arms grew from the arm plates at the top of the cup, dividing many times and producing hundreds of densely packed branches. The whole crown may have been as much as 80cm (31in) across. Some specimens have stems over 1m (3ft 3in) long. *Pentacrinites* is often found with fossil wood. It has been suggested that it attached itself to floating wood and died when the wood became waterlogged and sank. This mode of life is known as pseudoplanktonic.

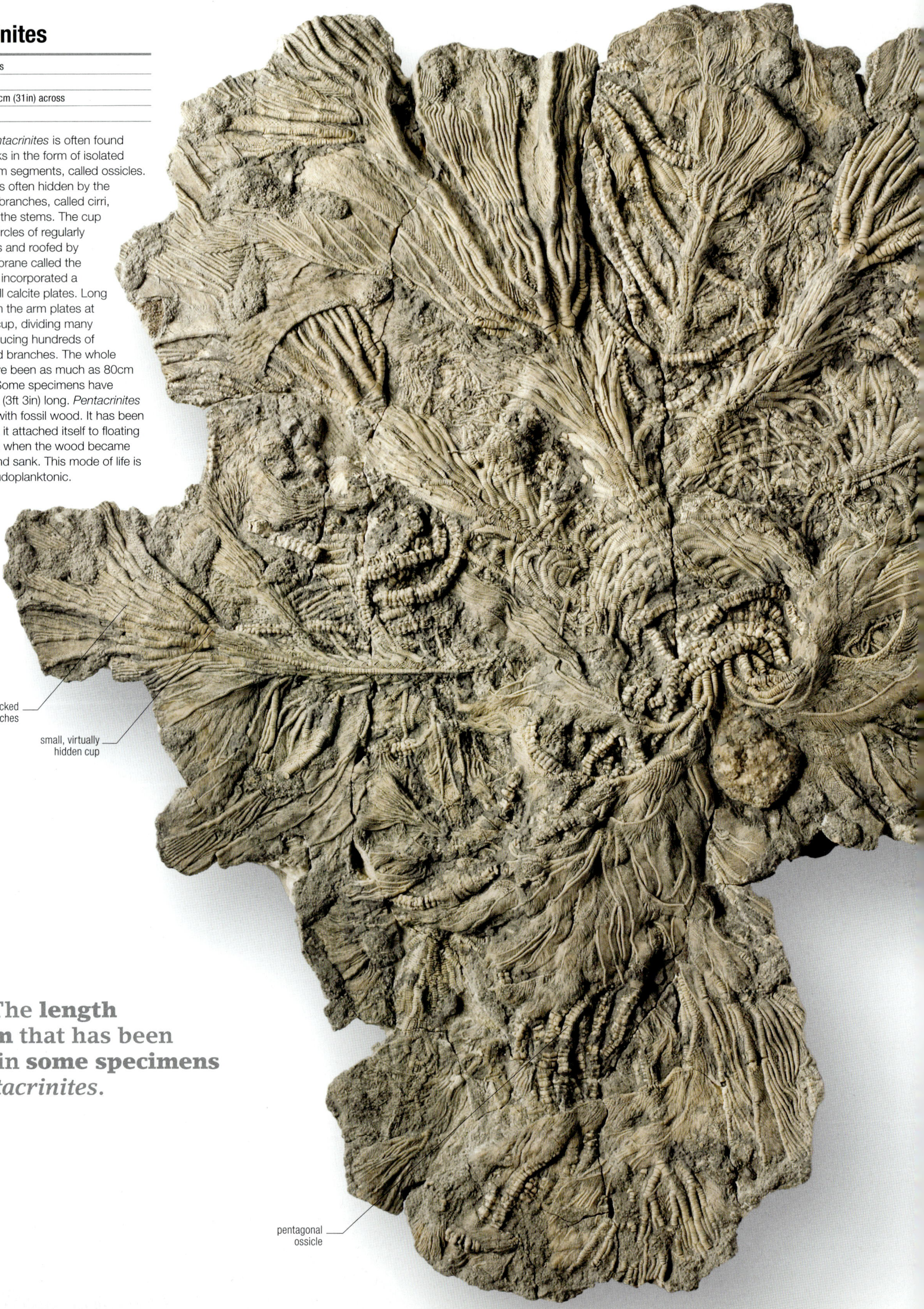

1m The **length of stem** that has been found in **some specimens** of *Pentacrinites*.

PENTACRINITES SP.
With its hundreds of densely packed arm branches, as well as the fact that it is often found with fossil wood, *Pentacrinites* resembles a type of beautiful, fibrous plant more than an invertebrate animal.

Apiocrinites

GROUP Echinoderms
DATE Middle to Late Jurassic
SIZE 30cm (12in) high, with stem; cup 3cm (1¼in) across
LOCATION Europe, Asia

Apiocrinites was a crinoid with a large cup. Its long stem was made up of individual segments called ossicles, and the cup was made up of two rings, each with five plates. Above these were rows of arm plates, from which arose ten arms used for gathering food. Each arm had one or two branches, which housed a groove along which food particles were transferred to the mouth within the cup. A leathery dome (tegmen) covered the oral surface. *Apiocrinites* attached itself to hard surfaces in clear, warm waters.

APIOCRINITES ELEGANS
This specimen shows the large cup and circular stem. Half of what seems to be the cup is really solid stem ossicles.

Pentasteria

GROUP Echinoderms
DATE Lower Jurassic to Lower Paleogene
SIZE Up to 12cm (4½in) across
LOCATION Europe

PENTASTERIA COTTESWOLDIAE
Pentasteria had five arms of roughly equal length.

Pentasteria was a typical Jurassic asteroid or starfish, and in most respects it differed little from starfish alive today. Like them, it had a double row of tube feet along the underside of each arm, which it used to move along with. Also like them, its mouth was positioned on its underside in the centre of its body. As its name suggests, *Pentasteria* had five arms, each of similar length, that projected outwards like the points of a star. Unlike many modern starfish, *Pentasteria* lacked suction discs on its tube feet, so it was unable to use them to prise open closed bivalve shells.

Palaeocoma

GROUP Echinoderms
DATE Early Jurassic
SIZE 5–10cm (2–4in) across
LOCATION Europe

Like all brittlestars, *Palaeocoma*'s body openings were located on the lower surface of its flat, disc-like body. These included a central, star-shaped mouth that also served as the anus, the opening into the water vascular system, and five pairs of gill slits found on either side of each arm near the disc's outer edge. As well as being used for respiration, the gill slits acted as release points for eggs or sperm. Food grooves with tiny, muscular tube feet ran along the underside of the arms and led to the mouth. The tube feet brought food towards the mouth and helped the organism to move.

PALAEOCOMA EGERTONI
Brittlestars are delicate and decay quickly. Only a swift burial in sand preserved this specimen intact.

Clypeus

GROUP Echinoderms
DATE Middle to Late Jurassic
SIZE 5–12cm (2–4½in) across
LOCATION Europe, Africa

Clypeus was a large, flattened sea urchin. On its upper surface, the ambulacra were shaped like petals; in the centre lay the apical disc, with four genital plates containing pores for the release of eggs or sperm. Its anal opening was in a groove at the rear of the upper surface. The mouth was located in the centre of the lower surface; the ambulacra on this side had large pores, and a single tube foot protruded through each pair of these. The tube feet were used for respiration, feeding, and locomotion, including burrowing.

CLYPEUS PLOTI
Large *Clypeus* fossils such as this are common in parts of the English Midlands, where they are often known as "poundstones" due to their size and weight.

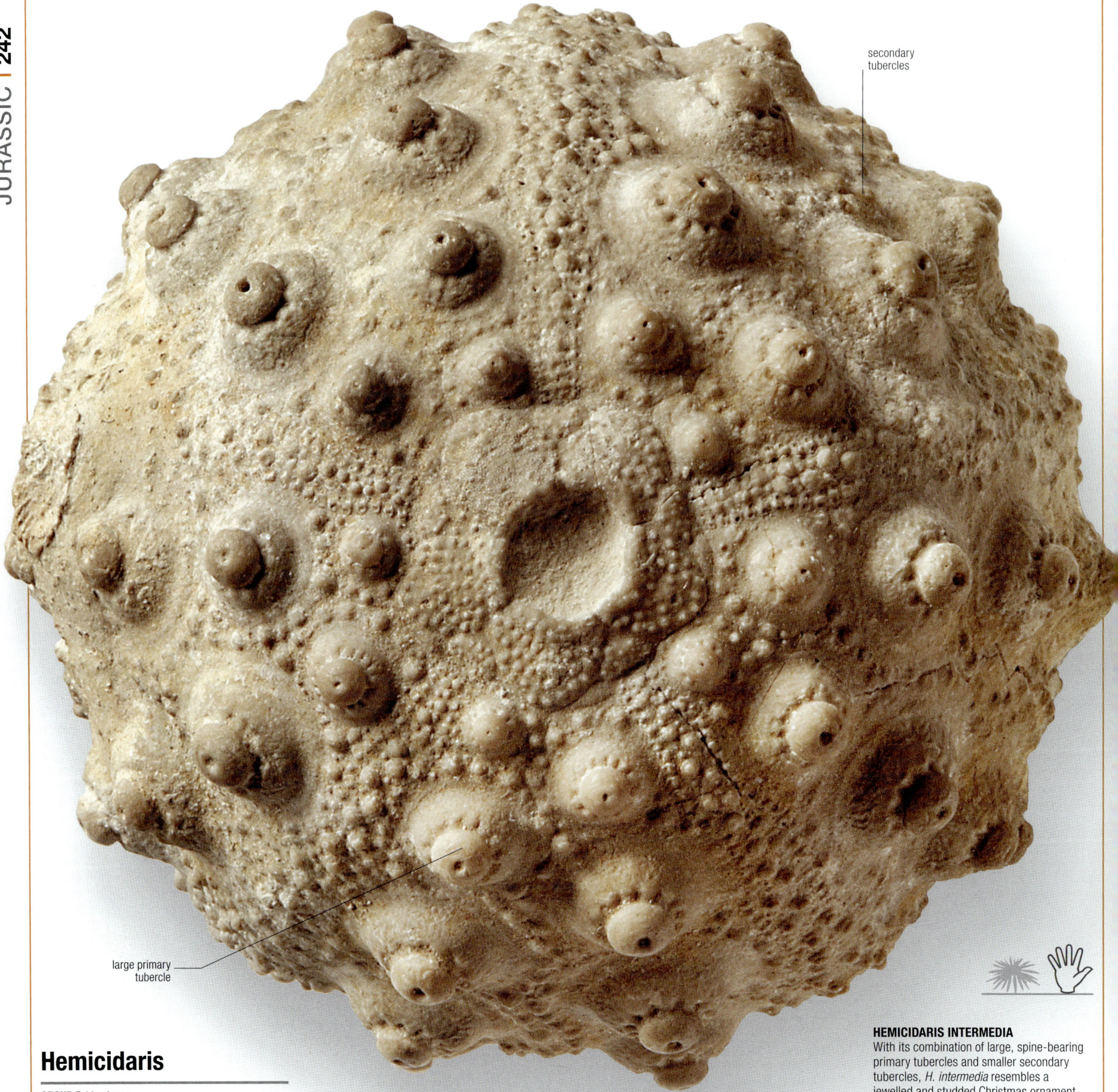

HEMICIDARIS INTERMEDIA
With its combination of large, spine-bearing primary tubercles and smaller secondary tubercles, *H. intermedia* resembles a jewelled and studded Christmas ornament.

Hemicidaris

GROUP Echinoderms
DATE Middle Jurassic to Late Cretaceous
SIZE Including spines, 20cm (8in) diameter, without spines, 2–4cm (¾–1½in) diameter
LOCATION England

Shaped like a flattened sphere, *Hemicidaris* was a medium-sized sea urchin. The most noticeable feature of this species are the large bumps, called tubercles, that ran along the body. In life, these tubercles carried large, tapering spines up to 8cm (3¼in) long. The sockets found at the end of the spines fitted onto the ball-shaped ends of the tubercles, and the muscles that moved the spines were attached to the relatively smooth area around these tubercle end points. The mouth of *Hemicidaris* was positioned at the centre of the animal's lower surface, and the outer margin of the membrane that housed it had ten notches that corresponded to the position of external gills. The anal opening was located almost at the centre at the top of the shell. Five genital plates with pores surrounded it; one of these five plates also served as a pressure sensor to the water vascular system. Between and outside these were five smaller ocular plates that bore light-sensitive tube feet. *Hemicidaris* would have lived on firm sea beds and used the sticky tube feet on its lower surface for locomotion as well as for feeding.

ANATOMY

BALL-AND-SOCKET JOINTS

Ball-and-socket joints are common in vertebrate skeletons. Echinoids have evolved a similar type of articulation for their spines. This is most readily seen in those with large tubercles on their shells; these are the "ball" part of the ball-and-socket joint. The smooth plate area surrounding the ball serves as a muscle-attachment area; muscles stretch from here to the base of the spine around the socket part of the joint, allowing controlled movement.

LIBELLULIUM SP.
With its two sets of wings, long, segmented body, and large head (not well preserved here), *Libellulium* differs little from modern dragonflies.

Libellulium

GROUP Arthropods
DATE Jurassic
SIZE Wingspan up to 14cm (5½in)
LOCATION Europe

Libellulium was a large, prehistoric dragonfly. Like modern dragonflies, its head had large, bulging eyes located at the front. Its thorax was short, expanding slightly immediately behind the head. The abdomen was composed of seven or eight long, narrow segments, with visible divisions between them. The limbs, not seen on this specimen, were situated at the front end of the thorax. The long, narrow, powerful paired wings originated from the thorax at the front end of the animal. The forewings were a little narrower and slightly longer than the hind wings, which slightly overlapped them. Some of the fine venation of the wings, including major and secondary veins, can readily be seen. Dragonflies (*Odonata*) have a long geological history, and the earliest appeared during the Carboniferous period.

LIVING RELATIVE

GIANT DRAGONFLY

Libellulium belonged to the same family of large living dragonfly species such as *Petalura gigantea*, commonly known as the giant dragonfly, which is found in New South Wales, Australia. Despite a wingspan of just under 14cm (5½in), it is actually a fairly poor flyer, and rarely moves far from the area in which it emerged.

Mesolimulus

GROUP Arthropods
DATE Late Jurassic
SIZE Up to 8–9cm (3¼–3½in) long, excluding telson
LOCATION Germany

Mesolimulus was a prehistoric horseshoe crab, very little changed from today's species. The large headshield was horseshoe shaped, and its eyes small and compound. The abdomen was covered by a single, semi-oval plate with a narrow, raised area flanked by raised lobes. Its broad marginal area was flat, and adorned on either side by seven long, pointed spines. A long, sharp-pointed barb or telson arose from the back of the abdomen.

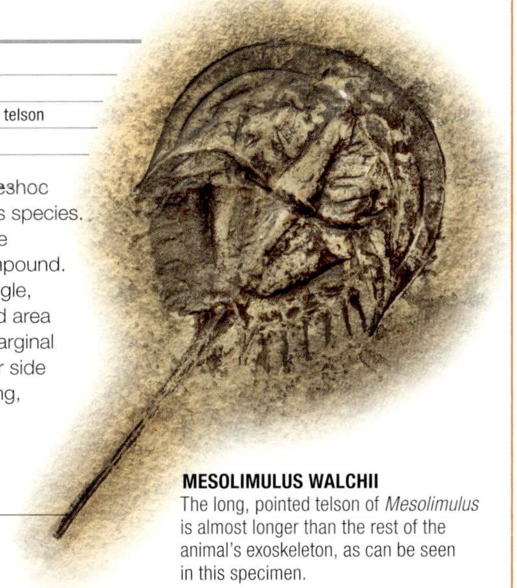

MESOLIMULUS WALCHII
The long, pointed telson of *Mesolimulus* is almost longer than the rest of the animal's exoskeleton, as can be seen in this specimen.

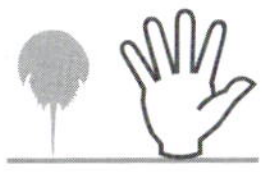

LIVING RELATIVE

HORSESHOE CRAB

The name "horseshoe crab" is derived from the shape of these animals' headshields; however, they are unrelated to true crabs. *Mesolimulus*, like other extinct forms, was smaller and flatter than its living relative *Limulus*, which is shown here, but both inhabited similar shallow marine environments. *Limulus* swims upside down, and this is likely to have been the case with its extinct cousins. *Limulus* is commonly found on the Atlantic coast of North America and in the Gulf of Mexico, and its lineage has a long pedigree. The oldest fossil horseshoe crabs have been found in rocks dating from the Permian period.

Eryma

GROUP Arthropods
DATE Early Jurassic to Late Cretaceous
SIZE 3.5cm (1½in) long
LOCATION Europe, E. Africa, Indonesia, North America

Eryma was a lobster. Its hard, upper shell (carapace) at the body's front end had a broadly oval outline and a prominent, backwards-curved groove in the area between the front limbs or chelipeds, which ended in large, lozenge-shaped pincers. Each pincer had an inner, moveable finger and an outer, rigidly attached one. There were four more long, slender, delicate walking limbs. Behind the carapace, the abdomen was narrower and composed of five broad segments. It terminated in the tail fan, made up of five, radiating flat lobes.

ERYMA LEPTODACTYLINA
This specimen, from limestone quarried in southern Germany's Solnhofen region, clearly shows the large claws and five-lobed tail fan.

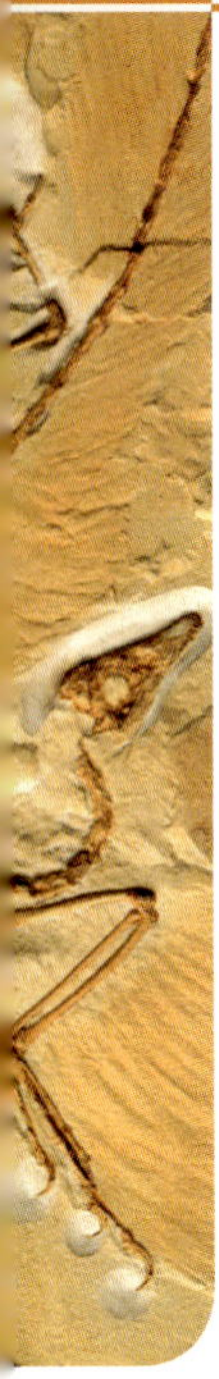

JURASSIC VERTEBRATES

Dinosaurs began to dominate in the Jurassic. Although dinosaurs originated about 30 million years earlier, it was only after the Triassic – Jurassic extinction of 200 million years ago that they became the top land vertebrates across the world. Meanwhile, in the seas, giant reptilian predators flourished.

The major subgroups of dinosaurs – theropods, sauropodomorphs, and ornithischians – originated during the Late Triassic. However, for the first 30 million years of their history, dinosaurs were only supporting actors on the world stage, and in many ecosystems were outnumbered and outmuscled by crurotarsans (crocodile-line archosaurs). Very suddenly, about 200 million years ago, all crurotarsans except for the crocodylomorphs disappeared in the midst of a global environmental meltdown, as volcanoes spewed toxic gas into the atmosphere and temperatures skyrocketed. This extinction was one of the “big five” mass extinctions in Earth’s history, and without it dinosaurs may have never risen to dominance.

Crocodylomorphs

Crocodylomorpha is the technical term for living crocodiles and alligators and their closest fossil kin. This remarkably successful group is the only surviving remnant of the great crurotarsan radiation of the Late Triassic, when crocodile-line archosaurs such as rauisuchians, phytosaurs, and aetosaurs dominated ecosystems around the globe. True crocodylomorphs evolved in the Late Triassic alongside their crurotarsan cousins. These proto-crocodiles looked nothing like the crocodiles of today, but instead were small, slender animals that walked upright and could run at high speeds. These animals, called

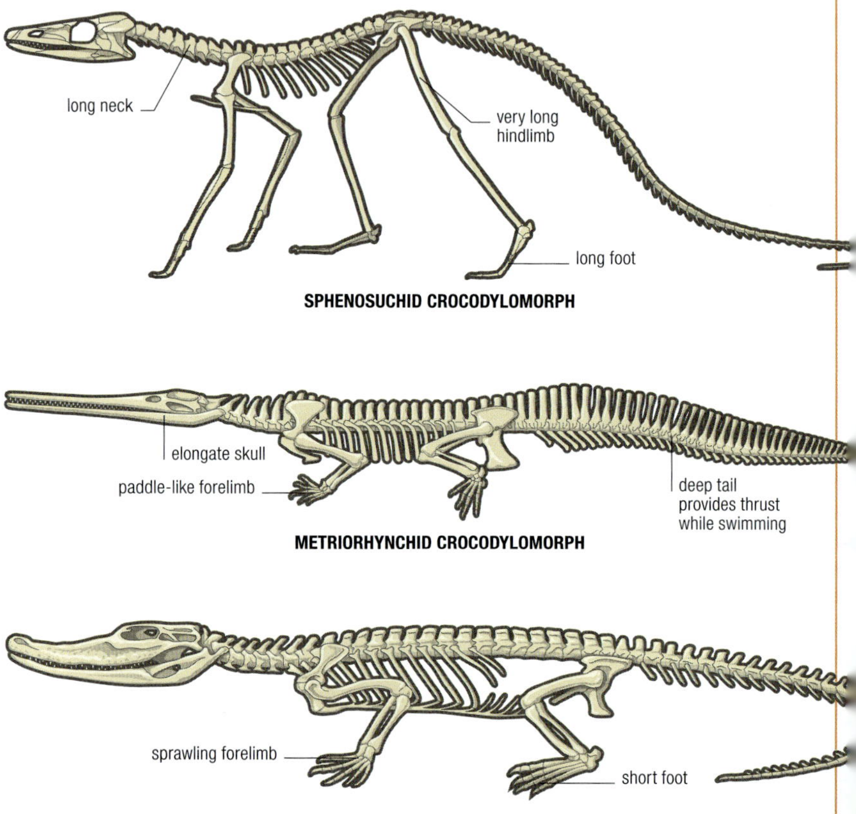

SKELETAL DIFFERENCES
Jurassic crocodylomorphs were a varied group that explored a wide range of skeletal designs. The earliest crocodylomorphs, the sphenosuchians, were small and graceful predators adapted for running. Metriorhynchids were colossal predators that lived in the ocean, with skeletons fine-tuned for their marine habitat. Modern alligators are slow, bulky creatures that lurk near the shore and ambush unsuspecting prey.

GROUPS OVERVIEW

The Jurassic was a time of reptilian dominance. Dinosaurs were the pre-eminent land vertebrates, and reptilian mega-predators, such as plesiosaurs, ichthyosaurs, and metriorhynchids ruled the oceans. However, other groups, such as mammals and amphibians, multiplied in the shadow of the dinosaurs.

TURTLES
Turtles are among the most unusual reptilian groups, and little is known about the early history of these animals. The oldest turtles come from the Middle Triassic, but these creatures only had a shell on their bellies. Some of the best-known primitive turtles lived in the Jurassic, such as *Kayentachelys* from southwestern USA, and had fully formed shells.

PLESIOSAURS
The long-necked, pot-bellied plesiosaurs look a bit like a sea-dwelling sauropod dinosaur. But unlike the peaceful, plant-eating sauropods, plesiosaurs were the most fearsome predators in most Jurassic oceans. Some of these creatures were over 15m (49ft) long – much longer than *Tyrannosaurus* – and could tackle the largest prey that the ocean had to offer.

ICHTHYOSAURS
The finned, streamlined ichthyosaurs resemble dolphins and tuna, but were actually ocean-living reptiles. Ichthyosaurs were enormously successful during the entire Mesozoic and fed on a variety of prey, from squid to large fish. The biggest ichthyosaurs were over 20–21m (66–69ft) long, making them the largest marine reptiles that have ever existed.

METRIORHYNCHIDS
Metriorhynchids are among the most unusual crocodylomorphs to ever live. Unlike most crocs, which are sluggish creatures that lurk at the water’s edge, metriorhynchids spent their entire lives in the open ocean. They were among the largest and fiercest predators of their day, and thrived worldwide before being replaced by plesiosaurs later in the Jurassic and Cretaceous.

50 tons The **weight** of *Brachiosaurus*, one of the **largest animals** ever to **walk** the Earth.

sphenosuchians, persisted until deep into the Jurassic. However, during the Jurassic, crocodylomorphs continued to diversify and many specialized subgroups evolved. These include not only sprawling species that lurk at the water's edge, such as living crocodiles, but also an unusual group called the metriorhynchids that lived entirely in the open ocean.

The age of giants

The Middle and Late Jurassic are often referred to as the "age of dinosaur giants". This fitting description pays homage to the evolution of several remarkable groups of colossal dinosaurs, which dominated terrestrial ecosystems across the planet. Foremost among these giants were the long-necked and plant-guzzling sauropods, such as *Brachiosaurus* and *Diplodocus*. The Late Jurassic was the peak of sauropod evolution.

FEARSOME PREDATOR
The powerful hunter *Allosaurus* was the most feared dinosaur of the Late Jurassic. It lived across North America and in Portugal, and used its size and strength to take down large prey, such as the sauropod giants.

THUNDER LIZARD
The Late-Jurassic *Brachiosaurus* is probably the most familiar of all of the long-necked, barrel-chested sauropods. This "thunder lizard" was one of the largest animals ever to walk the Earth. It approached lengths of 25m (82ft) and weighed as much as 50 tons.

At no other time were these thundering beasts so abundant and diverse. As many as 25 different sauropod species lived alongside each other in the Late Jurassic of North America, and today their fossils are common discoveries in the famous Morrison Formation (see p.40). Other colossal sauropods are known from Africa, China, and Portugal. Some of these creatures were among the largest animals to ever walk the Earth, and the ground literally quaked underneath their footsteps. Living alongside these herbivores, and probably preying on them, were many monstrous theropod carnivores. The poster child for Jurassic theropods is *Allosaurus*, a hatchet-skulled hunter that may have reached lengths of 12m (39ft) and masses of 1,800kg (4,000lb). Other nightmarish beasts lived alongside *Allosaurus*, including *Ceratosaurus* and *Torvosaurus*. Truly, the Jurassic was an age of giants.

PTEROSAURS
Often mistaken for dinosaurs, pterosaurs instead form their own unique subgroup of archosaurs. Pterosaurs arose during the Late Triassic and were the first vertebrate group to evolve flight. Most Triassic and Jurassic pterosaurs were small, and fed on fish, but some truly colossal species, such as *Quetzalcoatlus*, lived in the Cretaceous.

TETANURANS
The tetanurans are one of the most important and successful subgroups of theropod dinosaurs. Most Late Triassic and Early Jurassic theropods were small and had primitive skeletons, but in the Middle Jurassic, the much larger and more diverse tetanurans arose, including *Megalosaurus*, the "great lizard". Birds are a member of this subgroup.

SAUROPODOMORPHS
The sauropodomorph subgroup contains some of the most distinctive dinosaurs, such as the familiar long-necked herbivores *Brachiosaurus* and *Diplodocus*. Triassic and Early Jurassic sauropodomorphs were quite small and had unspecialized skeletons, but giant, quadrapedal, earth-shaking species with long, sinuous necks evolved in the Middle Jurassic.

ORNITHISCHIANS
Along with sauropodomorphs and theropods, ornithischians are one of the three major subgroups of dinosaurs. Ornithischians were rare in the Late Triassic, but became more common, and evolved into a diverse array of body forms, in the Jurassic. These include the plated stegosaurs, the armoured ankylosaurs, and the herbivorous iguanodontians.

MAMMALS
Mammals originated in the Late Triassic, around the same time as the dinosaurs. However, the furry mammals remained in the shadow of their giant reptilian contemporaries for most of the Mesozoic. But just because mammals were small does not mean they were unimportant. Mammals evolved into a range of different body types and ecological niches during the Jurassic.

Hybodus

GROUP Chondrichthyans
DATE Late Permian to Late Cretaceous
SIZE 2m (6½ft) long
LOCATION Europe, North America, Asia, Africa

The skeleton of *Hybodus* was very robust and more densely calcified than that of modern sharks. The jaws were solid and armed with rows of teeth that were continuously being replaced, as in modern sharks. The teeth measured up to 2cm (¾in) long, and varied in shape – some had a series of upright spikes or cusps, while others were low and stubby. By contrast to living sharks, in which the mouth opens on the underside of the head, the mouth of *Hybodus* opened at the end of the snout. In addition to pelvic claspers, used to insert sperm directly into the female, a male *Hybodus* also had specialized spines in the soft tissues of its head. These may have helped the male to grip the female during mating. Up to two pairs of these spines were situated just behind the eye.

STREAMLINED SHARK
The two dorsal fins were supported by long, ribbed fin spines that were embedded in the soft tissues. These probably helped the animal to cut through the water more efficiently.

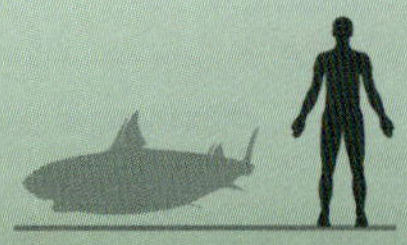

Ischyodus

GROUP Chondrichthyans
DATE Middle Jurassic to Miocene
SIZE 1.5m (5ft) long
LOCATION Europe, North America, Kazakhstan, Australia, New Zealand, Antarctica

Fish similar to *Ischyodus* are today restricted to the deep oceans, but in its time *Ischyodus* was a common bottom-dweller. It had a mainly cartilaginous skeleton, large eyes and pectoral (side) fins, and a long, whip-like tail. The first dorsal fin had a prominent spine that could be raised, and was probably used for defence. Instead of teeth, *Ischyodus* had dental plates that often protruded from its mouth. This is where it gets its common name, "rabbitfish". There were two pairs of plates in the upper jaw, and one pair in the lower jaw. They were used to crush mollusc shells and crustaceans. Dental plates are the most commonly preserved part of this fish, and are used to distinguish different species.

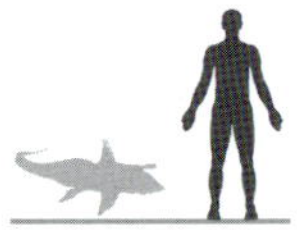

HEAD GEAR
Males had a curious head appendage, called a tenaculum. The tip was covered in sharp, tooth-like structures, and may have played a part during mating.

Leedsichthys

GROUP Actinopterygians
DATE Middle to Late Jurassic
SIZE 22m (72ft) long
LOCATION Europe, Chile

Possibly the largest fish that ever lived, *Leedsichthys* was first discovered near Peterborough, England in 1886. Despite its formidable size, *Leedsichthys* was a harmless filter feeder. Its huge mouth could take in large volumes of water in a single gulp. It then sifted micro-organisms from the sea water with gill rakers, or mesh plates at the back of its mouth, just as basking sharks and baleen whales do today. *Leedsichthys* is classified with the pachycormids: fish that have partly calcified skeletons. As cartilage does not fossilize well, this lack of calcification means that many of its bones have not been preserved and others are so thin that they are easily crushed. Teeth, however, are more robust than bones, and it has been possible to estimate that *Leedsichthys* had more than 40,000 of them.

GENTLE GIANT
A thrust from its powerful 5m (16½ft) tail was the only defense *Leedsichthys* had against attackers.

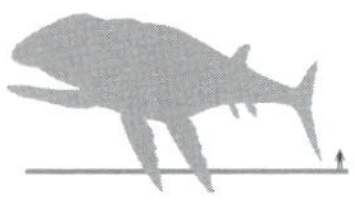

Lepidotes

GROUP Actinopterygians
DATE Late Jurassic to Early Cretaceous
SIZE 2m (6½ft) long
LOCATION Germany, Brazil

Lepidotes was a large bony fish with jaws that could be extended forwards to draw in food. At the time, this type of jaw mechanism was an evolutionary novelty, but it has since evolved independently several times and is present in many modern-day bony fish, including carp, perch, and herring. *Lepidotes* had batteries of peg-like teeth that it used to crush the shellfish on which it fed. Often found in isolation, individual teeth look like perfectly formed little stones. In medieval times they were called "toadstones" and in folklore were thought, among other things, to be an antidote for poison. *Lepidotes* had a single dorsal and anal fin, sited opposite each other, a symmetrical tail, and thick, enamelled scales that protected its body.

MARINE AND FRESHWATER
Lepidotes fossils have been found in ancient shallow marine and freshwater lake deposits and even in the stomach of the dinosaur *Baronyx*.

Eocaecilia

GROUP Amphibians
DATE Early Jurassic
SIZE 18cm (7in) long
LOCATION USA

This small, burrowing amphibian was discovered in Arizona, USA. It had a small head with heavy internal armour, a very long body, four tiny limbs, and a short tail. The skull of *Eocaecilia* is very like those of the group of living amphibians called caecilians, or "worm amphibians". *Eocaecilia* is believed to be the earliest known member of this group. Modern-day caecilians are burrowing creatures that have lost their limbs altogether. Although *Eocaecilia* still had limbs, its heavily built skull suggests that it used its head to plough through the ground like modern-day caecilians.

WORM AMPHIBIAN
Eocaecilia had a very elongated body, with around 50 vertebrae in the trunk section.

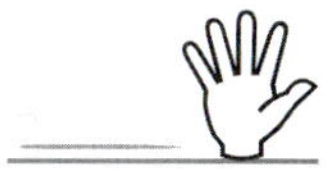

Prosalirus

GROUP Amphibians
DATE Early Jurassic
SIZE 6cm (2¼in) long
LOCATION USA

Only three fossilized skeletons of this small frog have ever been found. They were discovered in Arizona, USA, in the 1990s, and are the earliest frog skeletons to show the full set of jumping adaptations that characterize frogs today. These include long hip bones, long hind leg bones, and long ankle bones. Earlier ancestral frogs had been found in Madagascar and Poland, but these examples still had tails and did not have jumping adaptations, which meant they were not capable of frog-like leaping movements. Frogs began to diversify long after *Prosalirus* had died out, so it is not directly related to any one modern family. All the fossils were found in one small area, which probably represented the last puddle of water on a dried-up flood plain.

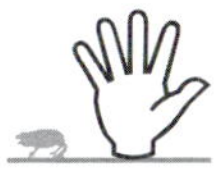

JUMPING FROG
This species had hind legs adapted for jumping. The name *Prosalirus* comes from the Latin verb prosalire, which means "to leap forwards".

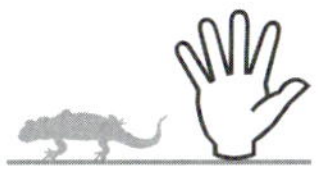

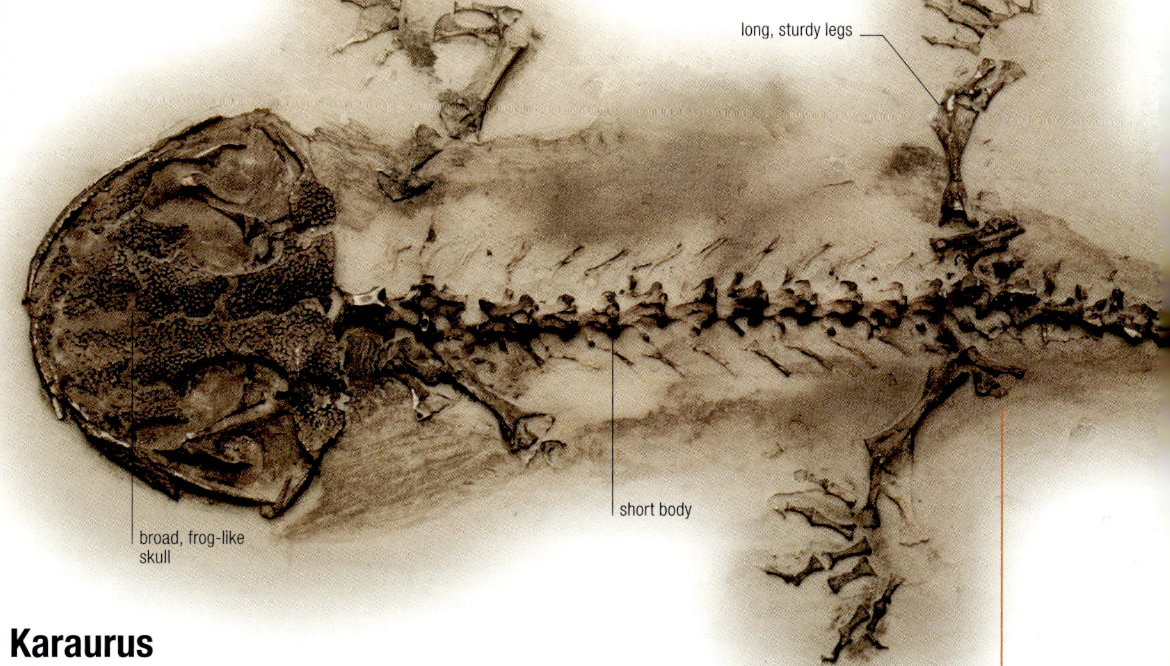

JUVENILE FOSSIL
In the only known skeleton of *Karaurus*, the limb bones are only partly developed, which suggests that this was a juvenile animal that would have grown larger.

Karaurus

GROUP Amphibians
DATE Late Jurassic
SIZE 19cm (7½in) long
LOCATION Kazakhstan

The only fossil of *Karaurus* was discovered in the 1970s in a fossil lake bed in Kazakhstan. For some years it was the earliest salamander known. The name *Karaurus* comes from a combination of "Kara" (after the Karatau Mountains where it was found) and "Urus", meaning tail. *Karaurus* was a medium-sized salamander, and was very heavily built with a broad, frog-like skull, a short body, and large legs. It was more primitive in many of its skeletal features than any living salamander, which has led scientists to believe that modern-day salamanders evolved from an animal that was more advanced than *Karaurus*. Since *Karaurus* was discovered, other Jurassic salamanders have been found in China, Europe, and North America, which suggests that they were widespread across the northern continents.

SHINY SCALES
This beautifully fossilized *Lepidotes* shows many of the characteristic features of this fish, in particular its thick, diamond-shaped scales. In life, each scale had an outer layer of ganoine, an enamel-like substance that reflected light and gave the fish a shiny appearance. Rows of peg-like teeth, once used for crushing shellfish, are also clearly visible.

Kayentachelys

GROUP Turtles
DATE Early Jurassic
SIZE 25–35cm (10–14in) long
LOCATION USA

Kayentachelys was one of the oldest turtles, and an important link between primitive Triassic turtles, such as *Odontochelys* (see p.208), and modern turtles. Like all modern turtles, *Kayentachelys* was sheathed in a box-like shell comprised of a carapace on top and a plastron underneath. Its skull was short and wide and was capped with a sharp beak. *Kayentachelys* lived in southwestern North America, alongside dinosaurs such as the therapod *Dilophosaurus* (see p.258) and early crocodylomorphs, such as *Protosuchus* (see p.254). The presence of large dinosaurs, crocodylomorphs, and turtles gave this ecosystem a modern character, in comparison with ecosystems of the Middle to Late Triassic.

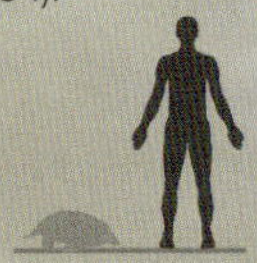

KEY SITE

THE KAYENTA FORMATION

The rusty red sandstones and shales of the Kayenta Formation cover broad swathes of western USA, including many of the most famous national parks, such as Grand Canyon and Zion. As well as producing stunning desert scenery, the Kayenta Formation has yielded many important fossils from the Early Jurassic period. During this time, about 190 million years ago, a varied fauna of early dinosaurs, mammals, crocodiles, and turtles thrived in the arid conditions of a region that at this time was located in the supercontinent Pangea. Many famous fossils, including *Kayentachelys*, have been found in the Kayenta Formation.

SMALL TURTLE
Kayentachelys was a small animal, no more than 35cm (14in) long, and weighing only a few kilograms. Its ecological habits are questionable, but it probably spent a lot of time on land.

Pleurosaurus

GROUP Lepidosaurs
DATE Late Jurassic
SIZE 50–70cm (19½–28in) long
LOCATION Germany

New Zealand's two tuatara species are the only living members of the once-diverse reptile order known as sphenodontia. Perhaps the most unique of these animals, *Pleurosaurus* was slender and streamlined, with a long tail. Its hands and feet were modified into paddles, making it highly manoeuvrable as it chased small fish.

SOLNHOFEN LIMESTONE
Pleurosaurus was found in the same German rock unit in which the first bird, *Archaeopteryx* (see p.264), was discovered.

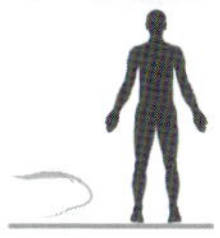

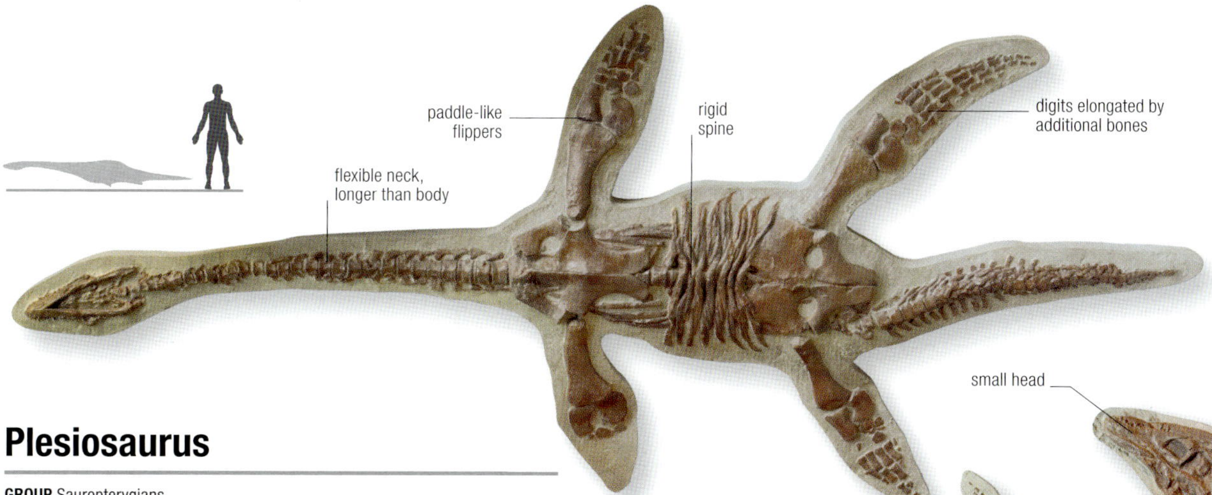

Plesiosaurus

GROUP Sauropterygians
DATE Early Jurassic
SIZE 3–5m (9¾–16½ft) long
LOCATION British Isles, Germany

Plesiosaurus was a large marine reptile that lived in the shallow seas of what is now Europe, during the Early Jurassic, about 190 million years ago. In many ways, it looked like an aquatic version of a dinosaur. However, it was only distantly related to the dinosaurs, and was more closely related to lizards and snakes. *Plesiosaurus* was a typical plesiosaur, with a stocky torso, four large flippers, an extremely long neck, and a tiny skull filled with many small teeth. This combination of features allowed *Plesiosaurus* to dominate its marine ecosystem. It was a highly successful predator of fish, squid, and other relatively small, fast-moving prey. *Plesiosaurus* was a fast and powerful swimmer that could strike at shoals of fish with great speed, using its muscular neck.

FIRST FOUND
Plesiosaurus was one of the first ancient reptiles to be discovered. It was first found by Mary Anning (see p.253) and described in 1821 by William Conybeare.

COMMON FOSSILS
Plesiosaurus fossils are common in the so-called Liassic rocks of southern England. These are well exposed along the coast of Dorset, particularly around the town of Lyme Regis.

POWERFUL PADDLES
Liopleurodon used its four huge paddles to propel itself through the water after prey. The size of the paddles suggests that it was a powerful swimmer capable of high speeds in short spurts.

CERVICAL RIB
The cervical ribs (or neck ribs) of plesiosaurs were short and robust, which allowed for the attachment of strong muscles, and great mobility.

Liopleurodon

GROUP Sauropterygians
DATE Middle to Late Jurassic
SIZE 7–10m (23–33ft) long
LOCATION British Isles, France, Russia, Germany

Liopleurodon was the apex predator of Europe's seas during the Middle to Late Jurassic. This monstrous reptile reached lengths of 10m (33ft), making it larger than a modern-day killer whale. *Liopleurodon* was a plesiosaur, a stout-chested carnivorous marine reptile. Most plesiosaurs were long-necked and fed on fish. However, *Liopleurodon* was one of the pliosaurids, a subset of fearsome, predatory plesiosaurs with shorter necks and larger heads. Like other pliosaurids, *Liopleurodon* used its powerful jaws, which were up to 1.5m (5ft) long and studded with conical teeth, to spear marine reptiles and large fish. Although hefty, its body was highly streamlined, and it used its four paddle-like limbs to drive through the water. *Liopleurodon* was an amazingly successful animal: it existed for around 10 million years and inhabited a wide belt of seas across the whole of ancient Europe.

FOSSIL VERTEBRA
Pliosaurids were massive animals. Their spines were made up of vertebrae the size of dinner plates.

Ichthyosaurus

GROUP Ichthyosaurs
DATE Early Jurassic
SIZE 2m (6½ft) long
LOCATION British Isles, Belgium, Germany

At first glance *Ichthyosaurus* looks rather like a dolphin or a fish, but it was a reptile closely related to lizards. Its streamlined body was designed for speed, with hands and feet modified into flippers, a dorsal fin, and a tall, paddle-like tail. *Ichthyosaurus*'s skull was adapted for hunting fast-moving, slippery prey – it was elongated and studded with rows of sharp teeth.

Stenopterygius

GROUP Ichthyosaur
DATE Early to Middle Jurassic
SIZE 2–4m (6½–13ft) long
LOCATION UK, France, Germany

Stenopterygius was a close relative of *Ichthyosaurus* (see p.251). Like its more famous cousin, *Stenopterygius* was a dolphin-like reptile that hunted schools of fish in the warm, shallow seas of Jurassic Europe. It grew to a slightly larger size than *Ichthyosaurus* but had a smaller skull, which was perfectly adapted for quickly striking, capturing, killing, and eating fish.

Stenopterygius's snout had the streamlined profile of a missile or submarine, and it could blast through the water like a projectile, rapidly striking fish with little or no warning. This creature's tail was also adapted for speed. The vertebral column in the tail bent downwards and supported a deep tail fin, similar to those seen in modern sharks. This fin shape would have powered *Stenopterygius*'s large body as it raced through the water in search of its next meal. One of the most interesting *Stenopterygius* fossils shows a mother giving birth – proving that ichthyosaurs, like the dolphins of today, bore live young, and that the young were born tail-first.

100kph The estimated maximum speed reached by *Stenopterygius,* based on comparison with modern-day tunas.

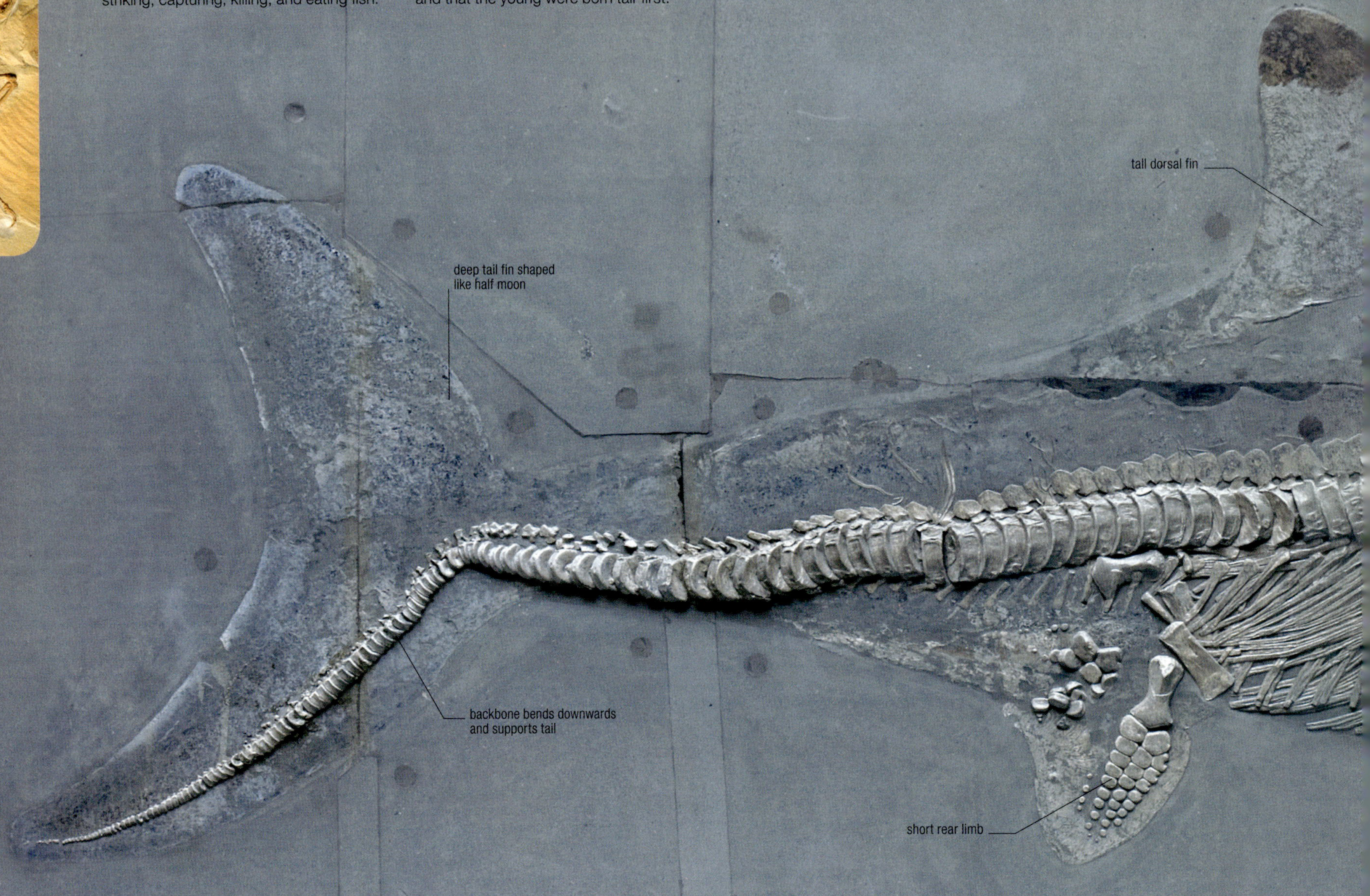

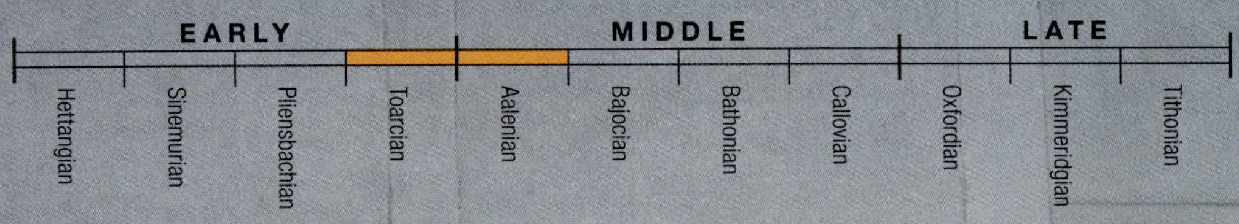

BIOGRAPHY

MARY ANNING

Mary Anning (1799–1847) came from a poor Dorset family that supplemented its income by selling fossils they found around Lyme Regis, on England's south coast. In the Early Jurassic period, this area was covered by a warm, shallow sea that was home to many ancient reptiles. At the age of 12, Mary found the first complete ichthyosaur fossil. She became an outstanding collector and accomplished naturalist, and was responsible for discovering the first plesiosaur in 1823 (see p.251). At a time when men dominated all aspects of society, she made a serious scientific contribution to research. An honorary member of the Geological Society of London, she died of breast cancer aged 47.

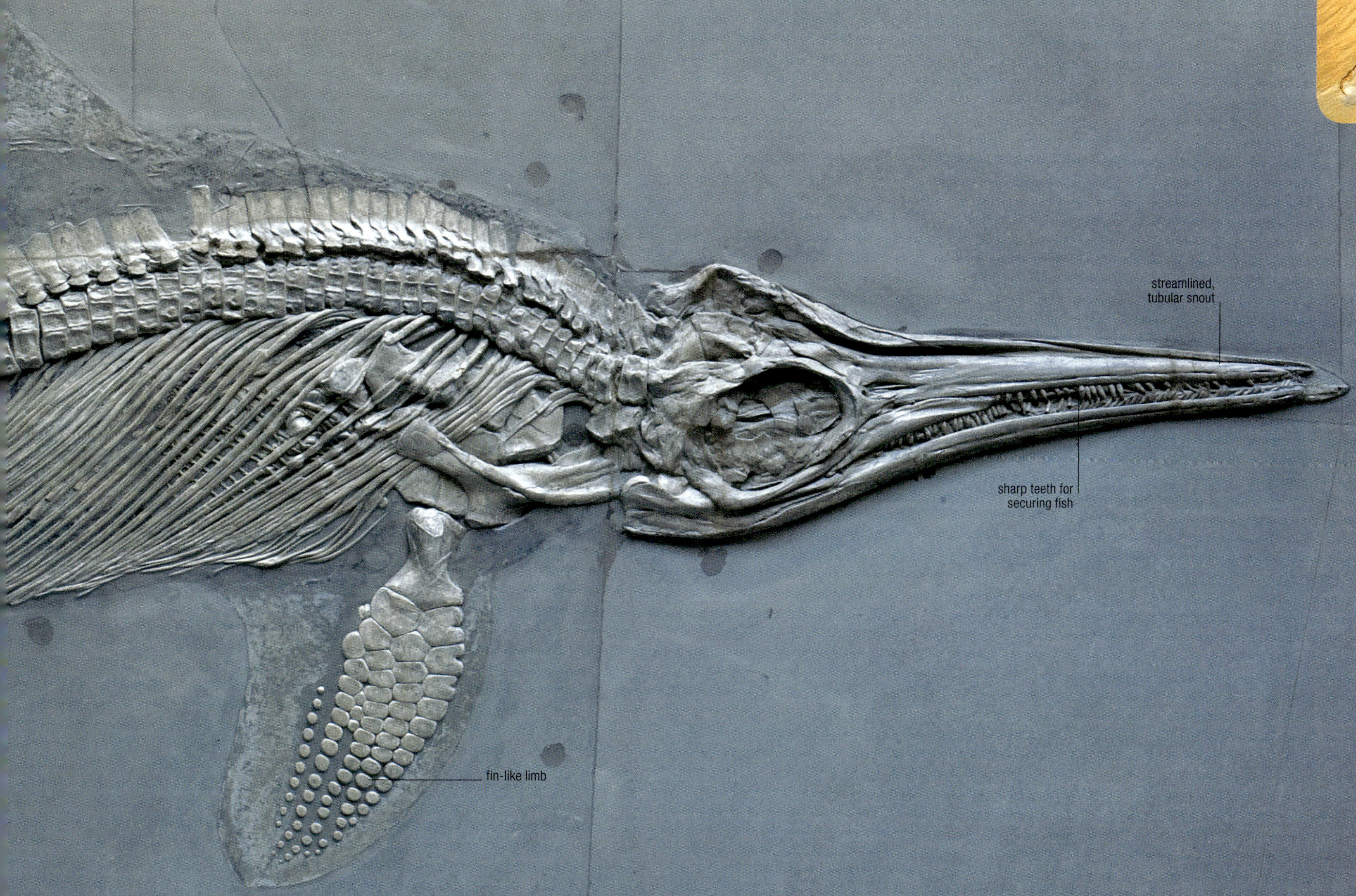

LASTING IMPRESSION
As seen in this specimen, ichthyosaur fossils are well known for preserving impressions of soft tissue next to bone. These carbonaceous impressions are thought to be produced either from residues of the animal's flesh, or by a film produced by bacteria.

Protosuchus

GROUP Crocodylomorphs
DATE Early Jurassic
SIZE 1m (3ft 3in) long
LOCATION Worldwide

Protosuchus is one of the earliest and most primitive relatives of living crocodiles. However, in many ways *Protosuchus* was very different from living crocodiles. It walked with an almost upright stance, its limbs were long and slender, and its feet were thin and capped with claws. These features suggest that *Protosuchus* was a strong runner and adapted for hunting on land rather than in the water. However, like living crocodiles, *Protosuchus* had a wide skull, powerful jaw muscles, and thick, conical teeth. It is no surprise that *Protosuchus* possessed a mixture of primitive and more modern physical features, as it lived towards the beginning of crocodylomorph evolutionary history. *Protosuchus* was first found in Arizona, USA, but more recently its fossils have turned up in many other places across the globe.

SMALL STATURE
At about 1m (3ft 3in) long, and weighing only 30–45kg (66–99lb), *Protosuchus* was a small animal that resembled a juvenile crocodile.

STENEOSAURUS SKULLS
These recognizably crocodilian *Steneosaurus* skulls were found in Oxfordshire, England.

Steneosaurus

GROUP Crocodylomorphs
DATE Early Jurassic to Early Cretaceous
SIZE 1–4m (3¼–13ft) long
LOCATION Europe, Africa

Steneosaurus was quite unlike most living crocodiles. It did not lurk around the edge of the water, but rather was a competent swimmer that could venture into estuaries and coastal waters to hunt fish. However, *Steneosaurus* was not fully aquatic. It still had a bony coat of armour and its limbs were not modified into flippers. This meant that it could hunt on land if necessary.

Sphenosuchus

GROUP Crocodylomorphs
DATE Early Jurassic
SIZE 1–1.5m (3¼–5ft) long
LOCATION South Africa

Around the same time that *Protosuchus* was running across the sand dunes of Arizona (see above), another crocodilian relative called *Sphenosuchus* was sprinting through the humid landscape of southern Africa. Although these two early crocodylomorphs lived at about the same time, they were quite different in their anatomy. *Protosuchus* was a quick runner that stood fairly upright, but *Sphenosuchus* had an even more upright stance and would have been much faster. It may have even reared up on two legs at times, and like its close cousin *Terrestrisuchus*, it greatly resembled theropod dinosaurs in its skeletal anatomy and lifestyle.

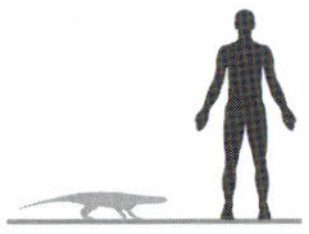

STRANGE CROCODILE
A fast runner with an upright stance, *Sphenosuchus* was quite a strange early crocodile with a lifestyle very different to that of modern crocodiles.

Dakosaurus

GROUP Crocodylomorphs
DATE Late Jurassic to Early Cretaceous
SIZE 4–5m (13–16½ft) long
LOCATION Worldwide

Dakosaurus was the largest and fiercest member of a group of marine reptiles distantly related to modern crocodiles. This behemoth reached sizes much larger than the other marine reptiles that, along with fish, formed most of its diet. It was well adapted to life in the open oceans, with a streamlined body and paddle-like limbs. However, unlike many of its relatives, which had tubular snouts, *Dakosaurus* had a deep skull that resembled those of predatory dinosaurs such as *Tyrannosaurus*. Its teeth were also similar to those of theropods, as well as modern killer whales. They were large, narrow, and covered with an array of sharp serrations, perfect for cutting flesh.

MONSTROUS SKULL
Dakosaurus had a monstrous skull full of huge teeth. One South American skull specimen measuring more than 1m (3ft 3in) long has been nicknamed "Godzilla" because of its size.

Geosaurus

GROUP Crocodylomorphs
DATE Late Jurassic to Early Cretaceous
SIZE 2–3m (6½–9¾ft) long
LOCATION Europe, North America, the Caribbean

Like *Dakosaurus* (above), *Geosaurus* is a member of the metriorhynchids, a group of distinctive crocodylomorphs that lived in the open oceans. Unlike its land-living relatives, *Geosaurus* had lost its protective coat of bony scales, making its body lighter and more streamlined for swimming. In addition, its hands and feet had evolved into flippers, its body was thin and elongated, and its tail had a deep fin. There is even evidence that *Geosaurus*'s skull housed a large salt gland – an organ present in some marine creatures that enables them to drink salty sea water and eat aquatic prey without dehydrating. A successful predator, it used its sharp teeth to plunder shoals of fish and squid.

Geosaurus, meaning "earth lizard", is a strange name for a creature that lived in the open oceans.

MEDIUM-SIZED
Geosaurus was a medium-sized metriorhynchid, reaching up to 3m (9¾ft) in length. It had a narrow, tapering snout full of fairly small, sharp teeth.

Dimorphodon

GROUP Pterosaurs
DATE Early Jurassic
SIZE 1m (3ft 3in) long
LOCATION British Isles

A relatively small pterosaur, *Dimorphodon* was only about 1m (3ft 3in) long but had a skull that was about 25cm (10in) long. Its body was surprisingly light, and its weight-saving adaptations for flight included hollow limb bones and a skull reduced to little more than a scaffolding around its large eye and face.

ON ALL FOURS
When not in the air, *Dimorphodon*, like other pterosaurs, would have walked – probably rather clumsily – on four legs and taken off from a crouched, quadrupedal stance.

Dimorphodon lived along the coasts of ancient seas and probably used its puffin-like skull and fang-like teeth to capture fish and rip them apart. This pterosaur would probably have caught fish at the surface by lunging for them while skimming low over the water. It is not thought to have entered the water or dived.

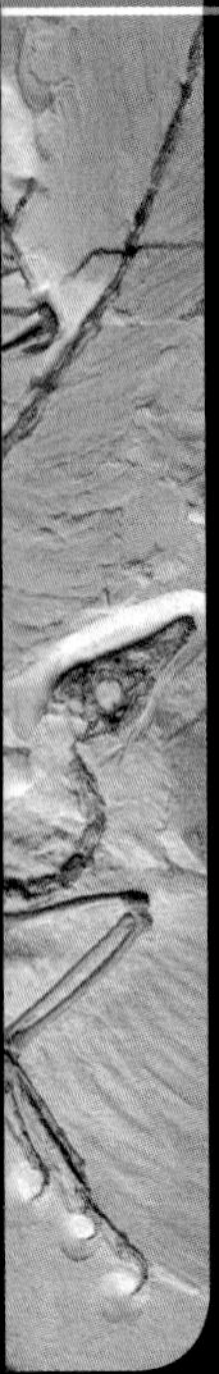

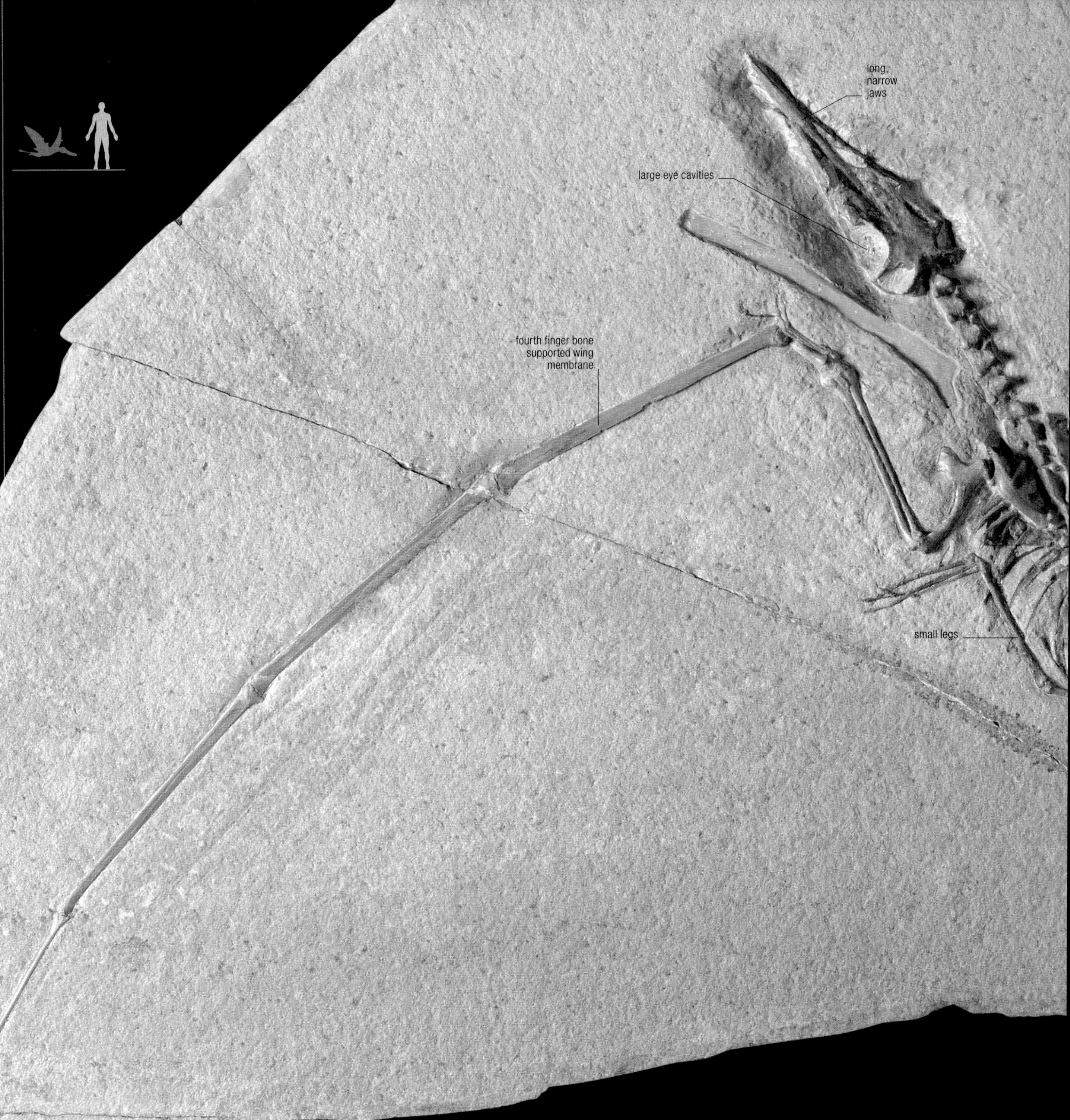

Rhamphorhynchus

GROUP Pterosaurs

DATE Middle to Late Jurassic

SIZE 1–1.3m (3ft 3in–4¼ft) long

LOCATION Europe, Africa

During the Late Jurassic, the pterosaur *Rhamphorhynchus* was the most common flying vertebrate in the UK and Spain. It had a wingspan of less than 2m (6½ft), and shared the skies with the first bird, *Archaeopteryx* (see p.264). Although it was small by pterosaur standards, *Rhamphorhynchus* was still several times the size of the birds that it flew alongside.

Like most pterosaurs *Rhamphorhynchus* lived in coastal environments and almost certainly ate a diet that was rich in fish. Its elongate skull had a series of nail-like teeth as well as a sharp beak, both of which were adaptations for catching prey. As *Rhamphorhynchus* skimmed over the waves, it immersed its head into water to scoop up fish in its beak. As the beak closed the teeth interlocked to secure the fish. *Rhamphorhynchus* was a powerful flier. Its long, thin wings provided thrust, and its slim, stiffened tail was used as a rudder for steering. Scans of its fossilized brain cavity indicate that *Rhamphorhynchus* had keen senses.

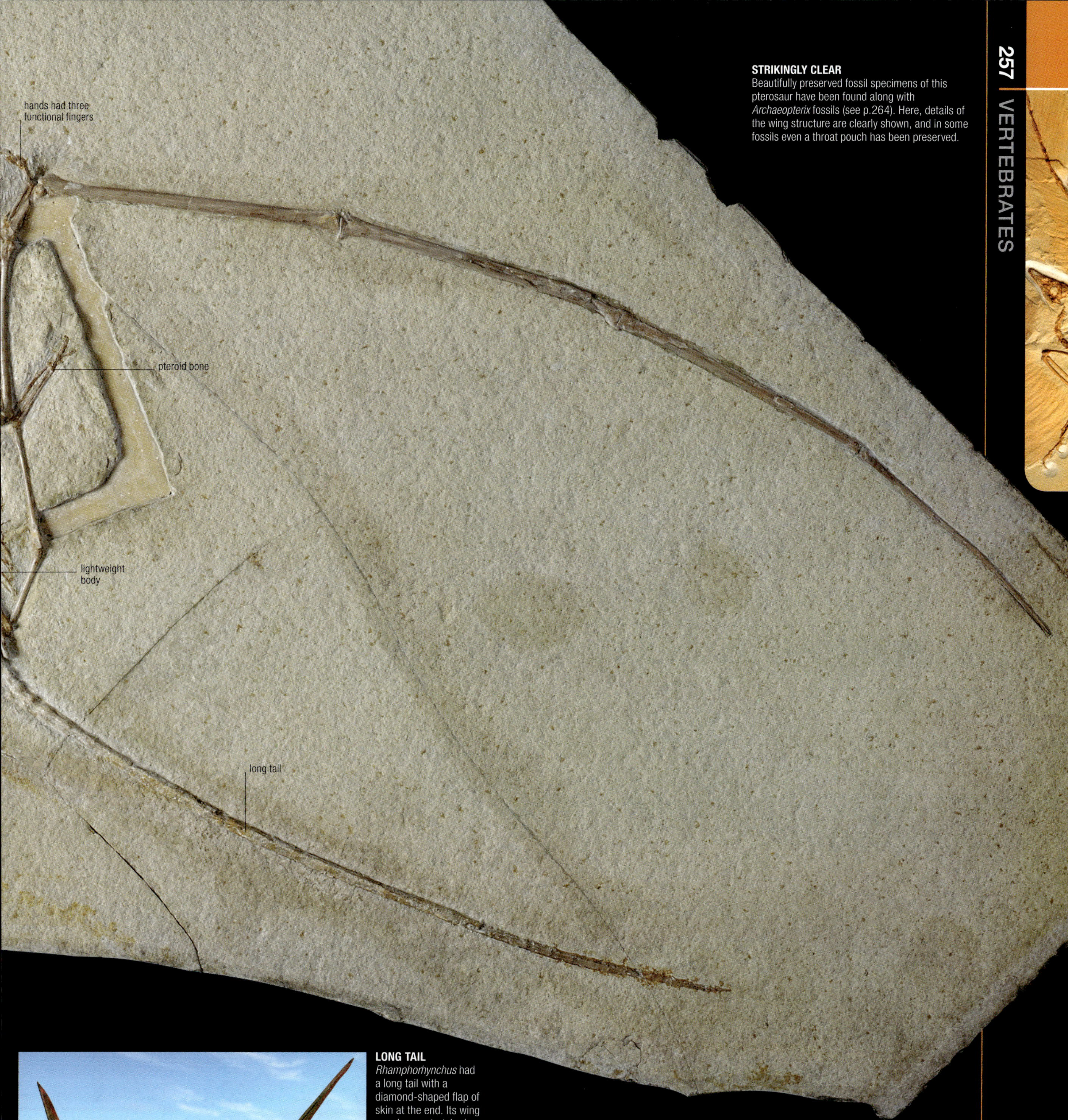

STRIKINGLY CLEAR
Beautifully preserved fossil specimens of this pterosaur have been found along with *Archaeopterix* fossils (see p.264). Here, details of the wing structure are clearly shown, and in some fossils even a throat pouch has been preserved.

LONG TAIL
Rhamphorhynchus had a long tail with a diamond-shaped flap of skin at the end. Its wing membrane stretched down to its ankle.

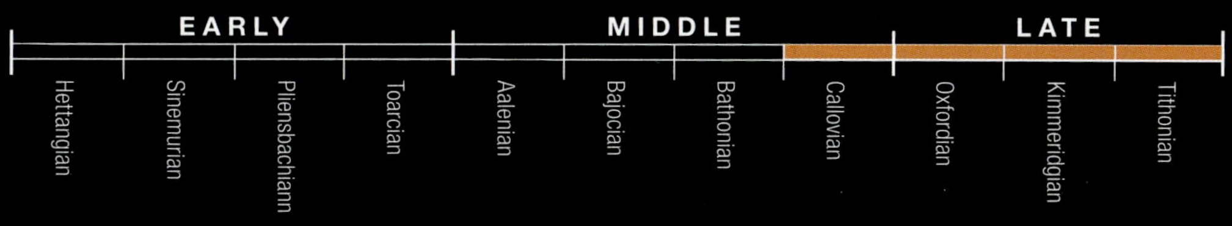

CRESTED HEAD
Recent fossil finds have shown that *Pterodactylus* had a small crest on its head that was probaby used for display.

Pterodactylus

GROUP Pterosaurs
DATE Late Jurassic
SIZE 1m (3ft 3in) long
LOCATION Germany

Pterodactylus, meaning "wing finger", is one of the best-known pterosaurs. Many complete skeletons have been found, from small juveniles to large adults. *Pterodactylus* was one of the first pterodactyloid pterosaurs. With shorter tails and longer necks, pterodactyloids were better adapted to flight than their earlier relatives. Unlike most other pterodactyloids, *Pterodactylus* was relatively small, with a wingspan of about 1m (3ft 3in). However there were a number of different *Pterodactylus* species, and it can be difficult to distinguish the young of a large species from the adults of a small species. *Pterodactylus* lived in the coastal areas of an inland sea that covered southern Germany in the Late Jurassic. It hunted small fish using its long jaws, which were full of small teeth.

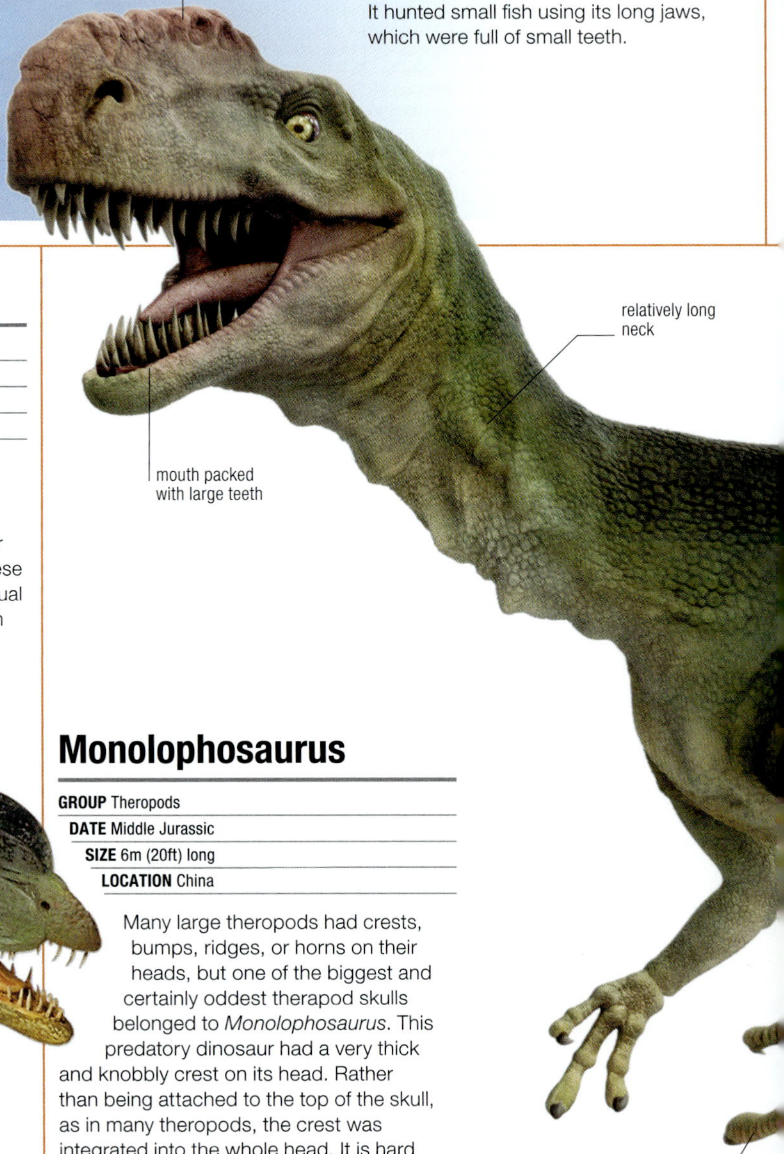

Dilophosaurus

GROUP Theropods
DATE Early Jurassic
SIZE 6m (20ft) long
LOCATION USA

One of the few known therapods from the Early Jurassic, *Dilophosaurus* was also one of the most striking. It was slender and lightly built, but is best known for its remarkable skull. This had two parallel, plate-like crests arranged along the upper surface of the snout. No one knows why *Dilophosaurus* had these crests, but the most popular idea is that they were used for sexual displays. Inside its mouth there was a prominent notch between the tip of the upper jaw and the rest of the teeth, and this may have helped *Dilophosaurus* to grab small prey. It was originally described as a new species of *Megalosaurus* (see opposite) and not assigned its own genus until 1970.

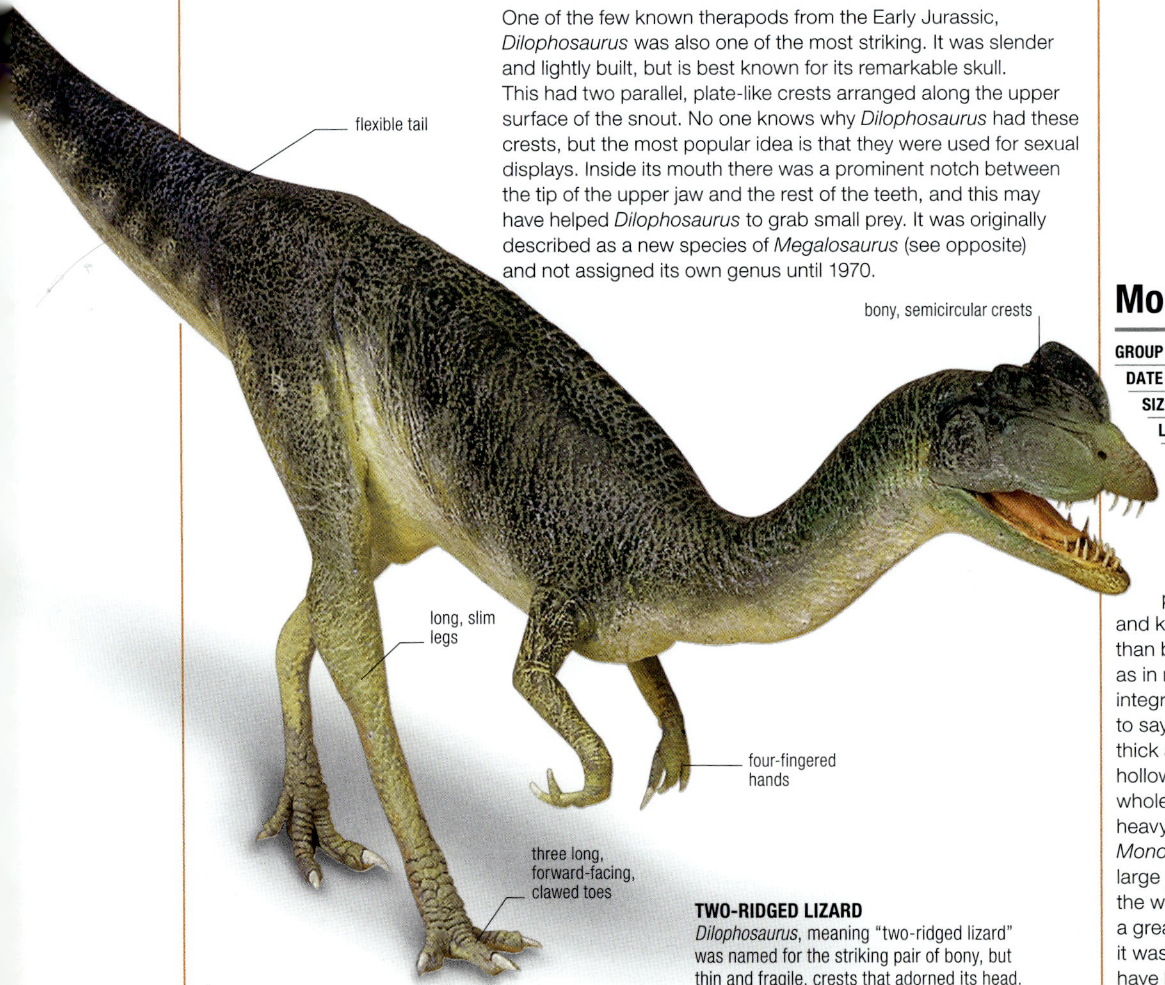

TWO-RIDGED LIZARD
Dilophosaurus, meaning "two-ridged lizard" was named for the striking pair of bony, but thin and fragile, crests that adorned its head.

Monolophosaurus

GROUP Theropods
DATE Middle Jurassic
SIZE 6m (20ft) long
LOCATION China

Many large theropods had crests, bumps, ridges, or horns on their heads, but one of the biggest and certainly oddest therapod skulls belonged to *Monolophosaurus*. This predatory dinosaur had a very thick and knobbly crest on its head. Rather than being attached to the top of the skull, as in many theropods, the crest was integrated into the whole head. It is hard to say how the crest functioned. It is very thick and strong at the top but it is thin and hollow lower down, suggesting that as a whole it was weak and could not resist heavy forces. Only one specimen of *Monolophosaurus* is known, and some large parts of this are missing – including the whole of the tail – so we do not know a great deal about this animal. However, it was certainly a top predator and would have been a capable killer.

Cryolophosaurus

GROUP Theropods
DATE Early Jurassic
SIZE 6.5m (21½ft) long
LOCATION Antarctica

Discovered in the Transantarctic Mountains during the 1990s, *Cryolophosaurus* was clearly a top predator. Fragments of ankylosaurs, ornithopods, and theropods had previously been found in Antarctica, but *Cryolophosaurus* is currently the only impressive dinosaur specimen known from the continent. *Cryolophosaurus* had a very unusual bony crest above its eyes. This was a thin, sheet-like structure that curved upwards and forwards over the top of the skull, and had parallel lines decorating the front and back. Perhaps a close relative of *Dilophosaurus* (see opposite), *Cryolophosaurus* had a shallow skull and long, slender proportions.

FORMIDABLE SIZE
With an estimated weight of more than 460kg (1,010lb), *Cryolophosaurus* is currently the largest known Early Jurassic theropod. As the only known specimen was not fully mature, adults were probably even larger than this.

SCIENCE
ANTARCTIC FIELDWORK

Cryolophosaurus is one of several dinosaurs that have recently been discovered in the rocks of the Transantarctic Mountains, Antarctica. Though covered by ice and snow today, the rocks on Antarctica surely contain many fossils of species that have yet to be discovered. Fieldwork in Antarctica presents special complications, as people and machinery must be able to withstand the cold, strong winds, and dangerous terrain. Furthermore, the rock containing *Cryolophosaurus* was particularly hard, so proper preparation of its bones could only be carried out in the laboratory.

This dinosaur's name means **"single-crested lizard"**. Despite the similar name, it is not a very close relative of Dilophosaurus, the **"double-crested lizard"**.

AGILE PREDATOR
A medium-sized therapod, *Monolophosaurus* had quite long, powerful legs and was probably a fairly agile predator that would have been a threat to other dinosaurs.

Megalosaurus

GROUP Theropods
DATE Middle Jurassic
SIZE 6m (20ft) long
LOCATION England

Megalosaurus, "great lizard", was described in 1824 after several large reptile bones were discovered in Oxfordshire, England. This makes it the very first dinosaur to be recognized by science. Because the fragmentary remains were superficially similar to the bones of large, living lizards, *Megalosaurus* was originally imagined to be a gigantic, four-footed reptile, similar to modern monitor lizards. Today we know that *Megalosaurus*, like other large theropods, was a two-legged predator with short arms. Despite the length of time since it was first discovered, *Megalosaurus* remains enigmatic due to the poor fossil record.

FOSSILIZED FOOTPRINTS
This is one of a group of fossilized *Megalosaurus* footprints found in a quarry near Oxford, England, in 1997.

SACRAL VERTEBRAE
These vertebrae were among the first *Megalosaurus* fossils to be found. They were identified by geologist William Buckland in 1824.

Gasosaurus

GROUP Theropods
DATE Middle Jurassic
SIZE 3.5m (11½ft) long
LOCATION China

The bones of this carnivore were discovered by chance after the rocks they were encased in were blown apart with dynamite. As a result, they are not in the best condition and many are missing. Nevertheless, *Gasosaurus* is an important and interesting animal because it comes from the Middle Jurassic, a time from which we have few dinosaur fossils. *Gasosaurus* reveals information about the evolution of some of the earlier predatory dinosaurs. It was one of the larger predators of its time, but was too small to threaten many of the giant herbivores that were its contemporaries.

***Gasosaurus* takes its name from the fact that it was found by a company searching for natural gas.**

TYPICAL THEROPOD
This species is thought to have had the typical theropod body shape, with a large head, long legs, and a thick tail. The few fossil fragments include arm, pelvis, and leg bones, but no skull.

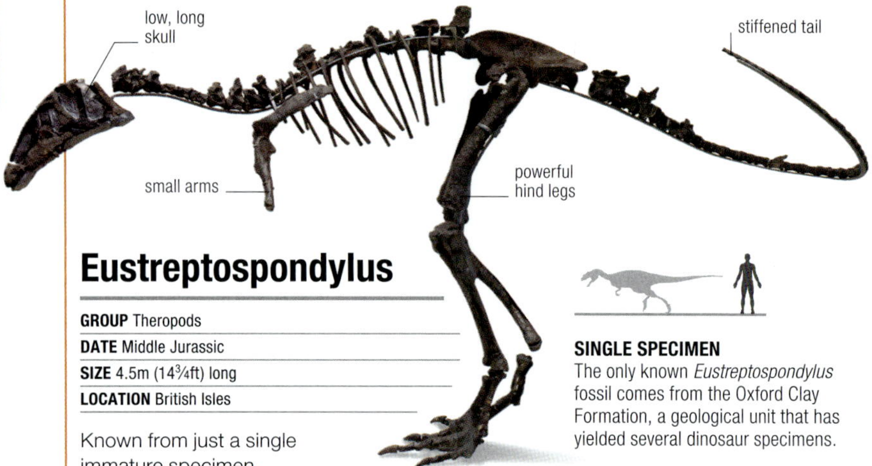

Eustreptospondylus

GROUP Theropods
DATE Middle Jurassic
SIZE 4.5m (14¾ft) long
LOCATION British Isles

Known from just a single immature specimen, *Eustreptospondylus* is one of Europe's best-preserved large theropods. It had a low, crestless skull, and a shallow notch in the upper jaw, close to the front of the snout, gave it curved margins to the edges of its mouth. The lower jaw was long and slender, and had a thickened and deepened tip. These features are like those developed to an extreme in the Cretaceous spinosaurids, which had skulls like those of modern crocodiles. Indeed, some experts have suggested that *Eustreptospondylus* may have foraged on shorelines for carcasses and marine life.

SINGLE SPECIMEN
The only known *Eustreptospondylus* fossil comes from the Oxford Clay Formation, a geological unit that has yielded several dinosaur specimens.

Ceratosaurus

GROUP Theropods
DATE Late Jurassic
SIZE 6m (20ft) long
LOCATION USA, Portugal

Ceratosaurus was a formidable and frightening predator. Its name – "horned reptile" – is a reference to its rounded nasal horn, but it also had tall triangular or rounded horns in front of its eyes. It had a large, deep skull that was well suited for a life of preying on large animals. Uniquely among theropods, *Ceratosaurus* had a continuous row of flat bony plates known as scutes running along its neck, back, and tail, which was deep but narrow.

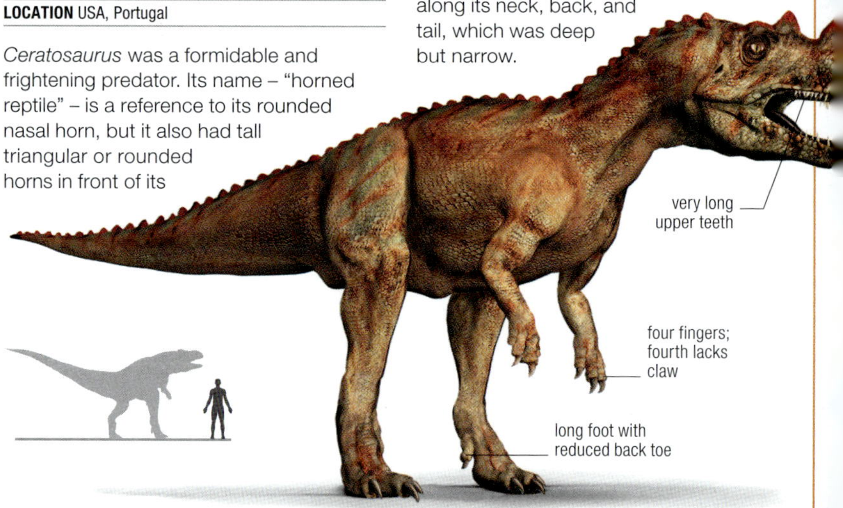

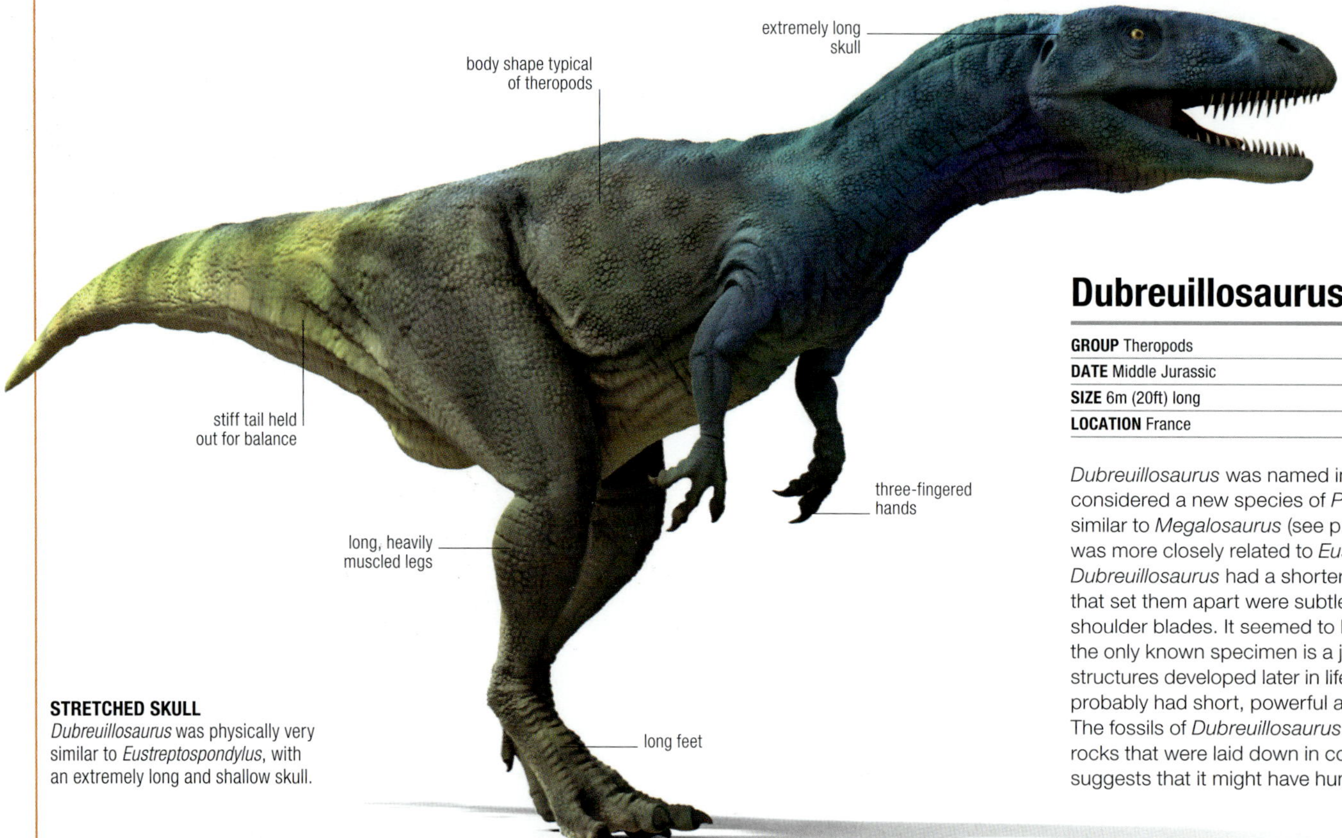

STRETCHED SKULL
Dubreuillosaurus was physically very similar to *Eustreptospondylus*, with an extremely long and shallow skull.

Dubreuillosaurus

GROUP Theropods
DATE Middle Jurassic
SIZE 6m (20ft) long
LOCATION France

Dubreuillosaurus was named in 2002, after first being mistakenly considered a new species of *Poekilopleuron,* a large theropod similar to *Megalosaurus* (see p.259). Later study showed that it was more closely related to *Eustreptospondylus* (see above). *Dubreuillosaurus* had a shorter snout but the only other features that set them apart were subtle differences in their vertebrae and shoulder blades. It seemed to lack any sort of crest or horns, but the only known specimen is a juvenile, and it is possible that these structures developed later in life. Like its relatives, *Dubreuillosaurus* probably had short, powerful arms with three-fingered hands. The fossils of *Dubreuillosaurus* were discovered in sedimentary rocks that were laid down in coastal, mangrove swamps. This suggests that it might have hunted fish and other marine prey.

Allosaurus

GROUP Theropods

DATE Late Jurassic

SIZE 7.5m (24½ft) long

LOCATION USA, Portugal

One of the most abundant of all Jurassic theropods, *Allosaurus* is known from many specimens discovered in the Morrison Formation, western USA. A formidable predator, its large skull and jaws held sharp, serrated teeth (see panel), and the three large claws on its hands may have been used to grip on to the sides of prey. Its powerful jaws and claws, combined with its considerable size, have led most scientists to imagine *Allosaurus* as a predator of stegosaurs, ornithopods, and perhaps sauropods. Evidence for the fact that *Allosaurus* fed on these large plant-eating dinosaurs comes from tooth marks preserved on their bones, though like all predators it would also have scavenged on dead dinosaurs. One *Allosaurus* specimen has a hole in one of its tail vertebrae exactly matching the size and shape of a stegosaur tail spike. On this occasion, it seems that the herbivore succeeded in causing a major injury to the predator.

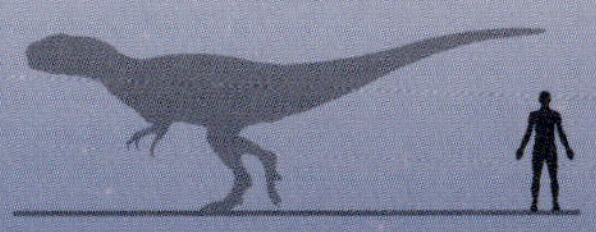

ANATOMY

DEADLY TEETH

Allosaurus had a massive, deep skull, and powerful jaws with blade-like teeth. Studies on skull function suggest that *Allosaurus* may have been able to open its jaws wide enough to deliver large slashing bites with the whole length of its long tooth rows, extracting huge mouthfuls of flesh.

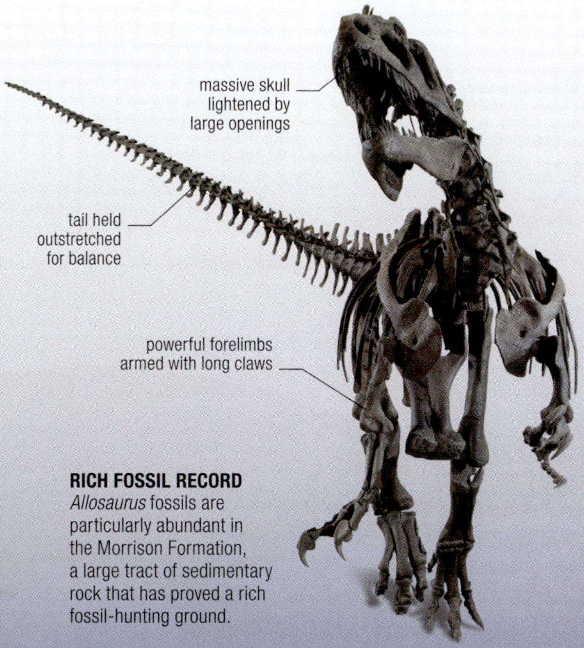

RICH FOSSIL RECORD
Allosaurus fossils are particularly abundant in the Morrison Formation, a large tract of sedimentary rock that has proved a rich fossil-hunting ground.

FAMILIAR PREDATOR
Allosaurus had low, parallel ridges running along the top of its snout, and distinctive triangular horns were present in front of its eyes. After *Tyrannosaurus* (see pp.322–23), *Allosaurus* is probably the best-known predatory dinosaur.

Sinraptor

GROUP Theropods
DATE Late Jurassic
SIZE 9m (30ft) long
LOCATION China

Sinraptor means "Chinese hunter", and so far all its fossils have been excavated in that country. It was a large predator that closely resembled its more famous relative, *Allosaurus* (see p.273). Several specimens are known from different parts of China, and more work is needed to establish exactly which specimens belong to which species. Some paleontologists think that several animals named as separate species of *Sinraptor* are actually different enough to be placed in their own genus. One thing that is known for certain is that *Sinraptor* occasionally fought with others of its kind. Scientists have found tooth marks on the skull and jaw of one fossil that appear to have been made by another *Sinraptor*.

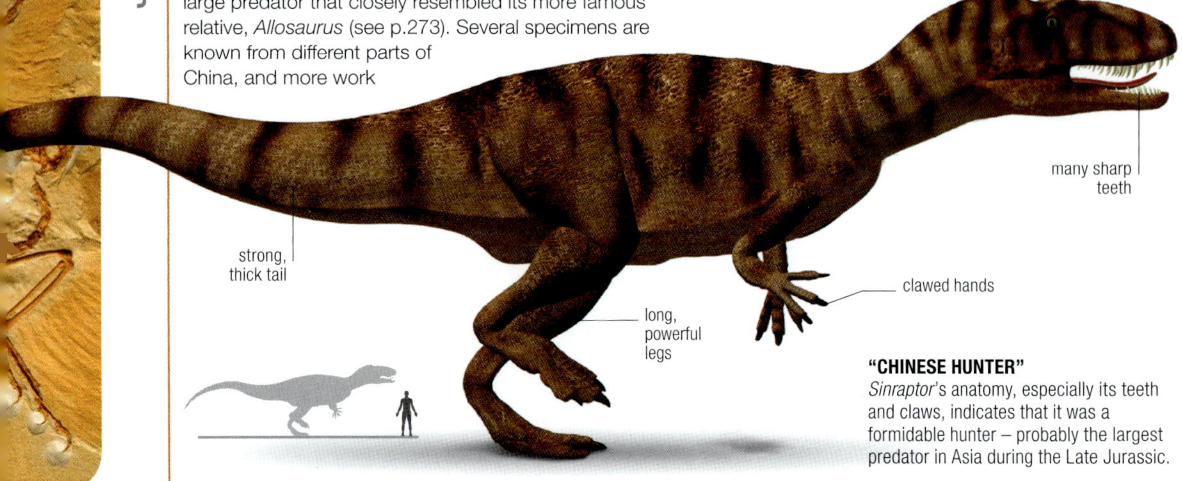

"CHINESE HUNTER"
Sinraptor's anatomy, especially its teeth and claws, indicates that it was a formidable hunter – probably the largest predator in Asia during the Late Jurassic.

Proceratosaurus

GROUP Theropods
DATE Early Jurassic
SIZE 2m (6½ft) long
LOCATION British Isles

Discovered in Gloucestershire, England, *Proceratosaurus* is known only from one well-preserved skull. At the tip of its snout the base of a crest is preserved. Because the top of the skull is missing, the shape of this crest is unknown, but it is thought to have been horn-like, similar to that of *Ceratosaurus* (see p.272). Alternatively it may have had a long crest that extended the whole length of its skull.

MISTAKEN IDENTITY
Due to its nasal horn, *Proceratosaurus* was named to reflect a close relationship to *Ceratosaurus*, but this relationship was later disproved.

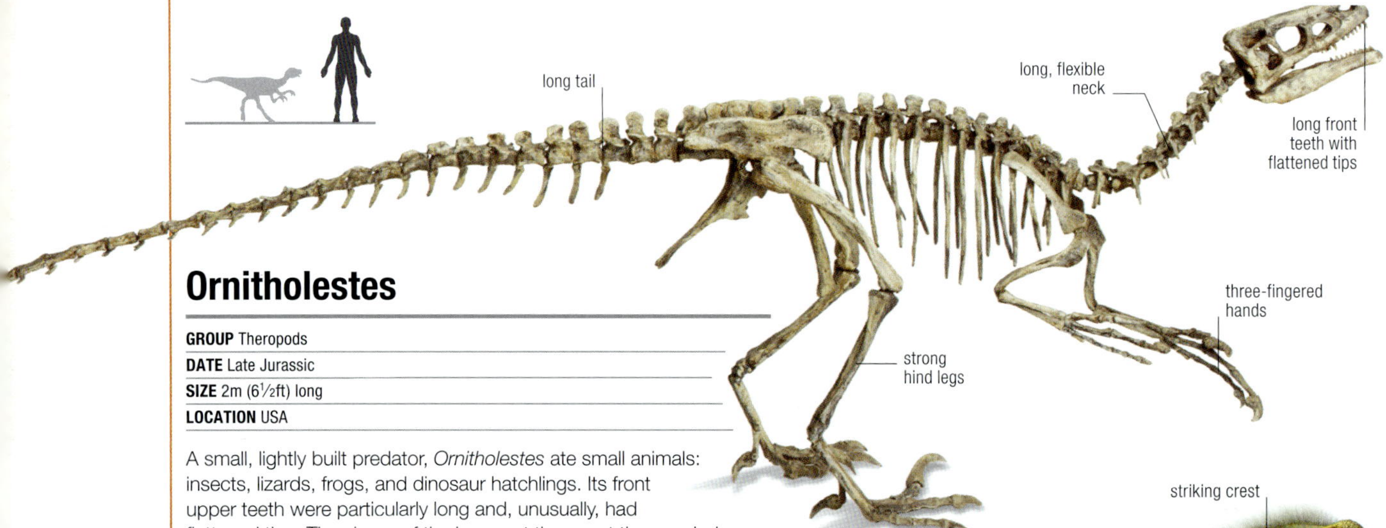

Ornitholestes

GROUP Theropods
DATE Late Jurassic
SIZE 2m (6½ft) long
LOCATION USA

A small, lightly built predator, *Ornitholestes* ate small animals: insects, lizards, frogs, and dinosaur hatchlings. Its front upper teeth were particularly long and, unusually, had flattened tips. The shape of the bones at the snout tip once led to the suggestion that it might have had a nasal horn, but new observations have shown that this was probably not the case. Its three-fingered hands were long and slender. *Ornitholestes* was almost certainly covered with filament-like feathers.

"BIRD THIEF"
The elegant shape of *Ornitholestes* made it a swift, efficient hunter. Its name means "bird thief".

BIOGRAPHY
HENRY F. OSBORN

Henry Fairfield Osborn (1857–1935) was one of America's most influential paleontologists. In 1891, he joined New York's American Museum of Natural History, where he assembled a team of expert fossil hunters, and named and described scores of new species, including *Ornitholestes* in 1903.

Guanlong

GROUP Theropods
DATE Middle Jurassic
SIZE 2.5m (8¼ft) long
LOCATION China

Guanlong is an early form of tyrannosaur, and although it may not look much like the famous *Tyrannosaurus*, it has much in common with its fiercer cousin. Among other things, it has relatively blunt teeth and the shape of the hips shows that there are evolutionary ties. *Guanlong* was a relatively small and lightly built predator that would have hunted other small dinosaurs. It was found in the fossil beds of western China, and we know of few other predators from that region. Despite its size, *Guanlong* may have been one of the top predators around. Its name means "crested" or "crowned dragon" in Chinese, and the large, thin crest on its head was made of bone. Analysis has shown that the crest may have played a part in the animal's biting and feeding mechanism, although it is thought that its primary function was one of display.

***Guanlong*'s striking crest ran from its nose to the back of its head, and was probably used mainly for display.**

PRIMITIVE TYRANNOSAURUS
Although small in comparison, *Guanlong* shared enough characteristics with *Tyrannosaurus* for scientists to be certain that the two were related.

Compsognathus

GROUP Theropods
DATE Late Jurassic
SIZE 1.3m (4¼ft) long
LOCATION Germany, France

This small dinosaur is known from only two fossils – one from France and a much smaller individual from southern Germany (see below). The German discovery changed how people thought about dinosaurs by showing that they could be small, and were not just giants like *Tyrannosaurus* and *Diplodocus*.

This dinosaur's close relatives, such as *Sinosauropteryx*, have fuzzy proto-feathers, so *Compsognathus* probably had them also, although none have yet been found on the specimens. Despite its small size, this species was a predator – this is suggested by its mouthful of sharp teeth. Its name means "elegant jaw".

DIMINUTIVE DINOSAUR
Compsognathus was a small dinosaur with an streamlined jaw, and may have been feathered.

COMPLETE SPECIMEN
This *Compsognathus* fossil, found in southern Germany in 1859, was one of the most complete fossils to have been found at the time. It was also the smallest known for many years.

FEATHERED FLYER
In addition to its wing and tail feathers, *Archaeopteryx* also possessed long feathers on its hind limbs, and its neck and body were fully feathered too.

Archaeopteryx **was found in the year that Charles Darwin published his *On the Origin of Species*, and provided powerful support for the *theory of evolution* by natural selection.**

Archaeopteryx

GROUP Theropods
DATE Late Jurassic
SIZE 30cm (12in) long
LOCATION Germany

Archaeopteryx, meaning "ancient wing", was discovered in the Jurassic limestone of Solnhofen in Germany in 1859. Perhaps the most famous fossil in the world, the first specimen – now known as the London Specimen – was disarticulated and incomplete. Ten skeleton specimens are now known, and some are amazingly well preserved. Like modern birds, *Archaeopteryx* had long wing and tail feathers, and this is why it has always been regarded as the first bird. However, we now know that complex feathers and other bird-like features were widespread among the meat-eating dinosaurs known as maniraptorans. While *Archaeopteryx* remains recognized as one of the earliest ancestors of modern birds, many maniraptorans were almost as bird-like.

SPECTACULAR FOSSIL
The most spectacular *Archaeopteryx* fossil, and the third of its kind to be found, the Berlin specimen exhibits strikingly clear feather impressions, including those of flight feathers on the wings.

Anchisaurus

GROUP Sauropodomorphs
DATE Early Jurassic
SIZE 2m (6½ft) long
LOCATION USA

Anchisaurus is North America's best known early sauropodomorph. Its skull was shallow, with a narrow snout, and the teeth at the front of its upper jaw pointed forwards. These features suggest that *Anchisaurus* was omnivorous, and one specimen contains a small reptile within its gut. *Anchisaurus* had broad hands, and its large thumb claws were strongly curved. Its feet are unusual compared to those of its relatives in that the claw on the first toe was smaller than those on the other toes.

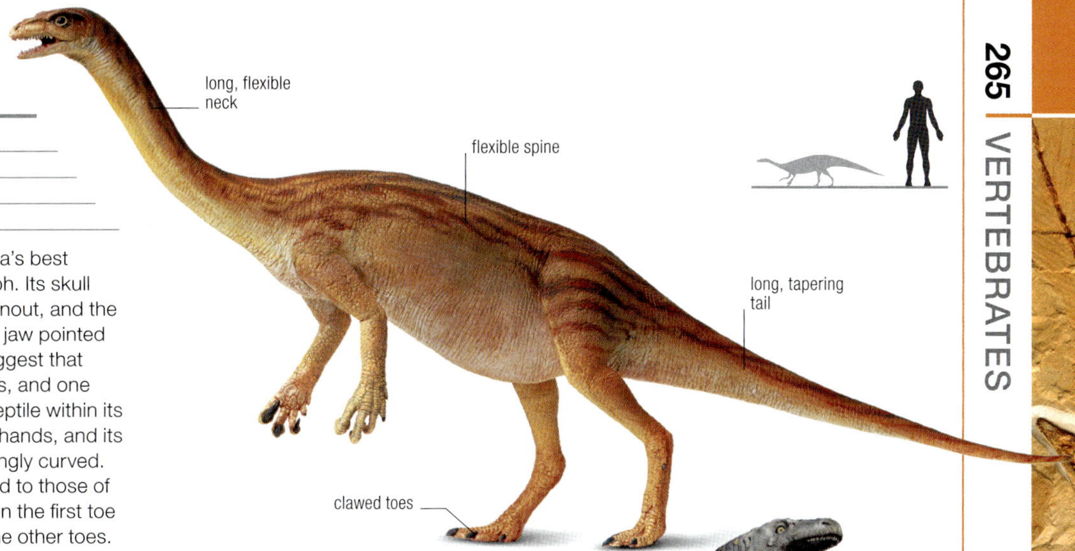

Lufengosaurus

GROUP Sauropodomorphs
DATE Early Jurassic
SIZE 5m (16½ft) long
LOCATION China

Lufengosaurus's snout was deep and broad, and it had distinctive bony bumps just behind its large nostrils and on its cheeks. A bony ridge on the side of its upper jaw might have helped anchor soft tissue. If so, then *Lufengosaurus* must have had larger cheeks than most other sauropodomorphs. Its closely spaced, serrated teeth suited a diet of leaves. *Lufengosaurus* was often thought to be very similar to *Plateosaurus* from Europe. However, new work has shown that the pair are quite different, and *Lufengosaurus* was most closely related to *Coloridasaurus* and *Massospondylus* (see below).

***Lufengosaurus* would have moved around on all fours, sometimes rearing up to feed.**

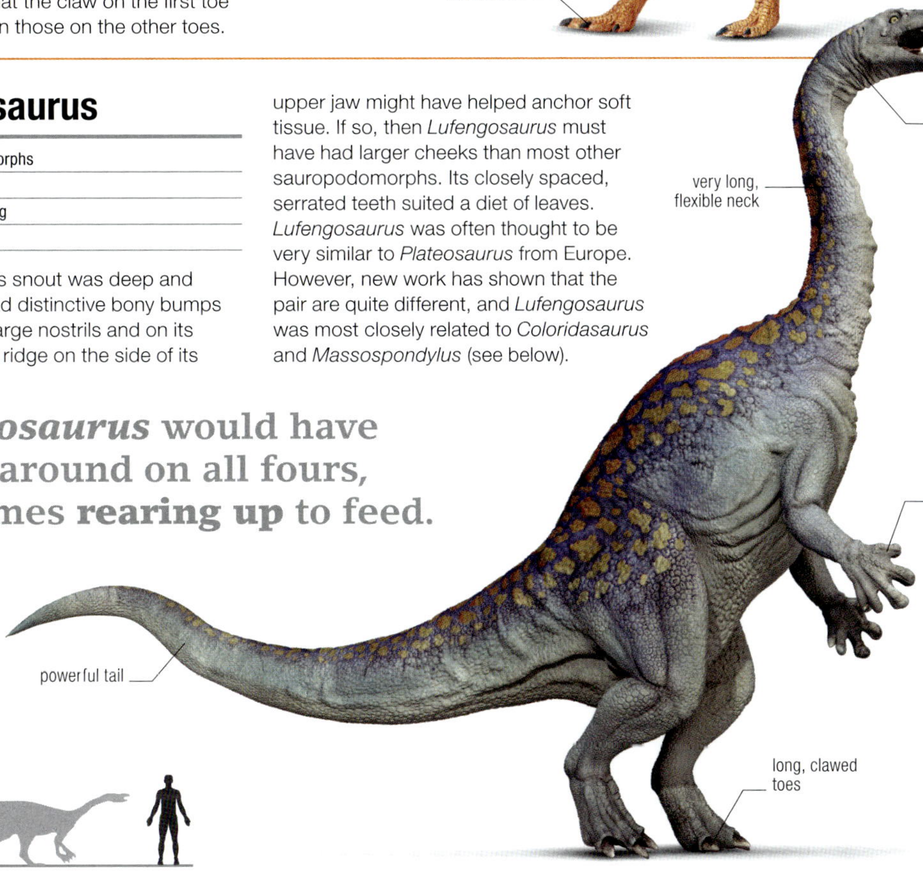

KEY FIND
CRUCIAL TIMING

Archaeopteryx was announced in 1861, two years after Charles Darwin published *On the Origin of Species*. It was to provide powerful support for Darwin's idea that organisms evolved over time, as here was a bird that had feathers (below) but still possessed obvious reptilian features. In the face of considerable doubt, *Archaeopteryx*'s long, bony tail, large hand claws, and toothed jaws convinced many that it was indeed a link between modern birds and their ancestors.

Massospondylus

GROUP Sauropodomorphs
DATE Early Jurassic
SIZE 5m (16½ft) long
LOCATION South Africa

Massospondylus is known from several complete skeletons and skulls and even some eggs containing embryos. It was a medium-sized, broad-skulled sauropodomorph with large eye sockets. Its broad, five-fingered hands had particularly large, curved thumb claws. A V-shaped bone, similar to the wishbone of birds, was present at the front of its ribcage. It had been thought that *Massospondylus* was able to walk on all fours. However, recent work on the mechanics of its short forearms suggests that it moved around, all of the time, on its hind legs.

EMBRYO FOSSIL
This *Massospondylus* embryo skeleton, measuring 15cm (6in) long, was found preserved inside its egg.

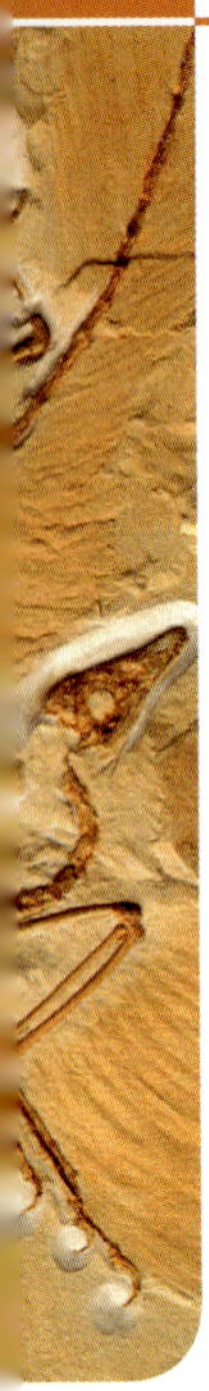

Vulcanodon

GROUP Sauropodomorphs
DATE Early Jurassic
SIZE 7m (23ft) long
LOCATION Zimbabwe

Vulcanodon was an early sauropod known from a single, partial skeleton without a skull. However, like other early sauropods it probably had a deep, blunt-snouted skull with leaf-shaped teeth. *Vulcanodon*'s feet were short, and large claws were present, at least on the inside toes. In contrast to later sauropods, the bones of its first toe were quite long. Two of the claws on each foot were broad and nail-like, not deep and narrow as they were in other sauropods. *Vulcanodon* was originally thought to have been a close relative of the sauropods, but not actually a sauropod in the true sense. However, it has now been established that it was indeed a true sauropod. Its closest relative is *Tazoudasaurus*, and both animals have been united in the family Vulcanodontidae.

EARLY SAUROPOD
An early sauropod, *Vulcanodon* was a specialized quadruped with long, column-like limbs. It was named after the volcanic rock in which the first skeleton was found.

Barapasaurus

GROUP Sauropodomorphs
DATE Early Jurassic
SIZE 18m (59ft) long
LOCATION India

Barapasaurus was originally named for just a sacrum (part of the backbone), but a large number of additional remains are now also thought to belong to it. Its skull is unknown, although isolated teeth have been found. These are broad at the tip but have a narrower base, and there are coarse serrations either side of the crown. *Barapasaurus* had particularly slender limbs for a sauropod. Its neck vertebrae were long but its body vertebrae were quite compressed. The shape of its vertebrae is unique among sauropods, leading some experts to suggest that it was an unusual side-branch in sauropod evolution.

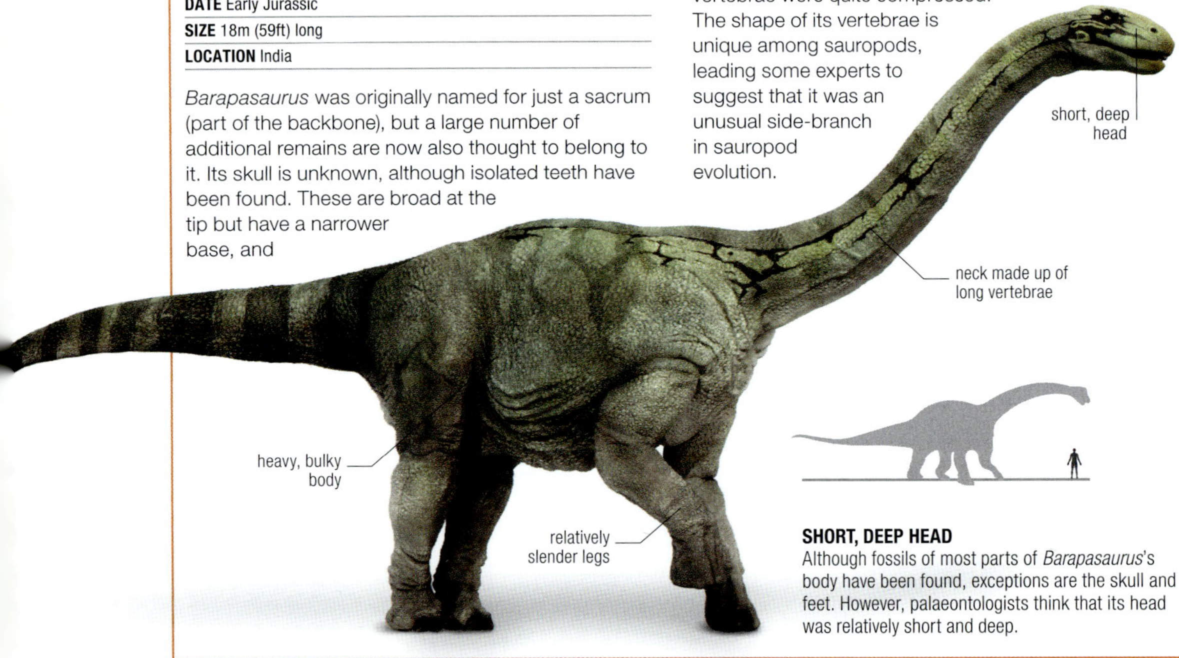

SHORT, DEEP HEAD
Although fossils of most parts of *Barapasaurus*'s body have been found, exceptions are the skull and feet. However, palaeontologists think that its head was relatively short and deep.

Shunosaurus

GROUP Sauropodomorphs
DATE Middle Jurassic
SIZE 12m (39ft) long
LOCATION China

Shunosaurus is the most abundant dinosaur in a fossil assemblage that has been named after it – the "*Shunosaurus* fauna". *Shunosaurus*'s neck was quite short compared with that of many other sauropods, but it does appear to have been quite flexible. Its skull was deep seen from the side but narrow looked at from above. *Shunosaurus* had more teeth than any other sauropod, with 25 or 26 being present in each half of the lower jaw.

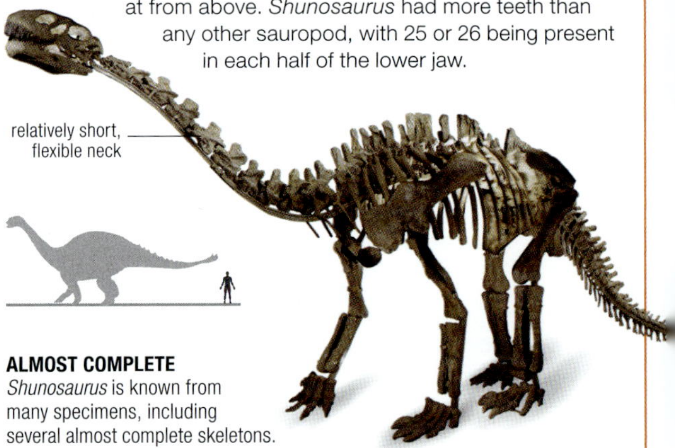

ALMOST COMPLETE
Shunosaurus is known from many specimens, including several almost complete skeletons.

This dinosaur had the longest neck of any known animal. One specimen had a neck more than 13m (43ft) long.

Mamenchisaurus

GROUP Sauropodomorphs
DATE Middle to Late Jurassic
SIZE 26m (85ft) long
LOCATION China

At first glance, one could mistake the front half of *Mamenchisaurus* with *Brachiosaurus* (see opposite), as they both have domed foreheads and enormously long necks. Both dinosaurs certainly used their necks to reach for food. However, the two are only distantly related and can be easily told apart: *Mamenchisaurus* had a pointed skull and its shoulders were lower and less massive than those of *Brachiosaurus*. *Mamenchisaurus*'s neck contained 19 vertebrae – more than any other dinosaur – and these were twice as long as those in its back. There is a large number of *Mamenchisaurus* species, and each stands out for its huge neck, even among other long-necked contemporaries.

CLUBLESS TAIL
Older reconstructions and illustrations of *Mamenchisaurus* sometimes show a club on the tail. This in now known to be a mistake.

Brachiosaurus

GROUP Sauropodomorphs
DATE Late Jurassic
SIZE 23m (75ft)
LOCATION USA, Tanzania

Unlike most sauropods, *Brachiosaurus* had very long front legs and a tall, erect neck. These features gave *Brachiosaurus* a huge reach. Without needing to stretch, a large individual could probably reach more than 15m (49ft) up into trees to feed. Despite what some early illustrations show, this herbivore probably could not rear up on its hind legs. Most of its weight was carried on its huge front limbs, so it would have been hard to take its full weight on its hind legs and stay balanced. In any case, *Brachiosaurus* was so tall it probably had little need to rear up. It could have reached higher to feed than almost any other dinosaur.

Brachiosaurus had a large skull, even when compared to similarly immense sauropods. It had a distinctive bar of bone in the middle of its forehead that created a large bulge on the top of its head. Inside the skull, this bar separates the two openings of the nostrils. It was once thought that the nostrils were large holes located high up on the skull, but recent research suggests that they were in fact relatively small and located closer to the front of the head.

ANATOMY

GIANT NUTRITIONAL DEMANDS

Sauropods were the largest animals to ever walk the planet. Scientists from the University of Bonn, led by Jurgen Hummel (pictured), have studied how these dinosaurs were able to grow to such large sizes by investigating the nutritional content of their diets. Using an artificial gut, they fermented plants that were present over 200 million years ago and found that sauropod diets were surprisingly high in nutritional value.

With its huge front legs and very long neck, *Brachiosaurus* could have reached higher to feed than almost any other dinosaur.

COLOSSAL BONES
This *Brachiosaurus* thigh bone measures 1.8m (6ft) from end to end, and was very thick, to support the dinosaur's considerable weight. One *Brachiosaurus* humerus (upper arm bone) fossil measures 2.1m (7ft) long.

DISTINCTIVE PROFILE
Due to its extremely long front legs and a long neck that it held erect, *Brachiosaurus* had a very distinctive asymmetrical profile. It takes its name – meaning "arm lizard" – from the fact that its front legs are longer than its back legs.

Diplodocus may have been able to move its long, **slender tail tip** at supersonic speed, creating a **whip-like crack.**

ELONGATED NECK
Diplodocus's neck contained 15 massive vertebrae and made up a considerable proportion of its total length. Whether the neck was held vertically or horizontally has been a subject of much debate.

DOUBLE BEAM
The long tail consisted of up to 80 vertebrae. The name *Diplodocus*, meaning "double beam", comes from the presence of chevron-shaped bones under the tail vertebrae.

Diplodocus

GROUP Sauropodomorph
DATE Late Jurassic
SIZE 25m (82ft) long
LOCATION USA

One of the best known of all dinosaurs, *Diplodocus* was like other diplodocids in having 15 neck vertebrae, proportionally short forelimbs, and a whip-like tip to its tail. Seen from above, its skull is rectangular and ends with a broad, squared-off mouth. *Diplodocus* had triangular spines along its back, and this was possibly the case in all sauropods. Studies of tooth wear suggest that *Diplodocus* used a feeding strategy known as unilateral branch stripping: a branch was gripped between its peg-like teeth, its head was pulled sharply upwards or downwards, and as a result either the upper or lower tooth row stripped the foliage off the branch. Three species of *Diplodocus* are currently recognized, and another diplodocid, originally named *Seismosaurus*, is now regarded as a fourth, and the largest, *Diplodocus* species.

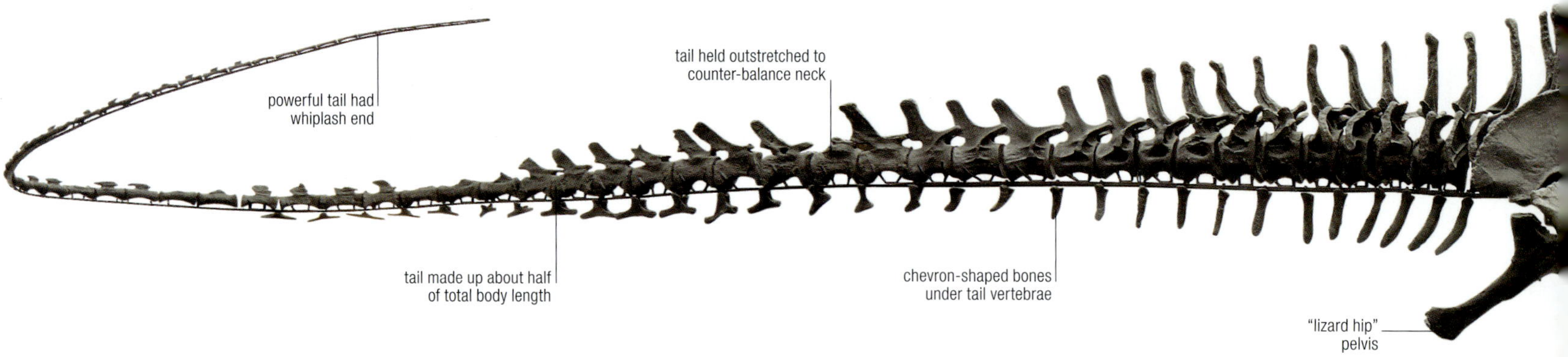

EARLY: Hettangian, Sinemurian, Pliensbachian, Toarcian
MIDDLE: Aalenian, Bajocian, Bathonian, Callovian
LATE: Oxfordian, Kimmeridgian, Tithonian

IMPRESSIVE PROPORTIONS
Diplodocus was one of the longest of all dinosaurs, possibly reaching up to 30m (98ft). Its tail, which made up about half its total length, counterbalanced an extremely long neck.

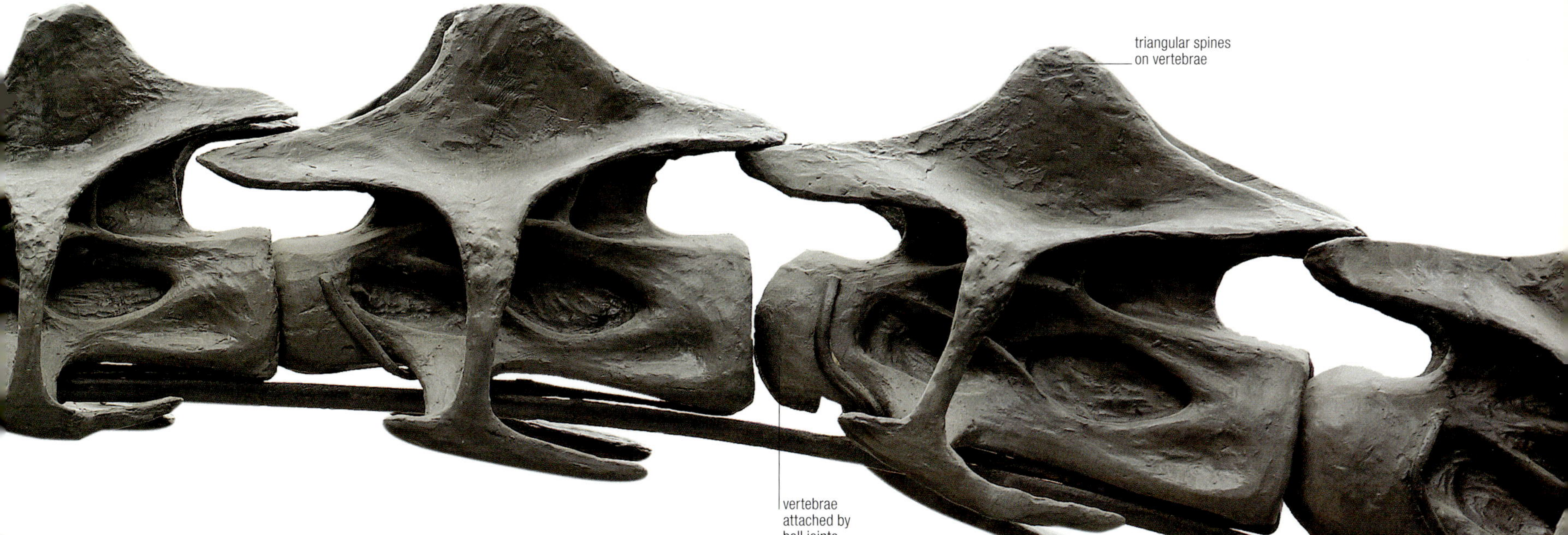

LIZARD HIP
As in other saurischians, *Diplodocus* had a "lizard hip" pelvic formation in which the pubis bone pointed forwards (see p.207). This formation distinguished the saurischians from the other order of dinosaurs, the ornithischians.

PEG-LIKE TEETH
Diplodocus's peg-like teeth, which were present only at the front of its jaws, were used to strip foliage from tree branches.

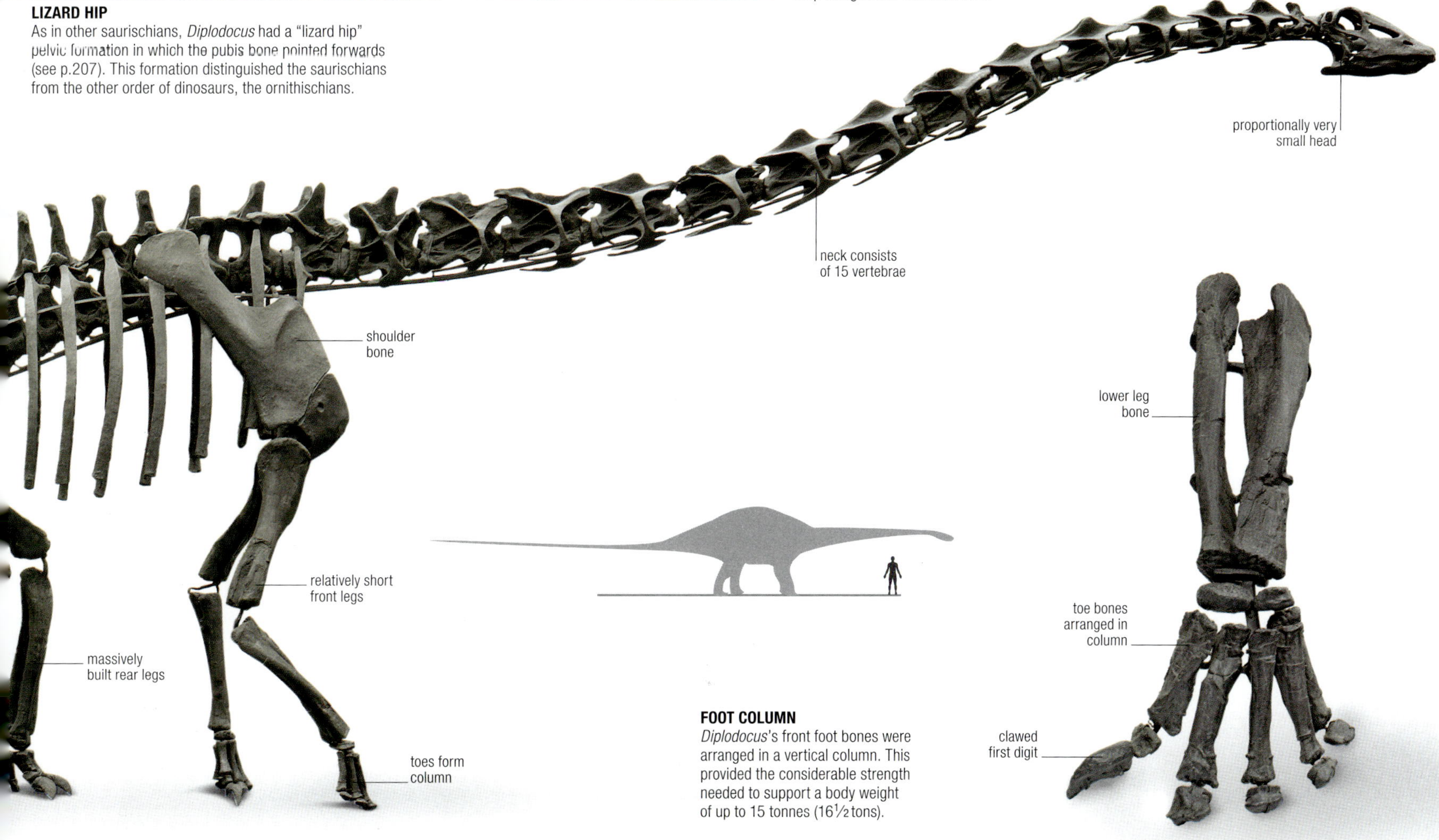

FOOT COLUMN
Diplodocus's front foot bones were arranged in a vertical column. This provided the considerable strength needed to support a body weight of up to 15 tonnes ($16\frac{1}{2}$ tons).

Barosaurus

GROUP Sauropodomorphs
DATE Late Jurassic
SIZE 28m (92ft) long
LOCATION USA

Barosaurus was similar to *Diplodocus* (see p.268) in many ways. Among other features, it shared distinctive hollows in its tail vertebrae and almost identical limbs. They mainly differed in the length of their neck vertebrae. *Barosaurus* probably had 16, which is one more than other diplodocids, and they were about a third longer than those of *Diplodocus*. This meant that *Barosaurus* probably had a larger feeding range than its shorter relative. For some time, fossils found in Tanzania were thought to belong to *Barosaurus*. It has now been confirmed that they are two different, but very similar, diplodocids: *Tornieria* and *Australodocus*.

LONG REACH
Thanks to its particularly long neck, *Barosaurus* was able to reach higher into trees and access more food than its close relatives.

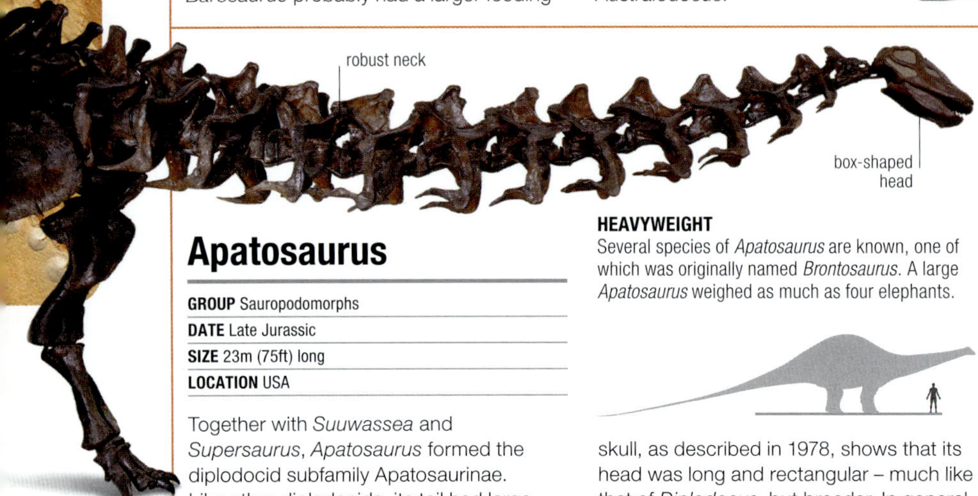

Apatosaurus

GROUP Sauropodomorphs
DATE Late Jurassic
SIZE 23m (75ft) long
LOCATION USA

Together with *Suuwassea* and *Supersaurus*, *Apatosaurus* formed the diplodocid subfamily Apatosaurinae. Like other diplodocids, its tail had large muscles at the base, while the tail tip was slender and whip-like. Many old restorations of *Apatosaurus* give it a box-shaped skull, but a true *Apatosaurus* skull, as described in 1978, shows that its head was long and rectangular – much like that of *Diplodocus*, but broader. In general, all apatosaurines had thicker legs and were stockier and more heavily constructed than other diplodocids.

HEAVYWEIGHT
Several species of *Apatosaurus* are known, one of which was originally named *Brontosaurus*. A large *Apatosaurus* weighed as much as four elephants.

Dicraeosaurus

GROUP Sauropodomorphs
DATE Late Jurassic
SIZE 12m (39ft) long
LOCATION Tanzania

Dicraeosaurus is a member of a group of diplodocoid sauropods called the dicraeosaurids. These dinosaurs were small, for sauropods, and had relatively short necks. *Dicraeosaurus*'s neck contained 12 unusually short vertebrae, which meant it could probably only browse vegetation from ground level to a height of around 3m (9¾ft).

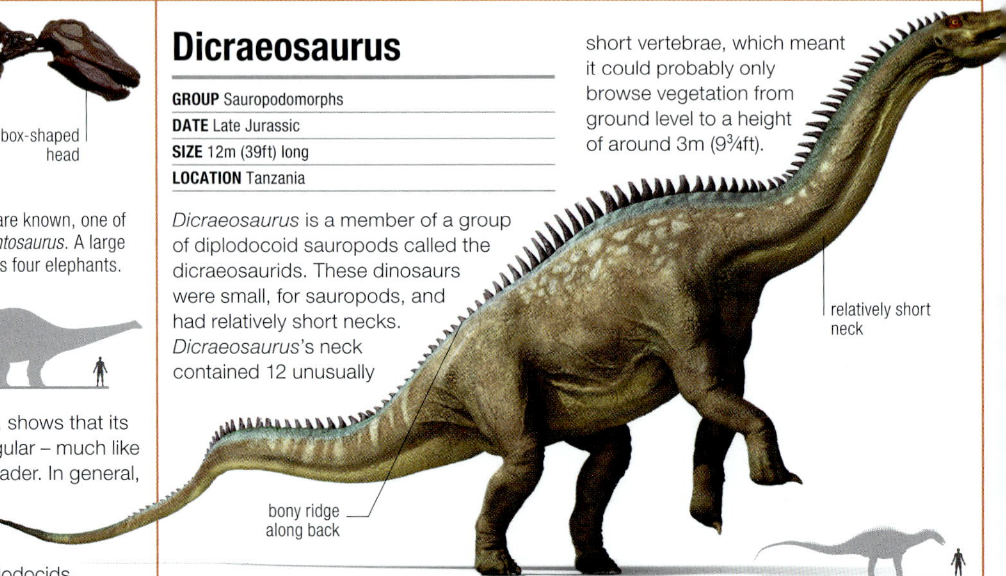

Camarasaurus

GROUP Sauropodomorphs
DATE Late Jurassic
SIZE 18m (59ft) long
LOCATION USA

Camarasaurus is the most common North American sauropod. Its neck was broad, but not as long as the necks of many sauropods. This means it probably fed on vegetation growing at various heights. It had a broad skull with short, strong jaws, and stout, spoon-shaped teeth that could have enabled it to feed on even the coarsest vegetation. *Camarasaurus* is one of the most primitive members of the macronarians, the sauropod group that also includes the brachiosaurs and titanosaurs. Like many macronarians, *Camarasaurus*'s skull had particularly large nostril openings. Its vertebrae also contained large openings that housed air sacs connected to its lungs. It is from these chambers that the dinosaur gets its name: "chambered lizard".

Scelidosaurus

GROUP Ornithischians
DATE Early Jurassic
SIZE 3m (9¾ft) long
LOCATION England

Discovered in the 1850s, *Scelidosaurus* was one of the first dinosaurs to be found as a complete skeleton. This original find is still one of the best preserved European dinosaurs. It had stout limbs and walked on all fours. Rows of oval armour plates ran along its neck, body, and limbs, and it had hooked spikes down the sides of its limbs. Because all the fossils have been found in marine rocks, it has been suggested that this dinosaur may have been an island-dweller or lived on the coast.

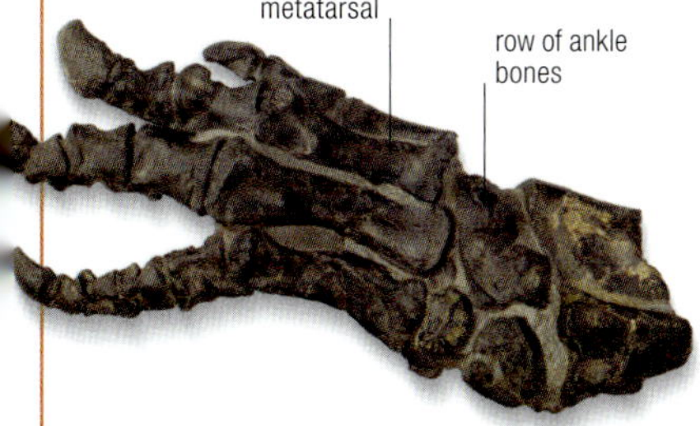

DINOSAUR TOES
Scelidosaurus had four toes, all tipped with blunt claws. Its toes were longer than those of the ankylosaurs and stegosaurs.

Lesothosaurus

GROUP Ornithischians
DATE Early Jurassic
SIZE 1m (3ft 3in) long
LOCATION Lesotho

Lesothosaurus is one of the earliest members of the ornithischians. It had long, slender hind limbs, and small forelimbs with hands that would not have been able to grasp properly. Like all ornithischians, the tips of *Lesothosaurus*'s upper and lower jaws were horny, forming a beak-like structure. Behind the beak, leaf-shaped teeth lined the jaws, and near the front of the upper jaw were 12 fang-like teeth. Analysis of its teeth have shown that *Lesothosaurus* sliced up plant material with its beak and was not able to chew its food.

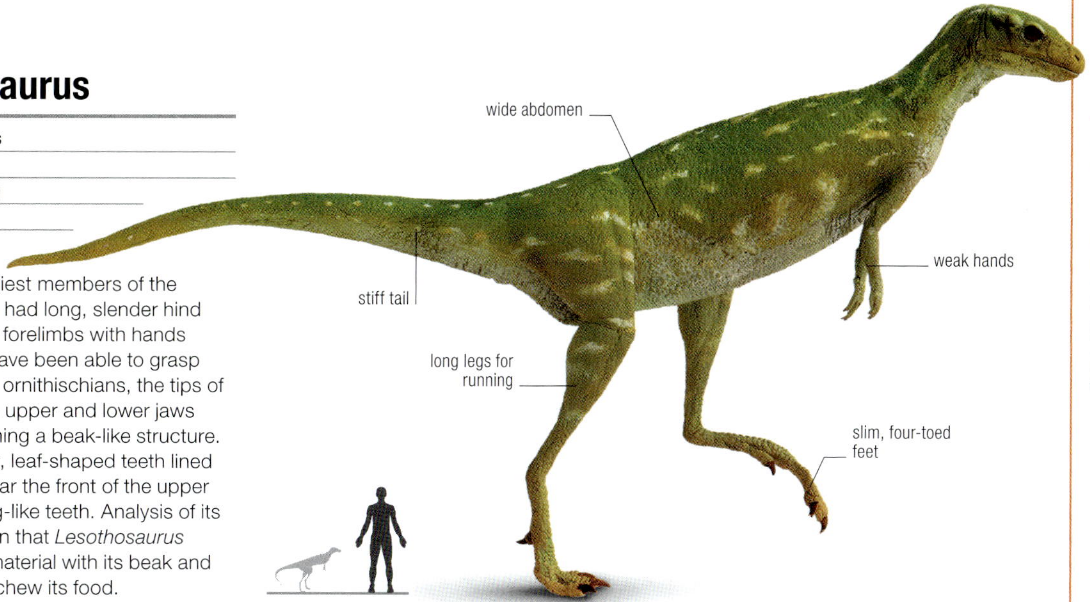

Heterodontosaurus

GROUP Ornithischians
DATE Early Jurassic
SIZE 1m (3ft 3in) long
LOCATION South Africa

Heterodontosaurus is a member of a small group of peculiar ornithischians called heterodontosaurids. Unlike virtually all other ornithischians, these animals had long, grasping hands, with strongly curved claws. *Heterodontosaurus*, meaning "different-toothed lizard", had three different tooth shapes. Small, incisor-like teeth were present at the front of the upper jaw, and blunter, chisel-shaped teeth were positioned further back. Most obvious were the large, fang-like teeth in both the upper and lower jaws. In addition, as in all ornithischians, it had a beak at the front of the upper and lower jaws. *Heterodontosaurus* was probably herbivorous, but its strong jaws, large fangs, and grasping hands suggest that it may also have eaten small animals. Its long, slender hind limbs also suggest that it was a fast runner.

ANATOMY
WORN TEETH

The top surfaces of *Heterodontosaurus*'s teeth are heavily worn down – clearly, it ate hard, abrasive foods. Other dinosaurs that experienced heavy tooth wear had numerous replacement teeth: as a tooth became worn down, a new one emerged from the jaw. Surprisingly, replacement teeth are usually absent in *Heterodontosaurus*, showing that tooth replacement occurred only very rarely.

BIPEDAL
This excellent, articulated skeleton shows that the hind limbs were much longer than the arms, which suggests it moved around on its hind legs.

The name "chambered lizard" comes from the large air sacs that were housed in the vertebrae of *Camarasaurus*.

ABUNDANT SAUROPOD
Camarasaurus was a robust, broad-skulled sauropod with spoon-shaped teeth. It was the most abundant North American sauropod and is better understood than any other dinosaur of this type.

Scutellosaurus

GROUP Ornithischians
DATE Early Jurassic
SIZE 1m (3ft 3in) long
LOCATION USA

Scutellosaurus is one of the oldest and most primitive thyreophorans, the ornithischian group that later included the armoured ankylosaurs and plated stegosaurs. It was lightly built, and probably capable of walking on its hind legs. Like other thyreophorans, *Scutellosaurus* had rows of armour plates along its body and tail. These formed parallel rows, with as many as five rows on each side. It also had a double row of scutes, or external plates, running neck to tail.

CHUBBY CHEEKS
Not much is known about the skull of *Scutellosaurus*, but like most other ornithischians it probably had fleshy cheeks, as shown here.

Stegosaurus

GROUP Ornithischian
DATE Late Jurassic
SIZE 9m (30ft) long
LOCATION USA, Portugal

Stegosaurus – the "plated lizard" – is the best-known stegosaur, and was the first member of that group to be named. Its bizarre, diamond-shaped plates have been the cause of much disagreement, but recent work has shown that they were arranged in two staggered rows that ran along the neck, back, and tail. In most other stegosaurs the plates were paired, so *Stegosaurus* was unusual in that respect. It has often been suggested that the plates were used in defence or in controlling body temperature. However, their position on the body makes a defensive role unlikely, while a role in temperature regulation also seems unlikely because the plates have the same anatomy as the armour plates that covered the bodies of other dinosaurs. The most likely role for the plates is that they were used in display. Small, rounded bones called ossicles covered the throat region of *Stegosaurus*, and two pairs of long spikes that projected from the tip of the tail were almost certainly used in defence.

DISTINCTIVE DINOSAUR
One of the most distinctive dinosaurs, *Stegosaurus* had a sweeping arched back, short forelegs, a small head held close to the ground, and a stiffened tail armed at the tip with long spikes.

The bizarre, **diamond-shaped plates** along the back of *Stegosaurus* were probably **used in display** rather than defence.

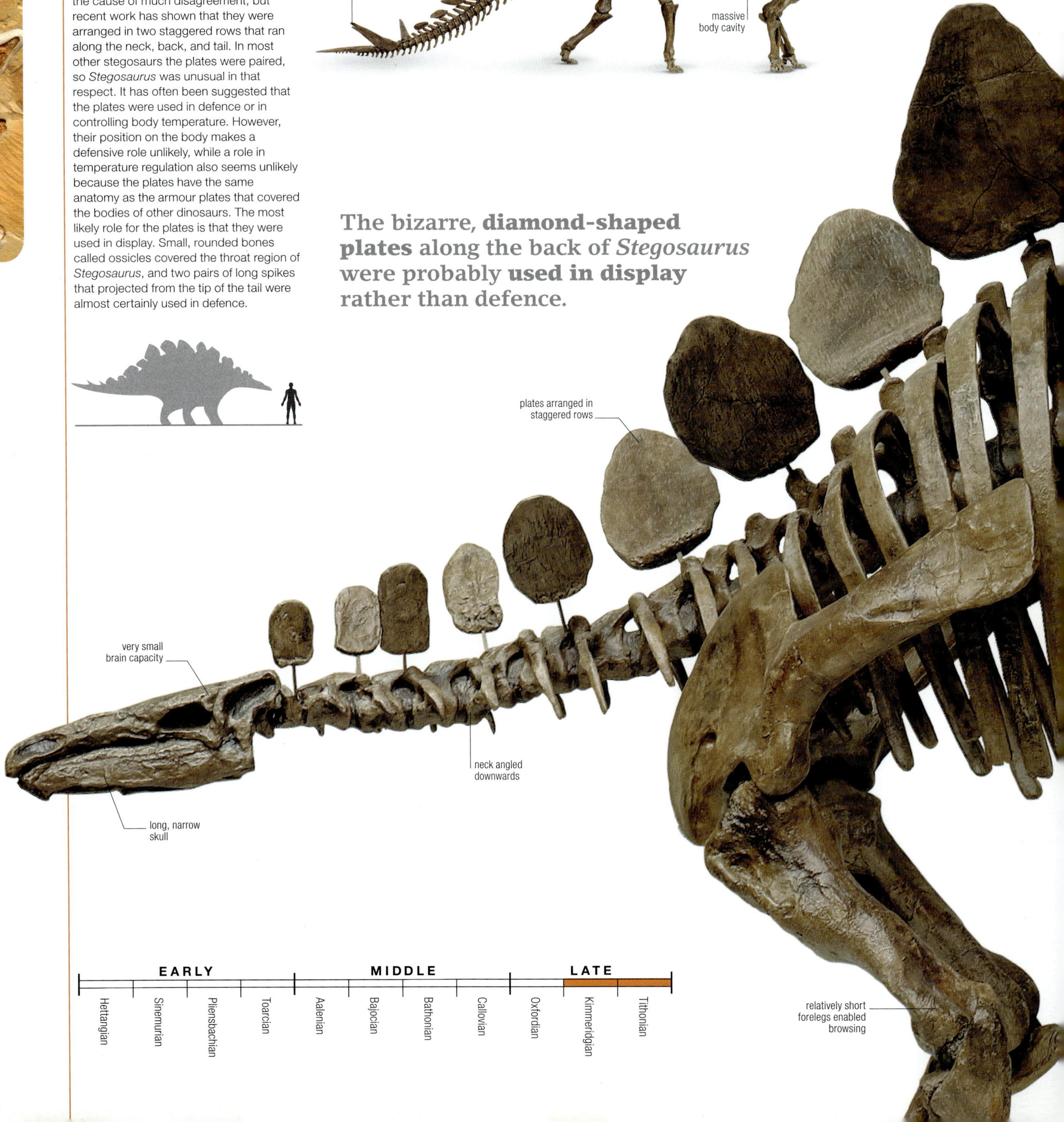

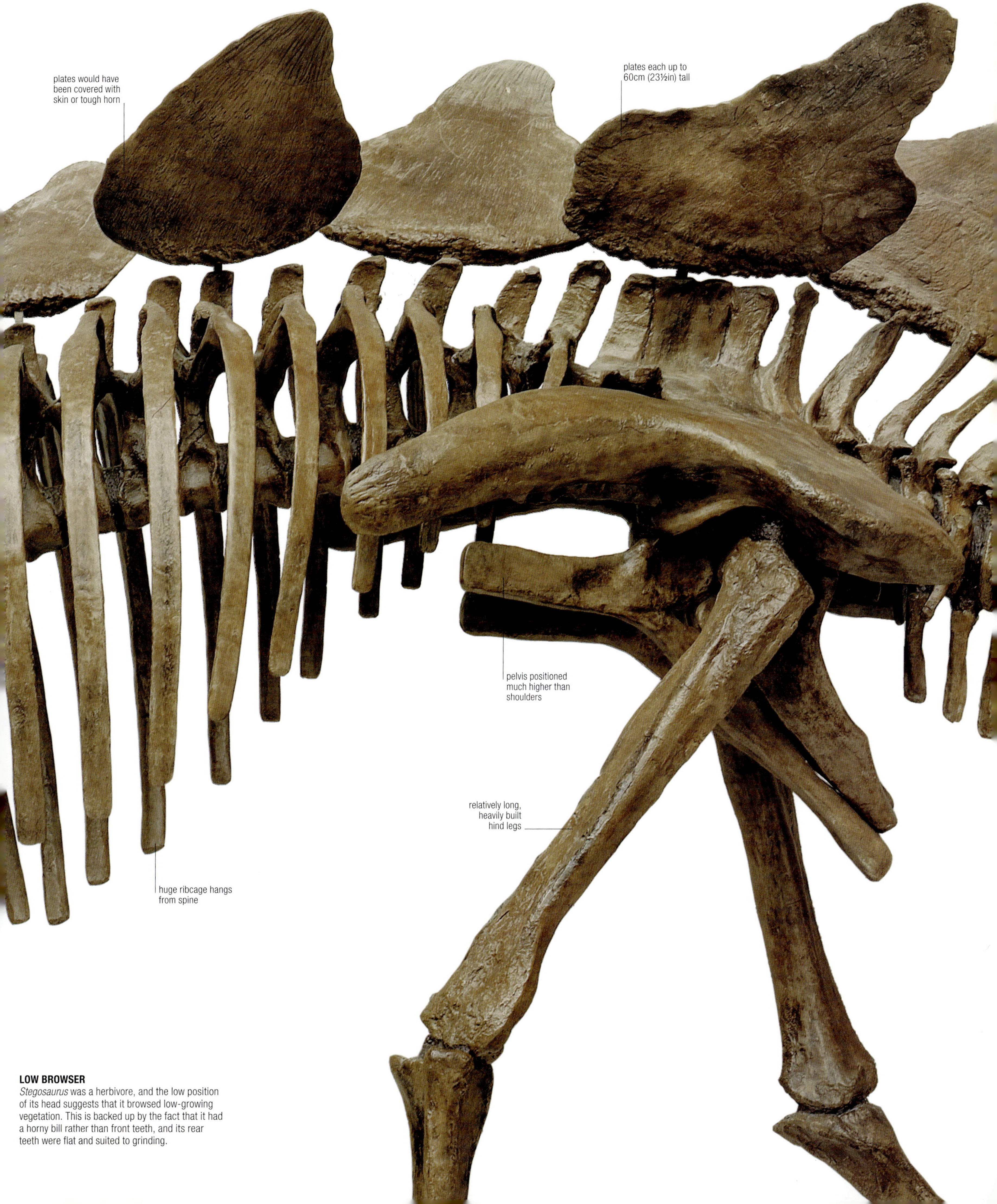

LOW BROWSER

Stegosaurus was a herbivore, and the low position of its head suggests that it browsed low-growing vegetation. This is backed up by the fact that it had a horny bill rather than front teeth, and its rear teeth were flat and suited to grinding.

Huayangosaurus

GROUP Ornithischians
DATE Middle Jurassic
SIZE 4m (13ft) long
LOCATION China

Huayangosaurus is one of the more primitive members of Stegosauria. It differed from the more advanced members of this group in having teeth at the front of the upper jaw. Its hips were also different from those of later stegosaurs, as was its snout – other stegosaurs had long, slim snouts, but the skull of *Huayangosaurus* was relatively short and broad. In some specimens, small horns were present above the eyes. However, these horns may also be absent in some *Huayangosaurus* skulls, so it is possible that they only appeared in adulthood, or were restricted to males or females. A large spike was present on each shoulder. Such spikes were typical for stegosaurs; the species that lacked them were unusual. In addition to the plates and spikes that ran along the backbone and tail of *Huayangosaurus*, several large armour plates were also arranged along both sides of the body for protection.

Huayangosaurus was **small** for a **stegosaur**.

PRIMITIVE STEGOSAUR
Although all stegosaurs were broadly similar, *Huayangosaurus* was probably more primitive looking. For example, its hind limbs were less column-like than those of other species.

KEY SITE
DASHANPU FORMATION

Huayangosaurus was discovered at Dashanpu Quarry in Sichuan Province, China. This site includes both Middle and Upper Jurassic sediments and is one of the most famous dinosaur-bearing locations in the world. Over 8,000 specimens have been collected from the site, and the discoveries have revolutionized our understanding of dinosaur diversity and evolution. In addition to *Huayangosaurus*, Dashanpu Quarry has yielded numerous sauropods, theropods, and small, bipedal ornithischians. Plesiosaurs, pterosaurs, and other fossil reptiles have also been discovered. Many of the fossils are carcasses that collected over the centuries in a gigantic lake.

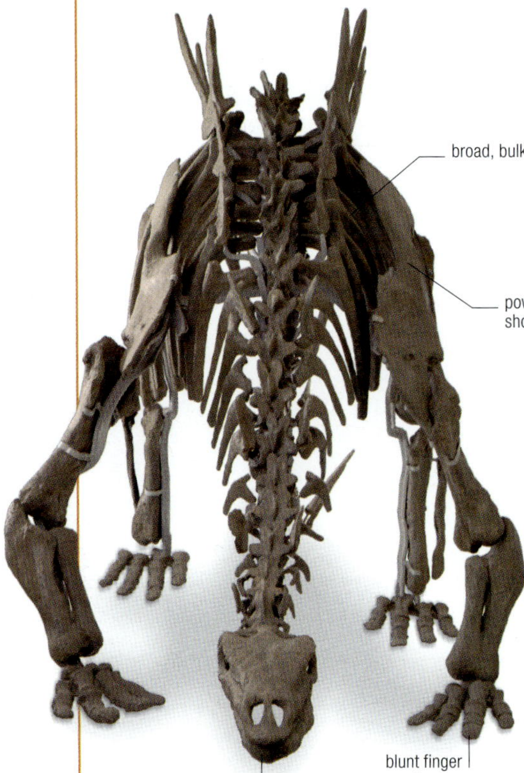

SPRAWLING LIMBS
This skeleton has been mounted so that the forelimbs sprawl out to the sides. This is almost certainly incorrect as it is thought stegosaur forelimbs were located almost directly beneath the body.

The group of **spikes** at the end of a **stegosaur tail** has been nicknamed the **thagomizer**.

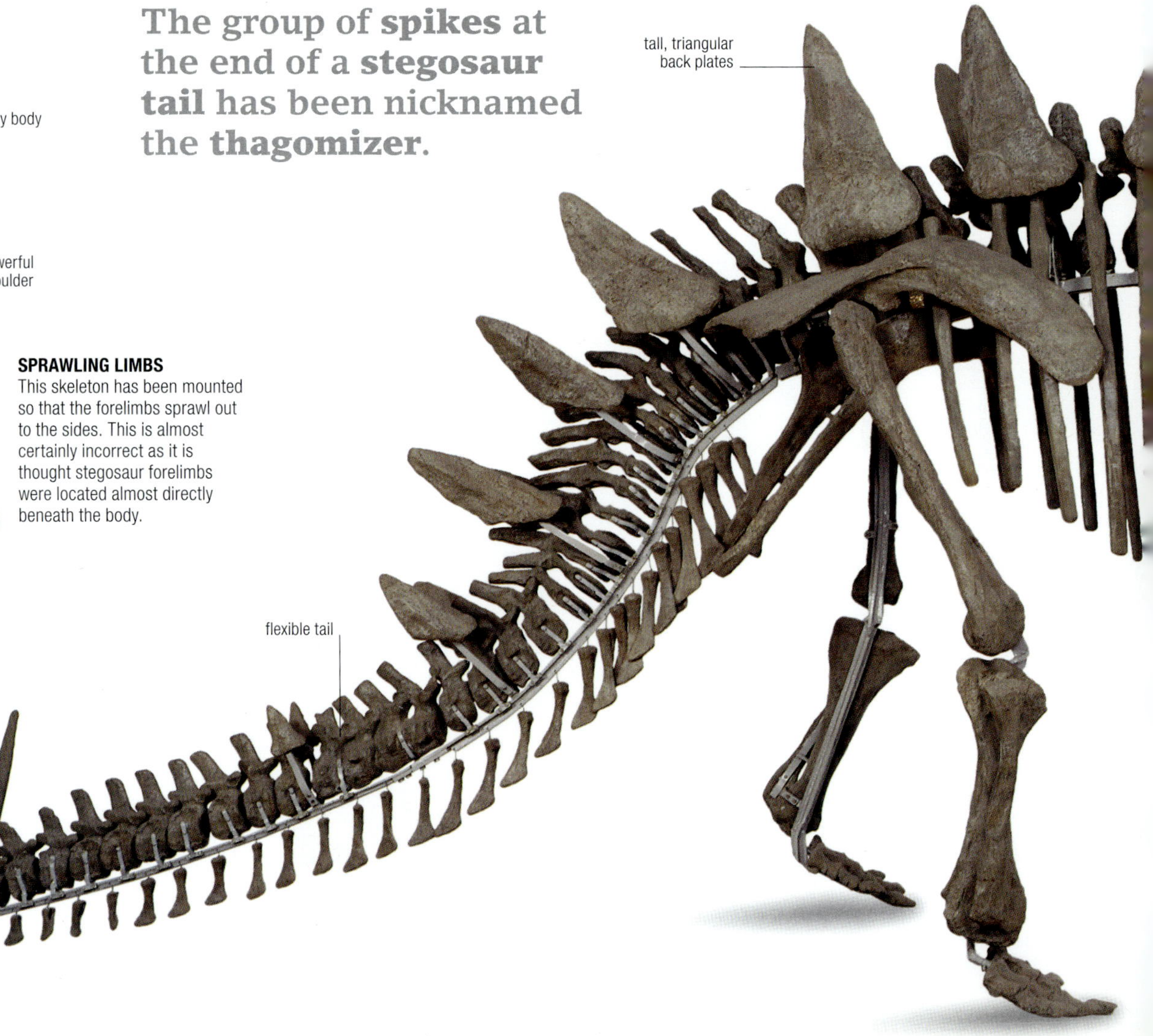

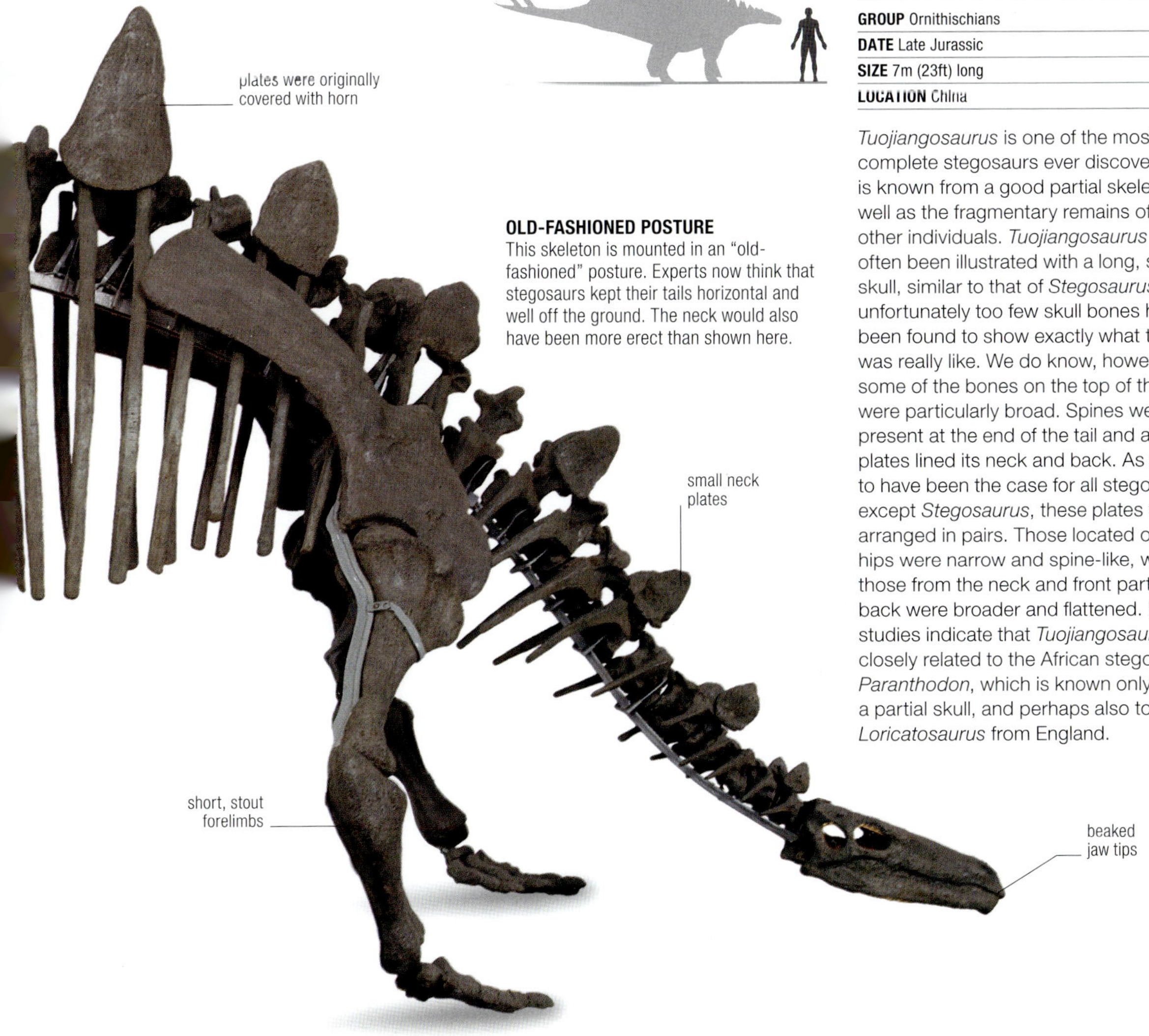

OLD-FASHIONED POSTURE
This skeleton is mounted in an "old-fashioned" posture. Experts now think that stegosaurs kept their tails horizontal and well off the ground. The neck would also have been more erect than shown here.

Tuojiangosaurus

GROUP Ornithischians
DATE Late Jurassic
SIZE 7m (23ft) long
LOCATION China

Tuojiangosaurus is one of the most complete stegosaurs ever discovered. It is known from a good partial skeleton, as well as the fragmentary remains of several other individuals. *Tuojiangosaurus* has often been illustrated with a long, shallow skull, similar to that of *Stegosaurus*, but unfortunately too few skull bones have been found to show exactly what the skull was really like. We do know, however, that some of the bones on the top of the skull were particularly broad. Spines were present at the end of the tail and armour plates lined its neck and back. As seems to have been the case for all stegosaurs, except *Stegosaurus*, these plates were arranged in pairs. Those located over the hips were narrow and spine-like, whereas those from the neck and front part of the back were broader and flattened. Recent studies indicate that *Tuojiangosaurus* is closely related to the African stegosaur *Paranthodon*, which is known only from a partial skull, and perhaps also to *Loricatosaurus* from England.

BIOGRAPHY
DONG ZHIMING

Tuojiangosaurus is one of almost 40 dinosaurs named and described by Dong Zhiming, one of China's most influential palaeontologists. Initially working under the great pioneer of Chinese vertebrate palaeontology Yang Zhong-jian (also known as C.C. Young), Dong was involved in the discovery of the Dashanpu Quarry in Sichuan Province in the 1970s, naming and describing many of the dinosaurs discovered there.

INCORRECT POSTURE
This *Kentrosaurus* skeleton is presented in a posture that has since been considered to be incorrect. Recent studies have suggested that its tail would have been held horizontally off the ground, the neck would have been held more erect, and the forelimbs would not have been positioned in a sprawling posture.

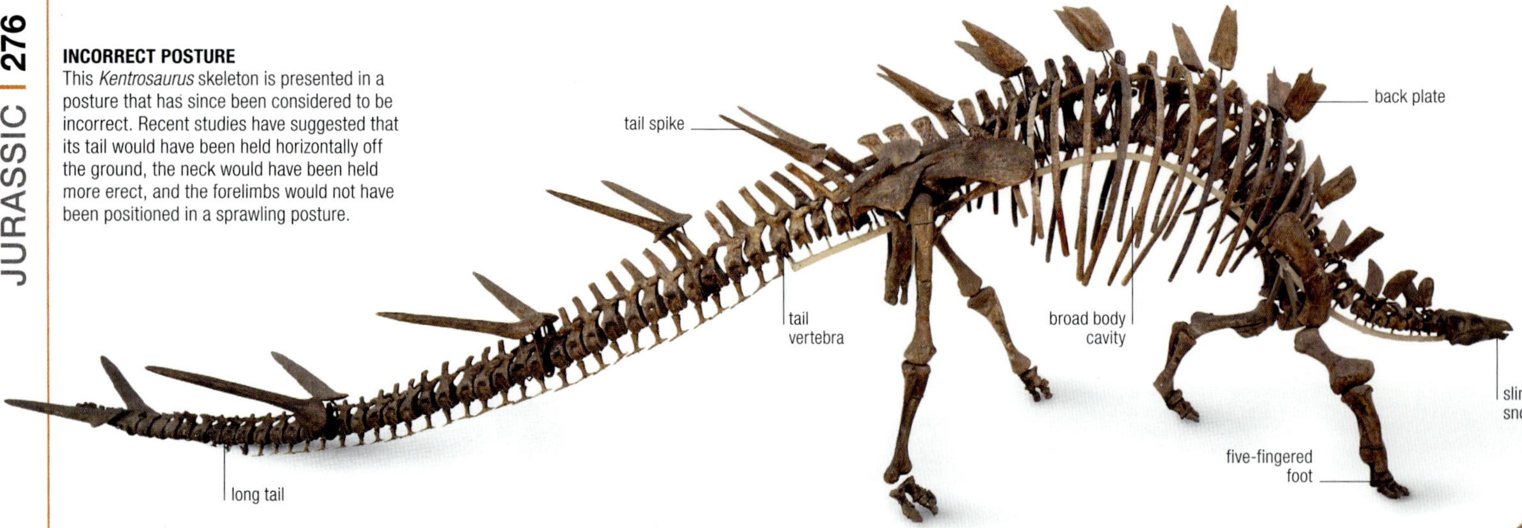

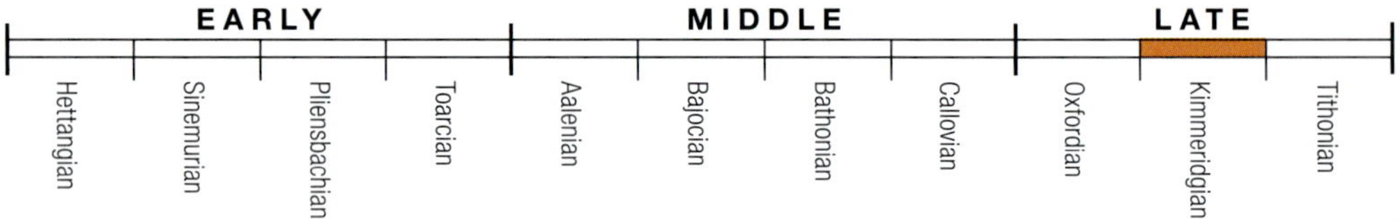

Over **900** ***Kentrosaurus*** **bones** were found at the **Tendaguru** fossil site in Tanzania.

SPIKED LIZARD
A relative and contemporary of the more famous *Stegosaurus* (see pp.272–73), *Kentrosaurus* was smaller but just as well protected. Its name means "spiked lizard", and its most prominent features were the plates that ran down its neck then gave way to sharp spikes that continued to the tip of its tail.

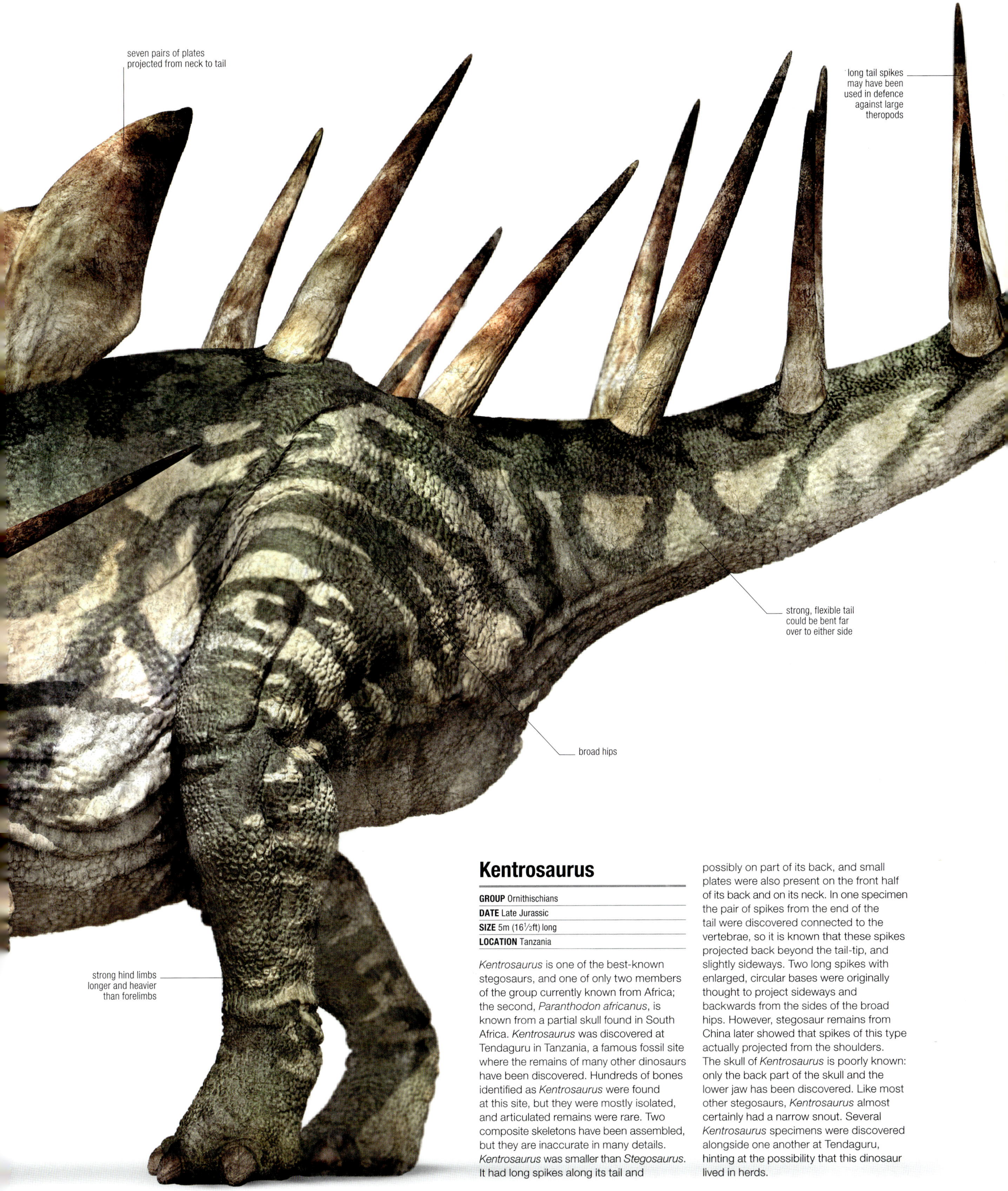

Kentrosaurus

GROUP Ornithischians
DATE Late Jurassic
SIZE 5m (16½ft) long
LOCATION Tanzania

Kentrosaurus is one of the best-known stegosaurs, and one of only two members of the group currently known from Africa; the second, *Paranthodon africanus*, is known from a partial skull found in South Africa. *Kentrosaurus* was discovered at Tendaguru in Tanzania, a famous fossil site where the remains of many other dinosaurs have been discovered. Hundreds of bones identified as *Kentrosaurus* were found at this site, but they were mostly isolated, and articulated remains were rare. Two composite skeletons have been assembled, but they are inaccurate in many details. *Kentrosaurus* was smaller than *Stegosaurus*. It had long spikes along its tail and possibly on part of its back, and small plates were also present on the front half of its back and on its neck. In one specimen the pair of spikes from the end of the tail were discovered connected to the vertebrae, so it is known that these spikes projected back beyond the tail-tip, and slightly sideways. Two long spikes with enlarged, circular bases were originally thought to project sideways and backwards from the sides of the broad hips. However, stegosaur remains from China later showed that spikes of this type actually projected from the shoulders. The skull of *Kentrosaurus* is poorly known: only the back part of the skull and the lower jaw has been discovered. Like most other stegosaurs, *Kentrosaurus* almost certainly had a narrow snout. Several *Kentrosaurus* specimens were discovered alongside one another at Tendaguru, hinting at the possibility that this dinosaur lived in herds.

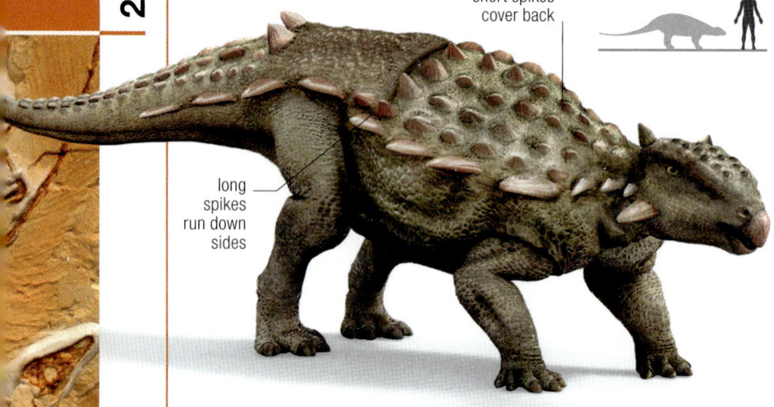

Gargoyleosaurus

GROUP Ornithischians
DATE Late Jurassic
SIZE 4m (13ft) long
LOCATION USA

Gargoyleosaurus was the earliest, and one of the smallest, members of the Ankylosauridae. This was a mostly Cretaceous group that included the giant *Ankylosaurus* (see p.335) and other species with tail clubs. Small bumps covered the upper surface of the skull, and four short, triangular horns projected from behind each eye and the cheeks. Unlike other ankylosaurids it had seven teeth in each of its premaxilla (bones forming the tip of the upper jaw). It also had simple, straight nasal passages instead of looping ones like otherankylosaurids.

Camptosaurus

GROUP Ornithischians
DATE Late Jurassic
SIZE 5m ($16^1/_2$ft) long
LOCATION USA

Camptosaurus was a member of the ornithopod dinosaur group Iguanodontia. Primitive iguanodontians were small, but *Camptosaurus* was one of the first to evolve into a larger animal. Like more primitive ornithopods, *Camptosaurus* was probably mostly bipedal. However, its short, stout fingers were also well-suited for supporting its weight, which suggests that it may have walked on four legs while it foraged. Later iguanodontians became increasingly quadrupedal. Most *Camptosaurus* fossils have been found in Late Jurassic rocks in North America. It shared its habitat with herbivores such as *Stegosaurus* (see pp.272–73) and *Diplodocus* (see pp.268–69) and it would have been preyed on by theropods such as *Allosaurus* (see p.261).

SCIENCE

MISTAKEN SKULL

Until recently, *Camptosaurus* was depicted with a deep, rectangular snout. This was because Othniel Marsh, the man who described *Camptosaurus*, assumed that a rectangular-snouted skull he also had belonged to this dinosaur. Recent work has shown that the original skull (shown here, on the left) does not belong to *Camptosaurus*, and that its skull was very differently shaped. The rectangular skull actually belonged to an entirely different animal, the recently named *Theiophytalia*.

Dryosaurus

GROUP Ornithischians
DATE Late Jurassic
SIZE 3m ($9^3/_4$ft) long
LOCATION USA

Dryosaurus was a medium-sized, bipedal ornithopod with short arms and tiny hands. Its skull was short and tall, with a sloping upper surface and a narrow beak, which suggests that it was a selective feeder that browsed on leaves. *Dryosaurus* is the best-known member of a small group of ornithopods called the dryosaurids. Named after *Dryosaurus*, their name means "tree lizards". Once thought to be close relatives of *Hypsilophodon* (see p.338), we now know that dryosaurids were members of the Iguanodontia, a group that also includes *Camptosaurus* (see above), *Iguanodon* (see p.338), and the hadrosaurs. Primitive, dryosaurid-like iguanodontians that had evolved a variety of body sizes and lifestyles prior to the evolution of the more familiar hadrosaurs have also been discovered.

The name *Dryosaurus*, meaning **"tree lizard"**, reflects the fact that the dinosaur lived in a **forest environment** and ate leaves.

LOW BROWSER
A low-browsing herbivore, *Camptosaurus* had a distinctive arched back, and arms long enough to enable it to move on all fours when feeding.

MISSING TOE
The first toe on the inner side of *Dryosaurus*'s foot (the hallux) was missing, its feet were long and slender, and its back legs were powerful. These features suggest that *Dryosaurus* was a fast runner.

SCIENCE
CROSSING CONTINENTS

In 1919, remains were found in Tanzania that were similar to *Dryosaurus* fossils discovered in the USA. They were so similar that the Tanzanian example was renamed as a second species of *Dryosaurus*, suggesting that the American and African continents were joined in the Late Jurassic. Research has shown that it was, in fact, a different animal, which matches geological evidence that America and Africa were not joined at this time.

Othnielosaurus

GROUP Ornithischians
DATE Late Jurassic
SIZE 2m (6½ft) long
LOCATION USA

Othnielosaurus was a small, bipedal ornithischian, with a short neck and forelimbs. Its hind limbs were much longer and had four elongated, clawed toes. A partial skeleton of *Othnielosaurus* was discovered in 1963. Unfortunately the skull, hands, and most of the tail were missing. Isolated teeth, thought to belong to *Othnielosaurus*, have also been identified. They are small, leaf-shaped, and equipped with numerous small points, or cusps, which made them well-suited for shredding leaves.

Megazostrodon

GROUP Placental mammals
DATE Early Jurassic
SIZE 10cm (4in) long
LOCATION South Africa

Megazostrodon's name means "large girdle tooth". Each cheek tooth possessed short, triangular points, or cusps, arranged in a line. They were simpler in shape than those of later mammals, especially in the way the cusps bit together, and were probably used for cutting up insects. The skeleton of *Megazostrodon* was unspecialized for any particular lifestyle, but it probably climbed, burrowed, and ran, much like modern-day rats and shrews. *Megazostrodon* shares several features with a number of other Late Triassic and Early Jurassic mammals. These early mammals are grouped together as the morganucodontids.

Sinoconodon

GROUP Placental mammals
DATE Early Jurassic
SIZE 30cm (12in) long
LOCATION China

Sinoconodon is one of the largest primitive mammals found to date. Several features of its teeth and jaws were unusual – the long gap between the canine and cheek teeth, the robust jaw joint, and the stout, strong chin. These features suggest that *Sinoconodon* had a powerful bite and might have preyed on large insects and small reptiles. It has also been suggested that *Sinoconodon* may have been closely related to *Megazostrodon* (above), but recent studies have shown that it was an even more primitive mammal – one of the earliest known.

CRETACEOUS

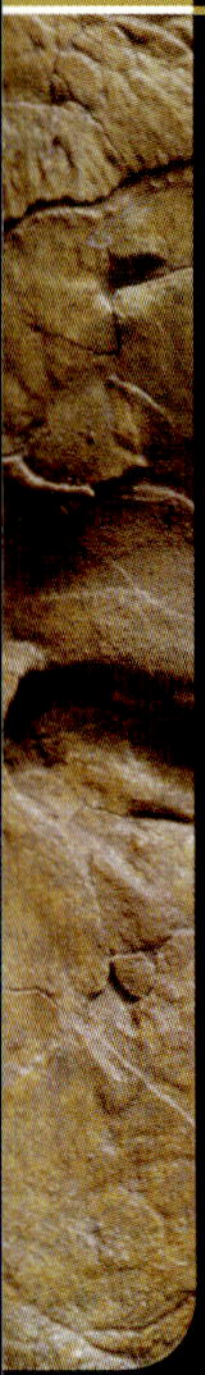

Cretaceous

The Cretaceous period in Earth's history was generally a greenhouse world, with high atmospheric carbon dioxide levels, high global temperatures, and sea levels about 200 to 300m (660 to 980ft) higher than today. Huge, shallow seas covered many of the present-day continents. It was a time when birds and flowering plants evolved and modern insect groups diversified.

WHITE CLIFFS OF DOVER
These iconic chalk cliffs on England's south coast are composed almost entirely of coccoliths, deposited in Cretaceous seas.

CALCAREOUS FOSSIL
This is a scanning electron micrograph of a coccolith, the fossilized skeleton of a coccolithophore (a single-celled marine alga).

Oceans and continents

The Cretaceous takes its name from its extensive chalk strata, composed of the skeletons of countless tiny algae in shallow seas. It was a time when today's oceans were beginning to take shape. Pangaea continued to disintegrate, with the South Atlantic Ocean opening up as Africa and South America rifted apart. Australia remained attached to Antarctica, but India separated from the western side of Australia during the Early Cretaceous and began to move in a relatively westerly direction. Later, India parted from Madagascar, then rotated to move northward, beginning its collision with the Asian continent at the end of the Cretaceous. This initially resulted in the outpouring of vast amounts of magma, covering much of India with lava that solidified as a basalt, known as the Deccan Traps. Some experts suggest that the End-Permian and Cretaceous mass extinctions might have been caused by this type of excessive volcanic activity. The Tethys Ocean contracted a little as the combined Eurasia, North China, South China, and Indochina blocks rotated clockwise, bringing southeast Asia closer to the Equator. Sea levels were very high across the globe in Late Cretaceous times, flooding North America and creating a huge seaway that extended from the Gulf of Mexico to the newly forming Arctic Ocean. The North Atlantic, like its southern counterpart, was expanding, but only in the south. Rock layers enriched with the rare element iridium indicate that Earth was hit by asteroids, one of which left an enormous impact crater at Chicxulub, on Mexico's Yucatán Peninsula. This event is one of several proposed as causes for the mass extinction of species at the end of the Cretaceous.

MONUMENT ROCKS
The widespread carbonates and other sedimentary deposits make the Cretaceous rock record especially fine – for example, the rich marine fossils of Smoky Hill Chalk Member in Kansas, USA.

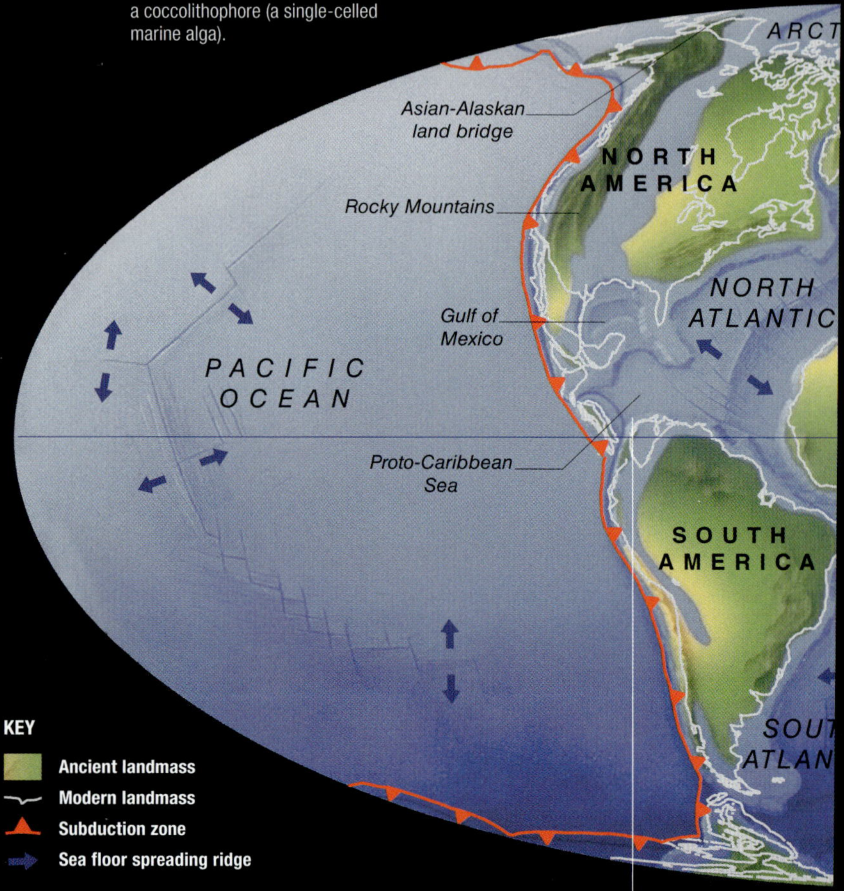

CRETACEOUS WORLD MAP
Pangaea continued to disintegrate, as Africa and South America parted. Oceans were rearranged, and Europe remained connected to North America.

Caribbean Sea opens up, moving North and South America apart

EARLY

MYA 140 · 120

MICROSCOPIC LIFE
- **140** Planktic foraminifers begin to diversify

PLANTS
- **125** First angiosperm (flowering) plants

BETULITES

INVERTEBRATES
- **140** First venerid bivalves
- **135** First tellinid bivalves
- **125** First mactrid, teredenid, and donacid bivalves
- **110** First limopsid, verticordiid, and thyasirid bivalves

TEREDINA

VERTEBRATES
- **140–100** Stegosaurs and high-browsing sauropod dinosaurs decrease in diversity; ornithopod and ankylosaur dinosaurs, and birds diversify
- **140** First ceratopsian dinosaurs
- **130** First freshwater pelomedusid turtles
- **125** First eutherian mammal (*Eomaia*); first marsupial mammal (*Sinodelphys*); first enantiornithine birds
- **115** First monotreme (egg-laying) mammals
- **110** First hesperornithiformes (diving birds)

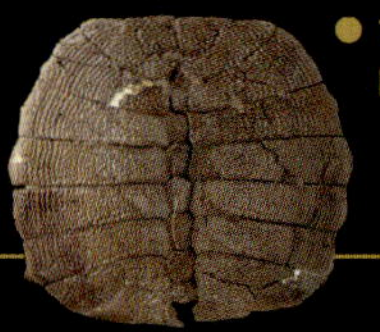

TRIONYX

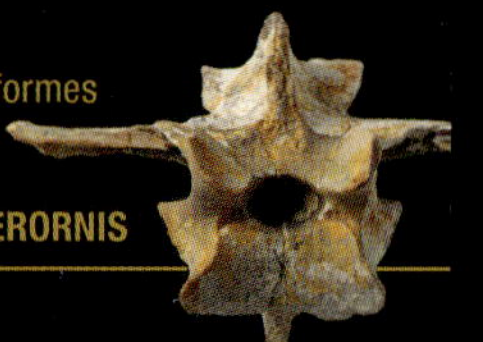

HESPERORNIS

MAGNOLIA TREE
Flowering plants (angiosperms) such as this magnolia, appeared and occupied new niches on the land. They co-evolved with pollinating insects, such as bees, and birds.

Climate

The Cretaceous period had a very variable climate, with oscillations between warm and cold about every 2 million years, and major swings in average temperature over a 20-million-year cycle. The Cretaceous is often described as a warm greenhouse world (see p.25), but this really only applies to part of it. At the outset, the Jurassic climate continued, with warm temperatures and no evidence for polar ice caps. Reptiles flourished, and flowering plants evolved and spread. However, by about 120 million years ago, global temperatures had reduced to the extent that east Antarctica was covered by ice sheets, and parts of Australia at latitudes above 65°S were glaciated. High-latitude temperatures in Australia averaged only about 12°C (54°F). Over the next 30 million years, global temperatures increased so that by 90 million years ago the Earth was experiencing one of its hottest periods. Globally, average surface temperatures were over 10°C (50°F) higher than today. Even in high latitudes they were between 22 and 28°C (72–82°F), and ocean temperatures in lower latitudes were as high as 36°C (97°F). The Late Cretaceous was much cooler, with ice sheets developing on the South Pole. Average global temperatures dropped from their previous high of 31° (88°F) to about 21° (70°F) near the end of the Cretaceous.

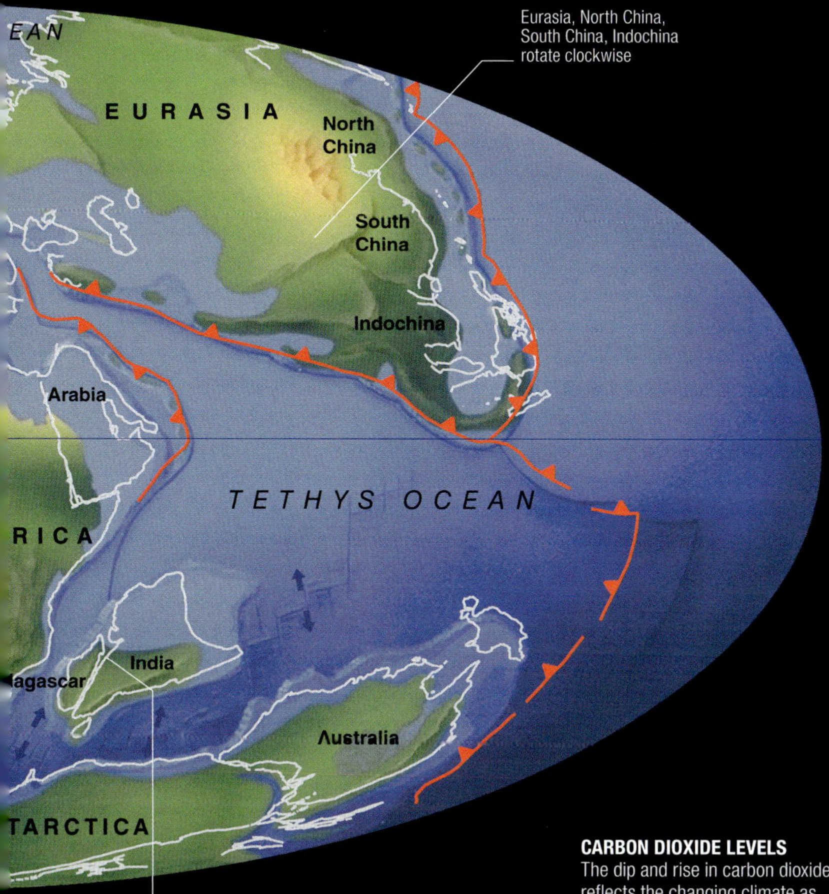

Eurasia, North China, South China, Indochina rotate clockwise

India separates from Madagascar and moves northwards

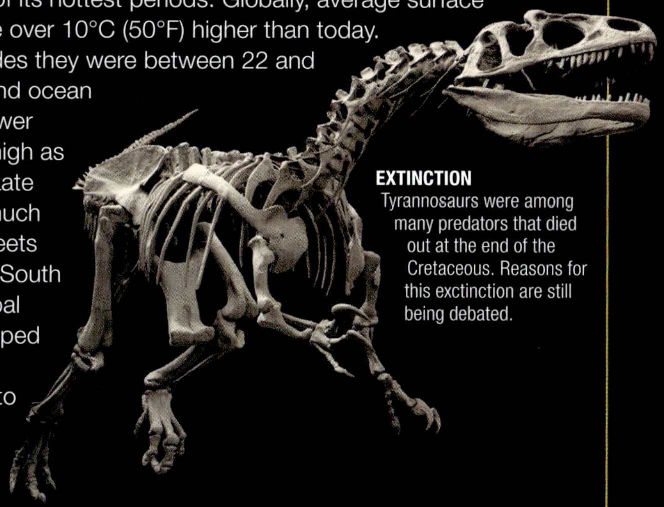

EXTINCTION
Tyrannosaurs were among many predators that died out at the end of the Cretaceous. Reasons for this exctinction are still being debated.

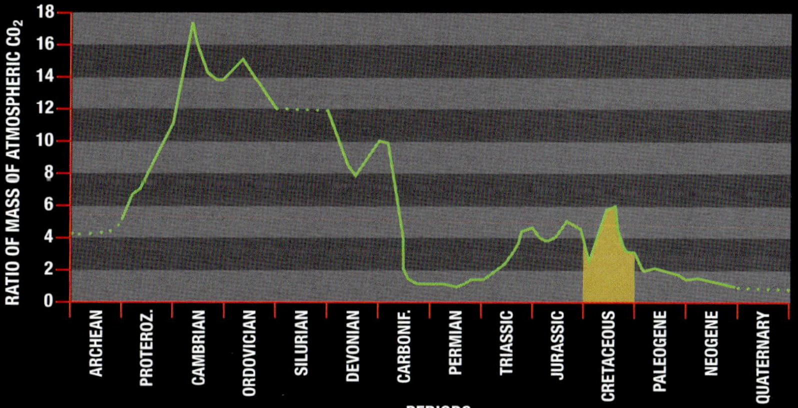

CARBON DIOXIDE LEVELS
The dip and rise in carbon dioxide reflects the changing climate as Earth moved from warm to cool to warm conditions. Gases from volcanic activity may have increased carbon dioxide levels.

LATE

80 | 60 | MYA

MICROSCOPIC LIFE
- **65** Extinction of most planktic foraminifers

PLANTS

NIPA

- **90** Large diversification in angiosperms (magnoliids, rosiids, hamamelidids), monocots e.g. palms (left) and ginger
- **65** Extinction of 60% of plant species

INVERTEBRATES
- First bees
- **90** First nuculanid bivalves
- **85** First ungulinid, kelliid, and chamid bivalves
- **80** First termites; first ants
- **70** First yoldiid bivalves
- **65** Extinction of ammonites, belemnites, inoceramid bivalves (right), rudist bivalves

VOLVICERAMUS

VERTEBRATES
- **90–65** Diversification of late ceratopsian and ornithopod dinosaurs
- **90** Extinction of ichthyosaurs; first snakes (right)

DINILYSIA

TRICERATOPS

- **70** Multituberculate mammals increase in diversity
- **65** Extinction of dinosaurs, mosasaurs, pterodactyls, plesiosaurs, pliosaurs

CRETACEOUS PLANTS

Angiosperms, also known as flowering plants, made their first appearance during the Early Cretaceous. At first they were rare and not diverse, but steadily they became increasingly prominent, and by the end of the Cretaceous, they dominated the vegetation in most parts of the world.

Fossil floras from the earliest part of the Cretaceous consist mainly of ferns, conifers, cycads, and groups of extinct seed plants – they look very similar to fossil floras from the Jurassic. By about 130 million years ago, traces of the earliest flowering plants provide the first hints of evolutionary changes to come.

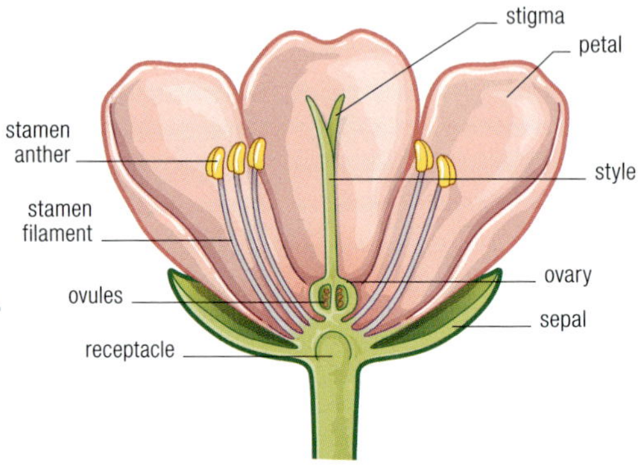

THE STRUCTURE OF A FLOWER
Typically, the flowers of angiosperms have a whorl of green, leaf-like structures on the outside that protect the flower. In the centre, the stamens, which produce the pollen, surround the carpels. The carpels ultimately develop into the fruit, which contains the seeds.

The earliest flowering plants

The earliest reliable fossils of flowering plants are pollen grains from the middle of the Early Cretaceous. They can be recognized as angiosperm pollen by their distinctive wall structure. Next, by about 125 million years ago, well-preserved fossils from eastern North America and Portugal record the first appearance of angiosperm leaves and the earliest fossil flowers. These flowers are all very small, but have the same basic structure seen in flowering plants today. Details of their pollen-producing structures, and of the organs in which the seeds develop, are also exactly like those of modern angiosperms, but different from the reproductive parts of other seed plants. Several of the earliest flowering plants from the Early Cretaceous were members of groups that are still alive today, such as the water lily and wintergreen families. The Chloranthaceae, an unusual family of living angiosperms, were especially prominent at this time. There are also fossils of early monocots related to modern aroids. By about 100 million years ago, the number of modern angiosperm groups that can be recognized had increased further. They included plants very similar to living magnolia and bay laurel, as well as early eudicots related to modern plane trees, sacred lotus, and box.

Towards modern vegetation

Flowering plants continued to diversify rapidly through the Late Cretaceous. By about 70 million years ago palms and gingers had evolved, which suggests that the variety of existing monocots had also increased. Among

DAKOTA SANDSTONE LEAF
The Middle Cretaceous Dakota Sandstone flora illustrates how much angiosperms had developed by about 100 million years ago. These fossils consist almost entirely of many different kinds of leaves from diverse angiosperms that grew on the banks of an ancient river system.

GROUPS OVERVIEW

The most obvious evolutionary change during the Cretaceous was the rise of flowering plants, but there was also expansion and decline of other groups. This resulted in ecosystems dominated by plants that are mostly familiar to us today. However, these ecosystems, populated by dinosaurs and other extinct reptiles, were still far from fully modern.

FERNS
The Cretaceous was a transitional time in the evolution of ferns. Ancient groups, such as Dipteridaceae, became less important, while Gleicheniaceae initially flourished but later declined. By the end of the Cretaceous, new groups of ferns were starting to diversify into the new habitats created by angiosperms.

HORSETAILS
Horsetails were common in many Early Cretaceous floras and they have a continuous fossil record up to the present day. Some seem to have been larger than their modern counterparts, but they grew in similar habitats – along the banks of streams and around the edges of lakes.

CONIFERS
The Cretaceous was a time of major modernization in the history of conifers, resulting in many of the living groups seen today. But some groups of conifers declined at this time. Cheirolepidiaceae, which were very prominent in the Jurassic and Early Cretaceous, were extinct by the end of the Cretaceous.

CYCADS
Cycads are one of the groups that appear to become less important during the Cretaceous. Some may have suffered in competition with angiosperms. However, there is evidence of diversification among some groups of cycads that survived up to the present.

There are several features that distinguish all angiosperms from other seed plants, but most obvious is the flower itself.

WATER FERNS
Many groups of plants that have a long fossil history continued to diversify through the Cretaceous. Water ferns are one of several new groups of ferns that first appear at this time.

eudicots, early dogwoods and witch hazels appeared, along with many forms related to heathers. Plants related to modern oaks, walnuts, and hazels appeared in eastern North America and western Europe, and southern beeches are known from the southern hemisphere at this time. In parallel with the modernization of angiosperms, other groups of plants also showed important changes. Among ferns, archaic forms from earlier in the Mesozoic became less important, as did cycads, bennettitaleans, and a key family of extinct conifers. At the same time, new groups appeared and plant life as a whole began to look more modern. For example, relatives of modern pines are known from the Early Cretaceous and true pines became widespread in the Late Cretaceous.

Evolution of flowers

Several features distinguish all angiosperms from other seed plants, but most obvious is the flower itself. Recently, beautifully preserved fossil flowers have been discovered in many parts of the world from several different stages in the Cretaceous. They provide important new insights into the evolution of flowers. The earliest angiosperm flowers were small – typically just a millimetre or two in diameter. Many were bisexual, but others were unisexual and produced only pollen or seeds. Very few had distinct petals. More typical showy petals appear later in the fossil record and are common in flowers from the Late Cretaceous. The structure of flowers became increasingly specialized in the Late Cretaceous. For example, petals became fused into a tube and the carpels were often fused into an ovary with a single pollen-receiving structure. Some of these changes probably indicate increasingly sophisticated interactions with insect pollinators. Several key groups of pollinators, such as bees and modern kinds of moths and butterflies, are first known from the Late Cretaceous.

MAGNOLIA
Among the several different modern groups of flowering plants that can be recognized by the Middle Cretaceous, there are forms clearly similar to modern magnolia. By the end of the Cretaceous the diversification of the magnolia family and its relatives was well underway.

GINKGOS
Ginkgo-like plants were widespread in both the northern and southern hemispheres during the Early Cretaceous. But by the end of the Cretaceous, they were less diverse and confined to the northern hemisphere. *Ginkgo* from the Early Paleogene is very similar to modern *Ginkgo biloba*.

GNETALES
Gnetales are a small group of living seed plants. Their distinctive pollen grains are first seen around the Permian–Triassic boundary. During the mid-Cretaceous, Gnetales underwent a major diversification. Fossil Gnetales similar to living *Welwitschia* (see pp.290-91) and *Ephedra* (see p.421) are known from this period.

BENNETTITALEANS
Bennettitaleans are a group of extinct seed plants that appeared in the Triassic. They were still important in Early Cretaceous floras, but were extinct by the Cretaceous–Paleogene boundary. In the Cretaceous, they are known mainly from *Cycadeoidea*, which were superficially cycad-like, but thought not to be related to cycads.

ANGIOSPERMS
Angiosperms radiated rapidly through the Cretaceous. Before the end of the Early Cretaceous, the three major groups of living flowering plant – eumagnoliids, monocots, and eudicots – existed. Today, eudicots include nearly three-quarters of all flowering plants, and monocots account for almost a quarter.

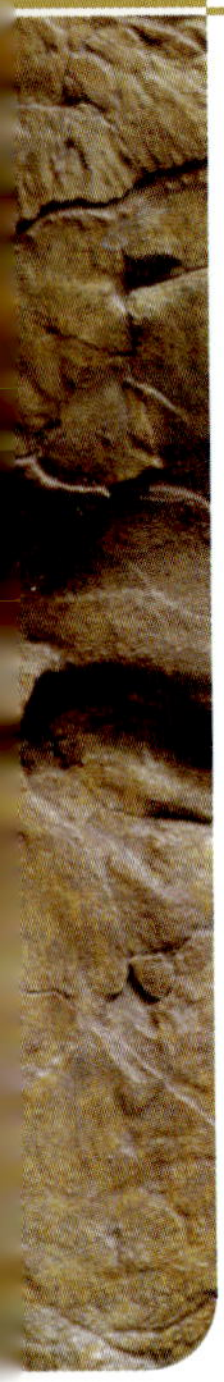

Nathorstiana

GROUP Lycophytes
DATE Cretaceous
SIZE Up to 4cm (1½in) long, 2cm (¾in) wide
LOCATION Germany

The fossil plant *Nathorstiana* is known only as a collection of moulds and casts taken from a single location in Germany. Its root-bearing base grew downwards, producing new roots and shedding older ones as it went. The root-growing tip was sunk in a central depression and covered with a membrane for protection. The radially symmetrical shape of the young bases divided into two and then four lobes as the plant matured; this maturation indicates age – not the very variable size shown by the fossils. *Nathorstiana*'s leaves and reproductive organs are unknown, but it is generally regarded as a lycophyte and part of a series of plants that gradually reduced in size from the Triassic *Pleuromeia* to the living *Isoetes*.

LIVING RELATIVE
QUILLWORT

Isoetes consists of about 150 mainly aquatic or semi-aquatic species. Narrow, quill-like leaves (hence the common name quillwort) are attached to a rooted, corm-like stem. The leaves' swollen bases have a large sporangium containing either megaspores or microspores, enclosed by a thin membrane called a vellum.

NATHORSTIANA SP.
This impression from Quelinburg in Germany shows the base of *Nathorstiana* with the lumps on the side marking the bases of the leaves.

Weichselia

GROUP Ferns
DATE Cretaceous
SIZE Leaf 1.5m (5ft) or more across
LOCATION Northern hemisphere

Weichselia was a successful genus of tree-like ferns that formed dense swards, rather like modern bracken. The plants had palmate leaves with 5–15 "fingers", or pinnae, each containing 14–20 spore-producing sporangia arranged in single rows on either side of the central rib. Climate change during the Middle Cretaceous seems to have eliminated *Weichselia*. It belonged to the family Matoniaceae, which died out throughout most of the world by the Late Cretaceous, perhaps out-competed by flowering plants. Today, only two living genera exist and are found in Malaysian Borneo.

WEICHSELIA RETICULATA
The fronds of *Weichselia* were similar to those of bracken.

LIVING RELATIVE
DIPTERIS

Dipteris is considered to be a broadleaf fern. Its leaves show a characteristic known as dichotomous venation, in which the veins branch off from one another like the branches of a tree. With its unequal leaf lobes arranged around the top of a vertical stalk, *Dipteris* resembles a fan or umbrella.

SPORE PRODUCTION
This close-up shows the single rows of sporangial clusters on one side of the central rib. These clusters produced the plant's reproductive structures (spores).

TEMPSKYA
With its huge, fibrous false trunk and covering of leaves, *Tempskya* probably resembled a young cedar or redwood rather than a species of tree fern.

Tempskya

GROUP Ferns
DATE Cretaceous
SIZE Up to 4.5m (14¾ft)
LOCATION Northern hemisphere

This very unusual tree fern – placed in a family of its own – has been described from well-preserved petrified specimens found in many locations in the northern hemisphere. What appears to be a trunk is actually a false trunk, made up of a large number of stems surrounded by a thick mat of fibrous, adventitious roots – that is, roots arising from a part of the plant other than a main root source. Sporelings grew vertically upwards and branched or dichotomized repeatedly, while giving off more adventitious roots that grew downwards and bound the stems together. The result is that the base of the trunk was almost entirely composed of adventitious root fibres. Although no leaves have been found attached, the fossilized leaf bases show that they were produced from the sides of the stems, so the upper part of the trunk would have been covered with leaves, rather than having a single crown of leaves on top as in other forms of tree ferns.

Salvinia

GROUP Ferns
DATE Cretaceous to present
SIZE Leaf 5–10mm ($^3/_{16}$–$^3/_8$in) long
LOCATION Tropical and subtropical regions worldwide

Salvinia is a living genus of small-leaved floating ferns. It has pairs of flat, oval, floating leaves attached to a rootless rhizome, and the leaves are covered with hairs to repel water. Living *Salvinia*, and the closely related genus *Azolla*, are two mainly tropical ferns that have fossil records dating from the Cretaceous period onwards. Both can propagate easily by growth and division. *Salvinia* also has highly dissected, fertile leaves hanging downwards into the water. They grow rounded, hard-coated sporocarps – specialized structures that produce and release spores. The sporocarps contain either a few female megasporangia, each with a single megaspore, or many male microsporangia, each with 64 microspores. Sporocarps enable the plants to survive periods of drought, opening to release their spores when water returns. Individual megaspores are covered with a foamy layer that makes them buoyant, allowing them to float on the water.

SALVINIA SP.
This specimen from Baden-Württemberg, Germany shows a length of *Salvinia* rhizome (the dark lines) with many leaves and the reproductive sporocarps (the dark oval areas).

SALVINIA FORMOSA
This magnified photograph of a fossilized *Salvinia* leaf shows the remains of the hair-like strands that grow from the tropical plant's leaves. These allow it to float on top of water.

Onychiopsis

GROUP Ferns
DATE Middle Jurassic to Early Cretaceous
SIZE Terminal segment 1cm ($^3/_8$in) long
LOCATION Japan, Europe, North America

Narrow, highly branched leaves, as shown on the fossil below, are usually included in the genus of ferns called *Onychiopsis*. The terminal segments on the sterile leaves were almost flat, while those on the fertile leaves were slightly enlarged, with sporangia on their lower surfaces that were covered by a protective membrane called an indusium. This fern's fertile segments were similar to those of the living fern *Onychium*. *Onychiopsis* was common in Middle Jurassic and Early Cretaceous Japan (where the genus was first named), Europe, and North America.

ONYCHIOPSIS PSILOTOIDES
The narrow, highly branched leaves of *Onychiopsis psilotoides* give it a fan-like appearance, similar to that of some species of the living fern *Onychium*.

Gleichenia

GROUP Ferns
DATE Cretaceous to present
SIZE Smallest leaflet about 2mm ($^1/_{16}$in) long
LOCATION Tropical and subtropical regions worldwide

This fossil fern is branched, with short, rounded leaflets. It is identical to the tropical *Gleichenia* that is found growing today in open places and on the edges of forests. It has a creeping rhizome (a constantly growing underground stem), and fronds that grow enormously large by continuously dividing their side branches. *Gleichenia* forms dense thickets by scrambling or climbing over other plants, and is a genus of the primitive Gleicheniaceae family, whose members have simple rhizomes covered with hairs. Its large sporangia are arranged in circles on the underside of the fronds. Gleicheniaceous ferns were most common in the Cretaceous; the earliest specimens were found in England.

Gleichenia **was much more scarce in the Cenozoic than in the Cretaceous, perhaps because of competition from flowering plants.**

GLEICHENIA SP.
The leaves of *Gleichenia* could reach enormous sizes by repeated growth and branching. In contrast, its ultimate leaflets (pinnules) are small.

Schizaeopsis

GROUP Ferns
DATE Cretaceous
SIZE Leaf segment 3–5cm (1¼–2in) long
LOCATION Worldwide

The leaves of this fern consist of repeatedly dividing, single-veined, narrow segments ending in rounded or pointed tips. When fertile, the narrowest segments have folded, pod-like structures at their end, within which are 8–12 sporangia containing spores that are about 80 micrometres in diameter. *Schizaeopsis* is very similar in structure to some living species of *Schizaea*, except in having different types of spores. It is very likely that *Schizaeopsis* represents an early stage in the evolution of *Schizaea*.

SCHIZAEOPSIS PLURIPARTITA
This small portion of leaf belongs to a family of living ferns called the Schizaeaceae.

Hausmannia

GROUP Ferns
DATE Triassic to Cretaceous
SIZE Leaf 5–8cm (2–3¼in) long
LOCATION Northern hemisphere

This group of ferns often had deeply lobed leaves with the major veins branching and running to the edge of the leaf-blade. Small sporangia were scattered across the lower, hairy surface of the leaves. *Hausmannia* is structurally very similar to the living tropical fern *Dipteris* and is therefore included in the family Dipteridaceae. Paleoecological studies of groups of fossils found together suggest that *Hausmannia* was a stream-side dweller. It was also a pioneer species, like the living *Dipteris*, in the way it invaded open areas.

H. DICHOTOMA
This leaf fragment shows the dividing veins. Between these veins, finer veins form polygonal meshes.

Cycadeoidea

GROUP Bennettitaleans
DATE Jurassic to Cretaceous
SIZE Stem up to 5cm (2in) across
LOCATION North America, Europe

These large, fossilized bennettitalean trunks are known from many places in North America and Europe. Most *Cycadeoidea* stems are short and barrel-shaped, with a dense covering of leaf bases. In life, they had a crown of pinnate leaves at the top of the stem. They were seed plants, but were unlike any other group in having their reproductive organs arranged in complex flower-like structures protected by stout, spirally arranged bracts. The size of these cones varied from species to species, and most species were bisexual. Their ovules were borne on a central swollen receptacle and separated by scales; the pollen organs surrounded the receptacle. It is likely that these flowers were self-pollinated, although it is possible that insects may have been involved.

One species of *Cycadeoidea* was **named from a stem** found as a **burial object in an Etruscan grave.**

AMONG FERNS
These slow-growing plants probably lived in open habitats, surrounded by many different species of ferns. The climate was probably strongly seasonal and the vegetation may have burnt during the dry season, similar to many modern savannas.

SPLITTING VEINS
This small fragment of a dipteridaceous fern leaf shows several veins branching and running to the edge of the leaf-blade.

Welwitschia

GROUP Gnetales
DATE Late Cretaceous to present
SIZE Leaf up to 9m (30ft) long
LOCATION Namibia

Welwitschia mirabilis is a highly specialized gymnosperm that grows in the Namib Desert in southwest Africa (see p.290). It has a very short stem, tapering down into a long taproot, and two leaves that grow continually from their bases. Individual plants produce cones that have either pollen organs or ovules, and pollination is probably achieved through insects. Pollen similar to that from this genus and the related, living *Ephedra* is known from the Late Triassic onwards. It is probable that *Welwitschia* and *Ephedra* diverged in the Late Cretaceous, especially as both *Welwitschia*-like and *Ephedra*-like fossils are known from this time.

Taeniopteris

GROUP Pentoxylales
DATE Early Cretaceous
SIZE Leaf up to 40cm ($15\frac{1}{2}$in) long
LOCATION Worldwide

Taeniopteris had elongated leaves with parallel margins and a leaf stalk that continued as a prominent midvein in the leaf blade. In some fossil floras, *Taeniopteris* leaves occurred with seed-bearing structures (*Carnoconites*) that were crowded on short shoots, pollen-bearing structures called *Sahnia*, and stems called *Pentoxlon*. These genera all belong to the Pentoxylales, which were an important part of the mid-Mesozoic Gondwana vegetation covering the conifer forest floors.

TAENIOPTERIS SPATULATA
The elongated leaves may be up to 40cm ($15\frac{1}{2}$in) long with parallel margins and blunt or sometimes pointed ends.

Araucaria

GROUP Conifers
DATE Jurassic to present
SIZE Cones 2.5–4.5cm (1–$1\frac{3}{4}$in) long and 2.5–4cm (1–$1\frac{1}{2}$in) across
LOCATION South America

These large, conifer trees from the Jurassic of Argentina bore very characteristic female cones with numerous, spirally arranged cone scales attached to a central axis. The corresponding pollen cones are not known. Each fertile cone scale had an ovule and a smaller bract scale. Immature cones contained ovules with three-layered skin (integument). Pollination would have been by wind, as in the living members of the genus. Mature cones contained seeds with embryos, apparently in a dormant state; some seeds were loose within the cone. It also appears likely that the seeds were not shed until the cones fell to the ground where they may have shattered on impact. The structure of the cone and the way that the seeds were dispersed is most similar to the living species *Araucaria bidwillii*, which is found in Southern Queensland, Australia.

PERFECTLY PRESERVED
Preserved cones such as the ones shown here have been over-collected in the past for museums and private collectors, so it is now forbidden to collect and export them from Argentina without permission.

LIVING RELATIVE
THE MONKEY-PUZZLE

There are 13 species of the evergreen conifer *Araucaria* living in the southern hemisphere. *Araucaria araucana* (the monkey-puzzle tree) is the hardiest species, and grows on the lower slopes of the Andes in southern Argentina and Chile. It reaches about 40m (130ft) in height and its branches grow in characteristic whorls that extend horizontally. The young stem and the branches are covered with tough triangular leaves with sharp edges and tips. The trees are unisexual, producing either large female cones or clusters of smaller, elongated male cones on the ends of the branches.

ARAUCARIA MIRABILIS
These anatomically perfectly preserved, silicified cones of *Araucaria mirabilis* come from the Cerro Cuadrado petrified forest in Patagonia.

WELWITSCHIA MIRABILIS
These plants may live for over 1,000 years. During its long life, the *Welwitschia*'s two leaves may split right down to their basal growing areas to produce numerous segments that appear to be separate leaves. Water condensing from coastal fogs is used by the plants, which helps them to survive in their harsh desert habitat (see p.289).

Brachyphyllum

GROUP Conifers
DATE Jurassic to Cretaceous
SIZE About 8mm (5/16in) wide
LOCATION Northern hemisphere

Fossil conifer shoots that are covered with spirals of short, stubby, scale-like leaves are called *Brachyphyllum*. Pollen-producing cones were found on the ends of these shoots. These cones had many scales, and each scale had three pollen sacs on its underside. The leaves of *Brachyphyllum* possessed distinctive stomata – the respiratory pores that allow plants to "breathe". *Brachyphyllum* leaves show a number of features that indicate that they would have grown in arid environments.

Pityostrobus

GROUP Conifers
DATE Cretaceous
SIZE Cone 5–6cm (2–2¼in) long
LOCATION North America, Europe

Pityostrobus was a seed-bearing conifer cone. In *Pityostrobus*, the seed-bearing scales were positioned in the axils – the angle between a leaf stem (or the bract in this case) and the main stem – of very small bracts. The seed-bearing (or ovuliferous) scales thinned towards their tips, unlike in modern pine cones. In mature cones, each ovuliferous scale had two winged seeds separated by a central ridge. The closest living relative to *Pityostrobus* is *Pinus*, the genus that contains pines and related conifers. However, *Pityostrobus* differs from *Pinus* in that it possessed separate bract and ovuliferous scales. In *Pinus* cones, the bract and ovuliferous scales were fused together and had a woody consistency.

PITYOSTROBUS DUNKERI
This carbonized cone from the Early Cretaceous is roughly three times longer than it is wide. In life the cones would have been woody.

Drepanolepis

GROUP Conifers
DATE Cretaceous
SIZE 10cm (4in) long
LOCATION Northern hemisphere

Drepanolepis was a genus of extinct conifers that produced seed-bearing shoots with a spiral arrangement of bracts. Each bract consisted of a single sickle-shaped, wing-like, seed-bearing scale with a single seed attached to its upper surface. The seeds themselves were curved, with their openings turned back towards the base of the scale. Each of these seed-bearing scales was positioned in the axil – the angle where the bract joined the stem – of a much longer bract. *Drepanolepis* was similar in structure to the Permian conifer *Ullmania*. It is also similar to the Araucariaceae, the family to which the somewhat primitive-looking but living monkey-puzzle tree belongs (see p.289). Some people think that *Drepanolepis* might be an evolutionary offshoot of early Araucariaceae. Like all conifers, *Drepanolepis* was a gymnosperm, meaning that its seeds were "naked" – as opposed to trees whose seeds are surrounded by fruit, which are known as angiosperms.

SPIRALLING STRUCTURE
The shoots of *Drepanolepis* featured an arrangement of seed-bearing bracts that spiralled around a central stem.

With its distinctive **bracts** and **bract-scales**, ***Drepanolepis*** has sometimes been compared to cones of the **Araucariaceae** family.

Sequoia

GROUP Conifers
DATE Cretaceous to present
SIZE 70m (230ft) tall
LOCATION Northern hemisphere

Sequoia is a genus of conifers of the bald cypress family. Also known as redwoods, they have both pollen and seed cones on the same tree. After pollination, the seed cones expand and become woody. The long, diamond-like shapes on the cone's surface consist of ovule-bearing scales (upper) and bract scales (lower). Such cones can remain closed for 25 years after falling from the trees and may not open until stimulated by fire.

SEQUOIA DAKOTENSIS
This impression of a fossil cone segment comes from the modern sequoia's Cretaceous ancestor.

LIVING RELATIVE
REDWOODS

There are two giant redwood evergreen conifers. The Californian coast redwood is *Sequoia sempervirens* and the giant redwood (or *Wellingtonia*) is *Sequoiadendron giganteum. Sequoia* is confined to a narrow coastal strip running from the Canadian border to southern California, USA. At up to 110m (360ft) high, these trees are the tallest in the world; some are over 2,000 years old. Their trunks can be up to 6m (20ft) in diameter and their 30cm (12in) thick bark protects them from animal damage and forest fires. The leaves on the main shoots are needle-like and arranged into one plane, but the leaves on the cone-bearing shoots are scale-like and set in a spiral.

DREPANOLEPIS ANGUSTIOR
The scales and ovules appear to depart at different angles from the axis. This is an illusion produced by the compression of the spiral during fossilization.

Geinitzia

GROUP Conifers
DATE Cretaceous
SIZE Shoot 8mm (5/16in) wide
LOCATION Northern hemisphere

This conifer's leafy shoot has long, outward-spreading, spirally arranged leaves. Each leaf is needle-shaped, as thick as it is broad, and merges below into a swollen leaf cushion. *Geinitzia*-like leaves are very similar to the young shoots of other conifers. Similar leafy shoots in the Permian are called *Walchia* and in the Triassic they are called *Voltzia*. According to some interpretations, *Geinitzia* is part of the extinct conifer family Cheirolepidiaceae. Seed cones of two *Geinitzia* species are known from Belgium and Germany.

GEINITZIA CRETACEA
This fossilized example shows a small portion of a branching, fertile, leafy shoot with two lateral female cones. The ovals on the cones mark the ends of the ovule-bearing scales.

Archaeanthus

GROUP Angiosperms
DATE Cretaceous
SIZE Cone 10cm (4in) long
LOCATION North America

This flower consisted of around 100 loosely packed, helically arranged, peapod-shaped follicles that were attached to a long, central axis. The flower was borne at the tip of a branch that bore leaves in an alternate arrangement. Leaves associated with *Archaeanthus* had a prominent midrib and are called *Liriophyllum*. *Archaeanthus* flowers are very similar to those of modern *Magnoliae*, which points to *Archaeanthus* as the earliest member of the family

ARCHAEANTHUS
With its large, *Magnolia*-like flower, petal-like perianth, and pod-like fruits, *Archaeanthus* shows all the characteristics that are traditionally associated with insect pollination.

Archaefructus

GROUP Angiosperms
DATE Cretaceous
SIZE 10cm (4in) long
LOCATION China

Archaefructus is an herbaceous, aquatic angiosperm. It does not have petals or sepals, but it does have carpels and stamens. These are attached to an elongated stem with the staminate (pollen-producing) flowers below, and pistillate (fruit-producing) flowers above. This ancient flower is similar in some respects to *Trithuria*, a peculiar living genus of Nymphaeales (waterlilies).

ARCHAEFRUCTUS LIAONINGENSIS
This fossil shows part of a reproductive axis with female carpels on short stalks. The oval outlines in the carpels are seeds.

Mauldinia

GROUP Angiosperms
DATE Cretaceous
SIZE Leaf 10cm (4in) long
LOCATION Europe, North America, Central Asia

The leaves of *Mauldinia* were typically simple, elongated angiosperm leaves. Flowers were radially symmetrical, bisexual, and about 3.5mm (1/8in) in length. The flower parts were arranged in groups of three. There were two whorls of three petals, surrounding three whorls of three stamens. The central ovary was rounded to triangular in cross section, and contained a single seed. Five individual flowers were grouped together on the upper surface of bilobed, scale-like structures. These scales were then spirally arranged on an elongated axis. The flower structure and the leaves of *Mauldinia* are characteristic of the living angiosperm family the Lauraceae (the laurels).

ARALIA SAPORTANA
This impression of a lobed leaf is from the classic Dakota Sandstone fossil flora from Kansas, USA.

Araliopsoides

GROUP Angiosperms
DATE Cretaceous to present
SIZE 10m (33ft) high
LOCATION North America, Europe, Asia

The palmate leaves of *Araliopsoides* had three distinct lobes. The wide base of the leaf stalks (petioles) suggest that the leaves were deciduous and shed seasonally. The plants themselves were small, shrubby, and probably grew in warm temperate to subtropical deciduous forests in middle- and high-level northern latitudes during the Late Cretaceous. *Araliopsoides* was a forerunner of the maples that came later.

ARALIOPSOIDES CRETACEA
This impression of a trilobed leaf from the Dakota Sandstone shows beautiful preservation of leaf venation.

Aralia

GROUP Angiosperms
DATE Cretaceous to present
SIZE Leaf 1m (3ft 3in) long
LOCATION North America

These lobed fossil leaves from North America are known as *Aralia*. Fossils such as this demonstrate the difficulties experienced in identifying leaves by comparing them to similar living species. In the past, the genus *Aralia* has been used loosely to describe fossil leaves that appear to have affinities with the living family Araliaceae. Consequently, it has become little more than a catch-all name that indicates a possible family relationship. For this reason, many Cretaceous and Paleogene records of genera of living angiosperms have had to be discounted.

Credneria

GROUP Angiosperms
DATE Cretaceous
SIZE Leaf 10cm (4in) long
LOCATION North America, Europe

Credneria is the name given to large angiosperm leaves produced by trees very similar to living plane trees. The leaves were originally attached to the shoots by stalks (petioles). Individual leaves were typically broadly oval or elliptical, with rounded bases. Their margins were smooth and they had pointed or blunt apexes. Leaf venation was usually pinnate, with a main vein and lateral veins. During the Middle Cretaceous, the leaves are very common in many fossil floras from all over the northern hemisphere, where they have been given a range of different names, such as *Aralia* and *Araliopsoides* (see opposite). The extinct plane trees that produced *Credneria* were especially common on ancient riverbanks. Living plane trees grow in similar places today.

CREDNERIA ZENKERI
Although there is no plant material remaining, the detailed outlines of these two leaves and their intricate pattern of veins are clearly impressed into the sandstone in which they are preserved.

Cercidiphyllum

GROUP Angiosperms
DATE Cretaceous to present
SIZE Leaf 6cm (2¼in) long
LOCATION Northern hemisphere

The fossilized leaves of this Cretaceous plant, together with other fossils of its fruits and seeds, are very similar to those of living *Cercidiphyllum*. *Cercidiphyllum* is known today from two very similar species that grow in China and Japan. Both of these are commonly called katsura. Male and female flowers are produced on separate plants. The leaves are simple, ovate, and wider than they are long. They have a midrib and one or two pairs of basal veins that curve out towards the leaf apex. Fossil *Cercidiphyllum*-like seedlings from the Paleocene of Canada have been found in their original growth positions. This discovery suggests that this species grew on open floodplains during the Late Cretaceous and the Tertiary.

CERCIDIPHYLLUM SP.
This fossil, from Alberta in Canada, shows impressions of simple, ovate leaves related to the modern katsura tree, *Cercidiphyllum japonicum*.

RADIATING VEINS
This close up of a fossil *Cercidiphyllum* leaf shows five main veins radiating from the base.

CRETACEOUS INVERTEBRATES

During the Cretaceous, the sea level stood higher than at any time in geological history. The continental shelves and adjacent lands were flooded, providing diverse marine habitats for invertebrate life. Yet this time of flourishing marine life ended in disaster for terrestrial animals as well.

Diversification and extinction

Cretaceous marine invertebrate faunas were not unlike those of the Jurassic, with ammonoids, bivalves, gastropods, brachiopods, echinoids, crinoids, and bryozoans. Yet many ammonoids developed strange shapes, some being uncoiled or partially re-coiled, or even twisted in a series of U-bends to form an unlikely tangle. The Late Cretaceous chalk of northern Europe is a source of beautifully preserved, though not especially common, marine invertebrates; this unique sediment is largely composed of calcareous plates of tiny algae (coccoliths). Marine life flourished, including sea-dwelling reptiles, while on land dinosaurs, flying reptiles, and birds proliferated. Yet all this was to change dramatically. At the very end of the Cretaceous, 65 million years ago, ammonites, belemnites, ichthyosaurs, and plesiosaurs abruptly vanished, as did dinosaurs and pterosaurs. An asteroid hit Earth, creating shock waves, dust clouds, forest fires, and acid rain; the collision site has been identified in Yucatán, buried by later sediment (see p.32). At the same time, huge volcanic eruptions in India created widespread devastation, and the effects of both together were catastrophic. Study of a fossil rainforest suggests that its recovery after this mass extinction took 1.5 million years.

NEOHIBOLITES MINIMUS
This small, slender mid-Cretaceous belemnite lived in vast numbers in the warm shelf-seas that existed during the Cretaceous, where it hunted small prey. It is a common fossil in Europe where it is often found in clay deposits.

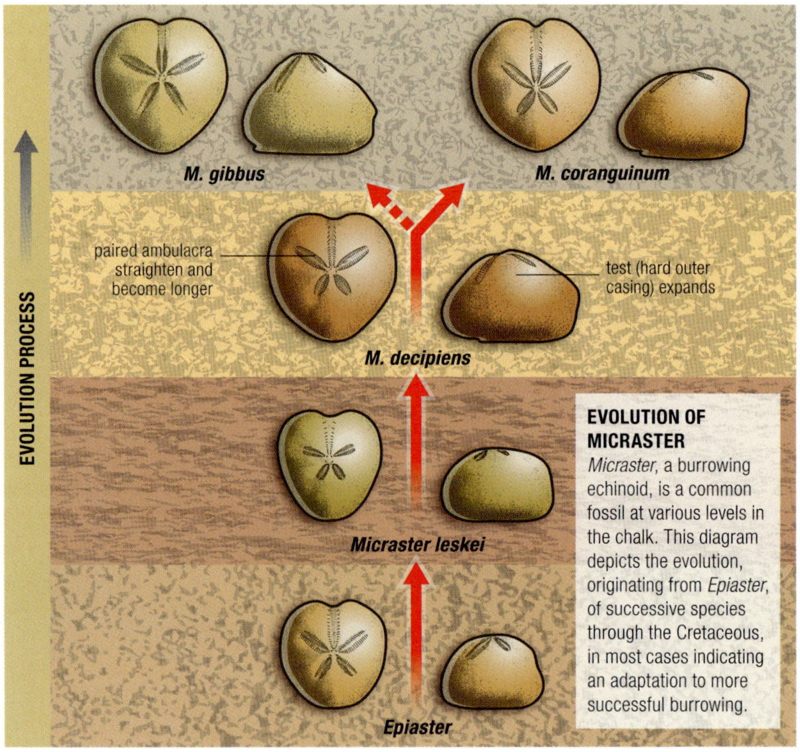

EVOLUTION OF MICRASTER
Micraster, a burrowing echinoid, is a common fossil at various levels in the chalk. This diagram depicts the evolution, originating from *Epiaster*, of successive species through the Cretaceous, in most cases indicating an adaptation to more successful burrowing.

GROUPS OVERVIEW

Many groups of invertebrates survived little changed from the Jurassic into the Cretaceous. Molluscs were abundant and diverse, as were cephalopods. Ammonites, belemnites, and nautiloids all evolved rapidly. Bivalves were abundant and became more diverse at low latitudes. Many new forms of echinoids also appeared.

AMMONITES
Cretaceous ammonites are of variable form, and include many differently shaped ones, such as helically coiled types, which look like gastropods. The helically coiled types may have lived like gastropods, creeping over the sea floor. Towards the end of the Cretaceous, ammonites declined, restricted to specific areas, before their final demise.

BELEMNITES
Belemnites are the internal shells of fossil, squid-like cephalopods, but they look like elongated bullets and are unlike the shells of modern squids or cuttlefish. They were common in the Jurassic and Cretaceous, and fossils are often found as mass-accumulations, or "belemnite battlefields", which, as with modern squids, may represent mass mortality after spawning.

BIVALVES
A very unusual group of bivalves, known as rudists, evolved at this time and was confined to the Cretaceous. Their external appearance is so modified that they are hardly recognizable as bivalves at all. Usually one valve is long and conical, the other sits on top of it like a lid. They grew very large and some species actually formed reefs.

ECHINOIDS
A trend towards deep burrowing, which began in the Jurassic, reached its greatest effectiveness in one group of Cretaceous echinoids, the spatangoids. Micraster (illustrated above) is one of these, and could burrow to many centimetres below the surface. Other unmodified echinoids, however, lived on the sea floor.

Siphonia

GROUP Sponges
DATE Cretaceous
SIZE Up to 6cm (2¼in) across
LOCATION W. Europe

Siphonia was a genus of unusual sponges that inhabited the western part of the Tethys Sea, a shallow body of water that lay over what is now western Europe. They were characterized by bulbous heads and were attached to the sea bed by a narrower stalk. Some forms resembled a partially inflated balloon. Others, such as *Siphonia tulipa*, looked more like tulips, hence their common name among fossil hunters: tulip sponges.

Fossil *Siphonia* had numerous tiny pores all over their outer surface. These were the openings of conical tubes that ran all the way through the sponge's body to its larger, central canal, called a cloaca. On the cloacal wall, the tube openings were larger. The cloacal surface was also marked with furrows. On the top of the outside of the sponge was a central, chimney-like opening to the cloaca. Like many sponges, *Siphonia*'s skeleton was composed of innumerable, branched, needle-like silica spicules of various sizes that were fused together.

SIPHONIA TULIPA
The bulbous head of *Siphonia* varied in shape, from distinctly tulip-like forms to those resembling inflated balloons.

PITTED SURFACE
No matter what their exact shape, all *Siphonia* fossils show this characteristic pitted surface. These pores were actually openings to tubes that ran through the sponge's body.

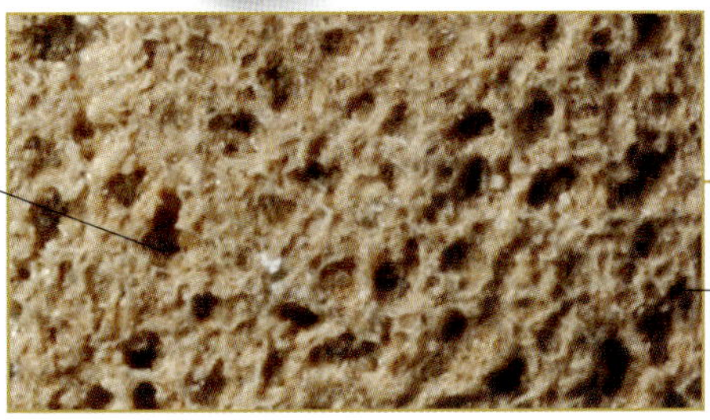

SPICULES
Siphonia's spicules are complex structures. Their edges were fused to those of adjoining spicules, and they were 0.5–1mm (1⁄32–1⁄16in) across.

Ventriculites

GROUP Sponges
DATE Cretaceous
SIZE Up to 12cm (4½in) high
LOCATION Europe, North America

Ventriculites was a glass sponge, with a skeleton of siliceous needles (spicules) that were fused into a rigid network to support the sponge's body. Between them were pores that dotted the surface. The skeleton of this fossil is vase-shaped, and siliceous fibres anchored it to the sea bed. In its centre was a cavity called the cloaca. Food and oxygen were removed as water passed through the body walls before exiting through the top opening.

VENTRICULITES SP.
Ventriculites occurred in many different shapes. Some were vase-like, such as the one above, while others varied from squat cups to tall cylinders.

LUNULITES SP.
Lunulites colonies were highly organized systems. A soft-bodied autozooid lived inside each of the chambers, which were known as autozooecia.

Lunulites

GROUP Bryozoans
DATE Late Cretaceous to present
SIZE 5–10mm (3⁄16–3⁄8in) across
LOCATION Worldwide

Lunulites was a cheilostome bryozoan that formed very small, circular colonies. Its lower surface was concave, with radial grooves. The colony originated at the centre, and radiating from this were rectangular to hexagonal autozooecia: chambers that housed soft-bodied, feeding autozooids – the animals that made up the colony. All autozooids possessed a lophophore or feeding structure consisting of a ring of tentacles surrounding the mouth, which could be withdrawn into the autozooecia for protection when not feeding. The autozooecia increased in size from the centre, and had wide openings. Modern relatives of *Lunulites* move across the sea floor by means of long, spiny appendages.

Crania

GROUP Brachiopods
DATE Cretaceous to present
SIZE Up to 1.5cm (½in) across
LOCATION Worldwide

Crania is a brachiopod that first appeared in the Cretaceous and still has relatives alive today. Scientists describe *Crania* as an inarticulate brachiopod, because the two halves of its shell are not "articulated" or joined together, but are held together by muscles alone. Unlike most brachiopods, *Crania* has no pedicle. It is commonly found in chalk deposits with one half of its shell cemented to a larger organism, such as a mollusc or echinoderm. Because its shell halves are held together by soft tissue, the bottom half is often the only part preserved, the upper half having been swept away before fossilization could take place.

CRANIA EGNABERGENSIS
Crania shells lack hinge teeth, and show tiny perforations across the shell. Inside are scars where muscles were attached.

Sellithyris

GROUP Brachiopods
DATE Cretaceous
SIZE Up to 3.5cm (1½in) long
LOCATION Europe

Sellithyris lived on the sea bed, attached by a flexible stalk or pedicle that passed through an opening below the beak of the larger shell (the pedicle valve); the smaller shell is called the brachial valve. *Sellithyris* was almost pentagonal in outline. Its outer surface was smooth but showed growth lines. All over the surface are tiny perforations (punctae) that housed bristles.

SELLITHYRIS SP.
The outer edge of *Sellithyris*'s shell is strongly folded, and there is an inward fold at the centre of the brachial valve, with two outward folds on either side.

Orbirhynchia

GROUP Brachiopods
DATE Cretaceous
SIZE Up to 2cm (¾in) long
LOCATION N.W. Europe

Orbirhynchia was a rounded rhynchonellid brachiopod with strongly convex valves. Each valve had a pronounced beak. In life, it would have attached itself to the sea bed via a fleshy stalk, or pedicle. This passed through the pedicle opening, visible on the specimen on the right, below the beak of the pedicle valve. The external shell had strong ribs, running from the edge of the beak to the shell's edge.

ORBIRHYNCHIA PARKINSONI
This internal mould shows the impressions of the adductor (closing) muscle scars.

CYMATOCERAS SP.
On this internal mould, the distinction between the ribs and the suture lines is clear.

Cymatoceras

GROUP Cephalopods
DATE Late Jurassic to Paleogene
SIZE Up to 30cm (12in) across
LOCATION Worldwide

Cymatoceras was a tightly coiled cephalopod mollusc that was similar in many respects to the modern *Nautilus*. The principal difference from *Nautilus*, apart from the broader shell, is the presence of fairly strong ribs, which are strongest on the outer part of the whorl and over the wide outer edge (venter). Like all cephalopods, *Cymatoceras* would have caught prey with its flexible tentacles. Prey would have then been dismembered by using its horny beak.

Deshayesites

GROUP Cephalopods
DATE Early Cretaceous
SIZE Up to 10cm (4in) across
LOCATION Europe, Greenland, Georgia

Deshayesites was an ammonite with a rather flattened, coiled shell. Like all ammonites, in life it would have had flexible, muscular tentacles around its mouth and emerging from its shell. *Deshayesites* has strong, sometimes branched, closely spaced ribs that have a slight forward bend in the middle, which points in the direction of the opening of the shell. The whorls themselves overlap each other very slightly, creating a wide, shallow depression, or umbilicus, at the centre.

DESHAYESITES SP.
In between the longer main ribs are shorter secondary ones, which begin halfway up the whorls.

MORTONICERAS ROSTRATUM
At the tops of the ribs on the outer whorl are prominent bumps, known as tubercles.

Mortoniceras

GROUP Cephalopods
DATE Early Cretaceous
SIZE Up to 30cm (12in) across
LOCATION Europe, Africa, India, North America, South America

Mortoniceras fossils are characterized by their square-section whorls and strong ornamentation, which is produced by a combination of ribs and tubercles. The whorls overlap only slightly, leading to a very shallow umbilicus – this coiling is described as moderately evolute. There is a pronounced ridge, known as the keel, on the outer edge (venter). In some mature specimens, a large horn is visible at the opening on the shell. This has not been associated with just one sex in this genus.

Scaphites

GROUP Cephalopods
DATE Late Cretaceous
SIZE Up to 20cm (8in) across
LOCATION Europe, Africa, India, North America, South America

Scaphites was an unusual ammonite. When young, it resembled many other ammonites, possessing a normal, spiral shell, but as it grew older the shape of its shell changed. Instead of continuing to spiral, it straightened, before starting to spiral once again, which created a hooked appearance. This left the shell opening (where the animal's head and tentacles emerged) facing upwards.

Fossils show that this opening was often constricted and thickened around its edges and positioned very close to its chambered shell (phragmocone). This raises the possibility that the final growth stage would have prevented the animal from actually feeding. It may be that, having reached maturity, the adult reproduced and then died. Like other ammonoids, *Scaphites* occurred in two different forms, which were slightly different in terms of size.

SCAPHITES AEQUALIS
The initial whorls are tightly coiled and the earliest ones are hidden by the succeeding ones. This is known as involute coiling.

The **final growth stage** may have **prevented the animal from actually feeding**. It may be that, when it reached maturity, the adult **reproduced and then died**.

BUOYANCY CONTROL
Like other ammonites, *Scaphites* would have been able to alter the amount of liquid and gas in its chambered shell to regulate buoyancy.

Baculites

GROUP Cephalopods

DATE Late Cretaceous

SIZE Up to 2m (6½ft) long

LOCATION Worldwide

Baculites was unlike most ammonoids; rather than having a spiral shell, like many of its relatives, it had a shell that was almost completely straight. Only on the very tip were there one or two minute whorls. These are rarely preserved and they do not appear on the specimen illustrated here. *Baculites* fossils are roughly egg-shaped in cross section. As they are long and straight, they fracture easily, and are often found as fragments rather than complete shells, breaking along the partitions between the shell chambers. The shape of the shell suggests that *Baculites* was a squid-like animal. In some places these fossils are found in large quantities, to the near-exclusion of other species.

BACULITES ANCEPS
This internal mould clearly shows *Baculites*' complex suture lines. The last two sutures are close together, suggesting that the animal was mature when it died.

The name *Baculites* means "walking stick rock", when translated literally.

Turrilites

GROUP Cephalopods

DATE Late Cretaceous

SIZE Up to 30cm (12in) long

LOCATION Europe, Africa, India, North America

At first sight a *Turrilites* fossil might be mistaken for a gastropod shell, but it actually belonged to an ammonoid. The shell was coiled helically, rather than forming the usual plane spiral, creating an elegant spire. Palaeontologists know that *Turrilites* was an ammonoid because of the shell's internal structure. As with all ammonoid shells, it was split into two distinct parts: the body chamber (where the living animal sat) and the phragmocone, a series of smaller, connected chambers.

TURRILITES COSTATUS
The tightly curled whorls on this shell have two rows of tubercles and ribs from mid-whorl tubercles to the joins.

Architectonica

GROUP Gastropods

DATE Early Cretaceous to present

SIZE Up to 4.5cm (1¾in) long

LOCATION Worldwide

Architectonica is a genus of rather striking and beautiful marine gastropods. The earliest examples are known from rocks dating back around 140 million years, and there are many species still alive today. *Architectonica* fossils have a very distinctive, ever-widening series of whorls. At the upper part of each whorl, immediately below the whorl suture, there are rows of small, tightly packed bumps, or tubercules. The shell is covered in strong spiralling ribs crossed by growth lines, which creates a fine network of ornament that covers the surface of each whorl and looks rather like the "engine-turning" seen on metalwork. On the underside of the last whorl the growth lines can only be seen on the edge of the umbilicus (the deep depression in the centre of the shell). In most fossil examples all trace of the original colour is lost. Modern species of *Architectonica* are carnivorous, feeding on sea anemones, coral polyps, and similar prey.

ARCHITECTONICA SP.
This specimen is preserved in its original aragonite but has lost the colour banding. Modern species of this genus include sundial shells.

Belemnitella

GROUP Cephalopods

DATE Late Cretaceous

SIZE Up to 13cm (5in) long

LOCATION Europe, North America

Belemnitella is a common belemnite fossil and several different species of the genus are known. Belemnites were squid-like in appearance, as a few rare fossils preserving their soft tissues show. The bullet-shaped hard part of the body, which forms most belemnite fossils, helped to keep the animal's body straight. The fossils are typically cylindrical with pointed tips. There is a distinctive slit in one side near to the pointed end. The large numbers of *Belemnitella* fossils found in some localities suggest that they lived in sizeable shoals.

BELEMNITELLA MUCRONATA
This specimen has a small, teat-like extension at the rear of the guard – a distinguishing feature of this genus.

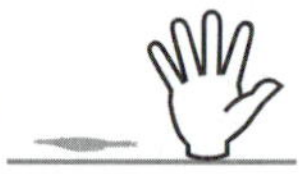

GERVILLELLA SUBLANCEOLATA
This fossilized shell displays the typical angled beak at the front end of the valve.

Gervillella

GROUP Bivalves
DATE Triassic to Cretaceous
SIZE Up to 15cm (6in) long
LOCATION Worldwide

Gervillella was a rather elongated bivalve mollusc, not unlike a modern mussel. The fossils show that one valve (half) of the shell was flatter than the other. There were two adductor (shell-closing) muscles, both of which left scars on the inside of each valve. One scar is much larger than the other, suggesting that the muscles were of unequal size. The hinge-plate has a few blade-like teeth at either end. The elastic ligament that pulled the shell open ran the full length of the hinge-line and the edges of both valves have a series of regular pits that show where it attached. Like a mussel, it lived its adult life attached to a rock or other hard surface by means of a byssus, or "beard", and fed by sifting plankton and other tiny particles of food.

Amphidonte

GROUP Bivalves
DATE Early Cretaceous
SIZE Up to 4cm ($1\frac{1}{2}$in) across
LOCATION Europe, North America

Amphidonte was a type of oyster that spent at least the early part of its life attached to the sea bed. Inside, the shell was pulled together by a single adductor muscle. The entire edge of one valve was dotted with tiny pits, while the edge of the other valve had corresponding lumps, or tubercles. This oyster lacked the hinge teeth seen in many other bivalves and in their absence the pits and tubercles helped the valves to fit together.

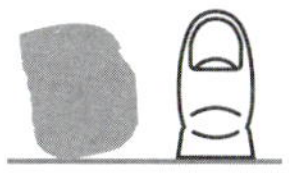

AMPHIDONTE OBLIQUATA
This species twisted as it grew, leaving its valves asymmetrical. One was strongly convex, often with a marked "shoulder", while the other was flat or even slightly concave in shape.

Spondylus

GROUP Bivalves
DATE Jurassic to present
SIZE Up to 9cm ($3\frac{1}{2}$in) across
LOCATION Worldwide in warm seas

Spondylus is an ancient genus of bivalve shellfish, with fossils that date back almost 200 million years. The genus is not extinct, however, and several species still exist in the world's warmer seas today. Both fossilized and modern shells vary greatly in appearance within and across species. *Spondylus* species live with the lower valve (half of the shell) cemented to the sea bed. This valve's beak region is often rather blunt because this is where the shell would have joined to the sea bed. The lower (right) valve is larger and more convex, whereas the upper (left) valve is flatter. Internally, there is a single large adductor (shell-closing) muscle, which leaves a single scar on the inside of each of the valves. The hinge has two strong teeth in each valve and a centrally placed ligament. Many fossils were originally described as having no hinge teeth, but it was later discovered that this absence was caused by the breakdown of the internal layers of the shell, which left only the outer layer intact. These often rather beautiful bivalve molluscs go by the collective common name of spiny oysters.

Neolithic Europeans used *Spondylus* shells in jewellery as far back as 5,000 years ago.

SPONDYLUS SPINOSUS
This fossilized lower valve is convex in shape and has strong radiating ribs, which have developed spines at irregular intervals. The upper valve (not shown here) would have been smaller and flatter.

LIVING RELATIVE
THORNY OYSTER

The Atlantic thorny oyster *(Spondylus americanus)* gives us some idea of what its fossil counterparts were like. Its mantle edge has brown, finger-like sensory papillae and a series of eyes on short stalks. Fossil *Spondylus* probably had these features too. As bivalves have no head, they evolved sensory organs on the mantle edge, where they are nearest to the outside environment.

Hippurites

GROUP Bivalves
DATE Late Cretaceous
SIZE 5–25cm (2–10in) tall
LOCATION S. Europe, N.E. Africa, S. Asia, American Antilles

Hippurites is a member of a very specialized group of bivalves, the rudists, which evolved in the Late Jurassic and died out at the end of the Cretaceous. The two valves of the shell are highly asymmetrical. The lower valve is conical and was attached to the sea bed. The upper valve is a lid-like, flattened cone, and is hinged to the lower valve by means of teeth and sockets in both the upper and lower valves. Two muscles were attached to the teeth in the upper valve.

HIPPURITES SP.
The outside of both the lower (seen here) and upper valve is ornamented by ridges crossed by growth increments.

Fissidentalium

GROUP Scaphopods
DATE Cretaceous to present
SIZE 7–10cm (2¾–4in) long
LOCATION Worldwide

Fissidentalium is a scaphopod, or tusk shell. The tapering shell is open at each end, with the larger opening at the front end. The shell lay buried in the sediment with the front opening almost horizontal and with the smaller (rear) end of the shell above the sea bed. The shell is ornamented by fine longitudinal ribs and by growth lines that cross these. The two sides of the animal were symmetrical and it fed on micro-organisms and organic particles in sea water. Respiration was carried out, not by gills, as in virtually all other groups of molluscs, but by the whole body surface. Scaphopods seem to have changed little through geological time. They first appeared in the Ordovician and were the last class of molluscs to evolve.

FISSIDENTALIUM SP.
This fossil shell has been preserved in its original aragonite. The ribs running along its length are crossed by growth lines.

Marsupites

GROUP Echinoids
DATE Late Cretaceous
SIZE Cup diameter up to 6cm (2¼in)
LOCATION Worldwide

Marsupites is a very large crinoid that has no stem. At the base of the cup is a large pentagonal plate where, in most crinoids, the stem attachment would be seen. Above this are three circlets of plates comprising the lower cup – they may be almost smooth or ornamented with ridges radiating from the centre of each plate. The upper plates have a marked semicircular depression at the centre of their upper edge where the arms join to the main body. It is believed that *Marsupites* lived on the sea bed or possibly with part, or even most, of its cup buried in the soft chalk sediment in which it is found. The arms would have been active in feeding, but on death they broke off and drifted away, so that only the base of the arms has ever been seen.

MARSUPITES TESTUDINARIUS
The arms are seen to branch once in well-preserved fossils, such as this, to give ten arms. It is likely that there were further branches of which we know nothing.

Temnocidaris

GROUP Echinoids
DATE Early Cretaceous to present
SIZE 4–7.5cm (1½–3in) in diameter (excluding spines)
LOCATION Worldwide

Temnocidaris is an echinoid with very strong differences between the ambulacral and interambulacral areas. The interambulacral areas are very wide and the plates are dominated by large spine bases, bearing long spines with smaller spines along their length. In the centre of each plate is a round boss that joined with a socket at the end of the long spines. Muscles that controlled the spines attached to the smooth area around the boss. The ambulacral areas are very narrow and sinuous, consisting of many smaller plates with perforations for the water supply to the tube feet. This supply enabled the tube feet to extend and retract.

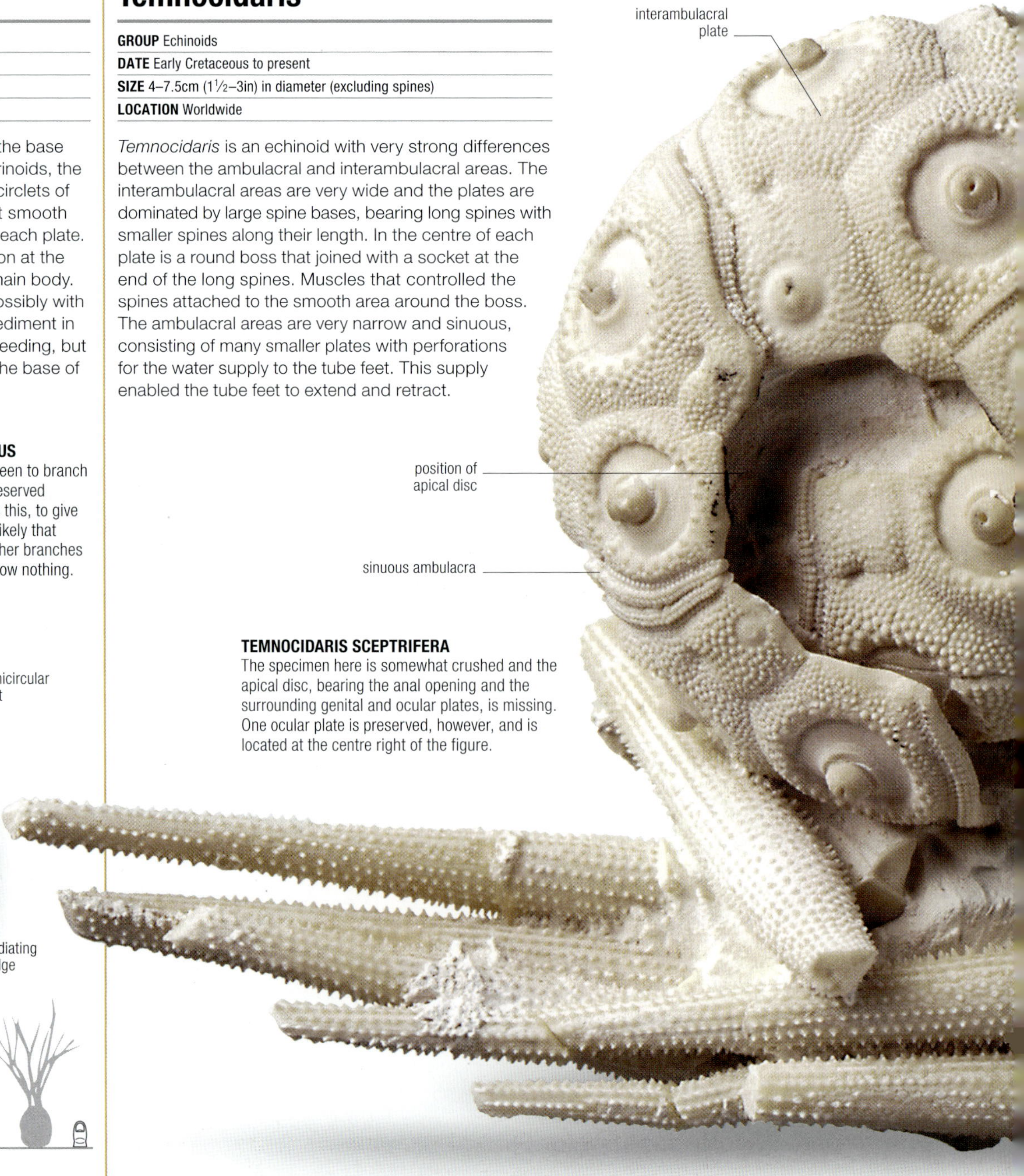

TEMNOCIDARIS SCEPTRIFERA
The specimen here is somewhat crushed and the apical disc, bearing the anal opening and the surrounding genital and ocular plates, is missing. One ocular plate is preserved, however, and is located at the centre right of the figure.

Micraster

GROUP Echinoids

DATE Late Cretaceous to Early Paleogene

SIZE 4.5–6.5cm ($1\frac{3}{4}$–$2\frac{1}{2}$in) long

LOCATION Europe, W. Asia

Micraster was a burrowing, heart-shaped echinoid with its five ambulacral areas forming a petal shape on the upper surface. The front ambulacra is in a groove that continues down to the mouth, which is found well to the front side of the oral surface. The anal opening is on the steep back part of the upper surface and beneath it is a smooth ring or fasciole on the shell (test) that in life would have carried hair-like structures (cilia) to convey waste material away from the anus. Most of the rest of the test is covered with small tubercles and during life these supported a dense cover of short, slender spines. In the centre of the upper surface the apical disc has four genital plates with pores for releasing eggs or sperm. A succession of species of this genus characterizes the chalk of northwest Europe and can be used to date it.

MICRASTER CORANGIUM
Known as the chalk heart urchin because of its distinctive shape, this specimen displays the characteristic deep anterior notch along which it would have channelled a current of water with food particles to its mouth.

This echinoid lived on a firm sea bed and used its rasping teeth to feed on algal films and other organic material.

Avitelmessus

GROUP Arthropods

DATE Late Cretaceous

SIZE Shell about 6cm ($2\frac{1}{4}$in) long

LOCATION S.E. USA

The shell or carapace of this Cretaceous crab is almost oval, with a rounded outline. The upper surface is marked with a pattern of broad, shallow grooves, which form two conjoined diamond shapes, with the larger towards the front. Smaller, subsidiary grooves run into either side of the larger diamond. The front edge of the carapace is fairly blunt, and has a very short rostrum in the centre. The orbits, where the eyes are situated in life, lie to either side of the rostrum. The first leg is modified to end in a pincer (cheliped); this comprises a moveable finger above and a fixed one below, which is an extension of the last segment on the leg. The surface of all the limbs is covered with fine granules. The oldest fossil crabs have been found in rocks of Early Jurassic age. By the Late Cretaceous, several lineages had evolved, but it was not until the Cenozoic that they became abundant and widespread.

AVITELMESSUS GRAPSOIDEUS
Only the upper parts of three of the four pairs of walking legs are preserved in this specimen. The three segments of the front leg, ending in a pincer, are intact.

CRETACEOUS VERTEBRATES

The Cretaceous was the pinnacle of dinosaur evolution, and at no point in their history were dinosaurs so abundant, diverse, and dominant. However, in the shadow of the dinosaurs, groups such as birds, mammals, and crocodiles also diversified, setting the stage for the modern world.

The dusty, sage-covered badlands of Montana and South Dakota preserve the best record of the final act of the age of dinosaurs. Here, the Hell Creek Formation yields spectacular fossils of *Tyrannosaurus*, *Triceratops*, and *Edmontosaurus*, some of the last dinosaurs to survive before a sudden asteroid impact 65 million years ago.

CAUDIPTERYX
The bizarre, feather-covered dinosaur *Caudipteryx* looks a bit like an ostrich, but it is a bona fide theropod, closely related to *Oviraptor*.

Early birds

One of the most extraordinary and important discoveries of modern palaeontology is that birds evolved from dinosaurs. The closest relatives of birds are theropods such as *Velociraptor* and *Deinonychus* – swift, sleek, keen predators with large brains and high metabolisms. More recently, even closer relatives of birds have been found. These include "raptors" such as *Microraptor*, which were small, lived in trees, and could probably even glide or fly.

Indeed, nowadays there is barely any distinction between the most bird-like dinosaurs and the earliest true birds – all of these animals were small, brainy, covered in feathers, and capable of flying. Put simply, the dinosaur–bird lineage is one of the best examples of a major evolutionary transition in the fossil record.

The oldest true bird is *Archaeopteryx*, a crow-sized flier known from a handful of spectacular fossils from the Late Jurassic of Germany. In many ways *Archaeopteryx* is half-bird and half-dinosaur: like dinosaurs it has teeth, claws, and a long tail, but like birds it has broad wings with asymmetric flight feathers and a brain adapted for powerful flight. Bird evolution kicked into high gear during the Cretaceous, as exemplified by the wonderful menagerie of feathered fossils from China's Liaoning Province. These fossils belong to an array of groups: primitive birds little-changed from *Archaeopteryx*, members

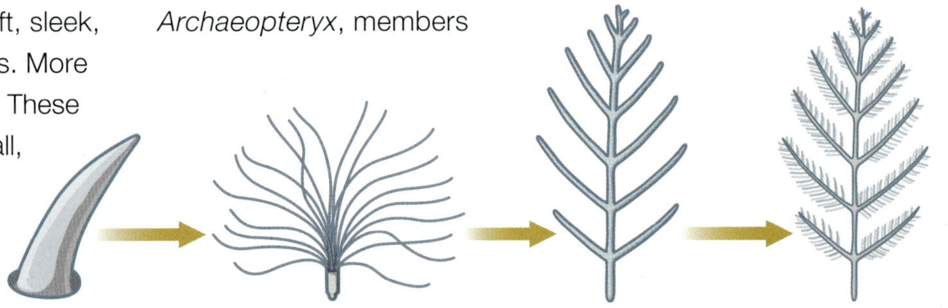

EVOLUTION OF A FEATHER
The earliest feathered dinosaurs, such as *Sinosauropteryx*, had simple strands that looked much like hair. Feathers subsequently became more complex, branching into individual barbs and later developing a central axis (rachis). In modern birds, flight feathers have a solid rachis and asymmetric barbs, both necessary to provide lift.

GROUPS OVERVIEW

The Cretaceous was a time of profound change on land and in the sea. Dinosaur groups continued to dominate most terrestrial ecosystems, and new subgroups (such as the tyrannosauroids and ceratopsians) became remarkably common. In the water, the reptilian mosasaurs ruled, but modern sharks were beginning to evolve.

CHONDRICHTHYANS
Chondrichthyans are one of the two great divisions of modern fish. This group includes sharks, rays, skates, and a variety of other animals with a skeleton made of cartilage (not bone, as in the osteichthyans). The first true sharks lived about 400 million years ago, but it was only in the Cretaceous that modern sharks first became abundant and diverse.

MOSASAURS
The bone-crushing mosasaurs were the most feared predators in most Cretaceous seas. They filled the same keystone predator niche as sharks do today. The largest mosasaurs were up to 18m (59ft) in length and could tackle any prey they desired. Like plesiosaurs and ichthyosaurs, mosasaurs were fully marine reptiles, and it is thought that they are close relatives of snakes.

ALLIGATOROIDS
The first crocodylomorphs evolved in the Late Triassic, but these small, sleek, fast-running animals were nothing like the crocodiles of today. One of the most important crocodylomorph subgroups, the alligatoroids, evolved in the Late Cretaceous. This group includes the familiar *Alligator* and its close relatives – animals that sprawl on four legs and lurk near the water's edge.

TYRANNOSAUROIDS
No dinosaur group inspires as much fear and curiosity as the tyrannosauroids. The most familiar member of this group is the colossal *Tyrannosaurus*, which was 12m (39ft) long, but several close relatives also approached this size. Small tyrannosauroids evolved in the Late Jurassic, but the giants dominated the Late Cretaceous world.

Several **recently discovered fossils** are of Cretaceous dinosaurs that were **covered in feathers**. There is barely any distinction between **these dinosaurs** and early **true birds**.

ARCHAEOPTERYX
There is no debate: *Archaeopteryx* is the most famous fossil of all time. Ten specimens of this, the oldest bird ever discovered, have been found in the limestones of Bavaria, Germany. Some of them preserve beautifully clear details of the feathers and skeleton.

of strange and wholly extinct groups, and even early representatives of the lineage leading to modern birds. Clearly, birds were the dominant flying creatures of the Cretaceous, and after the dinosaur extinction they continued to diversify.

The Mesozoic marine revolution

Marine ecosystems went through a remarkable period of change during the Mesozoic. This massive reorganization occurred on all levels, from primary producers all the way up to top predators. Large predators such as sharks and mosasaurs spread around the world, and bony fish became astonishingly common. The changes were not limited to vertebrates, and, in fact, many of the greatest changes involved invertebrates and microscopic organisms. The major modern groups of planktonic micro-organisms – the primary producers that form the base of every oceanic food web – originated and diversified during this time. Meanwhile, archaic invertebrates that were common during the Paleozoic, such as crinoids and brachiopods, were marginalized, and more modern groups such as clams, scallops, and heavily armoured gastropods exploded in abundance. The reasons for these changes are complex, but are likely to involve predation pressure from newly evolved giant predators (such as some sharks), as well as continental break-up and changes in sea level and ocean chemistry that are poorly understood.

ANCIENT SHARK
The ancient shark *Hybodus* was a small predator, only about 2m (6½ft) long, which thrived for over 100 million years. It is closely related to the modern sharks, but became extinct during the Late Cretaceous.

The end of the dinosaurs

The extinction of the dinosaurs has been one of the greatest mysteries of Earth's history. What could cause such a successful group to die off? Sixty-five million years ago a massive asteroid struck the Yucatán Peninsula of Mexico (see p.32). Most scientists think that this impact set in motion a chain reaction of environmental disturbances, such as tsunamis and acid rain, which spelled the end for the dinosaurs.

ORNITHOPODS
The ornithopods were some of the most successful ornithischian dinosaurs. Among this group of large, bulky plant-eaters are the iguanodontids and the duck-billed hadrosaurs, such as *Maiasaura*, which were the most common herbivores in most Late Cretaceous ecosystems. Hadrosaurs were probably the preferred prey of *Tyrannosaurus*.

CERATOPSIANS
Perhaps the most recognizable ornithischian subgroup is the ceratopsians, the horned-and-frilled herbivores that lumbered around on four legs. The three-horned *Triceratops* is the most familiar member of the group, but a variety of other species with even more bizarre horns also thundered across the North American plains during the Late Cretaceous.

AVES
Birds descended from theropods, and therefore represent a living subgroup of dinosaurs. The first bird, the beautifully feathered *Archaeopteryx*, is known from the Late Jurassic, about 150 million years ago. However, bird evolution jump-started in the Cretaceous, as several modern groups evolved and a variety of strange birds, now totally extinct, dominated the skies.

MARSUPIALS
The marsupials, which carry their young in a pouch, are one of the major subgroups of living mammals. Modern marsupials include animals such as kangaroos and opossums. The first marsupials must have evolved in the Jurassic, but their fossils become common in the Cretaceous. They were more geographically widespread than the marsupials of today.

EUTHERIANS
The other major subgroup of living mammals is the placentals, or eutherians, whose young develop inside the mother's womb. The oldest unquestionable eutherian is the tiny *Eomaia* from the Early Cretaceous of China. Throughout the rest of the Cretaceous, eutherians became larger and more common, and today live across the globe.

Hoplopteryx

GROUP Actinopterygians

DATE Late Cretaceous

SIZE 27cm (10½in)

LOCATION North America, Europe, North Africa, S.W. Asia

Hoplopteryx is an extinct relative of the modern slimeheads (see panel, below). Its remains have been found in chalk deposits, which suggests it inhabited a shallow-water environment. *Hoplopteryx* had large eyes, an upturned mouth, and jaws lined with small teeth. It was a deep-bodied fish with small pectoral fins and pelvic fins almost directly below them on the underside. The single anal fin had a straight rear edge, and the tail was deeply forked with equal-sized lobes. *Hoplopteryx* displayed an advance over other fish in the form of a series of bones called uroneurals in the upper lobe of the tail. These bones strengthened and supported the fin rays. Uroneurals in the tail fin allow fish to swim more powerfully, and their presence is considered a major advance in fish evolution. Fish with these skeletal elements are called teleosts.

Special tail-fin bones, called **uroneurals**, gave *Hoplopteryx* an **evolutionary advantage** over other fish of the time.

LIVING RELATIVE

SLENDER ROUGHY

A deep-sea reef fish from New Zealand, *Optivus elongatus* is commonly known as the "slender roughy". The roughys belong to a family known as "slimeheads", because of the numerous mucous canals in their heads. During the day it hides in crevices and cracks in the reef, but at night it emerges and hunts for prey, which it catches in its extensible jaws and swallows whole.

angular dorsal fin supported by nine unjointed fin rays increased the height of the fish
part of the powerful, forked tail fin
deep, flat body
small pectoral fin
small pelvic fin

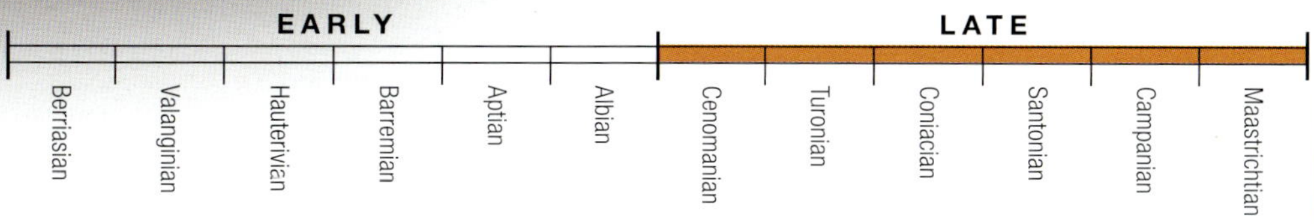
EARLY
LATE
Berriasian
Valanginian
Hauterivian
Barremian
Aptian
Albian
Cenomanian
Turonian
Coniacian
Santonian
Campanian
Maastrichtian

Xiphactinus

GROUP Actinopterygians
DATE Cretaceous
SIZE 6m (20ft) long
LOCATION North America

The largest-known bony fish, *Xiphactinus* was a formidable predator. It had a long, well-muscled body with a backbone of over 100 vertebrae. This, and the deeply forked tail, indicate it was a powerful swimmer and probably pursued its prey rather than ambushing it. The upturned lower jaw could gape widely, allowing it to consume large fish and possibly small marine reptiles – a 2m (6½ft) ichthyodectid fish has been found in the fossilized stomach contents of a 4m (13ft) *Xiphactinus*. In turn, *Xiphactinus* have been found in the fossilized stomach contents of a shark, suggesting that, despite its size and ferocity, it was not the top predator in its ancient ecosystem.

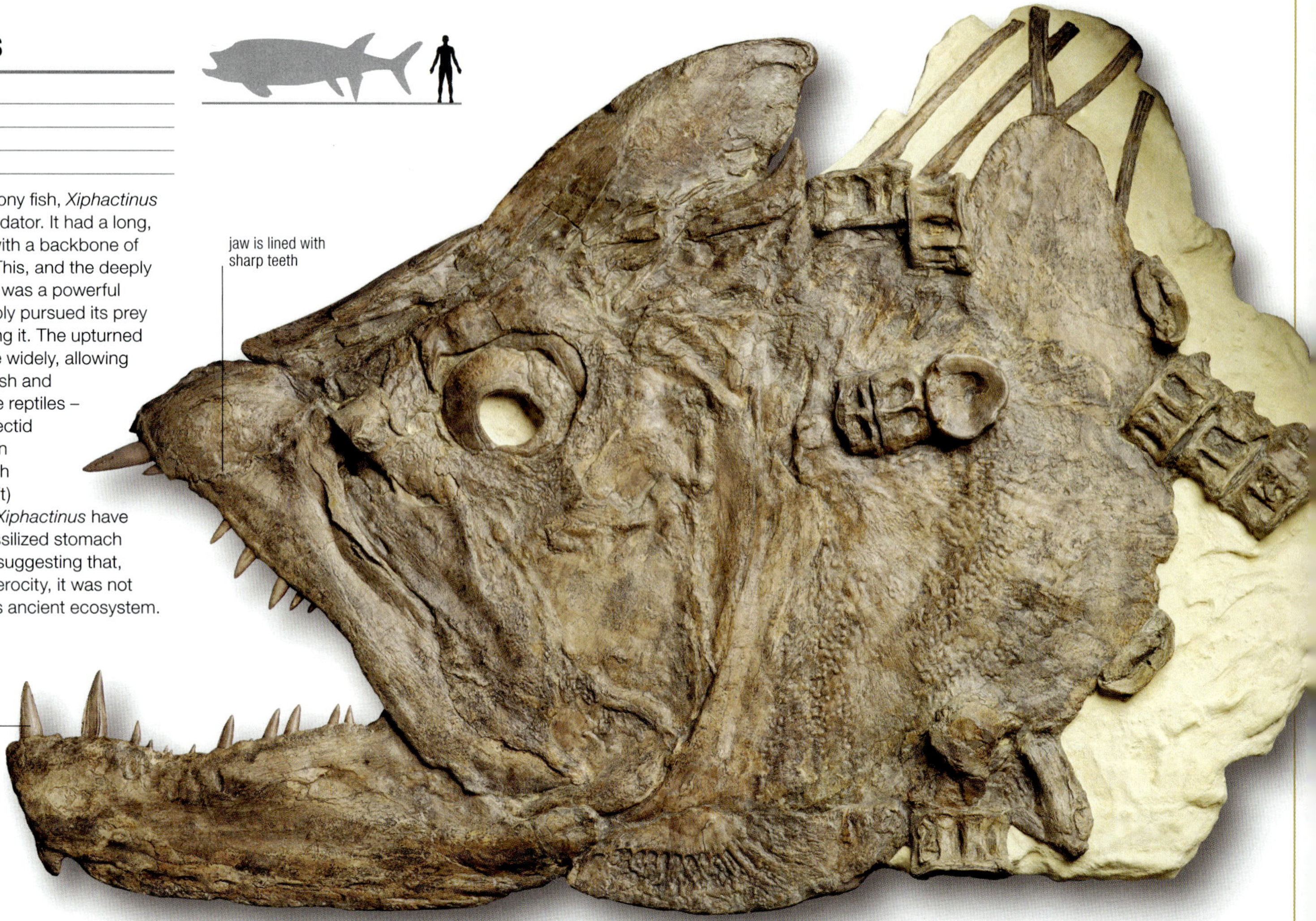

FANGED FISH
This species had fang-like front teeth for impaling, holding, and injuring their prey.

Squalicorax

GROUP Chondrichthyans
DATE Cretaceous
SIZE 5m (16½ft) long
LOCATION Europe, North America, South America, Africa, Near East, India, Japan, Australia, Russia

Squalicorax means "crow shark". Like *Cretoxyrhina*, it is an extinct member of the mackerel shark group. It had a typical shark body shape, and it's teeth were similar in outline to those of a modern tiger shark. It was a top predator and probably fed on mosasaurs, turtles, and fish. Isolated teeth have been found alongside a skeleton of *Cretoxyrhina*, which suggests it may also have scavenged on the carcasses of its larger cousin.

THORNY TOOTH
This tooth has a roughly rectangular root supporting a thorn-like crown, with a single, serrated, blade-like cusp.

Lepisosteus

GROUP Actinopterygians
DATE Eocene to present
SIZE 75cm (30in) long
LOCATION North and Central America, Cuba

Lepisosteus (gar pikes) first appeared about 110 million years ago. Today they are found in freshwater habitats in North and Central America and Cuba. They are similar in appearance to lizardfish, and it is possible they evolved in parallel to suit particular niches that existed in both saltwater (lizardfish) and freshwater (gar pike) environments. Modern gars, such as *L. oculatus*, have changed little from their Early Cretaceous ancestors, which makes them "living fossils".

LIVING RELATIVE
GAR PIKES

There are seven existing species of gar pikes, but they have changed little from their fossil ancestors. They are easily recognized by their long snouts with nostrils at the tip, and body scales with an enamel-like covering. All species are found in fresh and brackish water, except *Lepisosteus platostomus*, which is only found in brackish water.

DIAMOND SCALES
Lepisosteus had a long body, with dorsal and anal fins situated far down the back. It had a long snout, jaws lined with small, sharp teeth, and heavy, interlocking, diamond-shaped scales over its body.

Beelzebufo

GROUP Amphibians
DATE Late Cretaceous
SIZE 40cm (15½in) long
LOCATION Madagascar

Beelzebufo ("devil toad"), was only discovered in 2008. Its most remarkable feature is its size, which is far larger than any other known frog or toad, living or fossil. It would have co-existed with the last dinosaurs, being large enough to eat their newly hatched young. *Beelzebufo* is a close relative of the living South American horned frogs that ambush their prey, grabbing small animals as they pass with their huge mouths and the large, fang-like spikes of bone in their jaws. Its resemblance to modern South American frogs, rather than living Madagascan frogs, is yet more support for a past connection between South America, India, and Madagascar in the Cretaceous.

Trionyx

GROUP Turtles
DATE Cretaceous to present
SIZE 1m (3ft 3in) long
LOCATION Worldwide

Trionyx is a large, soft-shell turtle that survives today in Africa and the Middle East. Like other soft-shell turtles, it has a flat upper shell (carapace) that is covered with skin and lacks bony plates (scutes). The skulls of soft-shell turtles are unusual among living turtles as they are extremely long and very narrow, which is an ideal design for their ambush-hunting lifestyle. *Trionyx* has been around for a remarkably long time. Members of the genus have been in existence since the Cretaceous, surviving the devastating mass extinction 65 million years ago that wiped out the dinosaurs. Fossils of *Trionyx* are rare, but some exceptional specimens, such as the one above, have been found in the Eocene Green River Formation in Wyoming, USA (see p.375).

Protostega

GROUP Turtles
DATE Late Cretaceous
SIZE 3m (9¼ft) long
LOCATION USA

During the Late Cretaceous, North America was almost cut in two by a vast inland sea that stretched from the Arctic Ocean all the way to the present day Gulf of Mexico. A strange menagerie of prehistoric life inhabited these warm waters, including enormous fish, toothed birds, and a colossal turtle called *Protostega*. It was one of the largest turtles known to exist, reaching lengths of around 3m (9¼ft) and weights of several hundred kilograms, but it was only the second largest turtle of the Cretaceous, after *Archelon*. *Protostega* was fully marine and rarely ventured onto land. Its limbs functioned as effective paddles, and its thick shell was relatively light and streamlined so that it could travel quickly through the water. The skull of *Protostega* was characteristic of a turtle: it was short, wide, lacked teeth on its snout, and sported a sharp beak.

FAVOURITE FOOD
Although *Protostega* surely fed on fish, it is likely that it targeted jellyfish, squid, and other soft creatures as its favoured prey.

Mosasaurus

GROUP Lepidosaurs
DATE Late Cretaceous
SIZE 15m (49ft) long
LOCATION USA, Belgium, Japan, Netherlands, New Zealand, Morocco, Turkey

For about 20 million years during the Late Cretaceous, the oceans teemed with one of the most spectacular groups of predators to ever evolve: the mosasaurs. These were gigantic relatives of today's lizards and snakes that were adapted for life in the sea. *Mosasaurus* was a voracious, crocodile-like hunter that swam by undulating its long body. As a result, it was incapable of swimming fast over long distances, but it could accelerate very rapidly when required. It probably lived in the well-lit surface waters of the oceans, hunting slower-moving prey. Its bite marks have been found on ammonites and on the shells of large turtles, suggesting that it was capable of catching sizeable prey. The first *Mosasaurus* skull was discovered in 1774 in Maastricht, Holland in a limestone quarry.

FEARSOME LIZARD
Mosasaurs, including *Mosasaurus*, were some of the largest and most fearsome lizards in Earth's history. However, they only survived for a brief period of time, as they only appeared in the Late Cretaceous and were wiped out along with the dinosaurs in the Cretaceous extinction.

Elasmosaurus

GROUP Sauropterygians
DATE Late Cretaceous
SIZE 9m (30ft) long
LOCATION USA

Elasmosaurus was one of the last plesiosaurs. More than half its total body length was made up of the neck, which contained 71 vertebrae – more than any other animal that has ever existed. This was a source of great confusion for Edward Drinker Cope, who described the first specimen in 1869, as he placed the head on the end of the tail, which he had misidentified as the neck. *Elasmosaurus* had a relatively small head and probably ambushed fish by manoeuvring its neck. Its long, narrow teeth would have been perfect for piercing and trapping small, soft prey.

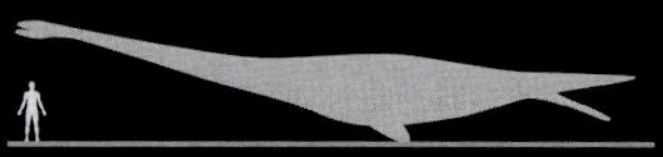

LONGEST NECK
Elasmosaurus belonged to a subgroup of plesiosaurans known as the plesiosauroids. All of them possessed small heads and very long necks, although none of them had a neck as long as *Elasmosaurus*.

BIOGRAPHY

EDWARD DRINKER COPE

Born in Philadelphia, USA, Edward Drinker Cope (1840– 97) was a palaeontologist, herpetologist, and evolutionist. He led numerous natural history surveys and fossil hunting expeditions across western North America, and described more than 1,000 vertebrate species, including a number of dinosaurs. Even so, Cope is most famous for his intense rivalry with fellow paleontologist Othniel Marsh, which has been dubbed the "Bone Wars".

Kronosaurus

GROUP Sauropterygians
DATE Late Cretaceous
SIZE 9m (30ft) long
LOCATION Australia, Colombia

Named after Kronos, a Titan of Greek legend, *Kronosaurus* was among the last of the giant pliosauroids, a group of plesiosaurians with large heads and short, stocky necks. However, recent work shows that *Kronosaurus* was not as enormous as was once thought and its estimated length has been downgraded from 12m (39ft) to 9m (30ft). Like other pliosauroids, it used four large flippers to swim, but whether it used these to "row" like the oars of a boat or "fly" underwater like a sea turtle is not known. It is possible that the answer is somewhere in-between.

POWERFUL JAWS
The powerful, crocodile-like jaws of *Kronosaurus* allowed it to attack other marine reptiles, but most of its diet probably consisted of large fish.

Plioplatecarpus

GROUP Lepidosaurs

DATE Late Cretaceous

SIZE 5–6m (16½–20ft) long

LOCATION North America, Europe

One of the best-known mosasaurs is *Plioplatecarpus*, a mid-sized predator that lived in the warm, shallow seas of North America and Europe approximately 80 million years ago. Its skull was long and powerful and studded with a series of thick, conical teeth like those of living crocodiles. Its body was elongated and streamlined, the hands and feet were modified into broad flippers, and the tail was deep and muscular. The jaws of this mosasaur were able to open very wide, enabling it to bite and swallow prey that was much larger than itself. This behaviour is also seen in living snakes and is one of the features that marks out the two groups as close relatives.

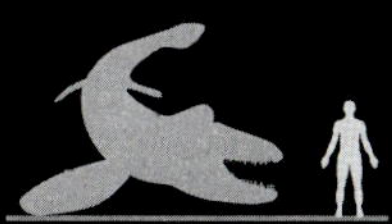

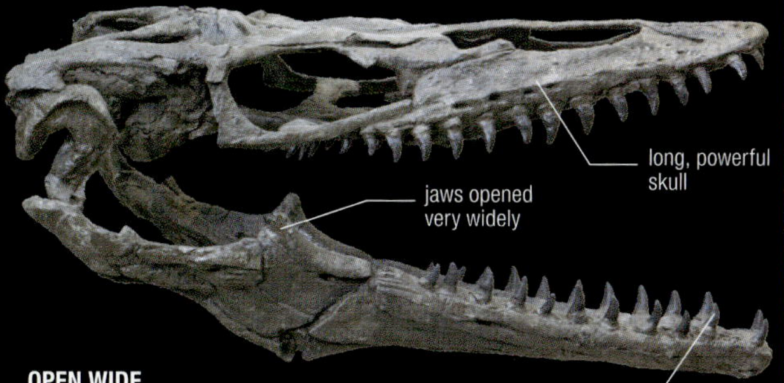

OPEN WIDE
Its thickset skull and massive jaws were able to open very wide, enabling it to catch large prey.

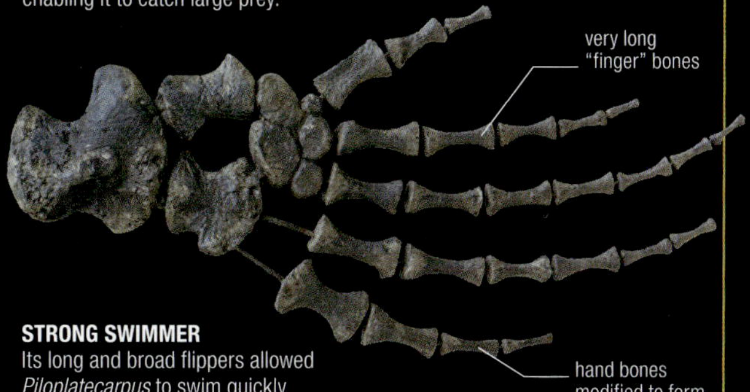

STRONG SWIMMER
Its long and broad flippers allowed *Plioplatecarpus* to swim quickly and powerfully.

Simosuchus

GROUP Crocodylomorphs
DATE Late Cretaceous
SIZE 1.5m (5ft) long
LOCATION Madagascar

Simosuchus is perhaps the strangest crocodylomorph to ever live. Most members of this group were powerful predators with strong, elongated skulls crammed full of sharp teeth. *Simosuchus*, however, had a short skull with a flattened face that resembled a pug dog. Not only that, but it also had a wide snout with a series of leaf-shaped teeth, perfect for shearing and chewing plants. This peculiar, herbivorous crocodile lived in Madagascar during the Late Cretaceous, about 70 million years ago. It lived alongside more characteristic crocodiles that ate meat, as well as many large theropod dinosaurs. *Simosuchus* may have avoided competition with these carnivores by evolving a plant-based diet. Other herbivorous crocodylomorphs also lived during the Cretaceous, although they were rare and none survive today. They shared ecosystems with herbivorous dinosaurs, which was a strange ecological pairing.

STRANGE CREATURE
Although the pug-nosed, plant-eating *Simosuchus* did not resemble the crocodiles of today, it was a close relative of living crocodiles, but with a comparatively bizarre lifestyle and strange body plan.

Deinosuchus

GROUP Crocodylomorphs
DATE Late Cretaceous
SIZE 12m (39ft) long
LOCATION USA, Mexico

The name *Deinosuchus* means "terror crocodile" – and for good reason. Along with *Sarcosuchus*, *Deinosuchus* was one of the largest crocodylomorphs to ever live, weighing up to 9 tonnes (10 tons). However, it lived much more recently than *Sarcosuchus*, and is a member of the alligatorid family that also includes the modern alligators. It was one of the most ferocious predators of the North American coastal regions, and in some areas overlapped with tyrannosaurids such as *Daspletosaurus* (see p.321). In these ecosystems it was *Deinosuchus*, not the tyrannosaurids, that was the largest and most powerful predator. Its anatomy and overall body plan was very similar to that of living crocodiles, so it is easy to imagine it was a giant version of living species. It probably hunted in a similar way to modern crocodiles, by lurking around the water's edge and preying on fish, marine reptiles, and the occasional land animal.

TERROR CROC
Deinosuchus was as long as *Tyrannosaurus*. It killed its prey in a similar way to modern crocodiles, by inflicting great wounds and dragging it under the water and drowning it.

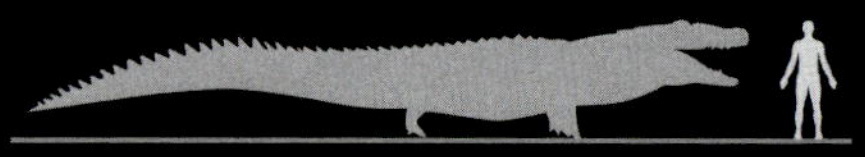

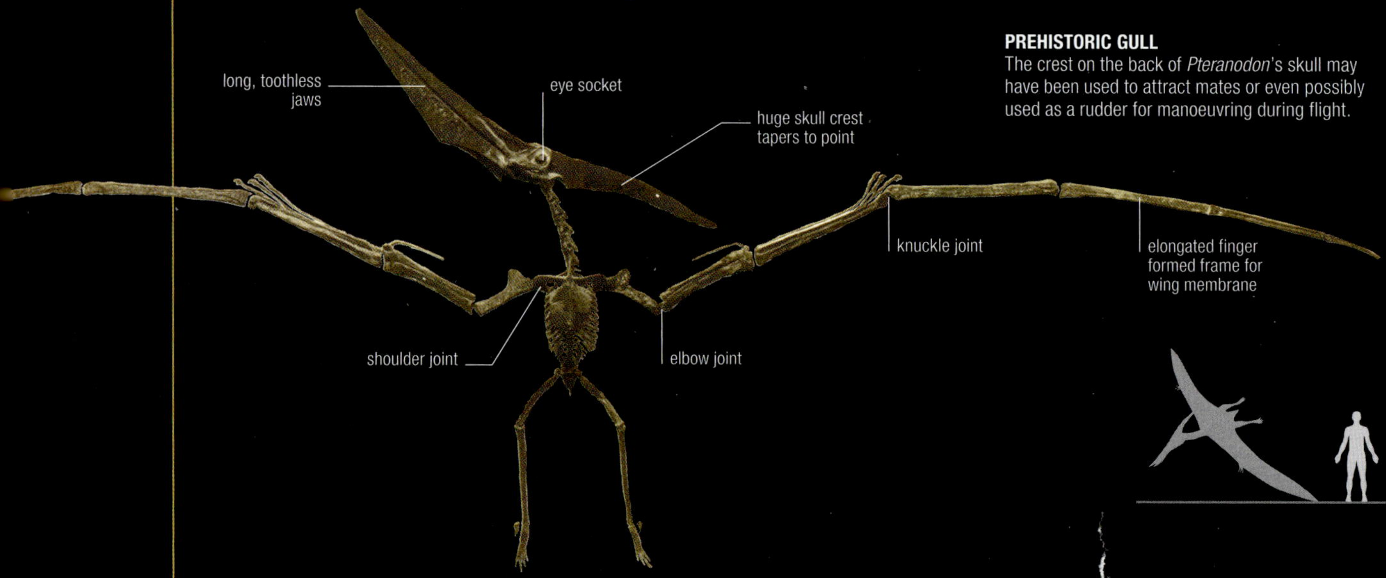

PREHISTORIC GULL
The crest on the back of *Pteranodon*'s skull may have been used to attract mates or even possibly used as a rudder for manoeuvring during flight.

Pteranodon

GROUP Pterosaurs
DATE Late Cretaceous
SIZE 1.8m (6ft) long
LOCATION USA

Pteranodon soared across the shallow seas of North America during the Late Cretaceous. It is likely that it flew and hunted in the same way as an albatross. Vast flocks of them probably glided over the ocean while looking for fish in the surface waters. It is certain that they ate fish, as fish bones have been found in the fossilized stomach of one specimen. The skull of *Pteranodon* was also well adapted for fishing, with long, toothless jaws and a streamlined skull for diving into the water.

Quetzalcoatlus

GROUP Pterosaurs
DATE Late Cretaceous
SIZE 12m (39ft) wingspan
LOCATION USA

The colossal Late Cretaceous pterosaur *Quetzalcoatlus* was the largest flying animal of all time. Its wingspan was larger than many small planes, and its 2.5m (8¼ft) skull would dwarf even the tallest basketball players. Despite its monstrous size, however, *Quetzalcoatlus* weighed no more than 250kg (550lb), thanks to a complex system of air sacs inside most of its bones. For a long time scientists thought that most pterosaurs ate fish and spent much of their time gliding over the sea, only venturing over land to catch small mammals and lizards. However, it is now believed that *Quetzalcoatlus* and its azhdarchid relatives may have spent most of their time flying over land and targeting large vertebrates as prey. It is likely, therefore, that *Quetzalcoatlus* fuelled its enormous metabolic needs by stalking and feeding on dinosaurs.

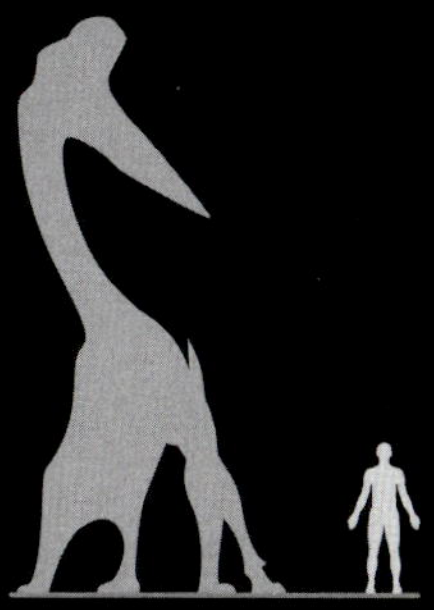

Quetzalcoatlus was **named after** the ancient Aztec **feathered serpent god** Quetzalcoatl, who was the patron god of Aztec priests.

MONSTER PTEROSAUR
Quetzalcoatlus is the most fantastic of all of the flying pterosaurs. This monstrous creature, named after an Aztec god, was larger than many small planes. While most other pterosaurs ate fish, *Quetzalcoatlus* was a fierce predator that hunted dinosaurs and other vertebrates.

Ornithocheirus

GROUP Pterosaurs

DATE Early Cretaceous

SIZE 8–10m (26–33ft) wingspan

LOCATION Europe, South America

In the Early Cretaceous, about 40 million years before the gigantic *Quetzalcoatlus* stalked dinosaurs across the North American plains (see p.313), another giant pterosaur ruled supreme in Europe and South America. This large beast, *Ornithocheirus*, is also sometimes referred to as *Tropeognathus*, an older name that is now considered incorrect. *Ornithocheirus* has long caused palaeontologists headaches, as it is only known from fragmentary fossils. However, these bones suggest that its wingspan may have approached 10m (33ft), about the same size as the more familiar *Quetzalcoatlus*. A close relative of *Ornithocheirus*, a smaller pterosaur from South America called *Anhanguera*, is known from exceptionally well-preserved fossils that have even allowed scientists to carefully study details of the bird's powerful brain. *Anhanguera* clearly had keen senses, and a strong sense of balance, all of which are necessary for flight, which is a complex and dangerous activity.

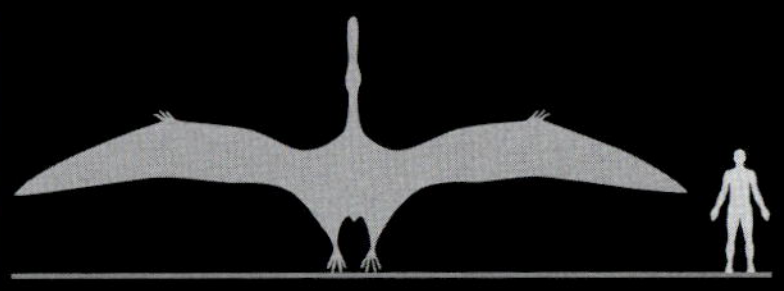

More than **ten species** have been assigned to the genus *Ornithocheirus*, but most of these are based on **fragmentary scraps** of fossils that are **difficult to study.**

SKY GIANTS
Ornithocheirus is one of the most mysterious and majestic pterosaurs, but it has also been a source of frustration for palaeontologists for many years. It is only known from small pieces of fossil, but these indicate that *Ornithocheirus* was one of the largest flying animals to ever live, and approached the size of a small aeroplane.

Suchomimus

GROUP Theropods
DATE Early Cretaceous
SIZE 9m (30ft) long
LOCATION Niger

Fragmentary fossils had long suggested that a close relative of *Baryonyx* (right) lived in Niger, in western Africa. This was confirmed in 1998 with the naming of *Suchomimus tenerensis*. A low, blade-like crest extended along the snout's upper surface, and a tall ridge ran along its back and probably its tail. *Suchomimus*'s arms were stout and the large muscle attachments indicate they were powerful.

CROCODILE MIMIC
Suchomimus means "crocodile mimic", so named because the animal had a very long, narrow, crocodile-like snout.

Baryonyx

GROUP Theropods
DATE Early Cretaceous
SIZE 9m (30ft) long
LOCATION British Isles, Spain, Portugal

Baryonyx walkeri was discovered in 1983 by an amateur palaeontologist, and it proved to be one of Europe's most interesting dinosaur fossils. Crocodile-like jaws and teeth were combined with a fairly typical theropod skeleton. *Baryonyx* is a spinosaurid and belongs to a family named after *Spinosaurus* (see p.316). *Spinosaurus* is known for its crest of tall spines running along its backbone. When first discovered, *Baryonyx* was thought to possess only short spines on the tops of its vertebrae. However, a new specimen has shown that these spines were longer, although still not as long as those of *Spinosaurus*. It seems that *Baryonyx*, and perhaps all spinosaurids, were specialized fish-eating dinosaurs that also ate other prey, including small dinosaurs.

FIRST DISCOVERY
The original *Baryonyx* specimen was found laying on its side. It died by the edge of a pool of water, and its remains were later buried by mud.

ANATOMY
LONG SKULL

Baryonyx had a long, crocodile-like skull. Its nostrils were positioned further back from the snout tip, and its upper jaw was curved. The skull shape suggests it was a fish-eating theropod that could plunge its long jaws into water.

Irritator

GROUP Theropods
DATE Early Cretaceous
SIZE 8m (26ft) long
LOCATION Brazil

A close relative of *Spinosaurus* (see p.316), *Irritator* was named in 1996 when a near-complete skull was found in the Cretaceous rocks of Brazil. The skull had been modified by its finder to look like a pterosaur skull. As a result, the scientists who studied the specimen were initially misled about its identity, and the name they chose reflects their irritation at having been deceived. As with *Spinosaurus*, a bony crest was present on the upper surface of *Irritator*'s snout. Unlike the crest of *Spinosaurus*, *Irritator*'s extended over the eye sockets. The nostril openings were well away from the tip of the snout, and the teeth had conical crowns. As no skeleton of *Irritator* has been found to date, little is known of the biology or behaviour of this animal. Like other spinosaurids, it might have caught fish. It could also have eaten carrion and land animals. Evidence for this feeding behaviour comes from the discovery of a spinosaurid tooth embedded within a pterosaur neck bone.

Spinosaurus

GROUP Theropods
DATE Late Cretaceous
SIZE 16m (52ft) long
LOCATION Morocco, Libya, Egypt

Spinosaurus is one of the most famous theropods and is also estimated to have been the largest. Unfortunately, the first and best specimen of *Spinosaurus* discovered was destroyed by an Allied bombing raid on Germany during the Second World War. This specimen included a lower jaw and well-preserved vertebrae with tall spines. Since that original discovery by Ernst Stromer (see panel, left), only a few articulated remains have been unearthed. However, a huge number of widespread, partial remains show that *Spinosaurus* was a relatively common animal in the Late Cretaceous. In many ways, *Spinosaurus* was typical of a large theropod. However, in two regions of anatomy – its skull and vertebrae – it was certainly not typical. The snout was long, like that of a crocodile, with the nostril openings placed well back from the snout tip and closer to the eyes than was usual for theropods. The tip of the upper jaw was expanded and rounded relative to the rest of the snout, and the teeth in this area radiated outwards like the spokes of a wheel. The teeth were rounded in cross section (whereas most other theropods had oval teeth), and serrations were absent from the teeths' keels. These features all suggest that *Spinosaurus* plunged its jaws into the water and caught fish. However, it was large and powerful enough to also prey on small- and medium-sized dinosaurs on land.

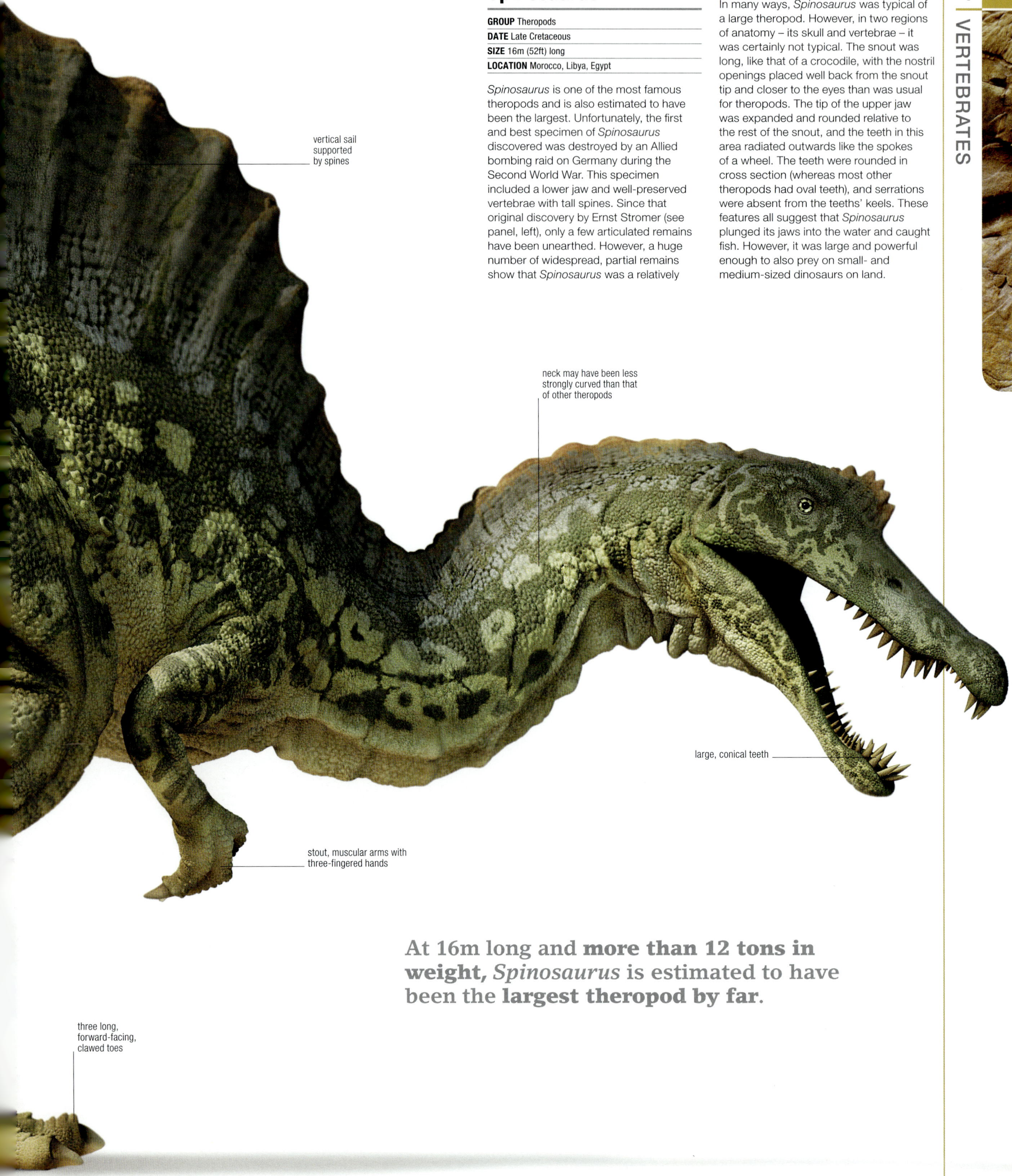

At 16m long and **more than 12 tons in weight,** ***Spinosaurus*** is estimated to have been the **largest theropod by far**.

SPINE LIZARD

This immense theropod had a sail-like structure running down its back. The sail was supported by tall, bony spines that grew upwards from the vertebrae. The function of the sail is unknown, but it might have been used in display or helped the animal control its body temperature.

BIOGRAPHY

ERNST STROMER VON REICHENBACH

Spinosaurus is one of several dinosaurs described by German palaeontologist Ernst Stromer von Reichenbach (1870–1952). Working together with fossil collector Richard Markgraf (1869–1916), Stromer explored the rocks of northern Africa in 1910 and 1911. He hoped to discover Eocene mammals, but instead discovered Cretaceous dinosaurs. In the Great Western Desert of Egypt, Stromer found numerous bones that he shipped back to Munich, Germany. These fossils were destroyed by a bombing raid during the Second World War. Stromer had made requests to the head of the museum where they were housed that the fossils be moved away somewhere safer, but this never happened. Tragically for Stromer, he also lost two of his sons in the war, while the third was captured by the Soviets.

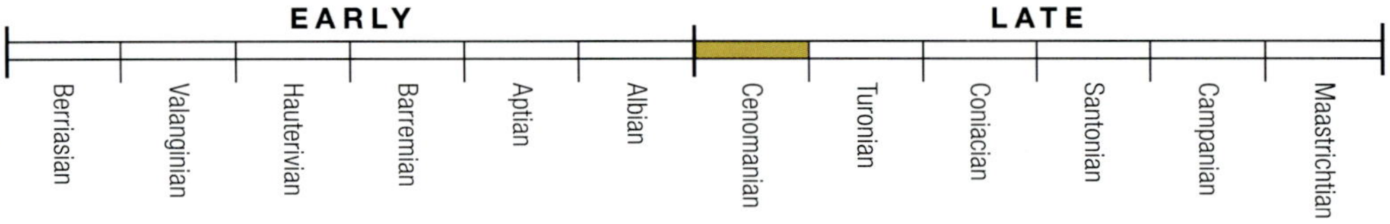

HEAVY CLAW
Baryonyx had a massive curved thumb claw – its name means "heavy claw" – and its upper arm bone (humerus) had attachment sites for huge muscles.

TWO LEGS
Experts first proposed that *Baryonyx* might have walked on all fours, but today we know that it was bipedal like other theropods.

TALL SAIL
It is assumed that, like *Spinosaurus*, *Irritator* had a tall sail along its back. However, this remains speculative as *Irritator* is only known from its skull.

Scientists who studied *Irritator* were **initially misled** about its identity. The name they chose reflects their **irritation** at having been deceived.

Carcharodontosaurus

GROUP Theropods
DATE Late Cretaceous
SIZE 11m (36ft) long
LOCATION Morocco, Tunisia, Egypt

Carcharodontosaurus was a giant African allosauroid theropod named in 1931. Its serrated teeth resemble those of *Carcharodon*, the great white shark, hence the dinosaur's name. *Carcharodontosaurus* had a narrow head that was taller at the back than at the front of the snout. A recently discovered partial skull of *Carcharodontosaurus* was more than 1.6m (5¼ft) long. Bony ridges on the top of the skull overhung the eyes, and the bones on the sides of its skull had a distinctive wrinkled texture. Using its massive jaws and long teeth, *Carcharodontosaurus* may have preyed on sauropods and other dinosaurs.

MASSIVE SKULL
With its huge frame and sharp teeth, *Carcharodontosaurus* was certainly an accomplished predator.

Giganotosaurus

GROUP Theropods
DATE Late Cretaceous
SIZE 12m (39ft) long
LOCATION Argentina

POWERFUL NECK
Remains suggest that the neck of *Gigantosaurus* was stout and powerful, and supported a large head.

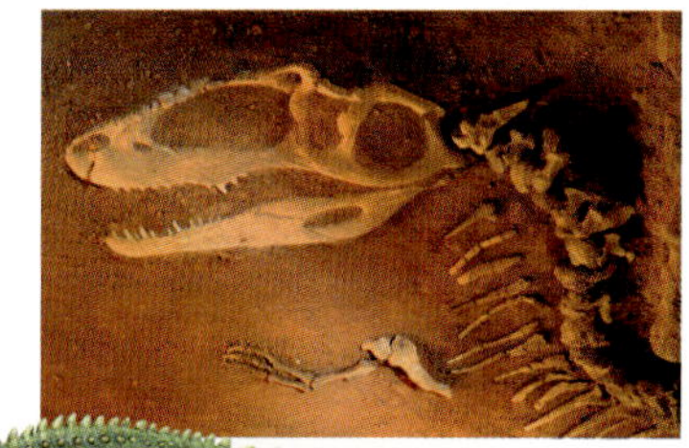

Giganotosaurus (meaning "giant southern lizard") was similar in size to the largest known individuals of *Tyrannosaurus* (see p.322). Its skull and skeleton also strongly resembled *Carcharodontosaurus* (above). The bones above and in front of the eye of *Giganotosaurus* had low, horn-like projections. *Giganotosaurus* lived alongside the sauropods *Limaysaurus*, *Andesaurus*, and *Argentinosaurus* (see p.332), and may have preyed on them.

DISTINCTIVE FACE
Shelf-like bony ridges featured above the eye socket of *Giganotosaurus*. Its lower jaw was squared-off with a distinct chin-like bony lump.

LETHAL JAWS
The jaws of *Acrocanthosaurus* were its primary killing weapon. Its skull was vaguely triangular, the eye sockets were narrow and a large bony boss was located over each eye.

Acrocanthosaurus

GROUP Theropods
DATE Early Cretaceous
SIZE 12m (39ft) long
LOCATION USA

Acrocanthosaurus was a giant North American allosauroid, closely related to the African *Carcharodontosaurus* (above). It had a wide, muscular ridge of tall spines running along its back that would have stiffened if *Acrocanthosaurus* had latched on to a large prey animal. This stiffening might have helped *Acrocanthosaurus* to anchor the weight of its body as it pulled and tore at its prey. Its powerful arms had a limited range of movement, but its fingers were equipped with massive curved claws that could withstand a lot of bending. This suggests that its hands might have grabbed prey, but it used its jaws for killing.

Aucasaurus

GROUP Theropods
DATE Early Cretaceous
SIZE 4m (13ft) long
LOCATION Argentina

Aucasaurus was closely related to *Carnotaurus* (see opposite) and they are united in a group called the Carnotaurini. Its short, deep-snouted skull was not as short or deep as that of *Carnotaurus*. Instead of horns, *Aucasaurus* had low swellings over the top of each eye. *Aucasaurus*'s small arms were also similar to those of its horned relative, but were proportionally longer, and the bones lacked some of the unusual proportions and extra bony processes present in *Carnotaurus*. The hand of *Aucasaurus* was unusual: four metacarpal bones were present, but the first and fourth lacked fingers. The second and third were attached to short fingers, but these lacked claws. *Aucasaurus* is known from findings in the Rio Colorado Formation – Late Cretaceous rocks in Argentina that have yielded many dinosaurs, including the theropods *Alvarezsaurus* and *Velocisaurus*, and the sauropod *Neuquensaurus*. Numerous sauropod eggs are also known from this deposit.

Known from a **beautifully preserved, near-complete skeleton** that is only **missing the end half of the tail,** *Aucasaurus* was named in 2002.

LIKELY PREDATOR
Little is known about the behaviour of *Aucasaurus*. However, like most large theropods, it was almost certainly a predator of other dinosaurs. Perhaps it preyed on smaller theropods and on ornithischians.

Carnotaurus

GROUP Theropods
DATE Late Cretaceous
SIZE 9m (30ft) long
LOCATION Argentina

Carnotaurus, which means "meat-eating bull", was named in 1985 from a well-preserved partial skeleton. Its skull was short and deep for a theropod, and prominent thick, blunt-tipped horns projected from the skull roof above the eyes, which may have been used in display or in fighting. The puny arms, which seemed to lack an obvious function, may have been used in display. Skin impressions preserved with the skeleton show that large, keeled scutes were arranged in rows on the neck and body. Until recently, *Carnotaurus* was the best-known member of the Abelisauridae family. However, the Madagascan *Majungasaurus* is now known from even more complete remains.

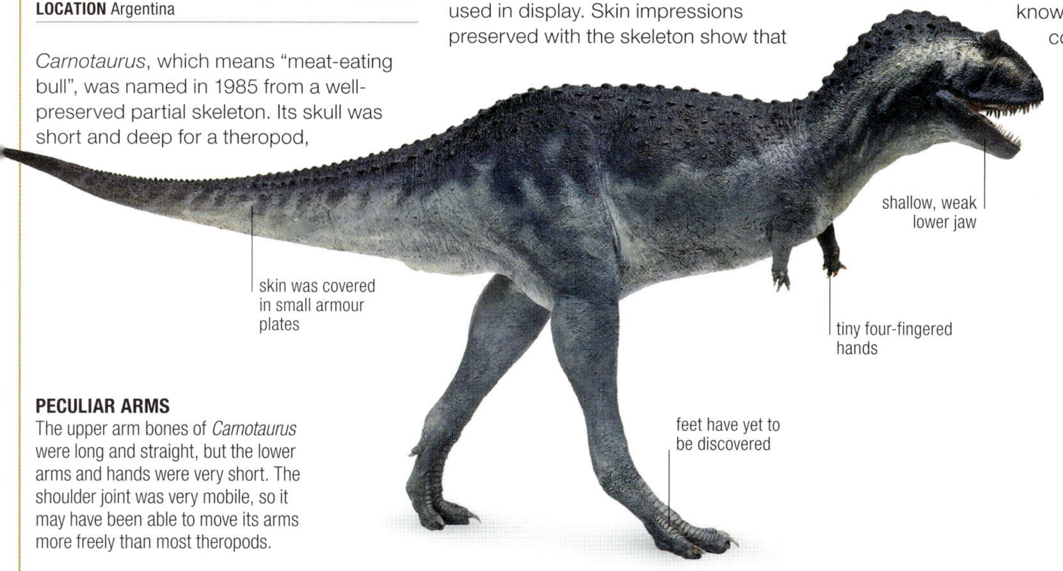

PECULIAR ARMS
The upper arm bones of *Carnotaurus* were long and straight, but the lower arms and hands were very short. The shoulder joint was very mobile, so it may have been able to move its arms more freely than most theropods.

ANATOMY
BLUNT HORNS

Carnotaurus is famous for possessing blunt horns over its eyes. In life, these were covered by horny sheaths, so their shape in the live animal might have been different from their shape in the fossil: they may have been longer, or more pointed, for example. It remains unknown what the horns were for. Some palaeontologists suggest that the horns could have been used to intimidate members of other theropod species. Others think that they were used in disputes between rival males.

Santanaraptor

GROUP Theropods
DATE Early Cretaceous
SIZE 3m (9¾ft) long
LOCATION Brazil

The single specimen of this small, poorly known theropod consists of bones from the pelvis, hind limbs, and tail. These provide little information on its overall appearance. However, it was definitely a coelurosaur, and a few of its details suggest that it might be an ancestor of the tyrannosaurids. It is presumed to be similar to animals such as *Dilong* and *Guanlong* (see p.262) and had long arms, three-fingered hands, and slim hind limbs. Fragments of preserved muscle and skin tissue were discovered, but unfortunately there was no trace of the skin's external covering.

Daspletosaurus

GROUP Theropods
DATE Late Cretaceous
SIZE 9m (30ft) long
LOCATION North America

Large and robust, *Daspletosaurus* was a close relative of *Tyrannosaurus*, but was geologically older. Its skull was proportionally larger and longer than that of most tyrannosaurids, although not as large, deep, or wide as that of *Tyrannosaurus*. Unlike *Tyrannosaurus*, it had a short, triangular hornlet above and in front of the eyes, a snout with a bumpy, gnarled upper surface, and broad cheeks. A blunt, horn-like lump projected sideways from beneath the eye socket. These features were typical of large tyrannosaurids.

STRONG JAW
The lower jaw of *Daspletosaurus* was powerfully built. The teeth were thick and robust, with long roots that were well embedded in the sturdy jaw bones.

Albertosaurus

GROUP Theropods
DATE Late Cretaceous
SIZE 9m (30ft) long
LOCATION Canada

Albertosaurus was closely related to another North American tyrannosaurid, *Gorgosaurus*. In the past, it was thought that the two were similar enough to be included in the same genus. However, they are currently regarded as distinct because they differ in various skull details. *Albertosaurus* also seems to have had more slender hind limbs and proportionally smaller forelimbs than *Gorgosaurus*, but both were similar in size overall. There is an *Albertosaurus* bone bed that contains the remains of numerous juvenile and adult individuals, a discovery that suggests that *Albertosaurus* was a social animal.

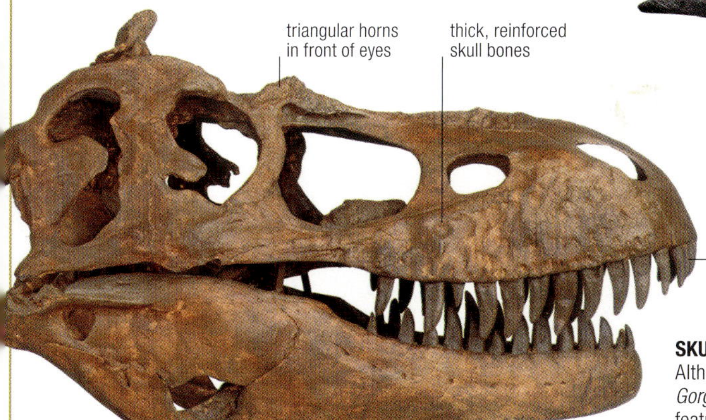

SKULL ANATOMY
Although *Albertosaurus* and *Gorgosaurus* share anatomical features, some skull anatomy of *Albertosaurus* is more like that of *Daspletosaurus* and *Tyrannosaurus*.

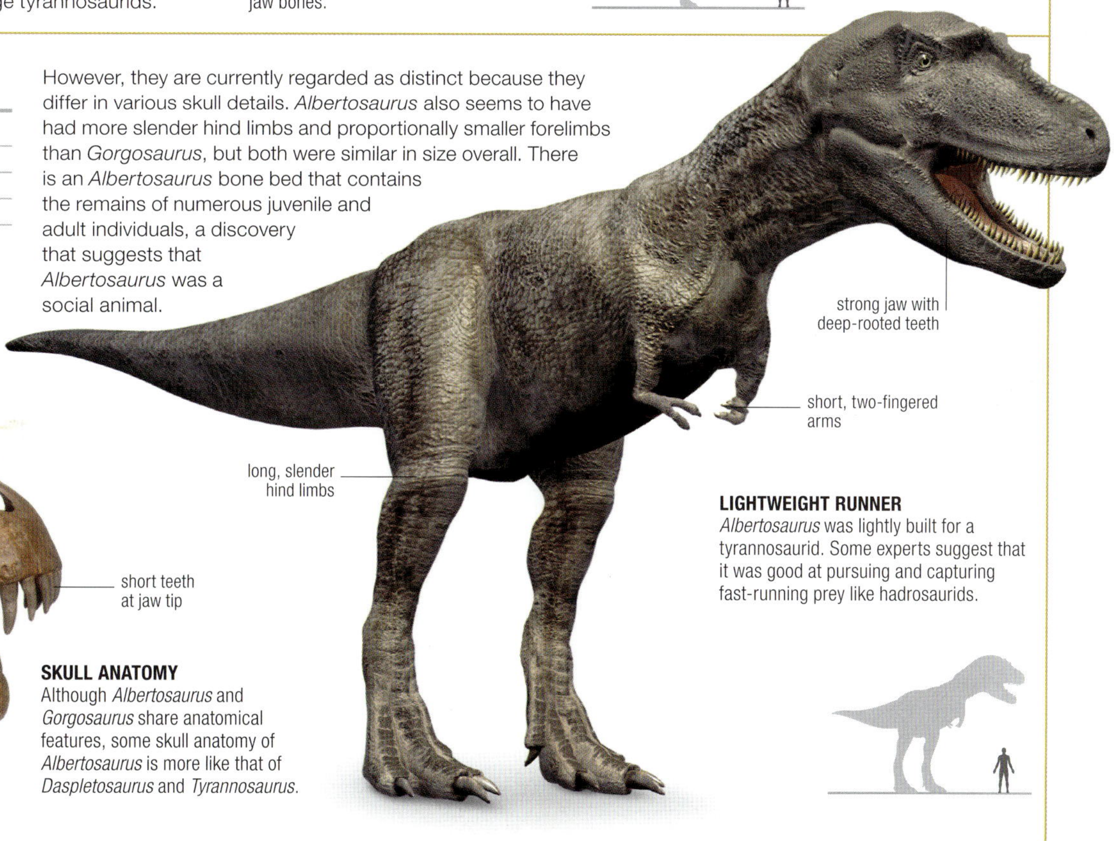

LIGHTWEIGHT RUNNER
Albertosaurus was lightly built for a tyrannosaurid. Some experts suggest that it was good at pursuing and capturing fast-running prey like hadrosaurids.

The strength of ***Tyrannosaurus's*** **bite** is estimated to be greater than that of **any other animal.**

BODY CAVITY
Although it had a very heavily built body, and a wide body cavity, some of *Tyrannosaurus*'s back vertebrae had holes to reduce their weight.

LIZARD HIPS
Tyrannosaurus was a saurischian, or "lizard-hipped" dinosaur, with a hip bone arrangement similar to that of most modern reptiles.

CLAWED FINGERS
Tyrannosaurus had very short arms, and hands with two prominent fingers and one vestigial (reduced) finger. The two larger fingers bore sharp, curved claws, although their function was probably very limited.

MASSIVE MOUTH
The massive mouth of *Tyrannosaurus* contained up to 58 serrated teeth. These were different sizes, but the longest were about 15cm (6in) long. Those at the front of the jaws were more closely packed in than those towards the back.

Tyrannosaurus

GROUP Theropods

DATE Late Cretaceous

SIZE 12m (39ft) long

LOCATION North America

The genus *Tyrannosaurus* contains the most famous dinosaur, *Tyrannosaurus rex*. Originally discovered in the Late Cretaceous Hell Creek Formation of Montana, USA, it was widespread throughout western North America. *Tyrannosaurus* was one of the largest ever terrestrial carnivores, but whether it was an active hunter or just a scavenger is a subject of much debate. Worn tooth tips and bone fragments preserved in its fossilized faeces indicate that it routinely crushed and swallowed bones. Some scientists have argued that this, along with the fact that its arms were too short to grip prey, and that it is believed to have had a highly developed sense of smell, support the idea that it was a scavenger. However, most palaeontologists believe that *Tyrannosaurus* was probably an opportunistic carnivore that was able to actively hunt for prey when it needed to.

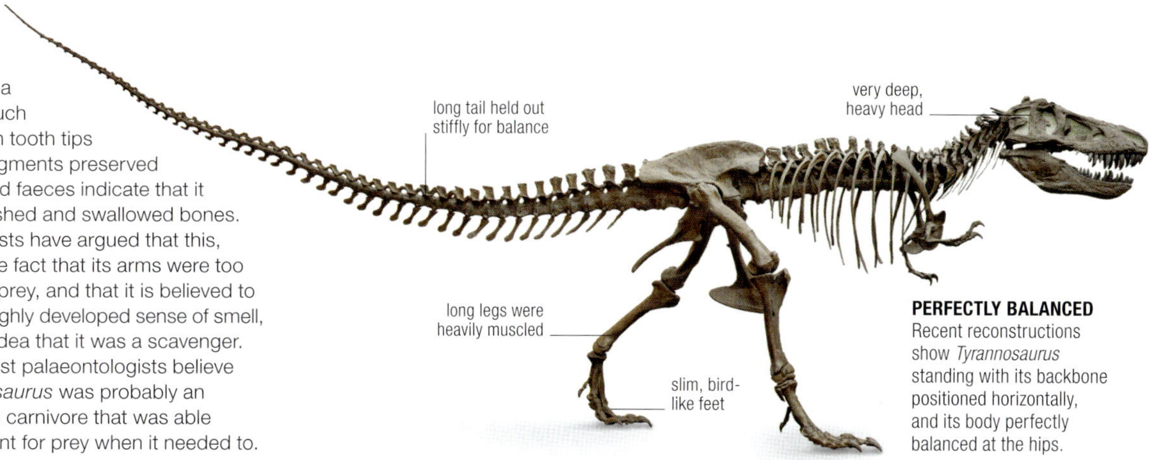

PERFECTLY BALANCED
Recent reconstructions show *Tyrannosaurus* standing with its backbone positioned horizontally, and its body perfectly balanced at the hips.

BULKY SKULL
Tyrannosaurus's snout and lower jaws were very deep, and the back of its skull was particularly broad, especially across the cheek region. The eye sockets faced forwards more than the eyes of any other tyrannosaurid, showing that *Tyrannosaurus* would have had acute binocular vision.

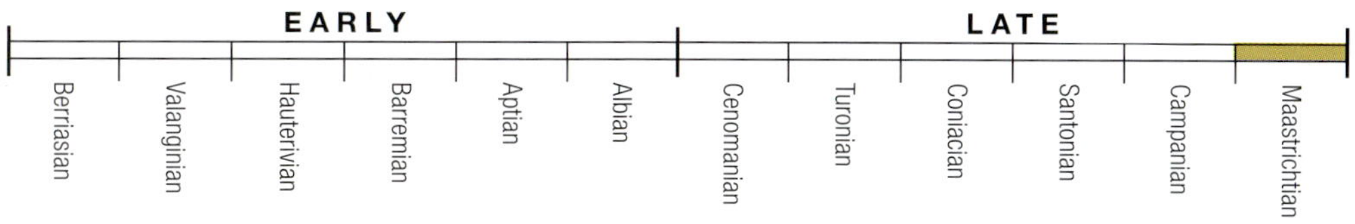

Tarbosaurus

GROUP Theropods
DATE Late Cretaceous
SIZE 12m (39ft) long
LOCATION Mongolia, China

Tarbosaurus was a large Asian tyrannosaurid, and a close relative of North America's *Tyrannosaurus* (see p.322). Indeed, several experts have proposed that they should be regarded as different species of the same genus. However, the two differ in many details, and *Tarbosaurus* could be more closely related to *Alioramus* and *Daspletosaurus* (see p.321). *Tarbosaurus* had a different skull and snout, and slightly more teeth than *Tyrannosaurus*. These differences might have evolved because the two predators relied on different prey. *Tyrannosaurus* lived alongside giant horned dinosaurs, while *Tarbosaurus* presumably hunted sauropods, hadrosaurs, and ankylosaurs. One *Tarbosaurus* fossil seems to preserve a throat sac beneath its lower jaw. This might have been used as an inflatable display structure during the breeding season.

POWERFUL PREDATOR
Tarbosaurus used its powerful jaws and teeth to attack herbivorous dinosaurs, such as the hadrosaur shown here. It may have attacked the flanks or thighs of larger prey, like sauropods.

The two-fingered forelimbs of *Tarbosaurus* were even smaller than those of its cousin *Tyrannosaurus*.

WEAKER SKULL
Tarbosaurus had a narrower, less strongly built skull than the similar *Tyrannosaurus*, and the bones along the top of its snout did not lock together in the same way.

Sinosauropteryx was the first "feathered" dinosaur to be discovered and caused a storm of interest when it was described in 1996.

Sinosauropteryx

GROUP Theropods
DATE Cretaceous
SIZE 60cm ($23\frac{1}{2}$in) long
LOCATION China

Like other members of the compsognathid family of theropods, *Sinosauropteryx* was a slender predator, with a long and pointed snout, that probably hunted for small vertebrate and invertebrate prey. What makes *Sinosauropteryx* fossils so significant is the discovery of a covering of fuzzy proto-feathers – an extremely important discovery because of the evolutionary position of compsognathids in relation to birds (see pp.304–05). Subsequent findings proved that some of the later theropods that shared a common ancestor with compsognathids inherited the same feathery features, meaning that even *Tyrannosaurus* (see p.322) and *Velociraptor* (see p.331) had feathers of some kind.

Alxasaurus

GROUP Theropods
DATE Early Cretaceous
SIZE 4m (13ft) long
LOCATION China

Alxasaurus is a therizinosaur, which were long-necked maniraptorans with leaf-shaped teeth, large hand claws, and broad feet with four toes. There is no direct evidence of how therizinosaurs lived, but it is likely that they used their hand claws to help hook leafy branches and stems in order to pull them closer to the mouth. The claws may also have been used in self-defence.

A full skull of *Alxasaurus* has never been found, but its lower jaw had a down-curved tip. Other therizinosaurs also had beaked jaw tips and may also have had cheek teeth at the back of the mouth.

LONG CLAWS
Alxasaurus had long, narrow hand claws, with strongly curved thumbs. These may have been used for feeding and possibly self-defence.

DISTINCTIVE FEATURES
Ornithomimus's distinctive features are its very large eye sockets, long legs, and its unusually similar-sized metacarpals and fingers.

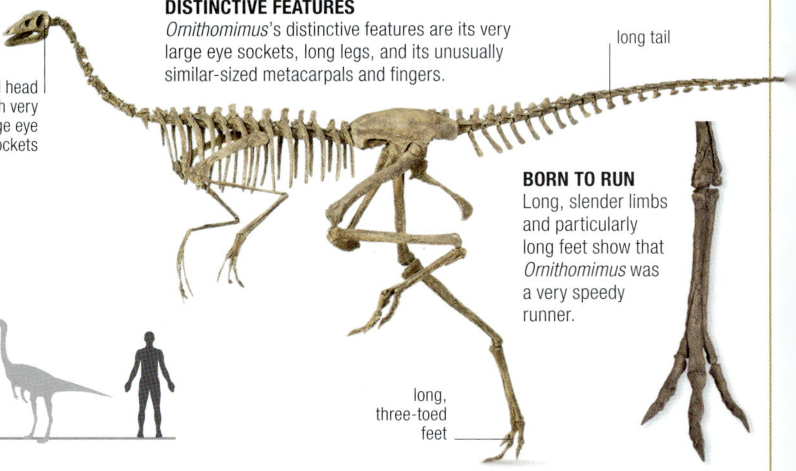

BORN TO RUN
Long, slender limbs and particularly long feet show that *Ornithomimus* was a very speedy runner.

Ornithomimus

GROUP Theropods
DATE Late Cretaceous
SIZE 3m ($9\frac{3}{4}$ft) long
LOCATION USA, Canada

Ornithomimus was the first of the ornithomimosaurs to be named, giving the group its name, which means "bird-mimic lizards". It was initially identified from just a hand and a foot, but is now known from many specimens. Like its close relatives *Struthiomimus* and *Gallimimus* (see opposite), it had a toothless beak and enormous eyes. In 2001, it was noted that vertical ridges on the beak's inner surface looked similar to structures used by ducks to filter food particles from water. However, these structures are also seen in other beaked animals that do not feed in this way.

ANATOMY

UNUSUAL FINGERS

Ornithomimus had unusually long, three-fingered hands. Its claws were slender and only gently curved. Its hands were also unusual in that the thumb was similar in length to the other two fingers. In other theropods, the thumb is usually shorter than the second finger. Because all three fingers were similar in length, it seems that they all worked in unison as a single, hook-like structure, possibly used to hook on to branches while feeding.

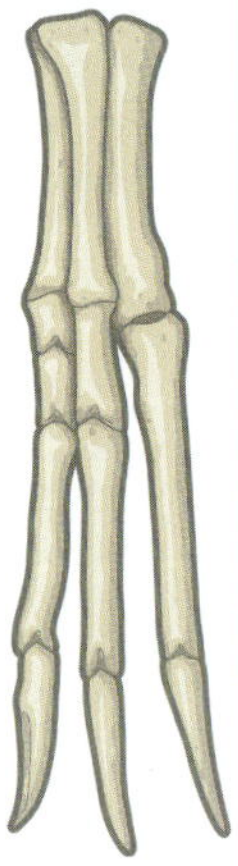

GROUNDBREAKING DISCOVERY
Sinosauropteryx was the first "feathered" dinosaur to be discovered and caused a storm of interest when it was described in 1996. Its proto-feathers seem to closely resemble the down feathers found on modern birds.

Struthiomimus

GROUP Theropods
DATE Late Cretaceous
SIZE 4.5m ($14\frac{3}{4}$ft) long
LOCATION Canada

Struthiomimus is one of several closely related ornithomimosaurs. The first *Struthiomimus* specimen was discovered in Alberta, Canada, and consisted of only pelvic and hind limb fragments. However, a far superior specimen, missing only the end of the tail and the skull roof, was discovered later on. *Struthiomimus* would have looked similar to *Ornithomimus* – its closest relative – but had a longer body and tail and shorter hind limbs. Its hands and hand claws were particularly long, and its thumbs did not oppose its fingers, reducing its grasping ability.

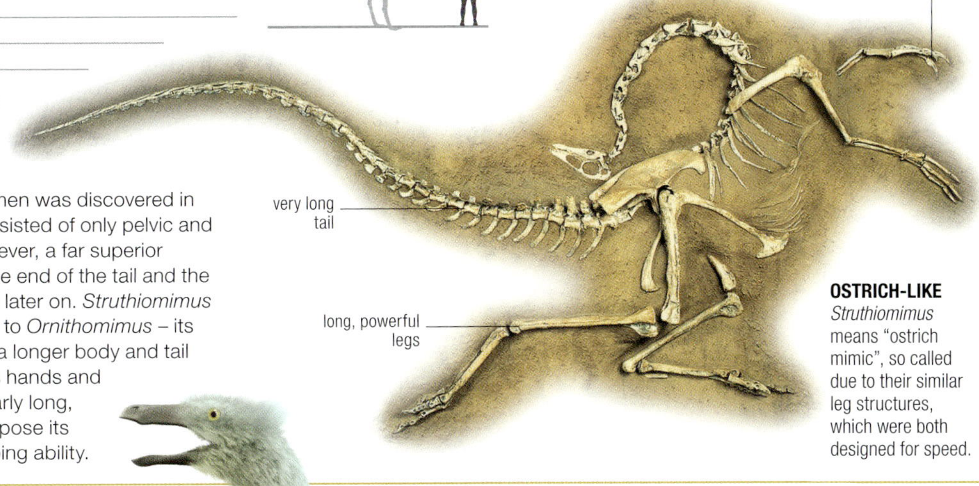

OSTRICH-LIKE
Struthiomimus means "ostrich mimic", so called due to their similar leg structures, which were both designed for speed.

Gallimimus

GROUP Theropods
DATE Late Cretaceous
SIZE 6m (20ft) long
LOCATION Mongolia

Gallimimus is one of the biggest and best-known ornithomimosaurs. It was originally thought to have a snout that curved upwards at its tip, but recent evidence suggests that it actually had a broad and blunt snout tip. Its lower jaw was deeper and shorter than that of other ornithomimosaurs. *Gallimimus* had proportionately shorter arms, smaller hands, and shorter hand claws than other members of this group, which suggests that it used its forelimbs differently to other ornithomimosaurs. It might have raked the ground to uncover food.

CHICKEN MIMIC
The anatomy of the neck in *Gallimimus* reminded its describers of the neck of a chicken, which explains why its name means "chicken mimic". Its slender legs and long tail suggest it was a fast runner.

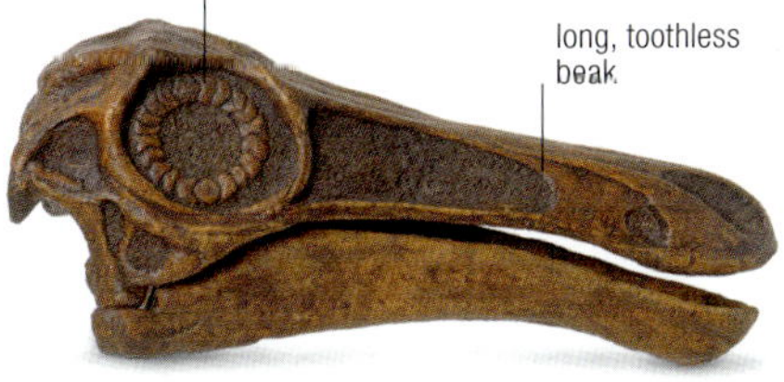

BLUNT BEAK
Gallimimus had a long, toothless beak that is noticeably blunt at the end, but it is not clear exactly what it ate. It had large eyes, but did not possess binocular vision.

Chirostenotes

GROUP Theropods
DATE Late Cretaceous
SIZE 4m (13ft) long
LOCATION USA, Canada

Chirostenotes was a large North American oviraptorosaur with a long skull and deep, rounded crest located over the top of its head. Its lower jaw was long and shallow, with an upturned, shovel-shaped tip. Two tooth-like prongs projected from the middle of the palate, but there were no true teeth. The claw on its second finger was straight in contrast to the curved finger shown on other oviraptosaurs.

UNIQUE TECHNIQUE
Chirostenotes had two curved claws and one straight one on each hand. It might have used these to probe under rocks, impaling small animals.

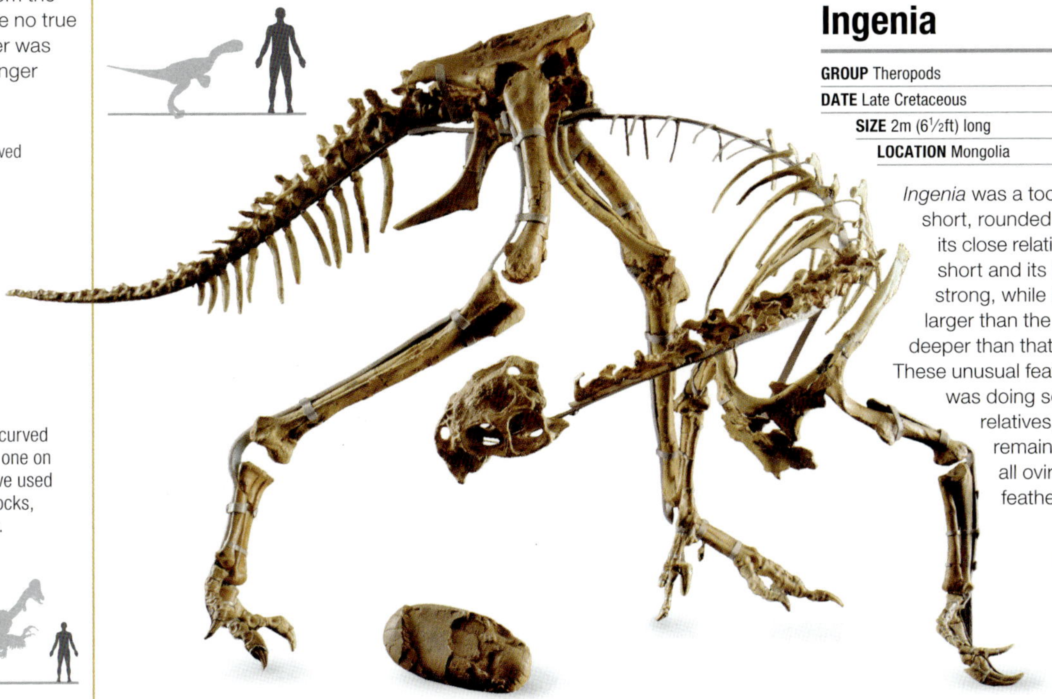

Ingenia

GROUP Theropods
DATE Late Cretaceous
SIZE 2m ($6\frac{1}{2}$ft) long
LOCATION Mongolia

Ingenia was a toothless oviraptorosaur with a short, rounded skull. Compared to those of its close relatives, its arms were particularly short and its hands were stout and strong, while its first finger was much larger than the other two. Also, its tail was deeper than that of other oviraptorosaurs. These unusual features all suggest that *Ingenia* was doing something quite different to its relatives, but its habits and lifestyle remain a mystery. However, like all oviraptorosaurs, *Ingenia* was feathered and bird-like.

UNKNOWN DIET
Although *Ingenia* was an oviraptorosaur, a name that suggests it ate eggs, there is no evidence for this and its true diet remains unknown.

***OVIRAPTOR* EGGS**
Discovered in the Gobi Desert, Mongolia, these fossilized dinosaur eggs were once thought to belong to *Protoceratops* (see p.350). The presence of an *Oviraptor* fossil found next to the nest was assumed to be evidence of its egg-stealing behaviour. However, further analysis has revealed that these were, in fact, *Oviraptor* eggs.

Caudipteryx

GROUP Theropods
DATE Early Cretaceous
SIZE 1m (3ft 3in) long
LOCATION China

Despite being a theropod dinosaur, which were typically carnivores, *Caudipteryx* used its large beak to eat plants and seeds, although it may have taken small animal and insect prey as well. Unlike other theropods, *Caudipteryx* did not have a bony crest on the top of its head. A number of complete fossils of this dinosaur have been found, giving us a good idea of how it looked. The abundance of remains also suggest that *Caudipteryx* was a common animal.

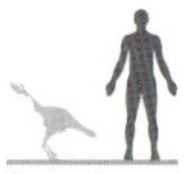

EXTENSIVE PLUMES
Fossil feathers show *Caudipteryx* bore large plumes on the arms and a large tail fan, but it was not a flying dinosaur.

Dromaeosaurus

GROUP Theropods
DATE Late Cretaceous
SIZE 2m (6½ft) long
LOCATION Canada

Dromaeosaurus was the first dromaeosaurid to be described. Ironically, however, it remains one of the most poorly known members of the group, and only a partial skull and a few bones from the hand and foot have been described. The skull was deep and broad for a dromaeosaurid, and the teeth at the tip of the upper jaw were wide. The lower jaw was also deep and robust compared to the far more shallow lower jaws of dromaeosaurids such as *Velociraptor* (opposite). These features suggest that *Dromaeosaurus* had a more powerful bite than other dromaeosaurid species.

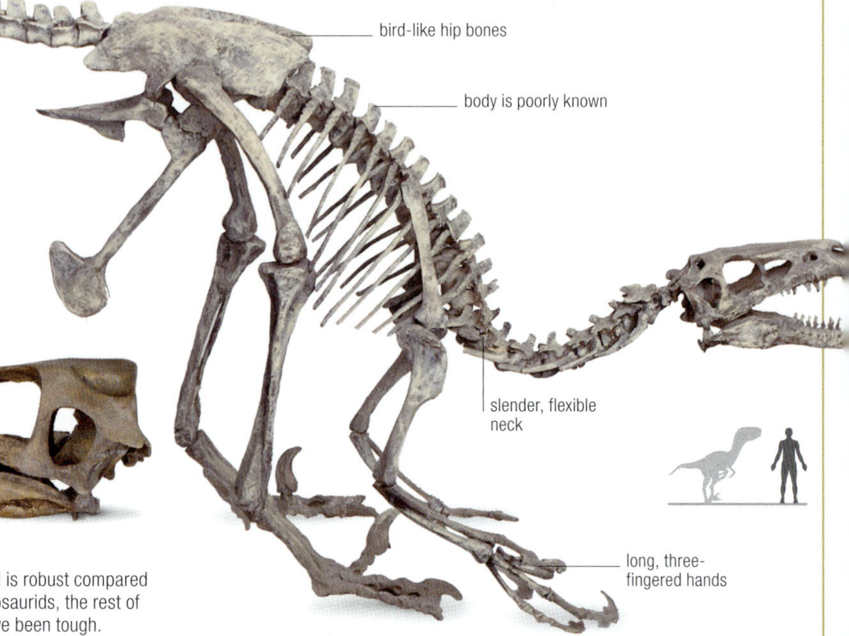

LONG FEATHERS
Dromaeosaurus would have been feathered, and particularly long feathers would have grown from its hands, arms, and tail.

ROBUST SKELETON
As *Dromaeosaurus*'s skull is robust compared to those of other dromaeosaurids, the rest of the skeleton may also have been tough.

Troodon

GROUP Theropods
DATE Late Cretaceous
SIZE 3m (9¾ft) long
LOCATION North America

Troodon was named for just a single tooth (the name means "wounding tooth") but we now know that skull and skeletal material originally named *Stenonychosaurus* belong to *Troodon*. *Troodon*'s teeth were coarsely serrated, so much so that some palaeontologists have suggested that it might have been able to shred leaves. It was probably mostly predatory, however, and could have preyed on animals ranging from small lizards and mammals to medium-sized ornithischians. In some places, large numbers of *Troodon* teeth are preserved alongside the bones of baby hadrosaurs. It might be that *Troodon* stayed close to hadrosaur colonies during the nesting season, grabbing unguarded young when it could. *Troodon* had large eyes and probably had well-developed binocular vision. Although often described as a brainy dinosaur, *Troodon*'s brain was proportionally about as big as that of an ostrich or emu. The bowl-shaped nests, eggs, and even some *Troodon* embryos have been discovered, as have preserved adults sitting on top of nests. A typical *Troodon* nest contained around 24 eggs.

SICKLE-SHAPED CLAW
Deinonychus's most famous feature was the enlarged claw on the second toe, which was probably used to disembowel prey in a series of raking kicks.

Deinonychus

GROUP Theropods
DATE Early Cretaceous
SIZE 3m (9$^{3}/_{4}$ft) long
LOCATION USA

When named in 1969, *Deinonychus* was used to promote the idea that dinosaurs were not slow and cumbersome animals destined for extinction, but were successful, often agile, and perhaps even warm-blooded. Its long fingers were equipped with three large curved claws. Like other maniraptorans, *Deinonychus* was almost certainly feathered, and long feathers called rectrices must have grown from the upper surfaces of its hands and second fingers.

BIOGRAPHY
JOHN OSTROM

Dinosaurs were considered unsuccessful until Ostrom argued in the 1960s that they were exciting and successful, with complex social lives. He described *Deinonychus*, showed that birds almost certainly descended from *Deinonychus*-type theropods, and studied the biology of hadrosaurs and horned dinosaurs. Many of his ideas and discoveries initiated a new phase in dinosaur research.

Velociraptor

GROUP Theropods
DATE Late Cretaceous
SIZE 2m (6$^{1}/_{2}$ft) long
LOCATION Mongolia

Velociraptor was discovered in the Gobi Desert during the 1920s and has become one of the most familiar dromaeosaurids. Its snout was long and narrow, with a concave upper border. Like other dromaeosaurids, *Velociraptor* had long hands, an enlarged claw on its second toe, and a fairly stiff, but lightly built tail. A second species of *Velociraptor*, *V. osmolskae*, was named in 2008. It differed from the other species, *V. mongoliensis*, in minor details of skull and tooth anatomy.

LUCKY FINDS
One spectacularly complete *Velociraptor* specimen is preserved locked in combat with a *Protoceratops*. Thanks to this and other specimens, we now know *Velociraptor*'s anatomy in detail.

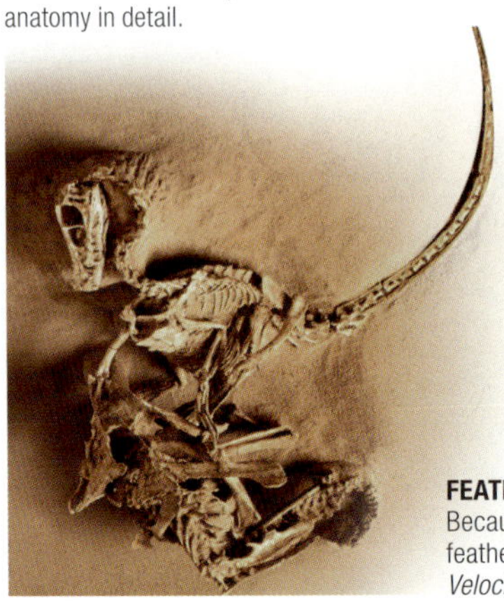

FEATHERED DINOSAUR
Because other dromaeosaurids were feathered, we can be confident that *Velociraptor* was feathered, too.

SCIENCE
MALE PARENTING

Small theropods, including *Troodon*, sat on egg-filled nests. *Troodon* folded its legs under its body and may have used its feathered arms to cover the eggs. Like female birds, female dinosaurs possessed a special type of bone, known as medullary bone, used in the production of eggshell. This is absent in the brooding dinosaur specimens, suggesting that they might be males. This supports the idea that both male and female *Troodons* contributed to nest care, much like modern ostriches.

OPPORTUNISTIC PREDATOR
Troodon was probably an opportunistic predator that ate a variety of animal prey. Its long, slim legs show that it was a fast runner. Like other small maniraptorans, it would have had a feathery coat.

Microraptor

GROUP Theropods
DATE Early Cretaceous
SIZE 1.2m (4ft) long
LOCATION China

This small, feathered dinosaur from China is related to the more famous *Velociraptor* (see p.331) and other dromaeosaurs. However, unlike those speedy runners, *Microraptor* was more at home in the trees, gliding from branch to branch. It spent most of its time hunting small prey, such as lizards and early mammals, and used its aerobatic skills to avoid predators. *Microraptor* did not fly, but glided like modern "flying" squirrels. Its feathered arms acted as glide surfaces, and its legs may have helped steer the animal through the air. *Microraptor* was not a direct ancestor of birds, but its lifestyle suggests a form of locomotion that may have been the forerunner of powered flight in birds.

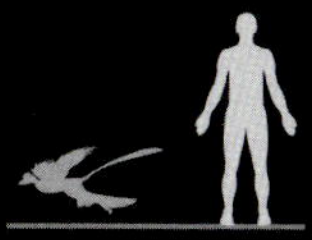

DINOSAUR FEATHERS
This fossilized *Microraptor* shows the long arms and associated wing feathers. The long feathers on the hind legs are also clearly visible, on the right below the stiff tail.

Argentinosaurus

GROUP Sauropodomorphs
DATE Late Cretaceous
SIZE 30m (98ft) long
LOCATION Argentina

One of the largest-known sauropods, *Argentinosaurus* was probably a primitive member of the group. Its broad vertebrae had small, peg-and-socket articulations that were located just above the spinal cord opening. Peg-and-socket structures were common in saurischians, and probably kept the animals' backbones rigid. These spinal features were absent in later titanosaurs (the lithostrotians), but it is unknown why they evolved more flexible backbones. *Argentinosaurus's* immense ribs were hollow, cylindrical tubes.

GREAT LEGS
The skull, neck, and tail of *Argentinosaurus* have never been found, but, based on one shin bone, the limbs would have been massive columns, with each hind limb being around 4.5m ($14\frac{3}{4}$ft) long.

Amargasaurus

GROUP Sauropodomorphs
DATE Early Cretaceous
SIZE 11m (36ft) long
LOCATION Argentina

Amargasaurus was an unusual-looking sauropod. It was relatively small and short-necked, with pairs of long spikes that projected from the top of its 12 neck vertebrae. The function of these spikes is as yet unknown. *Amargasaurus* may have fed on ground-level vegetation, while other, taller sauropods foraged for foliage higher up.

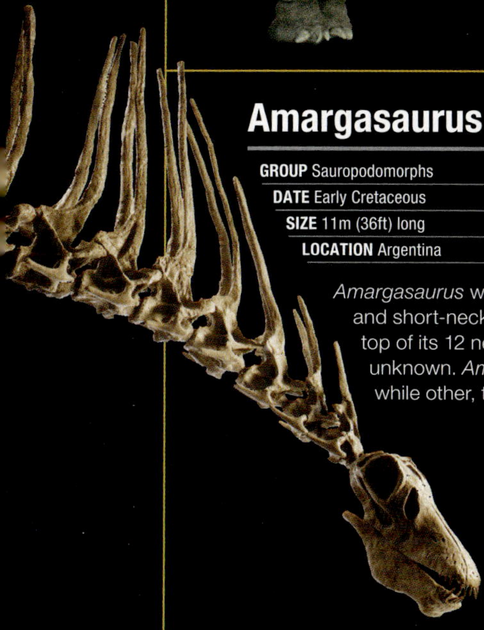

MYSTERIOUS SPIKES
Amargasaurus's spines may have supported skin sails or formed an array of horn-covered spikes.

FAMILY LIKENESS
This *Nemegtosaurus* has been modelled on *Rapetosaurus*, a close relative from Madagascar. All titanosaurs had wide bodies, and their flexible necks allowed them to feed high up in the trees.

Nemegtosaurus

GROUP Sauropodomorphs
DATE Late Cretaceous
SIZE 15m (49ft) long
LOCATION Mongolia

Only the skull of the Mongolian sauropod *Nemegtosaurus* has ever been found. At first glance, it resembles that of the Jurassic diplodocid *Dicraeosaurus*, and as a result *Nemegtosaurus* was originally thought to be a late-surviving diplodocid. More recent work, however, has shown that *Nemegtosaurus* was actually a

titanosaur, and therefore more closely related to animals such as *Saltasaurus* (see p.334). While some titanosaurs had stout, spoon-shaped teeth and a short skull, *Nemegtosaurus* had pencil-shaped teeth and a long snout. Some experts have reconstructed the dinosaur's head showing a rounded, bony hump on the top of its skull. If such a hump had existed, *Nemegtosaurus's* head would have looked very much like a brachiosaur's. However, it was more likely to have been low and subtle – the most recent studies show that the back of the skull was very tall compared to the snout, and that the whole skull was long and boxy. It is possible that *Opisthocoelicaudia*, a Mongolian titanosaur, could be a *Nemegtosaurus* specimen.

Whereas most **titanosaurs** have been found **without their heads**, only the skull of *Nemegtosaurus* has ever been found.

Saltasaurus

GROUP Sauropodomorphs
DATE Late Cretaceous
SIZE 12m (39ft) long
LOCATION Argentina

Saltasaurus is one of the best known of the titanosaurs. These were a group of sauropods traditionally thought to have been mostly restricted to the southern hemisphere but are now known to have been more widespread. Unlike other sauropods, some titanosaurs were armoured, and *Saltasaurus* was one of the first titanosaurs found that demonstrated this. The upper surface and sides of its body were covered with large, oval armour plates, some of which may have been tipped with spikes. Thousands of smaller, rounded bones covered the skin in between the large plates. As was the case in most other titanosaurs, *Saltasaurus* had very broad hips and its body would have been wide and rounded. Its limbs were stout and it had a flexible tail.

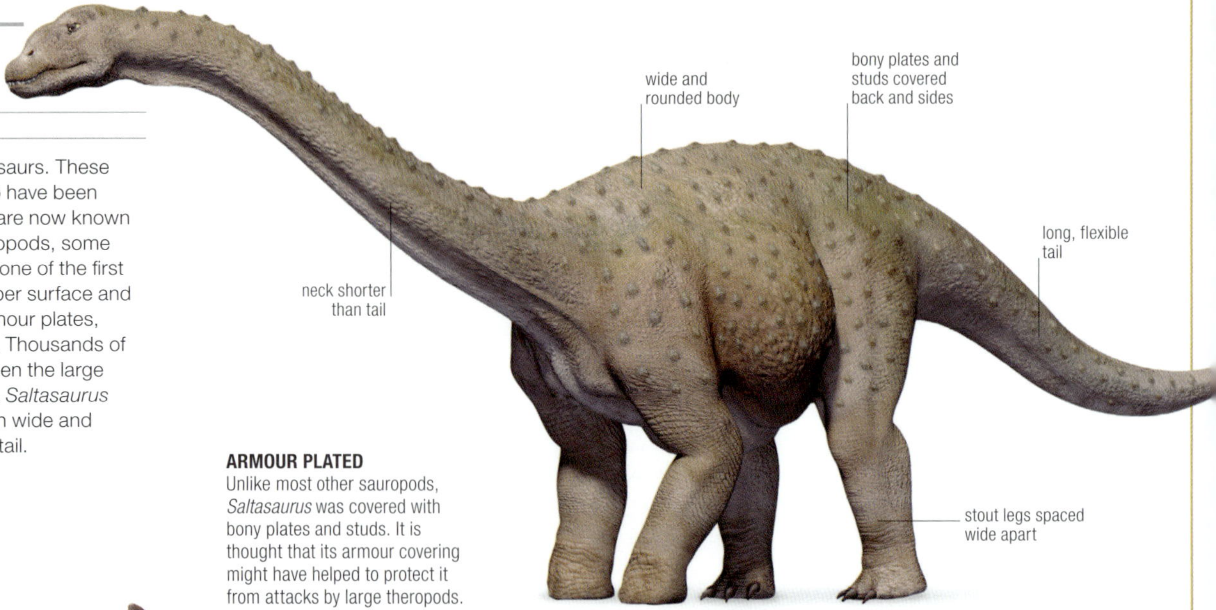

ARMOUR PLATED
Unlike most other sauropods, *Saltasaurus* was covered with bony plates and studs. It is thought that its armour covering might have helped to protect it from attacks by large theropods.

Minmi

GROUP Ornithischians
DATE Early Cretaceous
SIZE 3m (9¾ft) long
LOCATION Australia

Minmi is unique in having strange extra bones, called paravertebrae, along its back. These might have helped to provide increased support for its back muscles. Small, rounded armour plates covered its body, including its belly. One specimen includes a skull that is broad and deep, with a short, narrow snout and large eye sockets. Another *Minmi* specimen provides direct evidence of its diet: preserved gut contents show that it ate fruit.

SMALL SPECIMENS
All known *Minmi* specimens are relatively small, although this might be because all of those that have been discovered so far are juveniles.

IMPRESSIVE DEFENCE
Flattened, triangular spikes projected from the side of *Gastonia*'s body and tail, and long spines pointed upwards from the shoulder region. Smaller, oval plates were arranged across the animal's back.

Gastonia

GROUP Ornithischians
DATE Early Cretaceous
SIZE 4m (13ft) long
LOCATION USA

Gastonia is one of the world's best-known ankylosaurs, and most parts of its skeleton have been discovered. *Gastonia*'s skull was shallow and broad with a wide, square-tipped beak. As in nearly all ankylosaurs, its teeth were small and leaf-shaped. The bones forming the roof of the skull were thickened and domed, and a special joint around the bones that housed the brain may have provided a shock-absorbing function. Some experts speculate that these dinosaurs butted heads together when fighting.

Sauropelta

GROUP Ornithischians
DATE Early Cretaceous
SIZE 5m (16½ft) long
LOCATION USA

Sauropelta was a large, long-tailed, North American nodosaurid. Thanks to a well-preserved specimen that includes a near-complete skull, its anatomy is reasonably well known. *Sauropelta* had more teeth in its lower jaw than most other nodosaurids. The back of the skull was much broader than the snout, and the top of the skull was flat. The upper surfaces of its body and tail were covered in oval bone plates that formed a continuous armour covering. The name *Sauropelta* means "shield lizard". Long, conical spikes projected upwards and sideways from its neck and shoulders. Its relatively long tail was made up of more than 40 vertebrae.

FORMIDABLE SPIKES
It had been thought that only a single row of spikes was present on each side of *Sauropelta*'s neck, but new work has shown that there were two rows on each side. These spikes would have been formidable weapons.

Ankylosaurus

GROUP Ornithischians
DATE Late Cretaceous
SIZE 6m (20ft) long
LOCATION North America

The largest of the ankylosaurids, *Ankylosaurus* was a giant, club-tailed animal with large, triangular horns at the back of its skull. Its snout was short and broad. Small, leaf-shaped teeth lined the sides of its jaws, and at the front its toothless bill was broad and deep. The sides of the snout appeared to bulge outwards, and the nostrils were directed sideways. *Ankylosaurus* was very similar to its closest relative, *Euoplocephalus* (see pp.336–37).

BULGING SKULL
The bulging shape of *Ankylosaurus*'s snout was created by complicated air passages running through the skull. These were also present in several other ankylosaurids, but their function is unknown.

ARMED AND PROTECTED
Ankylosaurus was well protected and well armed. Its thick skin was studded with hundreds of bony plates of different sizes, and the great club on its tail could be swung with bone-shattering force.

Edmontonia

GROUP Ornithischians
DATE Late Cretaceous
SIZE 7m (23ft) long
LOCATION North America

Edmontonia was one of the largest and most widely distributed nodosaurids. One excellent specimen has allowed a good understanding of its anatomy and appearance. Bands of armour plates covered the upper surface of the neck and shoulder region, and smaller plates covered the rest of the back and tail. Several long spikes projected from each shoulder. The first two spikes pointed diagonally forwards, and the two further back pointed sideways. *Edmontonia*'s skull had a long, low snout and its eye sockets were placed far back.

LETHAL SHOULDER SPIKES
Edmontonia probably used its lethal shoulder spikes for fighting others of its own kind. The longest of the spikes had forked tips, which were more prominent in some individuals than in others.

mouth surrounded
by a horny beak
nose had a complicated
nasal structure, suggesting
Euoplocephalus had an
acute sense of smell
three toes on
hind feet
each toe tipped
with a blunt hoof

Euoplocephalus

GROUP Ornithischians
DATE Late Cretaceous
SIZE 7m (23ft) long
LOCATION North America

Euoplocephalus, which means "well armoured head", is one of the largest and best known of the ankylosaurids. Specimens have been found with almost all their thick plates of studded armour in place. *Euoplocephalus* was closely related to *Ankylosaurus* (see pp.335), and shared many of the same characteristics. The low-slung body and hips of *Euoplocephalus* were so broad that it must have been almost round in cross section. The limbs were short and stocky, and only three toes, tipped with blunt hooves, were present on each hind foot – other ankylosaurs had four toes. The vertebrae at the end of the tail were fused to form a stiffened, rod-like structure that helped to support the large, rounded club at the tip of the tail. The club was formed of four bony plates, held off the ground, and probably used in defence against predators. The base of the tail was almost certainly flexible enough to allow sideways movement, and the tail would have been quite muscular. *Euoplocephalus* was a herbivore, probably grazing on low vegetation and possibly even digging up roots and tubers. Remains are usually found singly, but the discovery of 22 young *Euoplocephalus* has raised the possibility that they may have lived in herds.

Mobile, bony shutters in its eyelids protected *Euoplocephalus*'s eyes.

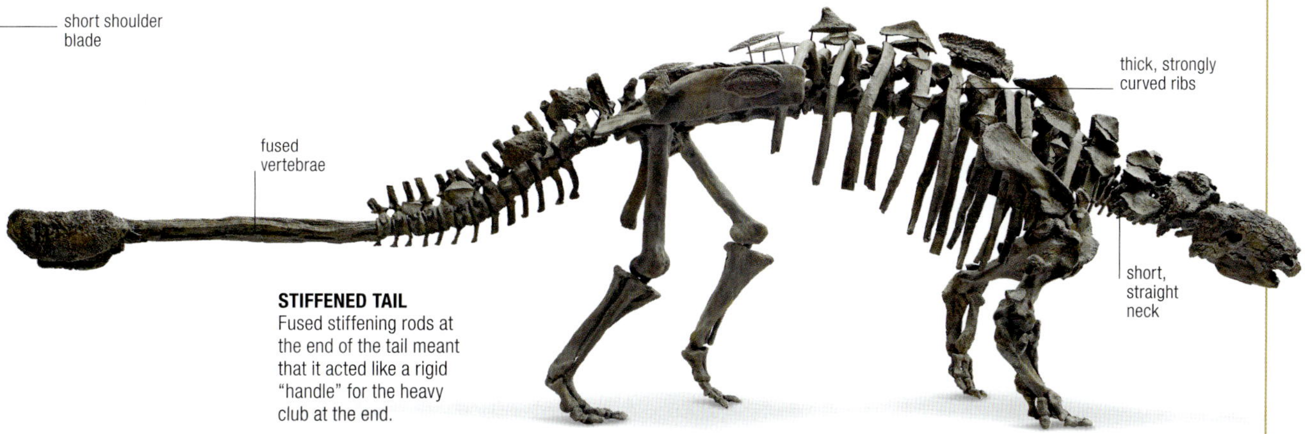

STIFFENED TAIL
Fused stiffening rods at the end of the tail meant that it acted like a rigid "handle" for the heavy club at the end.

HORNY PLATES
In life, the bony plates were covered by a protective horny covering. The plate shown here is from the side of the animal's hip.

WELL-ARMOURED HEAD
Small armour plates covered the surface of the skull, forming a mosaic-like pattern. They seem to be outgrowths of the skull bones themselves, and not separate bones.

TAIL CLUB
Euoplocephalus's tail club was wide and heavy. Studies have shown that it was tough enough to withstand substantial impacts.

PLATES WITH KEELS
Most of the armour plates that covered the body of *Euplocephalus* were oval, with keels, or ridges, that ran along their midline.

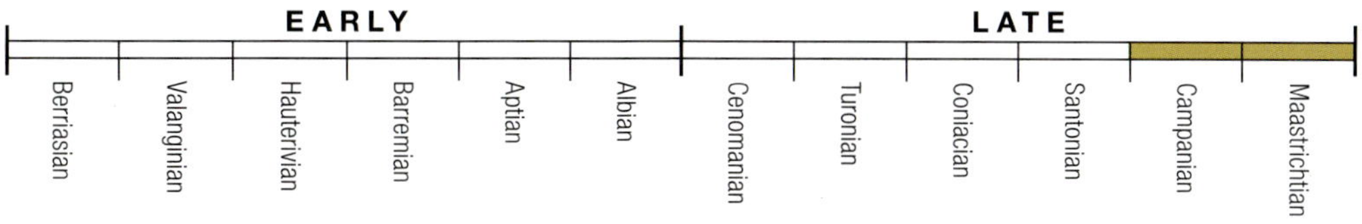

HAND FOSSIL
In *Iguanodon*'s hands, the middle three fingers were joined together, the fifth finger could curl to grasp food, and the thumb was armed with a vicious spike.

Iguanodon

GROUP Ornithischians

DATE Early Cretaceous

SIZE 9m (30ft) long

LOCATION Belgium, Germany, France, Spain, England

One of the most famous ornithopods, *Iguanodon* is best known for the many near-complete skeletons found in a Belgian coal mine. These were originally reconstructed standing upright in a kangaroo-style posture. However, it is now thought that *Iguanodon* walked mainly on all fours, with its body and tail held parallel to the ground. Its arms were long and robust, and well-adapted for bearing its weight.

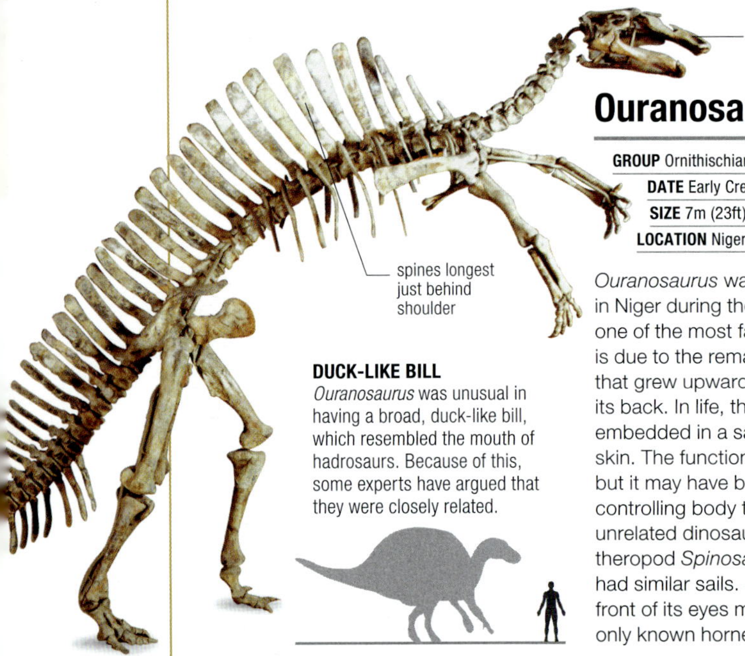

DUCK-LIKE BILL
Ouranosaurus was unusual in having a broad, duck-like bill, which resembled the mouth of hadrosaurs. Because of this, some experts have argued that they were closely related.

Ouranosaurus

GROUP Ornithischians

DATE Early Cretaceous

SIZE 7m (23ft) long

LOCATION Niger

Ouranosaurus was discovered in the desert in Niger during the 1960s, and has become one of the most famous ornithopods. This is due to the remarkably tall bony spines that grew upwards from the vertebrae on its back. In life, these would have been embedded in a sail formed of muscle and skin. The function of this sail is unknown, but it may have been used in display or in controlling body temperature. Many other unrelated dinosaurs, such as the giant theropod *Spinosaurus* (see pp.316–17), had similar sails. Small rounded horns in front of its eyes made *Ouranosaurus* the only known horned ornithopod.

The four **long, pointed claws** on its feet would have given *Tenontosaurus* a **dangerous kick.**

Leaellynasaura

GROUP Ornithischians

DATE Early Cretaceous

SIZE 1m (3ft 3in) long

LOCATION Australia

Leaellynasaura is one of several small ornithopods known from a famous fossil site in Victoria, Australia, called Dinosaur Cove. When *Leaellynasaura* was alive, southern Australia was within the Antarctic Circle, and although the polar regions were less cold than they are today, it would have been continually dark for several months a year. An internal cast of *Leaellynasaura*'s braincase shows that it had large optic lobes – the parts of the brain associated with eyesight. So it probably had large eyes and good sight for seeing in the dark.

STRIKING EYES
A small, bipedal herbivore, *Leaellynasaura*'s most striking feature was its very large eyes, used for seeing in the dark.

Hypsilophodon

GROUP Ornithischians

DATE Early Cretaceous

SIZE 2m ($6\frac{1}{2}$ft) long

LOCATION England, Spain

One of the best-known small ornithopods, *Hypsilophodon* is known from several near-complete skeletons. At one time, the misconception that it had grasping hands and a first toe that was directed backwards led some to think that it was a tree-climber. In fact, its rigid tail and long hind limbs and feet show that it was a fast-running ground animal. Leaf-shaped teeth show that, like other small ornithopods, it browsed on low-growing plants. Pointed teeth were present at the front of the upper jaw, and the beaked jaw tips were pointed.

VERTEBRAE FOSSIL
Hypsilophodon fossils, like this spine section, were first discovered on the Isle of Wight, England. Fossils from Spain also seem to belong to this dinosaur.

SPECIALIZED TENDONS
Tenontosaurus takes its name from the specialized tendons present in its back, hips, and tail – "tenon" is the Greek word for tendon. These enabled it to raise its tail when walking on all fours.

Tenontosaurus

GROUP Ornithischians
DATE Early Cretaceous
SIZE 7m (23ft) long
LOCATION USA

Tenontosaurus was a large and particularly long-tailed iguanodontian. Its long forelimbs and short, broad hands suggest that it walked on all fours, though it was probably able to rear on its back legs when feeding or fighting. Its skull was deep, and its nostril openings were long. Two species are known. In one, the front of the upper jaw was toothless, but in the other, teeth were present in the same region. The outer rim of the beak at the tip of the lower jaw was serrated. *Tenontosaurus* is best known for having been discovered alongside remains of the theropod *Deinonychus* (see p.331). This probably shows that the theropod preyed on *Tenontosaurus*. Groups of juvenile specimens have been discovered at two different locations, which suggests that young animals remained in groups after hatching.

HEAD STRUCTURE
Hypsilophodon's head was tall and short, with large eyes. Behind a bony beak, its mouth had cheek pouches that it used when chewing food.

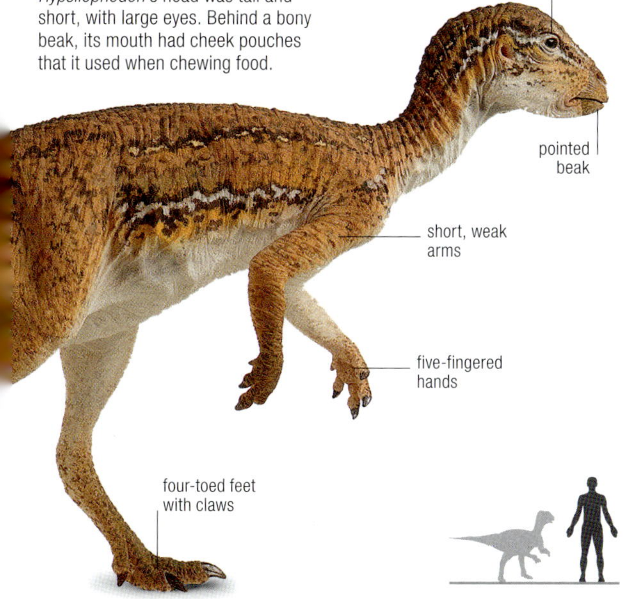

Muttaburrasaurus

GROUP Ornithischians
DATE Early Cretaceous
SIZE 7m (23ft) long
LOCATION Australia

Muttaburrasaurus was a large iguanodon-like ornithopod with a deep bony bump on the top surface of its snout. The skull bones beneath its eye socket were thick and strong. Some experts have suggested that these bones might show that *Muttaburrasaurus* was adapted for biting and chewing especially tough plants. It was considered to be related to *Iguanodon* (see opposite) – a spike-shaped bone was thought to be a thumb spike like that of *Iguanodon*. However, more recent studies have shown that *Muttaburrasaurus* was a far older species.

GIANT NOSE
Muttaburrasaurus might have used its giant nose lump to make loud, resonating calls. The size and shape of its nose differed between individuals, probably depending on sex or species variations.

1,600 The total number of teeth present in some *Edmontosaurus* skulls.

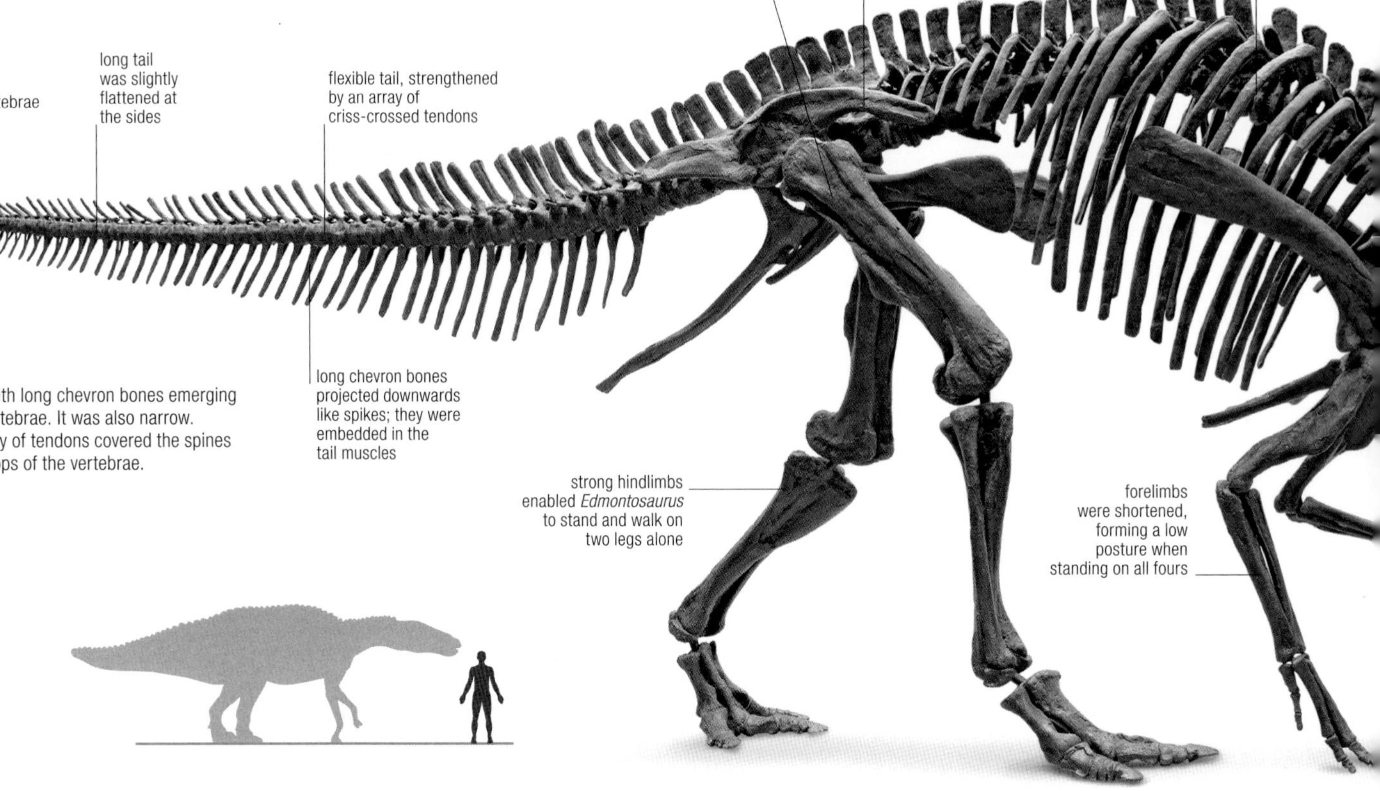

BALANCED TAIL
The tail was deep, with long chevron bones emerging from beneath the vertebrae. It was also narrow. A criss-crossing array of tendons covered the spines that grew from the tops of the vertebrae.

Edmontosaurus

GROUP Ornithischian
DATE Late Cretaceous
SIZE 13m (43ft)
LOCATION USA, Canada

One of the largest and best-known of the duck-billed dinosaurs, *Edmontosaurus* lived in North America during the late Cretaceous period. *Edmontosaurus*, meaning "Edmonton lizard", was named after Edmonton, Alberta, Canada, where the first fossils were found. A crestless hadrosaur, the bones at the front of its long jaws flared out sideways, forming a duck-like bill that was used to grab and crop large mouthfuls of vegetation. In fact, the term "duck-billed dinosaur" was originally coined for this species. As in other hadrosaurs, the jaw tips of *Edmontosaurus* were toothless, and its hundreds of teeth were arranged in tightly-filled batteries on both upper and lower jaws. Huge hollow areas surrounded the nostril openings. The function of these hollows is unknown, but they might have housed balloon-like sacs that the animal could inflate at will. These facial balloons would have enabled *Edmontosaurus* to make sounds, which may have been used to attract mates, signal to other members of the herd, or to threaten rivals.

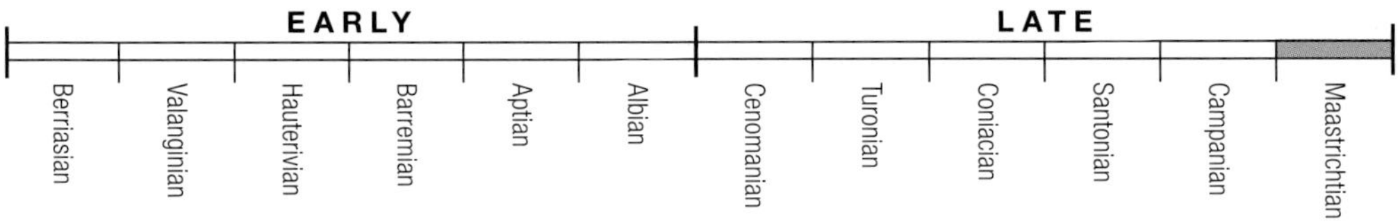

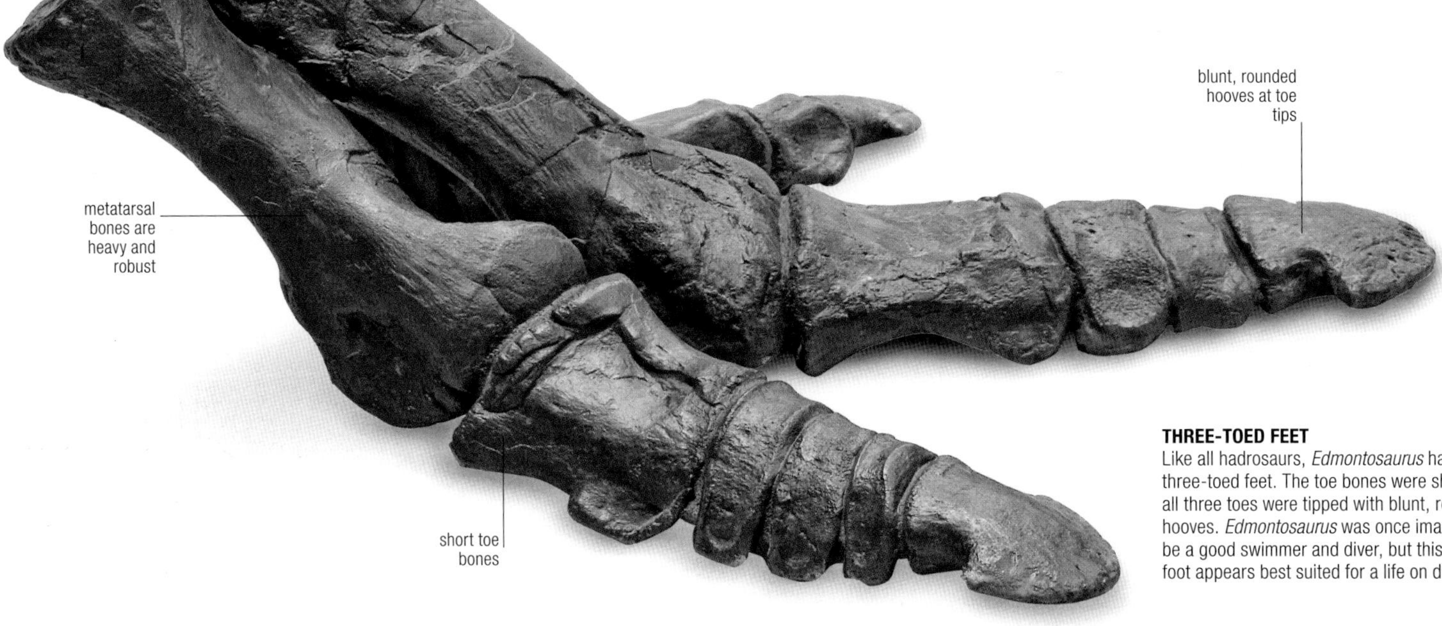

THREE-TOED FEET
Like all hadrosaurs, *Edmontosaurus* had three-toed feet. The toe bones were short and all three toes were tipped with blunt, rounded hooves. *Edmontosaurus* was once imagined to be a good swimmer and diver, but this sort of foot appears best suited for a life on dry land.

DUCK-BILLED SKULL
Viewed from the front, the skull really does look "duck-billed", thanks to its expanded jaw tips. The leading edge of the lower jaw is wide and scoop-like. A distinctive pitted bone texture shows where beak tissue covered the jaws; some hadrosaur specimens have beak tissue still preserved in place. The large eye sockets indicate that *Edmontosaurus* had a wide field of vision. The snout is dominated by the hollows surrounding the nostrils.

NECK VERTEBRAE
Edmontosaurus had 13 neck vertebrae. These formed a gently curved neck. Ball and socket-type articulations show that the neck was quite flexible.

GRINDING TEETH
The teeth were packed in vertical rows at the backs of the jaws. Only the teeth in the top row form the grinding surface.

PROJECTIONS ON SPINE
The bony spines on the tops of the vertebrae were short in *Edmontosaurus*. The ribcage was deep and narrow.

WEBBED HANDS
The four-fingered hand was slim and at least three of the fingers were tipped with blunt hooves. There was no thumb.

Brachylophosaurus

GROUP Ornithischians
DATE Late Cretaceous
SIZE 9m (30ft) long
LOCATION North America

Brachylophosaurus, whose name means "short-crested lizard", had a flattened, sheet-like crest that grew backwards from the snout, overhanging the back of the skull. An excellent *Brachylophosaurus canadensis* specimen was found in Alberta, Canada, after which it is named. Some individuals – presumably the males – were more heavily built than others, with deeper, lower jaws, stouter skulls, and wider crests that extended further along the skull. In 2000, an exceptionally complete specimen was discovered in Montana. It is covered in large amounts of preserved skin, and promises to provide much information on the appearance of this dinosaur.

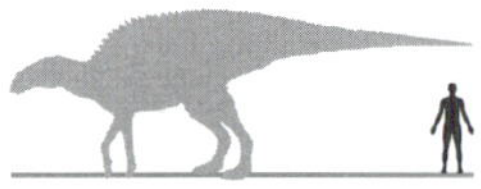

SKULL FEATURES
Brachylophosaurus had a particularly deep snout and a rectangular skull. Its nostril openings were enormous. Its jaw tips were stout and wide.

Maiasaura

GROUP Ornithischians
DATE Late Cretaceous
SIZE 9m (30ft) long
LOCATION USA

Maiasaura became world-famous thanks to the discovery of nests, eggshell fragments, and the remains of juveniles, all found alongside the skeletons of adults. As suggested by the generic name, which means "good mother lizard", the remains indicate that *Maiasaura* formed nesting colonies, with parents constructing crater-shaped nests, in which the hatchlings stayed for an extended period, being fed and looked after by their parents. It is possible that these behavioural traits were true of all hadrosaurids. *Maiasaura*'s skull possessed an expanded bill and a solid crest that extended across the top of the skull above the eyes. Several views have been proposed on the lineage of *Maiasaura*, but it shares some features with *Brachylophosaurus*, and the two appear to be close relatives.

COVERED NESTS
The rounded eggs of *Maiasaura* were covered by vegetation and sediment. On hatching, the babies may have dug their own way out, or they may have been helped by their parents.

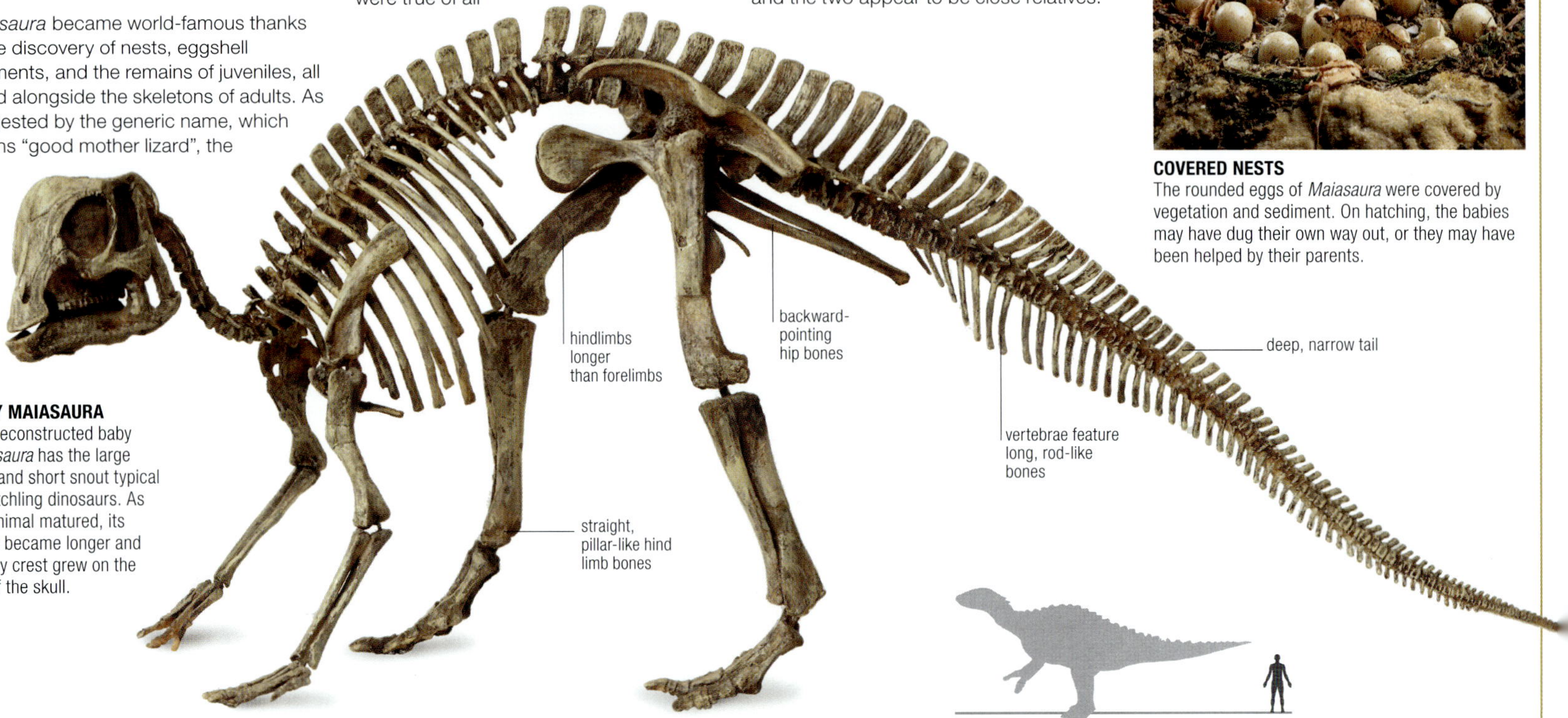

BABY MAIASAURA
This reconstructed baby *Maiasaura* has the large skull and short snout typical of hatchling dinosaurs. As the animal matured, its snout became longer and a bony crest grew on the top of the skull.

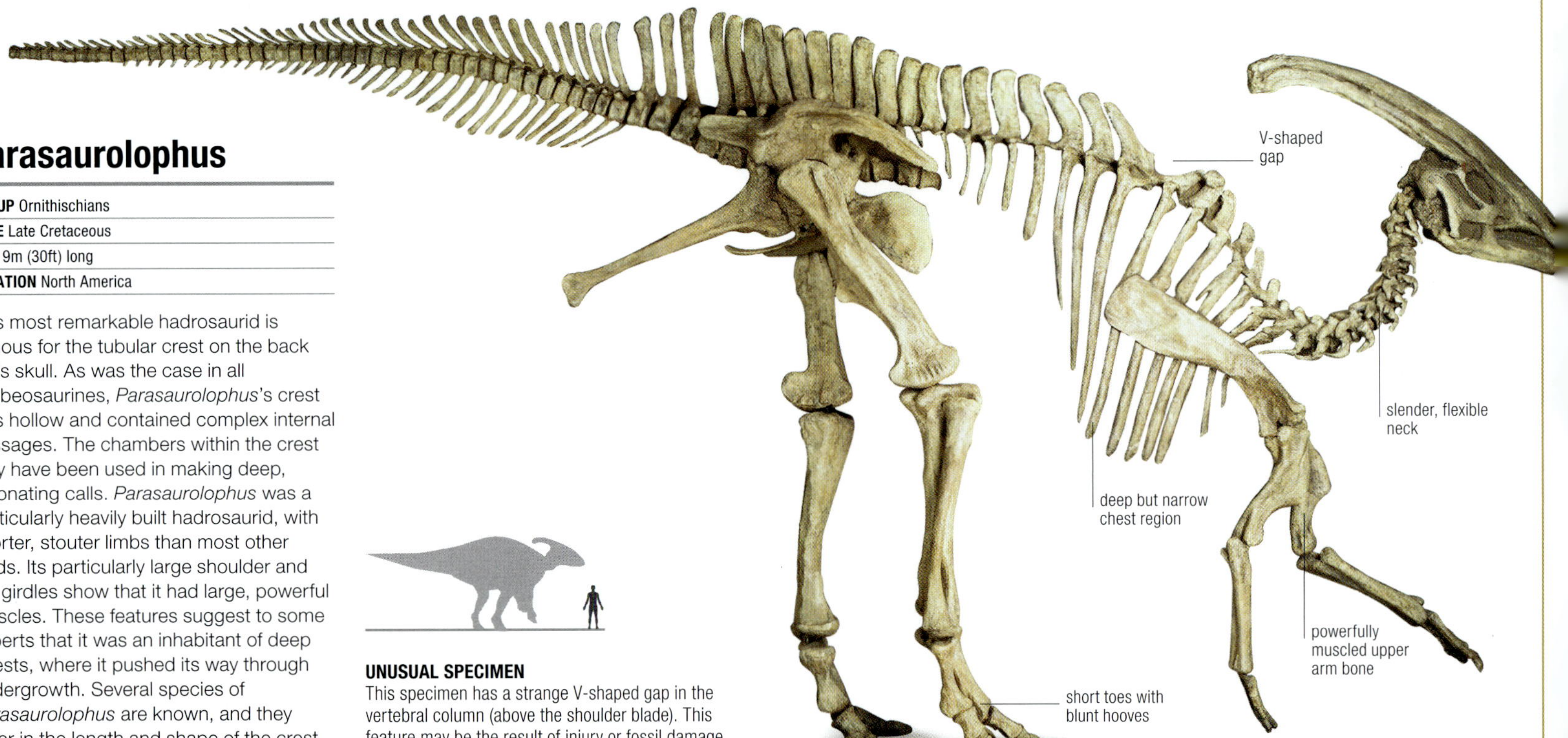

Parasaurolophus

GROUP Ornithischians
DATE Late Cretaceous
SIZE 9m (30ft) long
LOCATION North America

This most remarkable hadrosaurid is famous for the tubular crest on the back of its skull. As was the case in all lambeosaurines, *Parasaurolophus*'s crest was hollow and contained complex internal passages. The chambers within the crest may have been used in making deep, resonating calls. *Parasaurolophus* was a particularly heavily built hadrosaurid, with shorter, stouter limbs than most other kinds. Its particularly large shoulder and hip girdles show that it had large, powerful muscles. These features suggest to some experts that it was an inhabitant of deep forests, where it pushed its way through undergrowth. Several species of *Parasaurolophus* are known, and they differ in the length and shape of the crest.

UNUSUAL SPECIMEN
This specimen has a strange V-shaped gap in the vertebral column (above the shoulder blade). This feature may be the result of injury or fossil damage.

THE FUNCTION OF CRESTS
Lambeosaurines like *Lambeosaurus* probably used their bony crests as signalling devices. The crests helped distinguish the different species from one another, and were perhaps also used to demonstrate age and sex.

Lambeosaurus

GROUP Ornithischians
DATE Late Cretaceous
SIZE 9m (3ft) long
LOCATION Canada

Lambeosaurus was a large lambeosaurine that had a tall, rectangular crest extending forwards, overhanging the snout. It also possessed a shortened cheek region and a tall ridge along its backbone. Crest size and shape is variable in *Lambeosaurus* specimens, and as a result several species have been recognized. While some are probably distinct species, others might be males, females, or juveniles of one species. In *L. lambei*, the crest had a rectangular front portion that pointed upwards, and a spike at the back that was directed upwards and backwards. *L. clavinitialis* had a much shorter spike at the back and might be the female of *L. lambei*. In *L. magnicristatus*, there was no posterior spike, and the crest blade was large. Skin impressions show that small, non-overlapping nodule-like scales covered *Lambeosaurus*, and it seems to have lacked the large conical tubercles present on the underside of *Corythosaurus* (see pp.344–45).

Gryposaurus

GROUP Ornithischians
DATE Late Cretaceous
SIZE 9m (30ft) long
LOCATION North America

One of the "hook-nosed" hadrosaurines, *Gryposaurus* and its relatives had forelimbs that were about two-thirds the length of the hind limbs, unlike most hadrosaurids, which had forelimbs that were about half as long as the hind limbs. Why these hadrosaurids had such long arms is unknown, but it is likely that this adaptation allowed them to feed from higher up in the vegetation than the other herbivores that shared their environment. The best known *Gryposaurus* species is *G. notabilis*, but a second species, *G. latidens*, was named in 1992, and a third, *G. monumentensis*, in 2007. A very similar hadrosaurid – *Kritosaurus incurvimanus* – is regarded by some experts as another species of *Gryposaurus*. It has even been suggested that *K. incurvimanus* and *G. notabilis* might be the female and male of the same species.

HOOK-NOSED
The bones that formed the upper surface of *Gryposaurus*'s snout arched upwards, and a large depression surrounded the nostril. This enlarged nasal region may have been brightly coloured and used in display or to shove against rivals in disputes.

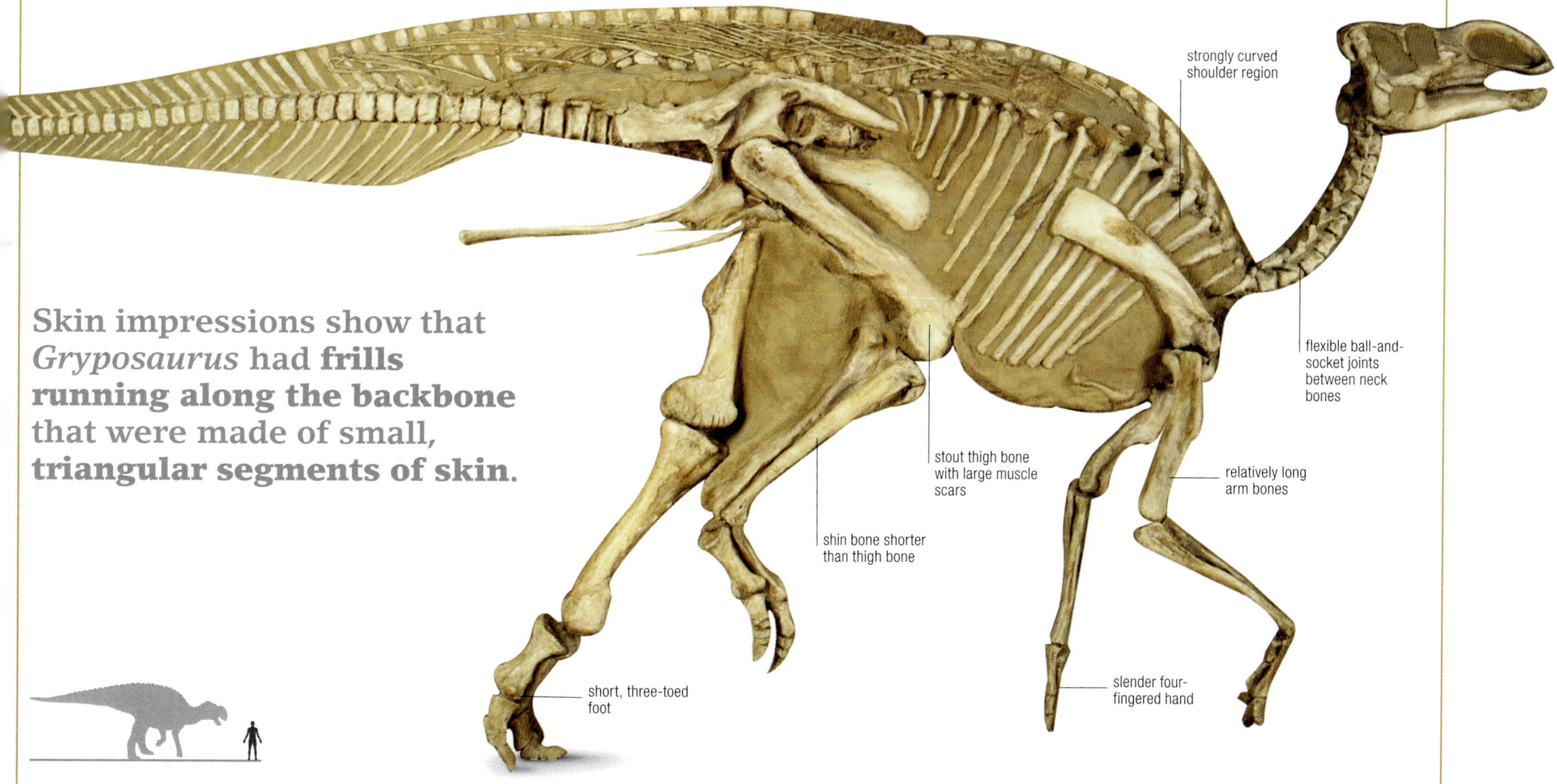

Skin impressions show that *Gryposaurus* had **frills running along the backbone** that were made of small, **triangular segments of skin.**

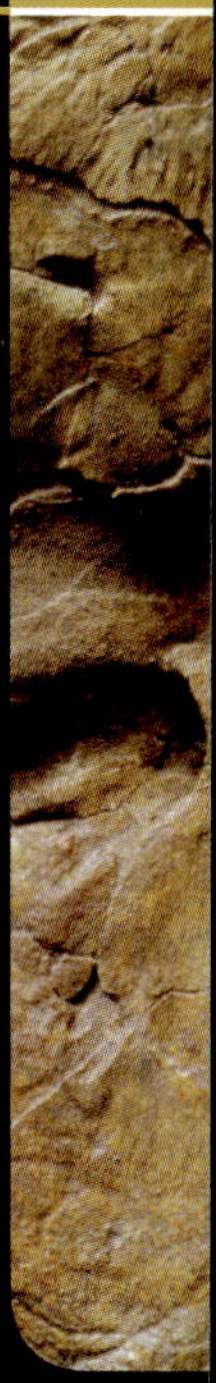

Corythosaurus

GROUP Ornithischians
DATE Late Cretaceous
SIZE 9m (30ft) long
LOCATION Canada

The name *Corythosaurus* means "helmet lizard". It was known for its hollow, plate-like crest, and was most closely related to *Velafrons* from Mexico, *Nipponosaurus* from Russia, and *Hypacrosaurus* from the USA. Together, these hadrosaurids are known as the fan-crested lambeosaurines. *Corythosaurus* was a large dinosaur with tall spines in a ridge along its back. Its snout was shallow and delicate, compared with that of many other hadrosaurids. This feature suggests that it may have been a selective feeder that browsed for the juiciest fruits and youngest leaves. Much is known about *Corythosaurus* as several complete specimens have been found. Some even have preserved skin impressions, which show that *Corythosaurus* had lines of conical tubercles along its underside and a continuous skin frill along the top of its backbone. The frill was attached to the rear of the head crest and was deepest in the gap between the shoulders and the back of the head.

FORAGERS
Although it may have wandered through swampy areas, *Corythosaurus* probably spent most of its time foraging for leaves in woodland habitats.

ANATOMY
CREST FOR CALLS

The crest of *Corythosaurus* was not solid bone – it contained tubes that were connected to the nostrils. All lambeosaurines had crests that differed according to species. It was once thought the crests were air tanks, but this was dismissed when it was realized the animals were not aquatic. It has also been proposed that the crests were lined with nasal tissue, to help provide a sensitive sense of smell. But the most popular theory is that they were used to make loud, resonating calls.

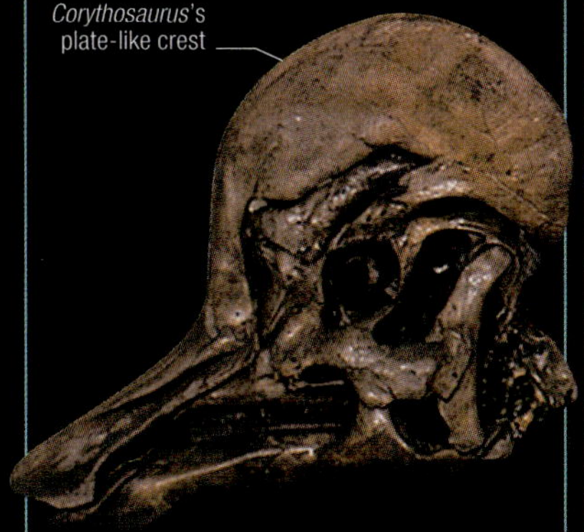

Psittacosaurus

GROUP Ornithischians
DATE Early Cretaceous
SIZE 2m (6½ft) long
LOCATION China, Mongolia

Psittacosaurus is one of the earliest members of Ceratopsia, the horned dinosaur group. It is also one of the best-known Mesozoic dinosaurs, owing to the number of specimens that have been discovered. This has enabled experts to identify 14 species according to skull shape, although not all experts agree with these classifications. Unlike later ceratopsians, *Psittacosaurus* was probably bipedal. It had four-fingered hands and long hind limbs. Its short, deep skull had a narrow, toothless beak, which gave *Psittacosaurus* its name, meaning "parrot lizard". Horn-like bony growths projected sideways from the cheeks, and differed in size and shape between the species. Young *Psittacosaurus* specimens have also been discovered. In one case, dozens of young were preserved along with the remains of an adult. One *Psittacosaurus* specimen was found with numerous, long filaments growing from the upper surface of its tail. Nothing like this has been seen in any other specimens, and the function of these bizarre structures remains a mystery.

PERFECT PRESERVATION
This spectacular *Psittacosaurus* fossil includes virtually the entire skeleton, as well as skin impressions and tail bristles. Bristle-like structures like this were present elsewhere in ornithischians, so perhaps they were widespread among these dinosaurs. The specimen is preserved lying on its back and its skull bones are jumbled.

NEW INFORMATION

The "bristly" *Psittacosaurus* specimen was only published in 2002, so many reconstructions, such as the model shown here, do not incorporate this new information. The filaments may have only occurred in certain species, or even just some individuals.

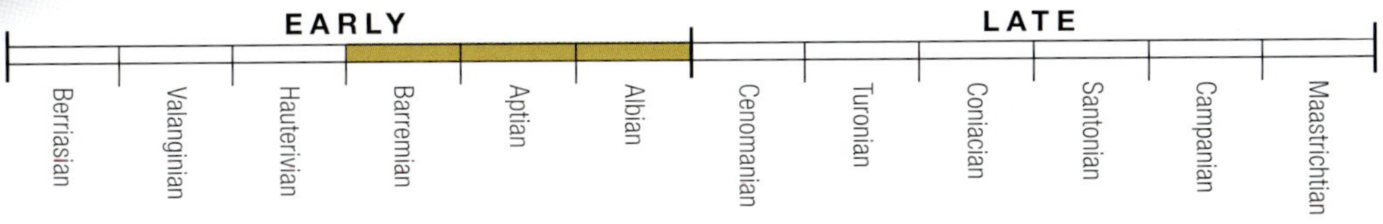

SOLID SKULL
Triceratops, which means "three-horned face", was one of the largest ceratopsians. As well as being very distinctive, with its neck frill and horns, its skull was also very solidly built. This has resulted in *Triceratops* skulls surviving the fossilization process better than most dinosaur skulls. About 50 *Triceratops* skulls have been found.

Triceratops

GROUP Ornithischians
DATE Late Cretaceous
SIZE 7m (23ft) long
LOCATION North America

One of the best-known of all dinosaurs, *Triceratops*'s anatomy is familiar to most people: it had a large neck frill, two long brow horns, and a short nose horn. The neck frill lacked the large openings that were normally present in that of most horned dinosaurs. Small, triangular bones lined the edges of some frills, but these were absent in many specimens. Injuries preserved on skulls show that individuals sometimes fought among themselves, perhaps over mates or territory. Some specimens preserve bite marks made by *Tyrannosaurus* (see pp.322–23), and one *Triceratops* even seems to have had one of its brow horns snapped off by a tyrannosaur.

About 15 *Triceratops* species have been named over the years, all of which were distinguished on the basis of differences in horn shape and size. However, these differences now seem to represent the sort of variation seen among individuals of the same species, and only two species are currently recognized.

HEAVILY BUILT HERBIVORE
This large, heavily built herbivore used its powerful beak to slice through tough plant stems. Its head was held low towards the ground, so it was probably a low browser.

The largest *Triceratops* skulls are **more than 2m long with 70cm-long brow horns.**

Protoceratops

GROUP Ornithischians
DATE Late Cretaceous
SIZE 2m (6½ft) long
LOCATION Mongolia, China

One of the most studied of all primitive ceratopsians, *Protoceratops* is known from many specimens collected in the Mongolian Gobi Desert. Some specimens have taller frills and deeper snouts than others, which may represent a difference between the sexes. *Protoceratops* most likely dug burrows where they had their young – one group of juvenile specimens was found preserved together in a burrow in the sand.

In 2001, a new *Protoceratops* species, *P. hellenikorhinus*, was named. Unlike the first species, *P. andrewsi*, it had two parallel nasal horns, lacked teeth at the front of the upper jaw, and its frill was orientated further forward.

BURROWING DEEP
The broad hands and wide, spade-like claws of *Protoceratops* suggest it was a capable digger.

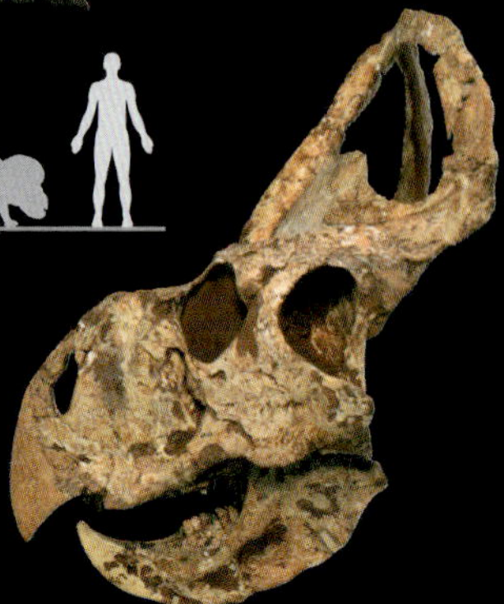

BIRD-LIKE HEAD
The face of *Protoceratops* was deep and its beak was narrow. It had two pairs of small, pointed teeth near the front of the upper jaw.

Pentaceratops

GROUP Ornithischians
DATE Late Cretaceous
SIZE 7m (23ft) long
LOCATION USA

Pentaceratops was a large chasmosaurine and a close relative of *Chasmosaurus* (below). The name of this dinosaur means “five-horned face” and reflects the particularly long horns (epijugals) that projected sideways from its cheeks. It had a deep snout, long brow horns, and an extremely tall frill above the neck. Six tongue-shaped bones (epoccipitals) were arranged around the frill border, and an additional smaller pair was present either side of the midline.

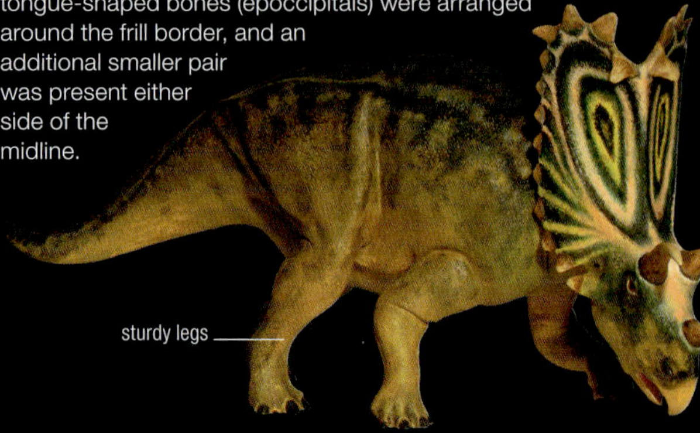

RECORD BREAKER
A *Pentaceratops* discovered in 1998 had a skull 3m (9¾ft) long, the longest of any land animal in history.

Chasmosaurus

GROUP Ornithischians
DATE Late Cretaceous
SIZE 5m (16½ft) long
LOCATION North America

Chasmosaurus is one of the best known of the chasmosaurine horned dinosaurs. Four species are currently recognized. It had a relatively long snout and a long, broad frill that contained two large holes. Conical bones (epoccipitals) were arranged around the sides of the frill, although not along its rear border. Some species, such as *C. irvinensis*, the most recently identified species, lacked brow horns entirely, while in others they were very long. In *C. belli*, some individuals had short, stubby brow horns while others had long, curved ones. The shape of the nose horn was also variable between the species. In some it was tall and obviously curved, while in others it was short, broad, and blunt.

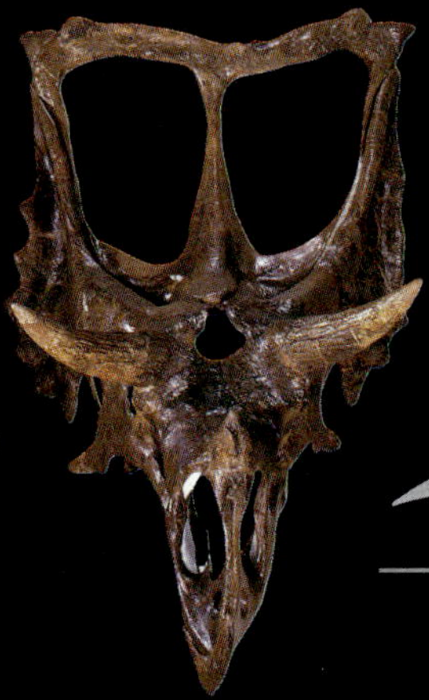

OPEN FRILL
Chasmosaurus’s name translates as “wide-opening lizard”, referring to the huge window-like holes in the frill. This was most probably used in display.

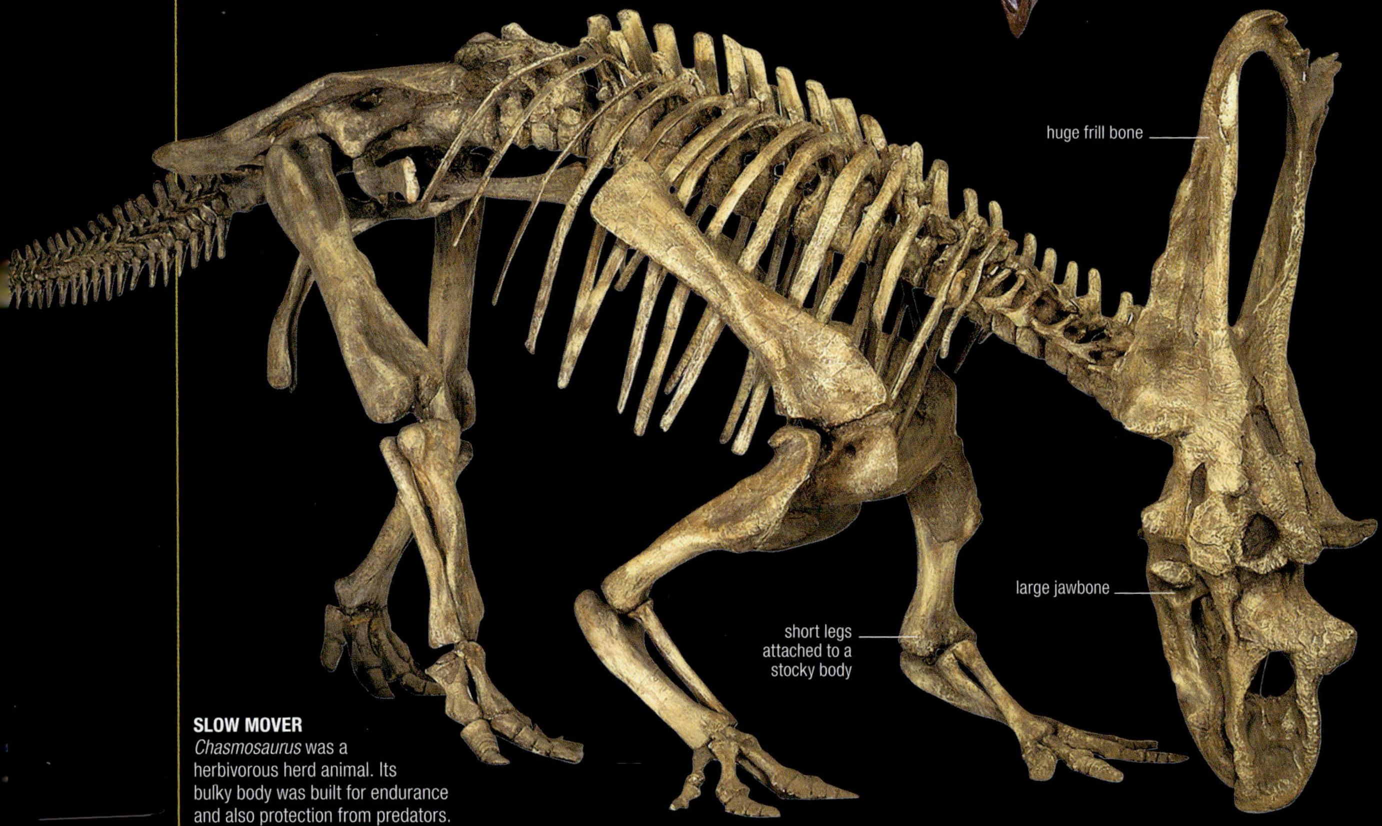

SLOW MOVER
Chasmosaurus was a herbivorous herd animal. Its bulky body was built for endurance and also protection from predators.

Centrosaurus

GROUP Ornithischians
DATE Late Cretaceous
SIZE 6m (20ft) long
LOCATION Canada

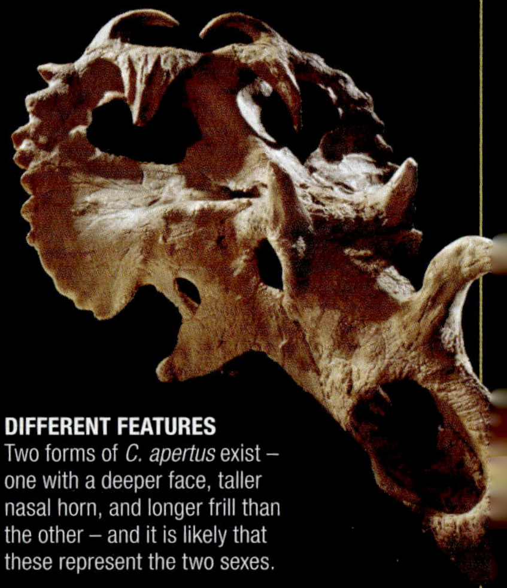

DIFFERENT FEATURES
Two forms of *C. apertus* exist – one with a deeper face, taller nasal horn, and longer frill than the other – and it is likely that these represent the two sexes.

Hundreds of *Centrosaurus* bones have been found together, suggesting that the animals lived in herds and underwent seasonal migrations, during which many drowned while crossing rivers. The shape of the nasal horn of *Centrosaurus* was variable. It could be short, tall, straight, or curved backwards or forwards. *Centrosaurus* means “spur lizard”, and the thick, upper border of the frill possessed inwardly curved, hook-like structures on either side of the midline. A second species, *C. brinkmani*, differs from *C. apertus*, in that the rear border of its frill was covered in small spikes.

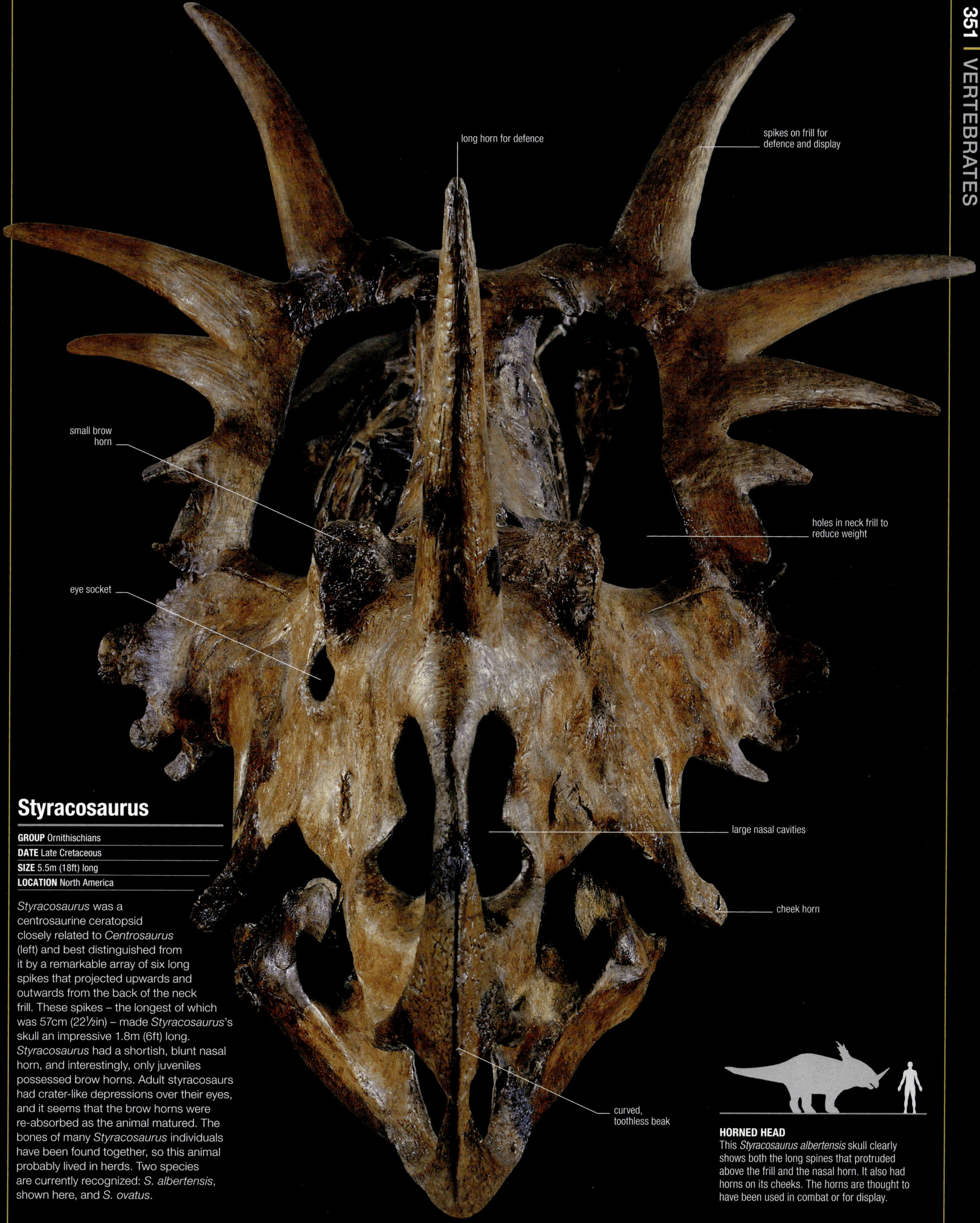

Styracosaurus

GROUP Ornithischians
DATE Late Cretaceous
SIZE 5.5m (18ft) long
LOCATION North America

Styracosaurus was a centrosaurine ceratopsid closely related to *Centrosaurus* (left) and best distinguished from it by a remarkable array of six long spikes that projected upwards and outwards from the back of the neck frill. These spikes – the longest of which was 57cm (22½in) – made *Styracosaurus*'s skull an impressive 1.8m (6ft) long. *Styracosaurus* had a shortish, blunt nasal horn, and interestingly, only juveniles possessed brow horns. Adult styracosaurs had crater-like depressions over their eyes, and it seems that the brow horns were re-absorbed as the animal matured. The bones of many *Styracosaurus* individuals have been found together, so this animal probably lived in herds. Two species are currently recognized: *S. albertensis*, shown here, and *S. ovatus*.

HORNED HEAD
This *Styracosaurus albertensis* skull clearly shows both the long spines that protruded above the frill and the nasal horn. It also had horns on its cheeks. The horns are thought to have been used in combat or for display.

Einiosaurus

GROUP Ornithischians
DATE Late Cretaceous
SIZE 6m (20ft) long
LOCATION USA

Einiosaurus was named in 1995. Like today's bison, these animals seemed to have lived in herds and they also had a curved horn on the snout. The one species' full name (*Einiosaurus procurvicornis*) translates as "buffalo lizard with forward-curving horn". The nasal horn of *Einiosaurus* was unique, being flattened from side-to-side and with a tip that curved over the end of the snout. In young animals, the horn was short and erect, and as the animal matured, it curved forwards. Like most other members of the centrosaurine group, *Einiosaurus* lacked brow horns and had either small lumps or rounded pits in their place. The margins of its frill were wavy, and two long spikes projected upwards from the frill's upper border. These spikes were straight, and it is difficult to imagine how they might have functioned in combat.

Einiosaurus appears to have been close to the ancestry of *Achelousaurus* and *Pachyrhinosaurus*, two centrosaurines that lacked nasal horns altogether and had large bony lumps on their noses instead.

The **distinctive** downwards curving **horn** was used both **for fighting** and **in display.**

GREEN EATER
Einiosaurus procurvicomis was a herbivore with teeth strong enough to consume a range of plants. It grazed in herds, much like modern bison.

Pachycephalosaurus

GROUP Ornithischians
DATE Late Cretaceous
SIZE 5m (16½ft) long
LOCATION North America

Pachycephalosaurus was the biggest of the pachycephalosaurs. It is best known from a single 60cm (23½in) long domed skull. Its name means "thick-headed lizard". It had large eyes, which suggests it had good vision, and small teeth, indicating it was either a herbivore or an omnivore. Two other North American pachycephalosaurs lived at the same time as *Pachycephalosaurus*: *Stygimoloch* and *Dracorex*. Both had smaller domes and larger horns than *Pachycephalosaurus*. However, some experts think that all these animals simply represent different growth stages of the same species.

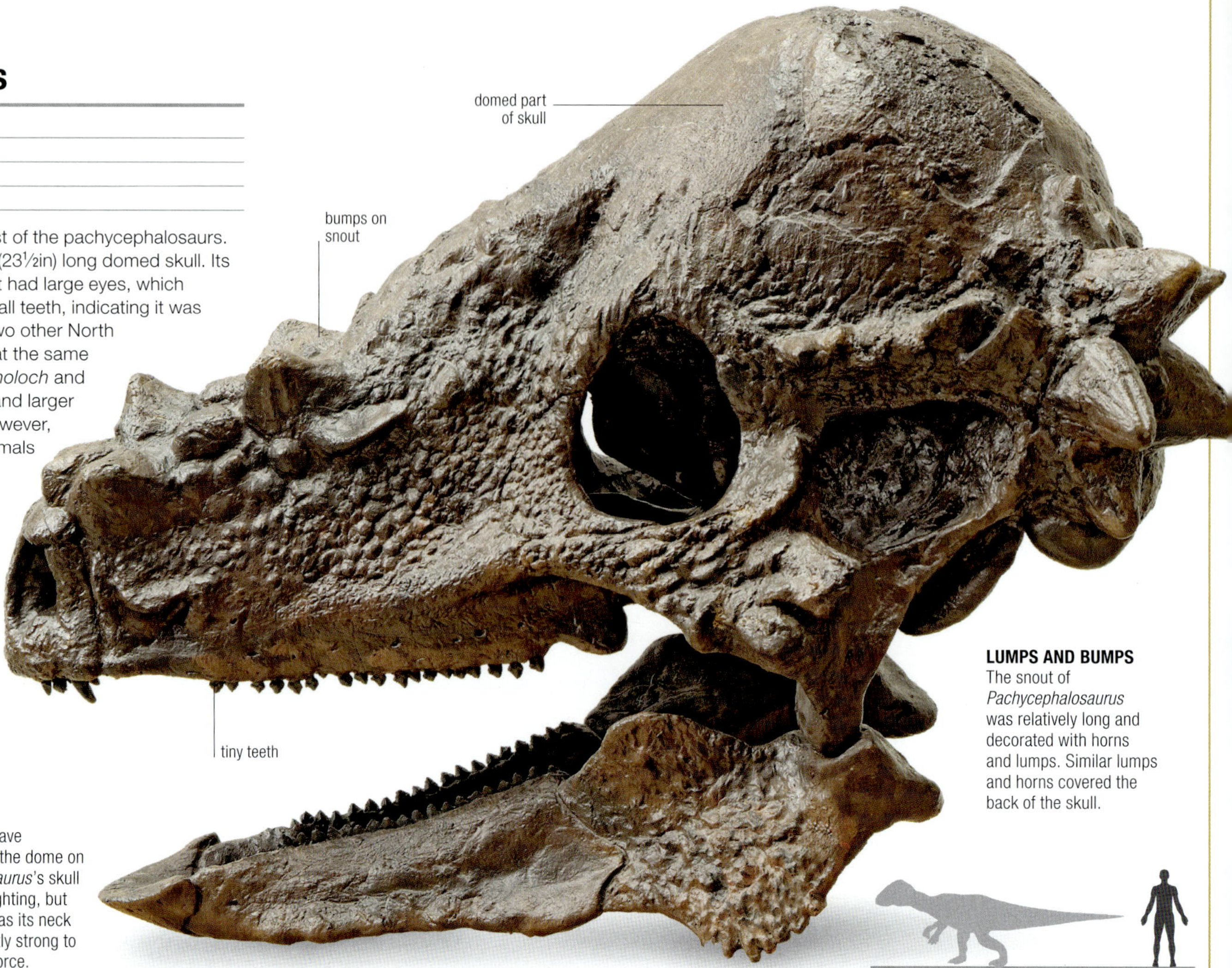

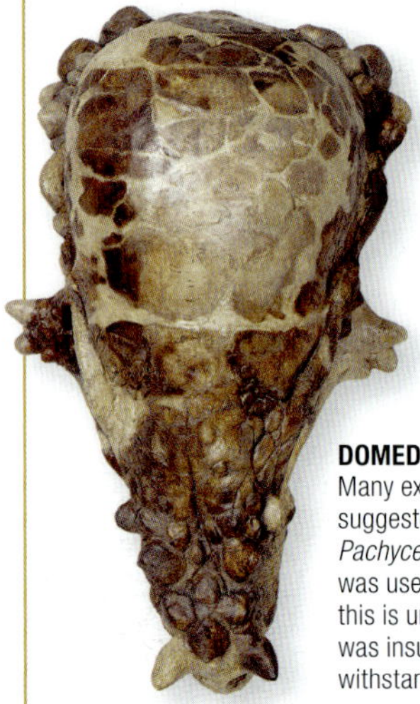

DOMED HEAD
Many experts have suggested that the dome on *Pachycephalosaurus*'s skull was used for fighting, but this is unlikely, as its neck was insufficiently strong to withstand the force.

LUMPS AND BUMPS
The snout of *Pachycephalosaurus* was relatively long and decorated with horns and lumps. Similar lumps and horns covered the back of the skull.

Stegoceras

GROUP Ornithischiansww
DATE Late Cretaceous
SIZE 2m (6½ft) long
LOCATION Canada

Stegoceras is one of the best known of the pachycephalosaurs. A prominent bony shelf decorated with bony knobs and spikes projected from the back of its domed skull roof – *Stegoceras* means "roofed horn". *Stegoceras* had a short face and, while its snout was narrow, its cheeks flared out. Its small, coarsely serrated teeth were most likely used in chewing and shredding leaves. Two species previously thought to belong to *Stegoceras* have recently been argued to be in separate genera. *Colepiocephale* lacked the bony shelf, as did *Hanssuesia*, which also had wider, flatter edges to its skull dome.

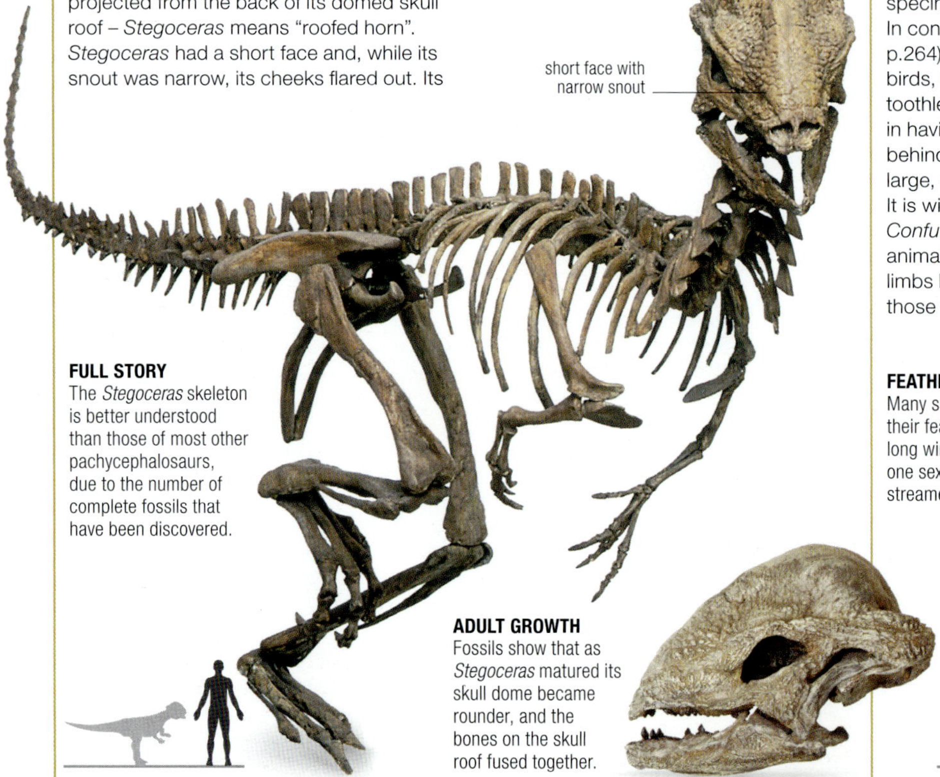

FULL STORY
The *Stegoceras* skeleton is better understood than those of most other pachycephalosaurs, due to the number of complete fossils that have been discovered.

ADULT GROWTH
Fossils show that as *Stegoceras* matured its skull dome became rounder, and the bones on the skull roof fused together.

Confuciusornis

GROUP Theropods
DATE Early Cretaceous
SIZE 30cm (12in) long
LOCATION China

Confuciusornis – the "Confucius bird" – is one of the best-known Mesozoic birds and hundreds of specimens have been discovered. In contrast to *Archaeopteryx* (see p.264) and other more primitive birds, *Confuciusornis* was entirely toothless. It was also unusual in having a robust, bony bar right behind the eye, and exceptionally large, curved thumb claws. It is widely thought that *Confuciusornis* preyed on aquatic animals. Its bill shape and hind limbs have been compared to those of modern kingfishers.

FEATHERY FRIENDS
Many specimens of *Confuciusornis* have their feathers still intact. The bird had long wing feathers, and possibly only one sex exhibited the two long, streamer-like tail feathers shown here.

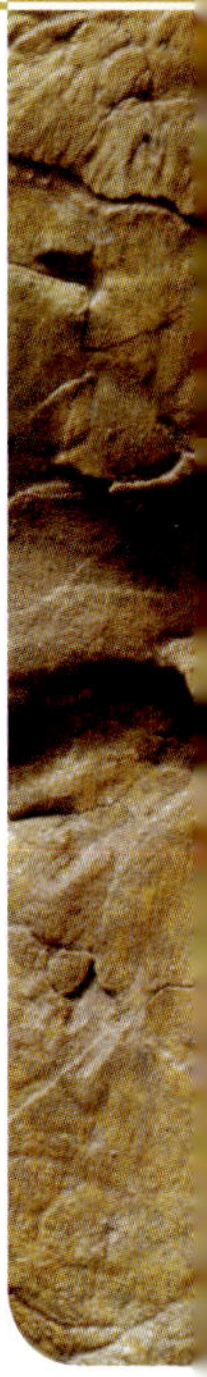

Iberomesornis

GROUP Theropods
DATE Early Cretaceous
SIZE 15cm (6in) long
LOCATION Spain

Originally described from a single, headless skeleton, *Iberomesornis* was a small, finch-sized Cretaceous bird. It is one of the most primitive known members of the entirely Cretaceous bird group known as the enantiornithines, or "opposite birds". They were closer to modern birds than they were to primitive species, in that they had shortened tails and large chest bones. However, like the much older *Archaeopteryx* (see p.264), they had teeth.

MODERN FEATURES
The wing bones of *Iberomesornis* show that it was capable of flight. Its curved claws suggest that it perched in trees like modern birds.

Gansus

GROUP Theropods
DATE Early Cretaceous
SIZE 20cm (8in) long
LOCATION China

Gansus was an amphibious diving bird with large feet. Fossilized skin impressions show that its feet were webbed all the way to the tips of the toes, so *Gansus* was probably similar to modern divers and grebes in its diving ability. Unfortunately, its skull remains unknown, so its feeding behaviour is still a mystery. *Gansus* does not have any close relatives. However, its features suggest that it was an early member of Ornithurae, the group that includes later Cretaceous forms, as well as all modern birds.

DIVING BIRD
This fossil, discovered along with a number of others in China in 2006, clearly shows *Gansus*'s long wing bones, indicating that it was a strong flier. Prior to this discovery, only one specimen was known.

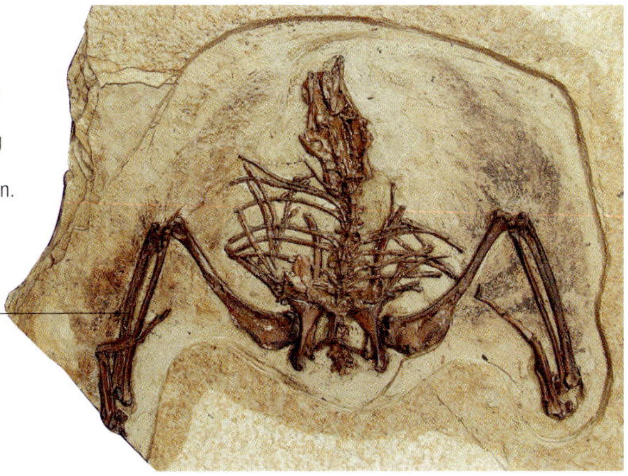

Hesperornis

GROUP Theropods
DATE Late Cretaceous
SIZE 1m (3½ft) long
LOCATION North America

Hesperornis (meaning "western bird") was a large, flightless sea bird with tiny wings, massive feet, and a toothed bill. First named in the 1870s, it is the best-known member of a group known as the hesperornithines, most of which were flightless. The wings on *Hesperornis* were so small that the hands and even lower arms were absent, and only a small, rod-shaped upper arm bone remained. The legs were placed far back on the body, as they are in modern birds that swim underwater. The toes were long and could close up tightly when the bird was pulling its leg back towards its body while diving. Skin impressions preserved with one specimen show that the toes were not webbed paddles but had large, fleshy lobes projecting from their sides.

WATER CREATURE
The flightless *Hesperornis* probably fed while in the water. It had small, conical teeth in its long, pointed bill, which was well-suited for catching fish.

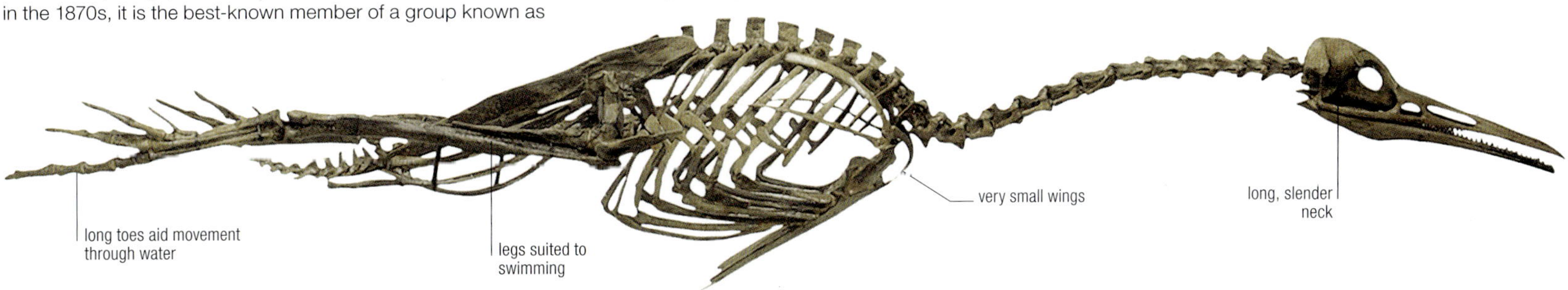

Ichthyornis

GROUP Theropods
DATE Late Cretaceous
SIZE 30cm (12in) long
LOCATION USA

Ichthyornis (the "fish bird") is one of the most famous of fossil birds. When originally described in 1872, it was one of only a handful of Mesozoic birds that seemed to bridge the gap between *Archaeopteryx* and modern birds. Today, many more Mesozoic birds are known. Studies have shown that *Ichthyornis* is more closely related to modern birds than were more archaic groups like the enantiornithines. However, unlike modern birds, *Ichthyornis* had teeth. These were small, smooth, and strongly curved, and were well suited for grabbing small and slippery prey, such as fish.

GULL-LIKE FEATURES
Ichthyornis probably looked much like a modern gull. Many of its fossils have been found in rocks in marine environments.

Vegavis

GROUP Theropods
DATE Late Cretaceous
SIZE 30cm (12in) long
LOCATION Antarctica

Discovered in 1992 on Vega Island in western Antarctica, *Vegavis* is a relative of *Presbyornis*, a fossil waterfowl from the Paleocene and Eocene of North America and part of the anseriform group Anatidae, which includes modern ducks, geese, and swans. The significant discovery of *Vegavis* shows that waterfowl – properly called anseriforms – were definitely alive during the Late Cretaceous. Furthermore, their closest relatives – the gallinaceous birds, or gamebirds – must have been present at this time as well, along with even earlier modern bird groups, such as the palaeognaths, which include ostriches.

MISSING LINK
Vegavis would have looked like a long-legged duck. Its skull is unknown, but like other fossil anseriforms, it probably had a duck-like bill.

Vincelestes

GROUP Early mammals
DATE Early Cretaceous
SIZE 30cm (12in) long
LOCATION Argentina

Compared to many Mesozoic mammals, *Vincelestes* is known from excellent fossils. Of nine specimens, six include skulls. These show that *Vincelestes* had large, stout canine teeth and fewer premolars and molars. It also had a short, deep snout. These features suggest that it was a predator, perhaps preying on smaller mammals, as well as reptiles and large insects. *Vincelestes* was large compared to most Mesozoic mammals, and its length was increased by its very long tail.

Repenomamus

GROUP Early mammals
DATE Early Cretaceous
SIZE 1m (3ft 3in) long
LOCATION China

Repenomamus, meaning "reptile-like mammal", is one of the most famous of Mesozoic mammals. Whereas most mammals from this time were the size of mice and shrews, *Repenomamus* was a giant in comparison. One species, *R. giganticus*, had a skull that was 16cm ($6^1/2$in) long and a total body length of about 1m (3ft 3in) long. Its jaws were stoutly constructed, and it was definitely a predator of smaller animals, including baby dinosaurs.

MESOZOIC DEVIL
Repenomamus fossils show that it had a long body and short, muscular limbs. It was similar in size, and possibly also in lifestyle, to the modern Tasmanian devil.

Volaticotherium

GROUP Placental mammals
DATE Middle Jurassic to Early Cretaceous
SIZE 20cm (8in) long
LOCATION China

It was once thought that Mesozoic mammals were all tiny, scurrying, shrew-like animals. However, new discoveries have shown that Mesozoic mammals were more diverse than once thought. *Volaticotherium* is especially surprising as it was clearly a glider – the only known Mesozoic gliding mammal. Large skin membranes stretched between its body and long limbs, and the shapes of its finger and toe bones, and its large claws show that *Volaticotherium* was a good climber. It was probably an insectivore.

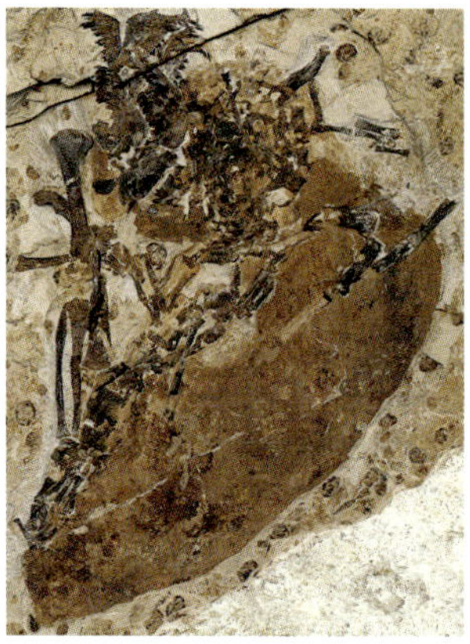

GLIDING MAMMAL
The skin membranes preserved in this *Volaticotherium* fossil suggest that it would have looked like a living flying squirrel.

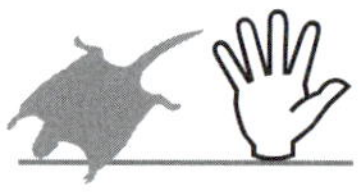

ANATOMY

WING MEMBRANES

Flying squirrels climb trees in pursuit of insects, and then spread their limbs to open up the membranes between them. They then glide from one tree to another. Like flying squirrels, *Volaticotherium* presumably had muscles within the membranes. These would have helped it control its gliding and allowed it to fold away the membranes.

Nemegtbaatar

GROUP Multituberculate mammals
DATE Late Cretaceous
SIZE 10cm (4in) long
LOCATION Mongolia

Nemegtbaatar is a member of an important group of Cretaceous mammals called the multituberculates, or "multis" for short. Multis were similar in size to modern mice and rats. They seem to have been mostly herbivorous and probably lived like rodents. However, they were not closely related to rodents, nor to any other group of placental mammal. *Nemegtbaatar* is one of many multis named since 1970. Its front incisors were large and protruding and it had no canine teeth. Instead, the cheek teeth were separated from the incisors by a short gap. *Nemegtbaatar* had a deep skull and a particularly wide snout, and its snout bones had small holes running through them that blood vessels passed through. This extra blood flow may have supplied a gland or a patch of sensitive skin on the top of its skull, the function of which is unknown.

MOLE-LIKE SKULL
Nemegtbaatar belonged to a Late Cretaceous Asian group of multis called Djadochtatherioidea. These mostly had short, deep skulls that superficially resembled those of modern voles.

Teinolophos

GROUP Egg-laying mammals
DATE Early Cretaceous
SIZE 10cm (4in) long
LOCATION Australia

Teinolophos is a poorly-understood Mesozoic mammal, currently known only from a few partial lower-jaw bones. Several jaw features demonstrate that *Teinolophos* was a monotreme – sometimes called egg-laying mammals. The platypus and the echidnas (or spiny anteaters) are the only living monotremes and are found only in Australasia. *Teinolophos* was originally thought to be an early monotreme, only distantly related to the platypus and echidnas. However, work published in 2008 showed that it possessed several features unique to platypuses, but not echidnas.

DISTINCTIVE JAW
Teinolophos's jaw shares many features of the living platypus, including a particularly large canal inside the bone of the lower jaw.

LIVING RELATIVE
DUCK-BILLED PLATYPUS

Teinolophos was similar in many ways to its modern relative, the duck-billed platypus of Australia. The platypus is an aquatic predator that forages along the bottoms of streams and pools for crustaceans and other prey. It finds these by using specialized sensory cells embedded in the rubbery skin of its soft, sensitive bill.

Sinodelphys

GROUP Marsupial mammals
DATE Early Cretaceous
SIZE 15cm (6in) long
LOCATION China

Sinodelphys is one of the oldest and earliest known members of Metatheria, the mammal group that includes marsupials and all of their fossil relatives. Discovered in the Yixian Formation in Jehol, China, its fossils have hair preserved around their bodies and limbs. *Sinodelphys* was small and slender and would have resembled opossums. It had a slender snout and dainty jaws, and its hands, wrists, ankles, and feet show that it was a good climber. This might indicate that the first metatherians lived in trees. That *Sinodelphys* is a metatherian suggests that the split between marsupial and placental mammals – eutherians – occurred during the Early Cretaceous. This is supported by the fact that *Eomaia* (see below), one of the most primitive eutherians known, is the same age as *Sinodelphys*.

FLEXIBLE ANKLES
Sinodelphys's ankle bones show that it was able to rotate its feet backwards when climbing down tree branches head first.

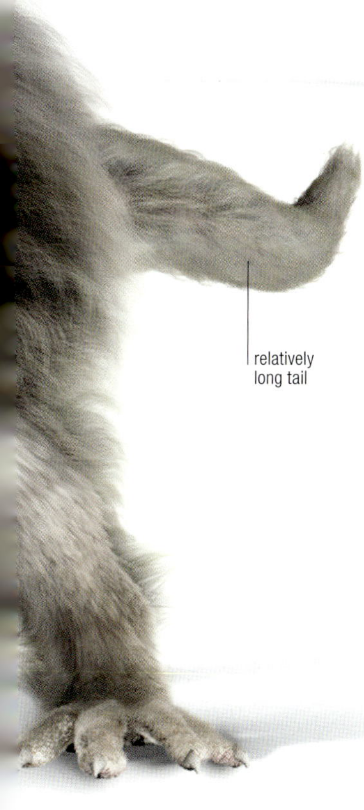

FUR COAT
Eomaia is known from a single, remarkable fossil that shows rarely preserved details including ears and a thick coat of fur.

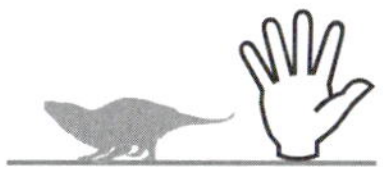

Eomaia

GROUP Placental mammals
DATE Early Cretaceous
SIZE 20cm (8in) long
LOCATION China

Mouse-sized *Eomaia* is currently one of the oldest and most primitive members of Eutheria, the mammal group that includes modern placental mammals and their fossil relatives. Its name means "dawn mother", reflecting its crucial position within our own family tree. *Eomaia* is from the Yixian Formation of Jehol, China. Like other fossils from this site, the outline of *Eomaia's* body has been preserved. Its bones are surrounded by a thick coat of fur, and its long tail is covered in short hairs. The hands and feet of *Eomaia* were similar to those of modern climbing mammals, such as opossums and dormice, so it is thought that *Eomaia* clambered about in bushes and trees. Tall, sharp points, or cusps, on its teeth suggest that it was a predator of insects and other small animals.

Zalambdalestes

GROUP Placental mammals
DATE Late Cretaceous
SIZE 20cm (8in) long
LOCATION Mongolia

A rat-sized mammal from the Late Cretaceous of Mongolia, *Zalambdalestes* is best known for its long, narrow snout and long, slender hind limbs. Its incisors at the tips of its jaws were long and grew continuously throughout its life – in the same way as a modern rodent's front teeth. A gap separated the front teeth from those further back. The tall, pointed teeth suggest

Zalambdalestes had a diet of insects and perhaps seeds. With its long hind legs and shorter front legs, it is thought that this animal probably hopped like a jerboa.

RODENT-LIKE APPEARANCE
Zalambdalestes's narrow, upturned snout and long tail gave it a rodent-like appearance, though it was not a rodent relative.

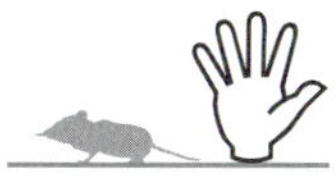

PALEOGENE

Paleogene

Following the major extinctions at the end of the Cretaceous, new forms of life emerged. Mammals, previously small and insignificant by comparison with dinosaurs, were now able to exploit many of the niches left vacant. Earth, after an interval of greenhouse conditions, began a long period of cooling that continued into the Quaternary ice age. Grasses, better able to cope with cooler conditions, evolved and spread.

CRETACEOUS–PALEOGENE BOUNDARY
A thin band of rocks containing iridium mark the Cretaceous–Paleogene boundary in both USA (above) and in Italy (right). It has been suggested that this is due to the collision of a meteorite with Earth at this time.

Oceans and continents

During the Paleogene, Earth's geography began to take on a more familiar shape, as Gondwana continued to break up. New oceans widened the gaps between continents, and ocean currents became established. South America and Africa spread further apart, expanding the South Atlantic Ocean. The North Atlantic Ocean also widened, and the ancestor of the powerful Atlantic current, the Gulf Stream, began to strengthen. The Rocky Mountains were forming in western North America, and the Himalayas and Tibetan Plateau were being folded and uplifted as India continued to plough northwards into the Asia plate. Antarctica remained attached to the southern tip of South America until the Late Eocene, when it finally separated and the Antarctic circumpolar current became established. In the early Paleogene, Australia had begun to separate from Antarctica and as it moved northwards, the Southern Ocean opened. As Africa began to also move in the same direction, the Tethys Ocean began to contract, forcing up the Alps. As continents separated, individual land masses, such as South America, Australia, and New Zealand, became isolated from each other, allowing their faunas and floras to develop independently. Large seaways in western Russia, such as the Turgai Strait that extended north of the Caspian Sea to the Arctic and separated Europe from Asia, also restricted some animal populations to certain areas. Marine life, devasted by the mass extinction at the end of the Cretaceous, slowly recovered and became more abundant by the end of the Paleocene. By the Mid-Paleogene, mammals had arrived in the oceans, in the form of whales.

ROCKY MOUNTAINS
New mountain ranges emerged as a result of collisions between lithospheric plates. The North American Rockies were the result of one such mountain-building episode.

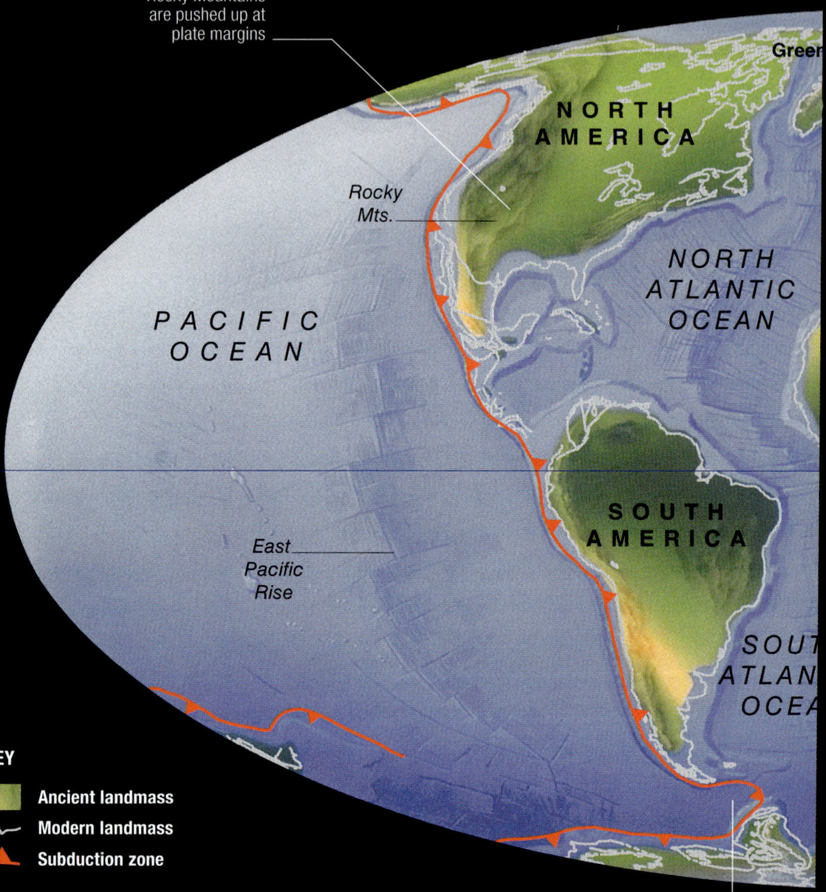

PALEOGENE WORLD MAP
North and South America were now separate, the Himalayas were created, and the Alps were formed by the closure of the Tethys Ocean.

PALEOCENE

MYA	70	60	50
MICROSCOPIC LIFE			50 Peak diversity of dinoflagellates and nanofossils
PLANTS	65–55 Dominance of conifers, gingkos in high latitudes	55 Angiosperms diversify, especially *Nothofagus* in southern hemisphere; first grasses	
INVERTEBRATES	65–55 Rapid diversification of ants; 65 First psammobiid bivalves	60 First semelid bivalves; 58 First clypeasteroid echinoids (right)	50 Increase in diversity of anomalodesmatan and heteroconch bivalves
VERTEBRATES	65–55 rapid diversification of mammals	60 Diversification of large, flightless birds (left); first primates; first edentates (anteaters/armadillos); first carnivorous mammals; first lipotyphlan mammals (hedgehogs, shrews, moles); first owls; 55 Modern bird groups diversify: first passerines (song birds), parrots, divers, swifts, woodpeckers; first whale (*Himalayocetus*); first rodents; first artiodactyl and perissodactyl mammals, including horses; first lagomorphs (rabbits); first armadillos; first proboscidea (elephants); first sirenians (dugongs)	52 First chiropterans (bats); 50 First brontotheres; first tapirs; first rhinos; first camels; diversification of primates

CHAMA

PLESIOLAMPAS

PACHYDYPTES

PALAEOCHIROPTERYX

SUBTROPICAL SEAS
Even in the High Arctic, sea surface temperatures were subtropical during the Paleocene/Eocene Thermal Maximum (PETM), a short period when there was extreme, widespread climate warming.

Climate

During the 40 million years of the Paleogene, Earth experienced dramatic climate change, with a number of warm and cold cycles lasting roughly 10 million years. After a drop in global temperatures at the end of the Cretaceous, they increased again in the early Paleogene, then soared dramatically about 55.4 million years ago. During a short interval of between just 10,000 and 30,000 years, sea-surface temperatures in the tropics rose by about 5°C (41°F), while at high latitudes they increased by about 9°C (48°F). Higher temperatures were accompanied by higher rainfall, with the result that much of the world was covered by tropical rainforests. These conditions persisted for about 170,000 years, until temperatures quickly dropped back to earlier levels, then began to rise again for a few million years during the later part of the Early Eocene epoch. After this a period of cooling began that was to lead to the onset of major glaciations some 50 million years later. This decline was slow and gradual, until at the Eocene–Oligocene boundary, the mean annual temperature fell substantially, by more than 8°C (46°F) in just 400,000 years. Mean annual temperatures dropped from about 20°C (68°F) to just 12°C (54°F). Falls in ocean temperatures were less, perhaps 2 to 3°C (35–37°F). As the climate became cooler, it also became drier.

ANTARCTIC ICE SHEET
By the end of the Paleogene, much of Antarctica was covered by an ice sheet. The temperature drop at the end of the Eocene Epoch marked Earth's transition from a greenhouse to an icehouse world.

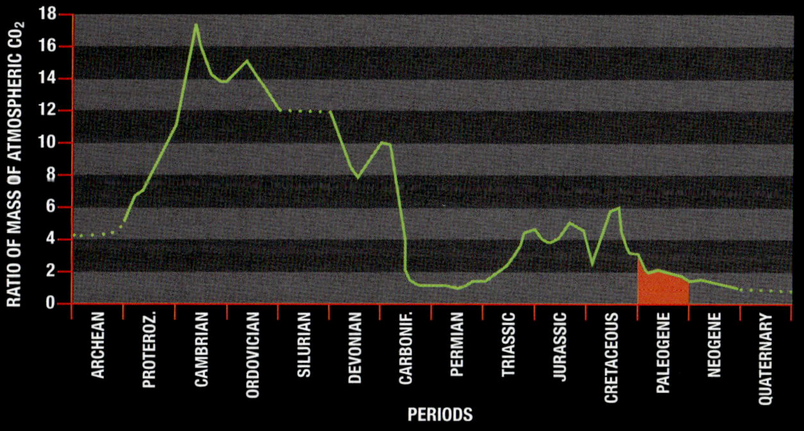

CARBON DIOXIDE LEVELS
Carbon dioxide levels gradually decreased through the Paleogene, while temperatures rose. This sudden global warming may have been a result of the release of methane from deep in the oceans.

OLIGOCENE

40 | 30 | 20 | MYA

MICROSCOPIC LIFE
- 40 Peak diversity of planktic foraminifers (left)
- 35 Increase in diversity of cold-tolerant ostracods and foraminifers

MMULITES

PLANTS
- 35 Beginning of expansion of grassland ecosystems
- 30 First eucalypts in Australia

INVERTEBRATES
- 40 Appearance of modern moths and butterflies
- 35 Major extinctions in gastropods
- 30 First balanids (barnacles)

VELATES

VERTEBRATES
- First embrithopod (extinct, hoceros-like) mammals
- 40 Major diversification of perissodactyl mammals
- 35 Major extinctions in reptiles and amphibians; toothed and baleen whales diversify; many modern mammal groups appear: glyptodons, ground sloths, dogs; first peccaries; first eagles and hawks
- 30 Extinction of embrithopod and brontothere mammals; first pigs; first cats
- 25 First deer

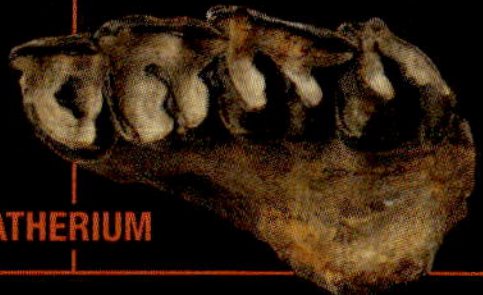
PARACERATHERIUM

PALEOGENE PLANTS

The mass extinction at the end of the Cretaceous had a less dramatic impact on plant life than other organisms. Most plants survived and some continued to diversify. The subsequent changes in climate and the evolution of new animals had more profound consequences for the world of plants.

The rapid evolution of different land animals during the Paleogene (see pp.368, 374) contributed to the development of ecosystems that were much more similar to those of today than had existed in the Cretaceous.

Tropical forests

An important new feature of Paleogene vegetation was the development of closed forests composed of large angiosperm trees. There is little evidence for these kinds of communities in the Cretaceous. By the time of the Eocene, very warm climates with abundant rainfall led to the extensive development of rainforests, not only in equatorial regions but also as far north as southern England. The Eocene London Clay flora of southern England contains the fruits and seeds of mangroves and many other plants that are characteristic of tropical Southeast Asia today. Eocene floras of a similar age from other parts of the world often contain a high proportion of large leaves, a characteristic of tropical vegetation today. At the height of the warm Eocene climate, trees also grew very close to the North Pole, well above the latitudes at which forests grow today.

MODERN TROPICAL RAINFORESTS
Tropical rainforests were not widespread until the Paleogene. They included many groups of plants that are characteristic of tropical vegetation today.

FLOWERING PLANTS
The legume family, illustrated here by the genus *Robinia*, is one of the many flowering plants that made its first appearance during the Paleogene.

Co-evolution with animals

During the Eocene, there is increased evidence of co-evolution between angiosperms and the new groups of vertebrates and insects that flourished after the Cretaceous–Paleogene boundary. Important insect pollinators, such as bees and moths, became more common in the Paleogene and some fossil flowers from the Eocene show specializations for insect pollination, such as the presence of oil glands and bilaterally symmetrical flowers. The average size of angiosperm fruit and seeds also increased in Paleogene floras, reflecting an increased reliance on mammals and birds for dispersal.

GROUPS OVERVIEW

Most of the important lineages of living angiosperms, as well as other groups of plants, had probably begun to differentiate by the Paleogene. Most of these Paleogene plants are similar to plants of today. There were also early indications of ecological associations among different groups of living plants that have persisted up to the present.

FERNS
The development of modern kinds of forests provided new opportunities in understory habitats. These were quickly exploited by the explosive evolution of several groups of ferns. Many different groups of ferns, as well as some lycopods, also flourished as epiphytes in the new canopy habitats created in these new forests.

CONIFERS
Swamp cypresses and their relatives were important in Paleogene floras. The dawn redwood and other conifers flourished in the cool climates close to the North Pole. Fossil plants preserved in ancient moutain lakes also show that the differentiation of modern pines and their relatives was well under way during the Paleogene.

MONOCOT ANGIOSPERMS
The fossil record of monocots increases dramatically during the Paleogene. Many different groups of living monocots appear for the first time during the Eocene. These include grasses, sedges, and pondweeds. Especially common are monocots characteristic of aquatic and low-lying habitats.

EUDICOT ANGIOSPERMS
The Paleogene saw the continuing diversification of eudicots that were established in the Cretaceous, as well as the first appearance of many others. The diversification of many groups of woody angiosperms, as well as the modernization of their pollination and dispersal biology, is a particular feature of the Paleogene.

Lygodium

GROUP Ferns
DATE Cretaceous (possibly Triassic) to present
SIZE Sterile leaflet up to 11cm (4½in) long
LOCATION Europe, North America, South America, China, Australia

Lygodium contains about 40 species of ferns, which are characterized by having leaves that continue growing to undetermined lengths, with slender twisting axes that allow them to climb. These climbing leaves are unique among ferns giving, at first sight, the wrong impression – that of a climbing leafy stem. The pinnules are either untoothed, toothed, or regularly lobed and there is a thin vein running down the middle of each lobe, which splits into secondary veins. Its fertile leaflets are also unusual in having very narrow leaf segments with their spore cases (sporophores) on cone-like extensions. It is the distinctive sterile and fertile foliage that permits the recognition of *Lygodium* in the fossil record. Leaves in the fossil record take *Lygodium* back to the Late Cretaceous, while spores similar to those of *Lygodium* are known from the Triassic onwards. Today *Lygodium* grows in tropical and subtropical regions on all continents.

UNEQUAL LOBES
Lygodium skottsbergii has lobed sterile leaflets, which usually have three, as here, or five lobes of unequal length.

LYGODIUM SKOTTSBERGII
This fossil from the Paleogene of Chile in South America shows the characteristic mass of fertile branched leaf axes with fruiting cones at the tips.

Osmunda

GROUP Ferns
DATE Permian to present
SIZE Fertile leaf up to 2m (6½ft) long
LOCATION Worldwide

The Osmundaceae appeared in the Late Permian of both the northern and southern hemispheres. Evolution of the family was rapid during the Late Paleozoic and Early Mesozoic, giving it the longest fossil record of any of the ferns. There are more than 150 extinct species, and a number of living genera, including *Osmunda*. Much of the geological history of the Osmundaceae is based upon mineralized trunks. These fossil stems are very similar to those of living members of the Osmundaceae. Some fossil leaves have been assigned to living genera when they are nearly identical to the foliage of living ones, and there are even Triassic specimens from Antarctica that are indistinguishable from the living species *Osmunda claytonensis*. The fossil record also shows that the family had produced different fertile and sterile fronds by about the Jurassic.

OSMUNDA SP.
Mineralized sections of *Osmunda* trunks, such as this specimen, show the stems to have had many vascular strands embedded in softer tissue surrounded by a mantle of leaf bases and roots.

LIVING RELATIVE
ROYAL FERN

Osmunda regalis, the royal fern, is one of about 12 species of deciduous ferns found almost worldwide. The sterile fronds have broad leaflet blades, while the sporangia-bearing fertile fronds, seen in the centre here, have very little, if any, leaflet blades.

Metasequoia

GROUP Conifers
DATE Cretaceous to present
SIZE Up to 40m (130ft) tall
LOCATION North temperate regions.

The earliest records of this genus are from the Cretaceous, but *Metasequoia* became one of the most abundant conifers in the Paleogene and early Neogene northern hemisphere, growing as far north as Arctic Canada. Large petrified trunks and stumps of *Metasequoia* have been found in North America. *Metasequoia* shoots can be easily recognized by the stems that branch simultaneously on either side, the two ranks of paired leaves flattened into one plane, and by the structure of the cones, if they are still attached to the shoots. *Metasequoia* has, like other conifers, both male pollen-producing cones and female seed cones. Differentiating *Metasequoia* species can be a problem as many are based on just a few fossil specimens. There are now four or five fossil species that are reliably distinguished by shoot and cone structures or by their seedlings. There are no known fossils of *Metasequoia* in the late Neogene and it was thought that they had died out until living plants were found in China (see panel, below).

PAIRED LEAVES
The distinguishing feature of *Metasequoia* is the way in which the leafy shoots are arranged in opposite pairs, and flattened onto a single plane.

BRANCHING STEMS
Metasequoia is deciduous. Its shoots branch freely, with only the smaller and more terminal stems bearing the leafy shoots that are shed at the end of the growing season.

LIVING RELATIVE
DAWN REDWOOD

Metasequoia glyptostroboides, like the ginkgo, is a living fossil. Amazingly, it was not discovered until 1944 when it was found growing in west central China – three years after it had been named as a fossil. *Metasequoia* is a fast-growing tree that is now commonly grown in temperate regions around the world. It can reach over 40m (130ft) in height and nearly 2m (6½ft) in diameter.

METASEQUOIA OCCIDENTALIS
This is the most common and best-known fossil foliage species of *Metasequoia*, and is the most similar species to the living *Metasequoia glyptostroboides* (see panel, left).

Picea

GROUP Conifers
DATE Paleogene to present
SIZE Up to 90m (295ft) tall
LOCATION North America, Europe, Asia, Japan

There are about 35 species of spruce living in the northern temperate and boreal regions today. They are tall evergreen trees with broadly spire-like crowns and shoots that can be identified by the woody peg-like structures at the bases of the leaves. *Picea* bear cylindrical, pendent seed-cones on leafy branchlets on their upper branches. Today *Picea* is found in montane, and sub-alpine forests in North America, Europe, and Asia.

PICEA SP.
The scales of cones, such as this one, are thin and fan shaped. Their winged seeds mature and are shed in the autumn.

LIVING RELATIVE
BRISTLECONE PINE

Pinus aristata is a slow-growing tree, 5–15m (16½–49ft) tall, and often contorted with strips of living bark separated by sun-bleached wood. It grows in the mountains of Colorado, New Mexico, and Arizona, where some can live up to nearly 2,500 years old.

Macginitea

GROUP Angiosperms
DATE Eocene
SIZE Leaves up to 35cm (14in) long
LOCATION W. North America

Macginitea is an extinct genus of plane tree that has been reconstructed by putting together different *Platanus*-like parts of plants that were repeatedly found next to each other at various localities in western North America. The leaves are large with five to seven palmately arranged lobes. The trees bore stalked globular masses of male flowers or female flowers. They were adapted for early colonization in open disturbed areas, especially close to water.

Podocarpus

GROUP Conifers
DATE Cretaceous to present
SIZE Up to 45m (148ft)
LOCATION North America, Australia, Asia, South America, Africa

These conifers have needle-shaped leaves that are flattened into one plane. Individual leaves are narrowed at the base and have rounded tips. Fossil *Podocarpus* can be distinguished from the similar shoots of *Taxodium*, *Taxites,* and *Sequoia* by anatomical details of their leaf surfaces. There are reports of *Podocarpus* pollen grains from the Cretaceous of North America, but by the Paleogene, the genus was restricted to the Mississippi Basin area. *Podocarpus* leafy shoots are known from the Paleogene of Australasia, South America, and the southern part of North America. The probable origin of the genus was in Australasia with subsequent migration to South Africa and across the Pacific from west to east. There are about 100 living species of *Podocarpus* in the warm temperate and subtropical regions of the southern hemisphere.

PODOCARPUS INOPINATUS
This shoot from the Paleogene of Chile shows the spirally developed flattened leaves arranged into one plane, so that all are orientated towards the light.

Platanus

GROUP Angiosperms
DATE Cretaceous to present
SIZE Trees up to 42m (138ft) tall
LOCATION Europe, Asia, North America

Most of the information on the early fossil history of the Platanaceae family comes from fossil leaves. The earliest come from the Cretaceous of Europe and North America, but they are far more common in the Paleogene and Neogene of Europe, Asia, and North America. Many of the leaves have leaf venation patterns and flower structures that exclude them from *Platanus* and refer them to other genera, such as *Macginitea* (see below, left). Leaves of *Platanus* and similar plants are distinctive in being large, stalked, and palmate with five or more lobes. The leaf margins are smooth or have serrations near the tips. There is a midrib in each lobe, either coming from a common point above the stalk or from extra branches a short distance above it. There are numerous secondary and tertiary veins, although the latter are hard to see. *Platanus* fruits have hairs to help them disperse on the wind. Similar plants, such as *Macginitea*, had hairless fruits.

LIVING RELATIVE
PLANE TREES

There are six species of large, spreading deciduous plane trees that are found in Southern Europe, western Asia to India, North America, and Mexico. They all have maple-like dissected leaves as shown here, and wind-pollinated flowers in spherical balls on long, pendulous stalks. After pollination the fruits hang on the trees through the winter. The London plane, *Platanus* x *hispanicus*, is a hybrid between two *Platanus* species: the European oriental plane and the American sycamore. This hybrid grows even in compacted soil and, being pollution resistant, is ideal for towns.

"PLATANUS"
This five-lobed leaf has clear midribs and secondary veins. The marks around the tips of the lobes on this specimen show where the overlying rock was removed to uncover the leaf.

Florissantia

GROUP Angiosperms
DATE Paleogene
SIZE Flower 2.5–5.5cm (1–2¼in) in diameter
LOCATION North America

There are three known fossil species of these flowers and fruits. The flowers are radially symmetrical and borne on long, slender stalks (pedicels). They have five petal-like organs (sepals), fused to at least halfway along their length, and a prominent radiating network of veins. Only one species has petals and these are free and smaller than the sepals. The projecting ovary has the bases of five pollen-producing stamens fused around it. After pollination and fertilization the ovaries swelled to form five-lobed nut-like fruits similar to those of linden trees, but they were still attached to the remains of the sepals and stamens suggesting that the whole flower was wind dispersed.

SCIENCE

INSECTS

Pollination of flowering plants can occur by means of wind, water, or animals, of which insects are the most common. This delicate insect wing can be seen on the fossil below. The development of nectar and more organized, specialized flower parts led to flowers and insects co-evolving into highly specialized forms (see p.28).

FLORISSANTIA QUILCHENSIS
The stalks of *Florissantia* species, such as this one, suggest that the flowers hung down, and dense, hairy areas at the base of the sepals may have produced nectar to attract insects or birds.

Banksia

GROUP Angiosperms
DATE Paleogene to present
SIZE Up to 30m (98ft) tall
LOCATION Australia

There are many species of fossil *Banksia* based on the leaves and the fruits from the Paleogene onwards in Australia. Records from outside of Australia are very debatable. Fossil leaves suggest that *Banksia* was well-adapted to limiting water loss as it can be seen that the breathing pores (stomata) are confined to the lower surfaces of the leaves and in depressions lined with surface hairs. In living *Banksia*, only a few of the flowers in each flower cluster ever develop and expand into the distinctive two-valved fruits. The valves of the fruits open in order to liberate the seeds for dispersal.

BANKSIA ARCHAEOCARPA
Fossil fruiting bodies, like the one illustrated here, show the positions of the spirally arranged former flowers and, in this case, three mature fruits.

BANKSIAEPHYLLUM SP.
These fossilized leaves, usually assigned to the genus *Banksiaephyllum*, show imprints of the midrib, secondary veins and a network of tertiary veins.

In living *Banksia* only a few of the flowers ever develop into the distinctive fruits.

BANKSIA COCCINEA
Commonly known as scarlet banksia, this shrub grows in southwestern Australia where it grows to about 8m (26ft) tall, producing spiked flowering heads about 8cm (3¼in) in height and width.

Nypa

GROUP Angiosperms
DATE Late Cretaceous to present
SIZE Fruit 8–12cm (3¼–4½in) long
LOCATION Northern hemisphere

The earliest palm fossils are leaves, stems, and pollen from the Late Cretaceous. Their remains were widespread and abundant in the Paleogene and fruits of the mangrove palm, *Nypa*, are common in European deposits. The size and number of fossil *Nypa* suggest that conditions in northwest Europe at that time were similar to present-day brackish mangroves in India and Southeast Asia.

NYPA SP.
This *Nypa* fruit eroded partially while drifting in the sea. The outer coat (epicarp) remains only at the top, exposing the inner coat (mesocarp) below.

Cyclocarya

GROUP Angiosperms
DATE Paleogene to present
SIZE Up to 30m (98ft) tall
LOCATION Europe, North America, Asia

Cyclocarya is a genus in the walnut family with only one living species. It is a deciduous tree growing to about 30m (98ft) tall with pinnately divided leaves and male and female catkins. Its distinctive fruits are surrounded by disc-like wings that help them disperse in the wind. Fossils of the walnut family are known from the Late Cretaceous of North America and Europe, but the first genus of living plants to appear is *Cyclocarya* in the Paleogene; a genus that then quickly spread from North America to Europe and Asia. The fossil assemblages where *Cyclocarya* has occurred have also included such plants as *Glyptostrobus*, *Metasequoia*, and *Liquidambar* – suggesting a warmer temperate to subtropical climate. Today, *Cyclocarya* is found only in China.

Sabalites

GROUP Angiosperms
DATE Cretaceous to present
SIZE Leaf about 1–2m (3¼–6½ft) long
LOCATION North America, Mexico, Europe

The earliest palms are found in the Late Cretaceous but none can be reliably assigned to modern genera. Palms are found in tropical and subtropical areas today; they are good indicators of warm climates. The sabal palm was one of a variety of palms with divided leaves, found in deposits in North America, Mexico, and Europe. It survived into the Paleogene, and its fossilized fruits have also been found in North America and southeast England. The name *Sabalites* was originally used for the leaves, but now refers to its seeds.

SABALITES SP.
In this fossil the stalk, or petiole, can be seen below the base of the sabal leaf. The fine lines indicate the leaf divisions.

PALEOGENE INVERTEBRATES

Following recovery from the Late Cretaceous extinction, the fifth to affect evolving life on Earth, marine invertebrate faunas similar to those of today occupied the available ecospace. A shell-littered shoreline of the Paleogene would look much like today's. Though warm, the first stages in global cooling had begun.

Establishment of coral reefs

One factor in the diversity of marine invertebrates during the Paleogene is the evolution of coral reefs. Scleractinian corals came into existence during the Triassic, and flourished through the Mesozoic, but there were no true coral reefs until the Paleogene. Living coral reefs are of three kinds: fringing reefs, barrier reefs, and atolls. Fringing reefs develop along shorelines, especially those of volcanic islands. If the island subsides, a circular or horseshoe-shaped atoll will form, and it has been shown by deep-drilling through Pacific atolls that coral reefs began in the Paleogene. Scleractinian corals have two advantages as frame-builders. Firstly, they can cement themselves to the substratum, unlike the Paleozoic corals that had no facility for attachment. Secondly, they have symbiotic algae in their tissues, which provide them with oxygen and carbohydrates, enabling the corals to grow rapidly. As a result, they have been able to build up a substantial framework, inhabited by many other living creatures – forming the basis for the most complex and productive ecosystems of all time.

CORAL REEF
This living Pacific coral reef contains colonies of many different species living side by side. Reefs provide support for other animal and plant life.

WORM WOOD
This fossilized log contains the preserved remains of the shipworm *Teredo*. It still exists today and lives in groups in either floating or submerged timber.

The Danian

Much of Denmark is underlain by Upper Cretaceous Chalk. On the shore of the island of Sjaelland, cliff sections reveal a grey, iridium-rich layer, with scattered fish bones, marking the Cretaceous mass-extinction event. Above this, chalky sediment continues well into the Paleogene. This is known as the Danian. The last ammonites and belemnites perished in the catastrophe. Nautiloids persisted, and in the Danian they take the place of the ammonites.

SPIRED SHELL
Turritella is a large, high-spired gastropod, typical of the Paleogene, and ornamented with ribs.

GROUPS OVERVIEW

Gastropods continued to evolve and spread during the Paleogene, with new forms appearing and adapting. Bivalves also appear to have diversified. Many echinoids emerged from the Late Cretaceous relatively unscathed. New forms of crustaceans appeared and quickly became established as highly effective predators.

GASTROPODS
During the Paleogene, gastropods continued to diversify and colonize new habitats. At this time, large – often highly ornamented – gastropods appeared, along with the pteropods, which are small gastropods with straight shells, adapted for a floating life in the plankton. Some freshwater Paleogene limestones may consist entirely of gastropods.

BIVALVES
Many of the common bivalve genera living today first appeared in the Paleogene. During this period bivalves appear to have evolved many different life habits, these include: byssal attachment (in which the animal is secured to a hard surface by a thread-like structure), bacterial feeding, and both deep and shallow burrowing.

ECHINOIDS
Most echinoid groups were not greatly affected by the late Cretaceous mass extinction, and both surface-living genera, such as *Echinocorys*, and deeper-burrowing genera, such as *Linthia*, continued as an important component of the Paleogene fauna.

CRUSTACEANS
Lobsters and crabs are the largest living crustaceans and they became highly abundant and diverse in the Paleogene. Their grasping pincers (unknown in Paleozoic crustaceans) were a major factor in their success.

ROTULARIA BOGNORIENSIS
This gathering of *Rotularia* were probably swept together by currents on the sea bed.

Rotularia

GROUP Worms
DATE Middle Jurassic to Paleogene
SIZE Up to 2cm (¾in) across
LOCATION Worldwide

Because of the soft nature of their bodies, worms are rarely fossilized and the most we know about many shell-less worms are the burrows they made into the sediments. There are some notable exceptions, namely the serpulid worms. These marine polychaete worms, which are relatives of the earthworm, secreted tubes composed of calcium carbonate around themselves to protect their soft bodies. Serpulid worms are therefore preserved as fossils in the same way as other shelled invertebrates. Unlike most surviving serpulid worms, which live cemented to the sea bed or to the shells of other invertebrates, *Rotularia*, although attached in its early stages of growth, was free-living. The tubular shell it lived in was at first coiled in a low spiral, but at maturity the last part of the shell straightened out and broke free from the coiled part.

Aturia

GROUP Cephalopods
DATE Paleogene to Early Neogene
SIZE Up to 15cm (6in) across
LOCATION Worldwide

Aturia was a nautiloid, similar in appearance to modern species belonging to the genus *Nautilus*. It had a very tightly coiled shell with heavily overlapping whorls and a very small umbilicus at the centre of the coil. This type of shell is said to be extremely involute. One of *Aturia*'s most distinctive features is its complex suture line with sharply angled folds (lobes). These V-shaped lobes face away from the shell's opening, or aperture. In cross section, the whorls are somewhat compressed, with flattened sides, but the venter (outer edge) is rounded. The shell is smooth, and ornamented only by very fine growth lines. *Aturia* is likely to have lived in fairly open water and probably fed on small fish and small crustaceans.

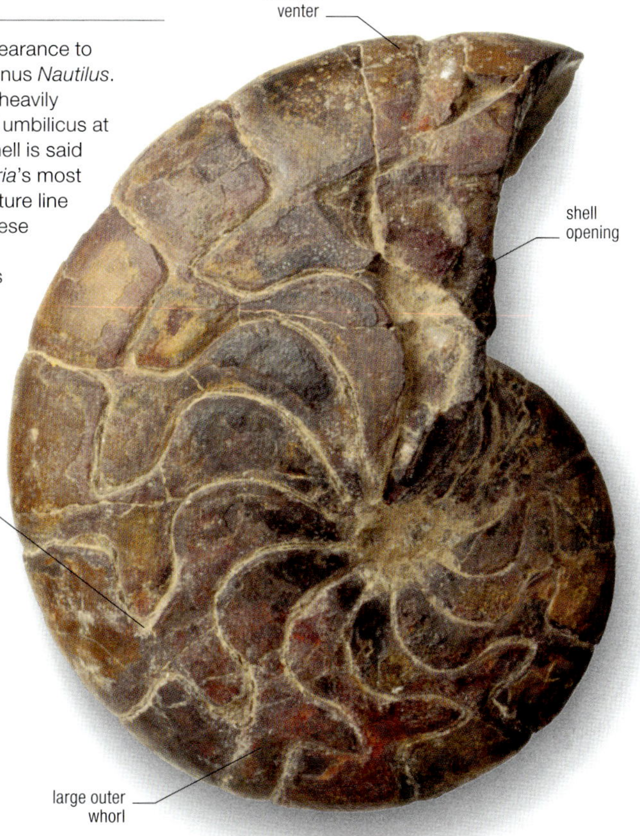

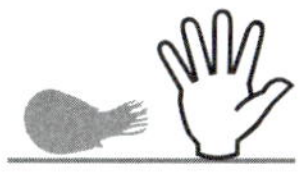

ATURIA PRAEZIGZAC
Clearly visible on this internal mould, the suture lines of *Aturia* are quite complex for a nautiloid. These superficially resemble the suture lines of some of the earlier ammonoids.

Xenophora

GROUP Gastropods
DATE Paleogene to present
SIZE Up to 4cm (1½in) long
LOCATION Worldwide

Xenophora is a genus of marine gastropods that still exist in the world's seas today. Among other distinctive features, it had a conical shell with a fairly low cone (spire), and very little overlap between successive whorls. There are strong growth markings on the shell, most clearly visible on its base. Again viewed from the underside, the central umbilicus is narrow and steep-sided. Unlike many gastropods, which have nodes on their shells, *Xenophora*'s bumpy outline was the result of shell fragments or rock particles embedded in its shell. The animal picked up pieces of shells and held them in the mantle margin against the outside of its soft body until other shell material was secreted around the foreign objects to fix them into place.

***Xenophora* means "bearer of foreigners"; the name was given because it picked up pieces of shells.**

SHELL OPENING
This view shows the underside of the shell. *Xenophora* had a rounded aperture, overhung in part by the shell. The lower lip of the shell's mouth is set back.

COSMETIC ENHANCEMENTS
Xenophora's shell had an irregular surface. This was because it would pick up pieces of shell from the sea bed and cement them into its own shell.

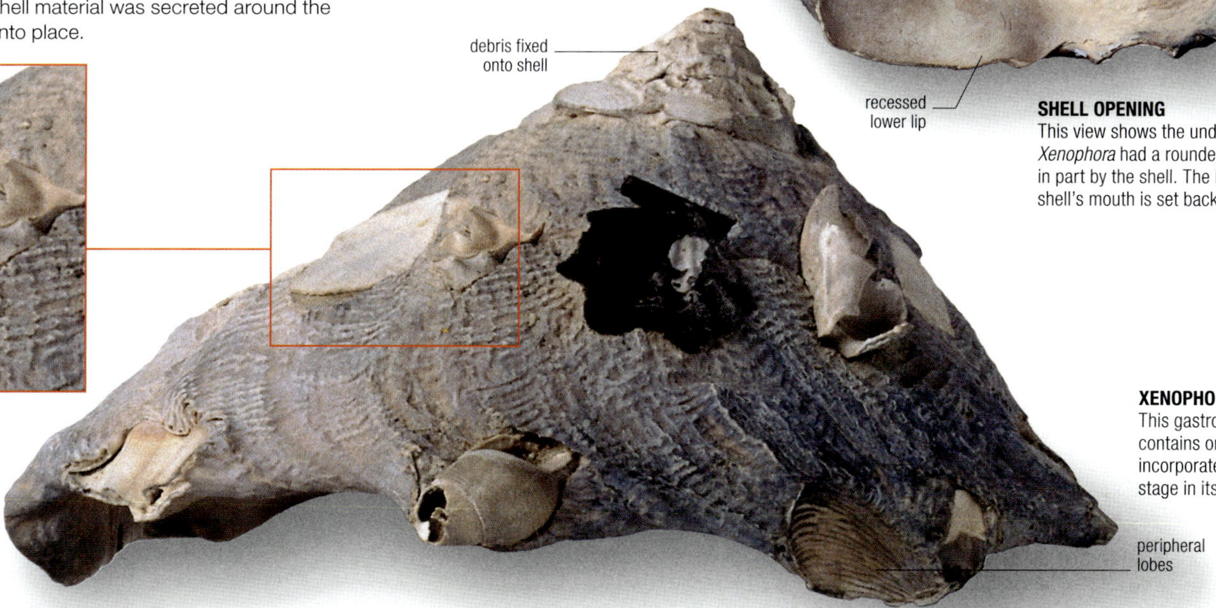

XENOPHORA CRISPA
This gastropod's apex (spire) contains one shell that was incorporated at an early stage in its life.

The shell of *Athleta* is **dominated** by the **pointed nodes** on the **shoulder of the whorl**, at the widest part of the shell.

Athleta

GROUP Gastropods

DATE Paleogene to present

SIZE 6.5–10cm (2½–4in) long

LOCATION Worldwide

Gastropods of the genus *Athleta* were predatory carnivores. *Athleta* fossils have a distinctive shape, with the tip of the shell forming a small, steep cone. As the animal matured, the whorls of its shell broadened. The most prominent parts of the shell's ornament are the spiny nodes on the shoulder of the outermost whorl at the widest part of the shell. From these nodes, ribs run up to the whorl suture, where the whorls join, and down the side of the outermost whorl. On the lower part of the outermost whorl there are strong spiral ribs, but because of the heavy overlap between whorls, these are not visible on the earlier parts of the shell. The mouth (aperture) of the shell is long, with a notch at the side for the inhalant siphon, a tubular structure used by the animal to take in water.

ATHLETA ATHLETA
Athleta can be distinguished by the spinous nodes on the shoulder of the outermost whorl. Parallel to the ribs that run downwards from these nodes are fine growth increments.

Clavilithes

GROUP Gastropods

DATE Paleogene to Neogene

SIZE 13cm (5in) long

LOCATION Europe, Africa, North America, South America

Clavilithes was a stout gastropod with a large shell, which had flat-sided whorls that barely overlapped each other. The whorl suture, where one whorl joined the next, was distinctly stepped, and the outside of the shell was covered with growth lines. The shell opening was large and oval in shape. Comparisons with modern, related forms suggest that *Clavilithes* was carnivorous.

CLAVILITHES MACROSPIRA
The shell of *Clavilithes* is covered with growth lines, which are deflected backwards in the mid-whorl position.

ANATOMY
INSIDE A SHELL

The soft body parts of a mature gastropod would only have occupied around the newest two whorls of the shell. The pillar-like structure in the centre is called the columella. It is formed by fusion of the wall of each whorl as it coils round the axis.

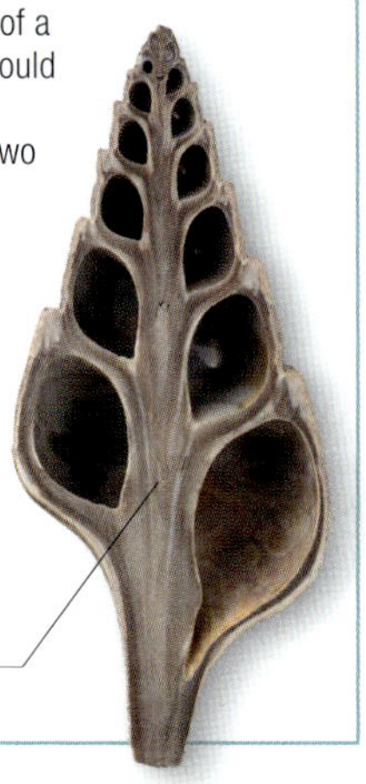

Venericor

GROUP Bivalves

DATE Paleogene

SIZE 2–8cm (¾–3¼in) long

LOCATION North America, Europe

Venericor was a filter-feeding bivalve. Its shell was a rounded triangular shape with forward-facing, curved beaks. The outside of the thick shell had strong radial ribs that became broader and flatter towards the ventral margin. *Venericor* had two teeth and sockets in each valve. Behind the teeth, a long, curved depression housed the ligament. The interior margins of the valves had small crenulations (notches).

VENERICOR PLANICOSTA
Variable growth lines cross the ribs, suggesting seasonality in growth rates.

PHOLADOMYA AMBIGUA
Pholadomya's ornamented shell helped it to grip against the soft sediment in its burrow. It could partly withdraw its siphons into a sinus at the posterior end of the shell.

Pholadomya

GROUP Bivalves
DATE Late Triassic to present
SIZE 7cm (2¾in) long
LOCATION Worldwide

Pholadomya is a burrowing bivalve with shells that gape open permanently at the posterior end. The shell is strongly inflated and its beaks fold inwards. It has concentric lines, parallel to the growth direction, and there are clearly marked radial ribs, which are strongest towards the middle part of the shell. The front and outer parts of each valve lack these radial ribs. There were no hinge teeth and the ligament was housed in folds of the outer margin of the shell, behind the beaks.

Crassatella

GROUP Bivalves
DATE Late Cretaceous to Early Neogene
SIZE 3.5cm (1½in) long
LOCATION Europe, North America

Crassatella was a bivalve with very strong concentric ribs parallel to the shell growth. The fossil shows a strong shoulder to each valve running from the beak to the ventral margin. Internally, it has deep, equally sized adductor muscle impressions, with a distinct pallial line running between them. The adductor muscles were responsible for closing the shell, and the pallial line marks where the animal's mantle would have been attached. The hinge has two teeth below the beak in each of its valves.

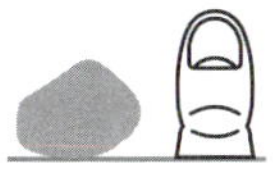

C. LAMELLOSA
Crassatella fossils are made up of two halves (valves) that are roughly symmetrical. On the outside of each valve there are very clear growth ribs.

Chama

GROUP Bivalves
DATE Paleogene to present
SIZE 4cm (1½in) long
LOCATION Europe, North America

Chama is a bivalve with a strongly asymmetrical shell. The animal is attached to the sea bed by its larger, more convex left valve throughout its life. The right valve is smaller and much flatter than the left, though still convex. Both valves had well-developed beaks. The valves grew spirally, along a horizontal plane. The surface of each valve shows concentric frills and radially arranged, flattened spines; these two features combine to give the shell a rather scaly appearance. Internally, there are two muscle scars, where the adductor muscles, which open the shell, attach in life. The pallial line, where the mantle attached, is smooth. On the hinge-line there is a single, rather blunt tooth in each valve, and above the hinge-plate there is a curved groove that housed the ligament.

C. CALCARTA
The spines of *Chama* might have attracted the growth of algae and small animals, which camouflaged it.

Glycymeris

GROUP Bivalves
DATE Early Cretaceous to present
SIZE 5cm (2in) long
LOCATION Worldwide

Glycymeris is a fairly thick-shelled bivalve with an almost circular outline. Its beaks are situated more or less centrally on its midline. Beneath the beaks, and extending both to the front and rear of them, there is a large triangular area which is marked with tiny grooves, where the ligament was housed. Beneath the ligamental area is the hinge-plate, which has numerous teeth and sockets that are smallest beneath the beaks, but become markedly larger and more curved as they follow the hinge margin round the outer curve. Unlike many of its arcoid relatives, *Glycymeris* was not attached by a bundle of fibres (byssus) in the adult stage.

G. GLYCYMERIS
The shell is ornamented with a combination of concentric growth lines, some stronger than others, and fine, radial ribbing.

GLYCYMERIS BREVIROSTRIS
Along the hinge area are numerous teeth and sockets. They are smallest beneath the beaks and become larger further along the hinge margin.

Teredina

GROUP Bivalves
DATE Late Cretaceous to Neogene
SIZE 15cm (6in) long
LOCATION Europe

Teredina was a very specialized type of bivalve mollusc. Its small, bivalved shell extended back into an irregularly shaped, long tube with a narrow opening. *Teredina* bored into wood or other soft material in marine or brackish water by moving its valves from side to side. This activity caused excessive wear in the beaks of the valves, so extra plates were secreted to counter this problem. When young, the animal possessed a foot, but as it burrowed and became permanently lodged in a piece of wood, the foot wasted away. *Teredina* is a relative of the modern shipworm *Teredo*, a type of clam that burrows into wood. Like the shipworm, *Teredina* was a filter feeder, drawing water into its body and removing minute particles of food using specialized gills. *Teredina* was long thought by fossil experts to be a type of tube worm, due to the long, tube-shaped end to its shell.

***Teredina* bored into wood in marine or brackish water by moving its valves from side to side.**

TEREDINA SP.
The specimen to the right shows a cross section of a log containing many *Teredina.* These worms bored their way towards the centre of the log with the valves at the top of their bodies.

Lovenia

GROUP Echinoderms
DATE Late Paleocene to present
SIZE Up to 3.5cm (1½in) long
LOCATION Indian Ocean, Pacific Ocean

Lovenia was a flattened, heart-shaped, burrowing echinoid that lived buried in the sand of the sea bed, usually in inshore waters. Some species of the genus, called heart urchins, still exist today. *Lovenia* fossils consist of the hard "test" or endoskeleton of the animal. Like all echinoderms, *Lovenia's* body was split into five different sections, which appears as a star-shape on the upper side of the fossilized test. This star-shape comprised the five ambulacral areas, which had large pores around their margins, each of which would have been the site of a tube foot. One of the ambulacral areas formed a deep furrow that continued down to the oral surface, where the mouth was situated. There were several fascioles – smooth areas of the test surface that carried tracts containing small, hair-like extensions (cilia).

LOVENIA SP.
In the centre of the upper surface of the test is the apical disc, made up of four small, circular pores through which eggs and sperm were released.

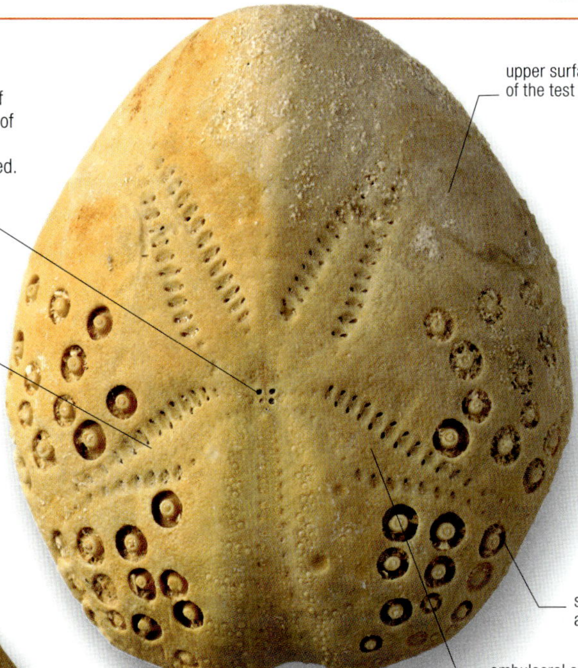

PROTECTIVE SPINES
There is a fine covering of bumps (tubercles) on both sides of the test. All the tubercles were sites of articulation for protective spines.

LIVING RELATIVE
HEART URCHIN

The heart urchin shown to the right *(Lovenia cordiformis)* is a relation of the fossil *Lovenia*. It is well adapted to life buried in the sediment, and relies upon the water currents that its small, hair-like extensions generate to bring food particles and oxygenated water down to its body. The earliest urchins had their mouth located in the centre of their underside, and the anal opening positioned in the middle of their upper side; such forms are still living (see *Temnocidaris* p.314). During the course of evolution, the anal opening migrated backwards, while the mouth moved forwards, as in the fossil *Lovenia*.

Palaeocarpillius

GROUP Arthropods

DATE Paleogene

SIZE Up to about 6cm ($2\frac{1}{4}$in) long

LOCATION Europe, Egypt, Somalia, India, Zanzibar, Java, Mariana Islands

The domed carapace (shell) of the fossil of this early crab is oval in outline and smooth, with spiny margins at the front and sides. The orbits, which housed the stalked eyes, are well developed. The pincers on the anterior limbs (chelipeds) are clearly visible; the right one is markedly larger and more robust than the one the left. The moveable finger on the right claw had a serrated inner edge, which contrasted with the comparatively slender left pincer. Although crabs began to diversify greatly during the Cretaceous, their numbers increased spectacularly during the Cenozoic Era. This highly successful group is more numerous today than it has been at any time during its history.

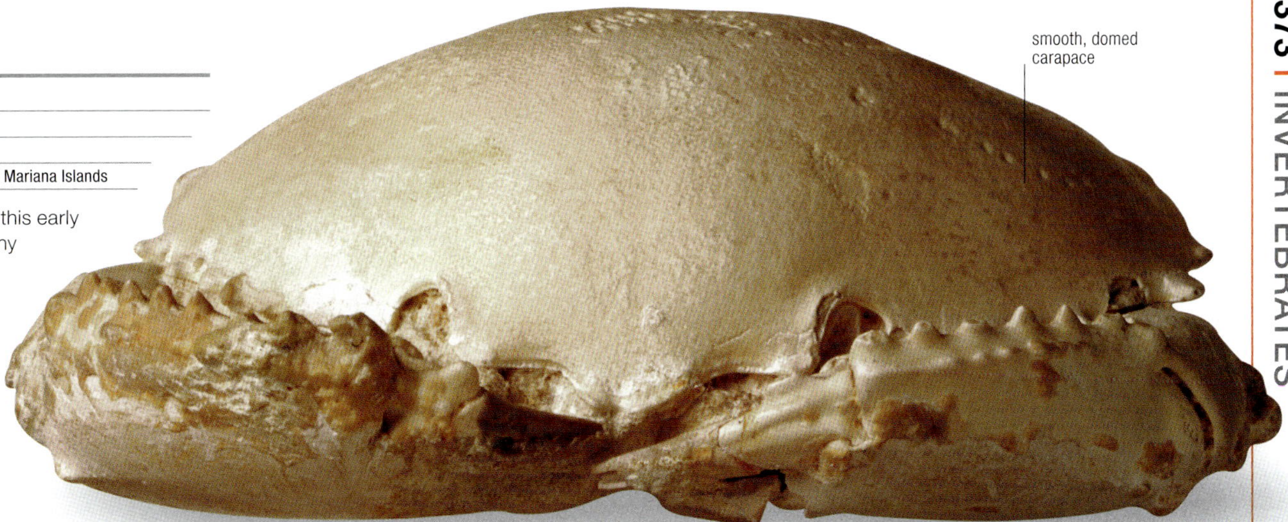

PALEOCARPILIUS AQUILINUS
The upper edge of both claws contain large bumps (tubercles). The moveable finger on the right claw has a serrated inner edge that the left pincer lacks.

Crabs are a highly successful group that are more numerous today than they have been at any time in their history.

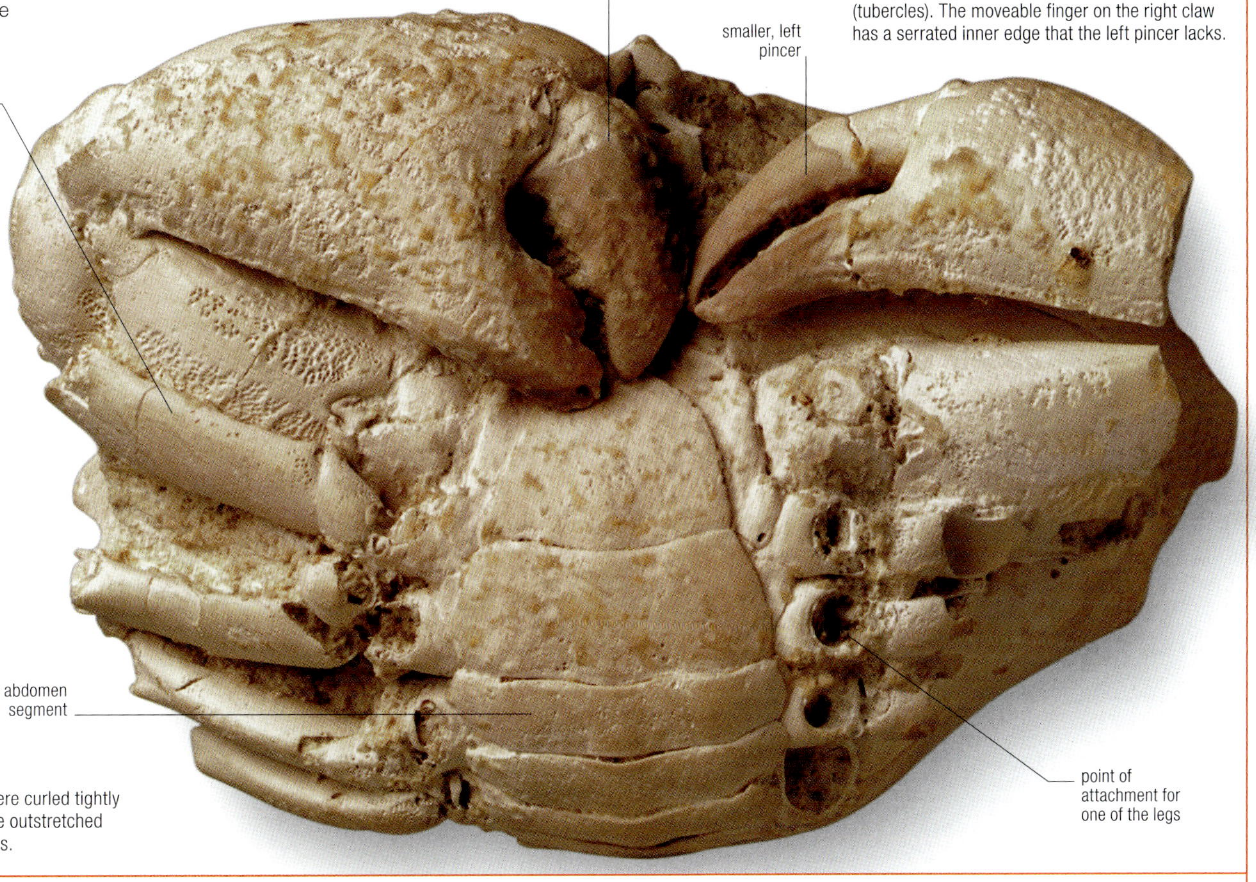

CURLED ABDOMEN
The segments of the abdomen, which were curled tightly under the carapace, are equivalent to the outstretched abdomen in modern lobsters and shrimps.

LINUPARUS EOCENICUS
Usually only the fossilized carapace of lobsters are found, but the side view of this example shows the limbs, which can be seen underneath the carapace.

IN THE GROOVE
The top of the carapace is divided into three sections. The front and central areas are divided by a V-shaped groove, and the central and rear sections are separated by a curved groove.

Linuparus

GROUP Arthropods

DATE Paleogene to present

SIZE Carapace about 5cm (2in) long

LOCATION Europe, West Africa, North America, Northeast Asia

The earliest, lobster-like crustaceans appeared during the Permian, but they are not abundant in the fossil record until the Paleogene. This lobster had an elongated shell (carapace) with a rounded margin at the front. The sides of the front and central parts of the carapace had a coarse, rough texture. *Linuparus*'s body was divided into three portions, which were defined by deep grooves. The abdomen was composed of five segments with smooth articulatory surfaces between them that were exposed when the abdomen was curled down. The abdomen formed a tail fan, which the animal used for propulsion through the water.

PALEOGENE VERTEBRATES

Following the demise of the dinosaurs at the end of the Cretaceous, mammals and birds began to radiate into diverse forms on land, underwater, and in the sky. Early in the Paleogene, fish and reptiles had taken on more recognizable forms and did not change very much.

At the start of the Paleocene epoch, 65 million years ago, mammals were small and insectivorous, with little limb or teeth specialization. They soon began to change in order to exploit the ecological niches left vacant by the dinosaurs – growing in size, evolving new modes of locomotion, and diversifying their diets to include plants and other vertebrates. The mammalian radiation continued during the Oligocene.

FOSSIL BAT
Fossil bats from the Messel oil shale pit in Germany are astonishingly well-preserved. The wings and skin of this specimen are clearly seen.

Taking to the air

Bats are sparsely represented in the fossil record because their light bones and delicate wing membranes do not fossilize well. For this reason, the precise time in evolutionary history that bats took to the air is not known. However, a few excellent bat fossils have been found at two locations dating from the Eocene epoch – an oil shale pit in Germany and the Green River Formation in Wyoming, USA. They show that flying bats existed at least 52.5 million years ago. Their arm and greatly elongated finger bones support a stretchy skin membrane that, powered by muscles, enables flapping flight. Bats probably evolved from a nocturnal insectivore that glided from tree to tree. The first fruit-eating bats appear in the Oligocene fossil record.

Carnivores and carnassials

At the beginning of the Paleogene, mammals had unspecialized teeth that could cope easily with their diet of insects. During the Paleocene epoch however, two major groups of mammals, the carnivores and the creodonts, evolved teeth suited to a diet of meat. Carnassial teeth – the defining feature of modern-day carnivores – are supremely adapted to shearing flesh from bone in a scissor-like action, while the large, pointed canines deliver the killing bite. Carnivore ancestors, the miacids, were small, weasel-sized mammals. Dogs appeared during the middle Eocene and true cats in the early Oligocene, in parallel with a cat-like group called nimravids (also known as false sabre-tooths).

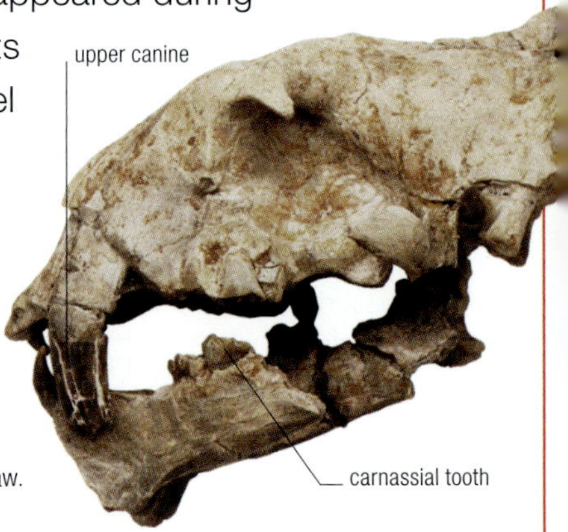

SLICING MOLARS
In primitive carnivores (shown here) and their modern descendants the carnassial teeth are the last upper premolar and the first lower molar. Creodont carnassial teeth were further back in the jaw.

GROUPS OVERVIEW

By 50 million years ago, only 15 million years after the end-Cretaceous extinction, primitive representatives of most mammalian orders had appeared. Early hoofed mammals (ungulates) were preyed upon by carnivores and creodonts, while whales took to the oceans and bats to the skies. The first primates, rabbits, and rodents also appeared.

CREODONTS
Creodonts were the dominant predators throughout the Eocene and Oligocene. They occupied many of the same niches as carnivores, which eventually replaced them. There were two families, *Oxyaenidae*, and the more widespread *Hyaenodontidae*, which included *Megistotherium* – one of the largest land predators to have ever lived.

PERISSODACTYLS
The first odd-toed ungulates appeared at the end of the Paleocene and radiated during the Eocene into primitive horses, rhinos, tapirs, and brontotheres. The first species browsed leaves and were relatively small and unspecialized. They gradually increased in size and further diversified throughout the remainder of the Paleogene.

ARTIODACTYLS
The first even-toed ungulates also appeared in the early Eocene and were small and long-legged. They soon radiated into numerous different forms, including the precursors of pigs, camels, and ruminants such as giraffes and cattle. All had crescent-shaped cheek teeth and a special double-pulley bone in the ankle characteristic of artiodactyls.

CETACEANS
Throughout the Eocene, a branch of hoofed mammals gradually returned to the water as primitive whales. Early species still had well-developed limbs and could walk as well as swim, probably entering the water to hunt. Toward the end of the epoch the first fully aquatic whales appeared. They had small hind limbs that were essentially useless.

Diplomystus

GROUP Actinopterygians
DATE Early Eocene to Miocene
SIZE 65cm (26in) long
LOCATION USA, Lebanon, Syria, South America, Africa

A distant relative of herrings and sardines, *Diplomystus* was a widespread freshwater fish. Many of the best specimens have been found in the Green River deposits of Wyoming, USA. It had a single dorsal fin, pairs of small pectoral and pelvic fins, and an anal fin that stretched back to the narrow base of the tail. The tail itself was deeply forked, with the upper and lower parts of the fin being of equal size. A number of smaller fish, such as *Knightia* (see below), have been found preserved in the stomachs of *Diplomystus* fossils. *Knightia* was once thought to be a relative as well as prey of *Diplomystus*. However, this is now known to be incorrect. *Diplomystus* can be distinguished from *Knightia* by two rows of modified scales. The first runs along the midline of the back, from the skull to the dorsal fin; the other row runs along the belly.

UPTURNED MOUTH
Diplomystus's mouth faced upwards, with the jaw set at an angle to the backbone, rather than being parallel to it. This suggests that it hunted smaller fish swimming just below the surface.

Knightia

GROUP Actinopterygians
DATE Middle Eocene
SIZE 25cm (10in) long
LOCATION USA

Knightia is thought to have been a freshwater relative of the modern herring. Every year, huge numbers of *Knightia* fossils are unearthed at quarries in the Green River region (see panel), and they are a common sight in rock shops around the world. Most *Knightia* specimens are small, but very well preserved. The sheer number of specimens suggests that *Knightia* lived in large shoals in the Green River. *Knightia*'s body size and abundance tells us that the fish was probably a secondary feeder, filtering out ostracods, diatoms, and other microscopic plankton in the lake water. Many *Knightia* skeletons have been found in the stomachs of larger fish.

OFFICIAL FOSSIL
In life *Knightia* swam in large shoals. It is so common in the Green River Shale that it has been designated as the official fossil of the US state of Wyoming.

KEY SITES
GREEN RIVER

Knightia is the most abundant fossil in the Green River Formation, an extensive series of lake deposits in Wyoming, Colorado, and Utah. Shales more than 3km (9,800ft) thick were deposited there between 53 and 46 million years ago. The Green River Shale is world famous for revealing hundreds of beautifully preserved skeletons complete with many fine details.

Sciaenurus

GROUP Actinopterygians
DATE Early Eocene
SIZE 20cm (8in) long
LOCATION England

The oldest known representative of the broam family, *Sciaenurus* was described by the scientist Louis Agassiz in 1845. The head had large eyes and a wide mouth. The body was about four times longer than it was deep. Many *Sciaenurus* specimens have been found in the London Clay deposits, which were formed by a subtropical sea that once covered the southeast of England.

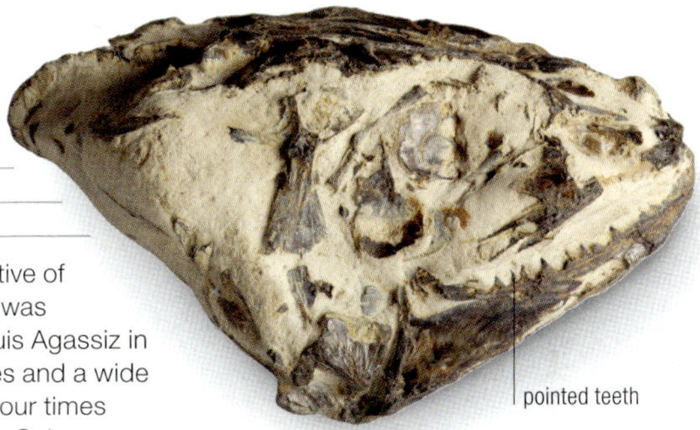

ACTIVE HUNTER
The lower jaw had a single row of sharp teeth, which suggests that *Sciaenurus* was an active hunter of other fish.

Mioplosus

GROUP Actinopterygians
DATE Eocene
SIZE 25cm (10in) long
LOCATION USA

A member of the perch family, *Mioplosus* was a freshwater lake fish. Its fossils are found singly, never in groups, and the fish is thought to have been a solitary hunter. It was also capable of attacking fish up to half its own size, using its many sharp teeth. A large tail, with symmetrical forks extended from a narrow base, suggests that *Mioplosus* was a powerful swimmer. It had two dorsal fins. The first one lay just behind the pectoral fins, and the rear dorsal fin was positioned above the anal fin.

CONSUMING PREY
As shown here, *Mioplosus* fossils are sometimes found in the process of consuming prey.

Mene

GROUP Actinopterygians
DATE Early Eocene to present
SIZE 25cm (10in) long
LOCATION Worldwide

Two species of *Mene* – *M. rhombea* and *M. oblonga* – have been found in significant numbers at Monte Bolca in northern Italy. This site, nicknamed "the fishbowl", has yielded a number of fish fossils with a remarkably high degree of preservation. It is even possible to tell what colour many of the specimens would have been in life. There is one surviving representative of the *Mene* genus: the moonfish, *Mene maculata*, from the Indian and Pacific Oceans. This living species is typical of the genus. It has large eyes and a protruding, upturned mouth. It is a marine fish with a very deep body, due mainly to downward expansion of its lower half. At the same time, the body is also highly compressed from side to side. *Mene's* shape suggests that its body was rather stiff and capable of only limited movement. It is likely that most of the propulsive force was generated by rapid movement of its tail.

DIFFERENT FINS
Mene had a small dorsal fin, a long anal fin, and small pectoral fins high on its side. The two pelvic fins were reduced to long spines that trailed in the water and may have extended beyond the tail.

LIVING RELATIVE

FRESHWATER EEL

Anguilla is the only surviving genus of the freshwater eel family Anguillidae. There are 20 species, found mainly in Europe, Asia, and North America. Eels spend their adult lives in fresh water, but return to the deep ocean to spawn. Their larvae (young) eat plankton as they drift on ocean currents towards the coasts, where they then enter river systems. Eels navigate along the river bottom, avoiding the stronger currents in the middle of the river.

Anguilla

GROUP Actinopterygians
DATE Middle Eocene to present
SIZE 1m (3ft 3in) long
LOCATION Worldwide

Anguilla is a living genus better known by the common name, eel. Eels are easily recognized by their distinctive, elongated bodies. The earliest *Anguilla* fossils are from Monte Bolca in Italy. Unlike today's forms (see panel, left), these early specimens were marine in origin. Only later did the genus begin to exploit the brackish and freshwater habitats that it occupies today.

UNIQUE FORM
Eels are fish like no others. The head is small, but the mouth can gape widely. The long, muscular body is the main source of power, and the fins are small and ineffective.

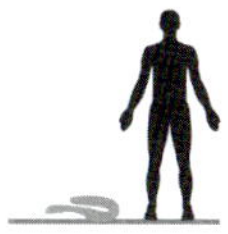

Heliobatis

GROUP Chondrichthyans
DATE Middle Eocene
SIZE 1m (3ft 3in) long
LOCATION USA

Heliobatis may have been an ancient relative of the stingray. Its fossils are common in the Green River Shale of Wyoming, USA (see p.375). When the long tail spine is included, *Heliobatis* was about 1m (3ft 3in) long, a similar size to many living stingrays. However, it was not as large as today's Chinese stingray, which is more than 5m (16½ft) long and weighs 1,500kg (3,300lb). Most modern stingrays live in sea water, but a few species inhabit river and lake habitats. The Green River deposits that contained *Heliobatis* specimens formed on the bottom of freshwater lakes.

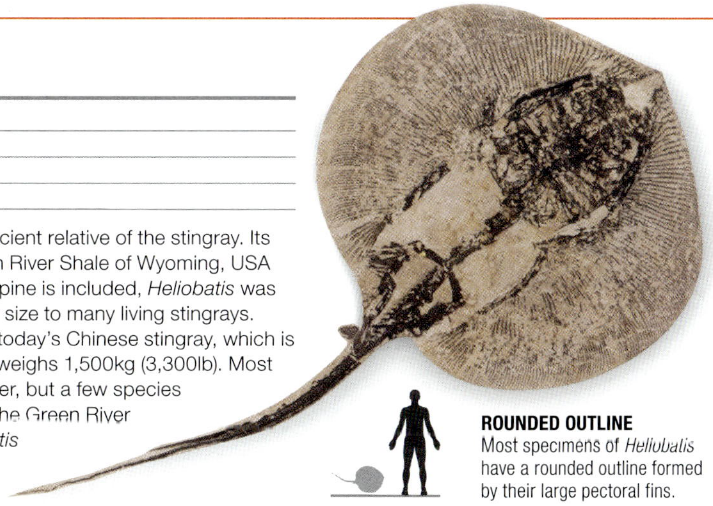

ROUNDED OUTLINE
Most specimens of *Heliobatis* have a rounded outline formed by their large pectoral fins.

Titanoboa

GROUP Lepidosaurs
DATE Middle Paleocene
SIZE 13m (43ft) long
LOCATION Colombia

Following the sudden demise of the dinosaurs, the snake *Titanoboa* was the largest land hunter of the Paleocene. It was also the largest snake ever known, being 30 per cent larger than today's longest species. Its back would have been 1m (3ft 3in) off the ground. *Titanoboa's* enormous vertebrae were found in coal deposits in Colombia, along with the remains of the crocodiles and turtles preyed upon by the giant snake.

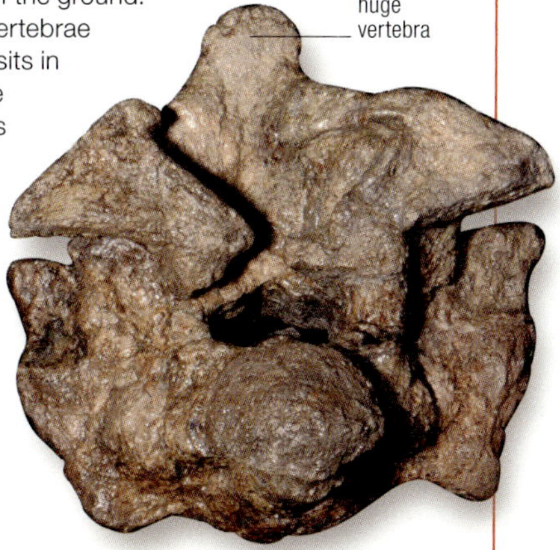

ENORMOUS FOSSIL
Scientists have estimated a length for the snake of 13m (43ft) from the enormous size of its vertebrae.

Puppigerus

GROUP Turtles
DATE Early to Middle Eocene
SIZE 90cm (35in) long
LOCATION USA, England, Belgium, Uzbekistan

Puppigerus was a sea turtle that lived in subtropical seas between the Early and Middle Eocene. Fossils found in the USA, England, and Belgium all show the same species, *P. camperi*. Until 2005, this was thought to be the only species in the genus. However, in that year, the discovery of *P. nessovi*, a new species from Uzbekistan, was announced. *Puppigerus* is an extinct genus in the Cheloniidae family. All living sea turtles belong to this family – except the leatherback. The cheloniids appeared in the Cretaceous, but *Puppigerus* most resembled modern forms. For example, the shell was completely ossified into solid bone. In addition, the pygal, the rearmost plate in the upper shell, or carapace, also lacked the notch seen in earlier cheloniids.

LOWER SHELL
The lower shell (plastron) of *Puppigerus* is small compared to that of land tortoises, so there is plenty of room for the paddle-shaped feet.

HUGE EYES
Puppigerus's eyes point sideways, not upwards like those of primitive sea turtles.

Primapus

GROUP Theropods
DATE Early Eocene
SIZE 15cm (6in) long
LOCATION England

Primapus is known from a single species whose specimens have been found in the London Clay deposits of southeast England. *Primapus* was an apodiform. Apodiformes is a surviving order of birds with living members that include swifts and hummingbirds. It is uncertain whether *Primapus* is an early member of the swift family, Apodidae, or belonged to a now extinct sister family called Aegialornithidae.

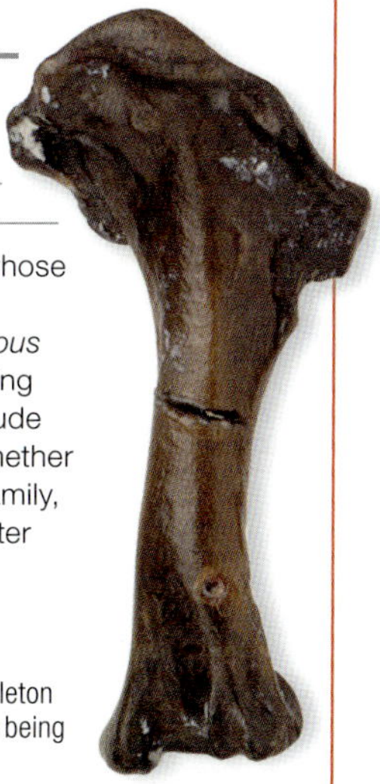

ARM BONE
The apodiform skeleton points to *Primapus* being an agile flyer.

Diatryma

GROUP Theropods
DATE Early to Middle Eocene
SIZE Over 2m (6½ft) tall
LOCATION USA

Diatryma was a huge, flightless hunting bird. It is known from complete skeletons found in Early and Middle Eocene rocks in the western United States that formed from a densely forested landscape. A very similar bird, *Gastornis*, lived in the forests of Europe at around the same time. Large *Diatryma* specimens are more than 2m (6½ft) tall and have a huge skull bearing a robust beak with a hooked tip. *Diatryma* also had stout legs with massive clawed feet, but the wings were tiny and vestigial, as the bird was too heavy to fly. *Diatryma* was an active hunter, and its powerful hooked beak was ideal for ripping flesh and crushing small bones. It has been calculated that the beak was strong enough to crack open a coconut, and *Diatryma* probably supplemented its diet by scavenging on dead bodies and foraging for fruits and other plant foods. *Diatryma* was well suited to feeding in the Eocene forests, making use of their high-density food supplies. However, as the global climate warmed, the forests turned to open plains. Despite its immensely powerful bite, *Diatryma* was no match for the speed and agility of the mammalian carnivores that evolved to hunt on the plains.

Diatryma was the top predator of its day, **considerably larger** than the mammalian carnivores of the early Eocene.

AMBUSH-HUNTER
These giant birds were too large and heavy to outrun prey, so they would have stalked them through the forest instead, and ambushed victims from among the dense undergrowth.

KEY FIND

MISTAKEN IDENTITY

The first complete *Diatryma* skeletons were found in Wyoming, USA, in the 1870s. A similar large bird, *Gastornis*, had been described in Europe in 1855. However, the European fossils were incomplete and incorrectly reconstructed, so few people noticed the similarities with *Diatryma*. In recent years, better fossils of *Gastornis* have been found, and they have shown that the European bird was almost identical to North America's *Diatryma*. Many scientists now think that the two names apply to the same bird. If this is accepted, then the first name, *Gastornis*, will be used, and the name *Diatryma* will be abandoned.

Presbyornis

GROUP Theropods
DATE Late Paleocene to Middle Eocene
SIZE 1m (3ft 3in) tall
LOCATION North America, South America, Europe

Presbyornis has been found in large numbers in the Green River Shale of Wyoming, USA, and in Eocene deposits representing shallow freshwater lakes. Eggs and nests have also been found in the same rocks, and *Presbyornis* probably lived in large flocks along the lake shores. It would have waded in the shallows and used its beak to filter out food from the water, as many ducks do today. It was one of the most successful species of its time, living for 20 million years.

DUCK-LIKE
With its long legs and neck, *Presbyornis* resembled a flamingo, but its wide bill and skull shape mark it out as a distant relative of ducks and geese.

Leptictis

GROUP Placental mammals
DATE Middle Eocene to Late Oligocene
SIZE 25cm (10in) long
LOCATION USA

Leptictis was an early relative of the insectivores: modern mammals that include hedgehogs, moles, and shrews. *Leptictis* would have stalked insects, amphibians, and lizards. It had a long snout filled with small teeth. These included simple, V-shaped molars like those of some modern insectivores. Along the top of the skull was a pair of long ridges. This was where strong jaw muscles would have been attached.

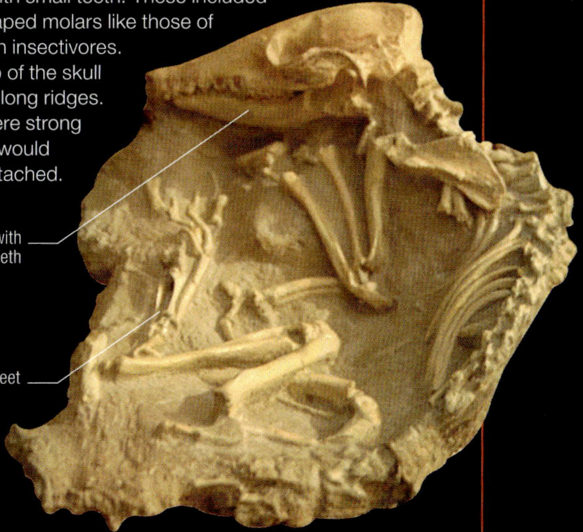

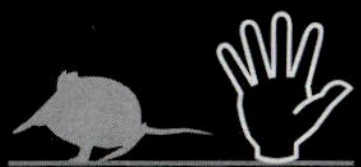

MANY SPECIMENS
Hundreds of *Leptictis* specimens, including some complete skeletons, have been found in rocks from the Middle Eocene to the Late Oligocene.

Plesiadapis

GROUP Placental mammals
DATE Middle Paleocene to Early Eocene
SIZE 18cm (7in) long
LOCATION North America, Europe, Asia

Plesiadapis looked like a ground squirrel in size and overall body plan, but was a relative of the primates. Even so, it had many rodent-like features: a pair of protruding incisors, a toothless gap between the front and back teeth, and eyes on the side of its head for spotting predators. *Plesiadapis* was so abundant that its species are used as index fossils for dating Paleocene deposits.

RODENT NICHE
Even though it occupied the rodent ecological niche before true rodents evolved, the teeth and skull of *Plesiadapis* show that it was a close cousin of the primates.

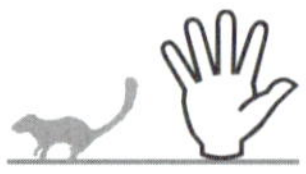

Uintatherium

GROUP Placental mammals
DATE Middle Eocene
SIZE 3.8m (12½ft) long
LOCATION North America, Asia

Uintatheres once lived **across North America** and Asia but vanished about **40 million years ago.**

Uintatherium was a giant horned mammal from the Middle Eocene. Most of its fossils have been found in Utah and Wyoming, USA, but it was widespread across North America and Asia, although probably in small numbers. It had huge upper canine tusks that protruded downwards and were protected by a bony flange on the lower jaw. The snout had a series of blunt, horn-like growths; like modern rhinoceroses, both sexes had horns. As is true of horned mammals today, these horns were probably used for display and species recognition. Large upper canines are also used in display of some small deer today. *Uintatherium* has no living descendants, and where it fits in the family tree of hoofed animals is a controversial subject.

HEAVY HERBIVORE
The size of a modern rhinoceros, *Uintatherium* was a heavy-limbed herbivore with massive bones and a bulky, barrel-shaped body. It was probably covered in a thick hide.

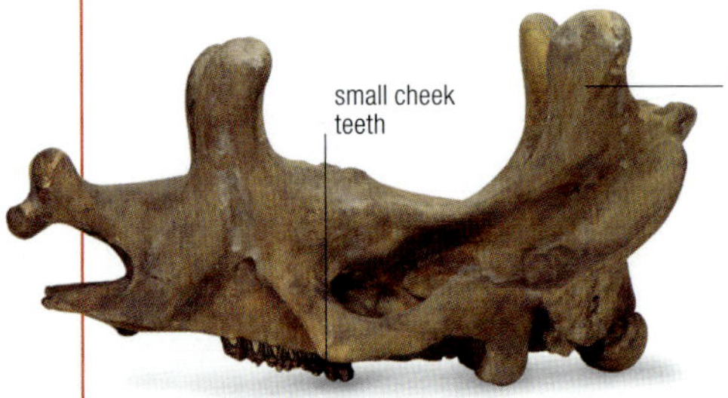

UPPER SKULL
At 1m (3ft 3in) long, *Uintatherium's* skull was similar to the size of a modern white rhino's. In the Middle Eocene, that made *Uintatherium* one of the largest mammals in the world.

BIOGRAPHY

ROY CHAPMAN ANDREWS

Andrewsarchus is named after the explorer Roy Chapman Andrews (1884–1960). Andrews started as a humble janitor at the American Museum of Natural History in New York, but rose to become its president. In the 1920s, he led expeditions to Mongolia, finding many important dinosaurs and the first known dinosaur eggs. He is thought to be the inspiration for the movie character Indiana Jones.

Andrewsarchus

GROUP Placental mammals
DATE Late Eocene
SIZE 3.7m (12¼ft) long
LOCATION Mongolia

Andrewsarchus was a giant predator that lived between about 40 and 37 million years ago. Most of what we know about this mammal comes from a single skull found in Mongolia. The skull is huge – more than 1m (3ft 3in) long – and it is estimated that the animal's total body length was about 3.7m (12¼ft), with a weight of about 250kg (550lb). That makes *Andrewsarchus* the largest predatory land mammal that has ever lived. *Andrewsarchus*'s pointed teeth are often worn and blunt. This suggests that it not only hunted large prey but was also a scavenger. Its enormous jaws would have crushed the bones of animal carcasses.

FORMIDABLE PREDATOR
The huge skull of this formidable predator suggests that it would have been larger than the largest brown bear species, which is the largest mammal predator on land today.

Hyaenodon

GROUP Placental mammals
DATE Late Eocene to Early Miocene
SIZE 0.3–3m (1–9¾ft) long
LOCATION Europe, Asia, Africa, North America

This species was a dog-like, predatory mammal that lived across the northern hemisphere in the Oligocene and persisted into the Early Miocene in Africa, finally dying out there 15 million years ago. There were many species of *Hyaenodon*, ranging in size from a weasel to a lion. It was probably the fastest predator of its time, but its legs were much shorter than those of a modern-day wolf.

LAST CREODONT
Although it looked like a modern hyena, with powerful jaws and sharp teeth for tearing flesh and breaking bones, *Hyaenodon* is not related to hyenas. It was the last surviving member of an extinct group of predators called creodonts.

Hesperocyon

GROUP Placental mammals
DATE Late Eocene to Late Oligocene
SIZE 80cm (31in) long
LOCATION USA

Hesperocyon is an extinct genus of dog that lived between the Late Eocene and Late Oligocene in North America. It is the earliest known member of the dog family, Canidae, and all modern species of dog, fox, and wolf are believed to have evolved from this small mammal over the past 30 million years. However, *Hesperocyon* looked very different to modern dogs. It had a long, flexible tail, like that of a coati, but relatively weak, short limbs, and a delicate skull capable of eating only small prey, such as birds and rodents. *Hesperocyon*'s teeth suggest that it was an omnivore, and probably foraged on the ground and in bushes for fruits and other plant items to supplement its diet of meat.

RACCOON-LIKE
Although it is related to dogs and foxes, the small skeleton of *Hesperocyon* resembles that of a modern raccoon, with a long tail and short legs.

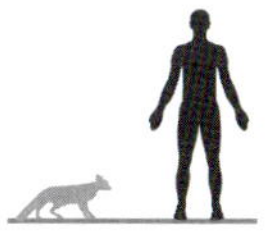

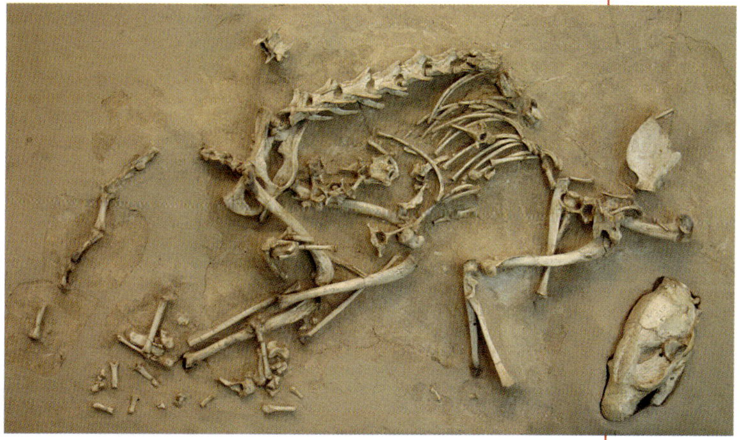

Icaronycteris

GROUP Placental mammals
DATE Middle Eocene
SIZE 14cm (5½in) long
LOCATION USA

Icaronycteris is one the earliest-known types of bat. It is known from a complete skeleton and a few other specimens from the Middle Eocene shales of Wyoming's Green River formation (see p.375). Like other insect-eating bats, *Icaronycteris* was small and a skilful flyer. It may have used an echolocation system to catch prey, although it was much more primitive than any living bat. It had a long tail that was not connected to the hind legs, and its first finger bore a claw and was not fused to the wing membrane as in modern bats. An even more primitive bat, *Onychonycteris*, recently found in the same deposits, lacks the inner ear features of modern bats indicative of echolocation.

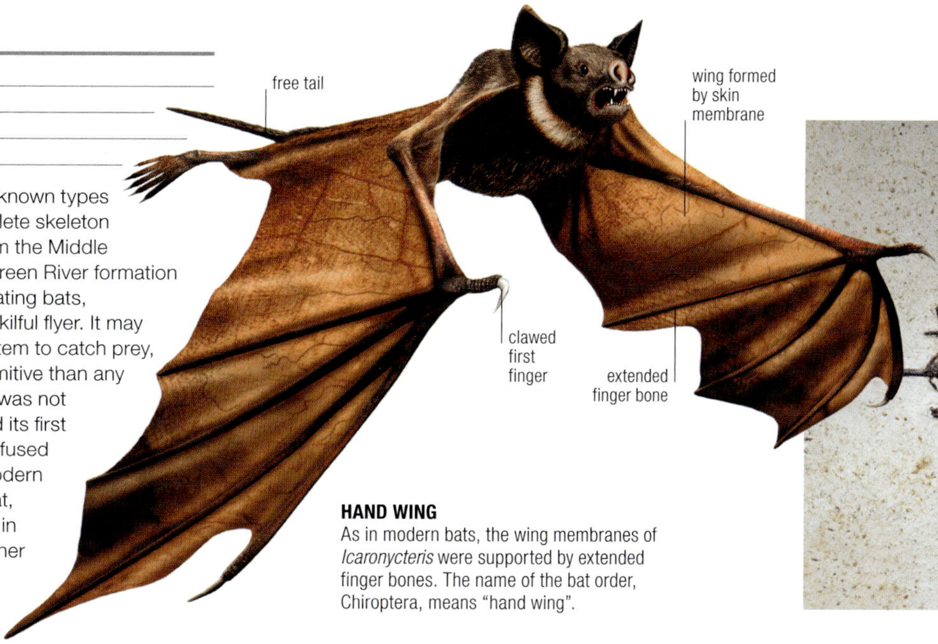

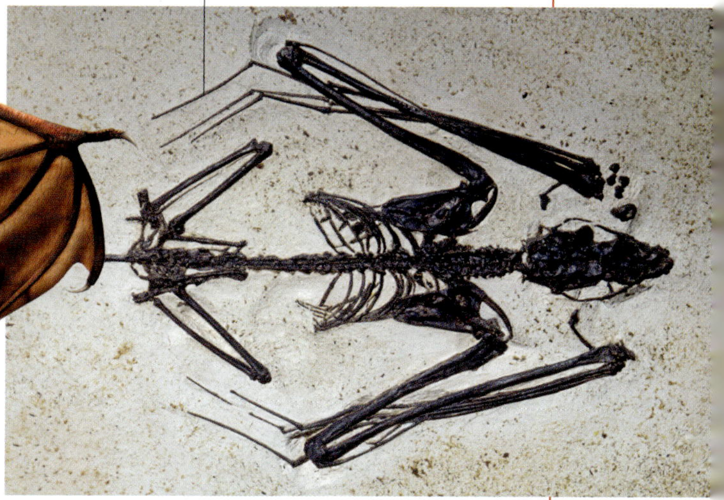

HAND WING
As in modern bats, the wing membranes of *Icaronycteris* were supported by extended finger bones. The name of the bat order, Chiroptera, means "hand wing".

Eurotamandua

GROUP Placental mammals
DATE Middle Eocene
SIZE 1m (3ft 3in) long
LOCATION Germany

One the earliest-known anteaters, *Eurotamandua* is known from a single complete skeleton fossil. It was originally thought to be related to tamanduas, tree-climbing anteaters from Central and South America (see panel, below). However, recent research has shown that *Eurotamandua* lacks the specialized vertebrae of anteaters, and their relatives, the sloths and armadillos. It may be related to modern pangolins, which are found in Africa and Asia. *Eurotamandua* was toothless and had a long tongue for licking up ants and termites. Large claws on its forefeet were used for ripping open insect nests. Like modern tamanduas, it had a long, flexible tail for gripping branches.

COMPLETE SKELETON
This skeleton was found in the Messel quarry in western Germany. The bones were preserved in bituminous shale, a mixture of mudstone and oily tar.

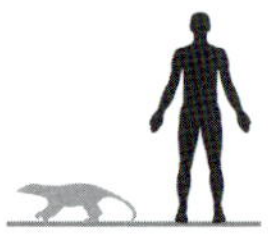

SCIENCE

TAMANDUA

Eurotamandua was named because it resembled the tree-climbing anteaters or tamanduas of Latin America. These animals have a similar appearance to the ground-dwelling anteaters, with a long, toothless snout and sticky tongue for lapping up ants and termites, long, sharp front claws for ripping their nests open, and short, bristly fur. However, they are only 1m (3ft 3in) long, weigh about 7kg (15lb), and have a long prehensile tail for climbing in trees.

Protorohippus

GROUP Placental mammals
DATE Early and Middle Eocene
SIZE 38cm (15in) tall
LOCATION USA

Protorohippus (formerly known as *Hyracotherium* or *Eohippus*) is one the earliest known horses. Its fossils were unearthed in western USA. It was about the size of a beagle or terrier dog, with short limbs bearing four toes on the front feet and three toes on the hind feet. The name *Hyracotherium* was once applied to early American horses, but recent research shows that this name applies to a horse-like European mammal, not a horse.

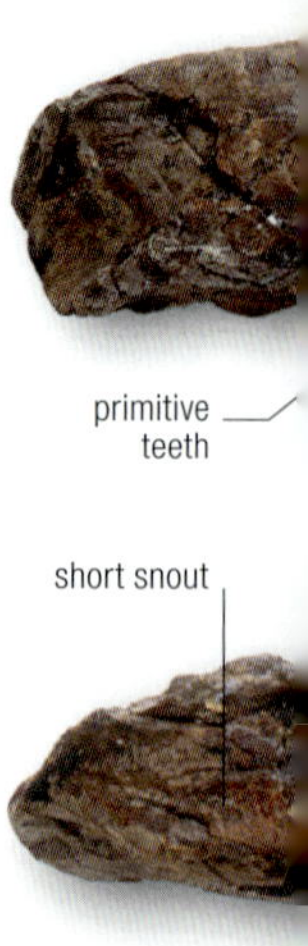

LOW-CROWNED MOLARS
The primitive, low-crowned molars of *Protorohippus* were very like the teeth of early tapiroids – the ancestors of modern tapirs.

Subhyracodon

GROUP Placental mammals
DATE Late Eocene to Late Oligocene
SIZE 2.5m (8¼ft) long
LOCATION USA

Subhyracodon was a hornless rhinoceros from the Late Eocene and Oligocene. It has been found in rocks across North America, especially in the Big Badlands of South Dakota, USA. *Subhyracodon* was not heavily armoured like modern rhinos, and avoided danger by running away on its relatively long, slender legs. The cheek teeth had low crowns suited for eating the leaves of trees and bushes, and they had the characteristic *π*-shaped pattern of crests that is typical of living rhinos.

At one time, many different species of *Subhyracodon* were named, although they have been reduced to just three valid species today. In the late Oligocene, *Subhyracodon* evolved into *Diceratherium*. This was a larger animal and it had paired bony ridges on its nose, which are thought to have supported short horns.

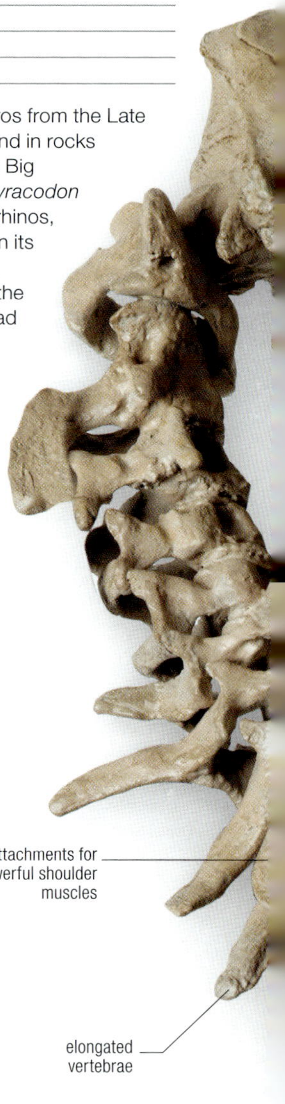

SLENDER LIMBS
Subhyracodon had more slender limbs than living rhinos, and was a little smaller – although it was still about the size of a modern cow. This specimen is the skeleton of a foetus found in Wyoming, USA.

Eomys

GROUP Placental mammals
DATE Late Oligocene
SIZE 25cm (10in) long
LOCATION France, Germany, Spain, Turkey

Eomys was a small, gliding rodent from the Late Oligocene. Several nearly complete skeletons show that it had a long skin membrane between its front and back legs, similar to that of living flying squirrels. Although the Eurasian *Eomys* had evolved a gliding lifestyle, dozens of other members of the Eomyidae family appear to have been more conventional ground or tree squirrels. The eomyids became extinct 2 million years ago. They are thought to have been close relatives of modern pocket gophers and pocket mice.

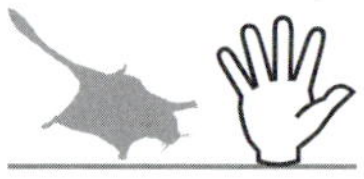

Palaeolagus

GROUP Placental mammals
DATE Late Eocene and Oligocene
SIZE 10cm (4in) long
LOCATION USA

Palaeolagus is one of the earliest-known fossil rabbits. The few complete skeletons discovered show us that, in most respects, *Palaeolagus* resembled modern rabbits. However, it had shorter hind legs, and its skull and teeth were also much more primitive than those of most later rabbits. At least eight *Palaeolagus* species are recognized. Some of them had specializations in the ear region indicative of acute hearing. *Palaeolagus* was descended from *Procaprolagus*, which in turn evolved from an earlier animal originating in Asia.

SCAMPERER
The short hind legs of *Palaeolagus* probably mean that it did not hop, but scampered more like a squirrel.

SMALL SKULL
The snout and braincase were smaller than those of later horses, and there was a smaller gap between the front and back teeth.

ANATOMY
HOOF EVOLUTION

Protorohippus had feet adapted for walking on rough terrain. There were short hooves on each of its four fingers and three toes. As horses evolved, their middle digits enlarged, and their side toes became smaller, leaving three fingers on each hand and three toes on each foot. In modern horses, the side toes are reduced to tiny splints, and all the weight is borne on the middle digit, making them very effective runners.

FOREFOOT OF PROTOROHIPPUS

Mesohippus

GROUP Placental mammals
DATE Late Eocene to Late Oligocene
SIZE 60cm (23½in) tall
LOCATION USA

An extinct horse about the size of a great dane dog, *Mesohippus* was found in the Big Badlands fossil beds of South Dakota, USA. Like modern horses, *Mesohippus* had a long snout with a gap between its front teeth and cheek teeth. Its teeth also show that *Mesohippus* was a specialized leaf-eater. During the Early Oligocene, there were more than a dozen *Mesohippus* species. By the end of that epoch *Mesohippus* lived alongside its later relative *Miohippus* in several places.

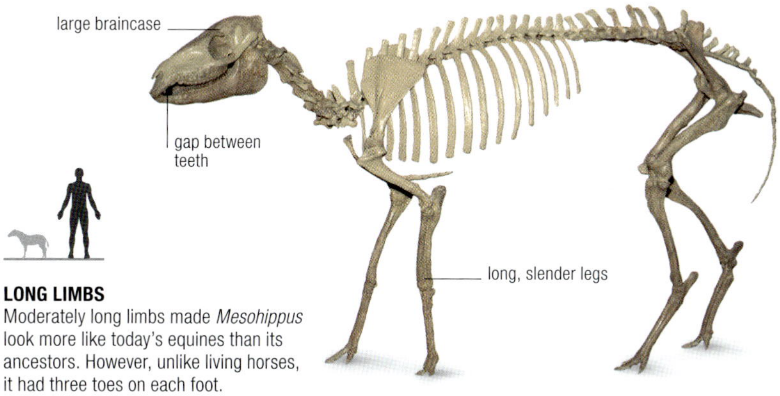

LONG LIMBS
Moderately long limbs made *Mesohippus* look more like today's equines than its ancestors. However, unlike living horses, it had three toes on each foot.

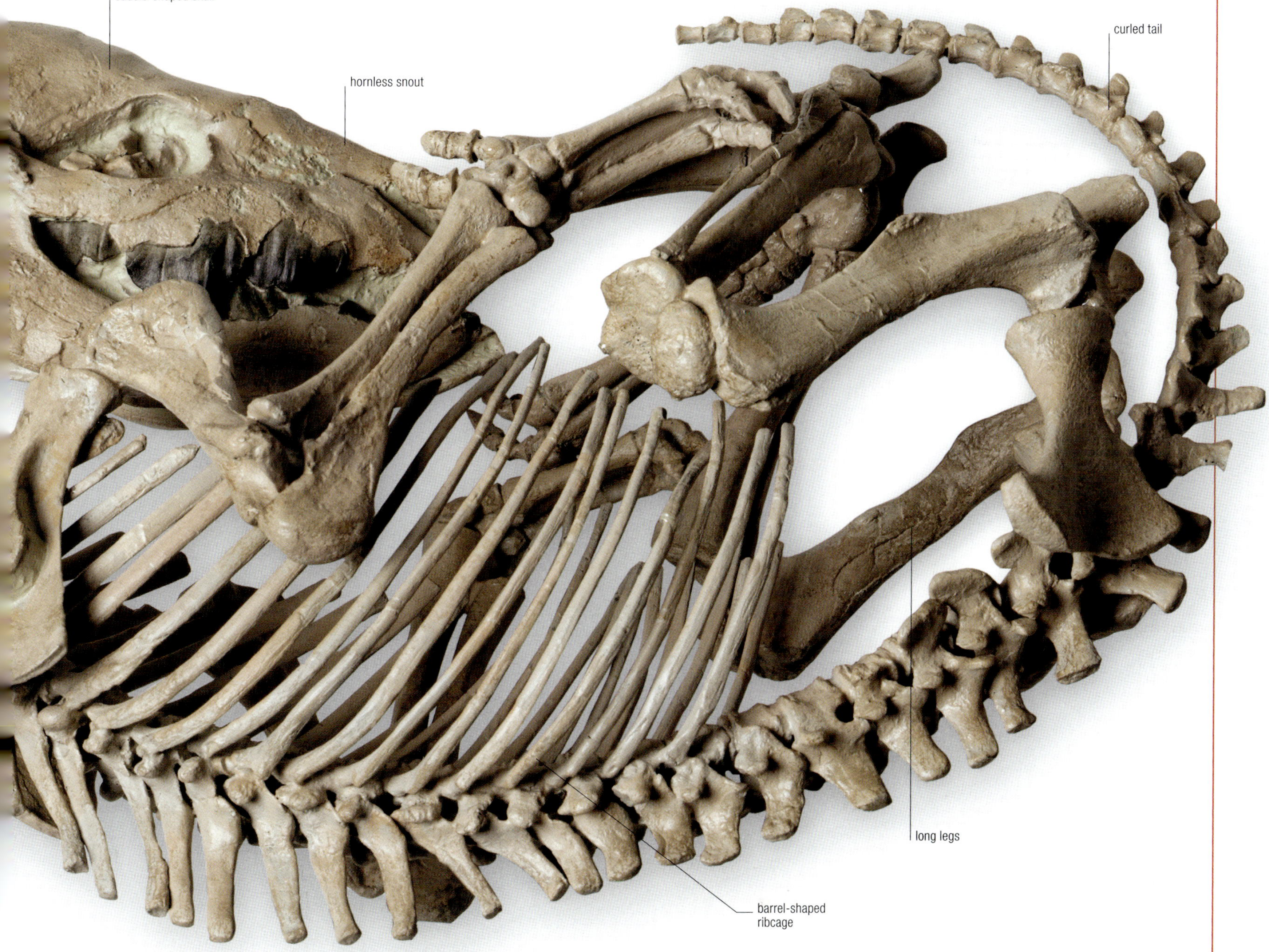

Megacerops

GROUP Placental mammals
DATE Late Eocene
SIZE 3m (9¾ft) long
LOCATION USA

This animal was the last and largest member of the brontotheres – meaning "thunder beasts". *Megacerops* was the size of an elephant and lived on the Great Plains of North America. It had a forked horn on its nose. The horn was relatively delicate and so was probably not used for head-to-head fighting.

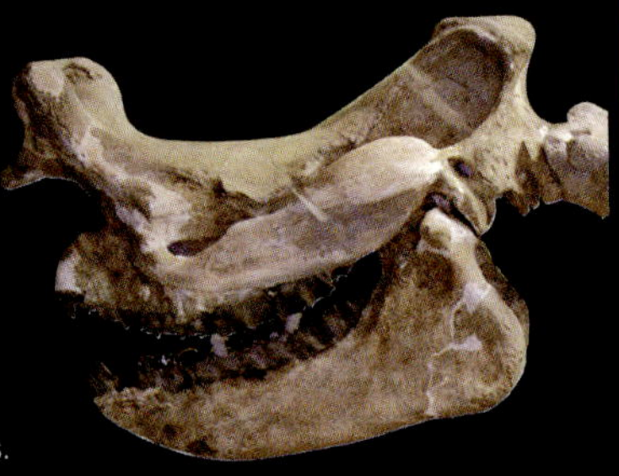

HERBIVORE TEETH
Megacerops had teeth with low crowns, showing that this mighty giant must have eaten the leaves of trees and bushes.

Mesoreodon

GROUP Placental mammals
DATE Late Oligocene
SIZE 1.3m (4¼ft) long
LOCATION USA

Mesoreodon was an oreodont, an extinct type of sheep-sized, even-toed hoofed mammal (artiodactyl). It had sharp canine teeth, used presumably for display and defence. Oreodonts were not very fast runners, and probably lived in herds for protection. Several complete *Mesoreodon* skeletons have been found, one of which had preserved vocal cords that show that it could make loud sounds like modern howler monkeys. These hooting calls may have warned the herd of attacks or frightened off predators.

UNSPECIALIZED ANATOMY
Mesoreodon's skeleton was not obviously specialized for a particular lifestyle. It had large eyes, a robust jaw, and cheek teeth with crescent-shaped ridges for eating leaves.

Leptomeryx

GROUP Placental mammals
DATE Late Eocene to Late Oligocene
SIZE 1m (3ft 3in) long
LOCATION USA

Leptomeryx was a hornless ruminant (cud-chewing animal) from the Late Eocene and Oligocene. Its fossils have been found in many parts of North America, and it is one of the most abundant fossils in the Big Badlands fossil beds of South Dakota, USA. *Leptomeryx* was about the size of a mouse deer or chevrotain. It had relatively slender limbs bearing two hoofed toes on each foot, which makes it an artiodactyl – an even-toed hoofed mammal. Although it is only distantly related to living chevrotains, *Leptomeryx* had many similar features, including a delicate stature, the absence of any antlers or horns, and, in the case of the males, enlarged upper canines that protruded as small tusks. Recent research has revealed that *Leptomeryx* was among the most primitive of ruminants, and therefore a distant relative of all ruminants, including deer, cattle, and camels.

During the Early Oligocene, *Leptomeryx* was one of the most common land mammals in North America.

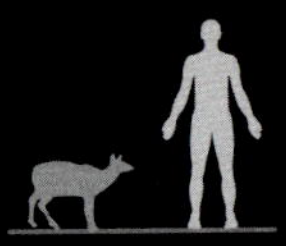

SIMILAR SKELETONS
Six species of *Leptomeryx* have been described. Their skeletons are too similar to tell them apart, so each species is identified by the shape of its teeth.

Darwinius

GROUP Placental mammals
DATE Middle Eocene
SIZE 90cm (35in) long
LOCATION Germany

This squirrel-sized primate, nicknamed "Ida", is remarkable for being the most complete primate fossil ever found – even its final meal of leaves and fruit is preserved in its stomach. *Darwinius*'s skeleton shows it to have been an agile but generalized climber, and some scientists think that this animal, while still technically a prosimian, is close to the origins of anthropoid primates, and thus a distant ancestor to humans.

REMARKABLE FOSSIL
First discovered in 1983, this fossil remained hidden in a private collection until it was revealed to the world as *Darwinius masillae* in 2009.

Eosimias

GROUP Placental mammals
DATE Middle Eocene
SIZE 5cm (2in) long
LOCATION China

Eosimias is one of the oldest known primitive anthropoids, the group that contains monkeys and apes. It is known from the Middle Eocene beds in Shanxi, China. *Eosimias* was a very small primate – some specimens were only about the size of a human thumb. Its grasping hands and long tail gave it the appearance of a tiny marmoset, but it is unrelated. Like marmosets, *Eosimias* probably caught insects to supplement its fruit diet. The presence of this animal, and some other species of primitive anthropoids, in the Eocene of Asia suggests that Asia was the birthplace of this group, rather than Africa as previously assumed.

LARGE EYES
Its large eyes suggest that *Eosimias* needed to see well in the dark and was possibly nocturnal – unlike other anthropoids, but like living prosimian primates such as lemurs and bushbabies.

Ambulocetus

GROUP Placental mammals
DATE Early Eocene
SIZE 3m (9¾ft) long
LOCATION Pakistan

An ancestor of modern whales, *Ambulocetus* retained strong legs and could walk as well as swim. It may have hunted like crocodiles, lurking in the shallows and then lunging forwards onto land to catch prey that wandered near the water. We know *Ambulocetus* was a relative of modern whales because of the shape of its skull and teeth. The similarity of its jaw and middle ear to that of modern whales suggests that, like them, it was specialized for underwater hearing and so might have lacked the external ear that picks up sounds on land. *Ambulocetus* is one of many transitional fossils that show how whales evolved from land mammals.

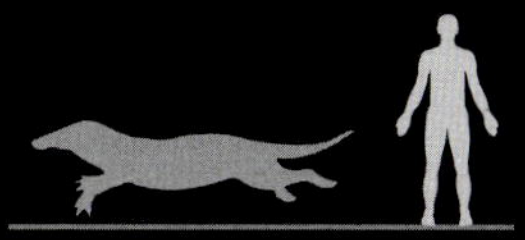

ABLE SWIMMER
Ambulocetus had webbed feet for paddling. It probably swam with an up-and-down flexing motion of the spine and tail, like otters do today.

Moeritherium

GROUP Placental mammals
DATE Late Eocene
SIZE 3m (9¾ft) long
LOCATION Egypt

Moeritherium was one of the earliest fossil relatives of the Proboscidae, the family that includes the elephants and mammoths. Its short proboscis made it look similar to a tapir. However, it also had short tusks like other members of the Proboscidae, and many other features of the skull and skeleton that show it is a proboscidean, not a relative of tapirs. The semi-aquatic nature of *Moeritherium* has lead scientists to speculate that the earliest proboscideans were semi-aquatic, and that their lineage became fully terrestrial later.

GRINDING TEETH
Moeritherium's teeth were adapted for eating leaves and soft vegetation.

NEOGENE

390 Plants

396 Invertebrates

404 Vertebrates

Neogene

By the end of the Neogene Period, the Earth had plunged into an ice age. More water circulating in the atmosphere, drawn from an Atlantic Ocean, resulted in high snowfall on Antarctica, and the ice sheet increased to its present size. As Earth cooled and became drier, grasslands spread at the expense of forests. Many large and small herbivorous mammals and their predators had to adapt to a demanding life on the open plains.

THE ALPS
Collisions between Earth's lithospheric plates threw up great mountain ranges, including the high alpine peaks in Italy, France, and Switzerland.

Oceans and continents

Neogene, meaning "new born", is a period when many modern animals evolved, including *Australopithecus,* one of our early ancestors. Major features of today's world geography were also taking shape. The Indian plate continued to plough beneath Asia, pushing the Himalayas even higher, and closing the great Tethys Ocean. Due to the northward movement of the African, Arabian, and Australian plates, Spain collided with France, forming the Pyrenees; Italy with France and Switzerland to form the Alps; Greece and Turkey with the Balkan region to create the Hellenide and Dinaride mountains; Arabia with Iran to form the Zagros mountains; and, more recently, Australia with the Asian plate to form Indonesia. In the process of mountain-building, the continental crust was compressed horizontally, as rocks were forced together by converging plates. The area occupied by the continents slightly decreased and the size of ocean basins increased, causing sea levels, on average, to fall. These changes may have all contributed to the cooling of the Earth, which transformed its ecosystems. As forests gave way to grasslands in mid-latitudes, hoofed mammals, including horses, developed teeth suitable for grazing, and long legs to gallop away from predators. The newly formed Panama Isthmus also affected the world's climate, by altering ocean current circulation (see p.24). It linked the American continents, allowing fauna and flora that were once separated to migrate freely. Cats, snakes, tapirs, wolves, and deer moved southwards, while opossums, armadillos, hummingbirds, and sloths travelled northwards.

PANAMA LAND BRIDGE
About three million years ago, the Panama Isthmus bridged the gap between the American continents, changing the routes of currents in the Pacific and Atlantic Oceans.

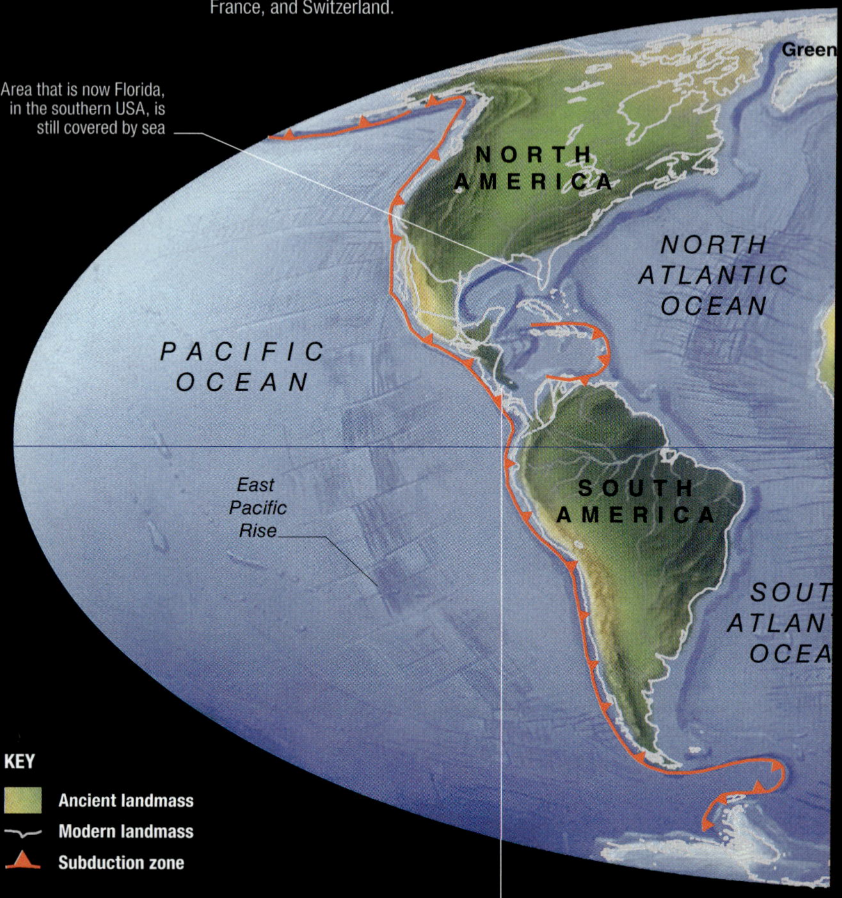

NEOGENE WORLD MAP
Familiar mountain chains have appeared, including the Himalayas, Alps, and Pyrenees. The Panama Isthmus links North and South America.

MIOCENE

MYA	25	20	15
PLANTS			15 Expansion of broad-leaved evergreen vegetation and conifers in northern hemisphere; reduction in broad-leaved forests
INVERTEBRATES			
VERTEBRATES		20 First giant anteaters; first giraffes; birds increase in diversity	15 First *Mastodon*; first bovids; firs kangaroos; Australian megafauna diversifies: thylacoleonids (below), thylacinids, diprotodontids, and vombatids

THYLACOLEO

SAVANNA VEGETATION
As savanna-like plants evolved in arid low-latitude areas, artiodactyls, including the giraffe, adapted to browse on and digest the tough vegetation in their grassland habitats.

Climate

During the early Neogene, the Antarctic Circumpolar Current became firmly established, isolating Antarctica as a cold region. The change from warm to cooler water is reflected in a change from carbonates to silica-rich sediments. By 12 to 14 million years ago, global temperatures had fallen and more ice built up in eastern Antarctica. While ocean temperatures dropped abruptly at the poles, they remained relatively warm in low latitudes, at about 22 to 24°C (72–75°F). Dry, savanna-type vegetation developed in low latitudes about 15 million years ago. By the end of the Miocene, the ice cap expanded greatly, becoming even larger than it is today. Two periods of glacial expansion, lasting about 15,000 years, occurred 5.2 and 4.8 million years ago, separated by a relatively warm phase. Yet, at the North Pole, ice had not started to form all year round. The colder conditions lowered sea levels by trapping water in the polar regions. Although the climate warmed for a short period in the early Pliocene, it then began to cool again, with large ice sheets forming in the northern hemisphere. This changing climate required new strategies for survival, such as herd behaviour and seasonal migrations in hoofed mammals, and burrowing and hibernation in early rodents.

STRAIT OF GIBRALTAR
The Mediterranean Basin was isolated 5.7 million years ago, when sea levels lowered. It partially dried out, leaving a huge salt pan, until 5 million years ago, when the Gibraltar Strait (above, left) opened.

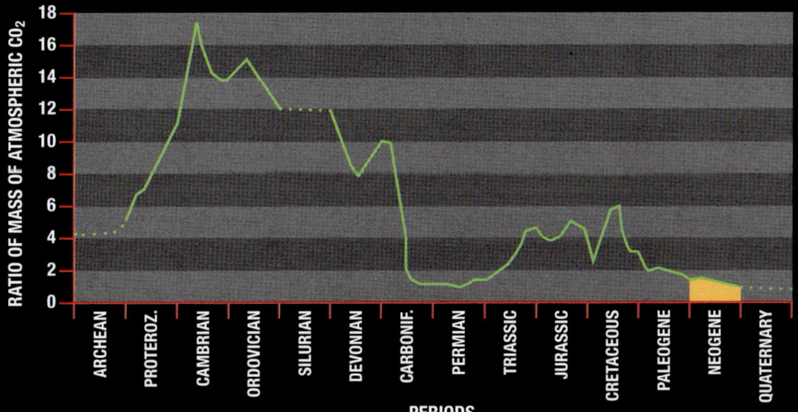

CARBON DIOXIDE LEVELS
Cores drilled by scientists in the Antarctic ice caps and the deep sea provide data about palaeoclimates. Atmospheric carbon dioxide was fairly stable during most of the Neogene, despite the changing climate.

PLIOCENE

10 | 5 | 0 | MYA

PLANTS

- **10** Savannas and grasslands become established

BIBIO

INVERTEBRATES

- **10** Insects (right) diversify, especially ants and termites

VERTEBRATES

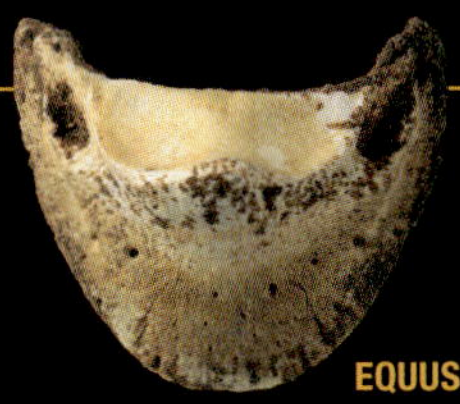

EQUUS

- **10** Small mammals (esp rodents) diversify; diversification of open vegetation herbivores, large mammalian carnivores, and snakes; horses increase in body size and develop high-crowned teeth
- **9** Increased diversity of artiodactyls, with larger body size, cursorality (active running), and grazing
- **6.5** First hominin (*Sahelanthropus*)
- **6** Australopithecines (*Orrorin, Ardipithecus*) diversify
- **5** First tree sloths and hippos; large, cursorial, grazing herbivores become more diverse; evolution of larger, faster carnivores (right); diversification of small burrowing rodents, birds, and small carnivores; very large vultures diversify; decrease in diversity of perissodactyls; diversification of kangaroos
- **4** First *Australopithecus*; largest known turtle, *Stupendemys*

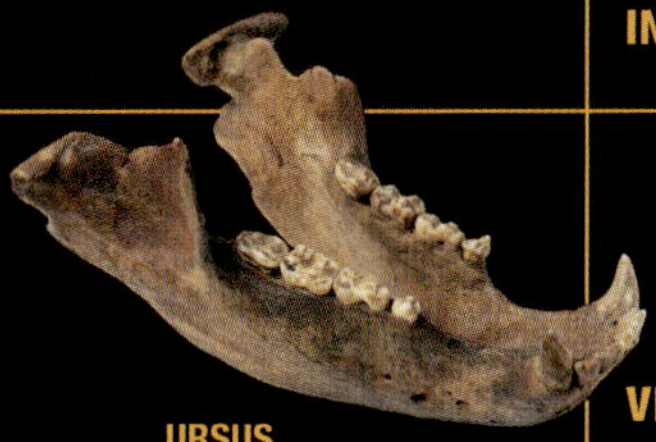

URSUS

NEOGENE PLANTS

The dramatic climatic cooling that began around the Paleogene–Neogene boundary, together with significant mountain building in Asia and the Americas, had a major impact on global vegetation. Increased separation of the continents also led to distinctive vegetation in different parts of the world.

The Neogene marks the debut of a truly modern world, with widely separated continents and a steep gradient of temperature from cold high-latitude areas to hot equatorial regions. As climates and geography became more similar to those of today, Neogene vegetation also became increasingly modern in appearance.

Seasonal forests

The cooler climates of the Neogene saw warmth-loving plants become increasingly restricted to the tropics and subtropics. In the northern hemisphere, to the north of the subtropics, areas with more strongly seasonal climates became dominated by trees such as oaks, maples, and beeches. A new flora of spring ephemerals became adapted to life below the canopy of these deciduous trees. Further north still, boreal forests developed, dominated by conifers. During the Neogene, temperate forests were widespread across the northern hemisphere. Their broadly similar composition probably reflects the relative ease of migration across the Bering Strait and northern Atlantic during the warmth of the Paleogene. Temperate forests developed much less extensively in the southern hemisphere, where there is much less land at the equivalent latitudes. Temperate forests in the southern hemisphere, for example in Chile, New Zealand, and Tasmania, remained dominated by southern beeches during the Neogene, as they had been since the latest Cretaceous.

NEW KINDS OF HABITATS
Drier climates, perhaps combined with increased pressure from grazing animals and an increased frequency of natural fires, created new habitats during the Neogene.

MAPLE FOREST
Modern kinds of temperate forest, dominated by familiar trees such as maples and oaks, became widespread for the first time as climates cooled during the Neogene.

Grasslands

In addition to global cooling, a key feature of the Neogene was drier climates in many parts of the world. In Australia, this drying trend was especially marked, but it also occured in Asia, and North and South America. It was accentuated by the development of rain shadows linked to the uplift of the Andes, the Rocky Mountains, and Himalayas. Grasses were the key group of angiosperms that exploited these new dry habitats effectively.

GROUPS OVERVIEW

Many new kinds of flowering plants, especially groups that are characteristic of open habitats, increased in diversity and became more prominent during the Neogene. The changing climatic conditions resulted in further diversification at species level in almost all groups of land plants during the period.

FERNS
Ferns have a relatively sparse fossil record during the Neogene, especially compared to their significance in many Mesozoic and earlier floras. However, the diversity of living ferns reveals that this group continued to diversify through the Cenozoic. Many living ferns show all the signs of groups that are still in a very active state of evolution.

CONIFERS
Changes in global climate during the Neogene resulted in the regional extinction of many conifer groups. Once widespread forms, such as redwoods, became restricted to a much smaller area. The exceptionally well-preserved Miocene fossil flora from Clarkia, USA, contains many conifers that are today restricted to eastern Asia.

MONOCOT ANGIOSPERMS
While most of the major groups of monocots were in place during the Paleogene, changing climates during the Neogene facilitated further diversification. In addition to the great Neogene explosion of grasses, the diversification of many groups of monocots in Mediterranean regions probably also dates from this time.

EUDICOT ANGIOSPERMS
The rapid diversification of eudicots continued through the Neogene. Many groups of herbaceous eudicots appear in the fossil record for the first time. These groups account for a very substantial proportion of the total diversity of living angiosperms, indicating that much of the diversity of living angiosperms is of relatively recent origin.

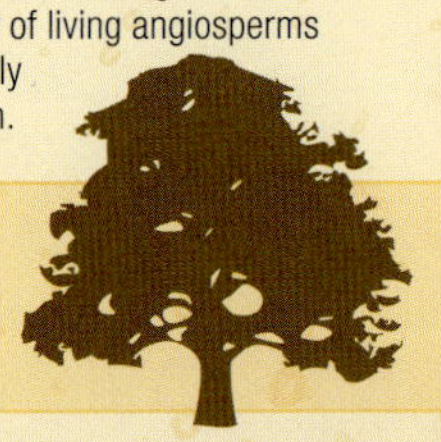

Taxodium

GROUP Conifers
DATE Cretaceous to present
SIZE Up to 40m (130ft) tall
LOCATION High northern latitudes

The early Tertiary northern flora that grew throughout what is now the Arctic belt began in the Late Cretaceous. As the climate cooled, the flora migrated south. Plant communities in these warmer, southern areas were dominated by *Taxodium* broad-leaved trees and shrubs. In areas where the flora could not migrate southwards, because of mountain ranges, the plants became extinct. The Cretaceous *Parataxodium* was the ancestor to *Taxodium*, *Sequoiadendron*, *Sequoia*, and probably *Metasequoia*. Living *Taxodium* are easy to recognize, but fossils can be hard to identify. Its shoots have two flat rows of needle-like leaves, but they are similar to those of *Metasequoia*. Its male and female cones are more distinctive, but far less common.

TAXODIUM DUBIUM
This conifer produces small male cones in large numbers off the sides of hanging axes.

SEED CONES
Large, ball-shaped seed cones occur on the ends of the leafy shoots.

LIVING RELATIVE
TAXODIUM

The three living species of *Taxodium* are confined to southern North America, Mexico, and Guatemala. The swamp cypress thrives around the Gulf of Mexico, where it grows to nearly 40m (130ft) tall. In wet areas it has a buttressed trunk and aerating roots. Its leaves turn brown and are shed in late autumn.

Taxites

GROUP Conifers
DATE Cretaceous to present
SIZE Shoot up to 12cm ($4\frac{1}{2}$in) long
LOCATION High northern latitudes

Taxites is a generic name that has sometimes been given to leafy shoots that are in fact the smaller "short shoots" of *Taxodium* (see left) that were naturally shed (abscized) from the trees. This happened either seasonally or annually. The linear leaves were spirally attached to the shoot. Because of their abscission, shoots are often found in large numbers, with hundreds on a single bedding-plane surface.

TAXITES LANGSDORFI
The linear, flattened leaves of *Taxites langsdorfi* are spirally attached to the axis; they are twisted and appear to be spread in two ranks.

Glyptostrobus

GROUP Conifers
DATE Cretaceous to present
SIZE Up to 35m (115ft) tall
LOCATION High latitudes in the northern hemisphere

The first records of *Glyptostrobus* are in the Late Cretaceous of Japan, North America, and Spitzbergen. There are many more Paleogene and Neogene records of shoots and their cones around the northern hemisphere and up to the present high-Arctic latitudes. The leaves were spirally arranged on the shoots, forming three ranks along them. Today, the only living species, the Chinese swamp cypress, grows in subtropical southeastern China and Vietnam.

GLYPTOSTROBUS SP.
The seed-bearing cones of this conifer were about 1.5–2cm ($\frac{1}{2}$–$\frac{3}{4}$in) long, and were borne on the ends of the leafy shoots.

MAGNOLIA LONGIPETIOLATA
Fossil leaves like these have often been assigned to living genera on the basis of superficial similarity to those of the living plants.

Magnolia

GROUP Angiosperms
DATE Cretaceous to present
SIZE Up to 45m (145ft) tall
LOCATION Temperate northern hemisphere

The characters that allow fossilized leaves of this genus to be assigned to the living genus *Magnolia* are their elongated shapes with curved edges, gradually tapering tips, stout stalks, obvious midribs, parallel secondary veins that depart at less than 45 degrees, and a meshwork of tertiary veins. Leaf size varies considerably on a single plant, so it is not a good feature for identifying a species. Fossil remains of *Magnolia* are known from the Paleogene and the Neogene. During these periods, *Magnolia* grew in slightly humid environments, either as understorey shrubs and trees or in openings along stream banks, and extended into higher ground on the fringes of more open habitats. Today, *Magnolia* lives naturally in the open forest in temperate and warm-temperate areas of eastern Asia.

LIVING RELATIVE
MAGNOLIA

Magnolia includes about 70 species of trees and shrubs with simple, leathery leaves and large, bisexual flowers. The showy, white or pink petal-like structures surround numerous stamens and a central mass of free carpels. *Magnolia* is deciduous and in many species the flowers emerge before the leaves.

Tectocarya

GROUP Angiosperms
DATE Middle Paleogene to Neogene
SIZE Fruit 2cm (3/4in) long
LOCATION C. Europe

Tectocarya was a member of the dogwood family. Their fruits had a fleshy outer layer that attracted animals and an inner, thickened, woody wall that was resistant to digestion. Fossils were formed when the part-digested seeds that had passed through animals were swept away and covered by sediments. In time, minerals infiltrated the seeds, sometimes preserving traces of the cell walls.

TECTOCARYA RHENANA
The woody, outer walls of these mineralized, nut-like seeds have water-worn surfaces walls.

Alnus

GROUP Angiosperms
DATE Cretaceous to present
SIZE Up to 39m (130ft) tall
LOCATION Northern hemisphere, South America

Alnus, or alders, first appeared in the Cretaceous. The leaves are mostly elliptical or ovate, and broadest in the middle or upper half. Some living alders have heart-shaped leaves. The trees have small catkins of male and female flowers. After pollination, the female catkins turn into overwintering, woody, cone-like structures that open the following year to release the seeds.

ALNUS CECROPIIFOLIA
Fossilized alder leaves, like this one, can be difficult to identify without associated male catkins and pollen, or their cone-like fruits.

Podocarpium

GROUP Angiosperms
DATE Paleogene to Neogene
SIZE Seed pod about 3cm (1 1/4in) long
LOCATION Europe, North America, China

Podocarpium was a member of the legume family. There are around 14,000 living species of legume, which include peas, beans, clover, peanuts, and lupins, and many of them are economically important. This family had its origins in the late Cretaceous in the southern continent, Gondwanaland, but by the Eocene, fossil leaves, flowers, seed pods, and pollen grains reveal that the three major groups of legumes had evolved and spread into the northern continents.

PODOCARDIUM SP.
The fruits of *Podocarpium* (above) were like small pea pods on a stalk. The compound leaves (left) had one terminal and several pairs of lateral leaflets.

Hymenaea

GROUP Angiosperms
DATE Neogene to present
SIZE Tree up to 25m (80ft) tall
LOCATION Tropical Central and South America, E. Africa

Amber-preserved specimens of *Hymenaea* were formed when flowers and leaves fell from large, evergreen, leguminous trees to become entombed in resin oozing from the trunk and branches of the same plants. The resin infiltrated the plant tissue before hardening and preserving cell walls, chloroplasts, xylem, and even nuclei and some cell membranes. *Hymenaea* had an extensive range in the Neogene of Central and South America and Africa. Today there are 13 tropical species in the Caribbean and from southern Mexico to Brazil, and one on the east coast of Africa and its islands.

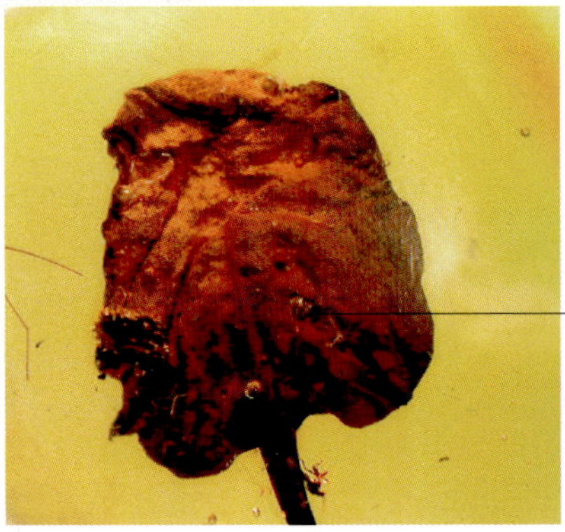

HYMENAEA SP.
The flower shown here is one of several species of *Hymenaea* that are preserved in Neogene Dominican amber.

Betula

GROUP Angiosperms
DATE End of the Cretaceous to present
SIZE Up to 24m (80ft) tall
LOCATION Northern temperate regions in Asia, Europe, North America

Betula, or birch, leaves are ovate to rounded or triangular in shape, with toothed margins. They are stalked and have a main vein with branching secondary veins. There are about 40 living species of birch and all are comparatively short-lived deciduous trees with both male and female catkins on the same tree. Some species grow to nearly 30m (98ft) tall and because they shed many of their small branches they are usually slender trees with few large branches. They all grow in open areas with poor soils and they are pioneer trees on disturbed or abandoned land. Today, birches are found in Europe, Asia, and North America.

BETULA ISLANDICA
These fossil leaves are preserved as impressions, but still show their *Betula* characters quite clearly.

LIVING RELATIVE
ULMUS MINOR

Many elm species hybridize with each other, especially when grown together. Artificial propagation can produce varieties of such hybrids that show desirable traits. *Ulmus minor* "dicksoni" is slow growing, with leaves that turn a bright golden yellow in the autumn.

Ulmus

GROUP Angiosperms
DATE Paleogene to present
SIZE Up to 36m (120ft) tall
LOCATION Northern temperate regions

Ulmus, or elm, leaves are elongated, broadest in the middle, with toothed margins, and an asymmetric rounded base; some species have an asymmetric apex as well. The leaves have a prominent midrib with equally prominent, straight, regularly spaced, and sometimes branched lateral veins that extend to the leaf margin, where they usually end in the teeth. These are all characteristics of elm leaves, but the problem, as in many other fossils, is identifying the exact species. There is a natural variation in the leaves on living elms, which should be taken into account when studying fossils. There are about 20 elm species living today in Europe, Asia, North Africa, and North America. All are large, deciduous trees with some species reaching 36m (120ft) in height. Small clusters of bisexual flowers are produced in the spring before the leaves open and winged seeds are shed either later in the spring and germinate immediately, or in the autumn, when they overwinter before germinating.

ULMUS SP.
Although this leaf shows many characteristics of an elm leaf, identifying the species is especially difficult if there is only one specimen to study.

Populus

GROUP Angiosperms
DATE Paleogene to present
SIZE Up to 60m (195ft) tall
LOCATION Northern temperate regions

Populus, or poplar, leaves are rounded with lobed or blunt-toothed margins, sometime leathery, and almost maple-like in appearance. They are stalked and the branching leaf veins are all of the same thickness, without much of a prominent midrib. Poplars today are fast-growing, deciduous trees that have their flowers in catkins. The flowers emerge in spring before the leaves, enabling the wind to carry the pollen to the female catkins relatively unhindered. Individual trees have either the male pollen-producing catkins or the female catkins. There are about 30 species spread across the northern temperate regions of Europe, Asia, and North America. The American west coast black cottonwood is the largest today, growing up to 60m (195ft) in height. A commonly planted tree in Europe and North America is the Lombardy poplar, which grows to 30m (98ft) tall and has an elegant, spire-like shape.

POPULUS LATIOR
These fossils reveal the fine network that the smallest leaf veins form between the major veins of a poplar leaf.

LIVING RELATIVES
GOLDEN POPLAR

This is a young plant of the vigorous hybrid cross between American cottonwood and the European black poplar. The particular variety shown here (*Populus* x *canadensis* "Serotina Aurea") is commonly called the golden poplar because its leaves turn golden yellow in autumn.

Zelkova

GROUP Angiosperms
DATE Paleogene to present
SIZE 40m (130ft) tall
LOCATION North America, Asia, Europe

There are five living species of *Zelkova* that grow in the Caucasus and eastern Asia. Its ancestors are known from the early Paleogene of North America. From there it migrated to Central Asia, and then into Central Europe. *Zelkova* survived in Europe into the Neogene, until it became extinct at the onset of the Quaternary.

Fagus

GROUP Angiosperms
DATE Late Cretaceous to present
SIZE Up to 45m (145ft) tall
LOCATION Northern temperate regions

Fagus comprises ten large, deciduous beech trees that are found throughout the northern hemisphere in Europe, Turkey, Japan, China, and eastern North America. The trees produce male flowers in hanging bunches and female ones in small clusters, with each flower producing two seeds. Leaves of *Fagus* are characteristic of many Neogene floras and are also known from the Paleogene. There are two basic forms of leaves: elongated and slender with 12–16 pairs of lateral veins in North America; and shorter and broader with 6–9 pairs of lateral veins in Europe. Intermediates occur in other parts of the world. The earliest assemblages of fossils showed all the variations, but the forms gradually became separated by distribution during the Neogene.

FAGUS GUSSONII
This leaf of *Fagus gussonii* was discovered in Greece, where the plant material has been compressed in layers of sediments.

Nothofagus

GROUP Angiosperms
DATE Late Cretaceous to present
SIZE Up to 45m (145ft) tall
LOCATION Australia, Papua New Guinea, New Zealand, South America, Antarctica

Nothofagus is known from fossil leaves, wood, seed cupules, and pollen. Its leaves are symmetrical and elliptical in outline, with broadly rounded tips and toothed margins. *Nothofagus* was the most significant tree in the temperate zone of the southern hemisphere throughout the Paleogene and Neogene. It is known as fossils from Antarctica, Australasia, and South America, and first appears in the Late Cretaceous. It invaded the earlier conifer forests and became the dominant tree. In the Neogene, its distribution was more limited in Australia, as the continent drifted north and became drier. *Nothofagus* became extinct in Antarctica, along with the rest of the vegetation, when the continent drifted southwards and became far colder.

NOTHOFAGUS SP.
This Antarctic fossil contains fragmentary remains of *Nothofagus* leaves. These have prominent midribs and parallel secondary veins that run to the toothed margins.

LIVING RELATIVE
SOUTHERN BEECH

Nothofagus, or southern beech, has about 20 living species of mostly evergreen, large trees native to Australia, New Zealand, and South America. All are temperate in distribution, except the subtropical Australian beech of New South Wales.

Quercus

GROUP Angiosperms
DATE Paleogene to present
SIZE 15–45m (49–145ft) tall
LOCATION Northern hemisphere

There are over 800 species of oak, or *Quercus*, living in the temperate northern hemisphere, the Mediterranean, and tropical and subtropical Asia. Most species are deciduous. Many are economically valuable for their timber, for example the bark of the cork oak is a major source of cork. Fossilized oak leaves and more rarely found acorns are very recognisable, and differences in the shape and vein patterns of the fossil leaves can be used to identify various species of *Quercus*.

FOSSILIZED TRUNK
The annual growth rings in this section of fossilized oak trunk vary according to the rate of growth of the tree, which can be affected by the environment.

QUERCUS FURUHJELMI
Oak leaves are elliptical to roughly oval, with distinctly lobed margins and stalks. They have a central vein and secondary veins that end in the leaf lobes.

Durania

GROUP Angiosperms
DATE Neogene to present
SIZE Fruit about 4cm ($1\frac{1}{2}$in) long
LOCATION North America

Some Paleogene river or lake deposits have large numbers of *Durania* seeds and fruits. Any original fleshy layer that was attractive to animals is usually missing, and their outer surface often shows signs of water erosion. *Durania* belongs to the sweetleaf family (Symplocaceae), with about 300 species living in the warm, temperate to tropical regions of eastern Asia and the Americas.

DURANIA ELOENBERGI
Durania seeds are elongate to ovoid in shape. They are sometimes curved and often have longitudinal ribs on their surface.

Comptonia

GROUP Angiosperms
DATE Paleogene to present
SIZE Up to 20m (66ft) tall
LOCATION Northern temperate regions, N. South America

Living *Comptonia* consists of a single species, *Comptonia peregrina*, which grows in eastern North America. *Comptonia* is closely related to *Myrica*, a genus that consists of about 35 to 50 species of medium-height trees – up to 20m (66ft) tall – and aromatic shrubs that grow up to 1m ($3\frac{1}{3}$ft) tall. These plants can tolerate drier soil, but some species prefer acid, boggy ground. *Myrica* leaves are simple, but the leaves of *Comptonia* are distinctively and deeply toothed. *Comptonia* is first recorded from the Eocene; it then spread to most temperate northern regions by the Neogene.

COMPTONIA ACUTILOBA
This is an impression of a leaf on fine-grained sediment. The leaf is long and oval in outline with a distinctly lobed margin.

Acer

GROUP Angiosperms
DATE Paleogene to present
SIZE 9–30m (30–98ft) tall
LOCATION North America, Europe, Asia

Living *Acers*, commonly known as maples, are medium to large trees. They have distinctive stalked leaves with toothed margins. Although most *Acer* leaves have five lobes, some fossils have just three. The extent of the incisions between the lobes, the size of the marginal teeth, and the size of the leaf vary between species. Sometimes, as in the living Japanese maple, the lobes are virtually separate and are themselves lobed. The fruits of maple trees are produced in pairs and, being winged, can travel some distance from the parent tree by spinning in the wind. Fossils of these winged fruits, very much like those of sycamores, are also known.

ACER TRILOBATUM
Maples have easily recognized, lobed leaves, with stalks and toothed margins.

DISTINCTIVE LEAVES
The fossil above shows a maple leaf with the usual five lobed arrangement. On the left is a species with only three lobes, but it is still a distinctive maple leaf shape.

LIVING RELATIVES
MAPLES

There are over 200 species of maples living in North America, Europe, and Asia today. The European sycamore and the North American sugar maple are the largest, growing to over 30m (98ft) tall. Sugar maples yield up to 2.5 litres (5¼ pints) of syrup a year. Many maple leaves turn red or yellow in the autumn, as shown here by this red maple.

Sapindus

GROUP Angiosperms
DATE Neogene to present
SIZE Up to about 30m (98ft) tall
LOCATION Worldwide tropics and subtropics

There are about 12 living species of this genus, living in warm temperate to tropical regions around the world. The plants have clusters of creamy white flowers that form fruits called soap nuts. These contain natural saponins and have been used for washing by various people, such as Native Americans, for thousands of years.

SAPINDUS FALSIFOLIUS
This is a pinnately divided *Sapindus* leaf, with the leaflets arranged alternately on the shoot.

Porana

GROUP Angiosperms
DATE Neogene
SIZE Climbing vine up to 20m (66ft) high
LOCATION China, Europe, North America

Porana was a climbing vine. Rare, fossilized *Porana* flowers are sometimes found – the earliest of these was discovered in the Early Neogene of China. Later occurrences of this plant, from the Middle Neogene of Europe and North America, did not survive the climatic cooling that occurred at the end of the Neogene.

PORANA SP.
This rare fossil is of one of the extinct, subtropical vine *Porana*'s trumpet-shaped flowers. The five free petals can be clearly seen.

Typha

GROUP Angiosperms
DATE Paleogene to present
SIZE Erect stem up to 3m (9¾ft) long
LOCATION Worldwide

Typha, or bullrush, is a genus of about 11 species of monocotyledonous plants that grow in wetlands around the world, but primarily in the northern hemisphere. *Typha* fossils may have originated in the Late Cretaceous, but are definitely known from the Paleogene of the USA and the Early Neogene of Europe. By the Quaternary, Typha was widespread and common around the world.

TYPHA SP.
Bullrush flowers are packed into dense cylindrical spikes. Fossil fruits and pollen of *Typha* are common, but flower spikes, such as this, are rare.

Phragmites

GROUP Angiosperms
DATE Paleogene to present
SIZE Up to 3m (9¾ft) tall in the flowering season
LOCATION Worldwide

Fossil leaves that can be assigned to the genus of living monocotyledonous grasses *Phragmites* have large numbers of parallel veins of the same thickness. This is unlike most dicotyledonous angiosperm leaves that have a series of main, secondary, and tertiary veins. *Phragmites* leaves are commonly found in freshwater deposits from the Paleogene onwards. Its remains are also commonly found in Quaternary peats, where a layer of *Phragmites* indicates an increase in the water table and a regrowth of the reed beds.

LIVING RELATIVE
COMMON REED

The common reed, *Phragmites australis*, is the only species of this genus of perennial grasses. It grows into extensive reed beds in wetlands in tropical and temperate regions, spreading by as much as 5m (16½ft) a year by means of its rooting horizontal runners.

PHRAGMITES ALASKANA
This fossil leaf of *Phragmites alaskana* shows its characteristic parallel veins.

Palmoxylon

GROUP Angiosperms
DATE Neogene
SIZE Up to 30m (98ft) tall
LOCATION Warm temperate to subtropical areas in northern hemisphere

Palms are monocotyledonous angiosperms. Their trunk structure differs from that of normal conifers and angiosperm trees, in that the "wood" has no growth rings and is made up of a central core of vascular bundles. *Palmoxylon* from the Late Cretaceous of Texas is one of the earliest known monocotyledons.

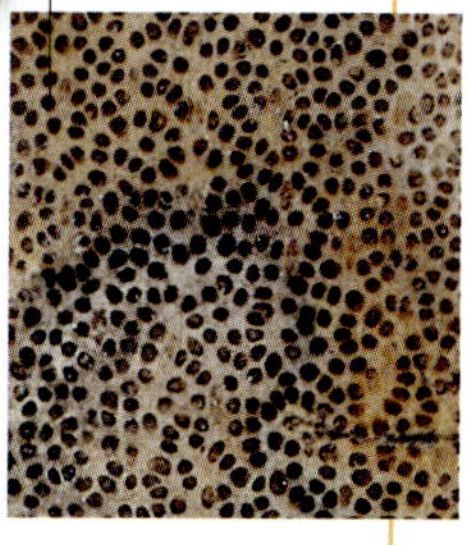

PALMOXYLON SP.
This slab of a fossilized palm trunk shows the detail of its internal anatomy, which consists of many individual vascular bundles.

NEOGENE INVERTEBRATES

This was a time of continued global cooling. The first ice sheets in Antarctica were beginning to spread about 25 million years ago, though this was followed by a warming event. By 14 million years ago, however, permanent ice sheets draped the Antarctic continent, and invertebrates adapted accordingly.

Continuing evolution

Neogene invertebrate faunas were similar to those of the Paleogene, with bivalves, gastropods, bryozoans, and echinoids as the main preservable components. The genera and species changed through time, and the original divisions of the Tertiary – Eocene, Oligocene, Miocene, and Pliocene – were based upon the ratios of the bivalves and gastropod faunas of these times to genera still living today. In Neogene sediments there are also crab, lobster, and fish remains. Neogene sediments are widespread in places: well-known samples are the "Crag deposits" exposed in low cliffs on the coast of East Anglia, Britain, replete with well-preserved fossils, which extend into the Quaternary. The unique Coralline Crag, exposed above the shoreline, is a shallow-water bank consisting almost entirely of the remains of bryozoans of many genera and species with attendant molluscs, echinoids, and giant barnacles. Similar banks can be found off south Australia today.

BIBIO MACULATUS
Known as the March fly, *Bibio* lived in grasslands. The delicate veins on the wings can be seen in this specimen.

COLEOPLEURUS PAUCITUBERCULATUS
This echinoid from the Neogene is preserved without its spines. *Coleopleurus* lived on hard, rocky substrates in shallow water.

Adaptations to cooling seas

During the Neogene many echinoids lived in cold Antarctic waters, where growth and development were slow, have developed "marsupiae" or brood pouches. This is an adaptation for incubating fertilized eggs, which develop directly without the free-swimming larval stages. The marsupiae form deep depressions on the surface of the echinoids. In the Paleogene and Neogene of Australia, marsupiate forms are abundant – Australia was originally situated nearer to the South Pole and gradually moved northwards. Marsupiate echinoids became progressively fewer during the Neogene and there are none in the warm Australian seas now.

MODERN FORM
This Neogene crustacean, *Archaeogeryon peruvianus*, was similar to crabs of the present day. It probably hunted in deep waters.

GROUPS OVERVIEW

Some invertebrate species that were prevalent during the Neogene are still living today, or they have some very modern-day equivalents. This is the case for bryozoans, bivalves, echinoids, and insects, which all continued to evolve and adapt to the changing habitats brought about by the gradual drop in global temperature throughout this period.

BRYOZOANS
The genus *Cupuladria*, still in existence today, is known from as far back as the early Neogene. Since the temperature limits of living colonies are well known, and assuming no change in habitat, *Cupuladria* can be used to gauge the temperature of the Neogene sediments in which its remains occur.

BIVALVES
Many of the species now found in cool–temperate waters, such as those found around the British coasts, are the same as, or very closely related to, those of the Neogene. There is a progressively increase in the number of cold-water forms through time.

ECHINOIDS
Both deep- and shallow-burrowing echinoids, as well as surface dwellers, continued as an abundant component of the marine fauna through the Neogene. These include the "living fossil" *Cidaris*, which has hardly changed since the Triassic.

INSECTS
Many insects are preserved in remarkable detail in amber; this was originally a sticky resin that oozed from conifers and other trees, trapping the insects for perpetuity. The most diverse faunas are from the Dominican Republic, and are probably early Neogene in age. The resin came from a now extinct plant, *Hymenaea protera*.

Meandrina

GROUP Corals
DATE Paleogene to present
SIZE Corallite 1–2cm ($\frac{3}{8}$–$\frac{3}{4}$in) across
LOCATION Worldwide

Meandrina is a colonial scleractinian coral that is often found in reefs. It is formed by budding from the initial, individual corallite, the skeleton of a polyp. The top of the colony is divided into meandering "ridges" and "valleys", and the polyps, the individual members of the colony, are housed in the cup-shaped depressions of the valleys. The partitions (septa) that divide the body cavity of the corallite are fairly straight and long, and those from adjacent corallites are often in line. The axial ends of the septa form the axial plate that runs along the centre of the horizontally-elongated corallites. Modern species of *Meandrina* include the brain corals, so-called because the meandering corallites resemble the surface of a human brain.

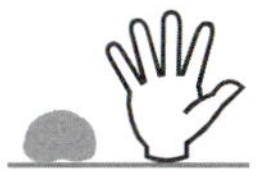

LONG PARTITIONS
The septa that divide each of *Meandrina*'s corallites are straight and elongated, and corallites are often arranged in lines, adding to the coral's folded appearance.

MEANDRINA SP.
In life, this coral would have been covered in a thin layer of colourful polyps, which collected particles of food from the water using tentacles.

LIVING RELATIVE
BRAIN CORAL

Brain coral is the common name given to modern corals that have prominent meandering grooves and intervening ridges on their outer surfaces, resembling a large human brain. Both *Meandrina* and *Diploria* have this morphology. These genera are slow-growing and well-cemented corals that provide stability in a coral reef because of their ability to withstand strong currents and storms. Like most other reef-building corals, their polyps live in a close symbiotic relationship with microscopic green algae. They occur both in the Atlantic and the Indo-Pacific oceans.

Sphenotrochus

GROUP Corals
DATE Eocene to present
SIZE 1cm ($\frac{3}{8}$in) high
LOCATION Worldwide

Sphenotrochus is a small, solitary, conical coral that in life was entirely covered by the polyp body. In cross section, the skeleton (corallite) is elliptical. The thin partitions or septa that divide the inside of the corallite are fused together around the outside to form the corallite wall. There are three cycles of septa, resulting in a total of 24 septa. A distinct plate in the centre of the corallite runs down its entire length, and is produced from the modified ends of the septa. *Sphenotrochus* survives to the present day. It usually lives in water around 20–275m (66–900ft) deep.

SPHENOTROCHUS INTERMEDIUS
The septa radiate out from the centre of the corallite and are fused around the outside to form the corallite wall.

Meandropora

GROUP Bryozoans
DATE Pliocene
SIZE Colony up to 9cm ($3\frac{1}{2}$in) in diameter
LOCATION Europe

The bryozoan *Meandropora* formed large, rounded colonies, which were composed of radiating cylindrical clusters (fascicles) of small tubular structures called autozooecia. These clusters divided and rejoined to produce a distinctive pattern on the surface of the colony. The autozooecia were very long with thin walls, and some were partitioned with diaphragms. They housed *Meandropora*'s soft-bodied, feeding individuals (autozooids). Each autozooid possessed a feeding organ called a lophophore, which consisted of a ring of tentacles surrounding the mouth. The autozooecia may have been arranged in clusters in order to provide a focus for channelling water currents. This would have enabled the colony to remove the sediment from the areas of the feeding autozooids and prevent it from harming them.

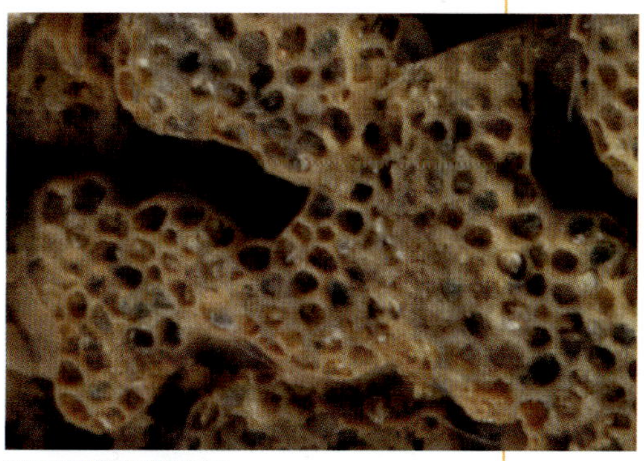

RADIATING CLUSTERS
The long, thin-walled, tubular autozooecia were arranged in clusters. This arrangement may have enabled *Meandropora* to channel water currents.

MEANDROPORA SP.
The clusters of autozooecia divide and rejoin to give the colony its distinctive surface pattern. The name "bryozoan" itself means "moss animal".

Biflustra

GROUP Bryozoans
DATE Cretaceous to present
SIZE Autozooecial openings 0.1–0.2mm wide
LOCATION Worldwide

Biflustra is a cheilostome bryozoan, with a hard skeleton that iss composed of calcium carbonate. It forms encrusting, flattened, or bifoliate (two-sided) colonies. Tubular chambers called autozooecia house the small, soft-bodied individuals (autozooids) of the colony. Each autozooid has a ring of tentacles around its mouth known as a lophophore. Small, beating, hair-like organs on the tentacles (cilia) create a flow of water to drive food particles towards the mouth. When the autozooids are not feeding, the tentacles are withdrawn into the autozooecia for protection.

There can be **thousands** of tiny individuals in a bryzoan **colony**.

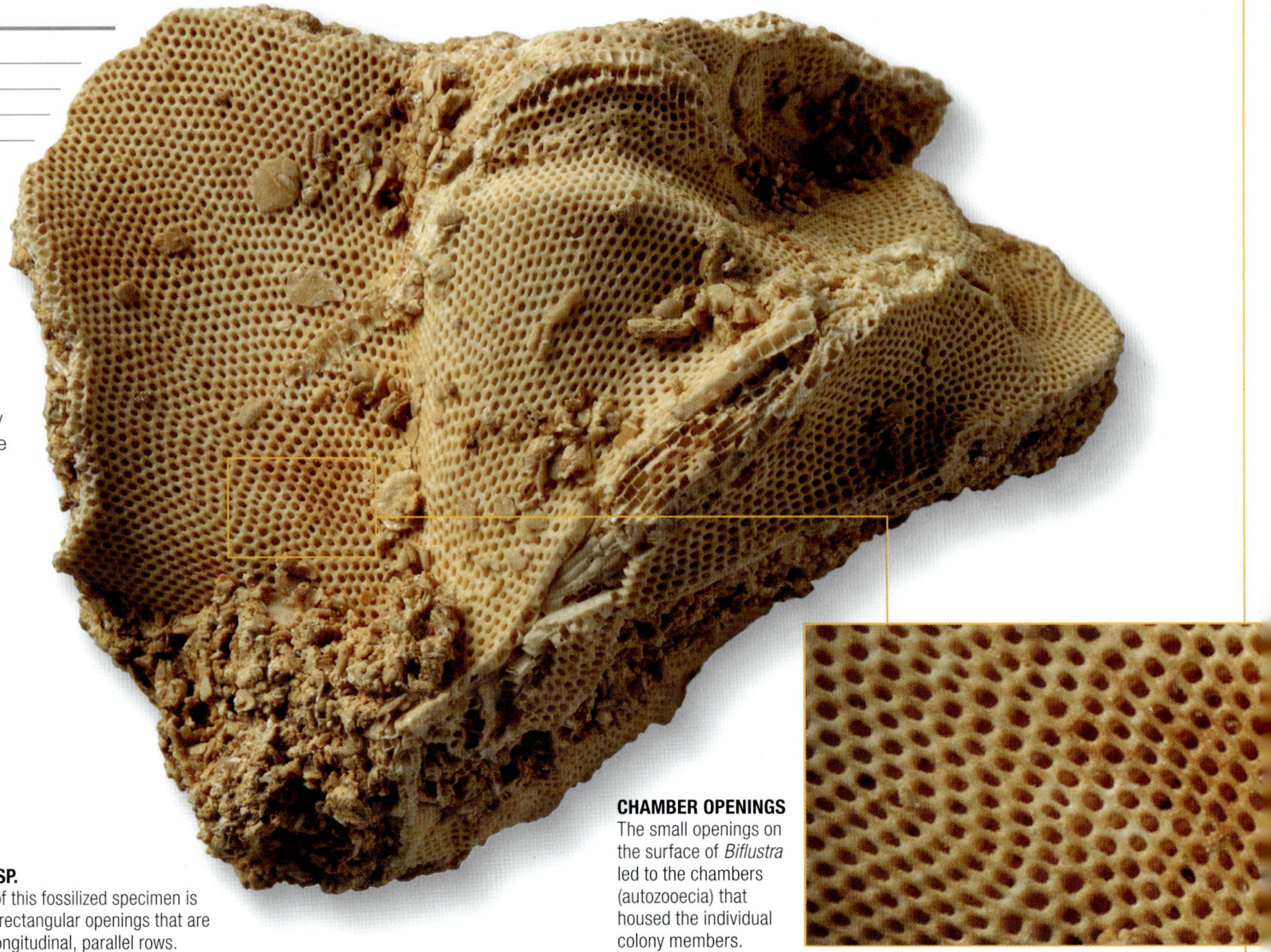

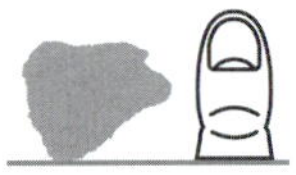

BIFLUSTRA SP.
The surface of this fossilized specimen is covered with rectangular openings that are arranged in longitudinal, parallel rows.

CHAMBER OPENINGS
The small openings on the surface of *Biflustra* led to the chambers (autozooecia) that housed the individual colony members.

Vasum

GROUP Gastropods
DATE Neogene to present
SIZE 4–10cm (1½–4in) long
LOCATION Tropics worldwide

This gastropod has a conical spire, with distinct spines that stick out around the shell and developed as part of its spiral ornament. There is one row of spines where the whorls meet each other and another below it at the shoulder of the whorl. Below the shoulder spines are five spiral ribs that have no spines. Lower down the whorl are four more rows of spirally developed spines. There are spines right back to the part of the shell that developed when the animal was a juvenile. The shell opening (aperture) is long and thin with a long anterior canal. This would have housed the animal's inhalant siphon, which brought water into the shell. Modern species of this genus predate marine worms and are found living in tropical zones worldwide.

VASUM SP.
Vasum's strong, thick, spirally ribbed shell would have offered good protection from most sea-floor predators.

Viviparus

GROUP Gastropods
DATE Jurassic to present
SIZE 1.5–3.5cm (½–1½in) long
LOCATION Worldwide

Viviparus is a freshwater gastropod. The whorls are rounded and convex, and have well-marked growth lines. Mature examples have some thickening on the outer lip, and some have similar earlier thickenings that suggest a slow down in the growth rate, perhaps due to environmental conditions. Unlike some other freshwater gastropods, it is not air-breathing. Instead, it has a single gill, which suggests it is derived from marine, gill-bearing ancestors. Females of modern *Viviparus* species keep their fertilized eggs in the space between the body and the shell wall (the mantle cavity) and give birth to live young.

VIVIPARUS SP.
A light, horny covering or operculum is attached to the foot and used to seal the shell opening.

LIVING RELATIVE

FRESHWATER SNAILS

Species of *Viviparus* are common and widely distributed across the globe. Some species display a burrowing habit, such as the European *V. viviparus* and the North American *V. intertextus* (shown here), which inhabits swamps. Both species have developed what is known as ciliary feeding, when the animal uses the leading edge of its single gill to collect food particles suspended in the water. This is similar to the filter-feeding method used among bivalves.

Terebra

GROUP Gastropods
DATE Neogene to present
SIZE 7–25cm (2¾–10in)
LOCATION Tropics worldwide

This pointed, multi-whorled gastropod is best described as "turreted". The shell is ornamented with sinuous, vertical ribs, and there is a faint spiral band about a third of the way down the side of each whorl. Modern species vary in appearance and size, with some species reaching 25cm (10in) long. Living species are all carnivores and often select a single species of invertebrate to feed on. Many bury themselves in soft sediment during the day and emerge to hunt at night. It is possible that the extinct Neogene species of *Terebra* behaved in a similar way.

TEREBRA FUSCATA
This specimen has around 15 whorls, which barely overlap each other.

Semicassis

GROUP Gastropods
DATE Neogene to present
SIZE 2–8cm (¾–3¼in) long
LOCATION Worldwide

The whorls of this gastropod are broad with each one overlapping about three-quarters of the following whorl. Some spiral ornamentation can be seen on the upper part of the outer whorl. The area surrounding the shell opening is called the mouth border, and there is a deep notch in this where the anterior inhalant siphon would have protruded to draw in water. Modern species of the genus are carnivores that feed on echinoids by boring into their shells with their tongue-like, toothed radula. It has been assumed that extinct species fed in a similar way.

SEMICASSIS SP.
This can be identified as a mature specimen because of the thickened lip that is partly visible at the opening to the shell.

Uzita

GROUP Gastropods
DATE Neogene
SIZE Up to 6cm (2¼in) long
LOCATION W. Europe

Uzita had stout whorls with convex sides, each overlapping about half the height of the preceding whorl. It had strong spiral ribs and grooves that were crossed by equally strong growth-parallel ribs, which created a pattern of small rectangles. The aperture was long with a well-marked anterior notch that housed the inhalant siphon. Living members of the same family as *Uzita* are scavengers and carrion feeders that feed on dead organisms on the sea bed, so it is probable that *Uzita* had the same feeding habits.

UZITA SP.
The mouth of the shell is large with a "lip" beneath the front of the largest whorl.

Ostrea

GROUP Bivalves
DATE Cretaceous to present
SIZE 7–15cm (2¾–6in) wide
LOCATION Worldwide

Neogene *Ostrea* were very similar to the living oysters that belong to the same genus. Like other members of the oyster group, the valves vary greatly in shape – the left valve is convex and often strongly ribbed, while the right valve is flat and lacks ribs. Unusually, *Ostrea* incubated its eggs within the mantle cavity, the space between the body and the body wall (see p.424). Protected in this way, when the young were finally released after 6–18 days, they already had a minute bivalved shell.

OSTREA VESPERTINA
Growth lines are visible on both valves. On the inner shell, a kidney-shaped impression has been left by the muscle that closes the valves.

Astarte

GROUP Bivalves
DATE Jurassic to present
SIZE 2–7cm (¾–2¾in) long
LOCATION Worldwide

Astarte is a triangular bivalve with forward-pointing beaks. It has two clear adductor muscle scars on the inside of the shell, and a well-marked pallial line. The pallial line runs between the two muscles scars and marks the attachment of the mantle (the external body wall) to the shell. The shell's exterior is marked with strong ribs parallel to its growth, and finer growth lines run between these. An external ligament, which opened the shell, was located behind the beaks, housed in a long groove. Internally, positioned under the beak in each valve, there are large, well-defined hinge teeth and sockets. These ensured accurate shell articulation. The family to which *Astarte* belongs can be traced back to the Devonian. This bivalve group evolved early and, due to its successful body-plan, it has changed very little over time.

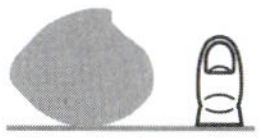

ASTARTE MUTABILIS
All *Astarte*'s relatives are characterized by the strong growth-parallel ribs over the whole surface of the shell.

Amusium

GROUP Bivalves
DATE Neogene to present
SIZE 5–12cm (2–4½in) wide
LOCATION Worldwide

Amusium is a large, smooth, almost circular bivalve from the same family as modern scallops. The main difference is that *Amusium's* valves are virtually smooth, whereas scallop valves are ribbed. *Amusium's* thin shell has two prominent, ear-like extensions on either side of the beak, with one slightly larger. Fine growth lines cover the shell surface, and faint radial striations are also visible.

AMUSIUM SP.
The virtually smooth valves of *Amusium* are the reason some species are referred to as "moon scallops".

Anadara

GROUP Bivalves
DATE Late Cretaceous to present
SIZE 2–6cm (¾–2¼in) long
LOCATION Worldwide

At first sight, *Anadara* looks like a modern cockle, with strong, radially arranged ribs crossed by growth lines. Inside, however, it differs greatly from modern cockles. Beneath the forward-facing beak of each valve is a long, striated groove that, in life, houses the long ligament responsible for opening the shell. Below the ligament groove is a long, straight hinge line, with many, almost vertical teeth and sockets that help the shell stay together when shut. There are two scars, where the muscles that controlled the movement of the valves would have been attached. An inconspicuous line between the scars is where the mantle was attached.

CLOSING MECHANISM
The hinge line, with its "teeth" and "grooves", helps *Anadara rustica* to stay shut when closed. Like most bivalves, it is a mainly sedentary filter-feeder.

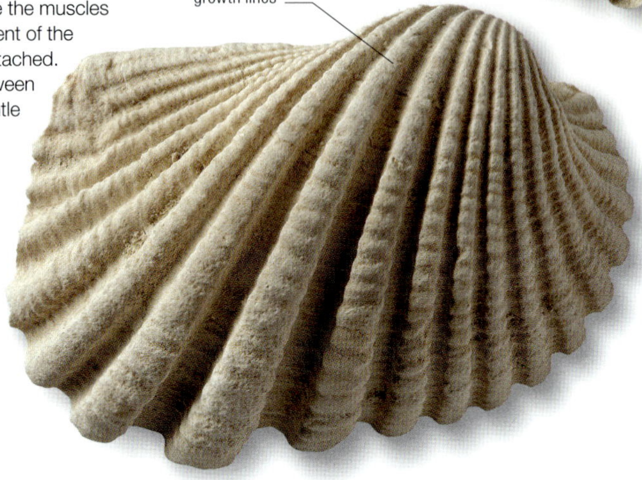

ANADARA RUSTICA
With its deep, radiating ribs, *Anadara* looks like a cockle, but it belongs to a group of bivalves called arcoids, which can be traced back to the Early Ordovician period.

Some species of *Anadara* have been used as a source of food by humans since prehistoric times.

Schizaster

GROUP Echinoderms
DATE Paleogene to present
SIZE 4.5–6.5cm (1¾–2½in) long
LOCATION Worldwide

Schizaster is a heart-shaped, burrowing sea urchin with five well-marked ambulacral areas or "petals" on its upper surface, where its tube feet protrude. The rear two petals are far shorter than the front three. The middle front ambulacral area is set in a deep groove that reaches down to the lower surface and the mouth. When viewed from the side, the shell is quite high, and there is a slight overhang at the rear above the anal opening.

SCHIZASTER SP.
Smooth fascioles are set around the petal-shaped upper ambulacra and beneath the anus. In life, hair-like cilia on these would have generated water currents.

Scutella

GROUP Echinoderms
DATE Late Paleogene to Early Neogene
SIZE 6–9cm (2¼–3½in) across
LOCATION Mediterranean

Scutella was a medium-sized, flattened, burrowing sea urchin, with internal "pillars" that strengthened its shell. There were five petal-shaped ambulacral areas on its lower surface where its many tube feet emerged. Around the edge of these areas were pairs of pores – each outer pore had a slit-like appearance. The mouth was situated in the centre of the animal's lower surface, and well-marked food grooves led directly to it. Where these food grooves crossed the junction between the upper and lower surfaces, they were as wide as the zones in between the ambulacral areas. *Scutella's* anal opening was also positioned underneath it, about midway between the back margin of the shell and the mouth. In the centre of the upper surface was the apical disc, a multi-plated area with four genital pores known as gonopores. *Scutella* is one of many echinoid fossils that are sometimes referred to as "sand dollars", although the sand dollars found by beachcombers today are usually the shells of modern echinoid species.

SCUTELLA SP.
With its five petal-like ambulacra and slightly domed shaped, *Scutella* has a pleasingly symmetrical appearance. Like modern echinoids, it was probably covered with tiny, hair-like spines.

CLYPEASTER SP.
This Neogene specimen is missing its central apical disc; there is a hole where the disc and gonopores resided.

Clypeaster

GROUP Echinoderms
DATE Paleogene to present
SIZE 5–15cm (2–6in) across
LOCATION Worldwide

Clypeaster is a medium to large burrowing echinoid with a variable shape that ranges from a flattened disc to bell-shaped forms, depending on the species. Sometimes, particularly in the flattened varieties, the shell wall is made up of two separate layers separated by supporting pillars, which gives the shell greater mechanical strength. There are five similarly shaped and prominent ambulacra (the petal-shaped areas where the tube feet protrude) on *Clypeaster's* upper surface, and these are far wider than the interambulacral areas. On its lower surface, these ambulacral areas are far less obvious. The mouth is located in the centre of the lower surface, and the anus is situated close to its edge. The disc-like structure located in the centre of the upper surface contains five genital pores. Some species of *Clypeaster* live in burrows up to 15cm (6in) below the sea bed. The shells of many modern species are often called "sea biscuits" by beachcombers.

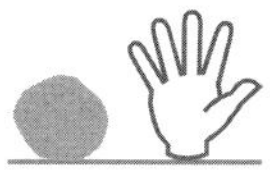

15cm The distance below the sea bed that some *Clypeaster* species dig their burrows.

PRESERVED FOR ETERNITY
Amber – fossilized tree resin – preserves organisms in exquisite detail. This example features a house centipede, the leggy *Scutigera* (centre), and above and to its left, two small flies. Trapped in the amber, the struggling flies probably attracted the predatory centipede, which in turn fell victim to the sticky resin. This all happened around 23 million years ago.

NEOGENE VERTEBRATES

From the start of the Neogene period 23 million years ago, the world's climate became gradually cooler and drier. This caused deserts and grasslands to advance across much of the continental land surface. Predators and prey either adapted to the new conditions or became extinct.

Many mammal groups colonized new continents during the Neogene, crossing land bridges created by reduced sea levels and the collision of landmasses. Ancestral camels originated in North America but crossed to Eurasia about 7 million years ago, where they later evolved into dromedaries and Bactrian camels. Movements to South America about 3 million years ago gave rise to llamas, guanacos, and vicunas.

Grasslands and grazers

The expansion of grasslands gave rise to a new mode of feeding among herbivores – grazing. The even-toed ungulates (hoofed mammals), especially ruminants, were better suited to eating the coarse, poorly nutritious grasses than the odd-toed ungulates because of their more highly developed stomachs. Ruminants have four-chambered stomachs containing microbes that digest cellulose. They also regurgitate and re-chew the tough plant material before swallowing it again. These adaptations led to a dramatic diversification of artiodactyls such as cattle, sheep, and antelopes during the Miocene at the expense of perissodactyls such as horses, rhinoceroses, and tapirs. The grazers began to herd together for safety while feeding and travelling on their seasonal migrations to find fresh pasture. They evolved to become bigger and faster to evade the predators that followed them. In turn, carnivores evolved strategies such as pack and pursuit hunting to help them catch their fast-moving prey. The advance of grassland in Africa also contributed to apes (hominoids) moving out of the forests and adopting a more upright posture (see pp.450–51).

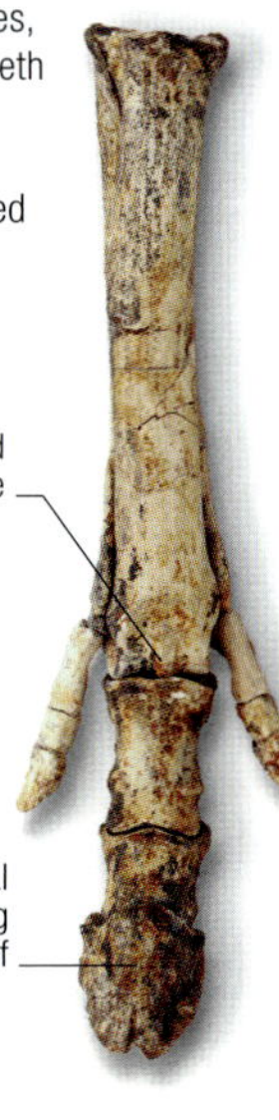

TEETH AND TOES
During the Miocene epoch, three-toed horses, such as *Hipparion*, evolved high-crowned teeth with grinding surfaces, enabling them to eat tough grasses. On each foot, the side toes became reduced and the central toe modified into a single weight-bearing hoof, allowing them to run fast across the plains.

Giants of the sea

Both modern lineages of whales – the mysticetes, or baleen whales, and the odontocetes, the toothed whales – existed at the start of the Neogene. Toward the end of the Miocene, representatives of several modern whale, dolphin, and porpoise families had appeared, alongside some families that are now extinct. The first true seals, sea lions, and walruses also evolved in the Miocene and the herbivorous sirenians (dugongs and manatees) diversified. All of these marine mammals were preyed upon by the gigantic shark *Carcharodon megalodon* (see opposite).

GROUPS OVERVIEW

Changing conditions during the Neogene prompted the appearance of new mammalian groups and the extinction of some older ones. The even-toed ungulates expanded at the expense of their odd-toed cousins, while carnivores took over from creodonts as dominant predators. Small mammals such as rabbits, rodents, and raccoons flourished.

PERISSODACTYLS
Horses continued to diversify during the Neogene as they exploited new grassland habitats. However, many species became extinct by the end of the Miocene around 5 million years ago leaving only one genus – *Equus*. Many Paleogene rhinos and tapirs also died out and the number of species continued to dwindle thoughout the Neogene.

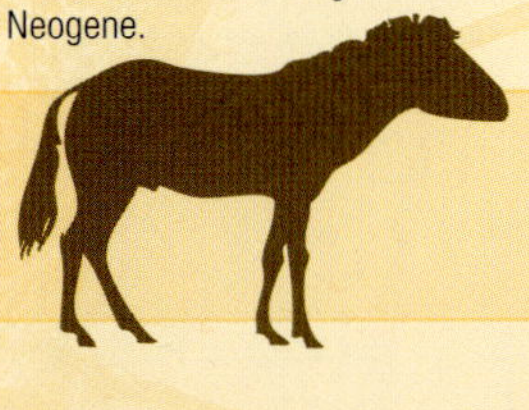

ARTIODACTYLS
Even-toed ungulates rose in prominence during the Neogene. Pigs, peccaries, and hippos were soon joined by the first giraffes, deer, and bovids such as gazelles and goats. Bovid evolution occurred primarily in Africa and Asia, but the Bering Strait land bridge allowed the ancestors of modern-day bison and bighorn sheep to move into North America.

RODENTS
Rodents did well during the Neogene, aided by their generalist habits and rapid reproduction rates. Modern-looking squirrels had appeared by the Miocene and murids, the family that includes mice, rats, gerbils, voles, lemmings, and hamsters, underwent a dramatic radiation in the Pliocene. The land bridge from Africa to Asia allowed porcupines to colonize Eurasia.

HOMINOIDS
The apes diverged from Old World monkeys about 30 million years ago. They evolved larger brains and bodies and became gradually more terrestrial. The lesser and great apes split from one another about 20 million years ago and species spread from Africa into Asia. The fossil genus *Sivapithecus* is considered to be the ancestor of orang-utans.

Carcharodon

GROUP Chondrichthyan
DATE Early Miocene to Pliocene
SIZE 18m (59ft) long
LOCATION Europe, North America, South America, Africa, Asia

Weighing in at around 50 tonnes (55 tons) or more, *Carcharodon* was the largest predatory shark that ever lived. The species – *megalodon* – is named after its huge triangular teeth, measuring up to 17cm (6½in) high. These teeth have serrations running down both cutting edges, ensuring an efficient slicing action. Cut marks in bones and isolated teeth lying next to fossil carcasses show that *Carcharodon* must have preyed upon a wide range of whales, dolphins and porpoises, seals, giant turtles, and fish. It probably ambushed its prey by attacking them at speed. Once disabled, the prey could be processed using its huge jaws and colossal bite force. The traditional view is that *Carcharodon* is a close relative of the modern great white shark (*Carcharodon carcharias*). There are certainly many similarities between the teeth of these two species, and in the absence of a complete specimen, *Carcharodon* is usually reconstructed using the great white as a model. However, some experts argue that these similarities evolved separately, and that *Carcharodon megalodon* belongs in a different genus – *Carcharocles*.

"BIG TOOTH"
This shark's species name – *megalodon* – is taken from the Greek for "big tooth". The size of its teeth has been used to estimate *Carcharodon*'s overall size.

STREAMLINED SHARK
Carcharodon is traditionally thought to have had the appearance of a scaled-up great white, with the classic, streamlined body shape, pointed snout, and large fins.

MASSIVE MOUTH
The open reconstructed jaws are the height of a man, and many times larger than those of the great white.

Myliobatis

GROUP Chondrichthyan
DATE Miocene to present
SIZE 1.5m (5ft) long
LOCATION USA, Belgium, Morocco

TOOTH PLATE
Myliobatis fossils are often identified by their tooth plates. These have one row of long teeth in the middle with rows of hexagonal teeth either side.

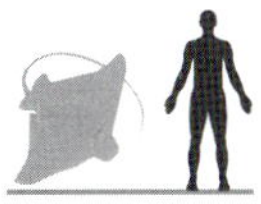

Myliobatis was a common stingray during the Neogene. At least 11 *Myliobatis* species survive today and are found worldwide – known as the eagle rays. Many specimens of extinct species have been unearthed in Pliocene deposits in North Carolina, USA. However, other finds in North America, Africa, and Europe show that *Myliobatis* was also around in the Miocene and perhaps extends back as far as the Eocene.

Gavialosuchus

GROUP Crocodylomorph
DATE Late Oligocene to Early Pliocene
SIZE 5.4m (17¾ft) long
LOCATION North America, Europe

Gavialosuchus was an extinct gharial (crocodile-like reptile) from the Late Oligocene to Early Pliocene of North America, and the Early Miocene of Europe. Like living gharials, it had a very long skull with a narrow snout that it used to catch fish. *Gavialosuchus* fossils have been found in coastal deposits, which suggests that it lived in estuaries or shallow seas, where it could have fed on a variety of different types of fish. Although living gharials are restricted to India and Southeast Asia, fossils of different *Gavialosuchus species* found in Florida, Austria, and Georgia suggest that the group was once distributed in swampy and coastal habitats all over the tropical regions of the world.

ELONGATED SKULL
Like modern-day crocodilians, *Gavialosuchus* had long jaws, full of sharp teeth.

Phalocrocorax

GROUP Aves
DATE Miocene to present
SIZE 45–100cm (17½in–39in) tall
LOCATION USA, France, Spain, Moldova, Bulgaria, Ukraine, Mexico, Mongolia, Australia

Modern-day members of the *Phalocrocorax* genus are the cormorants and shags. There are 36 living species of these medium-sized, fish-eating seabirds. A similar number of extinct species from all over the world have been described. *Phalocrocorax* hunted by diving and had a flexible neck and long bill for catching fish. The tip of its upper bill was slightly hooked to cling on to slippery prey. Historically, cormorants were thought to be the shoreline cousins of ravens. However, *Phalocrocorax* belongs to the Pelecaniformes, a group of water birds that includes pelicans and gannets.

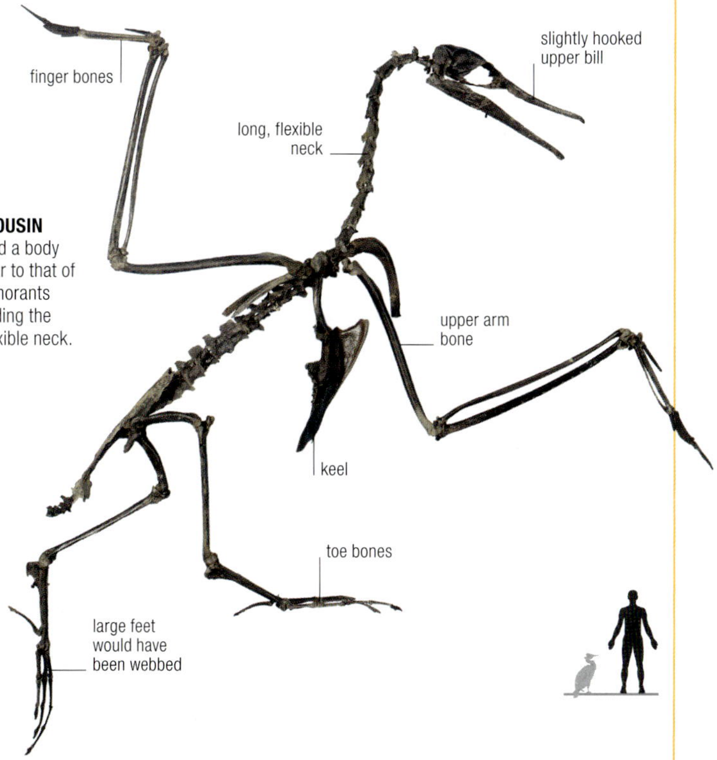

CORMORANT COUSIN
Phalocrocorax had a body shape very similar to that of modern-day cormorants and shags, including the characteristic flexible neck.

LIVING RELATIVE

ANDEAN CONDOR

The Andean condor has a 3m (9¾ft) wingspan and weighs 15kg (33lb). It soars at altitudes of up to 5,500m (18,00ft) above the Andes, looking for dead or dying animals, and may fly 200km (124 miles) a day in search of food. It usually follows other scavengers, such as small vultures or crows, to a carcass, then uses its strong bill to rip through the tough hide.

Argentavis

GROUP Theropods
DATE Late Miocene
SIZE 3.5m (11½ft) long
LOCATION Argentina

MASSIVE WINGS
With its wing area of almost 7 square m (75 square ft), *Argentavis* was able to soar effortlessly in the skies above Argentina, riding the thermals and updraughts.

Argentavis is by far the largest flying bird to have ever lived. It resembled a gigantic condor, with a wingspan of 8m (26ft); a large specimen would have weighed 80kg (176lb). This is more than double the wingspan and five times the weight of the Andean condor, the largest bird flying today (see panel, left). *Argentavis* had strong legs and wide feet, so it was able to walk easily. Its bill was long with a hooked end, like that of an eagle, so it could rip open almost any carcass. *Argentavis* was almost certainly a scavenger, searching for carrion as it soared in the air, before landing and driving off other large predators from their kills. This giant of the sky would have had to cover a vast territory to find enough food to sustain itself.

Thylacosmilus

GROUP Marsupial mammals
DATE Late Miocene to Early Pliocene
SIZE 1.5m (5ft) long
LOCATION South America

Thylacosmilus was a large predatory marsupial that was about the size of today's jaguar. Like the sabre-toothed cats such as *Smilodon* (see p.435), to which it was unrelated, it had long, sabre-shaped upper canine teeth. This is a classic example of convergent evolution. However, in *Thylacosmilus*, these canines grew continuously throughout its life. When the mouth was closed, the sabres were protected by a pair of scabbard-like flanges in the lower jaw. *Thylacosmilus* was the last of the line of marsupial carnivores that diversified in isolation in South America. When North and South America became connected via the Panama land bridge, around 3 million years ago, placental carnivores, including *Smilodon*, moved into South America and took their place.

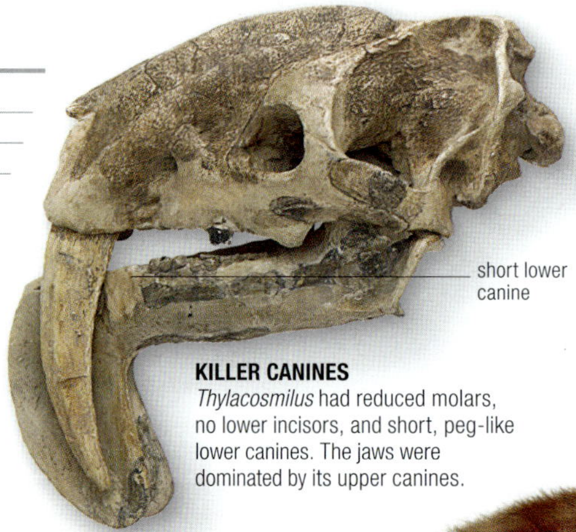

KILLER CANINES
Thylacosmilus had reduced molars, no lower incisors, and short, peg-like lower canines. The jaws were dominated by its upper canines.

***Thylacosmilus*'s upper canines were even longer than those of *Smilodon*, the sabre-toothed tiger.**

POWERFUL PREDATOR
The heavily muscled shoulders and neck would have given great force to *Thylacosmilus* as it stabbed downwards or slashed at its prey with its sabres.

Deinogalerix

GROUP Placental mammals
DATE Late Miocene
SIZE 60cm (23½in) long
LOCATION Italy

Deinogalerix was a giant, spineless, hairy hedgehog that lived on Gargano during the Late Miocene, when it was an island off the coast of Italy. Like today's hedgehogs, to which it was related, *Deinogalerix* fed on a mixture of insects, snails, and other invertebrates. However, it was huge for an insectivore – weighing as much as a large cat – and its large body size would have enabled it to hunt birds and small mammals as well. *Deinogalerix* is a classic example of how small mammals tend to grow large on islands where they do not face competition from larger predators, such as cats, dogs, and bears. However, the giant barn owl *Tyto gigantea* also lived on Gargano during the Late Miocene and may even have preyed upon *Deinogalerix*.

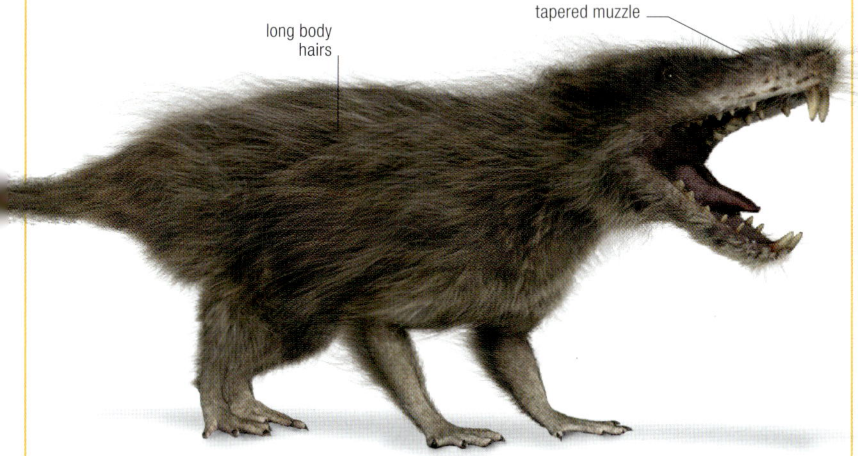

HAIRY HEDGEHOG
Deinogalerix means "terrible shrew", but it actually resembles a large moonrat, a primitive type of hedgehog found today in Southeast Asia, which has a long tail and fur instead of spines.

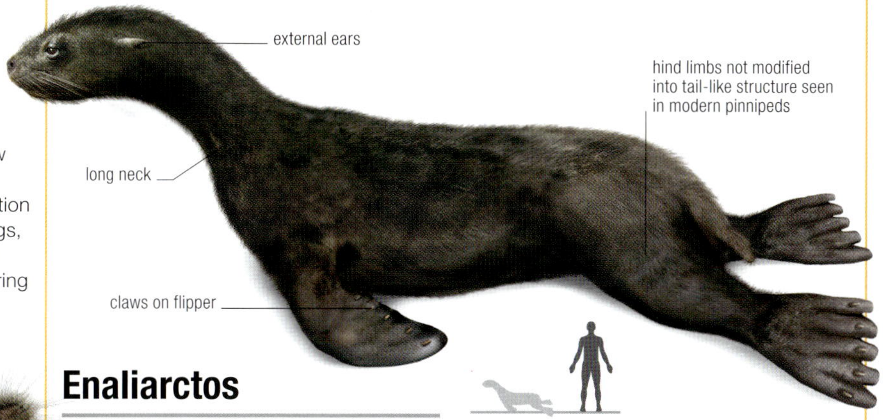

Enaliarctos

GROUP Placental mammals
DATE Early Miocene
SIZE 1.5m (5ft) long
LOCATION USA

PRIMITIVE PINNIPED
Its elongated and strongly muscled hind limbs indicate that *Enaliarctos* was more active on land than modern pinnipeds.

Enaliarctos is the earliest-known relative of the pinnipeds (seals, sea lions, and walruses). Its fossils have been found in Early Miocene rocks in California and Oregon, USA. Like later seals and sea lions, *Enaliarctos*'s limbs were modified into crude flippers, although they were not as highly specialized as those of modern pinnipeds. *Enaliarctos* used both front and hind limbs for swimming; modern sea lions use just their forelimbs for swimming, and all four flippers to move on land, while modern seals use their hind flippers for swimming and are less mobile on land. Although it had large eyes, sensitive whiskers, and good underwater hearing, like its living descendants, *Enaliarctos*'s teeth were more primitive and resembled those of its bear-like ancestors. It ate fish but, unlike seals, could not eat its catch while swimming – instead it had to drag its prey to the shore to tear it apart.

***Enaliarctos* had sensitive whiskers, large eyes, and good hearing.**

Allodesmus

GROUP Placental mammals
DATE Middle Miocene
SIZE 1.5m (5ft) long
LOCATION USA, Mexico, and Japan

Allodesmus was an early relative of sea lions and, like them, it swam with its forelimbs and could rotate its hind limbs forwards to help it walk in a shuffling fashion on land. It had a long, slender skull like that of a modern leopard seal, but its teeth were blunt with bulbous crowns, more suited for catching fish and squid, which it would then have swallowed whole as modern sea lions do. Its large eyes would have helped it see while hunting underwater. *Allodesmus* showed a wide range in sizes between males and females, suggesting that the large bulls, which weighed up to 360kg (794lb) would have guarded a harem of cows.

***Allodesmus* displayed sexual dimorphism, with males being much larger than females.**

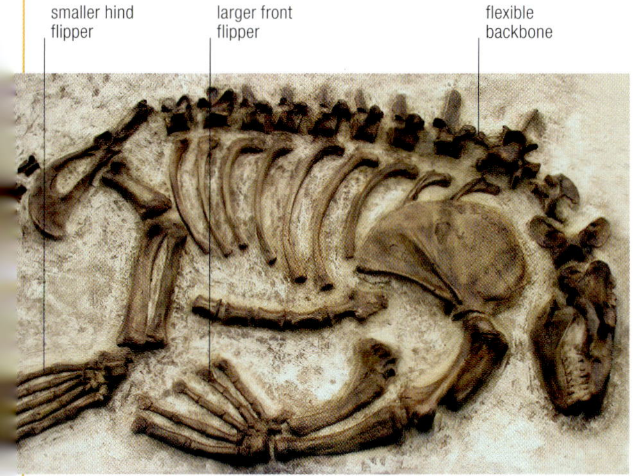

KEY SITE
SHARKTOOTH HILL

Sharktooth Hill, which lies northeast of Bakersfield, California, USA, is the preserved remains of a North Pacific near-shore oceanic ecosystem of the Middle Miocene. The sheer quantity and quality of its fossils were first documented by Swiss palaeontologist Louis Agassiz in 1856. In addition to marine molluscs, the fossils of a wide variety of fish have been found, including the teeth of 30 species of sharks (hence the site's name). Turtles, birds, and 17 species of whales and other marine mammals have also been discovered, including many *Allodesmus* fossils.

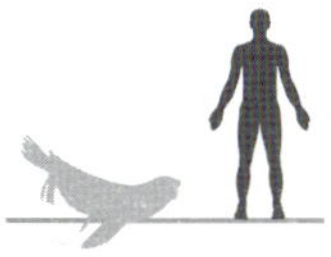

FORWARD PROPULSION
Allodesmus propelled itself through the water with its front flippers, which were larger and more powerful than its rear flippers.

Ceratogaulus

GROUP Placental mammals
DATE Middle to Late Miocene
SIZE 30cm (12in) long
LOCATION Canada, USA

Ceratogaulus was a burrowing rodent distinguished by a pair of straight horns on its nose. It was roughly the size of a modern marmot, but more closely resembled a gopher, with its strong forelimbs equipped with huge claws. The role of the horns has been much debated. It has been suggested that they were used for digging, but they were not in the correct position. Both males and females had horns, so they are unlikely to have been used for display during courtship. It is now thought that the horns were used as defensive weapons when the animal was out in the open, away from the safety of its burrow.

DESIGNED FOR DIGGING
Ceratogaulus used its large claws for digging. It fed on roots, bulbs, and other underground plant parts, but it also came above ground to forage.

Palaeocastor

GROUP Placental mammals
DATE Early Oligocene to Early Miocene
SIZE 40cm (15½in) long
LOCATION USA, Japan

Palaeocastor was one of the earliest known relatives of living beavers. It lived between the Early Oligocene and the Early Miocene. Fossils have been found in western USA and Japan. Although related to beavers, *Palaeocastor* was not an aquatic tree-cutter. Instead it was a terrestrial burrower, about the size and shape of a modern marmot.

BURROWING BEAVER
A member of the beaver group, *Palaeocastor* was a burrowing mammal that used its strong front teeth to dig into the ground.

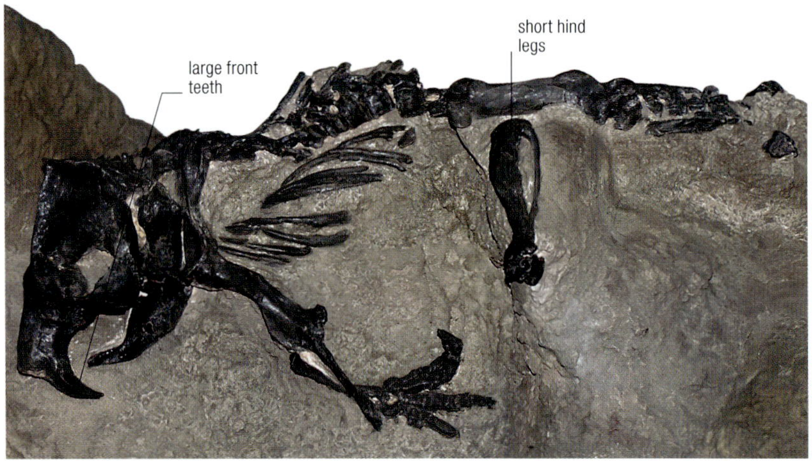

KEY FIND

DEVIL'S CORKSCREWS

Palaeocastor is most famous for making long spiral burrows that were found in Nebraska, USA. These spirals, known as "Daemonelix", are shaped like corkscrews and went as deep as 3m (9¾ft) below the surface. At first they were a mystery, until a specimen of *Palaeocastor* was found at the bottom of one of them. The scratch marks along the sides, long-attributed to the claws of a burrower, were actually cut by the chisel-like front teeth of the beaver as it dug with its mouth.

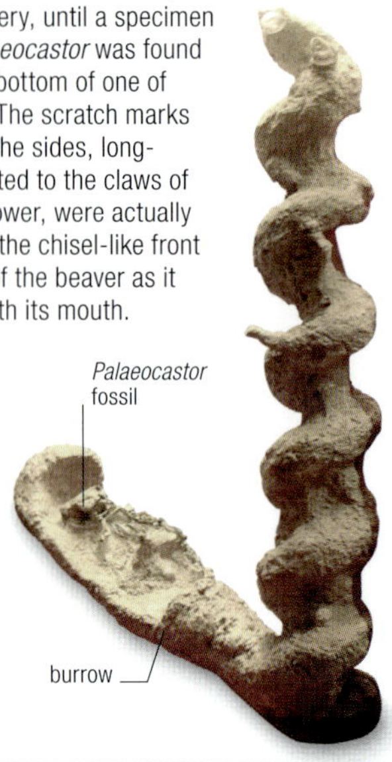

Teleoceras

GROUP Placental mammals
DATE Miocene
SIZE 4m (13ft) long
LOCATION USA

Teleoceras was a large rhinoceros with a single small horn on the tip of its nose. It comes from Miocene deposits of North America. Although it is a true rhinoceros, its body plan reflected that of a hippopotamus, with short, stumpy limbs, a massive barrel chest, and a skull with high-crowned cheek teeth. *Teleoceras* fossils are found in large numbers in many ancient river and pond deposits on the high plains of western North America. One such place is Ashfall Fossil Beds State Park (see panel, right). Here hundreds of complete *Teleoceras* skeletons are found. Some specimens have grass seeds fossilized in their throats, confirming that *Teleoceras* was a grass-eater.

Stenomylus

GROUP Placental mammals
DATE Late Oligocene to Early Miocene
SIZE 1.5m (5ft) tall at the shoulder
LOCATION USA

Stenomylus was a humpless camel that lived in North America between the Late Oligocene and the Early Miocene. Although *Stenomylus* was a true camel, it was small, reaching only 1.5m (5ft) tall at the shoulder. Its legs and body were delicate and slender, like those of a modern gazelle. Its most remarkable features were its elongated, high-crowned molars, which had deep roots that reached the base of the jaw and the top of the skull. These must have been used to eat very gritty plants, as they show signs of extreme wear during the life of the animal.

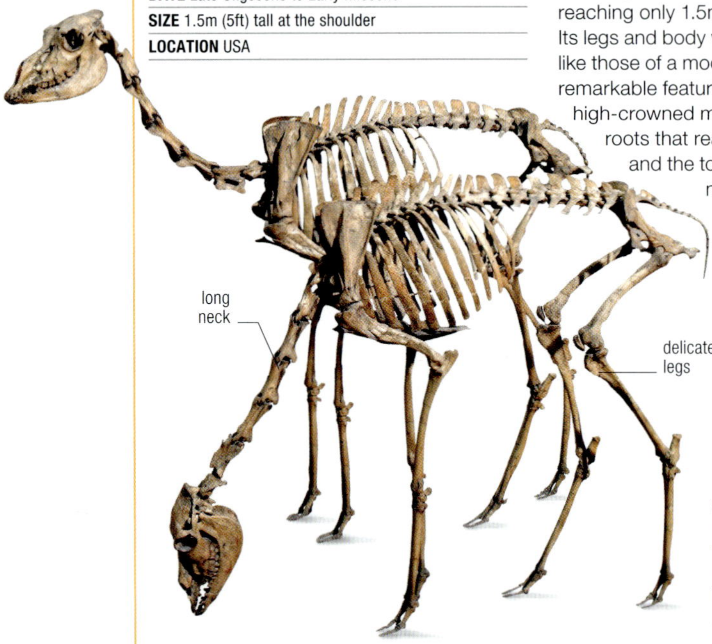

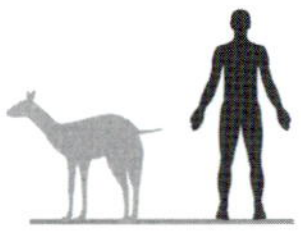

SOCIAL CAMELS
Stenomylus is best known from large quarry samples in western Nebraska, USA. These humpless camels apparently lived in large herds.

Menoceras

GROUP Placental mammals
DATE Early Miocene
SIZE 1.5m (5ft) long
LOCATION USA

Menoceras was a small rhinoceros that has been found in the early Miocene deposits of western North America, especially Nebraska and Wyoming. It was a true rhinoceros, but is more slender in its build than modern members of the group. It was also smaller and only reached the size of a large pig. Males had two bony horns positioned side by side on the tip of the nose, but the females were hornless.

TWIN-HORNED RHINOCEROS
Menoceras is the first-known rhino to develop horns. However, only the males developed a pair of horns on the tip of their nose.

The large number of *Teleoceras* fossils found at the Ashfall Fossil Beds State Park has led to the site being called "Rhino Pompeii".

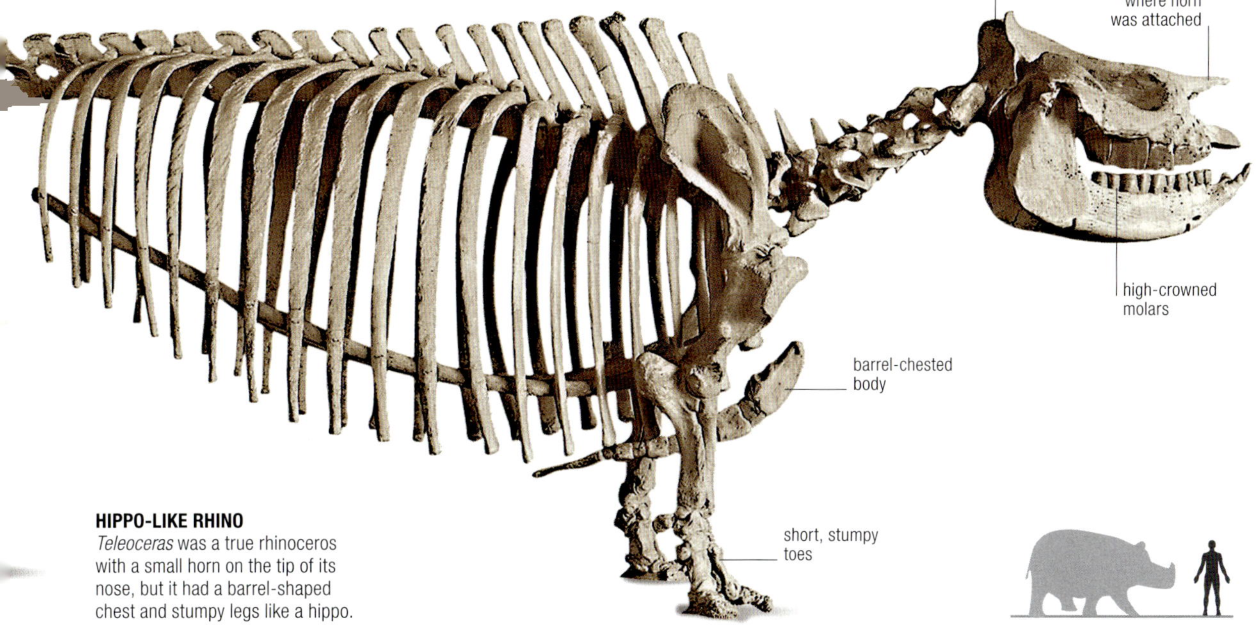

HIPPO-LIKE RHINO
Teleoceras was a true rhinoceros with a small horn on the tip of its nose, but it had a barrel-shaped chest and stumpy legs like a hippo.

KEY SITE

ASHFALL FOSSIL BEDS

Teleoceras is best known from Ashfall Fossil Beds State Park in northeast Nebraska. Here hundreds of complete skeletons of *Teleoceras* and other mammals and birds were fossilized as they died, suffocated by a volcanic ash fall that buried their waterhole 10 million years ago. Some specimens were removed, but most have been left excavated in the ground and can be viewed as they were found.

IMMENSE MAMMAL
Paraceratherium was the largest land mammal that ever lived, reaching 5.5m (18ft) at the shoulder, 8m (26ft) in length, and weighing about 15 tonnes (16½ tons).

Paraceratherium

GROUP Placental mammals
DATE Late Oligocene to Early Miocene
SIZE 8m (26ft) long
LOCATION Pakistan, Kazakhstan, India, Mongolia, China

Paraceratherium (formerly known as *Indricotherium*) was a gigantic, hornless rhinoceros from the Late Oligocene to Early Miocene of Asia. Its immense skull alone was 1.3m (4¼ft) long and had two conical upper tusks, flaring lower tusks, and retracted nasal openings suggesting a fleshy proboscis. Its immense size and long neck allowed *Paraceratherium* to feed on the tops of trees, like a giraffe does today. Despite its huge size, *Paraceratherium* still had the long toes and limbs of its ancestors, the hyracodonts, and not the stumpy toes found in similarly huge beasts, such as elephants.

ANATOMY

PREHENSILE LIP

Paraceratherium, like many rhinos and tapirs, probably had a long prehensile lip that could wrap around branches and strip leaves from the trees that it fed upon. The living black rhinoceros does this with its upper lip, and so do most tapirs. Tapirs and many rhinos have the nasal bones retracted far back on the skull, leaving a large nasal notch where the complex muscles attach to manipulate their lip or proboscis.

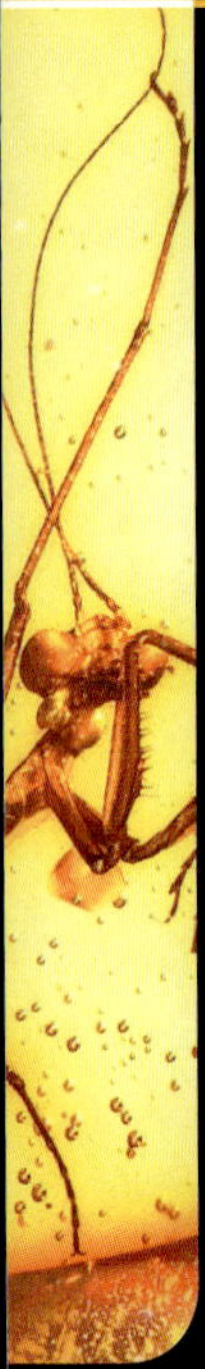

Chalicotherium

GROUP Placental mammals
DATE Late Oligocene to Early Pliocene
SIZE 2m (6½ft) long
LOCATION Europe, Asia, Africa

Chalicotherium was a perissodactyl (odd-toed, hoofed mammal) that lived between the Late Oligocene and the Early Pliocene in Europe, Asia, and Africa. It looked a little like a large horse, except that it had claws on its toes rather than hooves. The purpose of these claws on a hoofed mammal was a puzzle for some time, but it is now thought that they were used to hook branches and pull them. In addition, *Chalicotherium*'s front limbs were far longer than its back legs, and it had features on the hands and pelvic bones that suggested it knuckle-walked, much like a gorilla does today. When it was not on the move, *Chalicotherium* sat for long periods of time on its haunches feeding on vegetation. The relationships of chalicotheres have been a mystery for many years. Recent evidence has shown that they are cousins of the tapirs, but when the claws of *Chalicotherium* were first discovered they were thought to be a type of carnivore.

Chalicotherium means **"pebble beast"**, which refers to its **pebble-like molar teeth**. At maturity, its incisors and upper canines were shed.

RE-EVOLVED CLAWS
Although *Chalicotherium* was related to tapirs and other odd-toed hoofed mammals, its hooves had evolved into claws. This is the only known instance of a hoofed mammal reverting back to the ancestral clawed toe.

QUATERNARY

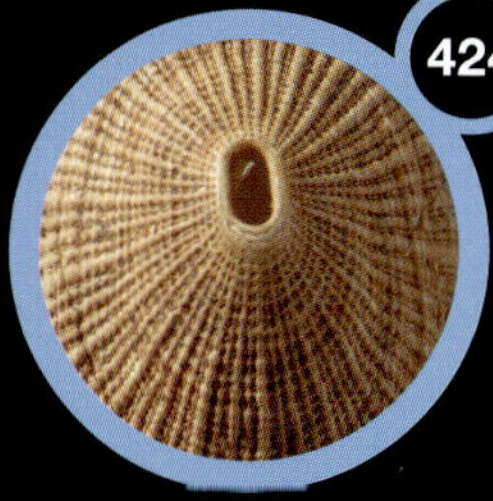

Quaternary

The Quaternary Period is part of a continuing ice age, during which intense periods of glaciation lasting nearly 100,000 years have been alternating with warmer interglacials 20,000 to 30,000 years long. Major changes in mammal fauna occurred during the last glacial and interglacial periods, with the extinction of the Pleistocene megafauna. Our own species may have played a significant role in their demise.

PO RIVER DELTA
Quaternary was the name given to one of four subdivisions of Earth's rocks made by early geologists, and typified by these alluvial deposits in Italy's Po River Valley.

Oceans and continents

Earth's oceans and continents have been more influenced by climate during the Quaternary than in any other period. While India continues to nudge into Asia, Australia into Indonesia, and Africa and Arabia into Europe and Asia, with associated and often catastrophic earthquakes, the oceans have risen and fallen like global soufflé. As the ice sheets advanced with a pincer movement from north and south, much of the water that otherwise would have been returned to the oceans became locked in the frozen continents and the sea levels fell substantially. In Europe, the rivers of the Rhine and Thames converged in a huge estuary that emptied into the North Sea off the northern coast of England, and the English channel did not exist. Rising oceans during the last 10,000 years of the Pleistocene brought dramatic changes to the geographies of these and many other parts of the world. During this period, humans, who had previously lived as hunter-gatherers, began to domesticate animals and establish permanent farming settlements. Our influence on the landscape has increased and we have come to dominate the environments in which we live, at the expense of many other living species. Yet the world geography we see today could be transient. Human activities seem to be causing global warming on a scale that threatens to flood Earth's terrestrial areas. No one can yet predict whether the Quaternary's natural cycles of glacial advance and retreat will continue and if they do, whether the next glaciation could overcome human impact on the Earth's temperatures. If it does, the size of the continents will increase, as once again there is a return to a windy, frozen world.

ENGLISH CHANNEL
The English Channel did not separate England and France 25,000 years ago. With much of Earth's water frozen in glacial ice, sea levels were up to 130m (425ft) lower than they are today.

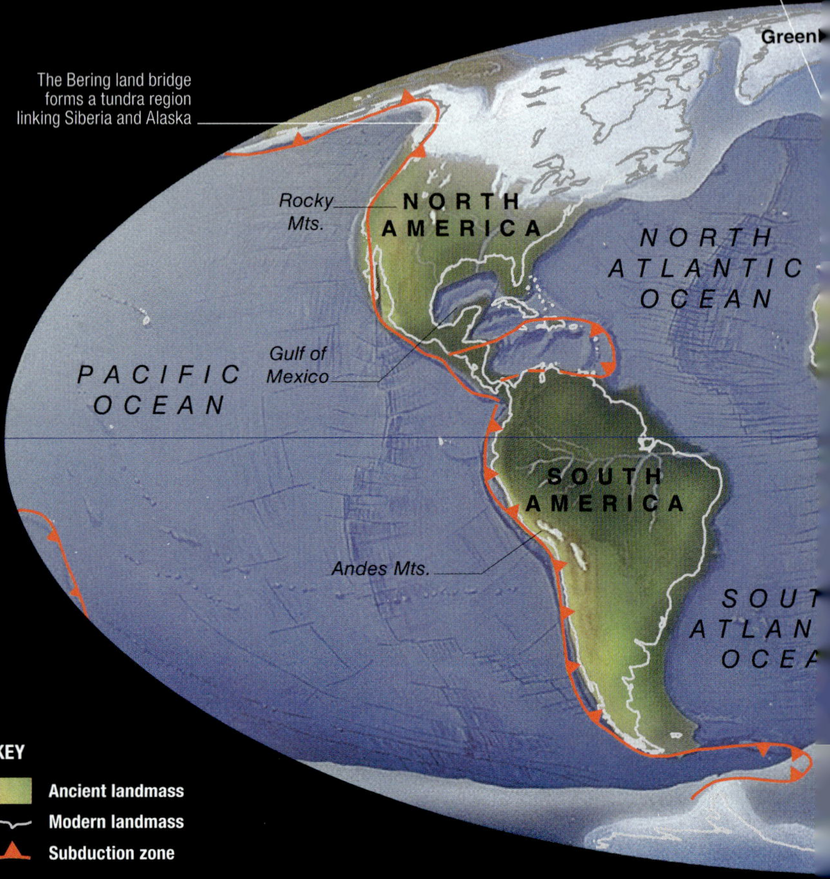

QUATERNARY WORLD MAP
With climatic fluctuations, the flooding of land masses alternated with the drying out of some shallow sea floors. At intervals, land bridges formed.

PLIOCENE

MYA	2	1.5
PLANTS	**2** Conifers at high latitudes become more diverse	
VERTEBRATES	**2** Extinction of australopithecines **2** First *Homo* **1.8** Evolution of *Homo erectus*; continued expansion of North American carnivores into South America; expansion of South American opossums, sloths, armadillos, and glyptodonts into North America	**1.2** Evolution of *Homo antecessor*

HOMO HABILIS

DIDELPHIS

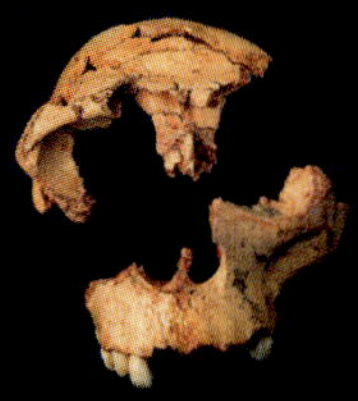

HOMO ANTECESSOR

FRESH WATER SUPPLY
Locked in the ice of the Perito Moreno Glacier, in Argentine Patagonia, is the world's third largest reserve of fresh water. Unlike many glaciers in the current interglacial period, it is not in retreat.

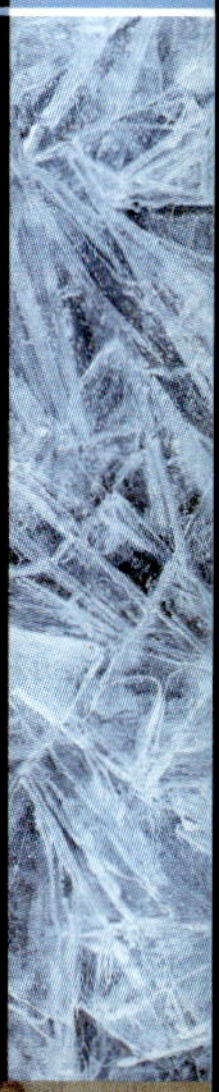

Climate

Earth's climate has fluctuated dramatically over the last 1.8 million years, and particularly during the last 600,000, while the planet has been locked in a series of cold glacial and warmer interglacial periods. Temperature, precipitation, and carbon dioxide levels have all varied cyclically. About 80 per cent of Quaternary climate change has been due to orbital forcing (see p.23) that has caused cyclical variations in the solar input to the Earth. From about 1.3 million years ago there were 30,000-year pulses of colder periods. Longer cold periods of nearly 100,000 years began about 900,000 years ago. During the cold phases ice sheets expanded enormously, especially in the last half a million years. With much of the world's water frozen and low evaporation, the climate was very dry, even in ice-free areas. Ice sheets covering Europe, Asia, and North America during the last glacial advance reached their maximum between about 15,000 to 20,000 years ago, and only started to retreat about 12,000 years ago. To cope with the cold, giant mammals developed thick protective pelts of fur or hair. Bison, woolly mammoths, and deer were hunted by large sabre-toothed tigers, cave bears, and by our ancestors – large-brained hominins that had begun to spread across the globe. Human impact on the climate and environment has been so extensive that some scientists have proposed a new epoch in Earth's history – the Anthropocene.

Large, permanent ice sheets cover Antarctica, advancing and retreating at intervals

FOSSILIZED MAMMOTH SKELETON
Many large mammals in Europe, including most mammoths, became extinct at the end of the Pleistocene. These large herbivores could have been exterminated by human hunting or habitat loss.

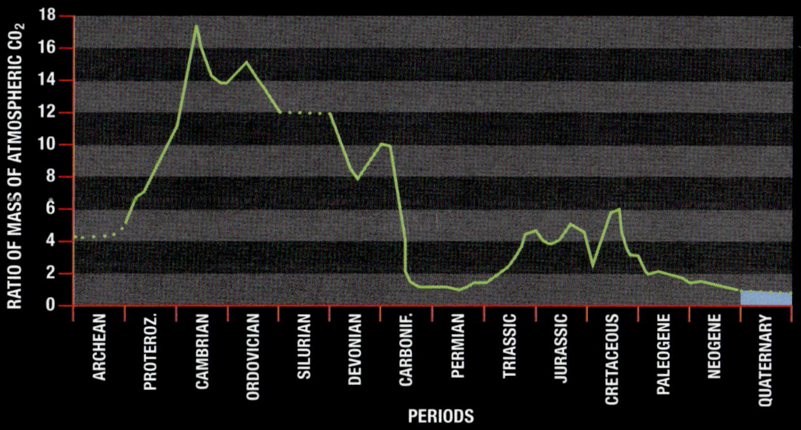

CARBON DIOXIDE LEVELS
Atmospheric carbon dioxide has fluctuated slightly for most of this period, but deforestation and the use of fossil fuels have increased levels by about 35 per cent since the Industrial Revolution began.

EISTOCENE | **HOLOCENE**

0.5 | 0 | **MYA**

PLANTS

- **0.9** Great expansion of steppe floras

VERTEBRATES

- **0.6** Evolution of *Homo heidelbergensis*

HOMO HEIDELBERGENSIS

- **0.35** Evolution of *Homo neanderthalensis*; giant flightless birds diversify (*Moa*, New Zealand; *Aepyornis*, Madagascar; *Genyornis*, Australia)

HOMO NEANDERTHALENSIS

- **0.15** Evolution of *Homo sapiens*
- **0.1** Major reduction in mammalian diversity in North America, especially mammoths, camels, perissodactyls, gomphotheres, sabre-toothed tigers, horses, sloths
- **0.01** Expansion of megafauna in Australia
- **0.05** Megafaunal extinctions in Australia
- **0.045** Arrival of *Homo sapiens* in Australia
- **0.03** Extinction of *Homo neanderthalensis*
- **0.012** Humans reach America
- **0.01** Megafaunal extinctions in Europe (e.g. mammoths)
- **0.0002** About 80% of small Australian mammals extinct
- **0.00007** Extinction in 1936 of last member of the family Thylacinidae (Tasmanian tiger)

QUATERNARY PLANTS

During the Quaternary, the geographic distribution of plants was shaped by dramatic changes in the global climate. The Quaternary marks the final phase in the evolution of land plants before the great modification of terrestrial environments caused by human activity over the past few thousand years.

The Quaternary was a time of rapid climatic change that had profound effects on global vegetation, and as the climate changed, it had a major impact on where new plants could establish themselves. The cumulative effect over hundreds, or thousands, of years is seen as "migration", with plants staying in step with the climatic conditions in which they grow best.

Regional extinction

Successive expansion and contraction of ice sheets and glaciers in the northern hemisphere during the Quaternary resulted in successive southerly and northerly migrations by different species. One effect of these migrations was significant regional extinction, especially in Europe, where east–west running mountain ranges partially blocked migration routes to the south. The strong similarities between the rich temperate forests of China and those of eastern North America largely reflect the elimination of many species from Europe during the Late Neogene and Quaternary. Fossil evidence shows that many of these plants, such as the tulip tree, flourished in Europe until relatively recently. There were also Late Neogene and Quaternary regional extinctions in western North America, most likely due to the increasingly dry climate.

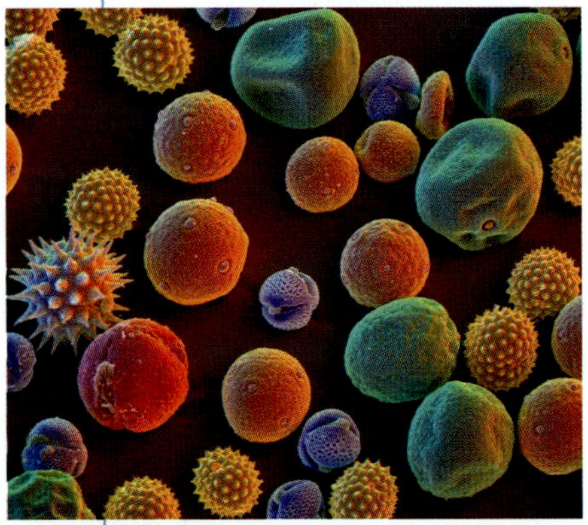

POLLEN GRAINS
By studying ancient pollen and spore assemblages it has been possible to reconstruct the main vegetational changes that occurred during the Quaternary.

TUNDRA
Distinctive tundra vegetation, often dominated by dwarf birches and willows, first became widespread during the successive expansions of ice sheets and glaciers that took place through the Quaternary.

Human landscapes

Many of our most familiar landscapes have been modified to a greater or lesser extent by human action. This trend is destined to continue, with the loss of much of the great diversity that has accumulated over more than 450 million years of land plant evolution.

HUMAN IMPACT
In the last few thousand years, and especially over the past few hundred years, human populations have exerted ever greater influence over the ecosystems that have grown up on land.

GROUPS OVERVIEW

There are probably more land plant species living today than at any time in the past. This extraordinary variety has arisen as group upon group has undergone bursts of diversification in response to changing environmental conditions. The fossil record provides ample evidence of extinction, but over time new species have arisen faster than they have disappeared.

CLUBMOSSES
Living clubmosses include forms very similar to ancient lycopods from the Early Devonian, alongside other plants that are the product of more recent bursts of diversification. The diversity of clubmosses is a wonderful mixture of ancient and modern forms that reflects more than 400 million years of plant evolution.

FERNS
Ferns are second only to angiosperms in the number of living species. Their success has been based mainly on exploiting the forested habitats that angiosperm trees created. However, among the great diversity of modern ferns there are also more archaic forms that have persisted largely unchanged since the Paleozoic and Mesozoic.

CONIFERS
With fewer than a thousand living species, conifers are nevertheless very important in the vegetation of many parts of the world. They have proved especially successful in the northern hemisphere, where they dominate the relatively uniform, cool temperate forests that extend across much of Asia, Europe, and northern North America.

ANGIOSPERMS
Despite their recent origin compared to the much longer evolutionary history of most other groups of land plants, angiosperms dominate the vegetation over most of Earth's surface. They exhibit bewildering diversity in almost every aspect of their structure and biology. There are about 350,000 species of living angiosperms.

Chara

GROUP Charales
DATE Silurian to present
SIZE Up to 1m (3⅓ft) long
LOCATION Worldwide fresh and brackish water

These fresh- or brackish water plants possess both algal and higher plant characteristics, which makes them an isolated group between the green algae and the bryophytes. Whole plant remains are extremely scarce, although their reproductive organs, called gyrogonites and utricles, are much more common. The plants are bushy and up to 1m (3⅓ft) long. Individual shoots are only 1 or 2mm (¹⁄₁₆in) wide and organized in short lengths called internodes with whorls of branches arising at the joints (nodes). The Charales has an extensive fossil record, first appearing in the Late Silurian and diversifying during the Devonian and the Carboniferous. The genus *Chara* first appeared during the Early Paleogene.

CHARA STEMS
Chara gyrogonites can be recovered from all kinds of sediments, but the shoots need to have been infiltrated and mineralized to survive. They may be preserved silicified or calcified in limestone, as in this rare specimen from China.

Marchantia

GROUP Bryophytes
DATE Quaternary to present
SIZE Thallus up to 10cm (4in) long
LOCATION Worldwide, except in arid regions

Marchantia are liverworts, which make up a major part of the bryophyte group. Some resemble leafy mosses, while others have a flattened body that is called a thallus. Spores similar to those of liverworts are in evidence from the Ordovician onwards, with the earliest fossils appearing in the Devonian. By the Late Triassic there were forms similar to *Marchantia* and many more are known from the Cretaceous and the Tertiary.

M. POLYMORPHA
This specimen shows impressions of thalli preserved in postglacial calcrete from Eskatorp, Sweden. The faint forking lines mark the centres of the thallus.

LIVING RELATIVE
COMMON LIVERWORT

The common liverwort (above) has dull green mats of branching thalli that can grow up to 10cm (4in) long and up to 1.4cm (½in) wide. The green, umbrella-like structures that form on the ends of the stalks have yellow bodies on their undersides, which carry female reproductive parts.

Sphagnum

GROUP Bryophytes
DATE Jurassic to present
SIZE Up to 30cm (12in) long
LOCATION Cool and wet regions worldwide

Commonly called peat moss, *Sphagnum* has between 150 and 300 species. It occurs mainly in the northern hemisphere, as far north as Svalbard, Norway, but also in the cool and wet regions of the southern hemisphere. Its leaves have a lattice of small, green, photosynthetic cells, and larger, dead cells that have a capacity to hold large quantities of water. *Sphagnum* can build up great thicknesses of peat through its ability to increase the acidity of its surroundings, which prevents decay by bacterial and fungal activity. When rainfall is high, *Sphagnum* bogs can grow into domes above the level of the water table. *Sphagnum*-like plants are known from the Russian Permian, but *Sphagnum* itself probably arose in the Jurassic.

SPHAGNUM SP.
This portion of Holocene *Sphagnum* peat contains long narrow leaves, possibly cottongrass, and fragments of larger leaves from trees such as alder or birch.

SCIENCE
PEAT MOSS

Sphagnum can absorb greater amounts of water than cotton wool, and it also has the ability to acidify its surroundings, which inhibits the growth of bacteria and fungi. For these reasons *Sphagnum* has been used for centuries as a dressing for wounds. It was, for example, used in both the First and Second World Wars.

Amblystegium

GROUP Bryophytes
DATE Neogene to present
SIZE Mat up to 20cm (8in) across
LOCATION Americas, Europe, Asia, Australasia, Pacific

This creeping moss has tapering leaves, 3–6mm (⅛–¼in) long, that end in fine points. It grows and divides irregularly and horizontally to form a mat of interlocking filaments. There are about 15 species and all prefer moist, temperate climates. The earliest record of fossil *Amblystegium* is in the Early Neogene of southern Germany and there are a number of records of it in the Quaternary.

AMBLYSTEGIUM SP.
This fossil shows a moss mat that has been infiltrated by calcium carbonate in mineral-rich rivers. It then crystallized, preserving the plant in three dimensions.

LIVING RELATIVE
FIR CLUBMOSS

Huperzia selago grows in northern grassland and rocky areas, and northern and more southerly mountainous regions above the tree line, in Europe, Russia, Asia, Himalayas, Japan, USA, and Canada. This distribution was a result of its migration away from advancing woodland as the climate warmed after the last ice age.

Huperzia

GROUP Lycophytes
DATE Cretaceous to present
SIZE Shoot up to 60cm (23½in) long
LOCATION Worldwide, except for arid areas

Huperzia is a simple clubmoss, and perhaps one of the most primitive vascular plants living today. It has small, spirally arranged leaves and instead of cones, like the other clubmosses, *Huperzia* has zones of spore-producing structures (sporangia) in the leaf axils. Nearly 400 species (about 85 per cent of all clubmosses) exist in temperate and tropical regions worldwide. *Huperzia* represents a lineage of living plants that stretches back to Devonian genera such as *Asteroxylon* (see p.118). The living species of *Huperzia* probably evolved, diversified, and spread during the Paleogene and Neogene.

Andreaea

GROUP Bryophytes
DATE Quaternary to present
SIZE Up to 1cm (3/8in) high
LOCATION Arctic regions, mountains throughout the northern hemisphere

Andreaea is one of two genera of mosses that are commonly found on granite rock faces in northern mountainous regions and the Arctic. Their hair-like, rooting rhizoids can penetrate small cracks in the rock and the plants form small, wind-resistant cushions. There are up to 100 species of *Andreaea*. Fossil mosses are known from the Carboniferous onwards, but there are no undisputed records of *Andreaea* before the Quaternary.

ANDREAEA ROTHII
Andreaea is protected against exposure to ultraviolet light by specialized compounds in its cell walls. The stalked red spore capsules release their spores through four longitudinal slits.

Marattia

GROUP Ferns
DATE Carboniferous to present
SIZE Frond up to 5m (16½ft) long
LOCATION Tropics worldwide

The group that includes *Marattia* is the most primitive of the living ferns, having large leaves (fronds) that are enclosed when young by two basal scales (stipules). Their sporangia are large and borne in clusters on the undersides of the fronds. The group can be traced back to numerous Carboniferous and Permian fossil stems and foliage. The best known of these is the tree fern called *Psaronius* (see p.148). The earliest fossils referred to as *Marattia* come from the Jurassic of Yorkshire, in northern England, where there are also specimens of another similar living genus, *Angiopteris*. The earliest of these ferns probably grew near streams, and they spread into other wetland and humid habitats as they underwent evolutionary changes and the habitats changed. The cooling of the climate during the Late Neogene and Quaternary caused their extinction in the temperate regions of the world, limiting the remaining 26 species to the tropics.

MARATTIA SALICINA
The king fern, with fronds up to 5m (16½ft) long, is native to New Zealand and the islands of the South Pacific. The starchy stem was once a traditional food for the Maori.

Molinia

GROUP Grasses
DATE Quaternary to present
SIZE Flowering head up to 2.5m (8¼ft) tall
LOCATION Europe, Iran, Siberia, E. Canada, N.E. USA

Molinia caerulea, purple moor-grass, is a densely tufted perennial grass that forms large swards in places that are at least seasonally wet. It is found on heaths, moors, bogs, fens, lake-shores, mountain grasslands, and cliffs. Its purple flower heads give it its common name. This grass grows best in the wettest areas and can dominate large parts of bogs and moors, where it contributes to the growth of peat. Sections through peat from both the Quaternary and today's bogs and moors reveal layers that are densely packed with leaves of this species.

MOLINIA CAERULEA
This compact, tufted grass is found in damp places, where it can cover large areas. Its stiff leaf-blades taper to form sharp points.

EPHEDRA MAJOR
All species of *Ephedra* are woody shrubs, with photosynthetic stems, and leaves that are reduced to brown scales.

Ephedra

GROUP Gnetales
DATE Triassic to present
SIZE Up to 2m (6½ft) tall
LOCATION Arid and semi-arid regions of Asia, Europe, N. Africa, W. North America, South America

The 35–45 living species of *Ephedra* are woody shrubs. It belongs to the same group of gymnosperms that includes *Gnetum* and *Welwitschia* (see pp.289, 290). *Ephedra* has seed-producing cones and pollen-producing cones. Its pollen is distinctive, with variable thickenings producing longitudinal ridges and grooves. Similar pollen first appeared in the Permian, but more reliable pollen is known from the Cretaceous.

Eriophorum

GROUP Angiosperms
DATE Quaternary to present
SIZE Flower head up to 70cm (28in) across
LOCATION Europe, Asia, North America, Arctic

Cottongrasses (*Eriophorum*) are actually sedges, and not grasses. *Eriophorum* is a genus with about 25 species. It is found in acid bog habitats throughout the temperate northern hemisphere, in Europe, Asia, North America, and particularly in the higher latitude Arctic tundra. They are herbaceous and perennial, spreading by underground stems. Their bisexual flowers are on stalks and produce masses of seeds that are covered in fluffy cotton, which allows the wind to carry them. The sedge family is known from the Early Neogene, where flower parts of *Carex* have been found in Nebraska, USA, but remains of *Eriophorum* are only found in Quaternary interglacial and postglacial peat. Layers of *Eriophorum* leaves and rhizomes occur in such peat, representing a time when the plants dominated the bog's surface.

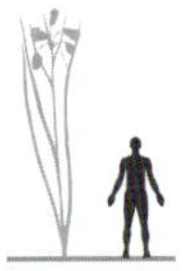

ERIOPHORUM VAGINATUM
The hare's-tail cottongrass can form tussocks that are difficult to walk over. It has only one cottony flower head.

SCIENCE
BOGS AND FENS

Swards of cottongrass grow on acid bogs and fens where there is shallow, standing water. In the foreground of this photograph is common cottongrass (*Eriophorum angustifolium*), which has several cottony heads on each flower, and leaves that are V-shaped in section. The bog's acidity prevents the plants rotting after death so an accumulation of cottongrass remains builds up. Water table or rainfall changes may favour the growth of other plants, such as *Sphagnum* (see p.420), which in turn dominate the bog's surface.

Davidia

GROUP Angiosperms
DATE Paleogene to present
SIZE Tree up to 18m (59ft) tall
LOCATION North America, Russia, China

Fossil leaves that are similar to those of living *Davidia* show that it was prominent in the Early Paleogene vegetation of western North America, and suggests that the family had its origins in the Cretaceous. Fossil fruits matured from flower heads that appear to have been borne between two large bracts, as in the living genus. *Davidia* spread to eastern Russia and China during the Paleogene and Neogene, and narrowly survived extinction during the Quaternary ice ages. There is only one living species in this genus.

LIVING RELATIVE
DOVE TREE

Davidia involucrata has large, green leaves with pink stalks. Its flowers are reddish, and clustered into tight heads that hang on the ends of drooping stalks, and enclosed by two enormous, petal-like, pure white bracts. The bracts give it another common name, "pocket handkerchief tree".

Corylus

GROUP Angiosperms
DATE Late Paleogene to present
SIZE Up to 24m (79ft) tall
LOCATION North America, Europe, Asia

There are about 15 species of *Corylus*, or hazel, from northern temperate regions. All are bushes, except the Turkish and Chinese hazels, which are trees. They are deciduous and have male flowers in catkins and female flowers enclosed within buds. *Corylus avellana* was an early colonizer after the retreat of the Quaternary glaciers.

CORYLUS AVELLANA
Broken nuts of a hazel from postglacial deposits in north Wales show the outer lignified shell, lined with fragments of the outer coat of the seed.

Trapa

GROUP Angiosperms
DATE Neogene to present
SIZE Fruit up to 4cm (1½in) wide
LOCATION Europe, Asia, Africa

The pollen and ornately shaped, nut-like fruits of *Trapa* have been found across the northern hemisphere, from Europe to Africa, India, and China. Evidence of the genus stretches back from the present to the Early Neogene. Each fruit has a single, very large, starchy seed that contains toxins. However, these toxins are destroyed during cooking, and *Trapa* seeds are known to have been used as a food source for thousands of years by early people in present day Europe and China, with the oldest recorded use being by the nomadic Maglemosian people. These people lived in northern Europe around 10,000 to 8,000 years ago.

TRAPA NATANS
Trapa natans has four sharp barbed spines on its fruits. Its generic name is derived from the Latin word for "thistle".

LIVING RELATIVE
WATER CHESTNUT

The water chestnut (*Trapa natans*) is an aquatic plant that grows naturally in slow-moving water in warm temperate regions of Europe, Asia, and Africa. It is used as a food source in China and India. It has also been introduced into parts of North America and Australia, where it has become an invasive weed.

Tilia

GROUP Angiosperms

DATE Paleogene to present

SIZE Up to 36m (120ft) tall, leaf up to 12.5cm (5in) long

LOCATION Temperate regions of northern hemisphere

Tilia (lime trees) are large, deciduous trees that are now spread across the temperate regions of the northern hemisphere, except in northwest America. Their fossils may be leaves, flower parts, wood, or pollen grains. The fossil leaves shown here are heart-shaped impressions in the surface of the rock, with no original plant material remaining. Some leaves landed the right way up, which resulted in their veins becoming embedded into the rock. Those that landed upside down show the smoother upper surface on the rock. The five veins that depart from the base of each leaf branch into secondary veins; a further transverse network of smaller veins is visible between them.

LIVING RELATIVE

COMMON LIME

Tilia x europaea is a cross between the large-leaved lime from southern continental Europe and the small-leaved lime from northern Europe. It grows larger than the two parents, but looks untidy because of the growth of small shoots around its base and on its larger branches.

LIME POLLEN
Lime pollen found in Quaternary peat reflects climatic and other factors. After the last ice age, it gradually increased, but then decreased when humans started to fell lowland forests.

TILIA EUROPAEA
These impressions of lime leaves on calcrete are from the postglacial deposits at Benestad, in southern Sweden. Differences in appearance reflect the orientation of the leaves.

QUATERNARY INVERTEBRATES

The start of the Quaternary is now taken as 2.6 million years ago. Temperatures dropped gradually, and many large mammals became extinct during the coldest time, but in general terms the extinctions on land and in the sea were considerably less severe than at the end of the Cretaceous.

Dominance of invertebrate groups

In the oceans, the same kinds of animals prevailed as in the Neogene, though their distribution changed as the ice-sheets advanced and retreated. The remains of both marine and land-living animals have been used to understand the climatic changes that took place, especially from cores drilled through Cenozoic and Quaternary marine sediments. Foraminiferids (fossilized plankton) are sensitive indicators of climatic change, as successive cold- and warm-water groupings change through time. The foraminiferids themselves preserve a record of stable oxygen isotopes, which can be used to measure the temperatures that existed when these microfossils were alive. Cold-water bivalves are different from those of warmer seas, and are used in a similar way. On land, the remains of beetles in peat and soil are used to understand past climatic changes. Many living species of beetles, the same as those of the Quaternary, have restricted temperature preferences. Successive groupings of warm- and cold-loving beetles in a peat core can be used to tell relative temperatures through time. This data all points to a dramatic temperature rise about 11,500 years ago.

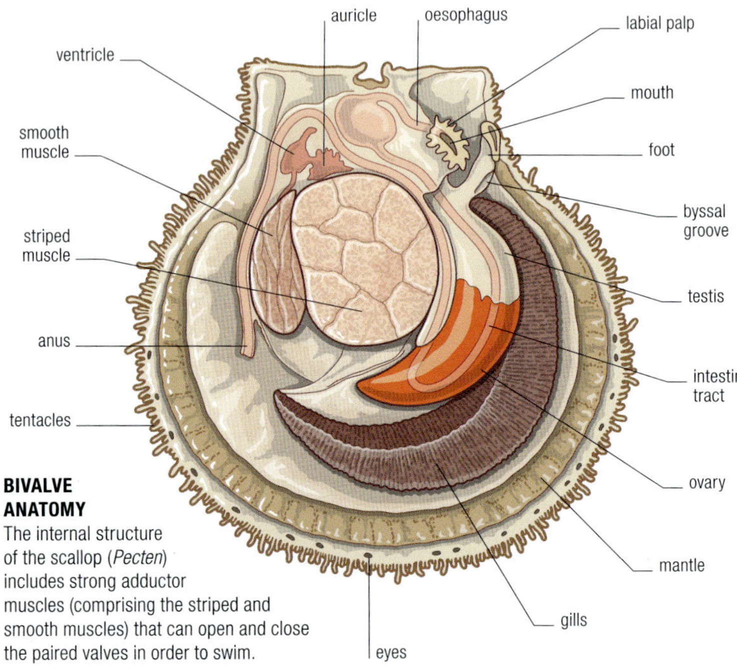

BIVALVE ANATOMY
The internal structure of the scallop (*Pecten*) includes strong adductor muscles (comprising the striped and smooth muscles) that can open and close the paired valves in order to swim.

LIVING MUSSELS
These mussels (*Mytilus*) and limpets (*Patella*) are cold-water molluscs, similar to those that existed during the Quaternary. They live on the shoreline in rock pools where they feed on algae.

GROUPS OVERVIEW

Invertebrate groups, both marine and land-dwelling, remained largely the same in the Quaternary as they were in the Neogene. However, the distribution of these groups was greatly affected by the changing climatic conditions. Gastropod, bivalve, and echinoid fossils all play a role in our understanding of how and where temperatures changed.

GASTROPODS
Cold-adapted gastropods, along with other specialized marine invertebrates, were first recognized as long ago as 1780, in Otto Fabricius' great work *Fauna groenlandica* (Greenland fauna). Several species of wholly Arctic whelks are known, for example, and the specific temperature preferences of many such gastropods are now known in detail.

BIVALVES
Arctic bivalves, quite distinct from those of temperate conditions, are well known from the Quaternary. The mussel *Yoldia arctica* for example, gave its name to the Yoldia Sea, a great area of cold brackish water that existed in what is now the Baltic region. The Yoldia Sea was formed by glacial meltwater from the retreating Scandinavian ice-sheet.

ECHINOIDS
These do not appear to have been adversely affected by the climatic changes of the Quaternary, the cold-adapted types tracking the edge of the ice sheets as these advanced and retreated. Echinoids survived to become an integral component of today's fauna, at various depths in the sea.

ARTHROPODS
The crustaceans (amphipods particularly) have temperature-restricted occurrence, with many forms confined to polar regions. These, however, have only a limited preservation potential, so are not of great use in tracking climate change.

Heterocyathus

GROUP Anthozoans
DATE Late Neogene to present
SIZE Up to 1cm (3/8in) long
LOCATION Indian and Pacific Oceans

Heterocyathus is a small tabular coral with a flat base and top, and roughly rounded sides. The centre of the coral has a series of slender, vertical pillars arranged opposite the ends of the dividing walls, which look like rings of small dots. Like all corals, *Heterocyathus* starts life as a planktonic larva. This eventually settles on the sea bed, usually attaching itself to a small shell, which it soon outgrows. Modern forms sometimes live in symbiosis with sipunculid worms (peanut worms), and it is likely that fossil forms did too. Living species can be found at depths of 10–550m (33–1,800ft).

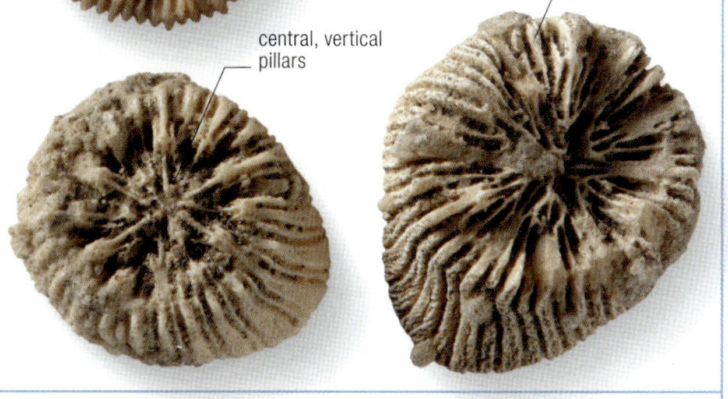

HETEROCYATHUS SP.
Radiating out from the centre of these coral fossils, are numerous vertical walls, or septa, which become thicker towards the edge.

Hippoporidra

GROUP Bryozoans
DATE Palaeogene to present
SIZE Can be over 10cm (4in) across
LOCATION Europe, Africa, North America

Colonies of the bryozoan *Hippoporidra* encrusted the surface of gastropod shells in many thick layers, sometimes forming large spiral extensions. On the colony surface, many circular features can be seen. These are the openings to structures called autozooecia, which contained the soft-bodied feeding animals (autozooids). The autozooids had a ring of tentacles surrounding the mouth, called a lophophore. When the zooid was not feeding the lophophore could be withdrawn into the autozooecia for protection. The frontal walls of the autozooecia are perforated with pores. Raised areas (monticules) can be observed on the colony surface, where the autozooecia are larger. Avicularia, which housed defensive, non-feeding zooids, can be found between the monticules.

HIPPOPORIDRA EDAX
This fossil suggests a symbiotic relationship between the bryozoan and the hermit crab, although the crab has not been preserved here.

***Hippoporidra* colonies have been found to live in symbiosis with hermit crabs.**

Magellania

GROUP Brachiopods
DATE Neogene to present
SIZE 2–3cm (3/4–1 1/4in) long
LOCATION Australia, South America, Antarctica

This is a typically shaped, weakly ornamented terebratulid brachiopod shell. Its pedicle opening lies close beneath the beak of the pedicle valve; in this example the opening has been slightly enlarged towards the beak by a small breakage. There is a well-marked triangular area beneath the pedicle opening that closes off its front side. The brachial valve houses the loop-shaped lophophore support, which, together with the fine perforations over the surface of the shell (punctae), are characteristic features of terebratulid brachiopods. The ornamentation consists of fine growth lines.

MAGELLANIA SP.
This specimen has a single, well-marked growth increment, possibly caused by a pause in growth. The growth lines are crossed by faint radial ribs that strengthen the front edge of the shell.

ANATOMY
PEDICLES

Most brachiopods, such as *Magellania,* have a fleshy stalk called the pedicle, which emerges through an opening in the larger pedicle valve and anchors them permanently. In some inarticulate brachiopods, such as *Lingula*, the pedicle is larger and emerges between the two valves to act as a movable anchor rather than a permanent fixture. In yet other Paleozoic brachiopods, the pedicle died away in the adult, its opening grew over, and the brachiopod was anchored by the shell's weight.

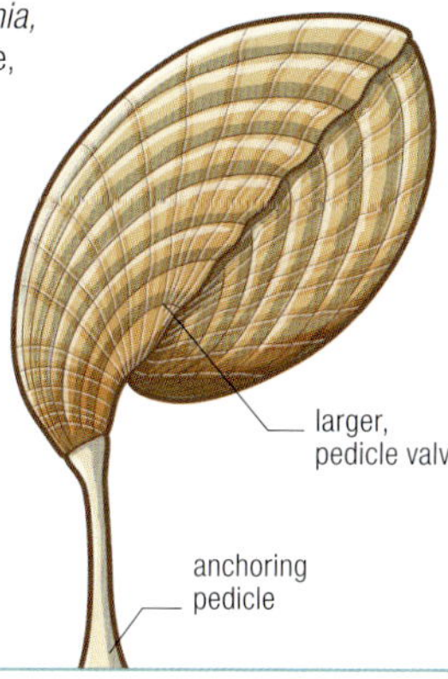

Capulus

GROUP Gastropods
DATE Paleogene to present
SIZE Up to 2.5cm (1in) across
LOCATION Atlantic and Mediterranean regions

Capulus is a cap-shaped gastropod with a tight, asymmetrical initial spiral. Modern *Capulus* are often found with bivalves, particularly *Pecten* (see p.428), *Chlamys*, and *Modiolus* (see p.239). They attach themselves by their large foot to the open edge of the bivalves' shells, close to the inhalant current that sucks in nutrient-rich water. In this position *Capulus* can use its long, extendable proboscis to collect food particles that the bivalve is drawing in. It may also remove food particles that are trapped on the edges of the bivalve's gills. This semi-parasitic mode of life does not appear to affect the bivalves adversely.

CAPULUS SP.
This fossil shell looks like a tiny elf's cap, with a wide opening, a brim, and a tip swept back in one direction.

Neptunea

GROUP Gastropods
DATE Neogene to present
SIZE Up to 8cm (3 1/4in) long
LOCATION North Atlantic Ocean, North Africa, Mediterranean Sea

Neptunea's shell has a tall spire, coiled so that each whorl overlaps the preceding one for about a third of its height. Fine spiral ribs are crossed by variably strong growth lines, which may represent seasonal variations in growth. *Neptunea* is essentially a marine scavenger, feeding on dead fish, crustaceans, and molluscs.

NEPTUNEA CONTRARIA
Unusually, this species has a sinistral shell, coiled so that the aperture is on the left side of the upright shell.

Lymnaea

GROUP Gastropods
DATE Paleogene to present
SIZE Up to 6cm (2¼in) long
LOCATION Worldwide

Lymnaea is a genus of freshwater snail. It includes the great pond snail (*Lymnaea stagnalis*), which is common across much of the northern hemisphere. The shells are pointed, with the whorls forming a long spire. *Lymnaea* snails have flattened feelers and curtain-like edges to the foot. All freshwater snails and land snails are hermaphrodite – individuals produce both eggs and sperm. Unlike marine gastropods, which have gills, *Lymnaea* snails are air-breathing and must surface to take in air.

LYMNAEA SP.
As in this specimen, the outside of the shell lacks ribs or any other noticeable structure. However, faint and closely spaced growth lines are often visible.

Littorina

GROUP Gastropods
DATE Neogene to present
SIZE Up to 5cm (2in) long
LOCATION N.E. Atlantic Ocean

Living species of *Littorina* include winkles and periwinkles. These molluscs graze algae and other micro-organisms from the surface of rocks in the intertidal zone. At low tide they may be found out of the water, stuck to rocks. When covered by the tide, they release their grip and search for food. Like all marine gastropods, *Littorina* breathe by removing oxygen from the water with their gills. When exposed by the tide they secrete a mucous plug to keep water in. They also have a "lid" (operculum) to their shells, with which they can seal the opening.

LITTORINA RUDIS
The surface of this shell displays typical spiral ribs, which are often crossed by faint growth lines.

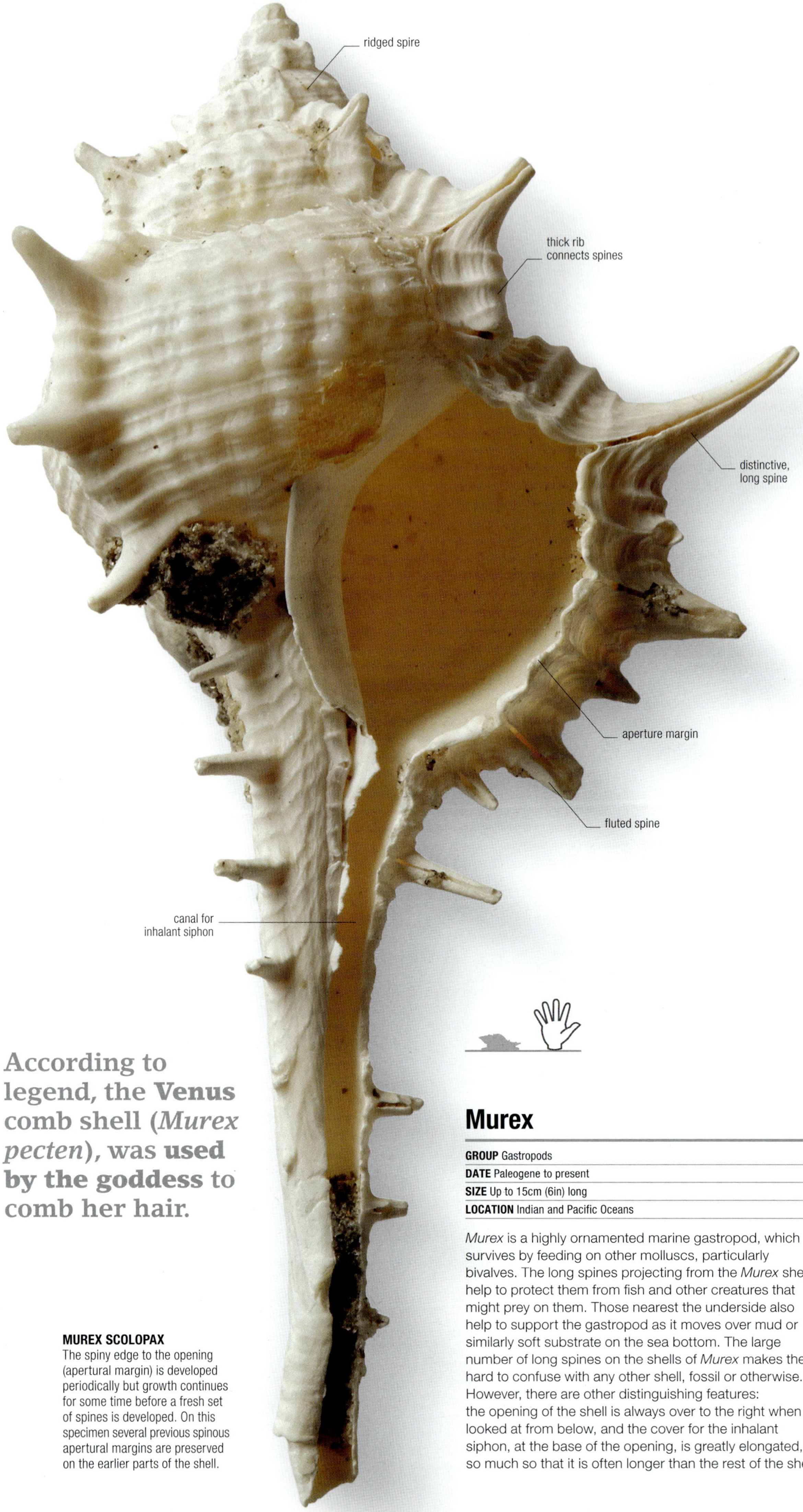

MUREX SCOLOPAX
The spiny edge to the opening (apertural margin) is developed periodically but growth continues for some time before a fresh set of spines is developed. On this specimen several previous spinous apertural margins are preserved on the earlier parts of the shell.

According to legend, the **Venus comb shell (*Murex pecten*), was used by the goddess** to comb her hair.

Murex

GROUP Gastropods
DATE Paleogene to present
SIZE Up to 15cm (6in) long
LOCATION Indian and Pacific Oceans

Murex is a highly ornamented marine gastropod, which survives by feeding on other molluscs, particularly bivalves. The long spines projecting from the *Murex* shell help to protect them from fish and other creatures that might prey on them. Those nearest the underside also help to support the gastropod as it moves over mud or similarly soft substrate on the sea bottom. The large number of long spines on the shells of *Murex* makes them hard to confuse with any other shell, fossil or otherwise. However, there are other distinguishing features: the opening of the shell is always over to the right when looked at from below, and the cover for the inhalant siphon, at the base of the opening, is greatly elongated, so much so that it is often longer than the rest of the shell.

Fusinus

GROUP Gastropods
DATE Paleogene to present
SIZE Up to 15cm (6in) long
LOCATION Widely distributed in warmer seas

This genus of gastropod molluscs contains a large number of species, many of which are still alive today. *Fusinus* shells vary in shape between species but all have certain features in common, such as a long, pointed spire and a slender canal extending down from the shell's opening. In life this acts as a cover for the long inhalant siphon, a tube that is used for sucking in water to pass over the gills. Living *Fusinus* species are known collectively as spindle shells. They are large, carnivorous sea snails from tropical and subtropical seas that feed on smaller gastropods.

FUSINUS CAROLINENSIS
Spindle shells, such as this, have spiralling ribs on the outer surface that are also visible on the inner surface. Although usually a predator, this specimen seems to have been prey itself – there is a clear drill hole in one of the whorls.

Trivia

GROUP Gastropods
DATE Paleogene to present
SIZE Up to 3cm (1¼in) long
LOCATION Worldwide

Trivia is a gastropod similar to the modern cowrie shell. Looked at from the outside, *Trivia*'s spiral shape is invisible. A cross section through the shell, however, shows it to be spiralled. As it grew, each whorl completely enveloped the preceding one, hiding it from view. Extending out from its long, narrow opening (aperture) are slender ridges. In life, its shell is at least partly covered by fleshy lobes extending from the body underneath. *Trivia* is a carnivore that feeds on sea squirts and other, similarly immobile, invertebrate animals.

TRIVIA AVELLANA
This fossil shows how the narrow opening runs the full length of the shell. Both its inner and outer edges are quite strongly ridged.

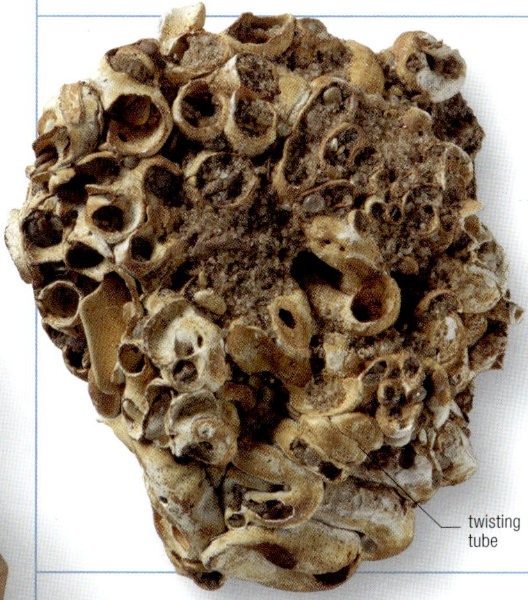

Vermetus

GROUP Gastropods
DATE Neogene to present
SIZE Tube up to 6mm (¼in) across
LOCATION Worldwide

Vermetus is a most unusual gastropod that looks more like a worm tube. The first part of its shell is coiled as in most gastropods, but then the spiral opens out to form an irregular, twisted tube. The tubular shell has flattened sides, which in places are slightly concave. It also has quite thick walls bearing finely spaced growth plates and thicker ribs parallel to the direction of shell growth. *Vermetus* often lives in clusters of several individuals.

VERMETUS INTORTUS
The holes bored into the tubes shown here were probably created by a predatory gastropod such as *Natica*.

Helix

GROUP Gastropods
DATE Neogene to present
SIZE Up to 4.5cm (1¾in) long
LOCATION Worldwide

Helix is a genus of land snail with many species, some living and others extinct. Each whorl of a *Helix* shell overlaps the preceding one to about half its height, giving a fairly high spire. At the mouth of the shell, in mature snails, there is usually a pronounced lip. This is absent in younger snails, where the shell is not fully grown. Like other air-breathing snails, *Helix* are hermaphrodite and their eggs develop into tiny adults without going through a larval stage. Land snails first appear in the fossil record in the Carboniferous, while freshwater snails do not appear until the Jurassic, suggesting that air-breathing gastropods were derived from marine ancestors, and freshwater forms from terrestrial snails that went back to the water.

HELIX ASPERSA
This shell displays the large rounded whorls and wide rounded opening typical of *Helix*. The lipped mouth indicates that this was a mature snail.

LIVING RELATIVE
VARIED COLOURS

The shell of *Helix* has an outer horny layer (periostracum), which protects the calcareous underlayers from dissolving in acidic rain. The variable colours of this layer has led zoologists to recognize different varieties, but in fossils colour banding is seldom preserved.

Land snails are hermaphrodite and have a variety of complex mating rituals.

Clinocardium

GROUP Bivalves
DATE Neogene to present
SIZE Up to 4cm (1½in) long
LOCATION North Pacific and North Atlantic coastal regions

Clinocardium is a small, cockle-like bivalve that still exists today. The shell has clear ribs radiating out from the beak to the rounded edges. These ribs are crossed by concentric growth increments, which are strong on some parts of the shell but very weak on others. Inside there are two teeth and sockets below the beak in each valve. There are also clear, equal-sized scars where the adductor muscles were attached.

C. INTERRUPTUM
This specimen shows three distinct growth ridges, suggesting that its growth was interrupted seasonally.

Corbicula

GROUP Bivalves
DATE Cretaceous to present
SIZE Up to 2.5cm (1in) long
LOCATION Worldwide

Corbicula is a small bivalve with a rounded triangular shape and a broad beak to each valve (half) of its shell. Inside, each valve has clear scars where the adductor muscles attached, and the hinge-line has prominent teeth. The central tooth below the beak of one valve fits in a central socket in the other. Outside the central teeth, there are lateral teeth and sockets that run parallel to the edge of the shell. The elastic ligament that opened the valves sat outside, behind the beaks.

CORBICULA FLUMINALIS
This shell was found with fresh-water fossils, so unlike some early forms it probably was not marine.

Pecten

GROUP Bivalves
DATE Paleogene to present
SIZE Up to 10cm (4in) along the hinge-line
LOCATION Worldwide

Both valves of a *Pecten* shell are ornamented with ribs that radiate out from the hinge end. The tops of the ribs are flat and the spaces between them narrower than the ribs themselves. The hinge-line of each valve is quite straight. Inside each valve the most obvious feature is the very large adductor muscle scar. As one might expect for a bivalve that is able to swim by opening and closing its shell, the adductor muscle, in life, is large and powerful. Most people know members of the genus *Pecten* as scallops.

LIVING RELATIVE

SCALLOP

Very few bivalves are free-swimming and *Pecten*'s ability to clap its valves together enables it to avoid many predators. It needs to sense their approach and has developed sophisticated eyes on the edges of its mantle. It also has chemoreceptors, enabling the bivalve to taste and smell the water, and statocysts to help determine its orientation.

PECTEN MAXIMUS
The valves of *Pecten* differ in shape; the valve above, which lay on the sea bed, is convex, while the one below is almost flat.

Cerastoderma

GROUP Bivalves
DATE Paleogene to present
SIZE Up to 4cm (1½in) long
LOCATION Europe

Cerastoderma is best known as the cockle commonly found on modern European beaches, but its history goes back to the Late Paleogene. In outline its shell is relatively square, with strongly radiating ribs crossed by raised riblets parallel to the direction of growth. The beak of each half of the shell (valve) is pointed and runs straight across the hinge-line. There are two teeth on the underside of the beak in each valve. Inside the shell there are two clear muscle scars where the adductor muscles (the muscles that pulled the shell closed) attached. In life, *Cerastoderma* lives in shallow burrows.

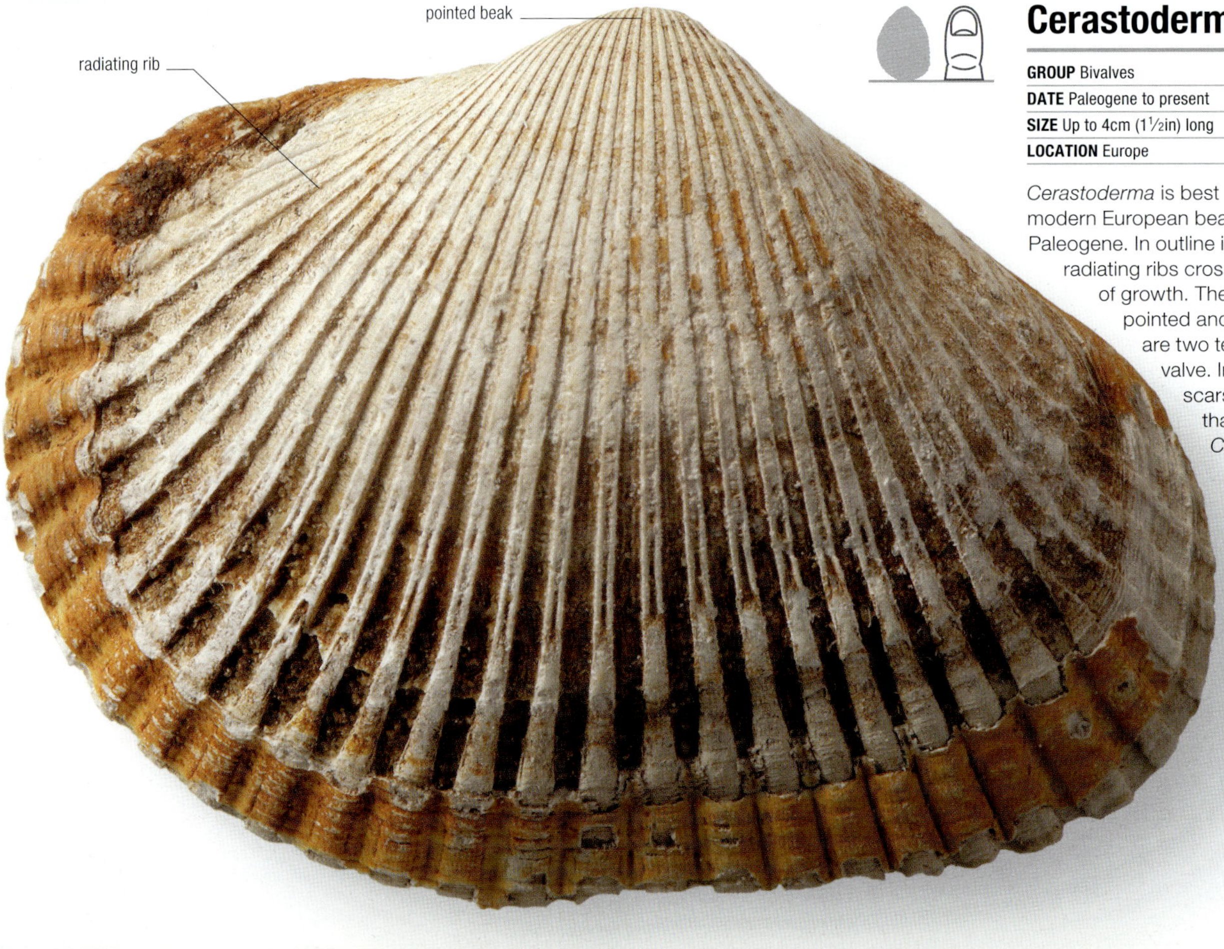

***Cerastoderma* feeds by extending its siphons up through the sea bed to suck in water and food.**

CERASTODERMA ANGUSTATUM
This fossil shell shows the strong, raised ribs fanning out from the beak, crossed by fine, parallel growth lines.

Venus

GROUP Bivalves
DATE Late Paleogene to present
SIZE Up to 3.5cm (1½in) across
LOCATION Europe, Africa, Indonesia, North America

The genus *Venus* includes both living and extinct bivalve species. Each shell has a beak that curves around to one side, behind which is a smooth, depressed area. On the inside of each valve (half of the shell), beneath the beak, there are three large teeth and sockets. Each valve also has two scars on the inside that are roughly equal in size: these mark the points where the adductor muscles, which closed the valves, attached. The living species are commonly known as Venus clams. Venus clams are filter feeders that live in burrows in the mud or sand of the sea bed. Like most other burrowing bivalves, they use their siphons to circulate water through the shell, using their gills to filter out food particles and for oxygen and passing the filtered water back to the sea bed.

Venus lived in burrows; its siphons kept it connected to the sea water.

VENUS VERRUCOSA
This fossilized upper valve shows that the outer surface is strongly ridged with curved ribs, which are crossed by less obvious radial ribs, running from the beak to the edge of the shell.

Echinocyamus

GROUP Echinoderms
DATE Late Cretaceous to present
SIZE Up to 1.5cm (½in) across
LOCATION Widely distributed in northern hemisphere

Echinocyamus is an extremely small echinoid with a rather flattened shape. As with most sea urchin fossils, those of *Echinocyamus* lack the spines and just show the hard "test", which in life contains all of the internal organs. This is roughly egg-shaped, flattened in profile, and has a thick, rounded edge. Both the mouth and the anus are on the underside, the latter smaller than the former and positioned behind it towards the edge of the test.

ECHINOCYAMUS PUSILLUS
This lower (oral) surface has numerous spine bases, and the test has small pores in the ambulacral areas that are less conspicuous than on the upper surface.

Balanus

GROUP Crustaceans
DATE Paleogene to present
SIZE Up to 2cm (¾in) across
LOCATION Worldwide

Members of the genus *Balanus* are commonly known as barnacles; this type, both fossilized and living, has a roughly conical form. They tend to be irregular in shape, especially when crowded together, as they often are. The walls of the shell are angled inwards, and are composed of four to six rigid plates, cemented together.

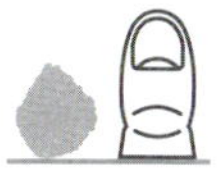

BALANUS CONCAVUS
Narrow, raised ridges can be seen on each plate of this fossil. In life, the plates are all cemented to a hard substrate.

QUATERNARY VERTEBRATES

Throughout the Pleistocene epoch, ice caps advanced and retreated around the globe, forcing vertebrates to move and adapt. Many species succumbed to extinction after the end of the last ice age about 10,000 years ago. Since then, modern humans have had a dramatic impact on the world's fauna.

Mammals abounded beyond the edges of the ice caps during the Pleistocene. Large animals were better able to conserve body heat and survive extreme cold than small animals. In addition, several northern species, such as woolly mammoths and woolly rhinos, sported thick fur coats to keep them warm. The ranges of many animals expanded and contracted according to the changing climate and vegetation. For example, reindeer moved southwards from the Arctic to southern Europe during glacial periods and, contrastingly, hippos moved northwards from Africa to central Europe during interglacials.

The rise and fall of megafauna

Giant mammals were ubiquitous during the Pleistocene. Elephant-like mastodons and car-sized armadillo-like glyptodonts lumbered around North America, while giant ground sloths, which were up to 6m (20ft) tall and weighed a metric ton, inhabited both North and South America. In Eurasia, there were giant rhinoceroses and a species of giant deer, *Megaloceros giganteus*, whose antlers measured over 3.5m (11½ft) from tip to tip. Australia was home to giant marsupials, including *Diprotodon*, a hippopotamus-sized relative of the wombat and koala, and *Procoptodon*, the largest kangaroo to have ever lived. Both of these were prey to the marsupial lion *Thylacoleo*. A combination of factors is likely to have contributed to the extinction of much of the prehistoric megafauna at the end of the last ice age. These included climate change, which caused many areas to dry out, and pressure imposed by the spread of modern humans, who hunted many of the species and also learned to use fire to clear vegetation and for cooking (see p.461).

CASTOROIDES

LITTLE AND LARGE
Giant beavers (*Castoroides*) inhabited North America during the Pleistocene. They were up to 2.5m (8ft) long and weighed as much as 100kg (220lb), dwarfing today's North American beaver (*Castor canadensis*), which is 1m (39in) long and weighs up to 35kg (77lb).

NORTH AMERICAN BEAVER

GROUPS OVERVIEW

Much of the world's existing vertebrate fauna appeared during the Quaternary, but many other species became extinct. Many seabirds, for example, died out, possibly due to competition with newly evolved marine mammals. The Hominoidea included *Gigantopithecus*, which, at 3m (9¾ft) in height, was the largest ape to have ever existed.

AVES
Birds grew large during the Pleistocene too. *Aepyornis*, the ostrich-like "elephant bird" of Madagascar weighed nearly 400kg (880lb), whereas the flightless moas of New Zealand reached a height of about 3.6m (12ft). They were preyed upon by the huge Haast's eagle, which weighed up to 15kg (33lb).

CARNIVORES
The largest terrestrial carnivore during the Pleistocene was the giant short-faced bear *Arctodus simus*, which weighed up to 900kg (2,000lb). Among the cat family, lions that were larger than today's African lions were found worldwide, and there was also the sabre-toothed cat *Smilodon*. The dire wolf, a close relative of today's grey wolf, lived in North America.

EQUIDS
Horses proliferated in North America during the Pleistocene. Using the land bridge to eastern Asia, they spread to Eurasia and Africa, where nine *Equus* species survive today, including two species of wild ass, three species of zebra, the kiang, and Przewalski's horse. No wild horses remained in North America after the end-Pleistocene mass extinction.

PROBOSCIDS
Mammoths and mastodons lived in North America and Eurasia during the Pleistocene but died out after the end of the last ice age. Not all proboscids were large – dwarf elephant fossils have been found on several Mediterranean islands. The last woolly mammoths lived as recently as 3,700 years ago on Wrangel Island in the Arctic Ocean.

Megalania

GROUP Reptilomorphs
DATE Pleistocene
SIZE 8m (26ft) long
LOCATION Australia

Megalania was a heavily built, giant monitor lizard that resembled the Komodo dragon. It was so impressively large and powerful that it was able to eat any of the large ice-age mammals of Australia, including the rhinoceros-sized wombats known as diprotodonts and the giant kangaroos. Its great size may have been due partly to the fact that there were few large mammalian predators in Australia. Only the much smaller marsupial lion, *Thylacoleo*, would have offered *Megalania* some competition. *Megalania* vanished about 40,000 years ago, at about the time that the first humans arrived in Australia, although there is no evidence that humans successfully hunted *Megalania*. It seems more likely that humans were the prey.

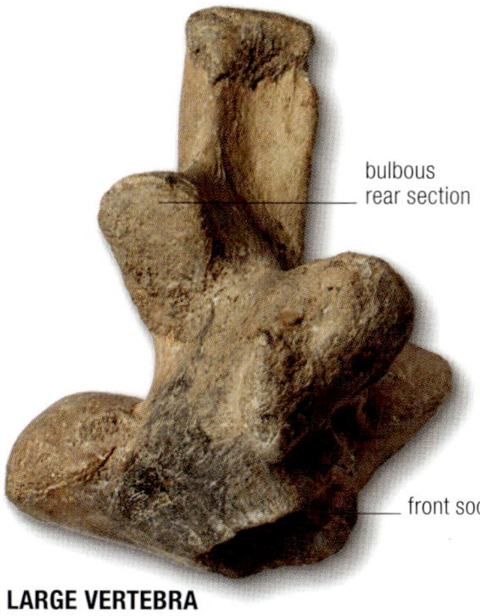

LARGE VERTEBRA
Megalania's spine was made up of 29 presacral vertebrae. Each had a strongly socketed front and a bulbous rear section.

LIVING RELATIVE
KOMODO DRAGON

The largest living lizard is the Komodo dragon, which lives only on the Indonesian island of Komodo. This impressive creature can be as long as 3m (9¾ft), and can weigh up to 70kg (155lb). It eats carrion and hunts large prey, such as deer and pigs, by ambushing them along game trails. Even if the catch escapes its clutches, the Komodo dragon's poisonous saliva eventually kills the animal.

FEARSOME PREDATOR
Megalania had a huge skull with large jaw muscles and long, sharp teeth. Compared to living monitor lizards, it was bulky, weighing 1,900kg (4,190lb).

Trilophosuchus

GROUP Crocodylomorphs
DATE Early Miocene
SIZE 1.5m (5ft) long
LOCATION Australia

Compared with modern crocodilians, *Trilolophosuchus* had a short snout, very large eyes, and three long ridges of bone running along the top of the skull, between the eyes. *Trilophosuchus* fossils have been found at Riversleigh in Queensland, Australia, where fossil beds have revealed the beautifully preserved remains of the inhabitants of a thick forest. Excavations show a rich variety of mammals, which could have provided *Trilolophosuchus* with ample prey. *Trilophosuchus* was a member of the crocodilian family Mekosuchinae, which first appeared in Australia in the Eocene. Members of this family were the dominant crocodilians in much of the southwest Pacific throughout most of the Cenozoic, until saltwater crocodiles arrived from Asia. The Mekosuchinae vanished at the end of the Pleistocene, probably due to human hunting, or the destruction of their habitat or food source.

"DROP CROC"
Due to its light build, some scientists suggest that *Trilophosuchus* captured prey by dropping out of trees – hence the nickname "drop croc".

Aepyornis

GROUP Theropods
DATE Early Pleistocene to Holocene
SIZE 3m (9¾ft) tall
LOCATION Madagascar

A giant, flightless, ostrich-like animal, *Aepyornis* was heavily built, weighing 400kg (880lbs), making it the heaviest bird that ever lived. It fed on a wide variety of fruits, nuts, and seeds. *Aepyornis* coexisted with humans for thousands of years, but became extinct soon after Europeans arrived in Madagascar. The reason for its extinction is unclear, but remains reveal that humans butchered *Aepyornis*, so over hunting may be blamed.

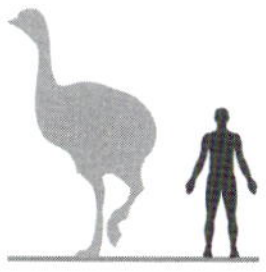

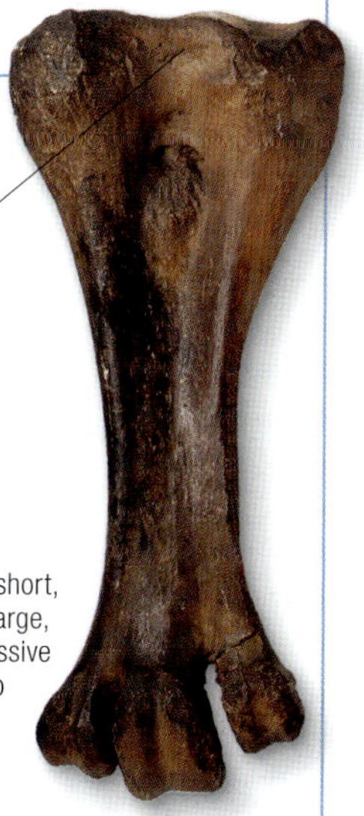

FOOT FOSSIL
Aepyornis had relatively short, but powerful, legs with large, three-toed feet. This massive metatarsus was joined to the ankle at the top and has toe articulations at the bottom.

SCIENCE
FOSSILIZED EGG

The huge fossilized eggs of *Aepyornis* are impressive by their sheer size – they have a circumference of more than 1m (3ft 3in) and a width of 34cm (13½in). Until recently, there was no way to see inside an intact egg without destroying its contents. However, high-resolution X-ray computed tomography (HRXCT) can now be used to photograph and digitally analyse the contents of the eggs. Images reveal nearly every bone of the skeleton as well as anatomical features that are only visible in embryos, disappearing as the bird becomes adult.

Dinornis

GROUP Theropods
DATE Pleistocene to Holocene
SIZE 3.6m (11¾ft) tall
LOCATION New Zealand

Dinornis was a moa, one of a group of giant flightless birds that lived in New Zealand from the early Pleistocene until they were wiped out by the Maori in the sixteenth century. It was the tallest bird that ever lived, and weighing in at 280kg (620lb), *Dinornis* resembled a gigantic, heavily built emu. Despite the relatively small size of its head and bill, it would have been able to eat almost any type of fruit, seed, or other plant food. It was covered in reddish brown, hair-like feathers, except on its leathery feet and legs and much of its head and neck. There were at least six genera of moas living in New Zealand when humans arrived. Their extinction was due to human hunting, combined with habitat loss as the Maori slashed and burned the forests.

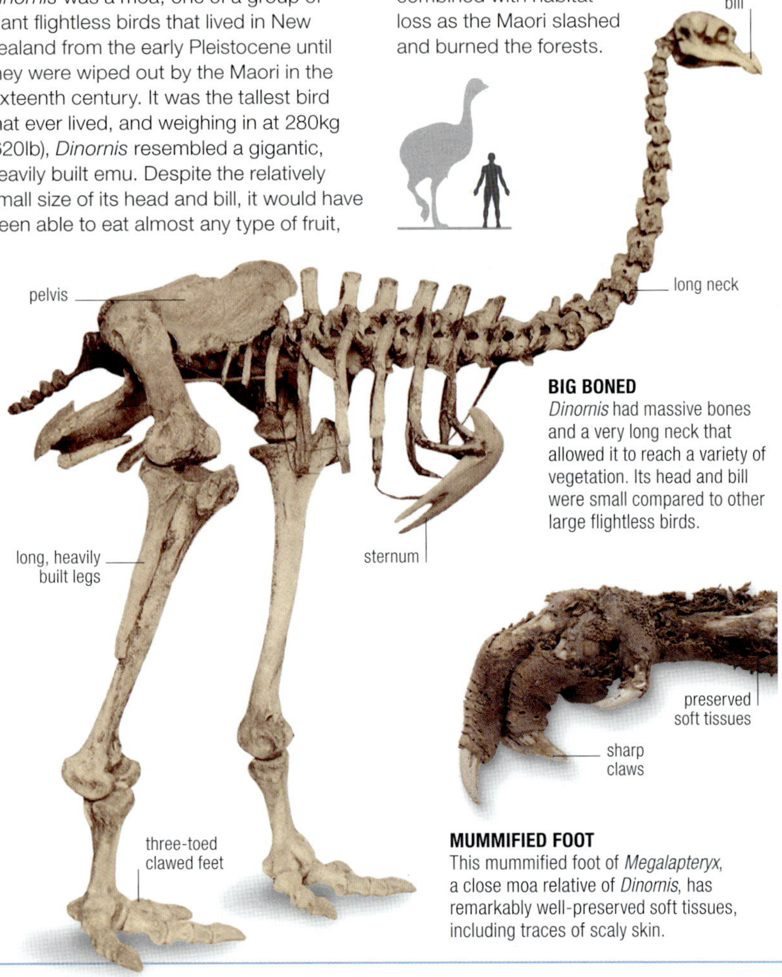

BIG BONED
Dinornis had massive bones and a very long neck that allowed it to reach a variety of vegetation. Its head and bill were small compared to other large flightless birds.

MUMMIFIED FOOT
This mummified foot of *Megalapteryx*, a close moa relative of *Dinornis*, has remarkably well-preserved soft tissues, including traces of scaly skin.

Toxodon

GROUP Placental mammals
DATE Late Pliocene to Pleistocene
SIZE 2.7m (8¾ft) long
LOCATION South America

Fossils of this rhinoceros-like mammal have been found in late Pliocene and Pleistocene deposits in Argentina and elsewhere in South America. It was a member of an extinct group of South American mammals known as the notoungulates. However, it was the size of a modern hippopotamus and looked remarkably like the rhinos of Eurasia and North America, making it a classic example of convergent evolution.

Toxodon had a short proboscis, a little like that of a tapir, and an enormous skeleton with short, stumpy limbs that had three toes on each foot. Its teeth were bow-shaped and suited to chewing a range of leafy vegetation. About 3.5 million years ago, the Panamanian land bridge connected North and South America, prompting the migration of many mammals between the two continents. *Toxodon* remained in South America and survived the arrival of northern mammals because no direct competitors made the journey. However, numerous remains embedded with arrowheads suggest that humans eventually hunted *Toxodon* to extinction.

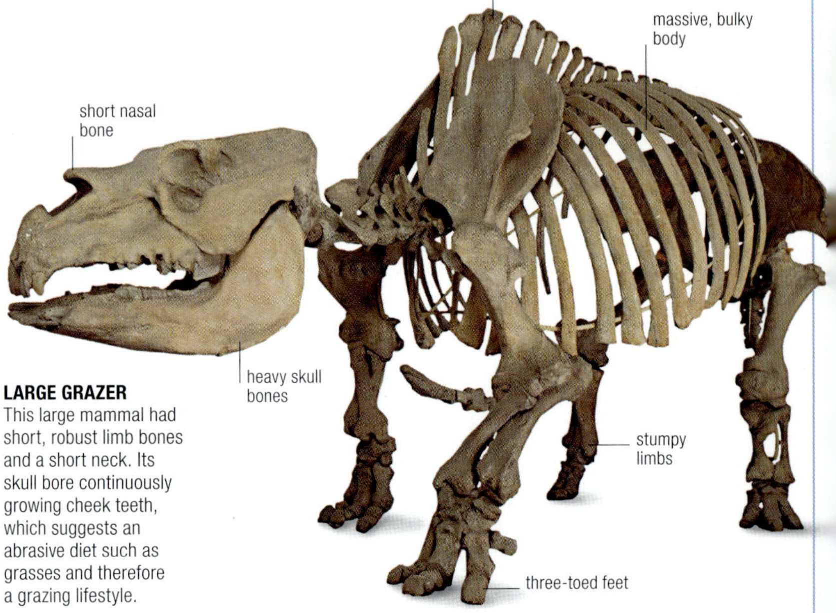

LARGE GRAZER
This large mammal had short, robust limb bones and a short neck. Its skull bore continuously growing cheek teeth, which suggests an abrasive diet such as grasses and therefore a grazing lifestyle.

Teratornis

GROUP Theropods
DATE Pleistocene
SIZE 75cm (30in) long
LOCATION North America

This huge bird is known from the vast number of bones preserved at La Brea tar pits in California (see p.434), but is also known from Pleistocene deposits in Nevada, Arizona, and Florida. It had stout legs and feet strong enough to hold down prey as it tore its food apart. Although *Teratornis* probably ate carrion, like vultures and condors, it also had features that suggest it was good at catching fish in the water. Large numbers of these birds may have been trapped in the tar pits at La Brea not only because they were attracted to carrion, but also because they may have tried to fish in the overlying film of water, or drink from the edge of the tar pools. *Teratornis* became extinct 10,000 years ago, perhaps due to the loss of carrion as many large American mammals vanished at the end of the last ice age.

CONDOR-LIKE
Although *Teratornis* resembled the giant condor, it was from a separate family. It was larger than the living Andean condor, and had a wingspan of up to 4m (13ft).

SCIENCE

HOW TERATORNIS FLEW

Like most modern condors and vultures, *Teratornis* depended on soaring using thermal updrafts so that it could cover large areas in search of carcasses to feed upon. Modern condors fly several hundred kilometres a day in this way. Most soaring birds have long, relatively large wings and small bodies that give them a low wing loading, enabling them to soar with minimal flapping. The wing dimensions of *Teratornis* are more like those of the California condor, suggesting that it could take off from the ground by jumping up with a few simple flaps of its wings, without needing to flap into the wind from a high point, as some vultures and condors must in order to take off.

Castoroides

GROUP Placental mammals
DATE Late Pliocene to Pleistocene
SIZE 2.5m (8¼ft) long
LOCATION North America

Castoroides was a giant ice-age beaver. Its remains are known from rocks of the late Pliocene and Pleistocene in North America. Fossils have been found primarily around the Great Lakes and in the Midwest, but they have also been dug up in Alaska, Canada, and Florida. *Castoroides* was the length of a modern bear and weighed up to 100kg (220lb). Its tremendous size was not the only difference with modern beavers – it did not have their simple chisel-like front teeth, or incisors. Instead, they were larger and broader, reaching to about 15cm (6in) long. In addition, *Castoroides*'s tail was longer and narrower than the broad, flat appendage of modern beavers. Its hind limbs were also shorter.

DAM LIFESTYLE
Although there is no firm evidence that *Castoroides* lived like modern beavers, it is likely that it built lodges and dams and gnawed on wood.

This **truly giant beaver** had teeth about 15cm long and **weighed as much as 100kg**.

Glyptodon

GROUP Placental mammals
DATE Late Pliocene to Pleistocene
SIZE 2.5m (8¼ft) long
LOCATION Argentina, Mexico, USA

Originating in South America, *Glyptodon* was a giant relative of the armadillo that migrated to southern North America when the Panamanian land bridge appeared 3.5 million years ago. It was a well-protected animal, being largely covered in skin-covered bones, known as osteoderms, which served as body armour. Even the tail was armoured, being constructed of a series of tapering bony rings made of osteoderms.

Glyptodon vanished at the end of the last ice age, possibly due to human hunting, although no direct evidence of human attacks on the genus has ever been found.

LONG IN THE TOOTH
The flat-crowned teeth grew continuously, so *Glyptodon* could eat a wide variety of tough vegetation, including gritty grasses.

WELL-ARMOURED
Glyptodon's shell was composed of over a thousand little tiles or osteoderms, each of which was around 2.5cm (1in) thick.

DOMED GIANT
Glyptodon weighed at least 1 tonne (1.1 ton) and had a huge, domed shell that made this creature about the shape and size of a Volkswagen Beetle.

LIVING RELATIVE
ARMADILLO

Armadillos are placental mammals that bear live young, which are nourished in the womb by a placenta. They have a leathery armoured shell, and a long, tubular snout. Armadillos are xenarthrans, related to sloths and anteaters. There are 10 genera and 20 species known throughout Latin America as far north as Texas. Their stout legs and claws allow them to dig burrows rapidly, or dig up ants and termites (their main foods) from their nests. Considerably smaller than their extinct cousin *Glyptodon*, their average length is about 75cm (30in), although the giant armadillo can grow to twice that length.

DEADLY CLAWS
Megatherium had long forelimbs that were equipped with impressive claws, used for grasping branches and fighting predators.

Megatherium

GROUP Placental mammals
DATE Pliocene to Pleistocene
SIZE 6m (20ft) long
LOCATION South America

A giant ground sloth, *Megatherium* weighed in at 4 tonnes (4.4 tons). It was about the size and weight of a bull elephant and was covered with long hair. It walked on the sides of its feet with its claws curled inwards, and could prop itself up on its back legs and massive tail. *Megatherium* vanished at the end of the last ice age, when humans arrived in South America, although there is no direct evidence of human attacks.

HIGH-CROWNED TEETH
Simple peg-like teeth made of dentin, with no enamel, were very high-crowned, allowing *Megatherium* to eat a variety of leaves and grasses.

Arctodus

GROUP Placental mammals
DATE Pleistocene
SIZE 3.4m (11¼ft) long
LOCATION Canada, USA, Mexico

This giant bear had a huge body that was covered with long hair, and a relatively short skull with very powerful jaws that were capable of crushing its prey – bones and all. Although *Arctodus* was very large, it had relatively long, slender limbs for a bear, and may have been a fast runner, ambushing its prey with a quick dash from cover, or covering long distances to wear down its prey. It could also have been a scavenger and thief, easily able to drive other hunters away from their kills. *Arctodus* vanished at the end of the last ice age, possibly due to competition from humans, although it was so large and fierce that humans probably did not try to hunt it down.

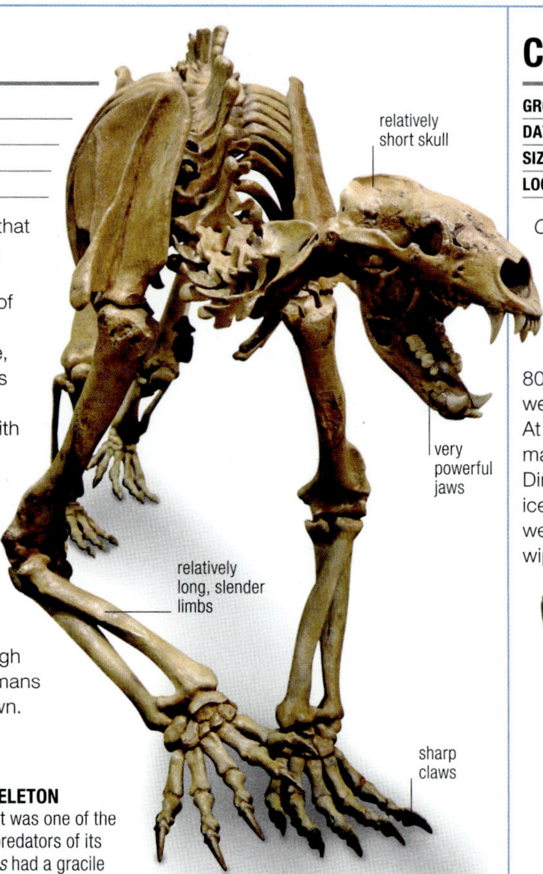

GRACILE SKELETON
Even though it was one of the largest land predators of its time, *Arctodus* had a gracile skeleton, which suggested it was an efficient runner.

900kg The approximate **weight** of *Arctodus*, one of the **largest land carnivores** that ever lived.

Canis

GROUP Placental mammals
DATE Late Pleistocene
SIZE 1.5m (5ft) long
LOCATION Canada, USA, Mexico

Canis dirus is the scientific name for the dire wolf, an extinct North American species. Its fossils have been found in rocks dating from the Late Pleistocene. It was the size of a modern grey wolf but was much heavier, weighing about 80kg (176lb). It is thought that dire wolves were scavengers, living like today's hyenas. At that time, there were no hyenas or other mammalian scavengers on the continent. Dire wolves died out at the end of the last ice age, probably because the first humans were spreading across the continent, wiping out the wolf's food sources.

KEY SITE
LA BREA TAR PITS

Dire-wolf fossils have been found at La Brea tar pits in California, the world's richest site for Pleistocene fossils. Tar from oil deposits, disguised by a layer of water, has trapped many animals over the past 38,000 years. Tens of thousands of fossils have been recovered, representing about 660 species.

COMPACT STRENGTH
About the size of a modern grey wolf, the dire wolf had a shorter, broader head, with sizeable teeth for crushing bone and shorter, more powerful limbs.

Crocuta

GROUP Placental mammals
DATE Pleistocene
SIZE 1.3m (4¼ft) long
LOCATION Africa, Europe, Asia

Crocuta is the genus that includes the living spotted hyena, an African scavenger and hunter that is well adapted to running long distances in pursuit of prey. The Pleistocene subspecies, *C. crocuta spelaea*, known as the "cave hyena", was considerably larger and more robust than its modern relative. It was apparently adapted for the colder climate along the edges of glaciers that covered Eurasia at that time. Although it is thought to have used caves, it probably hunted and scavenged in open areas. First appearing in India at the beginning of the Pleistocene, *Crocuta* had spread east to China, and also west to Europe and Africa, by the Middle Pleistocene. Eventually, most of its members died out, leaving only the lineages that still survive in Africa today.

POWERFUL JAW
Like modern hyenas, the cave hyena had extremely powerful jaws and teeth with which it crushed the bones of its prey.

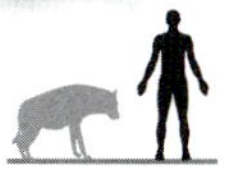

CANINE TOOTH
The cave hyena had two large canines like this in each jaw. As in other carnivores, these were primarily used to secure and rip apart prey.

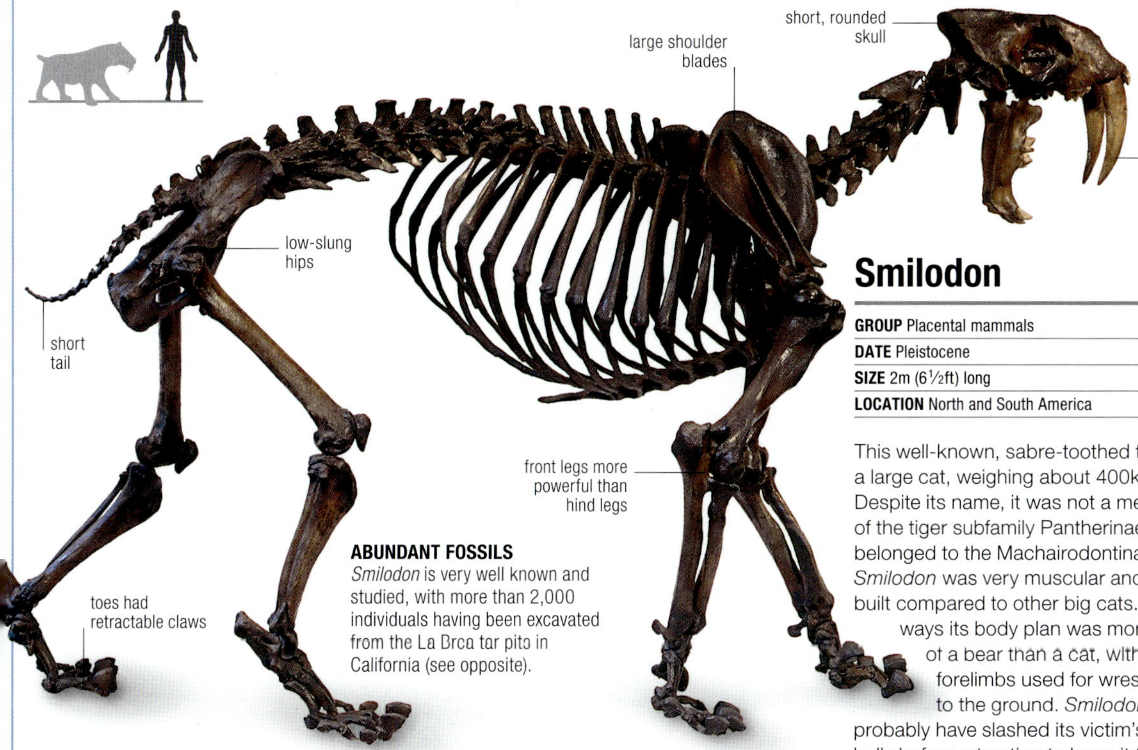

ABUNDANT FOSSILS
Smilodon is very well known and studied, with more than 2,000 individuals having been excavated from the La Brea tar pits in California (see opposite).

Smilodon

GROUP Placental mammals
DATE Pleistocene
SIZE 2m (6½ft) long
LOCATION North and South America

This well-known, sabre-toothed tiger was a large cat, weighing about 400kg (880lbs). Despite its name, it was not a member of the tiger subfamily Pantherinae, but belonged to the Machairodontinae. *Smilodon* was very muscular and heavily built compared to other big cats. In some ways its body plan was more like that of a bear than a cat, with powerful forelimbs used for wrestling prey to the ground. *Smilodon* would probably have slashed its victim's throat or belly before retreating to leave it to die of blood loss. Finds of multiple *Smilodon* fossils suggest that this cat, like modern-day lions, lived and hunted in large social groups. Like many other large mammals, *Smilodon* vanished at the end of the last ice age, possibly due to competition with humans.

ANATOMY
SABRE TEETH

Mammals exhibit a variety of specialized teeth, and the most spectacular are perhaps the long, blade-like canines. *Smilodon*'s canines were particularly impressive. They were very long, narrow recurved blades with serrated front and back edges, that were used for stabbing prey in the throat or belly, then ripping it open.

PREDATORY CARNIVORE
A formidable predator, *Smilodon* would probably have eaten a wide variety of large prey, including deer, horses, and mammoths.

RIDGED TOOTH
Mammoth teeth consisted of a series of enamel plates held together by a dental cement. Teeth erupted in the back of the jaw and moved forwards as they wore, to be replaced by other teeth from behind.

WOOLLY MAMMOTH
Mammuthus stood to about 5m (16½ft) at the shoulder and weighed 8 tonnes (8.8 tons). The species shown here, the wooly mammoth, was much like modern elephants in most features, except that it had smaller ears, was covered in thick hair, and had longer, curved tusks.

This genus became extinct in the Late Pleistocene, but one dwarf mammoth survived on Wrangel Island near Alaska, USA, until 4,500 years ago.

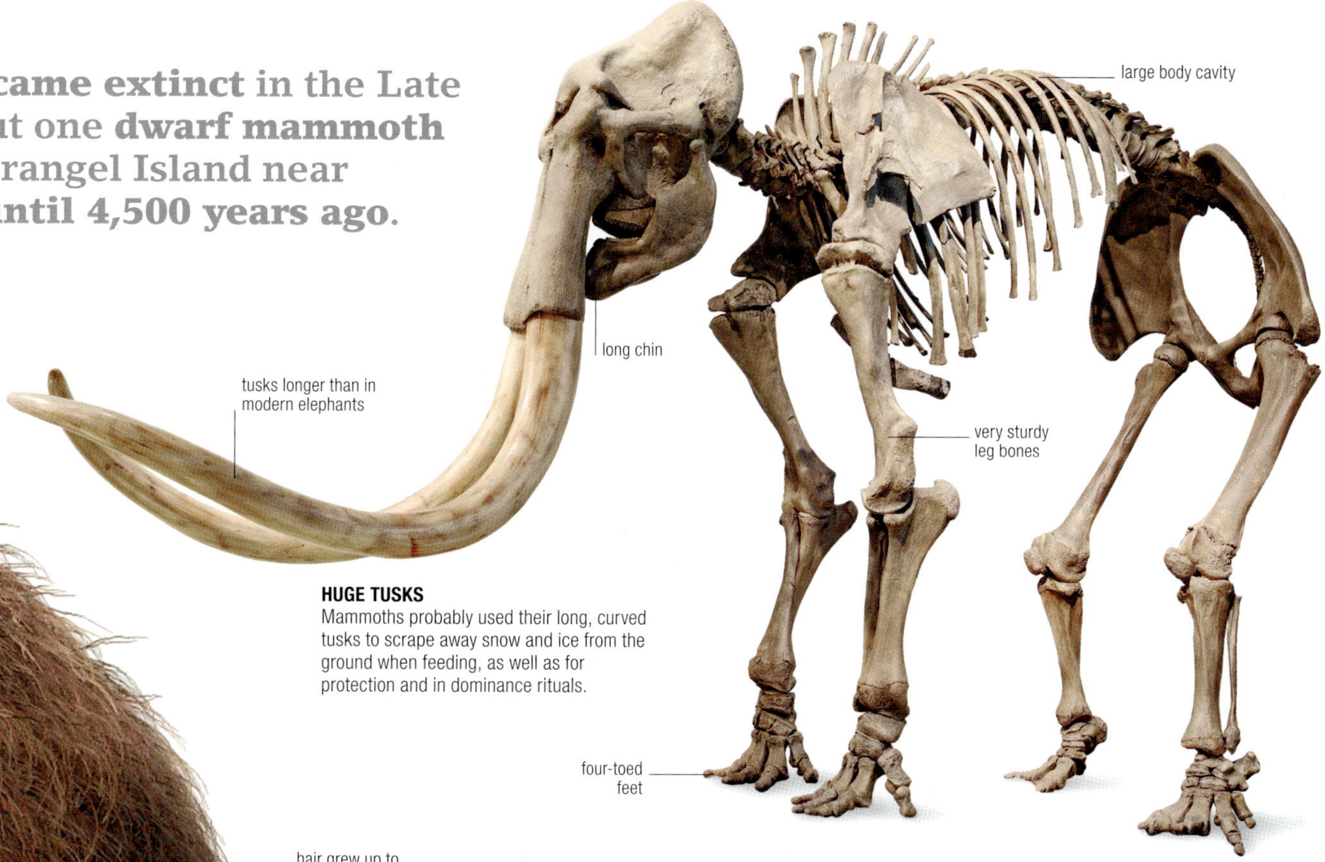

HUGE TUSKS
Mammoths probably used their long, curved tusks to scrape away snow and ice from the ground when feeding, as well as for protection and in dominance rituals.

Mammuthus

GROUP Placental mammals
DATE Pliocene to Late Pleistocene
SIZE 5m (16½ft) tall to the shoulder
LOCATION North America, Europe, Asia, Africa

A highly successful and widespread genus of mammoths, *Mammuthus* lived across the northern continents and Africa throughout the entire Pliocene and Pleistocene epochs. Species of *Mammuthus* include the famous woolly mammoth, which lived along the edges of the ice-age glaciers, the giant Imperial mammoth of milder regions, and a number of separate dwarf species, which lived on islands such as the Channel Islands of California, USA, and Sardinia, Italy.

Mammoth DNA, from many examples of freeze-dried and mummified specimens found in the Siberian permafrost, is almost identical to that of living elephants. Indeed mammoths were much like modern elephants in appearance, except that their tusks were considerably longer and curved inwards. One theory attributes their extinction to infectious disease. However, mammoths were hunted by humans all over the world and that, along with the effects of climate changes, probably contributed to their extinction.

KEY FIND
FROZEN BABY

Of the many specimens of freeze-dried mammoths found in the Siberian permafrost, most are incomplete. In 2008, a Siberian reindeer herder found a nearly complete baby mammoth, frozen in a near-perfect state. It was four months old when it died, about 30,000 years ago. It has been subjected to a detailed analysis, including CT scans, dissection, and chemical analysis of its tusks, which tell of the animal's health and the humidity and temperature of its environment.

QUATERNARY

Pliocene | Pleistocene | Holocene

Mammut

GROUP Placental mammals
DATE Pliocene to Pleistocene
SIZE 3m (9¾ft) tall
LOCATION North America, Europe, Asia

Mammut is perhaps better known as the mastodont, an extinct cousin of the elephants. It weighed up to 5 tonnes (5.5 tons) and had a long, flat skull with slightly curved tusks and a back that sloped down from the hump over the shoulders. It was covered in long, shaggy hair and was well adapted to living in dense conifer forests. Preserved stomach contents show that it ate spruce and other types of low-fibre vegetation. It seems that *Mammut* did not vanish when humans arrived in the Americas 10,000 years ago. Apparently, small populations survived until about 6,000 years ago in places like Utah and Michigan, USA.

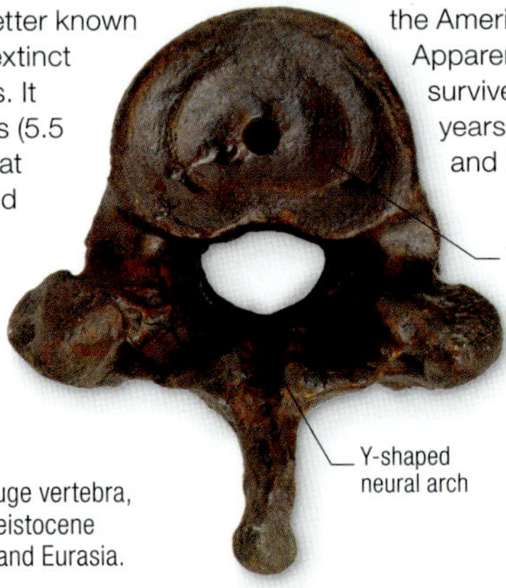

VERTEBRA
Mammut fossils, like this huge vertebra, come from Pliocene and Pleistocene deposits of North America and Eurasia.

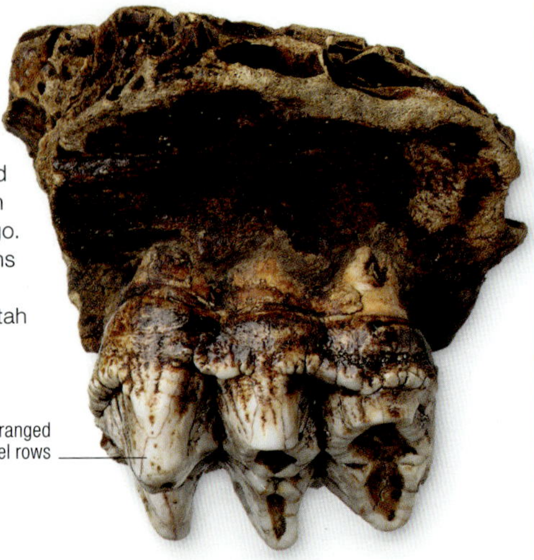

MAMMUT MOLARS
Mastodont means "breast tooth", which refers to the shape of the *Mammut's* molars, which had conical cusps shaped like breasts.

Stegomastodon

GROUP Placental mammals
DATE Pliocene to Early Pleistocene
SIZE 3m (9¾ft) tall
LOCATION North America, South America

Stegomastodon, a gomphothere mastodont, weighed over 6 tonnes (6.6 tons) and had a shortened skull and jaw compared to other gomphotheres, similar to those of mammoths and elephants. It had two long, incurved upper tusks that grew up to 3.5m (11½ft) long. Its huge molars had a series of cusps that looked like three-leaf clovers when worn. This form of tooth was well adapted for eating grass and other gritty vegetation. *Stegomastodon* originated in the grassy plains of North America. Several species crossed the Panamanian land bridge to South America, where *Stegomastodon* evolved into *Cuvieronius*.

Equus

GROUP Placental mammals
DATE Pliocene to present
SIZE 2.5m (8¼ft) tall
LOCATION Worldwide

Equus is the genus that includes today's horses, zebras, and donkeys. It first evolved from *Dinohippus* in the early Pliocene and still flourishes today in domestication, but most wild species of *Equus*, especially asses, onagers, and Przewalski's horse, are rare. In North America, *Equus* vanished with other large Pleistocene mammals about 10,000 years ago, possibly due to hunting by humans. Fossils from North America show a diverse range of species – for example, dwarf and giant forms, horses with robust legs, and others that were stilt-legged. The North American species appear to be related to African zebras and Eurasian asses, suggesting that horses once had a greater geographic spread.

GRAZER'S TEETH
This horse's skull has a combination of teeth designed for grazing. Incisors at the front cut vegetation, and high-crowned molars at the back ground it.

LIVING RELATIVE
PRZEWALSKI'S HORSE

Przewalski's horse is the species from which all domestic horses are descended. It looks like a short-legged version of a domesticated horse, with a brown coat, white belly, and dark brown mane. Originally found in the desert steppes of Mongolia, it was wiped out in the wild, surviving only in zoos, but has been reintroduced to its original habitat. There are about 1,500 alive today, both in the wild and in captivity.

Coelodonta

GROUP Placental mammals
DATE Pleistocene
SIZE 3.7m (12¼ft) long
LOCATION Europe, Asia

This ice-age woolly rhinoceros was about the size of a large white rhino, and bore a thick coat of long, shaggy hair. It had a pair of horns that were oval in cross section and curved backwards, which it probably used to shift snow when foraging. *Coelodonta's* cheek teeth had the characteristic π-shaped pattern of crests found in all rhinos, but were also highly ornate, with many additional ridges that increased the grinding surface. Many lines of evidence show that woolly rhinos were mostly grazers, although they could probably eat almost any vegetation.

Coelodonta vanished about 10,000 years ago at the end of the last ice age, probably due to the retreat of its glacial habitat caused by climate changes. As it had coexisted with humans for thousands of years, it is unlikely to have been hunted to extinction.

WOOLLY RHINO
Coelodonta lived across Eurasia during the ice ages of the Pliestocene. Several complete skeletons have been found, as well as mummified specimens.

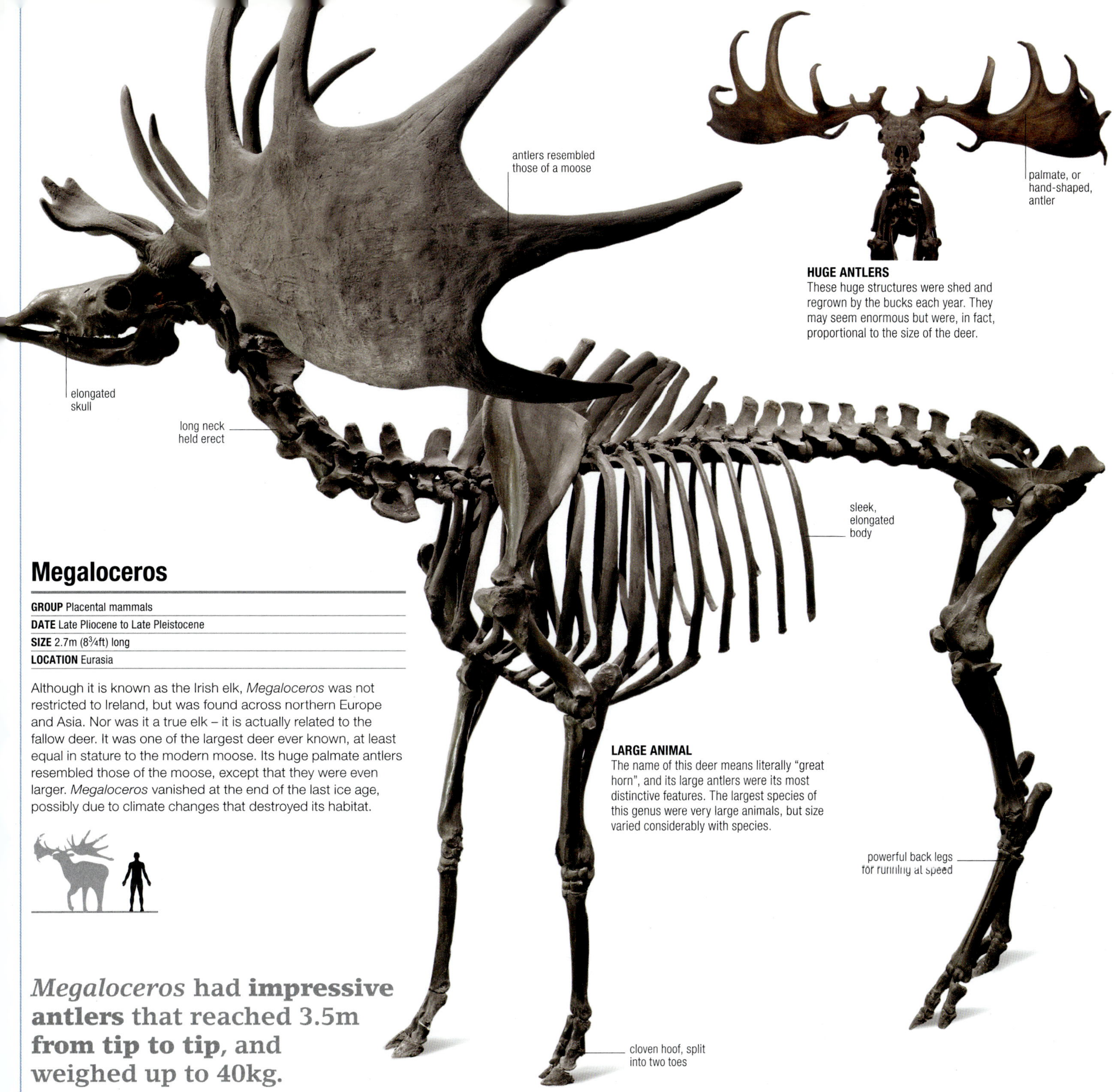

HUGE ANTLERS
These huge structures were shed and regrown by the bucks each year. They may seem enormous but were, in fact, proportional to the size of the deer.

LARGE ANIMAL
The name of this deer means literally "great horn", and its large antlers were its most distinctive features. The largest species of this genus were very large animals, but size varied considerably with species.

Megaloceros

GROUP Placental mammals
DATE Late Pliocene to Late Pleistocene
SIZE 2.7m ($8^3/_4$ft) long
LOCATION Eurasia

Although it is known as the Irish elk, *Megaloceros* was not restricted to Ireland, but was found across northern Europe and Asia. Nor was it a true elk – it is actually related to the fallow deer. It was one of the largest deer ever known, at least equal in stature to the modern moose. Its huge palmate antlers resembled those of the moose, except that they were even larger. *Megaloceros* vanished at the end of the last ice age, possibly due to climate changes that destroyed its habitat.

Megaloceros had **impressive antlers** that reached 3.5m **from tip to tip**, and weighed up to 40kg.

Gigantopithecus

GROUP Placental mammals
DATE Pleistocene
SIZE 3m ($9^3/_4$ft) tall
LOCATION China, Vietnam, India

No complete *Gigantopithecus* skeletons or skulls have been found, and most of what we know about this creature comes from its teeth and jaw bones. It was a huge ape that lived between 1 million and 300,000 years ago. It was built like a gorilla but was much bigger, weighing 540kg (1,190lb). Scientists speculate that the Bigfoot and Yeti myths may have been based on *Gigantopithecus*. However, there is no evidence to suggest that these animals survived beyond about 300,000 years ago – many thousands of years before humans lived in the area.

LIVING RELATIVE
ORANG-UTAN

Gigantopithecus's closest living relative, the orang-utan is the largest tree-dwelling mammal alive, weighing up to 120kg (260lb). Orang-utans have very long arms and quite short legs, and are covered with a coat of long reddish orange hair. They are found only in the vanishing forests of Borneo and Sumatra, but fossils have been found across Southeast Asia. Orang-utan means "person of the forest" in the Malay language, and indeed they look quite human at times. They spend all their lives up in trees, making nests of leaves and eating mostly fruit. According to recent studies, they are the most intelligent of non-human primates, able to solve problems that baffle chimpanzees.

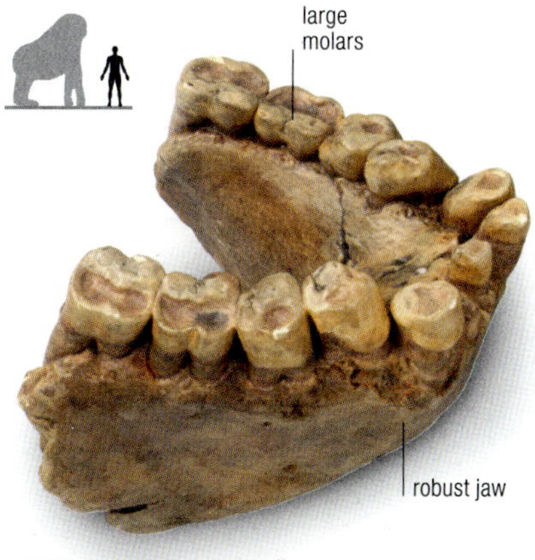

TEETH AND JAW BONES
Gigantopithecus's jaw bones were deep and robust, with low-crowned molars with thick enamel. Tooth-abrasion patterns suggest a diet of bamboo grass.

THE RISE OF

HUM

ANS

Human Relatives

Humans and apes are the living members of a primate "superfamily", called the hominoids, that share a common ancestor. While all hominoids exhibit features that distinguish them from other primates, humans have evolved an array of characteristics that make our species unique.

Ape origins

Primates diverged from other mammals at least 65 million years ago. Their evolved features include relatively large brains, eyesight that is superior to their sense of smell, and nails (rather than claws) on hands and feet that are adapted for grasping. The more primitive primates such as the lemurs, galagos, and lorises are known as prosimians.

Evolutionarily more recent primates include the New World monkeys and the Old World monkeys and apes. The earliest apes lived about 30 million years ago, and from these evolved an array of species, including *Proconsul* and *Pierolapithecus*, that may lay claim to being the earliest common ancestor of living apes and humans.

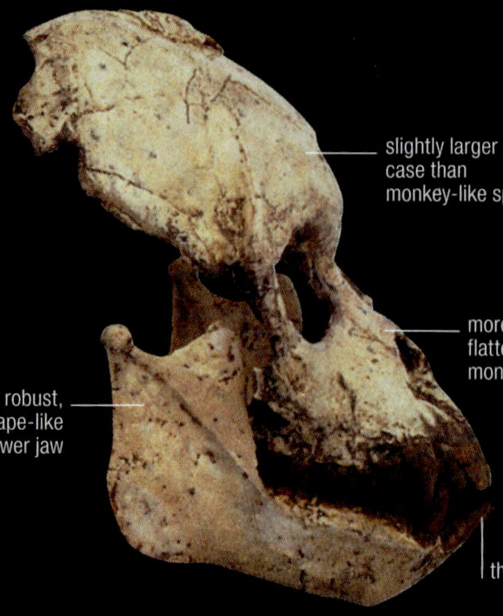

PROCONSUL
This species lived in Africa between 27 and 17 million years ago. It had the size and build of a monkey, but with a larger brain and no tail, like an ape.

KEY FINDS
APE ANCESTOR

In 2002, a team of Spanish paleoanthropologists discovered an unusually complete *Pierolapithecus* fossil that has since been dated to about 13 million years ago. Some researchers have claimed that *Pierolapithecus* is the ancestor of all great apes, including humans. Among the traits it shares with modern apes are its relatively upright posture and a flattish face. However, the skeleton also has some less-evolved, monkey-like traits such as relatively small hands and straight fingers. Also, while *Pierolapithecus* was found in Spain, all modern great ape species live in Africa or Southeast Asia.

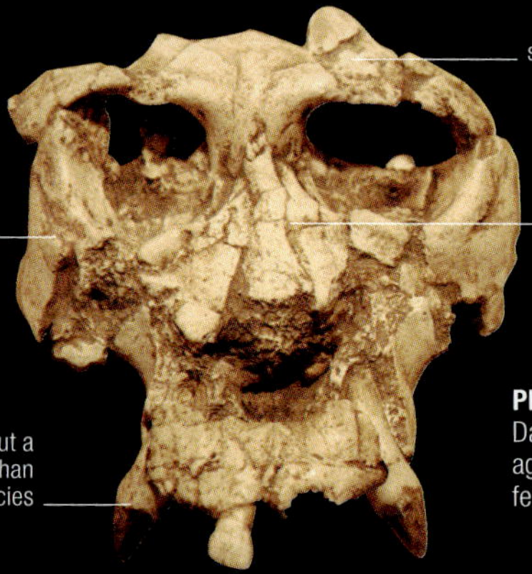

PIEROLAPITHECUS
Dating from about 13 million years ago, *Pierolapithecus* shares many features with modern great apes.

Primate family tree

The common ancestor of all primates is likely to have lived in the Cretaceous, with its descendants diverging into separate lineages over time.

Humans are most closely related to chimpanzees, from whom they diverged about 5–8 million years ago. The next closest relation is the gorilla, followed by the orang-utan.

The orang-utan lineage diverged earlier than the human-chimpanzee-gorilla line, with the fossil genus *Sivapithecus* its probable ancestor. The primates most distantly related to humans are the prosimians. They share a common ancestor with humans in the extinct species *Plesiadapis*, whose fossils date to more than 55 million years ago.

EXTENDED FAMILY
This diagram shows the primate family tree as currently understood mainly from genetic evidence. Humans, on the left hand side, are most closely related to chimpanzees and gorillas, and only distantly to lemurs and galagos.

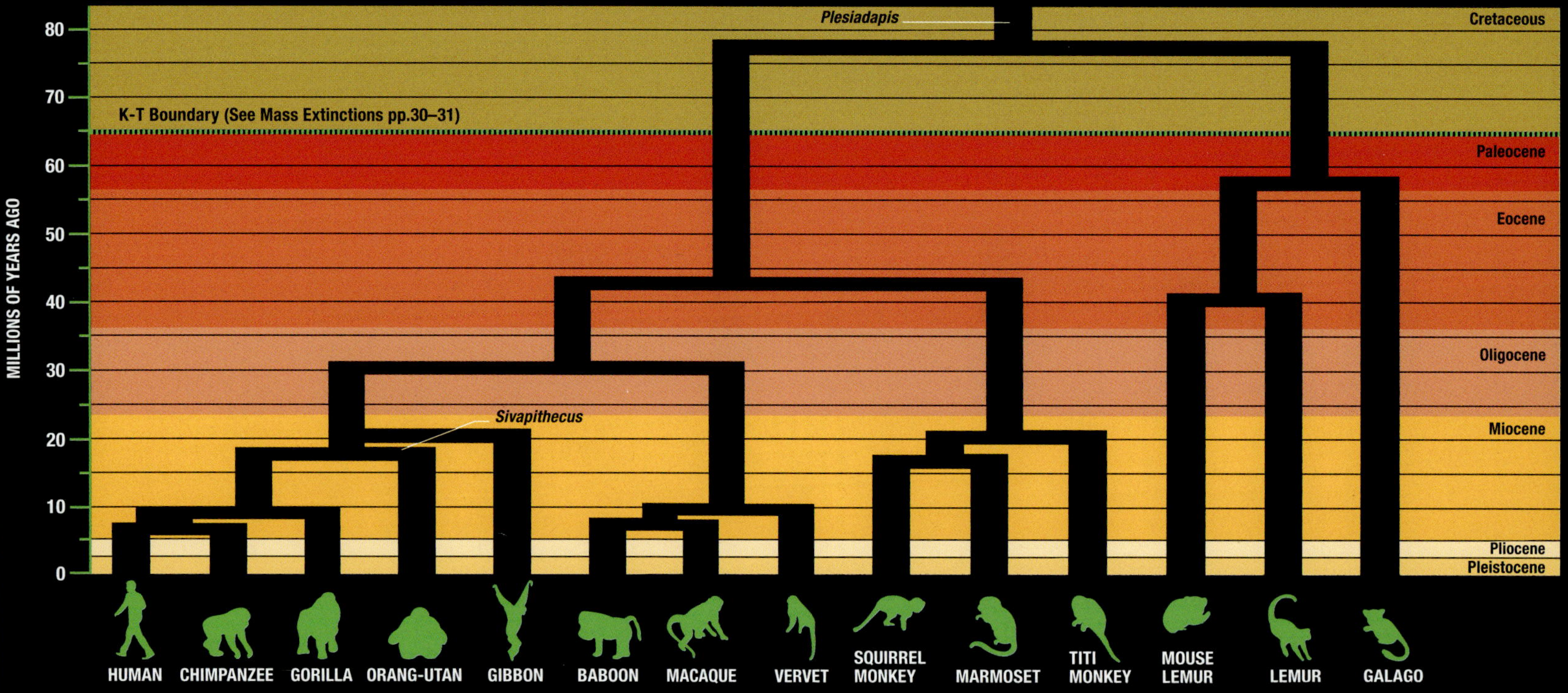

Great apes

It has long been recognized that modern humans have many similarities with chimpanzees and gorillas. However, we also have many differences, and the first scientists to look directly at the molecular differences between us were surprised to find how closely related we are. Most evidence demonstrates that human DNA differs from that of chimpanzees only by about 1.5 per cent, and from gorillas by 2 per cent. This suggests a relatively recent divergence from each (see panel, below). Therefore some researchers now prefer to call humans, chimpanzees, and gorillas "hominids" to distinguish them from the larger grouping of hominoids, which also includes the more distantly related orang-utans. Within the hominids, modern humans and our ancestors that evolved after our split with the chimpanzees and gorillas are now known as "hominins".

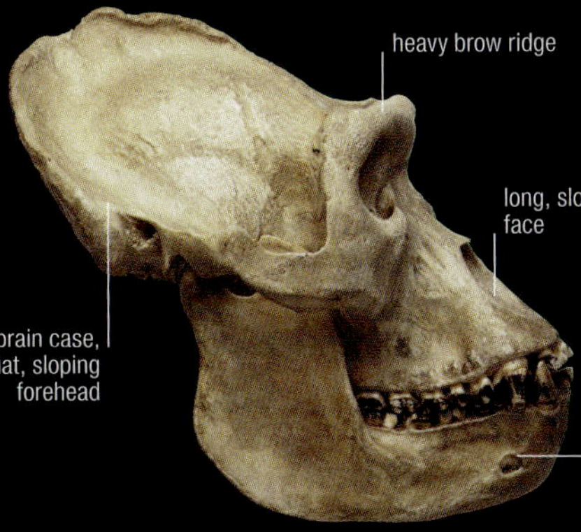

GORILLA SKULL
The gorilla skull is adapted for chewing tough leaves, with wide teeth set in powerful jaws that jut forwards, and broad cheekbones to accommodate large muscles for chewing.

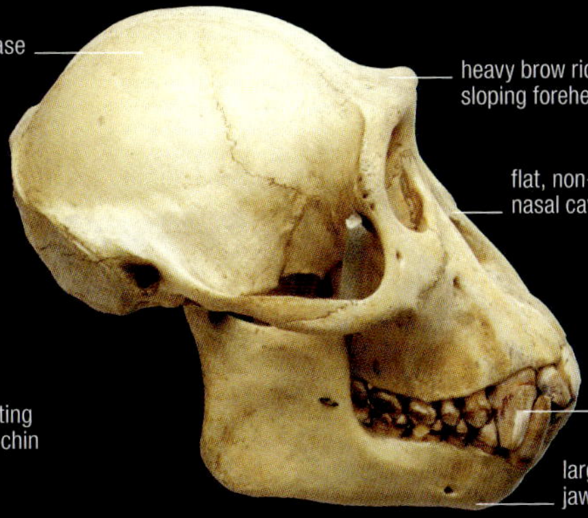

CHIMPANZEE SKULL
Chimpanzees have quite large jaws and powerful muscles for chewing, but their more omnivorous diet requires larger front teeth than those of a gorilla.

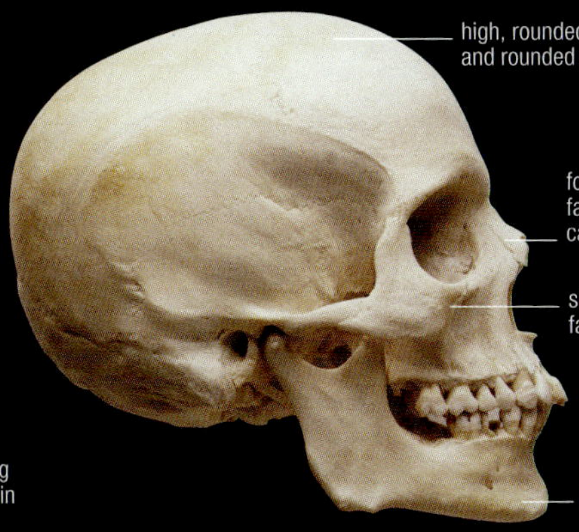

HUMAN SKULL
The human skull is higher and more rounded than those of other hominids. The face is flat, much smaller, and has smaller teeth, as a processed diet needs less chewing.

SCIENCE

DATING SPECIES DIVERGENCE

Recently, scientists have been able to investigate the relationships between species by looking at how they differ at molecular level, for example in their DNA and blood proteins. Because differences accumulate over time, the number of differences between species can be used to estimate how long has passed since species separated – the more differences, the longer ago the split. However, different tests produce slightly different results, and some species seem to accumulate more changes than others, suggesting rates of change vary. Scientists fine-tune the estimates provided by these "molecular clocks" by comparing them with the fossil record, itself patchy and incomplete in places. As more finds are uncovered and techniques improve, a more complete picture will emerge.

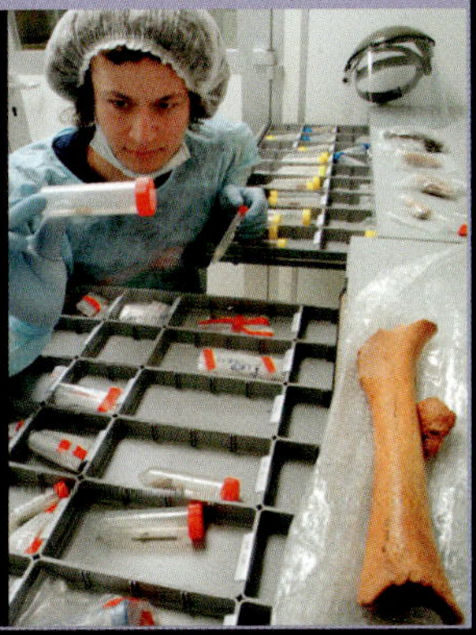

HUMAN SKELETON

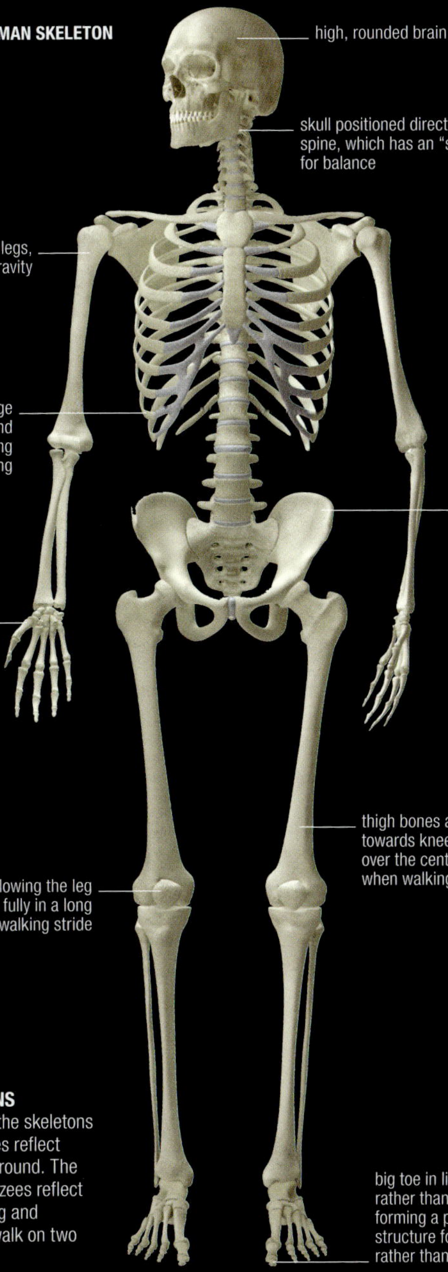

CHIMPANZEE SKELETON

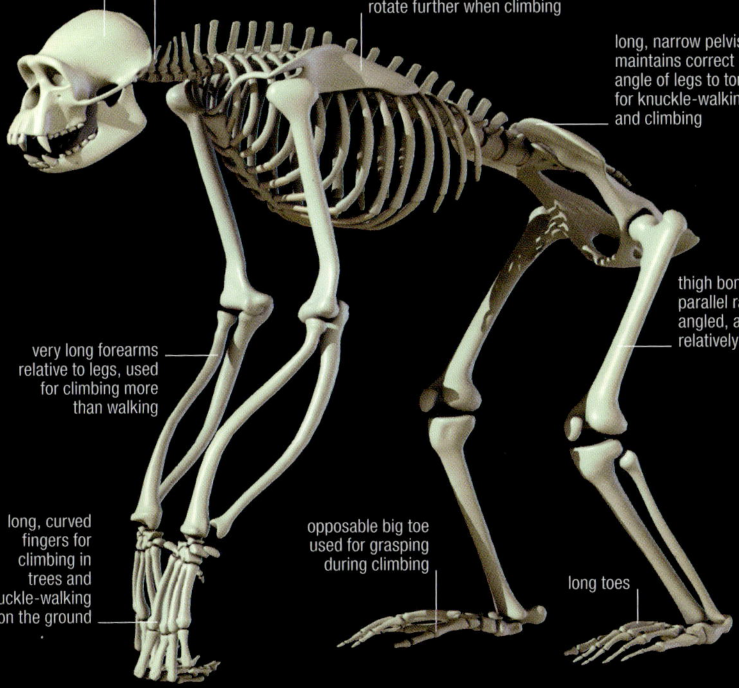

COMPARING APES AND HUMANS
Most of the differences between the skeletons of humans and those of other apes reflect differences in the way we move around. The skeletons of gorillas and chimpanzees reflect adaptations to time spent climbing and knuckle-walking, while humans walk on two legs (bipedally, see p.447).

Human Ancestors

Anthropologists have discovered thousands of fossils of our extinct ancestors. There is now a fairly good picture of the general pattern of our evolution since the separation of the human lineage from that of the chimpanzees some seven million years ago.

Hominin timeline

Until recently, it was believed that human evolution was a straightforward process, with a single line of species connecting modern humans to an ape-like ancestor. However, our history is actually much more complicated than this. Many species coexisted and overlapped, and some lineages became extinct along the way. There is no single "missing link" between our ancestors and those of other apes.

Of course, that is not to say that we know everything there is to know about human evolution: the fossil record contains only bones that ended up in the ideal conditions for preservation. Additionally, the individual members of any species – including modern humans – differ from one another to varying degrees, and different groups separate into distinct species only slowly, through a process of continual change. Many closely related but distinct species known today can be distinguished only by the colour or patterning of their skin or hair, or by the way they behave, clues that are not visible in the fossil record. This makes it difficult to tell whether individual specimens are different enough to be assigned to different species, particularly when fossils are damaged or fragmentary.

Anthropologists often disagree about what name to put to particular specimens; some prefer to place or "lump" fossil species together into large, variable groups and minimize the number of species, others "split" fossils with even slight differences into separate species. As a result, the exact number and characteristics of each fossil species is constantly being debated.

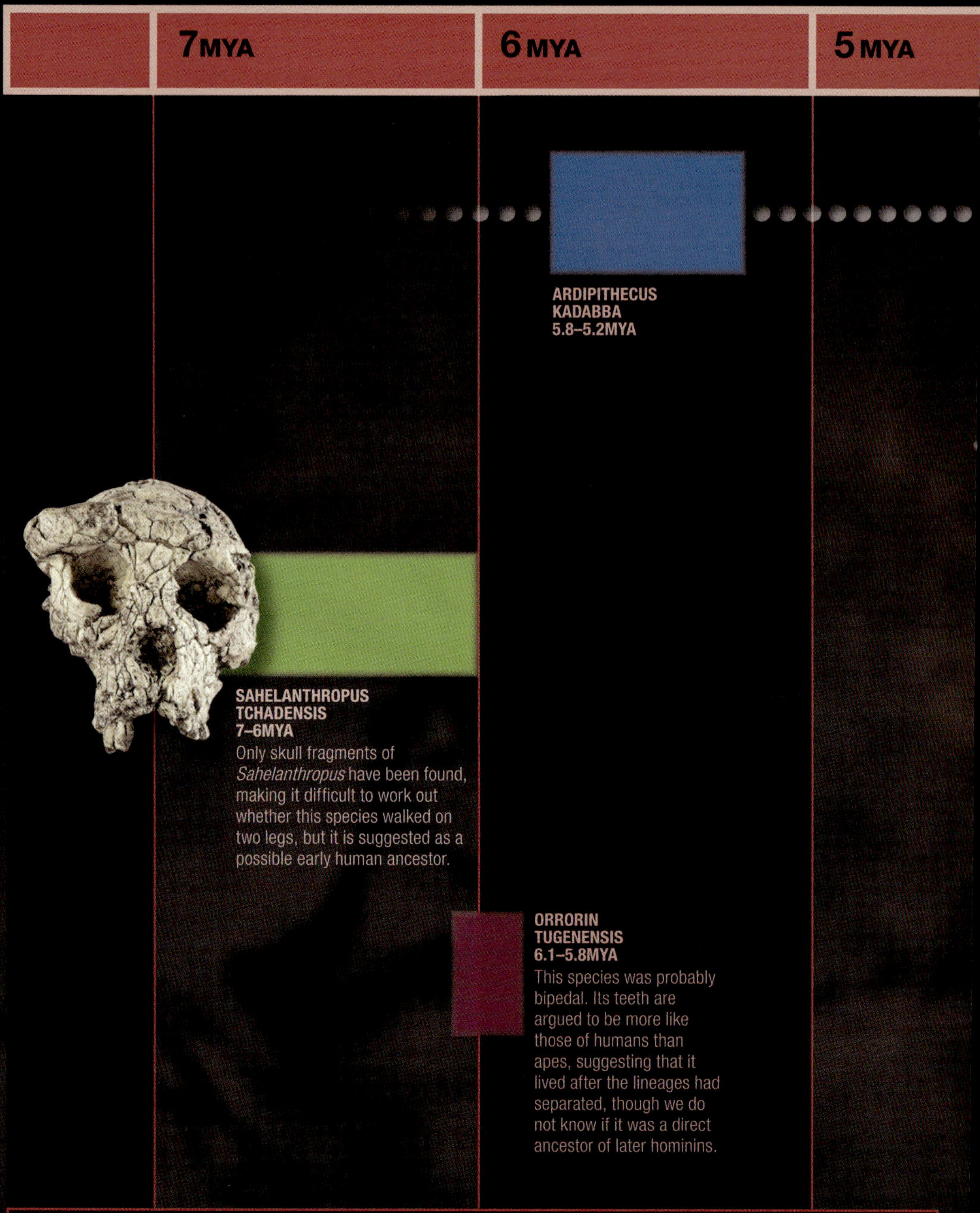

How this chart works

The hominin family tree is constantly being revised as more fossils are discovered and we learn more about how they relate to one another. This chart gives an overview of hominin evolution since the separation from chimpanzees, according to current knowledge. A coloured bar indicates the time range for each species. The colour of the bar indicates the family or genus each species belongs to.

Sahelanthropus tchadensis is perhaps the oldest known ancestor of all hominins. There is some suggestion that it may have walked on two legs, linking it more closely to hominins than apes.

The thigh bone of *Orrorin tugenensis* suggests that it was probably bipedal and more likely to be a hominin than an ape. It post-dates the split of hominins from chimpanzees.

Two species of *Ardipithecus* are currently known. Anthropologists do not agree on whether these early fossils are hominin; but they were definitely closely related to the common ancestors of hominins and chimpanzees.

The few specimens currently assigned to *Kenyanthropus* are badly distorted, making it difficult to decide where it should appear on the hominin family tree. It is currently thought to be a close relative of the australopithecines.

Several species of *Australopithecus* are known from across Africa. Like the earlier hominin ancestors, they were similar in size to modern chimpanzees, but with slightly larger brains.

The skulls and teeth of paranthropines are adapted for chewing and grinding tough vegetable foods; these specializations do not appear in the later fossil record, suggesting that they became extinct without leaving descendants.

Members of the genus *Homo*, including ourselves, have large brains and are tall, with longer legs than our earlier ancestors. All of these species were users of stone tools.

Dotted lines indicate the probable lines of descent among species. These are not necessarily direct relationships, but simply show the most likely patterns of human development.

OLDUVAI GORGE Olduvai Gorge is a huge ravine in northern Tanzania that forms part of the Rift Valley of eastern Africa. It was created by a stream cutting into layers of ancient volcanic ash, and fossils of many animals have been revealed here, preserved in the layers of sediment. Olduvai Gorge contains many of the earliest human fossil sites known, earning it the title "Cradle of Humankind".

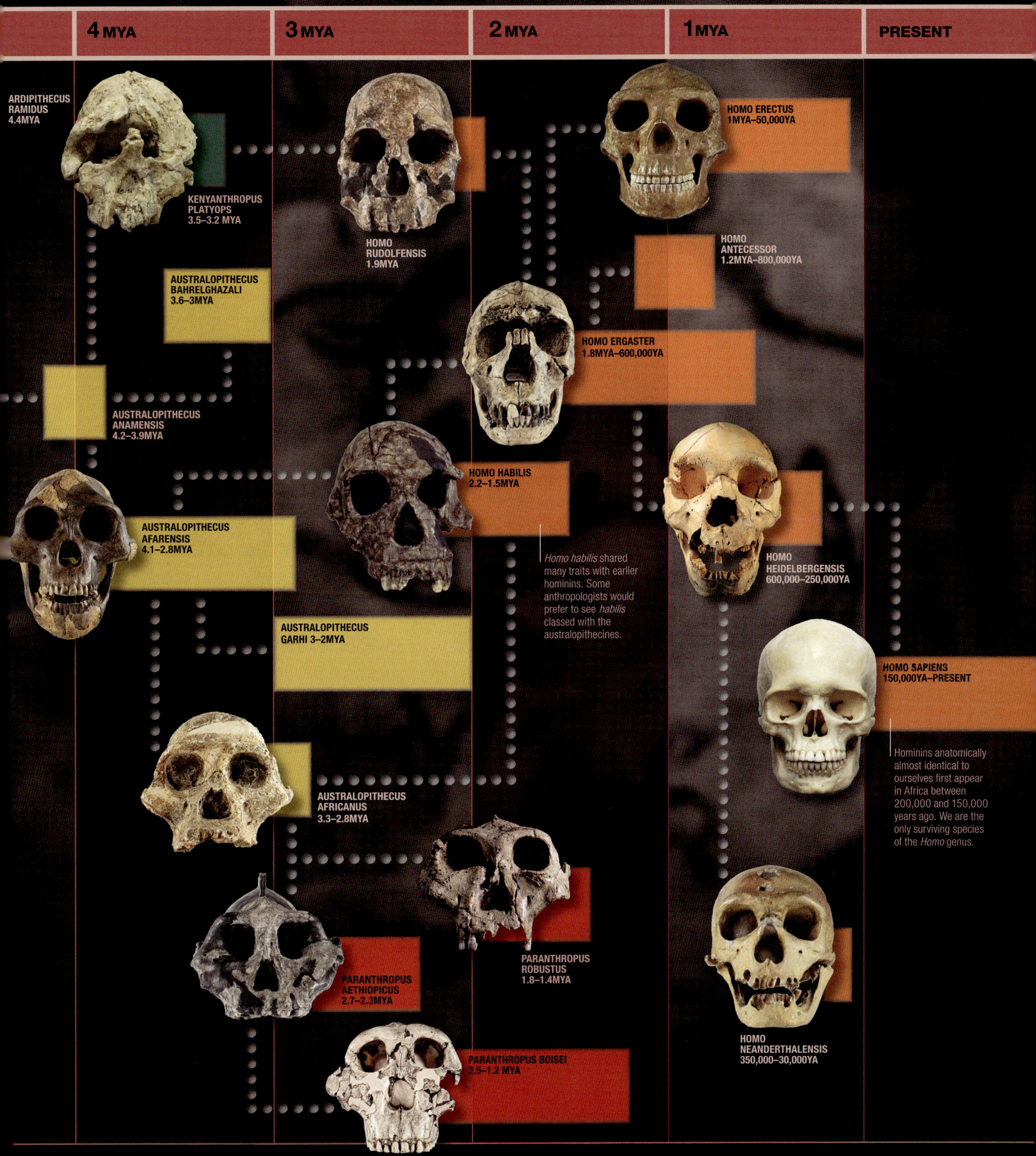
4 MYA
3 MYA
2 MYA
1 MYA
PRESENT
ARDIPITHECUS RAMIDUS 4.4MYA
KENYANTHROPUS PLATYOPS 3.5–3.2 MYA
HOMO RUDOLFENSIS 1.9MYA
HOMO ERECTUS 1MYA–50,000YA
HOMO ANTECESSOR 1.2MYA–800,000YA
AUSTRALOPITHECUS BAHRELGHAZALI 3.6–3MYA
HOMO ERGASTER 1.8MYA–600,000YA
AUSTRALOPITHECUS ANAMENSIS 4.2–3.9MYA
HOMO HABILIS 2.2–1.5MYA
AUSTRALOPITHECUS AFARENSIS 4.1–2.8MYA
Homo habilis shared many traits with earlier hominins. Some anthropologists would prefer to see *habilis* classed with the australopithecines.
HOMO HEIDELBERGENSIS 600,000–250,000YA
AUSTRALOPITHECUS GARHI 3–2MYA
HOMO SAPIENS 150,000YA–PRESENT
Hominins anatomically almost identical to ourselves first appear in Africa between 200,000 and 150,000 years ago. We are the only surviving species of the *Homo* genus.
AUSTRALOPITHECUS AFRICANUS 3.3–2.8MYA
PARANTHROPUS ROBUSTUS 1.8–1.4MYA
PARANTHROPUS AETHIOPICUS 2.7–2.3MYA
HOMO NEANDERTHALENSIS 350,000–30,000YA
PARANTHROPUS BOISEI 2.5–1.2 MYA

Sahelanthropus tchadensis

***Sahelanthropus* lived around the time of the last common ancestor of humans and other apes. Some scientists believe that it was the ancestor of both humans and chimpanzees, but few agree.**

Whether *Sahelanthropus* lived before or after the human and ape lineages separated is unknown, as is which of these lineages it was part of. Only a skull and teeth have been found, and while there are indications that it may have been bipedal, a lack of bones below the skull means anthropologists cannot be certain. If *Sahelanthropus* was proved to be bipedal, it would be a more likely candidate for human ancestry.

SMALL BRAIN
Sahelanthropus's skull is small, and relatively long and low, like those of apes. However, its face is shorter and more vertical, while its head would have been positioned on top of its spine, suggesting that the species may have walked on two legs (see opposite).

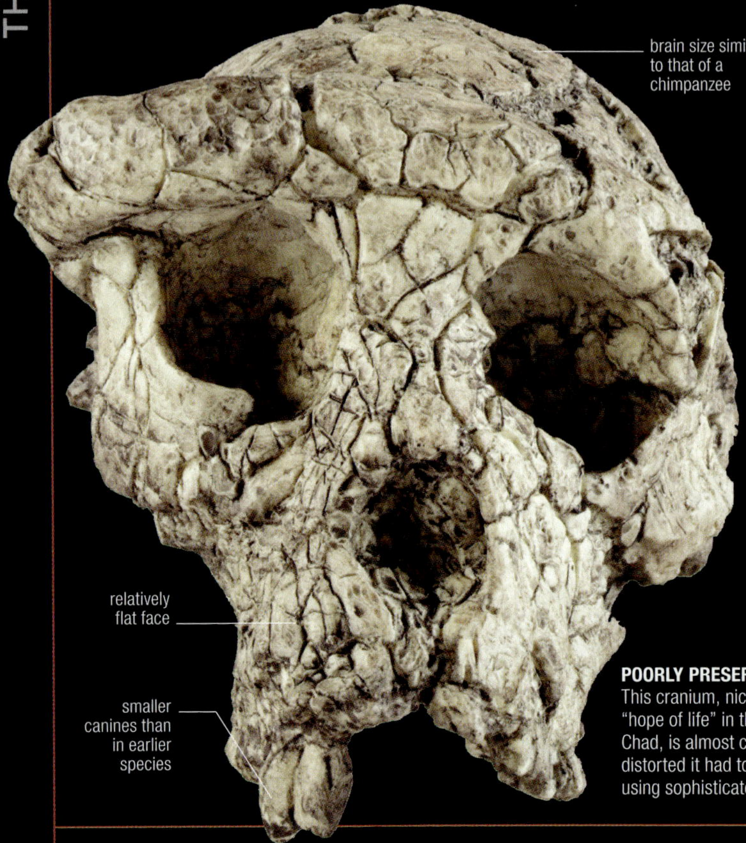

POORLY PRESERVED
This cranium, nicknamed "Toumai" or "hope of life" in the local language of Chad, is almost complete, but so badly distorted it had to be "straightened out" using sophisticated computer software.

Sahelanthropus tchadensis

DATES 7–6mya
BRAIN SIZE 320–380 cubic centimetres
HEIGHT Unknown
FINDS Single skull, pieces of jaw and teeth

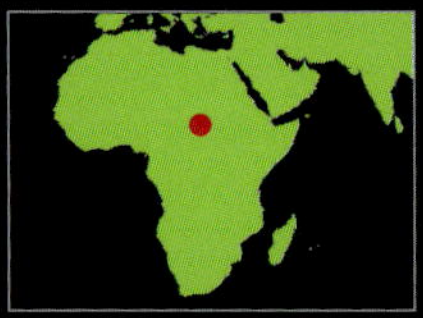

LOCATION
Djurab Desert, Chad

IN SITU
The *Sahelanthropus* fossils were found in Chad, in west Central Africa, an area not previously explored for hominin fossils. Further research here may well turn up more surprises.

Orrorin tugenensis

The fragmentary remains of *Orrorin tugenensis* suggest that this species was closely related to the common ancestor of both humans and chimpanzees, and it may be the oldest known biped.

The anthropologists who found *Orrorin tugenensis* claim that it belongs on the hominin side of the human-ape divergence. However, others would argue that while *Orrorin* definitely seems to have been closely related to the common ancestor of humans and chimpanzees, we cannot be sure on which side of the division it belongs. *Orrorin*'s features exhibit a mixture of ape-like and human-like traits. Like humans, it had teeth covered in thick enamel, although some of the teeth are more ape-like in shape and size. The shape of its thigh bones suggests that the hip joints carried most of its body weight and hence that it was bipedal. Also, marks on the bone that indicate where the hip muscles attached are located in similar places to those of later bipedal hominins. However, features of its arm bones suggest that it may also have climbed trees.

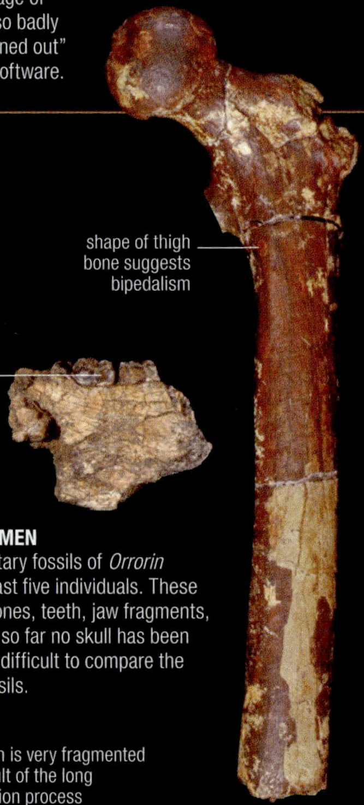

FRAGMENTED SPECIMEN
We have only fragmentary fossils of *Orrorin tugenensis*, from at least five individuals. These include leg and arm bones, teeth, jaw fragments, and a finger bone, but so far no skull has been found, which makes it difficult to compare the finds against other fossils.

TUGEN HILLS
At least five individual *Orrorins* were found in the Tugen Hills of Kenya in 2000, by Brigitte Senut and Martin Pickford. Orrorin is Tugen for "original man".

Orrorin tugenensis

DATES 6.1–5.8mya
BRAIN SIZE Unknown
HEIGHT Unknown
FINDS Jaw bone, teeth, fragmented arm and thigh bones, finger bone

LOCATION
Tugen Hills, Kenya

BIPEDALISM

Some of the most significant anatomical differences between humans and other primates are a direct result of our different ways of moving around. Other primates are quadrupeds and/or climbers, but modern humans walk upright on two legs. There are many theories about how and why this might have happened. For example, bipedalism frees the hands for carrying food, water, infants, stone tools, or weapons. It is also a highly effective way of moving across wide open spaces, wading through water, and minimizing exposure to the sun. However, the fossil record suggests that bipedalism developed before any tools or weapons are known, and even before our ancestors abandoned forested environments for more open ones. Because humans are the only primates who move around on two legs, it is probable that all biped fossils date to after the human lineage split from the apes.

GORILLA HUMAN

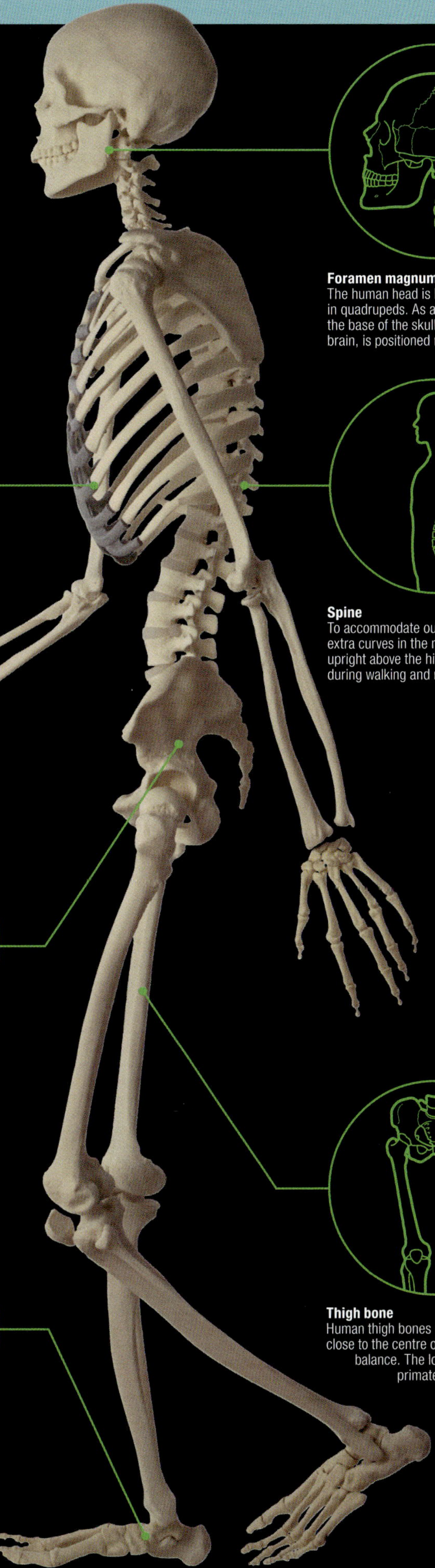

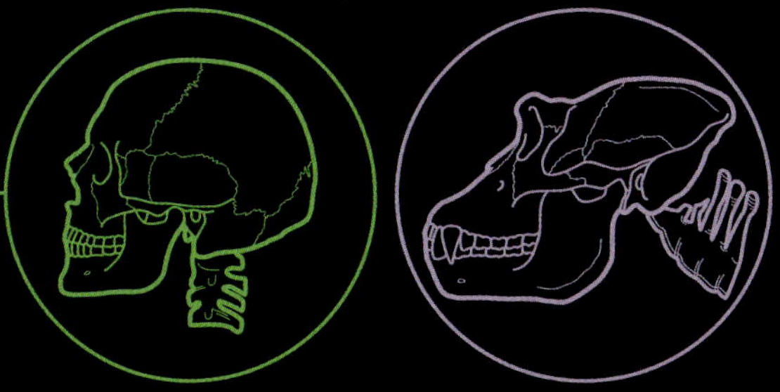

Foramen magnum
The human head is balanced above the spine rather than in front of it as in quadrupeds. As a result, the foramen magnum (literally "big hole") in the base of the skull, through which the spinal column passes into the brain, is positioned much further forwards than among other primates.

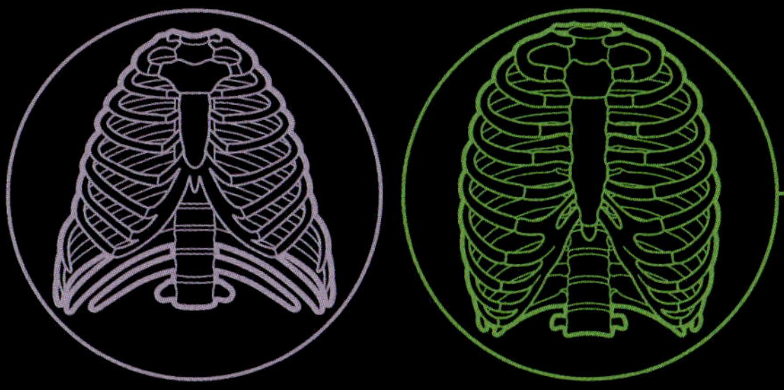

Ribcage
The cone-shaped ape ribcage allows increased flexibility of the shoulder joint, enabling them to reach above their heads when climbing. The barrel shape of the human ribcage allows the torso to flex and the arms to swing, which helps balance the body while walking on two legs.

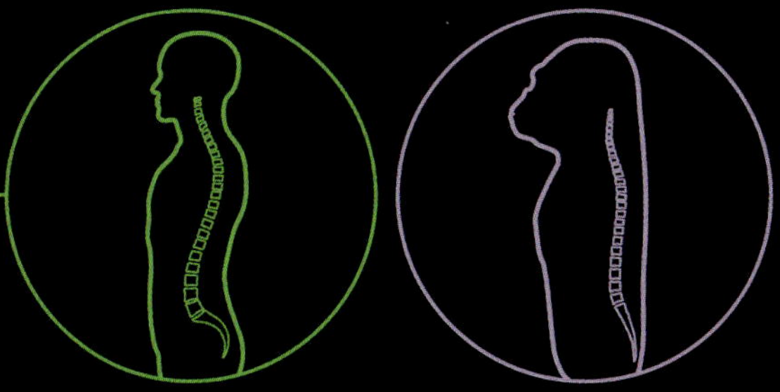

Spine
To accommodate our upright stance, the human spine has developed extra curves in the neck and lower back. These keep the torso balanced upright above the hips, and allow the spine to absorb the shock of impact during walking and running.

QUADRUPED COUSIN
Our closest modern relatives, the chimpanzees and gorillas, are "knuckle-walkers", resting on the knuckles, rather than palms or fingers, while moving on the ground.

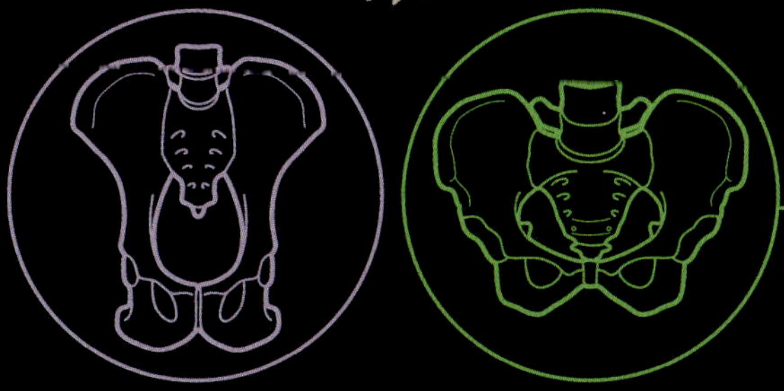

Pelvis
The human pelvis is lower and broader than that of other primates. This centres the torso over the hips, while strong ligaments and deep hip sockets stabilize the hip joints so they can support the full body weight.

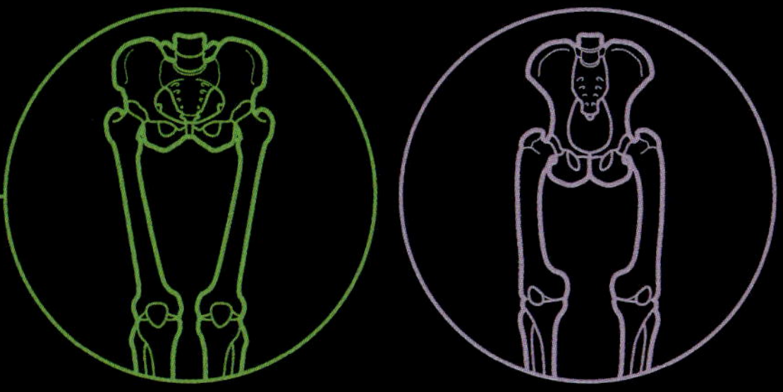

Thigh bone
Human thigh bones angle inwards from the hips to the knees, keeping the knees close to the centre of body weight, and the feet immediately beneath the hips for balance. The lower limbs are longer relative to the upper limbs than in other primates, and the leg can be fully extended, making walking very energy-efficient.

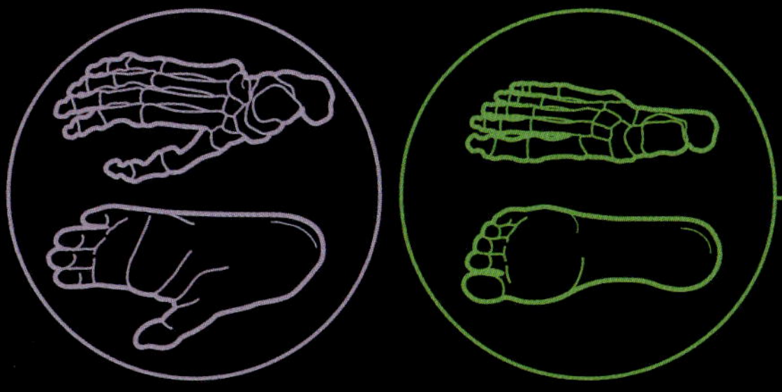

Feet
Primates' big toes are positioned to the side of the foot, for grasping branches with their feet while climbing trees. Uniquely among primates, human big toes are aligned with the other toes, and the heel is longer, creating a platform that supports our weight and allows us to walk.

Ardipithecus ramidus

The first member of the genus *Ardipithecus*, *ramidus* is thought to be an early part of the hominin lineage, and would have looked much like a modern chimpanzee.

This species was found in 1992 by the paleoanthropologist Tim White, and was initially identified as *Australopithecus ramidus* ("ramid" means "root" in Afar). It was only later on that it was assigned to an entirely separate genus, *Ardipithecus*, from the local Afar word for "ground" or "floor". Overall, *ramidus* seems to have been rather chimpanzee-like. However, several lines of evidence suggest that it was bipedal, and therefore probably an early hominin: its skull seems to have balanced squarely over the spine, and the arrangement of its toe bones suggests its feet were adapted for walking rather than grasping branches. Its teeth are chimpanzee-like, but with some hominin features: the canines are shaped more like those of later hominins; the front teeth are relatively small, like later hominins; and the cheek teeth are shaped differently to chimpanzees'. *Ramidus*'s arms show a mixture of features seen in climbing apes and in bipedal hominins.

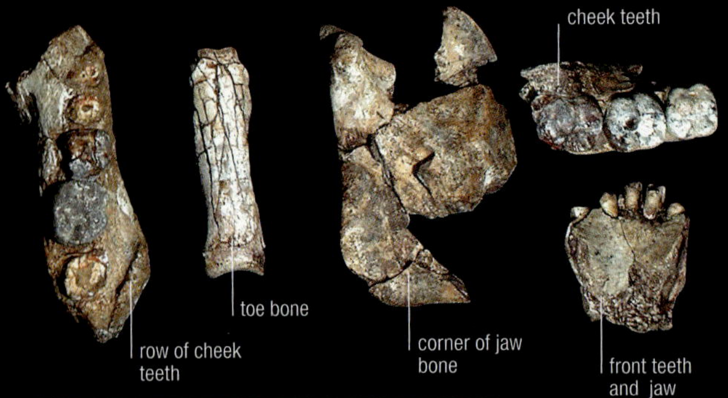

FRAGMENTED SKULL *Ardipithecus ramidus* is known only from fragments of skulls and jaws, isolated teeth, and toe and hand bones, as well as three almost complete bones from a single individual's arm.

JAW BONE Although it would have looked rather like a modern chimpanzee, *Ardipithecus ramidus*'s cheek teeth are similar to later species of australopithecine, and provide some of the strongest evidence that it was a hominin.

Ardipithecus ramidus

DATES 4.4–4mya
BRAIN SIZE Unknown
HEIGHT Unknown
FINDS Skull, jaw, teeth and arm bone fragments

LOCATION Afar, Ethiopia; Tabarin, Kenya

Ardipithecus kadabba

From the fossil record it is difficult to tell whether *Ardipithecus kadabba* was a hominin or chimpanzee ancestor.

The first *kadabba* fossils were discovered by the Ethiopian paleoanthropologist Yohannes Haile-Selassie in 1997. It was originally thought to be a subspecies of *Ardipithecus ramidus*, but, based on dental evidence, is now placed as a separate species of the same genus. *Kadabba* seems to have retained more ape-like characteristics than *ramidus*, such as large, pointed canines. Its other teeth and its hand bones are more similar to those of later hominins, while the toe bones appear intermediate between ape and hominin.

Ardipithecus kadabba

DATES 5.8–5.2mya
BRAIN SIZE Unknown
HEIGHT Unknown
FINDS Jaw, hand, foot, and arm bone fragments and collar bone fragments

LOCATION Middle Awash, Ethiopia

Australopithecus garhi

Most believe *Australopithecus garhi* only lived alongside our ancestors, while some believe we are its descendants.

Between 1996 and 1998, Berhane Asfaw and Tim White recovered a number of fossils from the same sediment layer in different locations, including skull, jaw, and teeth fragments, and a partial skeleton. The skulls and teeth are similar to those of *Australopithecus afarensis* (see pp.450–51) and *africanus*, though with some differences – the teeth are large – and the fossils were assigned to a distinct species. Stone tools are not found with the fossils, but are present in the same sediment layer, and animal bones found nearby bear cut marks.

Australopithecus garhi

DATES 3–2mya
BRAIN SIZE 450 cubic centimetres
HEIGHT Unknown
FINDS Skull and jaw fragments

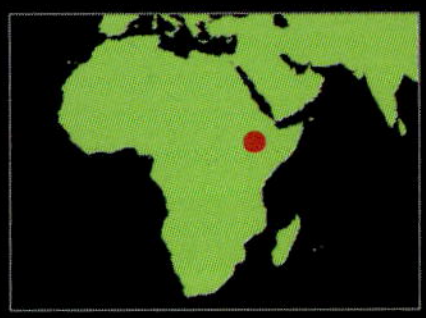

LOCATION Afar, Ethiopia

Australopithecus anamensis

***Australopithecus anamensis* was discovered in 1965, but was not immediately recognized as unique.**

These fossils were similar to those of later australopithecines, but also showed some ape-like features. It was not until 1995 that Maeve Leakey distinguished the 1965 find from fossils of *Australopithecus afarensis* (see pp. 450–51), and named it *anamensis*. The leg bones show that, like *afarensis*, *anamensis* was bipedal, but its skull and teeth are more primitive: the enamel is thicker, and the canines are larger. The rows of cheek teeth are also straight and parallel, like an ape's, rather than arched like those of later hominins.

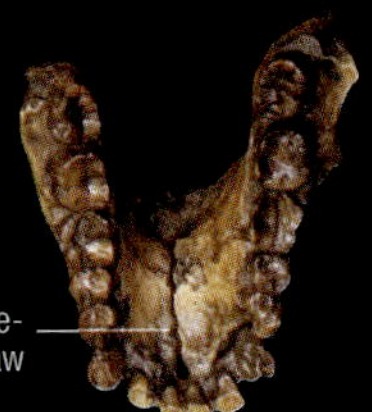

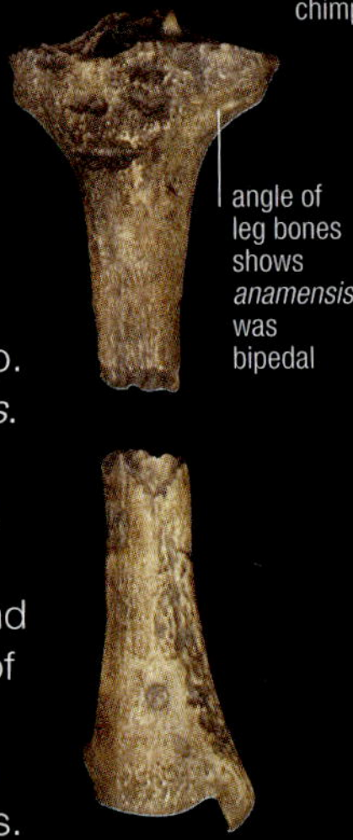

Australopithecus anamensis

DATES 4.2–3.9mya
BRAIN SIZE Unknown
HEIGHT Unknown
FINDS Teeth, jaw, arm, and leg bone fragments

LOCATION Eastern Turkana and Kanapoi, Kenya; Middle Awash, Ethiopia

Australopithecus bahrelghazali

***Australopithecus bahrelghazali* is a poorly understood species, and unusual among australopithecines in that its fossils were found in Chad, not in the Rift Valley.**

Australopithecus bahrelghazali is known only from a fragment of jaw and a few teeth, found by Michel Brunet in 1993. The jaw is also known as "Abel" after Brunet's deceased colleague Abel Brillanceau. Many anthropologists believe the fossil may represent a geographical variant of *Australopithecus afarensis,* and not a species in its own right. Even if further discoveries confirm this, the fossils are still important, as they tell us that australopithecines were more widespread throughout Africa than was initially believed.

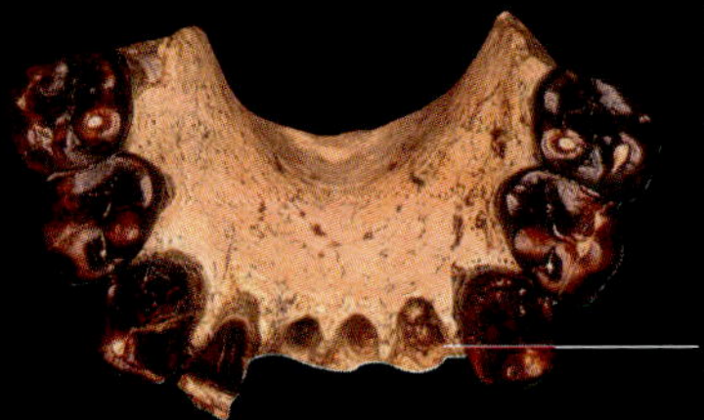

Australopithecus bahrelghazali

DATES 3.6–3mya
BRAIN SIZE Unknown
HEIGHT Unknown
FINDS Jaw fragments and a few teeth

LOCATION Bahr el Ghazali, Chad

Australopithecus africanus

***Australopithecus africanus* was the first early hominin species to be identified. It was ape-like but bipedal, with a small brain and human-like teeth.**

In 1924, Raymond Dart (see panel, right) identified a fossilized partial skull as belonging to a species intermediate between apes and humans. The claim met with considerable scepticism – at the time most anthropologists believed that the earliest human ancestors would be found outside Africa, and that they would have large brains but would not be bipedal. Dart's skull seemed to show exactly the opposite, and his claims were not widely accepted until more hominin skulls were discovered. *Australopithecus africanus* was very similar to *A. afarensis* (see pp.450–51) but its cheek teeth and jaw were larger and it had longer arms relative to its legs, suggesting it climbed more. It may also have had a more ape-like big toe better adapted to grasping rather than walking.

Australopithecus africanus

DATES 3.5–2mya

BRAIN SIZE 428–625 cubic centimetres

HEIGHT 1.4m (4ft 6in)

FINDS A number of jaw bones, several partial skulls, various skeleton fragments

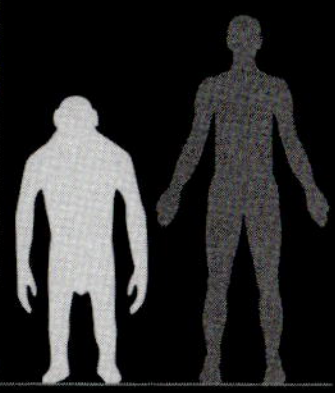

LOCATION
Various sites in South Africa

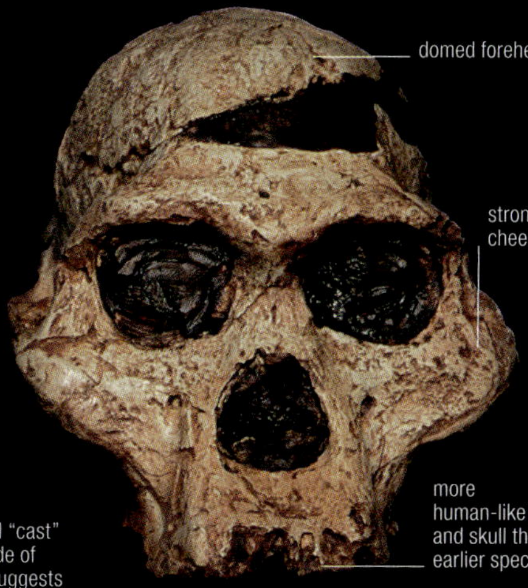

MRS PLES
The *africanus* skull known as "Mrs Ples" was found by Robert Broom in 1947, in Sterkfontein Cave in South Africa. The nickname was created by a press distortion of Broom's original classification, *Plesianthropus transvaalensis*.

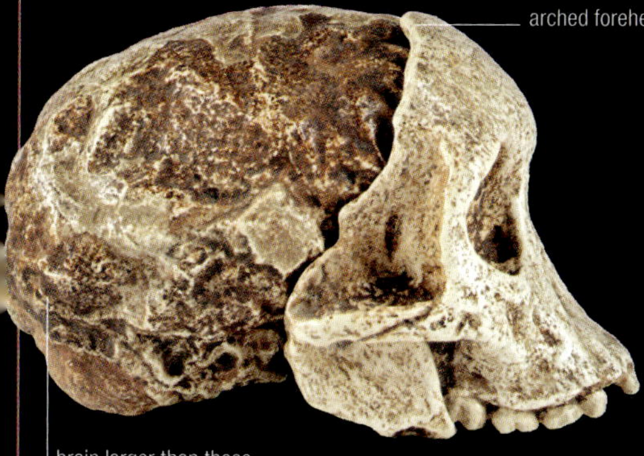

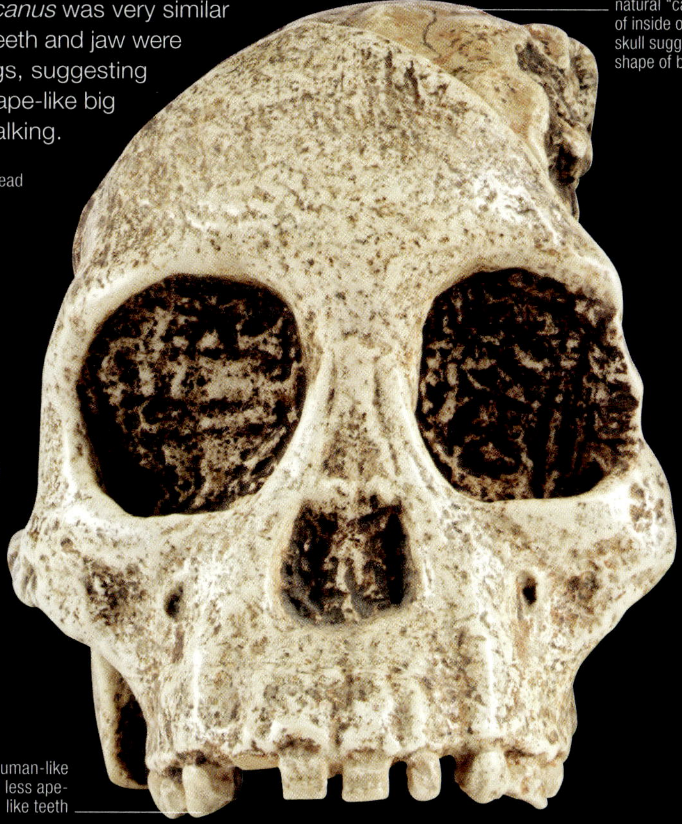

TAUNG CHILD
The Taung Child was the first hominin fossil to be found. A limestone cast of its brain was formed in the fossilization process, and while the skull was missing, the bones of the face fitted neaty onto the brain cast. Marks on the bones have led some anthropologists to suggest that the Taung Child was killed by an eagle.

BIOGRAPHY

RAYMOND DART

The Australian anthropologist Raymond Dart was working at the University of Witwatersrand, South Africa, when he identified fossils in a box of rocks as a new hominid species *Australopithecus africanus*. Although his discovery was first greeted with great criticism, Dart's ideas were eventually accepted, and his reputation established. In 1956, the Institute for the Study of Man in Africa was founded at the University of Witwatersrand in Dart's honour. Three years later, he published an account of his discovery, *Adventures With the Missing Link*. Dart died in 1988 at the age of 95.

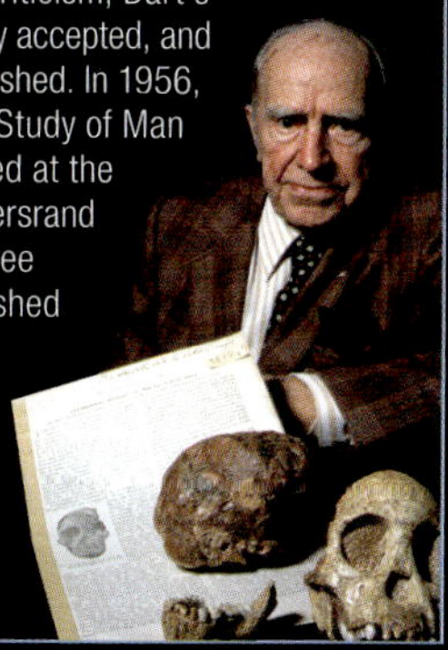

Kenyanthropus platyops

A deformed skull found by Maeve Leakey in 1999 has been assigned not only to a new species, but to a new genus: *Kenyanthropus*.

Kenyanthropus platyops looks less like its East African contemporary *Australopithecus afarensis* (see pp.450–51), and more like the much later *Homo rudolfensis* (see p.455). Its skull is a similar size to those of australopithecines and paranthropines (see pp.452–53), but although it shares some traits common to both, it lacks the distinctive characteristics of either. The significance of it not fitting any contemporary genus lies in the fact that it would prove that another type of hominin lived alongside australopithecines. This would disprove the notion that a single line of hominins evolved from ape-like ancestors to modern humans and that instead several types of hominin coexisted. However, some argue that the fossil is too deformed to be assigned to a new genus.

Kenyanthropus platyops

DATES 3.5–3.2mya

BRAIN SIZE Unknown

HEIGHT Unknown

FINDS Single skull and dental fragments

LOCATION
Lake Turkana, Kenya

LAKE TURKANA
Two to three million years ago, Lake Turkana was much larger than it is now, and the area around it was very fertile. It was an ideal environment for early hominins.

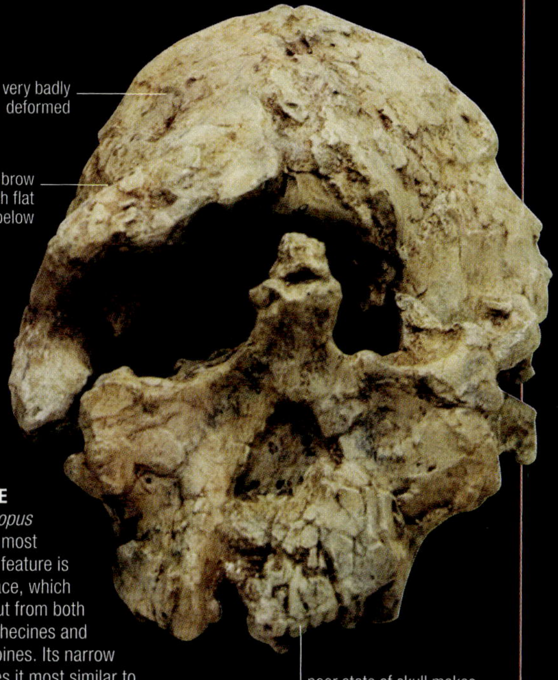

FLAT FACE
Kenyanthropus platyops's most distinctive feature is its "flat" face, which marks it out from both australopithecines and paranthropines. Its narrow nose makes it most similar to *Homo rudolfensis* (see p.455).

Australopithecus afarensis

This species is thought to be the ancestor of the genus *Homo*, to which modern humans (*Homo sapiens*) belong. It is one of the best-known early hominins, with a number of significant finds, including "Lucy", giving us a good idea of what the species looked like and how it behaved.

Significant finds

In 1973, the paleoanthropologist Donald Johanson found a 3-million-year-old knee fossil that clearly belonged to a hominin that had walked upright. The following year the famous "Lucy" was discovered, and more *Australopithecus afarensis* finds have been made since, including the "first family" and the "Dikika baby". As a result of these notable finds, we know quite a lot about *afarensis*'s physical characteristics and lifestyle. The species was clearly mainly bipedal, though it may have climbed trees occasionally. This was a surprise to many people: it was originally thought that our ancestors' brains grew before they began to walk upright. However, *afarensis*'s brain is not much bigger than that of a modern-day chimpanzee, and as far as we know it did not make stone tools. Like other australopithecines, *afarensis* was much smaller than modern humans and probably also matured faster and died younger, much as apes do today. Males were considerably larger than females, suggesting that they competed for female attention and did not associate with them for long periods.

THE FIRST FAMILY
Donald Johanson (left) displays the "first family", the remains of around 13 individuals who are thought to have died at the same time, perhaps killed by a flash flood. This discovery was made in 1975, very close to where Lucy was found, and consists of more than 200 bones of a mixed social group, including both sexes and a wide age range.

LUCY
The fossil find known as Lucy was named after the Beatles' song "Lucy in the Sky with Diamonds", which was playing while the archeologists celebrated their discovery. Until the discovery of the Dikika baby (see opposite), this was the most complete prehistoric hominin skeleton known, with almost 40 per cent of her bones recovered. Identified as female from the shape of her pelvis, Lucy is the best-known fossil of the species *Australopithecus afarensis*.

RECONSTRUCTED SKULL
The skull shown here has been built using the existing fragments of Lucy's skull as a guide. These are identified in the lighter colour.

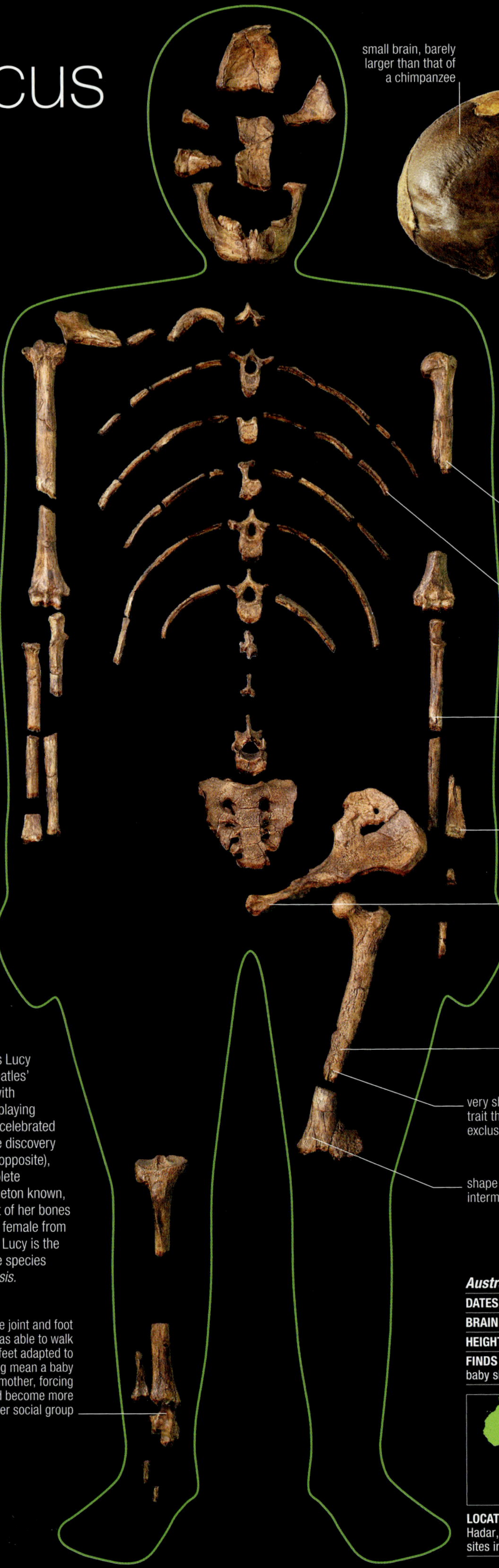

Australopithecus afarensis

DATES 4.1–2mya

BRAIN SIZE 380–485 cubic centimetres

HEIGHT 1.05–1.51m (3ft 5in–5ft)

FINDS Partially complete adult skeleton; almost complete baby skeleton; complete knee joint; many other fragments

LOCATION
Hadar, Ethiopia, and other sites in Ethiopia and Kenya

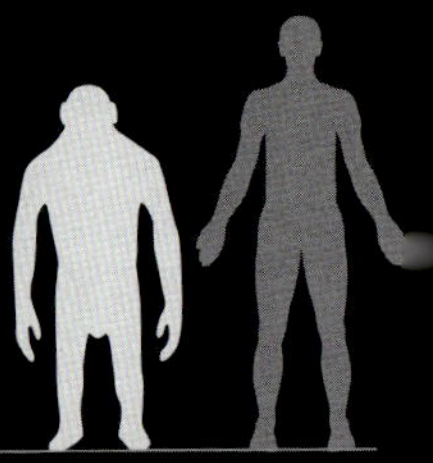

Walking tall

Archeologists disagree over the extent to which *afarensis* walked on two legs like modern humans, or climbed in trees like primates. Certainly, many features of *afarensis*'s skeleton suggest that they were moving towards a fully bipedal lifestyle. Their skulls were balanced squarely over their spines and their pelvis above their hips, while the structure of their knees allowed them to balance better while moving around on two legs. Marks are left on bones where ligaments attach, and these are in similar positions to those on modern human skeletons. Computer simulations of the way *afarensis*'s bones and muscles worked together suggest that they were efficient bipeds, while analysis of their preserved tracks suggests that they walked at a similar speed to modern humans. Their ankle joints and feet are also more similar to modern humans' than those of primates. However, as in primates, *afarensis*'s fingers are long and curved and their wrist bones are more like an ape's than a human's. The Dikika baby's shoulder blades are also more like a gorilla's than a human's, allowing a greater range of movement, which suggests that *afarensis* may have been partially reliant on climbing.

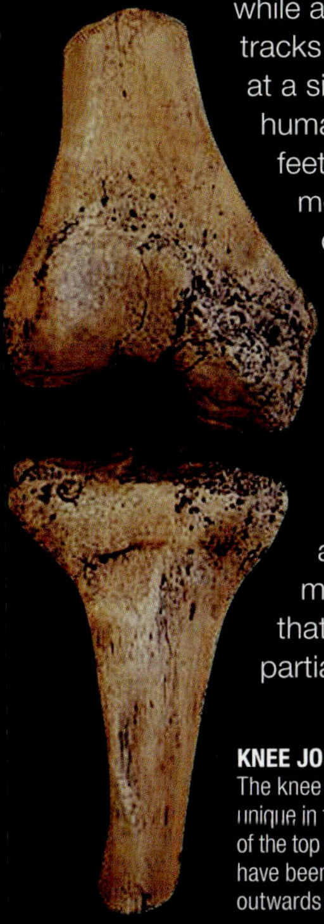

KNEE JOINT
The knee joint discovered by Donald Johanson is unique in that both parts have been found. The angle of the top piece demonstrates that *afarensis* would have been able to walk upright, as it angles outwards to the hips, as in modern humans.

KEY FINDS
DIKIKA BABY

The Dikika baby was found in 2000 by paleoanthropologist Zeresenay Alemseged in Ethiopia's Rift Valley, having been buried 3.3 million years ago by a rapid flood. The find is named Selam, meaning "peace" in Ethiopian. It is an important discovery because Selam is the most complete hominin fossil from this period, with a very well preserved head and face, including milk teeth. The skeleton was extracted very slowly from the sandstone that preserved it, using a dentist's drill to avoid damage. Even the fragile hyoid bone, which helps to enable speech, survived, giving us important information about *afarensis*'s throat – Selam's hyoid is more similar to a primate's than a human's, suggesting speech was not yet possible. Because Selam was only around three years old, the find has a great deal to tell us about how *afarensis* grew. Its brain seems to have been growing at the same rate as a chimp's, but because an adult *afarensis* brain is a little larger, it would have taken slightly longer to mature, implying a longer childhood spent around elder members of the group.

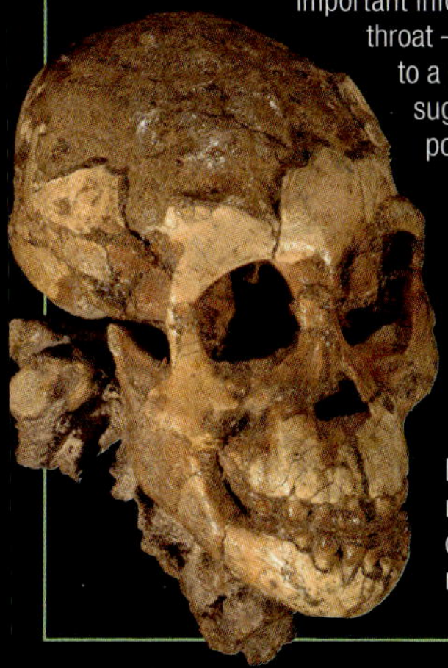

STEPS BACK IN TIME
These tracks were found in Laetoli, Tanzania, preserved in hardened volcanic ash that has been dated to an eruption 3.6 million years ago. There are three sets of footprints – two differently sized adults and a third, smaller individual following, stepping inside the largest set of footprints.

Paranthropus aethiopicus

Paranthropus aethiopicus **lived at the same time as some of the early species of the genus *Homo*, but seems to have been a side-branch of the family tree that has left no descendants. The word "paranthropus" means "alongside humans".**

Of the few *Paranthropus aethiopicus* fossils that exist, the most significant is the "Black Skull", which turned a very dark colour due to the high levels of manganese in the soil in which it was found. Like the other paranthropines, *aethiopicus* was probably descended from *Australopithecus afarensis* (see pp.450–51), but its relationship to the other species is unclear. It is sometimes thought that *aethiopicus* may have been an ancestor of *boisei* and *robustus*, but their visual similarities may be due to eating similar foods rather than being closely related. Although the paranthropines had small brains and forward-projecting lower faces in common with australopithecines, they also had huge jaws adapted for chewing and grinding tough seeds and vegetables. These features do not appear in members of the genus *Homo* or in modern humans, and anthropologists think that the paranthropines represented an unsuccessful path of evolution parallel to that followed by our direct ancestors.

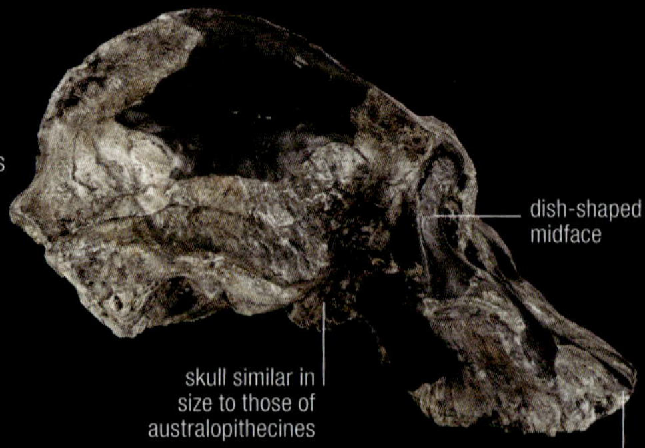

DISH FACE
Aethiopicus's face has a characteristically concave profile, or "dish" shape, when seen from the side.

MADE TO CHEW
Aethiopicus had large back teeth for grinding tough foods. These required powerful chewing muscles, which passed under the wide cheek bones and then attached to the bony crest on the top of the skull.

Paranthropus aethiopicus

DATES 2.5mya

BRAIN SIZE 410 cubic centimetres

HEIGHT Unknown

FINDS Two skulls, sections of jaw, and numerous teeth fragments

LOCATION Lake Turkana, Kenya; Omo, Ethiopia

Paranthropus robustus

Paranthropus robustus **was the first paranthropine discovered – by schoolboy Gert Terblanche in 1938. More than 130 individuals of this species are now known.**

Like other paranthropines, *Paranthropus robustus* had a massive jaw and teeth compared with the australopithecines and the early species of *Homo*. *Robustus*'s fossils are very similar to those of *boisei*, which lived at the same time but in East Africa. It is possible that further finds may demonstrate that the two were actually geographical variants of the same species. Although *robustus* does not seem to have made stone tools, some of *robustus*'s fossils were found alongside animal bones bearing the type of marks seen on the sticks that modern hunter-gatherers use to break into termite mounds.

OVERSIZED TEETH
Paranthropus robustus's teeth were particularly large, and huge jaw muscles were attached to its wide cheekbones and the bony ridge on top of its skull. However, although its teeth and jaws were larger than those of modern gorillas, *robustus* was still quite small-bodied.

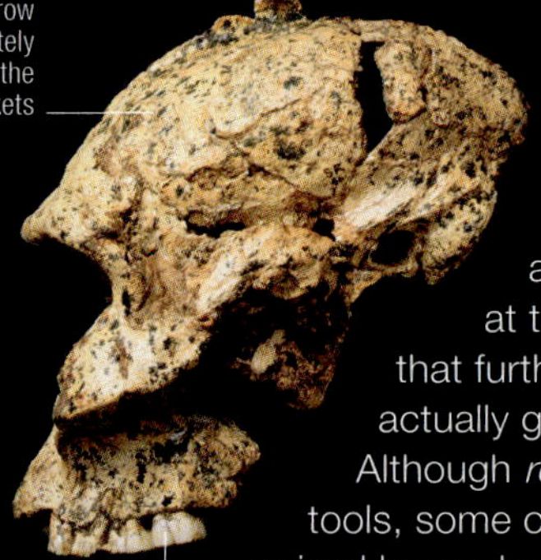

FLAT FACE
Paranthropus robustus has a flat, wide face. Its brain is only slightly larger than that of a modern chimpanzee.

Paranthropus robustus

DATES 2–1mya

BRAIN SIZE 530 cubic centimetres

HEIGHT 1.1–1.3m (3ft 6in–4ft 2in)

FINDS Various skulls and fragments of jaws and teeth; various skeleton fragments

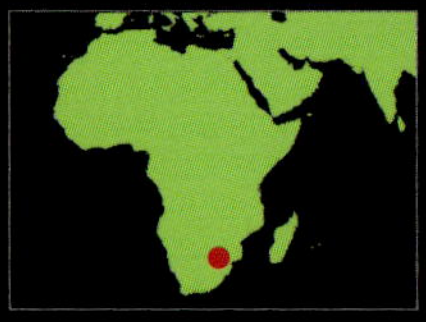

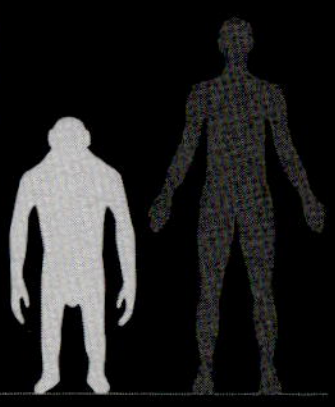

LOCATION
Various sites in South Africa

SCIENCE

DISPUTED DIET

The type of food that a species eats is usually reflected by the size and shape of its teeth. The paranthropines' huge back teeth suggest that they mainly ate tough foods such as seeds and raw vegetables. However, new research methods have led researchers to believe that paranthropines ate a much broader range of foods – including some protein-rich meat and insects – and reverted to tougher seeds and vegetables only when times were hard. These methods include studying patterns of wear on paranthropines' teeth – different foods leave different patterns – and the analysis of "isotopic signatures" present in fossil material (see p.466). These signatures are produced by traces of certain elements present in different types of food.

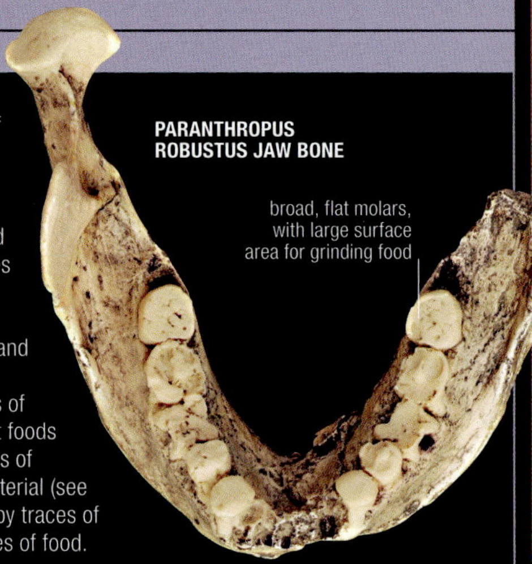

PARANTHROPUS ROBUSTUS JAW BONE

Paranthropus boisei

***Paranthropus boisei*, nicknamed "nutcracker man", was the largest and most specialized of the paranthropines, with the most powerful chewing apparatus, largest cheek teeth, and thickest tooth enamel of any hominin.**

Boisei's specialized chewing adaptations and large teeth originally led its finders, Louis and Mary Leakey, to assign the fossils to a new genus, *Zinjanthropus*. They named the species *boisei* after Charles Boise, who was funding their research. However, other anthropologists pointed out similarities with the fossils assigned to the genus *Paranthropus* by Raymond Dart (see p.449), and after much debate, most experts now agree that *aethiopicus*, *robustus*, and *boisei* should be linked together in a separate genus. All the fossils we have of *boisei* are skull fragments or teeth, but given the fossils' similarities to those of *robustus*, anthropologists have assumed that *boisei* was probably also rather ape-like in overall body shape, despite its larger brain. At some sites, fossils of *boisei* are found near stone tools, but anthropologists are not convinced that the paranthropines were toolmakers, as in most cases the fossils of early members of the genus *Homo* were recovered from the same sites.

BIOGRAPHY

LOUIS AND MARY LEAKEY

Louis was a Kenyan archeologist and naturalist whose work helped establish that the earliest human ancestors lived in Africa. In 1936, he married Mary Nichol, a British archeologist who became a noted paleoanthropologist herself, first working alongside Louis at Olduvai. It was Mary who discovered *boisei* at Olduvai, and who developed a system for classifying the stone tools found there, while Louis promoted research on African primates and established the Leakey Foundation.

FIRST FIND
A plaque marks a spot in Olduvai Gorge, Tanzania, where Mary Leakey found the first fossil of *boisei*.

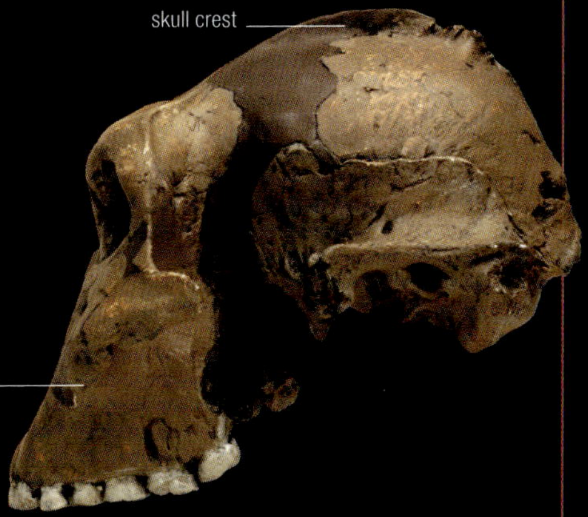

PROJECTING JAW
Boisei's face was deeper and flatter than that of *aethiopicus*, but its lower face and powerful upper jaw also projected forwards.

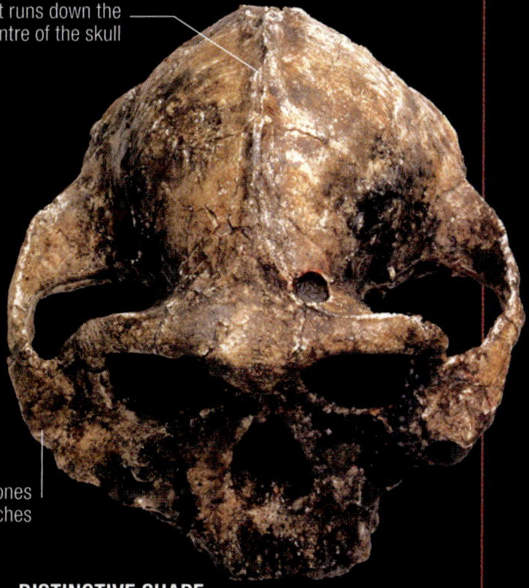

DISTINCTIVE SHAPE
Seen from above, *boisei*'s huge, flared cheek bones and narrow brain case (especially just behind the eye sockets) are very apparent.

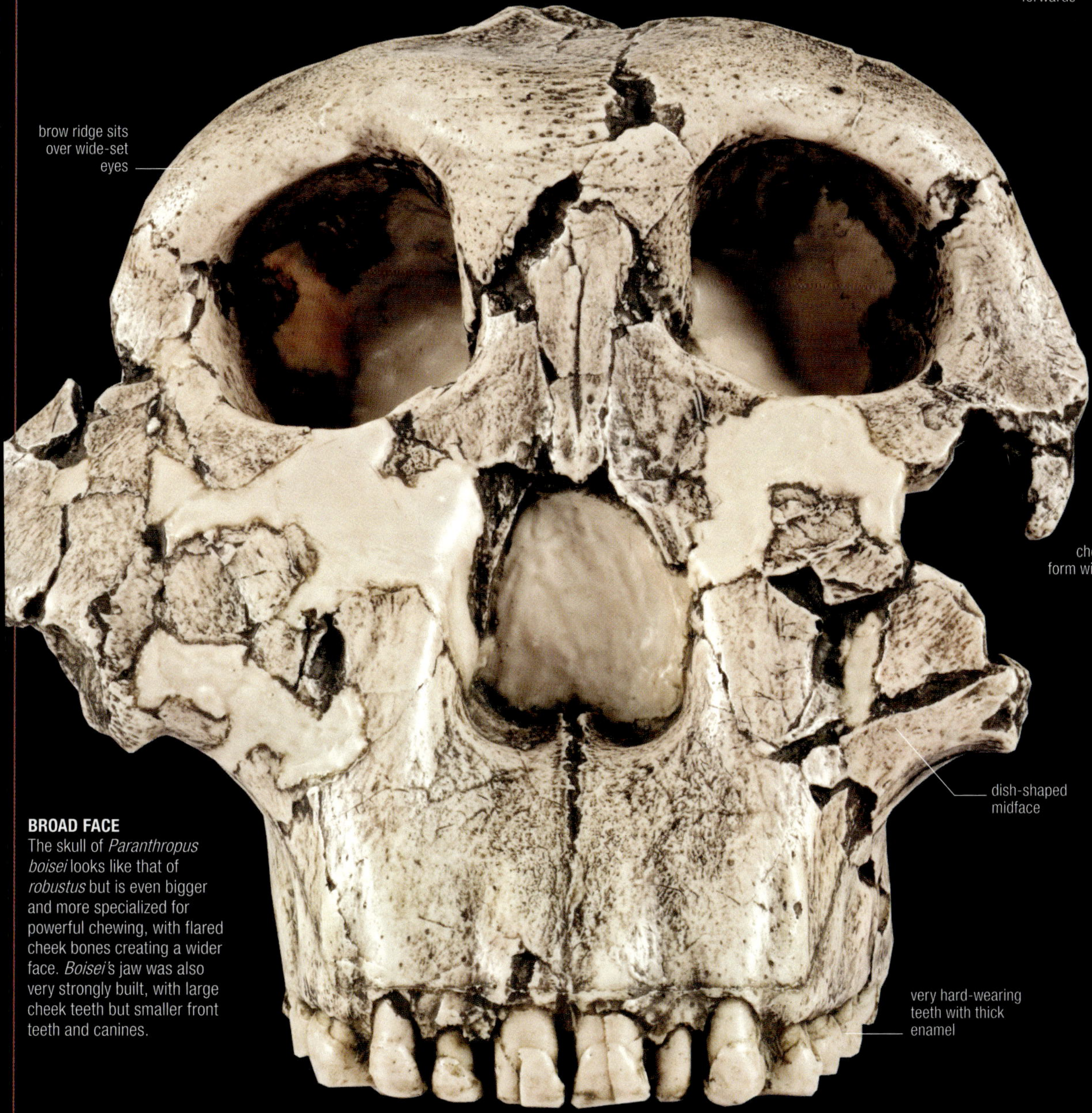

BROAD FACE
The skull of *Paranthropus boisei* looks like that of *robustus* but is even bigger and more specialized for powerful chewing, with flared cheek bones creating a wider face. *Boisei*'s jaw was also very strongly built, with large cheek teeth but smaller front teeth and canines.

Paranthropus boisei

DATES 2.5–1.2mya

BRAIN SIZE 410–550 cubic centimetres

HEIGHT 1.2–1.4m (4ft–4ft 5in)

FINDS Various skulls and fragments of jaws and teeth

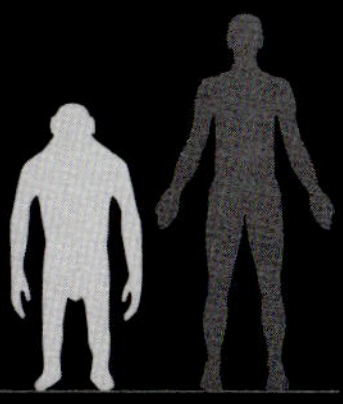

LOCATION
Various Rift Valley sites in Kenya and Ethiopia

Homo habilis

***Homo habilis* was the first member of the genus *Homo*, to which modern humans belong. The species was named *Homo habilis*, meaning "handyman", because some of its fossils are associated with early stone tools. *Homo habilis* is thought to be a direct ancestor of all later hominins.**

A new line

Homo habilis has a larger brain and smaller cheek teeth than the earlier australopithecines (see pp.448–51). The fact that its fossils have been found with early stone tools suggests that *habilis* had developed innovative tool-using behaviours and was consuming a diet that was less centred on tough seeds and vegetables – a characteristic of all later hominins, including ourselves. However, as more discoveries have been made, the fossils assigned to *habilis* have begun to fall into two groups: one with larger brains and teeth, now known as *Homo rudolfensis* (see opposite), and one that remained quite similar to its australopithecine forebears. As a result, some experts now argue that *Homo habilis* should be called *Australopithecus habilis*. Either way, *habilis* is still believed to be directly ancestral to later species belonging to the *Homo* genus.

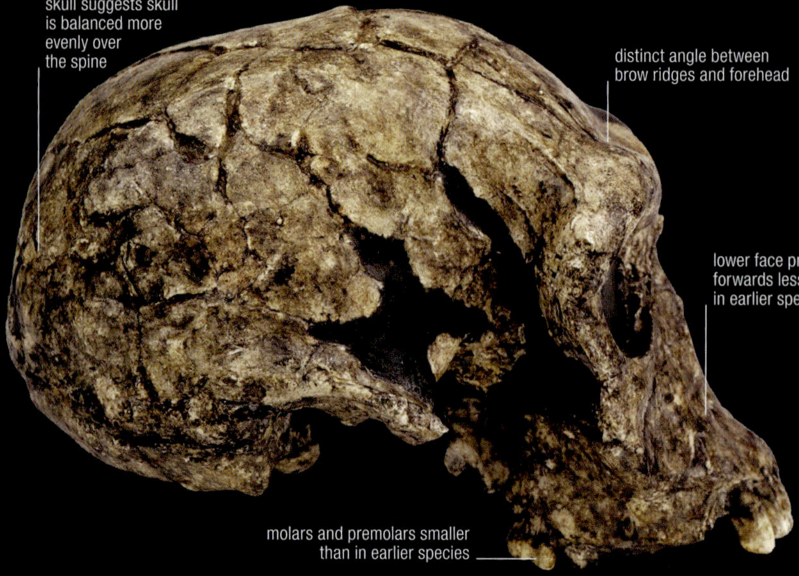

LARGER BRAINS
The use of stone tools allowed *Homo habilis* to access new foods such as tough buried tubers, nuts, meat, and bone marrow, which are richer in energy and protein than the grasses and seeds eaten by earlier species. This allowed *habilis* and its descendants to develop larger brains, while the massive cheek teeth so important for grinding tough vegetable foods became less essential.

FAMILIAR FACE
Although *habilis* was similar to its australopithecine ancestors in its body size and shape, its face looked slightly more human, as its smaller cheek teeth and less massive chewing muscles mean that its jaw did not project so far forwards from its face.

Homo habilis

DATE 2.2–1.6mya

BRAIN SIZE 500–650 cubic centimetres

HEIGHT 1–1.3m (3ft 3in–4ft 2in)

FINDS Various skulls and fragments of cranium; fragments of lower body, including feet and hands

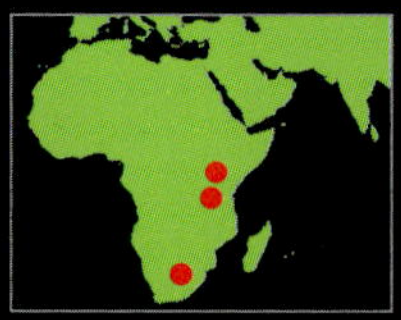

LOCATION Olduvai Gorge, Tanzania; East Turkana, Kenya; Sterkfontein, South Africa

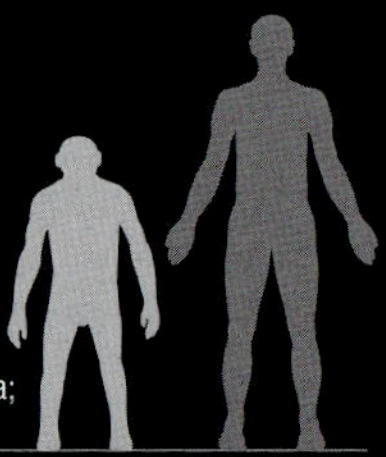

The original "handyman"

Homo habilis may have been the first hominin to make and use stone tools. This distinction is now contested, because even earlier tools have been found associated with *Australopithecus garhi* (see p.448). The tools associated with *Homo habilis* are known as "Oldowan" tools. Although simple, they required skill to make, suggesting that *habilis* had a good understanding of how rocks fracture. Tools were not designed for specific purposes – the aim was simply to produce a sharp edge. *Habilis* was particular about which stone it worked – there is evidence that it carried stones several kilometres from their sources. Tools were probably used for scavenging rather than hunting, with sharp flakes to cut meat from carcasses, and hammer stones to break open nuts or shatter bones to get at the marrow. Skulls of *Homo habilis* have been found with toothmarks in them, suggesting it was prey, not predator.

EXCAVATION
These paleontologists are working in the Olduvai Gorge, in Tanzania, Africa, a site which gives its name to Oldowan tools. All sites must be excavated slowly and carefully, with any finds plotted in three dimensions.

OLDOWAN TOOLS
The earliest tools are mainly made from cobbles of rough stone such as basalt, quartz, and quartzite, struck with a "hammerstone" to produce cores and sharp flakes for cutting.

FLAKE **FLAKED COBBLE** **CHOPPER**

SCIENCE
LEARNED BEHAVIOUR

The techniques required to make stone tools are not determined by our genes, but are learned from others. Humans are not the only species who can learn, or even the only species who can learn to use and make tools. Several other primate species also use tools, including chimpanzees. The one pictured here is using a stone as a hammer to open a nut. Some species pass methods of manufacture on through the generations, producing distinctive "tool kits". However, most primate tools are simply unmodified stones and sticks, and even the most accomplished primate stone toolmaker yet known, the chimpanzee Kanzi, has never learnt to produce any tools as complex as those of even the very earliest Oldowan stone tools, despite every effort to teach him.

Homo rudolfensis

***Homo rudolfensis* is a controversial species known essentially from a single fossil skull. As yet, it is not certain where *rudolfensis* sits in the human family tree, but it is not thought to be a direct ancestor of any later *Homo* species.**

The *Homo rudolfensis* skull was found in 1972 by fossil-hunter Bernard Ngeneo, at Koobi Fora near Lake Turkana (then called Lake Rudolf) in Kenya. We know very little about *rudolfensis*, particularly about its body, due to the lack of fossil material. Even the skull was damaged in the process of fossilization, and experts disagree about how it should be reconstructed. From the little evidence we have, *rudolfensis* seems to have been very similar to *habilis*. Its brain was relatively large compared to that of the australopithecines, perhaps larger even than that of *habilis*. Otherwise it had much in common with australopithecines, and like *habilis* may not belong in the genus *Homo*, particularly because its jaws and teeth are rather large. It is difficult to establish this species's relationships with other hominins — it was clearly closely related to *habilis*, but could have been either its ancestor or a fellow descendant of the australopithecines.

ONE OF A KIND
This skull is one of very few fossils that can be assigned to *rudolfensis* with any certainty. Experts are still debating whether a number of leg bones found nearby also belong to the same species, together with a few other isolated skull fragments, teeth, and jaw fossils from elsewhere in Kenya as well as Ethiopia and Malawi.

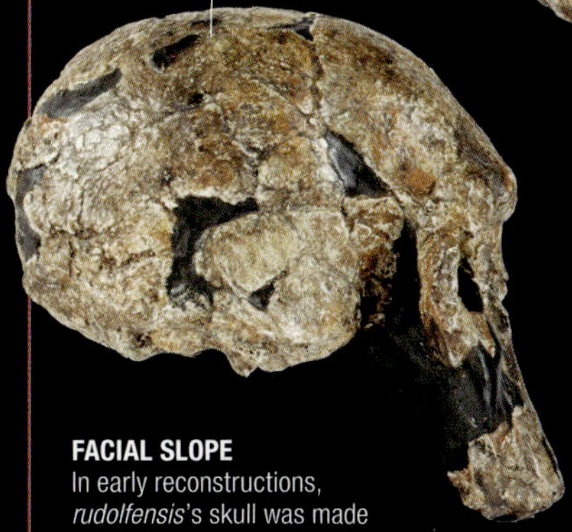

FACIAL SLOPE
In early reconstructions, *rudolfensis*'s skull was made relatively large-brained, with a steeply sloping face like a modern human. More recent versions have a smaller brain and a more projecting, ape-like face.

Homo rudolfensis

DATE 2.4–1mya
BRAIN SIZE 600–800 cubic centimetres
HEIGHT 1.5–1.6m (5ft–5ft 2in)
FINDS Single skull

LOCATION
Lake Turkana, Kenya, East Africa.

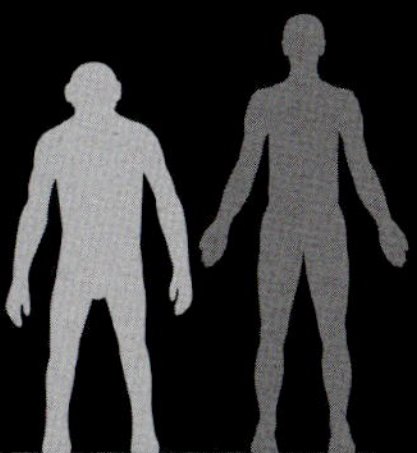

Homo ergaster

***Homo ergaster*, meaning "workman", takes its name from its association with a stone tool culture that was more complex than any that had come before (for early tools see p.455). Fossils of the species, such as the famous "Turkana Boy", look very different from those of their ape-like ancestors, being as tall as modern humans and with a similar build. *Homo ergaster* is probably the ancestor of all later species of *Homo*, including our own species, *Homo sapiens*.**

Turkana Boy

In 1984, Kamoya Kimeu, one of a team of archeologists led by Richard Leakey working at Lake Turkana, Kenya, found a virtually complete *Homo ergaster* skeleton. It is now known as "Turkana Boy" or "Nariokotome Boy". The skeleton is that of a male – distinguished by the shape of his pelvis – of between 9 and 13 years of age, dating from 1.5 million years ago. Turkana Boy is the most complete hominin fossil known from this time, and provides valuable information about the size, form, and proportions of the body of *Homo ergaster*. Some experts believe that Turkana Boy is actually a *Homo erectus* fossil, as the two species are closely related (see pp.460–61).

RICH FOSSIL SITE
The research centre at Koobi Fora, shown here, is on the opposite side of Lake Turkana from where Turkana Boy was found. Koobi Fora is a rich fossil site, and is now part of Sibiloi National Park, a UNESCO world heritage site.

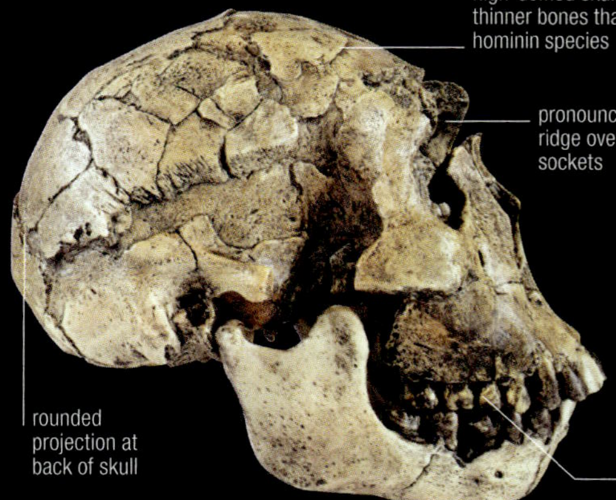

PROJECTING NOSE
Homo ergaster's skull displays features that are characteristic of the later hominin species. Its cranium is distinctly high and rounded and projects backwards, and its short, flat face has a forward-facing nasal aperture distinctive of modern humans.

SCIENCE

TURKANA BOY'S AGE

The age of Turkana Boy is a matter of considerable debate. Judging by the general growth of his bones, he would appear to have been as old as 13 years of age. However, when the limited development of his teeth is taken into consideration, he may well have been several years younger. The confusion might suggest that *Homo ergaster* matured more quickly than modern humans, but not as quickly as primates. Perhaps Turkana Boy was so tall, despite being so young, because his species had not yet developed the characteristic modern human growth pattern of slow early childhood growth followed by a growth spurt at adolescence.

REMARKABLE FOSSIL
Although few *Homo ergaster* fossils have been discovered, Turkana Boy is so complete that we can deduce a great deal about what his species looked like and how it lived. *Homo ergaster* was the first hominin to achieve the stature of modern humans. This individual was remarkably tall – although he was young, he stood about 1.6m (5ft 2in) tall and might have grown as tall as 1.85m (6ft 1in).

Homo ergaster

DATES 1.8–0.6mya

BRAIN SIZE 600–910 cubic centimetres

HEIGHT Up to 1.85m (6ft 1in)

FINDS One nearly complete skeleton; also a few fragments of skull, jaw, pelvis, and limb bones

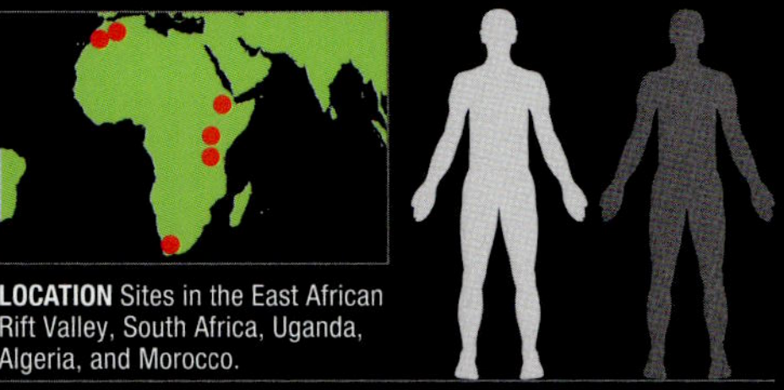

LOCATION Sites in the East African Rift Valley, South Africa, Uganda, Algeria, and Morocco.

BIOGRAPHY

RICHARD LEAKEY

Richard Leakey is the son of renowned anthropologists Louis and Mary Leakey (see p.461), and Professor of Anthropology at Stony Brook University, New York. Like his parents, he has made many significant hominin discoveries, digging in Omo, Ethiopia, as well as in his native Kenya. His finds include *Australopithecus boisei* fossils, important *Homo habilis* and *Homo erectus* skulls, some of the earliest *Homo sapiens* fossils, and Turkana Boy. He has been the Administrative Director of the National Museum of Kenya as well as the Head of the Wildlife Conservation and Management Department, appointed with the remit of reducing the poaching of elephants. In 2004, he founded WildlifeDirect, a Kenya-based charity providing support to conservationists in Africa. He has been very active in Kenyan politics throughout his life.

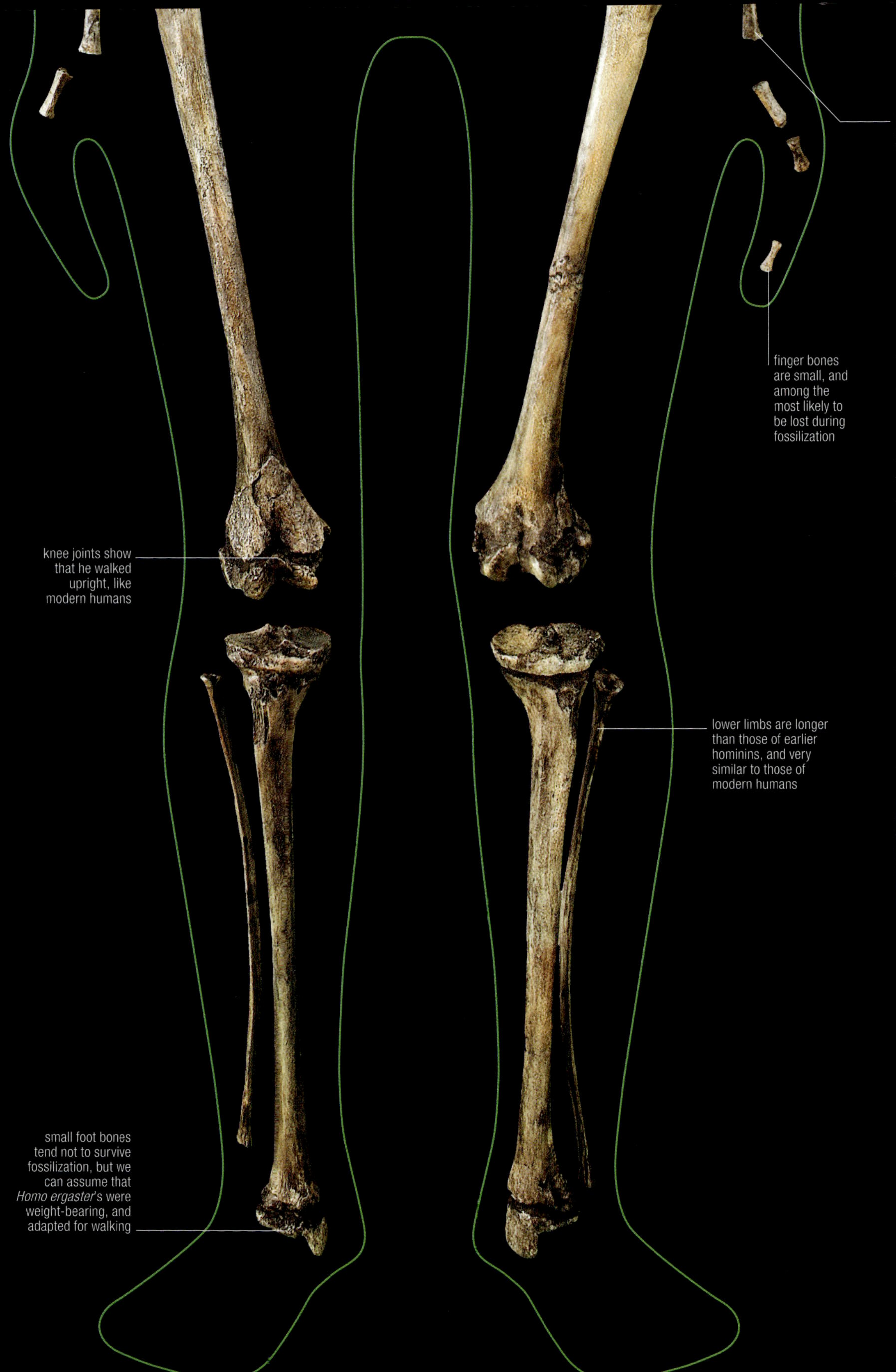

Slender stature

Homo ergaster's tall, slender body frame seems to have been an adaptation to the hot, dry savanna environments in which it lived. Tall bodies and long limbs maximize the surface area of the body relative to its volume, making sweating an effective way of keeping cool. As sweating is much more efficient when skin is bare, *Homo ergaster* may also have been the first hominin without much body hair. It also seems to have shown less sexual dimorphism than in earlier hominins, meaning males were only slightly bigger than females. This suggests that males did not dominate females in order to reproduce, and that couples may have co-operated in finding food and caring for their children.

ADAPTED TO HEAT
Modern humans living in savanna environments, such as the Masai of Kenya, tend to have the same tall and slender body proportions as *Homo ergaster*. This helps their bodies deal better with the hot and dry conditions.

First Human Migration

The very earliest hominin species were found only in Africa – the continent on which they evolved. It is only less than 2 million years ago, after the evolution of the genus *Homo*, that we start to find evidence of the movement of hominins out of Africa and into Europe and Asia.

Continental shift

Early hominins lived on the African savannas and were well-adapted to their environment. To move further north into Europe and Asia, they had to adjust to new environments, and very different types of animals and plants. In particular, northerly latitudes meant a more pronounced contrast between seasons, forcing adaptation to the cold (see panel, below). Archaeologists have had some difficulty in determining when hominins first left Africa. The arrows on this map show probable migration routes and times. The sites marked are those with fossil evidence of the hominins themselves, but there are many sites at which only stone tools have been discovered. It seems that groups of early hominins lived in parts of Europe and Asia in low population densities, that left little trace in the archaeological record, from around 1.8 million years ago. After around 600,000 years ago, hominins had become better adapted to living in these areas, with the effect that their populations expanded and the fossil evidence grew.

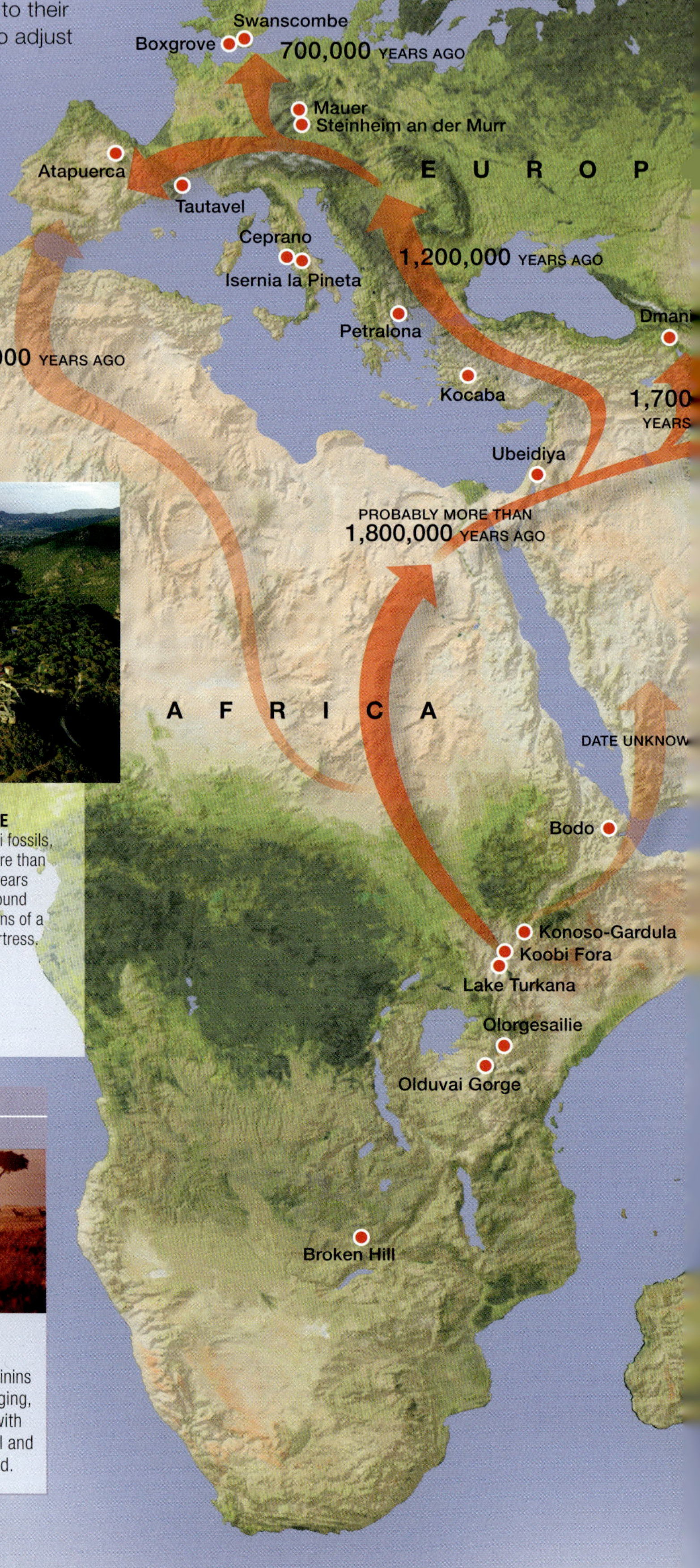

The earliest Europeans

The question of when hominins first arrived in Europe is complicated by much of the evidence coming from river sites, making it hard to tell whether stone objects were created by hominins or shaped by water. Only recently have securely dated finds at sites such as Dmanisi in Georgia and Sima del Elefante in Spain (see p.462) proved that hominins were present in Europe more than 1 million years ago. While many of these new fossils have been assigned to particular species, most of them resemble *Homo ergaster* (see pp.456–57), and probably represent geographical variants of that species.

GEORGIAN LANDSCAPE
The Dmanisi fossils, dated to more than 1.7 million years ago, were found amid the ruins of a medieval fortress.

DMANISI SKULL
The hominin fossils found in Dmanisi, Georgia, have been assigned to the species *Homo georgicus*. This might have been the first hominin species to settle in Europe.

SCIENCE

HOW HOMININS HEADED NORTH

Just before hominins first appeared outside Africa, Earth's climate cooled, and while more cold-adapted animals, such as mammoths, thrived, many large carnivores became extinct. It is thought that hominins were able to take advantage of the reduced competition for scavenging, and, with their large brains, they may have been equipped to deal with new challenges. In particular, it is thought they became more social and so better able to share information and work together to source food.

The Far East

The earliest Chinese hominin finds were from Longghushan, or "Dragon Bone Hill", near Zhoukoudian, southwest of Beijing. Between 1921 and 1937 the remains of more than 40 individuals were excavated. Originally named *Sinanthropus pekinensis*, or "Peking Man", the fossils were only later recognized as *Homo erectus* (see pp.460–61). These fossils could date to as early as 800,000 years ago, and stone tools from several sites west of Beijing are dated to 1.6 million years ago. However, these very early dates and the relatively small number of specimens mean that some archaeologists feel they need more reliably dated discoveries before they can say for certain when hominins arrived in the Far East.

MISSING SPECIMENS
The fossils from Zhoukoudian were lost at the start of the Second World War, while being sent to safety. Luckily, detailed casts and descriptions of the specimens survive.

RECENT DISCOVERIES
Archaeologists are continuing to make discoveries at the site in Zhoukoudian. In 2003, the remains of humans dating to 23,000 years ago were found.

SCIENCE

THE MOVIUS LINE

In 1948 archaeologist Hallam Movius mapped the distribution of early hominin sites with and without Acheulean handaxe technology (see p.461), and drew a line separating the two areas. A clear division was evident: across Africa and much of Europe, hominins had the Acheulean hand axe, but it was absent from large areas of Asia (although other types of stone tools were present). There are various explanations for the lack of hand axes in Asia. The first hominins to leave Africa may have carried Oldowan-style tools (see p.455), which occur at early sites in Europe. Alternatively, areas without hand axes may not have had the quality of stone needed, and groups might have lost their skills over time, or used alternate materials such as bamboo, which have not been preserved.

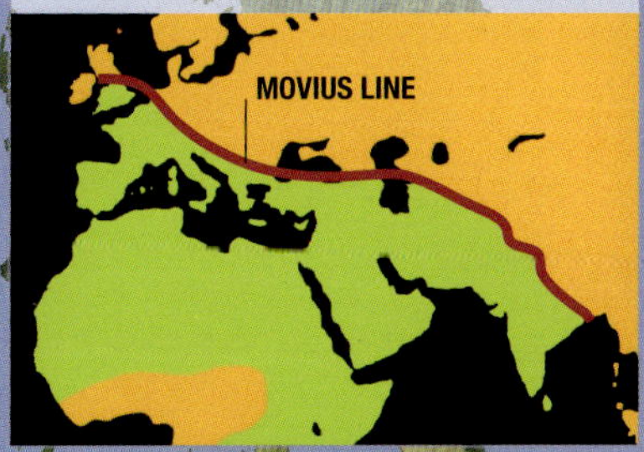

Java Man

A skull-cap from Trinil in Java was one of the first hominin fossils ever found. Other Javan discoveries include the skull of a child from Mojokerto, and several finds from the site of Sangiran (see right). All the fossils are of *Homo erectus*, but were not all excavated to modern standards, so dating them precisely is difficult. Java's geology is also complicated, making the dating of sediments problematic. As a result, some archaeologists do not believe the dates of 1.8 million years ago given to the fossils and think that hominins arrived in Indonesia much later.

SANGIRAN SITE
Archaeologists have unearthed many fossils of prehistoric hominins and of other animals from the site of Sangiran, Java, since its discovery in 1934.

SOLO RIVER VALLEY
The fossils from Sangiran and many other Indonesian Homo erectus finds come from along the Solo River of central Java.

EARLY SKULL
This skull from Sangiran is one of the most complete examples of *Homo erectus* known, and could date from as early as 1.6 million years ago.

Homo erectus

Although very different to modern humans, *Homo erectus* is much more "modern" looking than its ancestors. While the origins of this species are unclear – it probably evolved in Africa – it is especially well known from Asia, and may have survived there until about 50,000 years ago.

The first widespread species

Homo erectus shares a number of features with *Homo ergaster* (see pp.456–57) and the two species are clearly closely related, although anthropologists disagree to what extent. Some of the African fossils assigned to *ergaster* may belong to *erectus*, which would date *erectus*'s origins to much earlier. However, *erectus* seems to show more specialized features, such as a thicker skull and more massive jaws, that gradually disappear later in hominin evolution. Many anthropologists now believe that *Homo erectus* was a descendant of *Homo ergaster* that spread into Asia, or that evolved in Asia and later returned to Africa. Fossils from sites in China and Indonesia (see p.459) suggest that Asian populations remained isolated, without changing much until at least 300,000, and possibly less than 100,000, years ago.

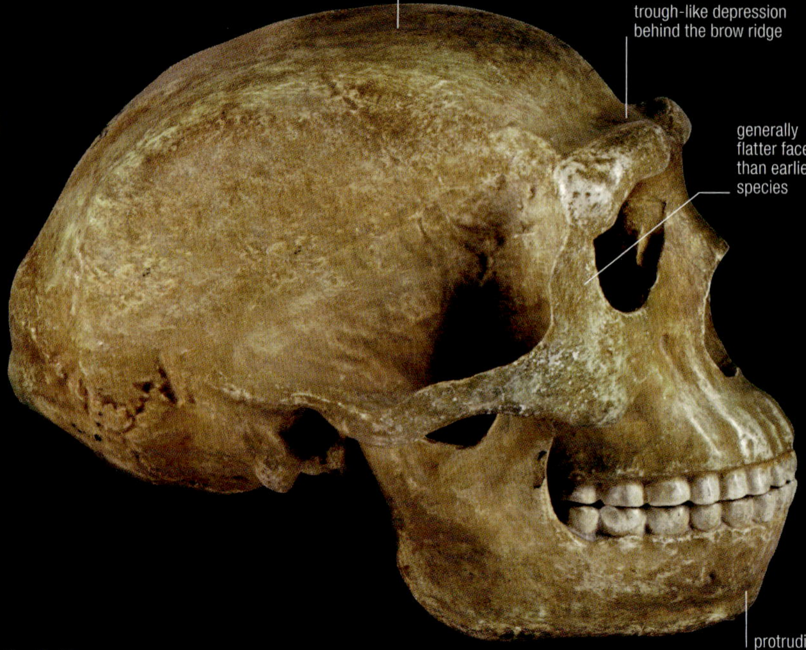

LARGER BRAIN
One of the most distinctive features of *Homo erectus* and *ergaster* is a much larger brain size than in earlier hominins, suggested by the increased size of their brain cases. They were the first species to live in environments outside their African savanna homelands, and some anthropologists have suggested that they may even have had some form of simple language.

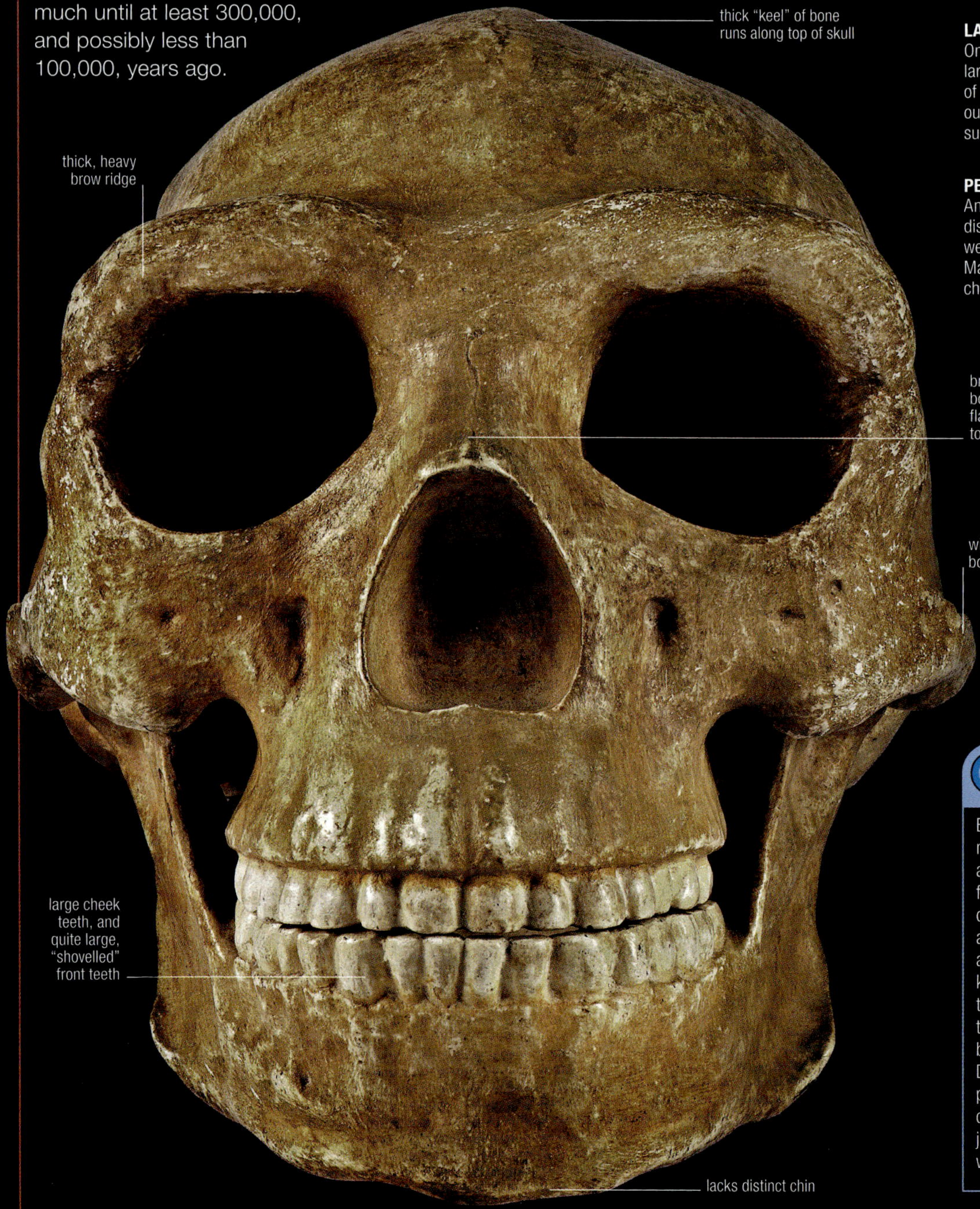

PEKING MAN
Anthropologists working on the African fossils disagree about how exactly to distinguish *Homo erectus* from *Homo ergaster*, which suggests that the two were very closely related. The Asian fossils, such as this skull of "Peking Man" from Zhoukoudian in China (see p.459), exhibit features that characterize *Homo erectus*.

Homo erectus

DATES Possibly 1.8mya–50,000ya

BRAIN SIZE 750–1300 cubic centimetres

HEIGHT 1.6–1.8m (5ft 3in–6ft)

FINDS Several relatively complete crania, as well as teeth and jawbones; a few limb bones

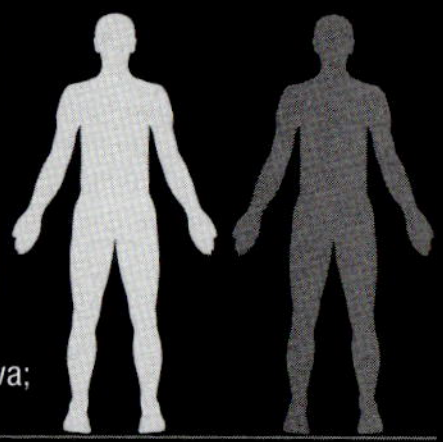

LOCATION
Various sites in China and Java; African sites are contested

BIOGRAPHY

EUGENE DUBOIS

Believing that Asia, rather than Africa, was man's ancestral homeland, in 1889, Dutch anatomist Eugene Dubois set out to search for fossils in what is now Indonesia. Within only two years he found a hominin skull-cap and thigh bone at Trinil, Central Java, and assigned them to *Pithecanthropus erectus*, later known as Java Man. Subsequently known as the "type specimens" for *Homo erectus*, these were the first hominin fossils to be found outside Africa and Europe. Dubois struggled to convince his peers of the significance of his discoveries, and they were only just beginning to be accepted when he died in 1940.

Brain development

One of the characteristics that really distinguishes modern humans from other primates is our remarkably large brain. Our bodies are also relatively large, but our brains are proportionately larger than would be expected for our size. Some parts of our brain have grown more than others – notably the neocortex, which deals with complex thought processes. Complex behaviour is required for manufacturing tools, foraging, and hunting, but increasing brain size seems to be most closely linked to our social behaviour, and particularly the size of the groups we live in. It would seem that what we really need a big brain for is maintaining group relationships. When we plot the size of hominins' brains over time, *Homo erectus* and later species seem to mark an acceleration of brain development, presumably as the size of the groups they associated in grew.

CHANGING CAPACITY
This graph shows the size of fossil hominins' brains increasing over time. *Homo erectus* shows one of the largest increases.

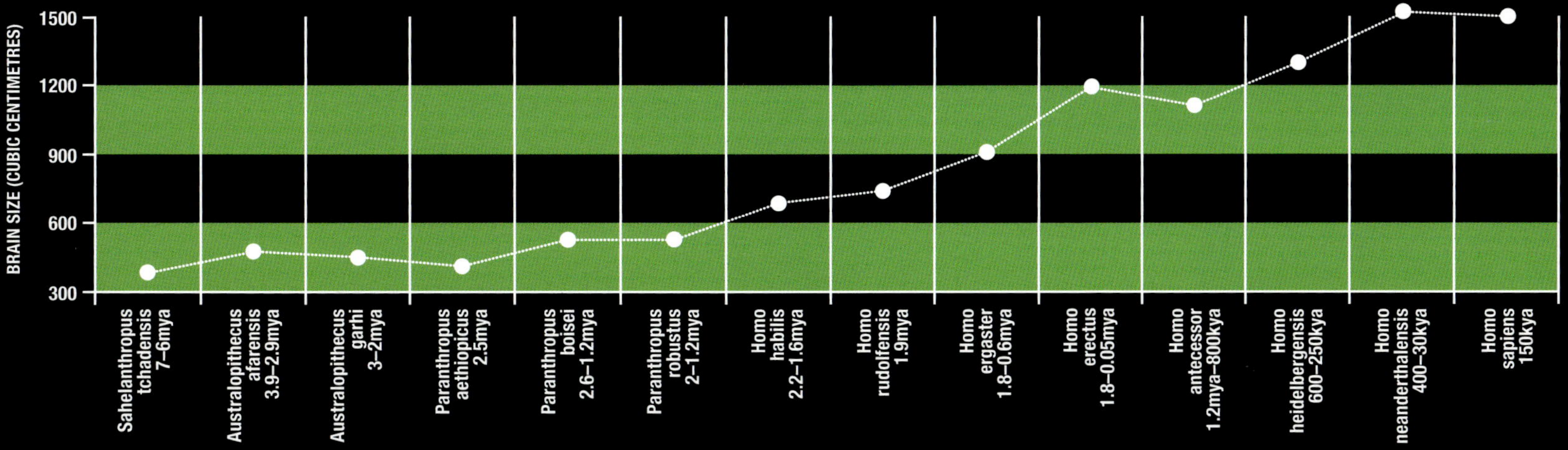

Advanced stone tools

The stone tool technology that developed from the Oldowan (see p.455) is known as "Acheulean". Its most distinctive tool type is the handaxe, which first appeared about 1.65 million years ago. Acheulean industry is first associated with early *Homo erectus* (or *ergaster*), though its use is continued by later species. Microscopic damage on the edges of some handaxes suggests that they were the "Swiss Army Knives" of stone tools, used for a huge variety of tasks including cutting wood, butchering animals, and scraping hides. However, the care taken in making them has led some archaeologists to think that handaxes were not just tools, but represented symbolic displays of strength and skill, or of personal or group identity. Their symmetry and the forethought involved in their creation has led some to argue that their appearance marks a significant step forward in hominin intelligence.

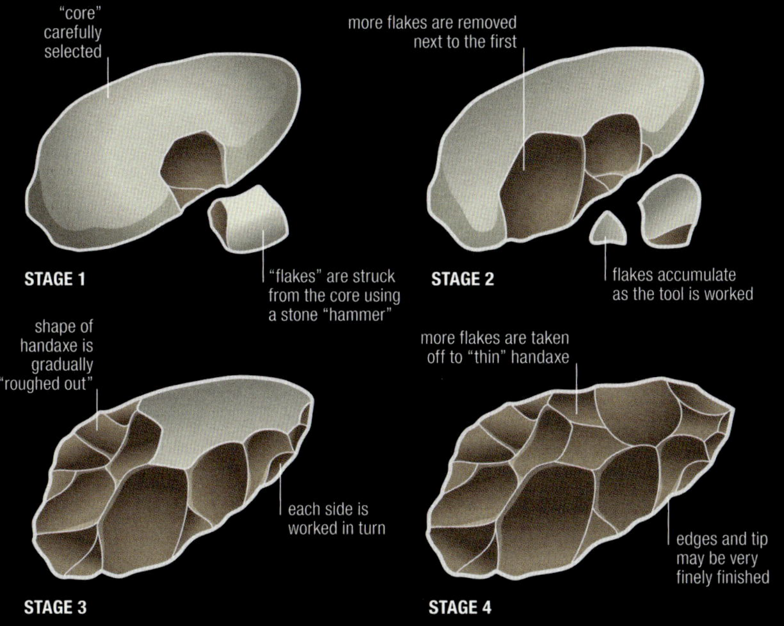

MAKING A HANDAXE
It takes much more skill to make an Acheulean handaxe than an Oldowan tool. A knapper would have to think several steps ahead when selecting a suitable piece of stone, and then prepare and place each strike precisely. Many handaxes were finished to a very high level of symmetry, sometimes using several different tools, including stone hammers and pieces of antler.

HOW WE KNOW
THE USE OF FIRE

At several African *Homo ergaster* or *erectus* sites, patches of baked earth, burned bones, ash, and charcoal have been argued to represent hominin use of fire. However, it has also been suggested that these are the result of naturally created wild fires. The earliest definite evidence for the use of fire by hominins occurs in the form of "hearths", in Chinese and European sites. Using fire may have been necessary in the hominin colonization of Eurasia, both to deal with the cold of winter and to cook meat – the most plentiful source of food further north – making it easier to digest.

TOOL DISTRIBUTION
Handaxes are found across large parts of Africa and Eurasia for well over a million years. They show only very small variations across the entire area and time period.

MASSIVE COLLECTION
At some Acheulean sites such as this one in Kenya, hundreds of handaxes lie scattered across the ground, many too large for practical purposes and apparently unused.

Homo antecessor

Homo antecessor is the second oldest fossil hominin that has been found in Europe. It is unknown whether _antecessor_ represents an early, small-scale, and unsuccessful colonization of Europe, or if it survived and evolved into later European species.

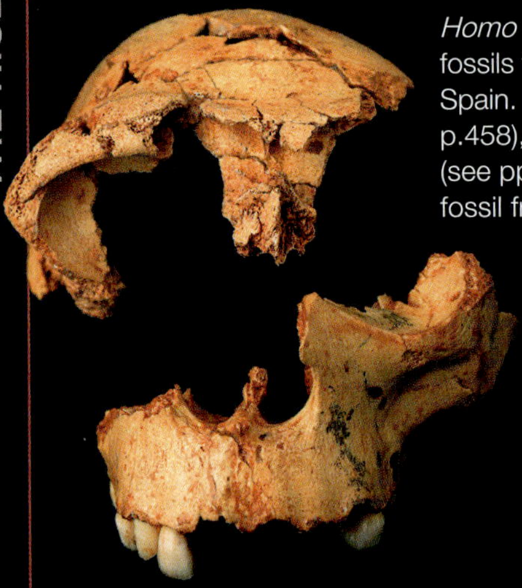

Homo antecessor is known from just a few fossils from two sites in Atapuerca, northern Spain. Like some of the Dmanisi fossils (see p.458), it is similar to African *Homo ergaster* (see pp.456–57). However, a similarly early fossil from Italy looks more like Asian *Homo erectus* (see pp.460–61). This diversity suggests that several groups of hominins may have colonized Europe before one eventually survived long-term.

SPARSE KNOWLEDGE
Most of the Homo *antecessor* fossils discovered by Juan Luis Arsuaga and his team at Atapuerca are fragments of the skull, jaw, and teeth, so little is known about the rest of their bodies, except that they were powerfully built. Their skulls were rather primitive, with low foreheads and brains only slightly larger than *Homo ergaster*.

Homo antecessor

DATES 1.2–0.8mya
BRAIN SIZE 1,000 cubic centimetres
HEIGHT 1.6–1.8m (5ft 3in–6ft)
FINDS Teeth and bone fragments from skull and body

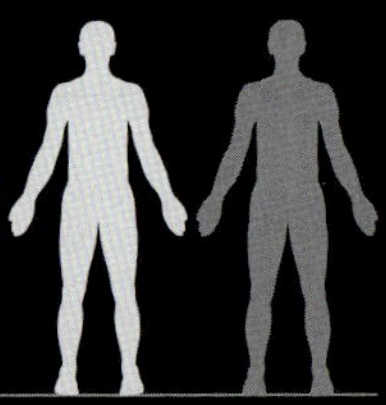

LOCATION
Atapuerca, northern Spain

GRAN DOLINA
The Gran Dolina site in Atapuerca is dated to around 800,000 years ago. About 80 fragments of fossil hominins have been found there, in a limestone hill riddled with caves. The fossils represent at least six individuals, all of whom were aged between 3 and 18 years old. With the fossils were found around 200 stone tools, as well as animal bones that bear the marks of having been butchered for meat.

KEY FIND
EVIDENCE OF CANNIBALISM

Many of the hominin bones found at Gran Dolina bear cut-marks that show that the muscle and flesh was stripped from the bone. Also, the way that the bones have fractured suggests that they were deliberately broken open to get to the marrow inside. While some modern human groups practise cannibalism for ritual purposes, the evidence from Atapuerca suggests that the individuals that lived here butchered and ate these bodies for food. This implies that it was difficult for these early pioneer populations to find other kinds of food. This may be why only a few hominin fossils are known from Europe at this time – these early pioneers still found it difficult to survive in their new and unfamiliar environments, which kept their numbers low.

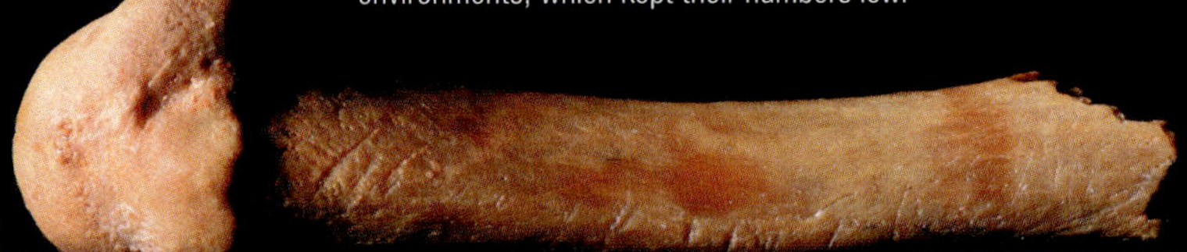

Homo heidelbergensis

This species was probably the last common ancestor of Neanderthals and modern humans. Fossils display a mixture of features inherited from earlier hominins, as well as more modern or "evolved" traits.

Changing taxonomy

The fossils from the African record of this time were initially assigned to a variety of species, including *Homo rhodesiensis*. However, as a group they are so similar to the European *heidelbergensis* that many anthropologists now consider them to be members of the same widespread species. Individuals were tall and well muscled, with thick, heavy limb bones, suggesting that they were engaging in tough physical activity. Marks on animal bones found with fossils tell us that these African hominins were capable scavengers, and possibly good hunters.

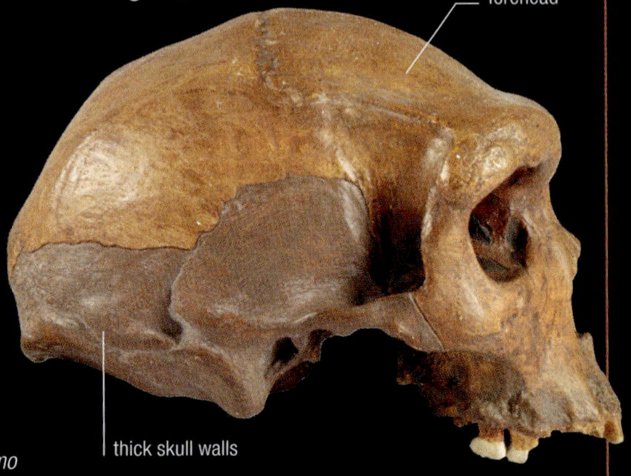

KEY CHARACTERISTICS
Heidelbergensis' large, projecting jaw, lack of a chin, thick skull bones, large brow ridges, and robust lower limbs were inherited from earlier hominins, and its immediate ancestor was probably *Homo ergaster*. Its more "modern" features include a larger brain, and a higher, broader skull.

EVOLVING FEATURES
This skull from Kabwé or Broken Hill, Zambia, was originally assigned to *Homo rhodesiensis*, but is now considered to be *heidelbergensis* by many anthropologists. Over time the African fossils become more like modern humans while European fossils become more like Neanderthals (see opposite).

Ancestors of Neanderthals

European fossils of *Homo heidelbergensis* can clearly be seen to evolve over time into Neanderthals, *Homo neanderthalensis* (see pp.464–67). Earlier specimens look similar to *ergaster* and *erectus*, with robust limb bones and large, heavy, protruding jaws. Early specimens with these features are found across Europe, and a mandible from Mauer, France, is the "type fossil" used to define *Homo heidelbergensis*. The relative isolation of the European population and the particular demands of its cold climates were what caused *heidelbergensis* to develop the distinctive anatomy of the Neanderthals, over hundreds of thousands of years. Later *heidelbergensis* specimens are much more Neanderthal-like, to the extent that some of these fossils display traits otherwise known only among the Neanderthals themselves.These include a depression in the back of the skull, a forward-projecting nasal area, and large front teeth compared to the back teeth. Three-dimensional scanning of the Atapuerca *heidelbergensis* specimens suggests that the inner ear was well-adapted for picking out the frequencies of human speech, which implies that it may have had some form of language.

SIMA DE LOS HUESOS
The "pit of bones" is a small cave in the Atapuerca complex, which can only be reached by clambering down a 13m (43ft) shaft. Archaeologists can only excavate for short periods, due to the need to conserve oxygen. Excavators believe that the high concentration of bones at Sima de los Huesos represents the deliberate disposal of bodies, but it is uncertain whether this was a ritual act or simply an easy method of disposal.

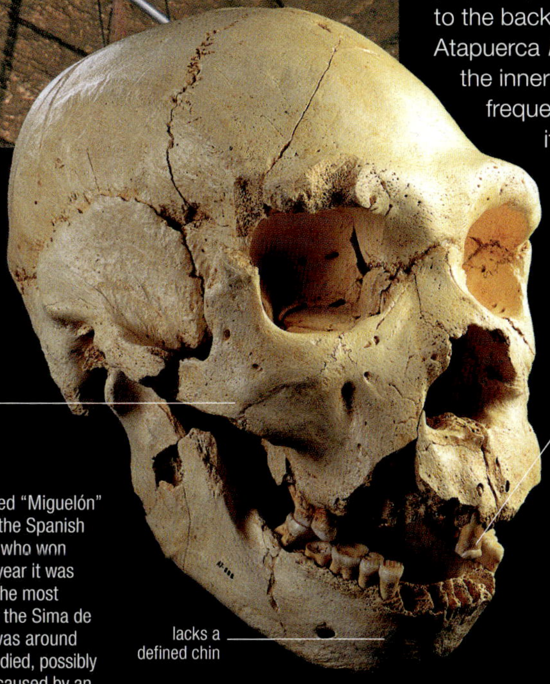

"MIGUELÓN"
This skull was nicknamed "Miguelón" by its excavators, after the Spanish cyclist Miguel Indurain, who won the Tour de France the year it was discovered. It is one of the most complete finds made at the Sima de los Huesos. Miguelón was around thirty years old when he died, possibly due to blood poisoning caused by an infection in a broken tooth. The skull is one of the most primitive-looking of the Atapuerca finds.

CLACTON SPEAR
This wooden spear-point, from Clacton-on-Sea, England, is 450,000 years old and was probably made by *heidelbergensis*. It is unusual because wooden artefacts survive only when preservation conditions are exceptionally good. Similar spears with fire-hardened tips are also known from Schöningen and Lehringen in Germany.

WIDE RANGE OF FOSSILS
The finds from Sima de los Huesos represent at least 32 individuals, with 5,500 fossils from nearly all parts of the body. They probably came from a single population, giving us valuable information about *heidelbergensis* society.

Homo heidelbergensis

DATES 600,000–250,000ya

BRAIN SIZE 1,100–1,400 cubic centimetres

HEIGHT Up to 1.8m (6ft)

FINDS Bone fragments from nearly all parts of the body, including a complete cranium and pelvis

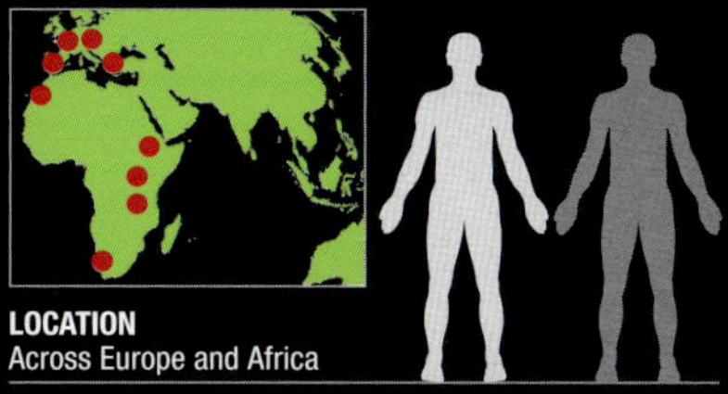

LOCATION
Across Europe and Africa

MYSTERIOUS HANDAXE
This handaxe, made of pink quartzite, is the only stone tool that has been found in the pit of bones. Named "Excalibur", after the sword of King Arthur, some archaeologists have suggested that it was thrown down the pit after the bodies as a ritual offering, implying that *heidelbergensis* believed in an afterlife.

KEY SITE

HEIDELBERGENSIS HANDYWORK

The site of Boxgrove, located in a quarry near Chichester in the south of England, presents a remarkable record of toolmaking and animal processing. Excavated between 1983 and 1999, the site is astonishingly well preserved, having been covered with fine silts washed over them by gently flowing water. The entire sequence of the manufacture and use of stone tools is preserved across a wide area, showing where flint was taken from the nearby cliff and made into handaxes. The shape of a stoneworker's knee can still be seen in the pattern left by waste flint on the ground. The handaxes were used to butcher elephants, rhinos, horses, and bison, whose cut-marked bones are also found alongside a *Homo heidelbergensis* leg bone and a couple of front teeth.

Homo neanderthalensis

The best-known of all fossil hominins, the Neanderthals thrived in Europe for over 300,000 years before modern humans arrived. The question of their interaction has been the subject of much debate – not just among archeologists but also in popular books and films.

Strong fossil record

The remains of more than 275 individual Neanderthals are known, from over 70 sites across Europe and western Asia. Most of these finds are fragmentary, but some are virtually complete skeletons, recovered from what may be intentional graves. As a result, we have a very good understanding of Neanderthal anatomy, and many sites from which to learn about their lives. The major debate concerning the Neanderthals is about their relationship to modern humans. Some believe that their anatomical adaptations were so specialized and so distinct from our own that they are a distinct species, *Homo neanderthalensis*, representing an evolutionary "dead end" that was completely replaced by modern humans. Others argue that they are similar enough to us to be a subspecies of *Homo sapiens*. Recent advances in DNA analysis (see panel, right) have begun to add more weight to the former view. However, anatomy and genetics tell us little about Neanderthal behaviour. The archaeological record shows that Neanderthals were excellent toolmakers and proficient hunters, although evidence for characteristic "human" behaviours such as language, art, and religion is less clear. Even so, Neanderthals thrived for more than 300,000 years, often through harsh climatic conditions. They were superbly adapted to their world, and were clearly a highly successful species.

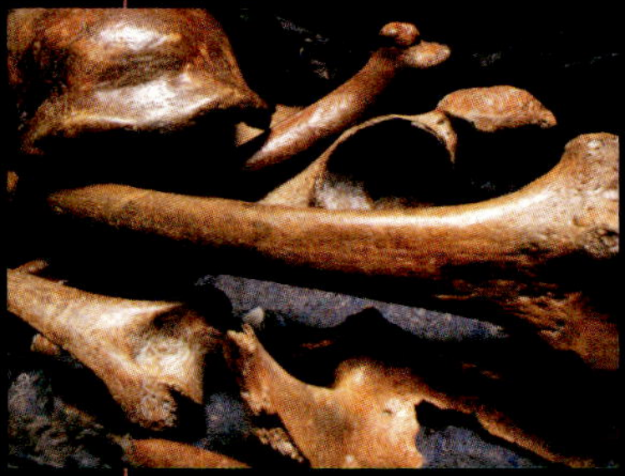

TYPE SPECIMEN
"Neanderthal 1", a partial skeleton found in 1856 at Feldhofer Cave in Germany's Neander valley, gave its name to the species.

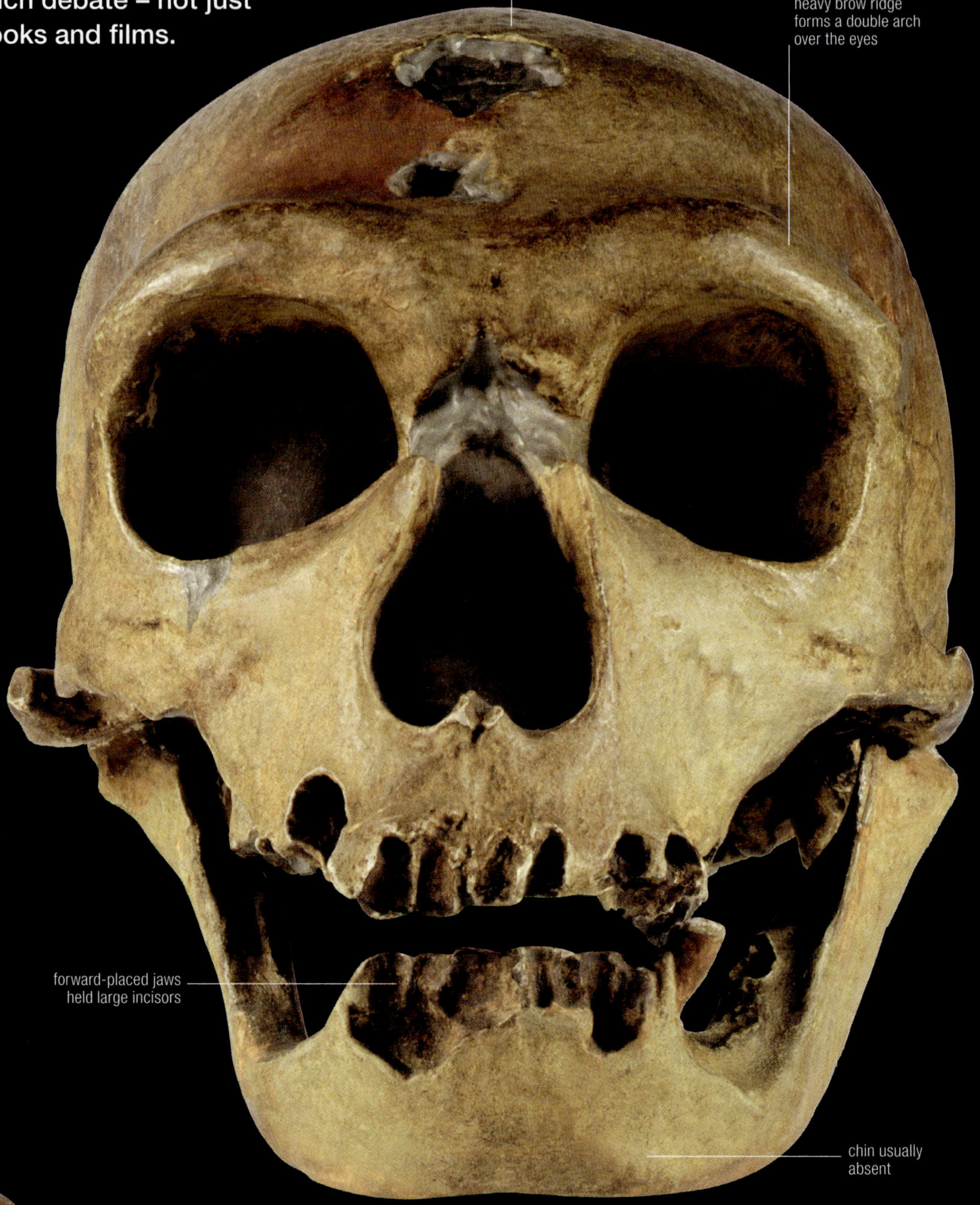

TEETH AS TOOLS
The most prominent part of a Neanderthal's face was their large nose. However, their jaws also project far forward, perhaps because they used their front teeth to grip things such as meat, hide, or sinew when cutting.

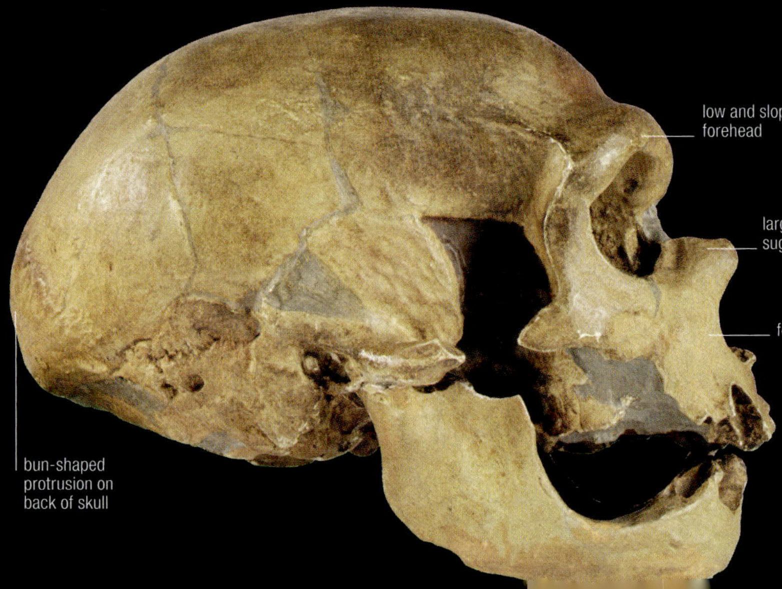

LONG, LOW SKULL
Neanderthal braincases may have been larger than those of modern humans, but they were shaped differently, with a low, sloping forehead and a long, low cranium.

Homo neanderthalensis

DATES 350,000–30,000ya

BRAIN SIZE 1,412 cubic centimetres

HEIGHT 1.52–1.68m (5ft–5ft 6in)

FINDS A number of entire skeletons, plus various bone fragments from more than 275 individuals

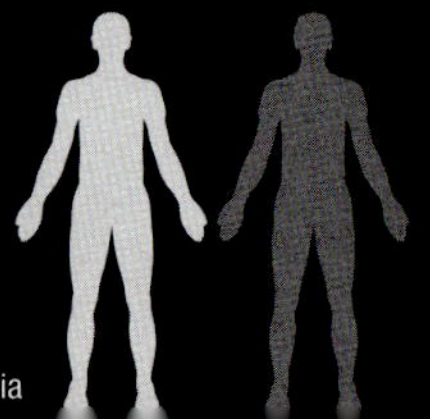

LOCATION
Across Europe and southwest Asia

Robust anatomy

Neanderthals were well adapted to the cold environment of Europe, and had lifestyles that involved intense physical activity. This probably explains their distinctive anatomy. Large brains were more heat-efficient, and big noses would have warmed the air before it reached the lungs. Neanderthal bodies were robust and heavily muscled, and the areas where their muscles attached to the bones are very marked, showing that they constantly put their muscles under a lot of stress. Their shoulders and arms were adapted for a powerful grip, and had highly developed throwing and striking muscles. The thickness and shape of the leg bones and joints reflect a great deal of use, and wide foot bones seem to be adapted for prolonged walking over rough terrain. The pelvis shows some interesting differences from modern humans, but anthropologists are not sure whether this suggests that Neanderthals were pregnant for longer, and gave birth to more mature babies, or whether it was because of their different stature and the intense physical stress they put on their hip joints.

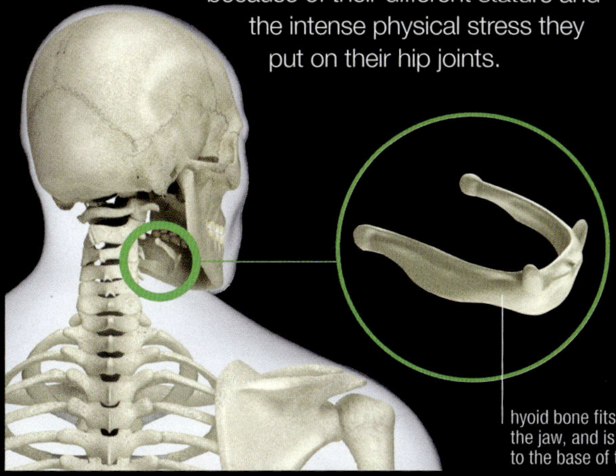

HYOID BONE
The hyoid bone is fragile, and rarely survives preservation. However, a complete Neanderthal hyoid found in Israel is virtually identical to that of modern humans. This suggests that Neanderthals might have had some form of language.

SCIENCE
DNA ANALYSIS

The DNA in bones degrades over time, and it is only recently that scientists have developed techniques for dealing with samples of low-quality DNA. This has allowed its extraction from increasingly older fossils, including those of Neanderthals. Sampling has to be done with care, to avoid contamination. Anthropologists are particularly interested in mitochondrial DNA, which is passed down from mother to child without combining with male DNA, meaning that changes can be tracked more easily. So far, results suggest that Neanderthal DNA is quite different to that of modern humans. The lineages are estimated to have separated about 500,000 years ago, and it does not seem likely that Neanderthals interbred with modern humans.

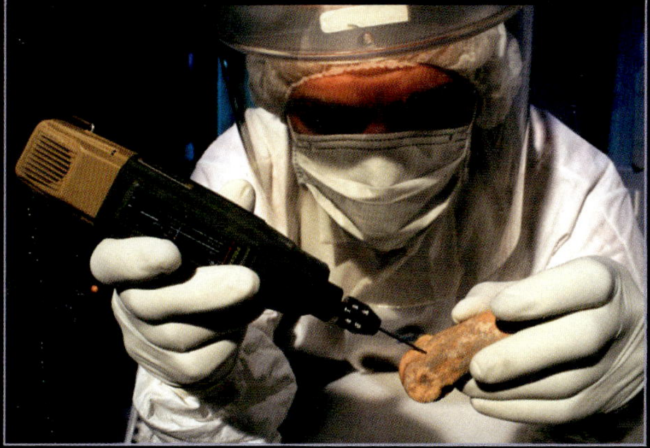

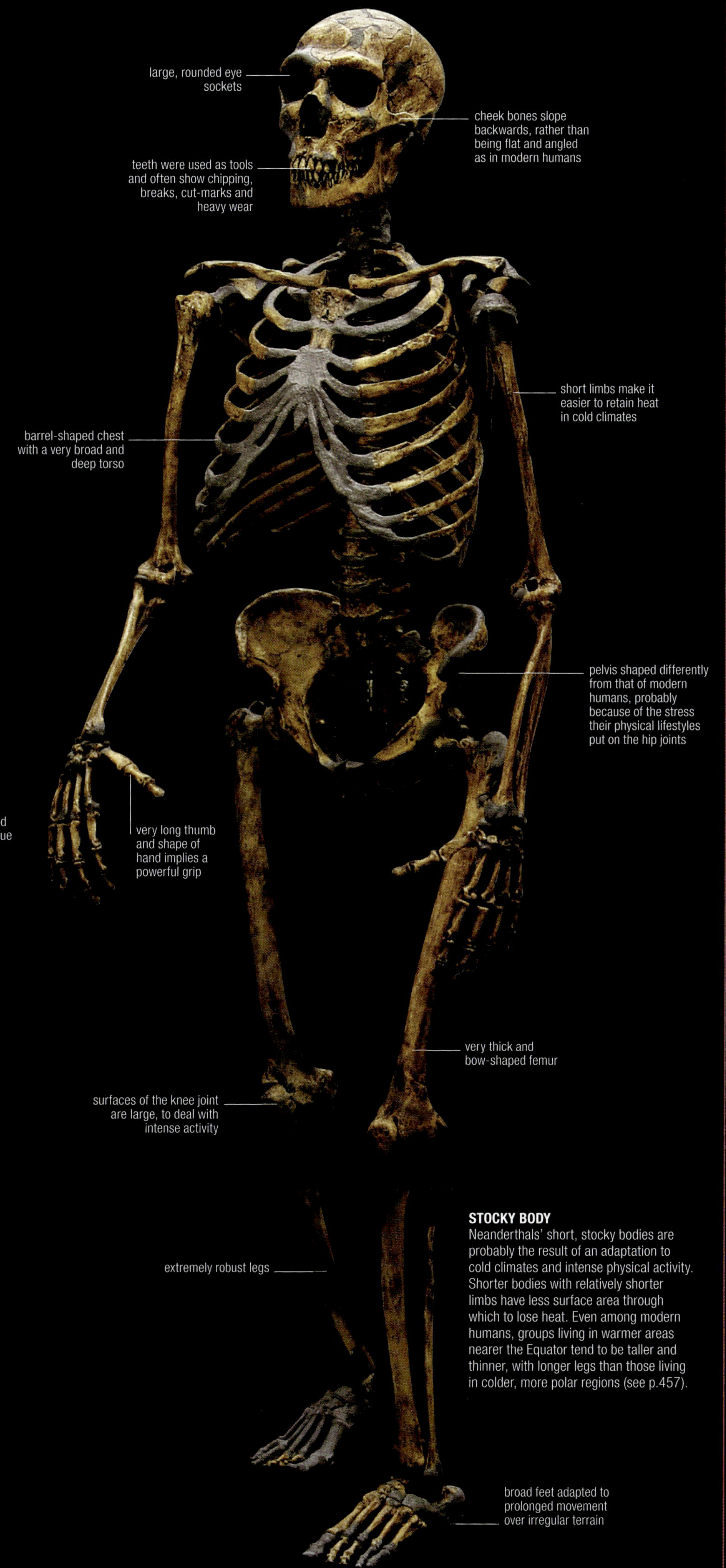

STOCKY BODY
Neanderthals' short, stocky bodies are probably the result of an adaptation to cold climates and intense physical activity. Shorter bodies with relatively shorter limbs have less surface area through which to lose heat. Even among modern humans, groups living in warmer areas nearer the Equator tend to be taller and thinner, with longer legs than those living in colder, more polar regions (see p.457).

Hunting and eating

Neanderthal stone tools are often found with butchered animal bones. However, archaeologists have debated whether Neanderthals obtained their meat by hunting or by scavenging the kills of other predators. Studying the parts of animal skeletons represented at these butchery sites gives us some clues. If Neanderthals scavenged, they would only have accessed carcasses after predators, who would have already eaten the easily accessible parts. Neanderthals would have had to focus on what was left, using stone tools to break into bones such as the toes and the skull, which contain nutritious marrow and brains.

The age of the animal prey is also important – older and younger animals are more likely to have been killed by non-human predators, then scavenged by Neanderthals, while animals in their prime are regularly hunted only by humans. Types of species are also significant: scavengers have little control over what carcasses they find, while hunters can choose their prey. Analysis of all these factors suggests that Neanderthal feeding strategies varied – there are many examples of scavenging sites, but also some sites with evidence of hunting. It would seem that while Neanderthals usually scavenged, they were also effective hunters when necessary.

IBEX
Tool-marked ibex bones were found in Gorham's Cave, Gibraltar, alongside Neanderthal tools, suggesting that they have been eaten by Neanderthals.

WILD BOAR
Wild boar would have been very dangerous to hunt without projectile weapons, though their bones have been found with Neanderthal tools.

FREQUENT INJURIES
The injuries suffered by Neanderthals are similar to those found among modern-day rodeo riders. This suggests Neanderthals had to get close to large animals during hunting, making the daily search for food a risky business.

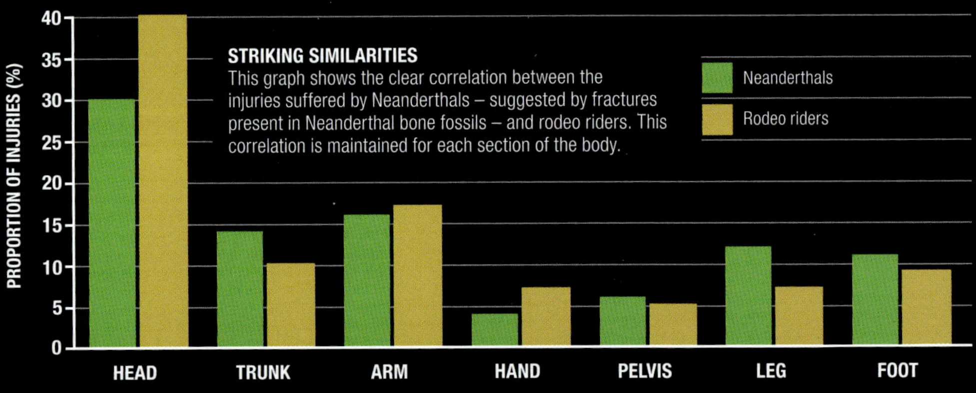

STRIKING SIMILARITIES
This graph shows the clear correlation between the injuries suffered by Neanderthals – suggested by fractures present in Neanderthal bone fossils – and rodeo riders. This correlation is maintained for each section of the body.

MAMMOTH GRAVEYARD
At La Cotte de St Brelade, on the island of Jersey, stone tools made by Neanderthals are found with the butchered remains of numerous mammoths and rhinos that Neanderthals drove off the cliffs above.

SCIENCE

ISOTOPIC ANALYSIS

Analyzing the chemical make-up of prehistoric bones can teach us much about the types of food that extinct species ate. This technique is known as isotopic analysis. During life, bones absorb some chemical elements from food, and ratios of different elements present in a fossil's bones can give us information about the kind of foods they ate. For example, by studying the different types of nitrogen present in a bone we can tell what proportions of a species' diet consisted of plants, meat, or marine life. While earlier hominin species have been shown to eat a combination of foods, including many different plants and seeds, analysis of Neanderthal bones has demonstrated that they ate a great deal of meat – even more than top carnivores such as hyenas. It has been suggested that meat comprised as much as 90 per cent of the Neanderthal diet.

Tools and culture

Neanderthals made stone tools using "prepared core technology". This involves preparing a stone core by giving it a particular shape before striking it, so that the shape of the flake produced can be pre-determined, and flakes detach with their edges pre-sharpened by the earlier work. Prepared core technology involved more skill and forethought than earlier handaxe manufacture, and also produced points that could be attached to spear shafts for hunting. This distinctive European form of flint-knapping technology is known as "Mousterian". Other evidence for Neanderthals having developed a more advanced culture than earlier hominins includes the presence of intentional graves. These suggest some element of ritualized behaviour and perhaps the development of social conscience and ideologies.

MOUSTERIAN TOOLS
The types of Mousterian tools present at different sites vary. Different groups of Neanderthals may have produced different kinds of tools; the variations might result from the types of tasks performed at each site; or the kinds of stone available.

Fate of Neanderthals

There is no evidence that modern humans and Neanderthals ever fought. In fact, there is little evidence to suggest they even met. Some archaeologists have claimed that the Neanderthals did not die out, but interbred with the new arrivals. They believe that a fossil from Lagar Velho, Portugal, represents a "hybrid", but this is controversial. More likely is that modern humans were better at exploiting their environment than the Neanderthals. Art, jewellery, and ritual is largely restricted to *Homo sapiens* sites, and it is believed that the skills and thought processes involved in these behaviours allowed modern humans to plan and organize their hunting more effectively, and to maintain relationships between groups over a much larger area. The two species seem to have coexisted for no more than 10,000 years in some areas, suggesting only limited competition between them. The European climate at the time was highly unstable, and though Neanderthals were adapted to this, rapid fluctuations may have put them under considerable stress. Faced with new competition for increasingly limited and unpredictable resources, even a small advantage for modern humans may have been enough to seal the Neanderthals' fate.

BURYING THE DEAD
Finds of complete or near-complete skeletons, such as this one from Kebara Cave, Israel, are more common for Neanderthals than earlier species. It is uncertain whether this means that Neanderthals deliberately buried their dead – although some sites may also preserve evidence of ritual burials.

GORHAM'S CAVE
This site on the rock of Gibraltar has yielded evidence of one of the last-known Neanderthal living sites. Although there are no fossils, Mousterian stone tools found here may date to less than 28,000 years ago.

"Two factors were predominantly responsible for **Neanderthal extinction** in western Europe: the arrival of competing human populations, and the **unstable climate.**"

CHRIS STRINGER, ANTHROPOLOGIST

Early Homo sapiens

While the Neanderthals thrived in Europe, *Homo sapiens* was evolving in Africa. Although poorly preserved, fossils of these early humans display a mixture of primitive and modern traits, with modern human features gradually becoming better established.

African origins

Most African hominins from between about 400,000 and 250,000 years ago have the long, low skulls of earlier species, with large faces, heavy brow ridges, a bony skull crest, and thick bones. These early specimens are variously known as *Homo rhodesiensis*, *Homo leakeyi*, and "archaic *Homo sapiens*". Fossils from between about 250,000 and 125,000 years ago still have some of the earlier features, but also exhibit more modern characteristics such as a chin, smaller brow ridges, and a rounder back to the skull. After around 125,000 years ago, fossils are recognizable as modern *Homo sapiens*, with much smaller faces, brow ridges, and teeth, and higher, more rounded braincases.

MIX OF FEATURES
This skull, from Jebel Irhoud in Morocco, is dated to about 160,000 years ago. It shows the mix of features of many specimens from this time period, with a flatter face than European Neanderthals.

KLASIES RIVER CAVE
The South African Klasies River Mouth cave sites have been excavated since the 1960s. They provide evidence of complex early human behaviour, particularly hunting, among groups of *Homo sapiens* around 125,000 years ago.

SUBSPECIES
This skull, discovered by anthropologist Tim White and his team in Ethiopia in 1997, is 160,000 years old. It has some primitive features, but the shape of its braincase places it in a subspecies of modern humans, *Homo sapiens idaltu*.

Homo sapiens

DATES About 150,000ya–present

BRAIN SIZE 1,000–2,000 cubic centimetres

HEIGHT Up to about 1.85m (6ft 1in)

FINDS Early *Homo sapiens* fossils include complete skulls, and various skull and skeleton fragments

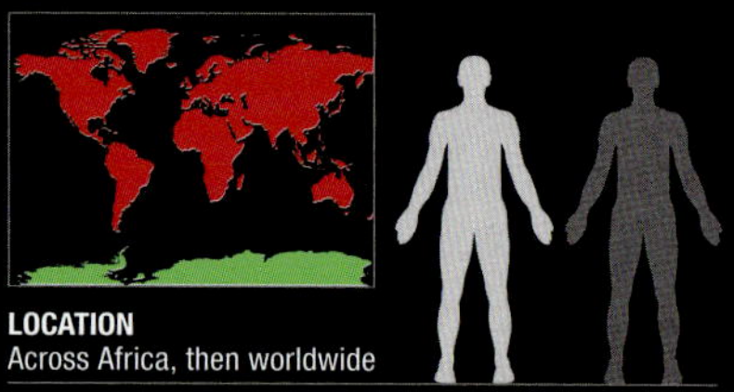

LOCATION
Across Africa, then worldwide

KEY FINDS

EVIDENCE OF EARLY CULTURE

Excavations at Blombos Cave, on the Southern Cape Coast in South Africa, have recovered early evidence for advanced behaviour by early *Homo sapiens*, dating to more than 73,000 years ago. Finds include two pieces of red ochre engraved with geometric patterns (right), many beads made of pierced shells, and finely made bone tools. The Blombos finds are important because it was previously thought that humans had only begun to exhibit such "modern" cultural behaviour much later, after they arrived in Europe.

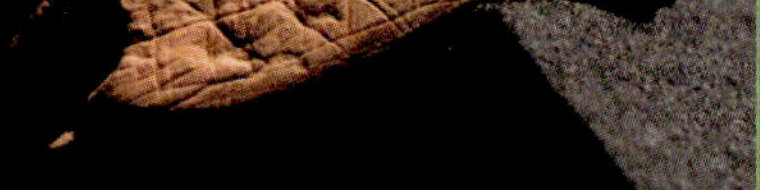

African exit

The earliest *Homo sapiens* fossils to be found outside Africa are from the Levant area of far-southwest Asia. At Skhul and Qafzeh in Israel, fossils of more than 20 individuals have been dated to around 130,000–100,000 years ago. However, these migrants do not necessarily represent the start of a continual migration from Africa, as Neanderthal fossils dated to between 60,000 and 45,000 years ago are also known from nearby sites. This suggests that the *Homo sapiens* fossils represent a brief foray into new territories, or that the two species somehow overlapped (see pp.464–67).

Modern Europeans

The first prehistoric hominin fossils to be discovered came from Europe. These fossils were often accompanied by sophisticated tools, art, and jewellery, and so it was long believed that the genus *Homo* first evolved in Europe. Increasing numbers of finds gradually established that our earliest ancestors evolved in Africa, and evidence for a long and gradual development of "modern" behaviour outside Europe has begun to be accepted. The apparently sudden appearance of "modern" behaviour in Europe does not mean that these practices began there, but rather that they were imported with the arrival of *Homo sapiens* (see p.472).

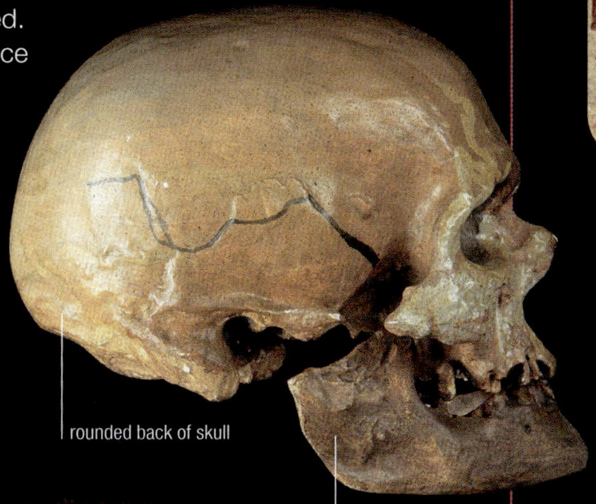

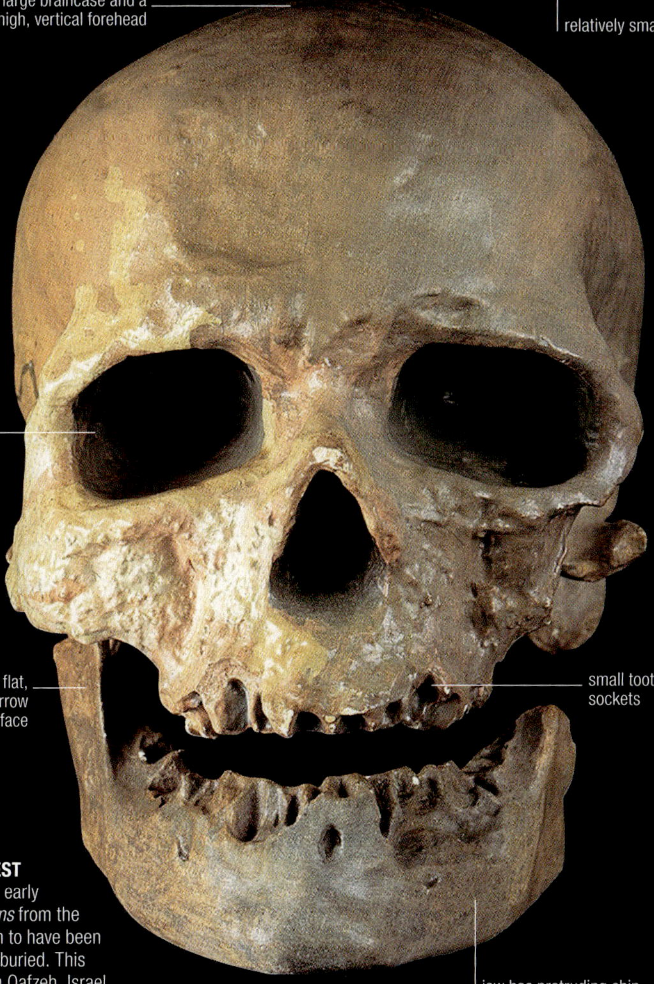

THE CRO-MAGNONS
Some of the earliest European *Homo sapiens* fossils were found in 1868 at the Cro-Magnon rock shelter in France. The "Cro-Magnons" were particularly large and muscular modern humans.

LAID TO REST
Some of the early *Homo sapiens* from the Levant seem to have been deliberately buried. This woman from Qafzeh, Israel, was buried with her child.

Out of Africa

For years, archaeologists had not been able to agree whether *Homo sapiens* evolved recently in Africa, or much more slowly from hominins that had spread across Europe and Asia much earlier (see pp.458–59). However, skeletal evidence and the results of DNA analysis now indicate that modern humans evolved in Africa before populating every continent.

Conflicting theories

The geographical origin of modern humans is one of the most controversial topics in archaeology. Two theories have dominated the debate: the theory of "multiregional" evolution, championed by Milford Wolpoff, and the theory of recent human ancestry, or "Out of Africa", which includes Chris Stringer among its proponents (see panel, opposite). The multiregional theory argues that *Homo sapiens* evolved gradually from different populations of *Homo erectus* scattered across the globe (see pp.458–61), with constant contact and interbreeding between the regions making them evolve in a similar way. The Out of Africa theory suggests that the origin of modern humans can be traced back to Africa, from where they spread and displaced existing hominid populations around the world. DNA evidence supports this theory, though there were also some early dispersals and retreats by small groups before the final dispersion around 60,000–40,000 years ago. Other researchers believe that incoming humans may also have interbred with local populations, and the fossil evidence from Asia and Australia is often presented as evidence of interbreeding between *Homo sapiens* and earlier groups. *Homo erectus* certainly survives late in Asia, and Australia's earliest colonizers have often been compared to *Homo erectus*. However, the fossil evidence from both areas is extremely poor. True "transitional" fossils and, arguably, truly "transitional" behaviour – are currently known only in Africa.

MIGRATION ROUTES
This map shows the routes apparently taken by *Homo sapiens* on leaving Africa, with the dates shown for the dispersals taken from find sites. Overall, the pattern is one where the earliest dates are in the areas closest to Africa, with the oldest sites in East Africa – traditionally viewed as the birthplace of many different hominin species. The last places reached by modern humans were those furthest away from Africa. Many colonization routes probably followed coastal habitats that humans already knew how to exploit. However, many coasts have since been drowned by rising sea levels, making evidence difficult to find.

SCIENCE

DNA ANCESTRY EVIDENCE

Modern human DNA has a lot to tell us about how and where we evolved. All living humans have very similar DNA, which suggests that we have not existed long enough for many differences to occur. The highest levels of DNA differences are found among African groups. This implies that, in order for the changes to have accumulated, these groups have been established longest. Assuming that changes occur at a constant rate, we can estimate that the ancestor of all modern humans lived between 270,000 and 200,000 years ago. This supports the Out of Africa theory of hominin dispersal. Multiregionalists argue that the estimated rates may be wrong, putting the date and location of human evolution in doubt.

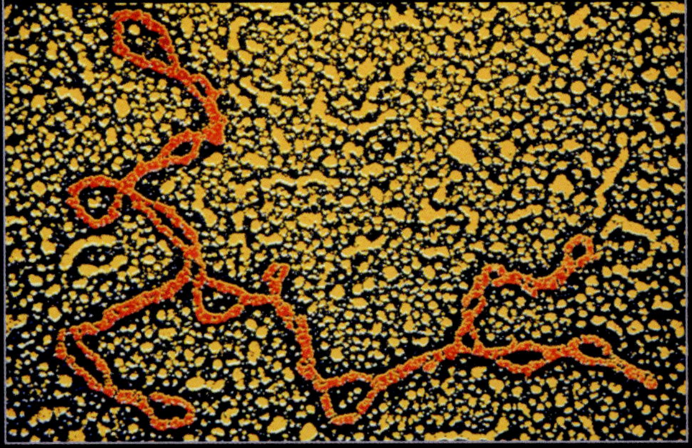

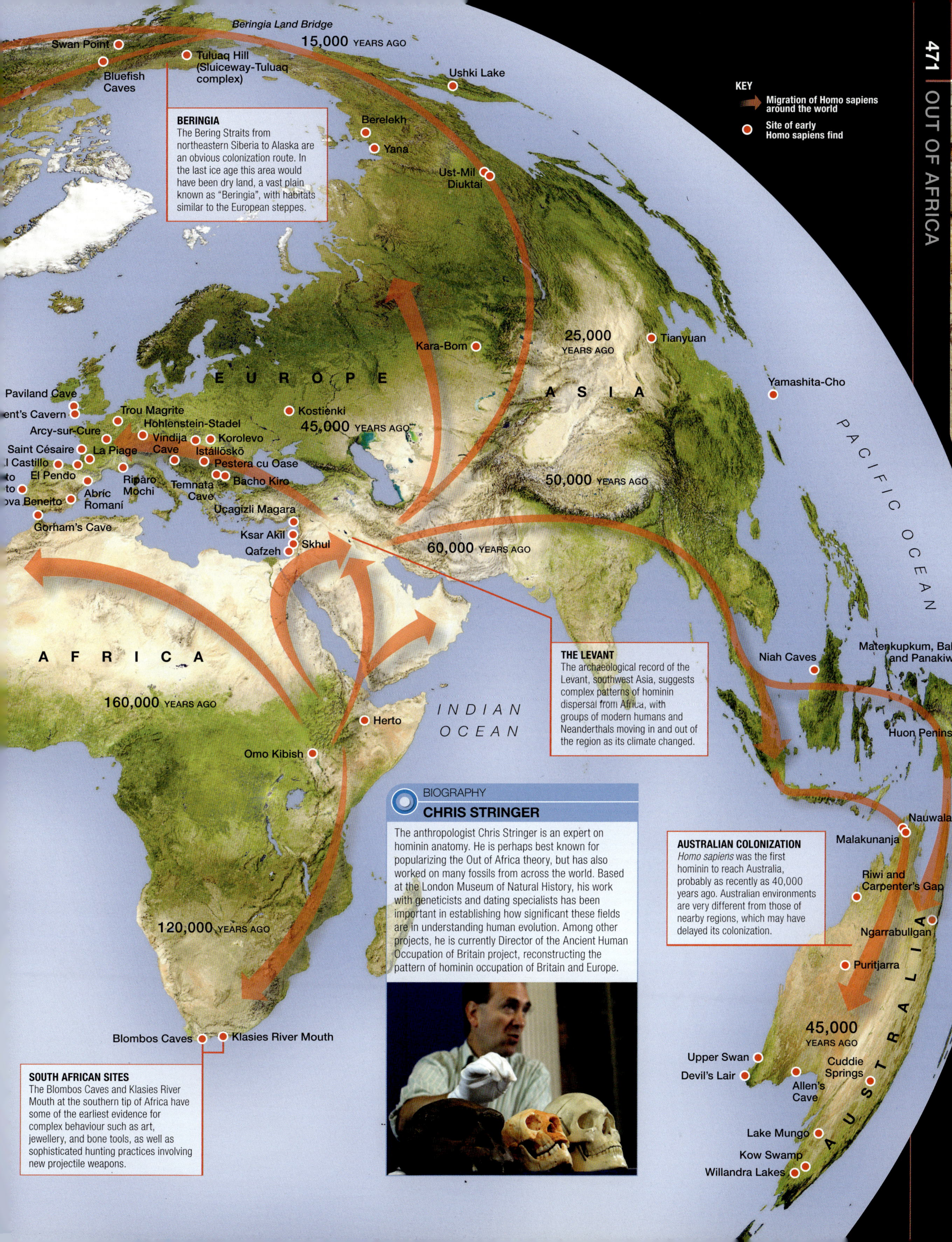

BERINGIA

The Bering Straits from northeastern Siberia to Alaska are an obvious colonization route. In the last ice age this area would have been dry land, a vast plain known as "Beringia", with habitats similar to the European steppes.

THE LEVANT

The archaeological record of the Levant, southwest Asia, suggests complex patterns of hominin dispersal from Africa, with groups of modern humans and Neanderthals moving in and out of the region as its climate changed.

AUSTRALIAN COLONIZATION

Homo sapiens was the first hominin to reach Australia, probably as recently as 40,000 years ago. Australian environments are very different from those of nearby regions, which may have delayed its colonization.

SOUTH AFRICAN SITES

The Blombos Caves and Klasies River Mouth at the southern tip of Africa have some of the earliest evidence for complex behaviour such as art, jewellery, and bone tools, as well as sophisticated hunting practices involving new projectile weapons.

BIOGRAPHY

CHRIS STRINGER

The anthropologist Chris Stringer is an expert on hominin anatomy. He is perhaps best known for popularizing the Out of Africa theory, but has also worked on many fossils from across the world. Based at the London Museum of Natural History, his work with geneticists and dating specialists has been important in establishing how significant these fields are in understanding human evolution. Among other projects, he is currently Director of the Ancient Human Occupation of Britain project, reconstructing the pattern of hominin occupation of Britain and Europe.

Near East

Populations of Neanderthals and modern humans in the Near East between 100,000 and 50,000 years ago varied, suggesting that groups of both species were moving in and out of the region as climates changed. It is likely Neanderthals lived there when harsher climatic conditions made environments more "European", and modern humans lived there when warmer climates extended "African" habitats. The modern humans are clearly *Homo sapiens*, although large and robust, but their stone tools are not as advanced as those we see at later European sites. Both Neanderthals and modern humans seem to be making and using very similar tools until about 50,000 years ago, when modern humans began to produce projectiles.

MOUNT CARMEL
The Carmel Caves in Israel have produced some of the most important hominin fossils. The caves of Skhul, Tabun, and El Wad have produced key Stone Age finds, including fossils of Neanderthals who continued to live in the area after *Homo sapiens* arrived.

RITUAL BURIAL
This early *Homo sapiens* skeleton found at the Qafzeh Cave site near Nazareth, Israel, was buried with a set of antlers laid across its body. It dates to 100,000–90,000 years ago, and may represent a form of ritual associated with burial, potentially indicating a belief in an afterlife. Although Neanderthal burials are generally accepted, evidence for Neanderthal ritual is rare, but common for modern humans, perhaps reflecting a greater capacity for symbolic thought.

Europe

A lack of early fossils makes it difficult for us to understand the spread of *Homo sapiens* across Europe. Archaeologists mostly track the arrival of modern humans by their distinctive stone-tool and worked-bone technology, called the "Aurignacian". However, some archaeologists have questioned this system, as occasionally similar stone tools seem to have been made by Neanderthals. Similarly, in western Europe a number of Neanderthal "Châtelperronian" sites appear to demonstrate that Neanderthals in the area were making supposedly "modern", complex artefacts such as beads. These Châtelperronian sites date to after the arrival of modern humans in the regions, and some archaeologists believe that Neanderthals may have copied modern human behaviours, or traded with them. The true Aurignacian technology, with characteristic bone and antler tools, jewellery, and art, is only ever found with *Homo sapiens*. When using it as a dating device, we can track the spread of modern humans across Europe from a number of sites, all dating to around 40,000 years ago. Aurignacian artefacts date some French sites to 35,000–30,000 years ago.

LION MAN STATUETTE
This ivory figure from Germany dates to about 32,000 years ago. Half human and half lion, some archaeologists believe that it may depict some kind of deity. At the very least, it demonstrates an entirely modern human imagination.

MAMMOTH-TOOTH CARVING
This carving of a lion, dated to around 32,000 years ago, was found near Stetten, southern Germany, at one of the earliest modern human sites in Europe. Such carvings are thought to represent "hunting magic" talismans for improving hunting success or appeasing the spirits of animals killed. No such carvings are known from any Neanderthal site.

GALGENBERG "VENUS"
This figurine of a dancing woman is from Galgenberg, Austria, and dates to around 30,000 years ago – the earliest phase in modern humans' colonization of western Europe. It is thought to have played a role in ritual or religion, and may be an early version of the many female Venus figurines that appear in Europe slightly later (see p.474).

Eastern Asia

Understanding the fossil record in eastern Asia is complicated by the poor dating of many fossils and sites. *Homo erectus* seems to have survived late here – perhaps even until 55,000–27,000 years ago in Java – while dates of more than 100,000 years ago from *Homo sapiens* sites in China suggest an overlap or even interbreeding between the two species. However, the dates from these early sites do not come from the hominin fossils, but from animal fossils in the same layer, making the dates less reliable. An interesting early site is at Tianyuan, where hominin fossils may show evidence of interbreeding between *erectus* and *sapiens* around 40,000 years ago. The earliest widely accepted *Homo sapiens* finds date to 33,000–24,000 years ago in China and Japan, but solid evidence for contact or interbreeding remains to be found.

KEY FINDS

FLORES MAN

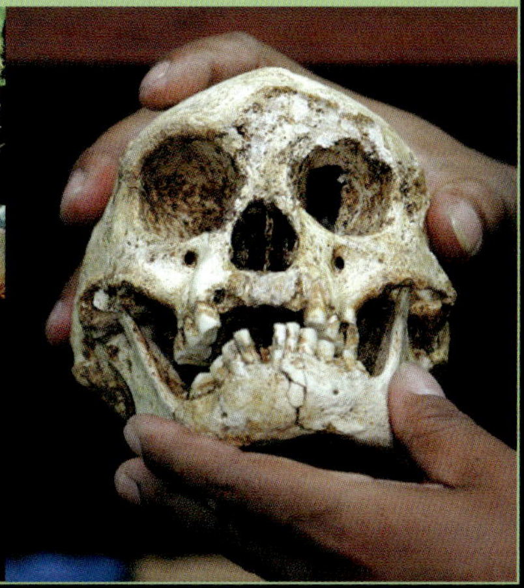

In 2003, the smallest hominin ever found was discovered on the island of Flores in Indonesia. Named *Homo floresiensis*, it is barely 1m (39in) high, and has a brain smaller than the earliest hominins. The fossils have been dated to less than 38,000 years ago. Their discovery has created significant debate: the excavators believe that they belong to a *Homo erectus* group that survived in isolation on the island until relatively recently. Other archaeologists have suggested that the bones belonged to modern humans who suffered from a growth-stunting medical condition. The evolution of dwarf forms of many animal species on islands is well documented, including on Flores. Debate continues, but at the very least the finds clearly demonstrate that hominin evolution was no simple, linear process.

ANCIENT FOOTPRINTS
At Willandra Lakes, southeastern Australia, 450 footprints have been found imprinted in a lake bed that was once a swamp. There are more than 22 trackways made by children, adults, and adolescents, which date to between 23,000 and 19,000 years ago.

Australia

Fossils from Australia have been claimed to represent colonization by both *Homo erectus* and *Homo sapiens*. It is now believed that all the fossils are those of *sapiens*, though they include some very heavily built specimens. Site dates have also been revised, with the earliest now around 40,000 years ago. This does not rule out earlier occupations, but any pioneer arrivals left few archaeological traces. Any colonization would have been demanding. Low sea levels during the ice age meant that the islands of Indonesia were a single land mass ("Sundaland"), as were Australia and New Guinea ("Sahul"), but sea crossings would always have been necessary to travel between them. Australia also boasts a combination of unique habitats, which may have been different to any the humans had previously experienced.

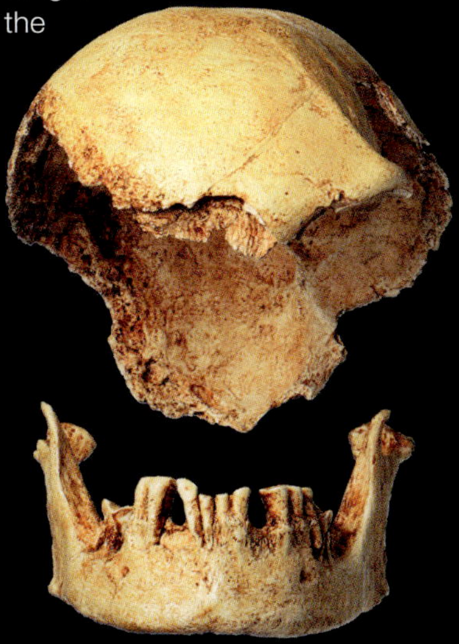

LAKE MUNGO SKULL
Fossil remains from Lake Mungo, southeastern Australia, were once thought to represent the evolution of *Homo erectus* into *Homo sapiens*. Now, most anthropologists believe that the Lake Mungo specimens are robust modern humans.

America

Dating the human colonization of America is difficult, with a range of dates given by different sites. "Clovis" spear points found across North America date human colonization to around 12,000 years ago. Other sites seem to pre-date this by several thousand years, while genetic evidence suggests colonization happened 18,000–16,000 years ago. All dating methods suggest Siberia as the likely origin of early populations, with Siberian Diuktai tools the likely forerunners of "Clovis" points. The Clovis sites probably represent the earliest widespread occupation of the continent, while smaller, more mobile groups could have arrived before, without leaving as many traces.

PEDRA FURADA
The site at Pedra Furada in Brazil has been dated to 50,000 years ago. This very early date has caused much controversy, with many archaeologists discounting its validity.

CLOVIS SPEAR POINTS

PROJECTILE TIPS
Clovis spear points appear across North America after 12,000 years ago, and probably represent the first widespread settling of the continent.

SCIENCE

MEGAFAUNAL EXTINCTIONS

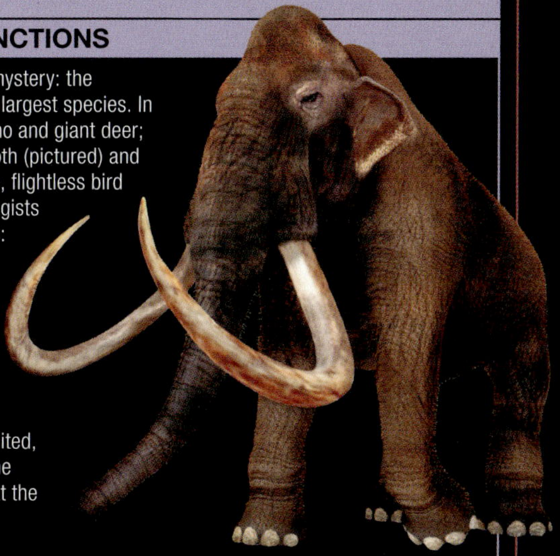

Archaeologists face a global murder mystery: the disappearance of many of the world's largest species. In Europe, victims include the woolly rhino and giant deer; in North America the imperial mammoth (pictured) and the mastodon; and in Australia a huge, flightless bird and giant marsupials. Many archaeologists believe the culprits were early humans: the "overkill" hypothesis. The disappearance of many of these animals appears to coincide with the arrival of modern humans in the same areas, and at some sites stone points are found embedded in animal bones. However, such direct evidence for human involvement is limited, and others believe the real cause of the extinctions was the unstable climate at the end of the Pleistocene.

European Hunter-gatherers

The early modern human hunter-gatherers of Europe lived through some of the harshest environments created by the last ice age. Nevertheless, this was a time of rapid technological innovation and social change as well as artistic expression, as groups battled to deal with the extreme conditions.

The early ice age

Between about 28,000 and 21,000 years ago, early modern human technologies developed into a series of more varied local forms. These are collectively known as the "Gravettian", after the French site of La Gravette, and are distinguished by their numerous small, narrow blades. These often have carefully shaped "tangs" or tabs for hafting them onto spears. At many sites the bones of fish and other small animals are found alongside those of larger species. This wide range of foodstuffs may explain why Gravettian sites seem to have been occupied for several months of the year, rather than being the short-term camps of mobile groups. At some of the more northeasterly Gravettian sites, pits have been found dug into frozen lower "permafrost" levels of the ground, creating natural freezers to store food. People living in these more enduring communities may have had more complex relationships with one another, which were expressed in the many burials containing finely-made jewellery and other grave offerings.

POSSIBLE PORTRAIT
This tiny mammoth ivory head is from Dolní Vestonice in the Czech Republic. Its face is uneven, and some archaeologists have suggested it depicts a skeleton found nearby.

MAMMOTH-BONE HOUSE
Mammoth-bone houses such as this one from Mezhirich have been found at several sites in the Ukraine. The bones used were probably scavenged rather than hunted.

GRAVE ADORNMENT
This skeleton of a young adult male found at the Gravettian site of Arene Candide, northwestern Italy, is adorned with a shell hat and necklace.

ABSTRACT FIGURES
These objects from Dolní Vestonice, in the Czech Republic, are made of mammoth ivory. They are believed to be highly stylized representations of women, related to the less abstract "Venus" figurines (see panel below).

AMULET

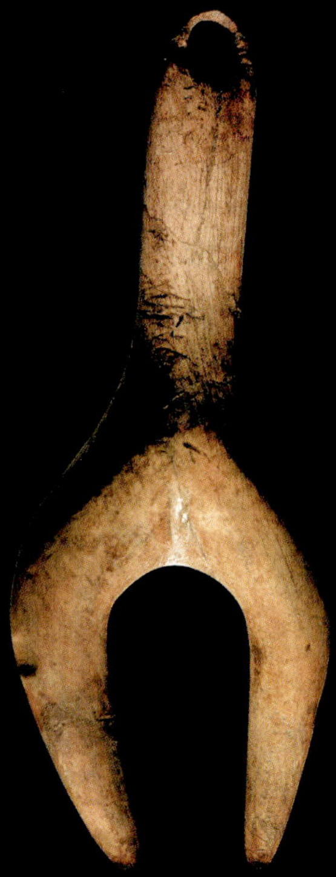

IDOL

KEY FINDS

VENUS FIGURINES

More than a hundred "Venus" figurines are known from across Europe, mainly from the Gravettian period. Each is unique, carved from soft stone or ivory, or sculpted in fired clay. All represent women, though often in a highly stylized way. Many seem to be obese or pregnant, and bellies, hips, and breasts are exaggerated, while arms, feet, and heads are given less emphasis. The figurines have been argued to be symbols or charms relating to fertility and childbirth, signs of social status, depictions of a Great Mother goddess, or items intended for exchange between groups. Given their wide distribution and timescales it is unlikely that they had only a single function.

In the shadow of the ice

After about 25,000 years ago, European climates began to cool significantly, and temperatures dropped to the coldest phase of the ice age. This period is known as the Last Glacial Maximum (see panel, right). Those living in more northerly and higher-altitude areas died, or were forced out by the expanding ice. The human occupation of Europe at the height of the ice age was largely restricted to the southwest, although some groups held on in areas such as the river valleys of the Ukraine. These people were skilled hunters, who spent much time crafting their weapons. They organized themselves to respond to the demands of the harsh environment by predicting and intercepting herd movements, thus ensuring the hunting success that was the difference between life and death.

ROCK CARVINGS
These animal carvings from the Roc de Sers, western France, are some of the few artworks reliably dated to the Last Glacial Maximum. They form a single frieze almost 10m (33ft) long.

SOLUTREAN INDUSTRY
"Solutrean" points are shaped like long, slender leaves. They were finely worked with a bone or antler tool or even a fingernail, giving the maker a great deal of control.

SCIENCE
LAST GLACIAL MAXIMUM

The global climate reached its coldest point of the ice age between about 21,000 and 18,000 years ago. Temperatures dropped significantly – by as much as 10°C – and the ice sheets expanded to their greatest extent. The period is known as the "Last Glacial Maximum". European environments would have changed dramatically, becoming almost entirely treeless, heather-rich heathland. However, the vast new open grasslands proved to be a rich environment for many large land mammals, such as mammoths, wild cattle, and reindeer.

The reindeer-hunters

After around 18,000 years ago, climates started to become warmer, and the areas of Europe that had been abandoned by humans as the ice sheets spread were re-colonized. Sites from this period are known as "Magdalenian" after the rock shelter of La Madeleine in France. Magdalenian tools are known for their sophisticated bone and antler points, needles, and harpoons. Although warmer than before, environments were still cold, and Magdalenians are often known as "reindeer-hunters" because of their focus on migrating reindeer. Many other species were also hunted, however, particularly freshwater fish. Most of the art of the Upper Palaeolithic dates to this period, and sites like Lascaux and Altamira, with their extraordinary cave paintings (see p.476–77), may have been places where people gathered at certain times of the year.

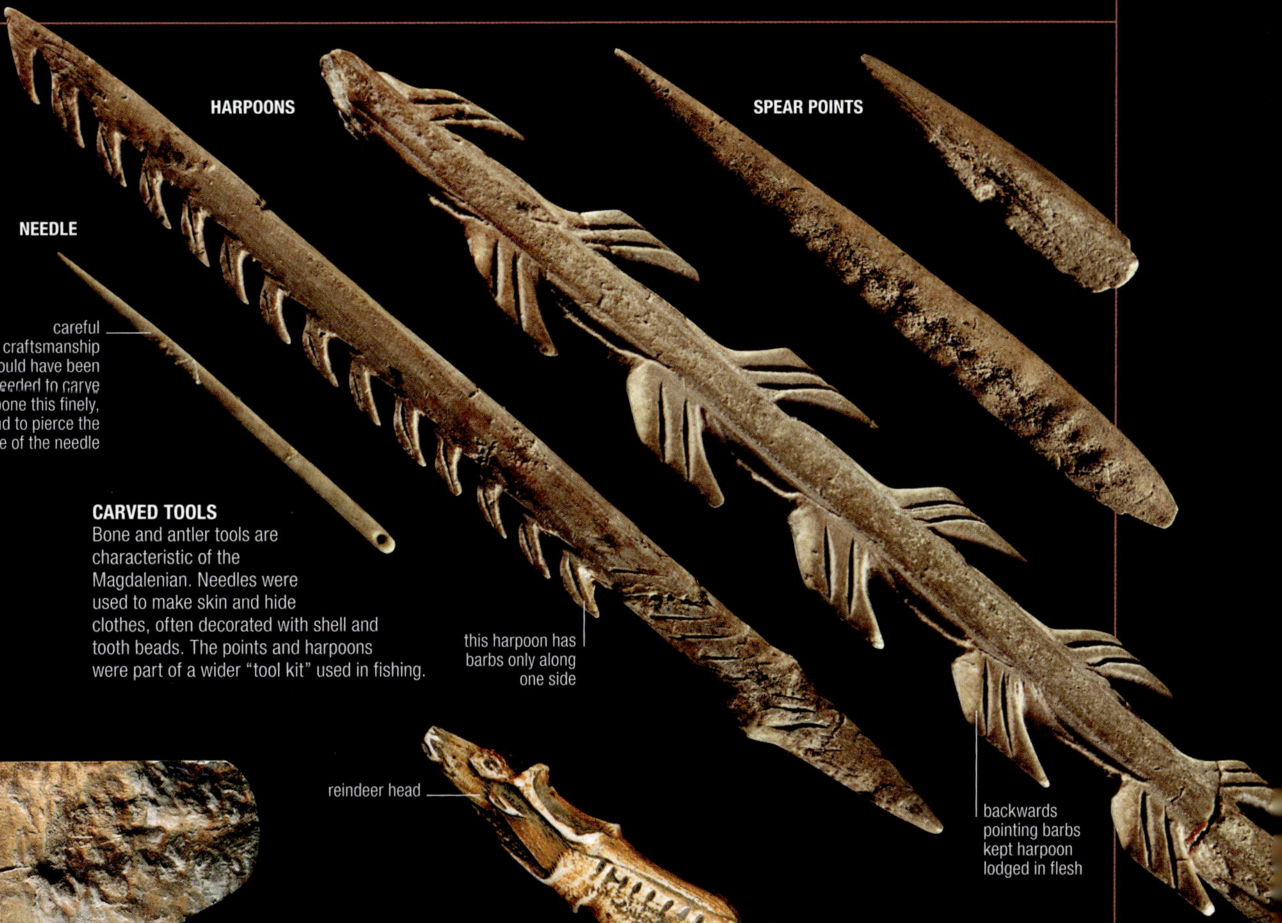

CARVED TOOLS
Bone and antler tools are characteristic of the Magdalenian. Needles were used to make skin and hide clothes, often decorated with shell and tooth beads. The points and harpoons were part of a wider "tool kit" used in fishing.

INTRICATE DESIGN
Art made on portable objects is common in the Magdalenian, and often depicts common prey animals, such as the reindeer and fish engraved on this piece of bone.

SPEAR-THROWER
This finely-engraved ivory object found at Bruniquel, France, is dated to around 14,000 years ago, and depicts two reindeer – a favourite prey of the Magdalenians. It is an "atlatl", a spear-throwing device that allowed a hunter to throw spears up to 100m (328ft), and thus tackle larger, more dangerous animals safely.

Paleolithic Cave Art

More than 350 painted caves have been discovered in Europe, mainly in France and Spain but also in Italy, the UK, and the Russian Urals. Most seem to have been painted around 18,000 years ago, but some may date to very soon after modern humans arrived in Europe, 40,000 years ago.

Altamira

The cave of Altamira is near the town of Santillana del Mar in northern Spain. It was discovered in 1879 by the nine-year-old daughter of Marcelino Sanz de Sautuola, an amateur archeologist, and is known for its plentiful images of bison. The paintings were so well conserved that many initially believed they were forgeries; they were not accepted as genuine until 1902. The great multitude of visitors to the cave threatened to damage the art, and in 2001 a replica was opened nearby.

◀ HUNTING MAGIC
The detailed depiction of prey animals such as this hind have led some archeologists to suggest the cave paintings may have been a form of "hunting magic".

▲ BEAUTIFUL BISON
Artists used natural pigment such as ochre, haematite, and charcoal to create their images, often utilizing the shape of the cave wall to create a more lifelike effect.

Chauvet

Chauvet Cave, in southern France, was discovered by cavers in 1994. It contains what may be the oldest cave art known, potentially dating to more than 30,000 years ago. Paintings include pictures of lions (see left), mammoths, and rhinos, as well as a famous half woman, half bison, which has contributed to some archeologists' view that cave paintings were created by shamans for magic or ritual purposes. Others argue that cave paintings were produced less exclusively, by people of both sexes and all ages, over a time span of more than 20,000 years. It is unlikely that there is a single explanation for cave art, and it is probable that explanations vary between caves.

▲ PANEL OF LIONS
Predators like the lions depicted here are less common in cave paintings than prey such as bison or horses. Chauvet, however, is well known for its depictions of predators, including bears, panthers, and hyenas.

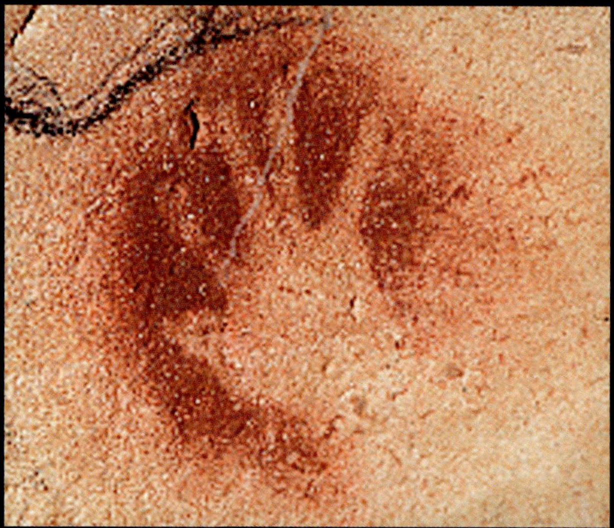

HAND STENCILS ▶
Stencilled handprints are common among the paintings at Chauvet. They were probably made by the artist placing a hand on the cave wall, then using a blowpipe to spray liquid pigment around it.

Lascaux

The famous site of Lascaux, containing around 2,000 paintings, is in the Dordogne region of southwestern France, and appears to date to the Magdalenian, around 16,000 years ago. It was found in 1940 by four teenagers and their dog, and opened to the public in 1948. However, as at Altamira, the sheer number of visitors damaged the paintings, and in 1983 a replica cave was created. Today few archeologists are allowed to enter the actual cave, and only for short periods, to protect the paintings from further damage.

◀ PROLIFIC PAINTERS
The Great Hall of the Bulls shows just how crowded the walls of Lascaux are. The stags above are visible by the head of the figure in the centre of this photograph.

▲ OCHRE STAGS
Most cave art is of individual animals, each representing separate artistic episodes, even when the animals are close together.

ON A GRAND SCALE
The Great Hall of the Bulls at Lascaux has some of the largest known cave art figures. Marks on the walls show where scaffolding was erected to allow the artist to reach high parts of the cave.

After the Ice

The Mesolithic – "middle stone age" – was the period after the end of the ice age and before the spread of farming. Many radical and exciting developments in human societies occurred during this time.

A changing world

The Mesolithic saw rapidly changing environmental conditions. Sea levels rose as the ice sheets retreated, and human populations spread northwards in their wake. Large land mammals were still present, but as forests expanded human diets shifted to concentrate on fish, seafood, waterfowl, and plant foods. Some areas were occupied year-round, and were close to coasts or inland lakes, where fragments of wicker fish traps are preserved, along with nets, textiles, twine, and even parts of wooden boats and skis. Large cemeteries from this period have been discovered, and while some of the burials contain many carefully made goods, others are much plainer. This suggests that Mesolithic groups may have become hierarchical, with some individuals enjoying higher status than others.

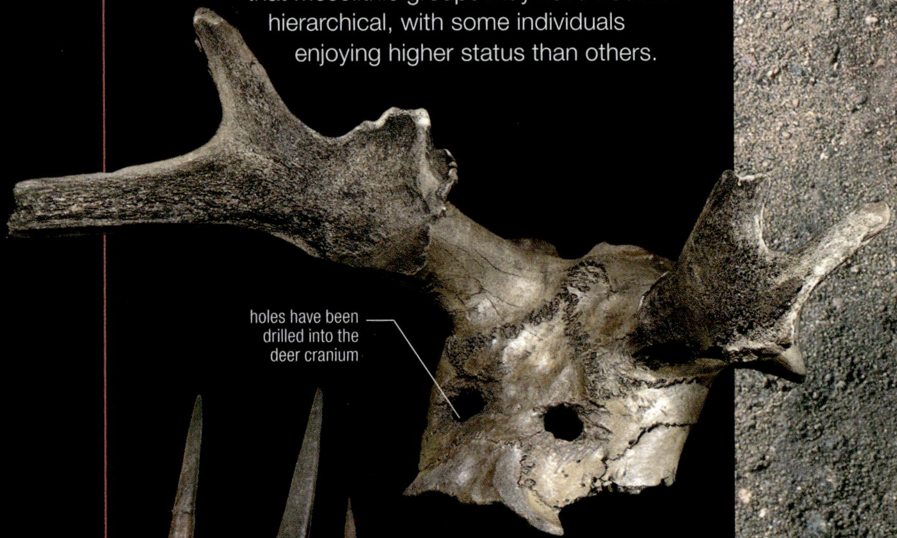

DEER SKULL-CAP
This red deer skull-cap comes from Star Carr in northern England. Holes have been bored in its skull to allow it to be attached toa human head. It has been suggested that this was either used by hunters to creep up on their prey, or for a ritual to improve hunting or appease the spirits of killed deer.

BARBED POINTS
Mesolithic technologies are based on tiny stone blades, bone and antler points, and hooks and harpoons, such as these barbed points from Star Carr. These were standardized, and attached to spears or arrows to make fishing equipment or hunting weapons, including bows and arrows. The bones of red and roe deer, elk, and wild boar have been found at Star Carr, and at some sites bones have been found with tiny stone tools still embedded in them.

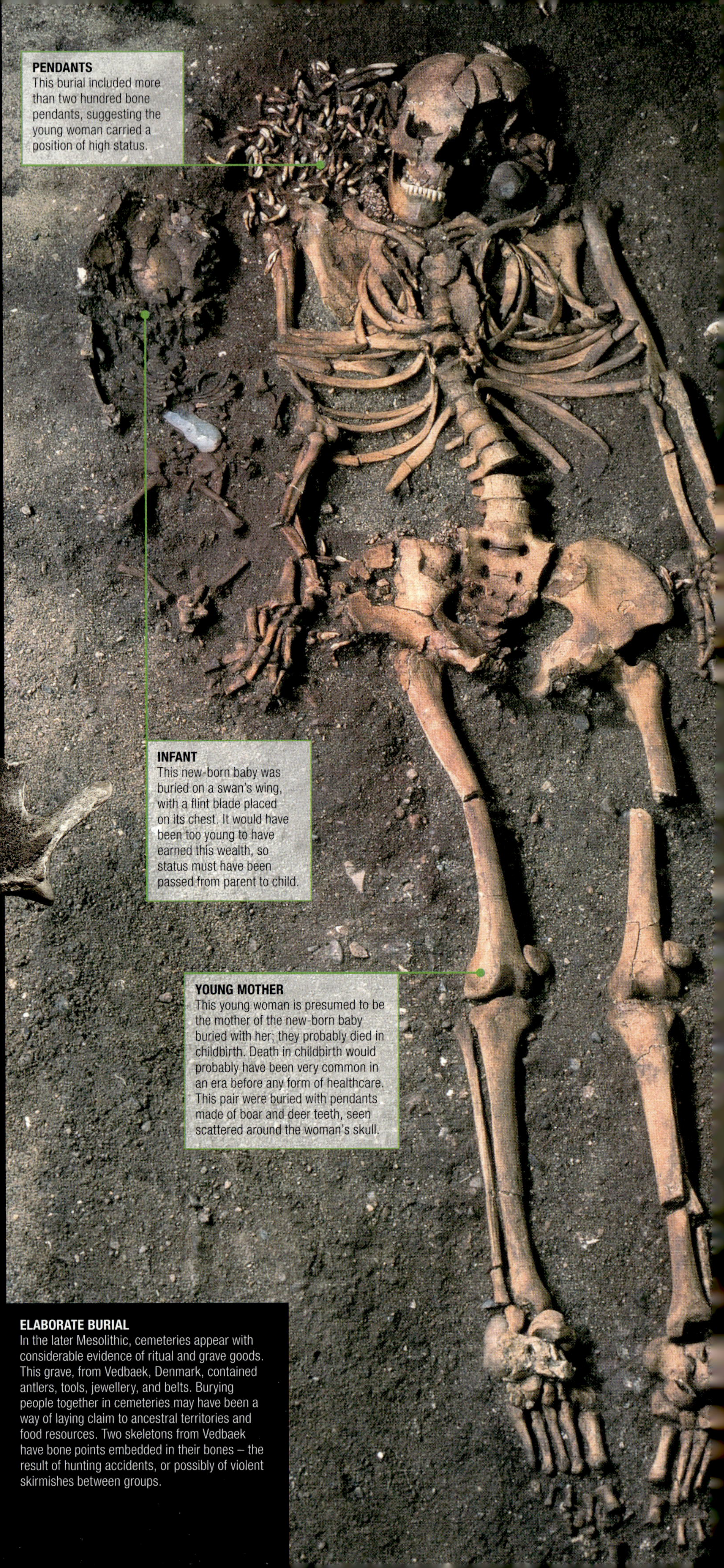

ELABORATE BURIAL
In the later Mesolithic, cemeteries appear with considerable evidence of ritual and grave goods. This grave, from Vedbaek, Denmark, contained antlers, tools, jewellery, and belts. Burying people together in cemeteries may have been a way of laying claim to ancestral territories and food resources. Two skeletons from Vedbaek have bone points embedded in their bones – the result of hunting accidents, or possibly of violent skirmishes between groups.

MESOLITHIC SHELL MIDDENS
The accumulation of Mesolithic shell middens, like this one near Glesborg, eastern Denmark, suggest that during this period groups had begun to occupy sites for longer and form communities. Shell middens consist mainly of huge heaps of oyster, mussel, and snail shells mixed with animal bones and objects made of bone, antler, and flint. These sites seem to have been kitchen refuse dumps, and so tell archaeologists a great deal about how Mesolithic people lived.

KEY SITES

LEPENSKI VIR

The site of Lepenski Vir on the Danube River in Serbia was a substantial village, occupied from about 8,000 to 6,000 years ago. It is one of few sites where it is possible to see a gradual transition from a hunter-gatherer lifestyle to a farming economy. The village consisted of carefully constructed wooden houses, with polished lime floors sometimes coloured red with ochre. These were built to a standard trapezoid shape, and each had a hearth of large stone blocks sunk into its centre. For food the villagers appear to have relied on fish from the river, as a bend in the Danube near the site creates ideal conditions for fishing. It has been suggested that this may have been the basis for a religion.

STRANGE SCULPTURES
In many of the houses, sculptures of fish or fish-human heads were positioned near the hearths. Their widespread distribution may suggest some kind of organized religion. The sculptures are thought to represent "river gods", petitioned for fishing success.

The earliest villagers

In the Near East, the Mesolithic period is more usually known as the "Epipaleolithic". The people of the Epipaleolithic hunted and gathered food, but they remained in the same place for much of the year, living in single-roomed, round houses dug partly into the ground. Over time, as villagers relied increasingly on wild cereals in the region, settlements became more substantial, and much more like those of the later Neolithic culture, but with little or no fired clay – the period is known as the "pre-pottery Neolithic". The buildings are rectangular rather than round, larger – sometimes two storeys high – and coated in thick lime plaster. Some are specialized workshops for crafts such as bead-making. Other buildings have basins, platforms, or engraved "standing stones", which may have been used in religious rites. These are the first permanent farming villages ever known, and some housed several thousand inhabitants. They mark the start of an urban life that continues to this day.

WALLS OF JERICHO
These walls and tower at Jericho, Israel, were built about 11,000 years ago. The tower was at least 8.5m (28ft) high and 8m (26ft) wide. It had been thought that they were built to defend the settlement, but it is now believed the wall protected against floods, while the tower was a shrine.

BIOGRAPHY

KATHLEEN KENYON

Kathleen Kenyon (1906–78) is best known for her excavations at Jericho that were very important in establishing the significance of the Epipaleolithic. She helped found London University's Institute of Archaeology, and later became its Director. In 1951 she became Honorary Director of the British School of Archaeology in Jerusalem, and, in 1973, was made a Dame.

DECORATED SKULL
Elaborate burial treatment of the dead is a hallmark of the pre-pottery Neolithic. Many people were buried under houses, and some had their skulls removed for plastering, painting, or coating in substances such as asphalt, as shown in this skull from Jericho, Israel. Decorated skulls seem to have been displayed together in small groups or "nests" that may have served to honour dead ancestors.

PLASTERED FIGURE
At 'Ain Ghazal, Jordan, a number of 9,000-year-old statues have been found, carefully buried in pits under the floors of abandoned houses. Some are of the head and shoulders only, some are full-figure, while some have two heads. They are about half life-size, and made of reed frameworks coated in plaster. Highly stylized, they have carefully painted facial details. Similar statues are known from Jericho, and may have been displayed ritually to represent ancestors or gods.

Glossary

abdomen
The belly of a vertebrate or the hind part of the body of an arthropod.

aboral surface
The body surface (e.g. in echinoids) opposite to where the mouth is located. *See also* oral surface.

acanthodians
An extinct group of jawed fish that existed from the Silurian (and possibly Ordovician) period, and died out during the Permian.

actinopterygians
A major group of bony fish that includes the great majority of living fish species. Also called ray-finned fish, their earliest known fossils date from the Silurian period. *See also* osteichthyans, teleosts.

adductor muscle
A muscle that pulls two structures together, such as the two shells of a bivalve.

aetosaurs
A group of plant-eating reptiles of the Triassic period, possibly related to crocodilians, that had backs protected by armoured plates and spines.

Afrotheria
A broad group of mammals originally thought to be unrelated, but now believed to have evolved from a single ancestor in Africa at a time when it was isolated from other continents. Afrotherians include elephants, manatees, aardvarks, and diverse smaller forms.

agnathans
Jawless fish, including living lampreys and many extinct forms.

algae (sing. alga)
Any of a wide variety of simple, mainly aquatic plant-like organisms, not all closely related, that produce their own food by photosynthesis. They include the single-celled "plants" of the plankton, as well as larger seaweeds. *See also* photosynthesis, cyanobacteria, protists.

ambulacra
The five narrower segments on the outer surface of an echinoid, through which structures of the water vascular system project. *See also* water vascular system.

amino acid
Any of about 20 different small molecules that are the "building blocks" of proteins, and thus are essential to all life.

ammonoids
An extinct group of swimming cephalopods with external, usually coiled, chambered shells. Includes ammonites and nautiloids. *See also* cephalopods, nautiloids.

amniotes
Land vertebrates (tetrapods) that lay shelled eggs, or whose ancestors laid shelled eggs. They include reptiles, birds, and mammals. *See also* amphibians, tetrapods.

amniotic egg
The type of egg laid by reptiles, birds, and early mammals.

amphibians
(1) The group of vertebrates that includes present-day frogs and salamanders, and their direct ancestors. (2) More broadly, all groups of tetrapods that never evolved amniotic eggs and so had to return to water to breed. In this broader sense there are a great variety of extinct species of amphibians. *See also* lepospondyls, temnospondyls, amniotes, tetrapods.

anal fin
A small fin near the anal opening in some fish.

anapsids
Reptiles characterized by having no skull opening in the region behind the eyes.

angiosperms
The plant group that contains all flowering plants. They are distinguished from other seed plants by the fact that angiosperm seeds grow hidden and protected in a flower structure called a carpel that matures into a fruit. *See also* eudicots, eumagnoliids, monocots.

anoxic
Lacking in oxygen.

anterior
Towards the front.

anthozoans
The group of cnidarians that includes sea anemones and most living corals. *See also* cnidarians, coral.

apical disc
Part of the outer skeleton of an echinoid, usually on the upper surface and containing the anal opening.

appendage
A limb-like structure, especially of an arthropod. Appendages may be modified as legs, gills, swimming organs, or in other ways.

arachnids
The group of chelicerate arthropods that includes spiders, scorpions, and relatives

Archaea
Microscopic single-celled prokaryotic organisms that were once classified with bacteria but are now treated as a separate kingdom in their own right. They are noted for flourishing in extreme environments. *See also* bacteria, prokaryotes.

archosaurs
A major group of reptiles that includes dinosaurs, pterosaurs, and crocodilians, as well as birds. They first evolved in the Triassic period. *Compare* lepidosaurs.

arthropods
A major group (phylum) of invertebrate animals with jointed legs and a hard outer skeleton. They include crustaceans, insects, chelicerates, and trilobites.

artiodactyls
Even-toed hoofed mammals, including pigs, camels, hippos, giraffes, deer, cattle, sheep, and antelopes. *See also* ruminants, perissodactyls.

asexual
Relating to reproduction not involving sex. Examples of asexual reproduction include creating new individuals by splitting or fragmenting the body, and forming specialized non-sexual structures to aid in dispersal (such as asexual spores).

australopithecine
Any member of the genus *Australopithecus* of extinct hominins. Several species are known dating from between 4.2 and 2 mya.

axil
The upper angle formed between a part of a plant, such as a leaf, and the stem that it grows from.

bacteria
Microscopic single-celled organisms that are prokaryotes. They abound in all habitats and may be free-living or parasitic. *See also* prokaryotes, Archaea.

barinophytes
An extinct subgroup of zosterophylls distinguished by the production of spores of two different sizes in a single sporangium. *See also* zosterophylls, spores, sporangium.

bedding
The laying down of sedimentary rock, or the manner in which a particular rock has been laid down.

belemnites
Extinct relatives of squids. Their long, often cigar-shaped internal shells are common fossils in Jurassic and Cretaceous rocks. *See also* cephalopods.

bennettitaleans
An extinct group of seed plants superficially resembling cycads in their leaves and their growth form but not closely related to them. *See also* seed plants.

biodiversity
The variety of different kinds of organisms found at a given time or location. Often calculated in terms of number of species, but can be measured in other ways.

bipedal
Walking on two legs.

bivalves
Molluscs, such as clams, mussels, and oysters, that have a shell made up of two hinged halves. Most species move slowly or not at all, and are filter feeders. *See also* molluscs.

blastoids
An extinct group of Paleozoic filter-feeding echinoderms. *See also* echinoderms.

bony fish
See osteichthyans.

brachial
Relating to the arms.

brachiation
A method of movement characteristic of apes such as gibbons, in which the body is swung through the trees by the arms.

brachiopods
A major group (phylum) of marine invertebrates, also known as lamp shells. Their two-part shell makes them look superficially similar to bivalve molluscs, but they are unrelated. There are some living forms, but they were far more abundant and diverse in the Paleozoic and Mesozoic.

brackish
Saltier than fresh water, but less salty than typical ocean water.

bract
A modified leaf at the base of a flower or flower head. May be small and scale-like, large and petal-like, or like normal foliage.

branchial
Relating to the gills.

bryophytes
A general term covering three groups of mainly small land plants: mosses, liverworts, and hornworts. These lack specialized water-conducting (vascular) tissue. *See also* mosses, liverworts, hornworts, vascular plants.

bryozoans
Also called moss animals, a group of filter-feeding invertebrate animals consisting of colonies of small immobile individuals.

budding
In zoology, new individuals growing directly from the flesh of existing individuals.

calcareous
Consisting of or containing the mineral calcium carbonate.

calice
The upper part of a coral skeleton.

carapace
The upper shell of a turtle; also the protective outer covering of some other animals, such as crabs and lobsters.

carbon dioxide
An atmospheric gas (carbon combined with two oxygens), released by burning, volcanic activity, and when organisms respire. Conversely, plants remove it from the atmosphere during photosynthesis. *See also* respiration.

carnivore
(1) Any flesh-eating animal. (2) A member of the order of mammals that includes cats, dogs, bears, weasels, hyenas, racoons, and mongooses.

carpel
The female part of a flower, containing and

protecting the unfertilized seeds, and also the seeds themselves after fertilization.

cartilage
A tough tissue formed mainly of protein that is found in the bodies of vertebrates. In sharks and relatives the skeleton is made of cartilage; in other vertebrates cartilage occurs in joints and other locations.

cartilaginous fish
See chondrichthyans.

cast (the filling of a mould)
The material filling a mould.

Caytoniales
An extinct group of seed plants with distinctive four-parted palmate leaves and small, fleshy, fruit-like structures that contain seeds.

cellulose
A carbohydrate that is the main structural material of plants.

cephalochordates
Small marine invertebrates that are of interest in understanding how vertebrates evolved. The living species, called amphioxus or lancelets, are filter feeders that burrow in shallow sea floors, but also swim in a fish-like way. Soft-bodied, they rarely fossilize. *See also* chordates.

cephalopods
The group of advanced molluscs that includes living squids and octopuses, the mainly extinct nautiloids, and the entirely extinct ammonites and belemnites.

cephalothorax
The front part of the body of some arthropods when not obviously divided into a separate head and thorax.

cetaceans
The group of mammals that includes whales and dolphins. They evolved from artiodactyl ancestors during the Palaeogene. *See also* artiodactyls.

chelicerates
A major subgroup of arthropods including arachnids (spiders and relatives) and many other forms, including the extinct eurypterids. They are characterized by joined pincers called chelicerae.

chemosynthesis
The phenomenon by which some specialized micro-organisms are able to make their own food using the energy from naturally occurring chemicals such as hydrogen sulphide.

chitin
A complex substance consisting mainly of carbohydrate that forms the tough outer covering (exoskeleton) of insects and other arthropods. It is also found in fungi.

chloroplast
A structure within a plant or algal cell within which photosynthesis takes place.

chondrichthyans
The major group of fish that includes the present-day sharks and rays, plus many extinct forms. Their skeleton is made of cartilage, not bone.

chordates
A large group (phylum) of animals that includes all the vertebrates as well as some related invertebrate forms such as cephalochordates. They are named for a rod-like support of cartilage called a notochord that runs down their backs. *See also* cephalochordates.

cilia (sing. cilium)
Microscopic beating hair-like structures on the surfaces of some cells, used to aid movement in small organisms, or to create water currents.

clade
A group of species consisting of all the evolutionary descendants of a given ancestor. Mammals are a clade, for example. Reptiles, however, are not a clade unless birds are included within them, since birds are descended from dinosaurs, which were reptiles. Cladistics is the approach to classification that seeks to use only clades to classify organisms, and to avoid using other terms.

cladistics
See clade.

claspers
Appendages found in some male insects and chondrichthyans that are used for grasping the female during mating.

clubmosses
See lycophytes.

cnidarians
A major group (phylum) of mainly marine invertebrates with simple bodies bearing stinging tentacles that surround a single opening (mouth). *See also* anthozoans, scyphozoans, hydrozoans, coral.

coevolution
The evolution of two or more different organisms together, so that they become increasingly adapted to one another. The mutual adaptations of flowers and the bees that pollinate them are an example.

colony
A colony of animals can consist of separate individuals working together, as in ant colonies, or animals joined by strands of living tissue, as with many corals. Individuals may be specialized for different roles, such as feeding, reproduction, and defence, in which case the colony may behave more like a single animal.

common ancestor
Any original species from which later species under discussion are all believed to have descended.

compound eye
An eye formed as a mosaic of smaller eyes, as in some arthropods.

compression fossil
A fossil where the original form has been squashed flat, destroying evidence of external structure.

condylarths
An extinct group of mammals known mainly from the Paleogene. They included plant-eating forms similar to today's hoofed mammals, although not closely related.

conifers
The group of non-flowering seed plants, mainly trees, that includes pines, firs, redwoods, cypresses, and other living and extinct species.

conodont animals
Extinct eel-like marine animals whose hard parts (small internal tooth-like structures called conodonts) were known and studied long before fossils of the whole animals were discovered from the 1980s onwards. Dating from the Cambrian to the Triassic, they are now thought to be chordates and possibly even primitive vertebrates. *See also* chordates.

continental shelf
A region where the sea floor is relatively shallow around the margins of a continent. At its outer edge the sea floor steepens until it reaches the depth of the deep ocean floor.

convergent evolution
The phenomenon in which unrelated organisms have evolved to look similar as a result of being adapted to similar environments or ecological niches. For example, marsupial mice in Australia look superficially similar to mice found elsewhere, but are not related.

coral
Any of various cnidarians that live fixed to the sea bed or to reef structures. The true corals are members of the class Anthozoa, secrete a calcareous skeleton, and may be solitary or colonial. *See also* cnidarians.

corallite
In corals, the skeleton of an individual coral animal.

corallum
The skeleton of a colonial coral, composed of corallites. *See also* corallite.

correlation (of rocks)
The matching of rocks from different regions of the world (e.g. by comparing their fossils) so that rocks of the same age in different places can be recognized.

cranial
Relating to the brain case, or to the skull as a whole.

creodonts
An extinct group of flesh-eating mammals with a similar ecological role to modern carnivores. They were dominant predators in the Eocene and Oligocene epochs.

crinoids
Stalked filter-feeding echinoderms, also called sea lilies. Living crinoids mainly live in deep water, except for stalkless forms called feather stars. *See also* echinoderms.

crocodilians
The group that includes living crocodiles and alligators and their immediate ancestors. They and their extinct relatives (together called the crocodylomorphs) belong to the archosaur group of reptiles. *See also* archosaurs.

crocodylomorphs
See crocodilians.

crustaceans
The diverse group of mainly marine arthropods that includes crabs, lobsters, shrimps, barnacles, and many smaller forms, including members of the plankton.

cusp
The sharp apex where two curved surfaces meet, especially in the context of the surface of a tooth.

cyanobacteria
Microscopic photosynthetic relatives of bacteria; also called blue-green algae.

cycads
A group of non-angiosperm seed plants that are superficially similar to palms but are reproductively very different and produce seeds and pollen in cones.

cynodonts
A group of advanced synapsids that arose in the Late Permian. *See also* synapsids.

czekanowskialeans
An extinct group of Mesozoic seed plants with distinctive finely divided, linear leaves often borne on short shoots.

decapods
(1) The group of cephalopods that includes squid and cuttlefish. (2) The group of crustaceans that includes crabs, lobsters, and relatives. The name literally means "ten feet", referring to the number of squid tentacles and crab/lobster legs respectively.

denticle
A small tooth-like projection in the skin of sharks and some other fish.

deuterostomes
Members of a major division of the animal kingdom, defined by the second opening formed in the developing embryo turning into the mouth. (The first opening becomes the anus.) They are thought to represent a single related "super-group" and include echinoderms, chordates (including vertebrates), and graptolites. *See also* protostomes.

diadectomorphs
A group of extinct tetrapods thought either to be the earliest amniotes or closely related to them. *See also* amniotes.

diapsids
A major division of reptiles whose name refers to the two openings on each side of the skull that are characteristic of the group. Two main groups within the diapsids are the archosaurs (including dinosaurs, birds, crocodilians, and pterosaurs) and the lepidosaurs (including lizards, snakes, and plesiosaurs, among others).

dicynodonts
A group of herbivorous therapsids with two tusks and a blunt beak. *See also* therapsids.

dinosaurs
A group of reptiles that evolved during the Triassic and dominated life on land in the Jurassic until their extinction (with the exception of their descendants, the birds) at the end of the Cretaceous. *See also* ornithischians, saurischians, archosaurs.

DNA
Short for deoxyribonucleic acid, a very long molecule made up of small individual units. DNA is found in the cells of all living things; the order of the small units "spells out" the genetic instructions (genes) of the individual.

dorsal
Relating to the back or upper surface of an animal. *See also* ventral.

echinoderms
A major group (phylum) of marine invertebrates that includes starfish, echinoids, crinoids, and many other extinct forms. They have calcareous protective plates under their skin, and living forms display radial symmetry.

echinoids
A group of living and extinct echinoderms. Typical echinoids, also called sea urchins, have a globular body, long spines, and a downward-facing mouth to graze algae. There are also many burrowing echinoids.

ecosystem
Any community of organisms considered together with the interactions between them and their associated physical environment.

edentates
An unusual group of eutherian mammals living mainly in South America and including armadillos, sloths, and some anteaters. ("Edentate" literally means "having no teeth".)

Ediacaran fauna
Soft-bodied invertebrates living before the Cambrian period and originally discovered fossilized in the Ediacara Hills in Australia.

embolomeres
A group of large aquatic amphibians from the Carboniferous, which may be the ancestors of amniotes. *See also* amniotes.

endemic
A species of organism native to only one restricted region of the world. (Also used as an adjective.)

enzyme
Any of a large variety of different molecules (nearly always proteins) that promote particular chemical reaction in the body.

Eocene
The second epoch of the Paleogene period, between the Paleocene and the Oligocene. It lasted from 55.8 to 33.9 mya.

eon
The largest and highest-level unit into which geological time is divided.

epidermis
The outermost layer of skin in animals; the outermost cell layer of a plant.

epithelium
The covering of cells that forms the surface of organs and tissues in the body.

epoch
A unit of geological time that is a division of a period – for example, the Eocene epoch of the Paleogene period.

era
A unit of geological time that is a division of an eon and is itself divided into periods – for example, the Paleozoic era of the Phanerozoic eon.

eudicots
Also called eudicotyledons, this is the largest group within the flowering plants (angiosperms). It contains many familiar flowers and tree species. *See also* angiosperms.

eukaryotes
Living things whose cells have a complex structure, including a distinct nucleus that contains the genetic material and is bounded by a membrane. Eukaryotes include all animals, plants, and fungi, as well as many microscopic single-celled species. *See also* prokaryotes, protists.

eumagnoliids
A group of angiosperms thought to have branched off early in angiosperm evolution, and that retains some of the features of primitive angiosperms. The group's members include magnolias and laurels.

euphyllophytes
One of the two main subgroups of vascular plants, including all land plants having complex leaves with many veins. Euphyllophytes include all seed plants, ferns, and horsetails. The term primitive euphyllophytes, as used in this book, refers to early lineages of euphyllophytes that do not fall into any of the main subcategories within the group. *See also* lycophytes, vascular plants.

eurypterids
Extinct aquatic predatory arthropods, sometimes called sea scorpions. They occurred form the Ordovician to the Permian. *See also* chelicerates.

eutherians
Mammals in which the foetus grows to a relatively advanced stage in the mother's womb, nourished by a placenta. Includes all living mammals except marsupials and monotremes.

evaporite
A sedimentary deposit, usually made up of salts of various kinds, that has been created by the evaporation of salty water.

evolution
In its straightforward modern definition, evolution is simply any change in the average genetic makeup of a population of organisms of the same species between one generation and the next. What is often called "the theory of evolution" is based on the idea, supported by various lines of evidence, that such genetic change is not random but is largely the result of natural selection, and that the operation of such processes over time can account for the huge variety of species found on Earth. *See also* natural selection.

exhalent siphon
The projecting tube in many bivalves through which circulating water leaves the animal. *See also* inhalent siphon.

exoskeleton
A skeleton located on the outside of an animal. Exoskeletons combine support with body protection.

fasciole
A band on the outside of the test of some echinoids, with minute granules to which in life very fine spines were attached that were specialized to create water currents.

fenestra
A hole or opening in the side of the skull, or in an individual bone.

ferns
A group of non-flowering vascular plants that reproduce by spores rather than seeds. They most commonly produce their leaves from underground stems, but some are large enough to form trees. *See also* euphyllophytes.

filter feeding
Feeding by collecting and separating small food particles from the environment. When the food particles are suspended in water, it is also called suspension feeding. Collecting and sieving small particles from mud and sand is called deposit feeding.

flagellum (pl. flagella)
A microscopic beating hair-like structure borne by some cells. Flagella are similar to cilia but much longer.

form genus
A scientific name given not to a whole organism, but to a particular part of one (such as a seed) that is regularly found in the fossil record. Form genera are convenient in classifying plant fossils, which are often found in fragments that are difficult to relate to each other.

fungi
A huge and diverse kingdom of living things that includes mushrooms, moulds, and numerous other species. The typical fungus "body" is a network of small threads. Structures such as mushrooms are specially produced "fruiting bodies" to disperse spores. *See also* lichens.

gastropods
The diverse group of molluscs that includes marine, freshwater, and land snails and slugs. *See also* molluscs.

genital plate
A section of the outer skeleton of an echinoid that contains small pores through which eggs or sperms are released.

genome
The complete set of genes of a given species of animal or other living thing.

Ginkgoales/ginkgos
A group of non-flowering seed plants with a single living species (*Ginkgo biloba*), but many extinct species. Designated as "gymnosperms" in traditional classification.

Gnetales
A group of living and extinct non-flowering seed plants of varying appearance. They are designated as "gymnosperms" in traditional classification.

gracile
In palaeontological contexts: smaller and/or more lightly built.

graptolites
An extinct group of colonial planktonic invertebrates, typically growing in the form of long narrow colonies of small individuals, supported by a hard skeleton.

green algae
A group of relatively simple water-living plants, including some microscopic plankton and some seaweeds. Many species live in fresh water. *See also* plants, algae.

greenhouse effect
Phenomenon in which heat radiating from the Earth's surface is absorbed by certain atmospheric gases, including water vapour, carbon dioxide, and methane, causing average temperatures to be warmer than otherwise.

guard cell
A cell whose swelling and shrinkage acts to open and close stomata in plants. *See also* stomata.

gymnosperms
A general term for several groups of seed plants that are not true angiosperms. The term literally means "naked seeds" in that their seeds are not fully protected within a structure called a carpel, in contrast to angiosperms. *See also* conifers, cycads, Ginkgoales, Gnetales, progymnosperms.

Holocene
The most recent epoch of the Neogene period, lasting from 11,700 years ago to the present. It is also designated as part of the Quaternary period. *See also* Pleistocene.

hominids
The subgroup of primates that includes humans, chimpanzees, and gorillas, together with their last common ancestor and related extinct forms. Does not include orang-utans, gibbons, or monkeys.

hominins
The subgroup of hominids that includes humans and their extinct ancestors and relatives, taken back to the point at which the lineage leading to humans split from that leading to chimpanzees.

hornworts
A group of small, low-growing land plants, living mainly in moist environments. They were traditionally classified with two other groups of "non-vascular" plants – mosses and liverworts – as "bryophytes". *See also* bryophytes, vascular plants.

hydrothermal vent
A fissure in a volcanically active region of the ocean floor from which superheated, chemical-laden water emerges. Found mainly at mid-ocean ridges.

hydrozoans
A group of cnidarians that includes many small colonial species with tiny polyps, some significant in coral reef formation, and also floating forms such as the Portuguese man-of-war. *See also* cnidarians, polyp.

ice age
Any episode in which the Earth's surface temperatures were much lower than today and ice cover more extensive.

ichthyosaurs
A group of marine reptiles that first appeared in the Triassic. Ichthyosaurs were predators, and most had streamlined bodies for fast pursuit, similar to present-day dolphins or tuna. They became extinct before the end of the Cretaceous.

igneous rock
Rock that originates from the solidification of magma rising from deep in the Earth.

inarticulate
Having no joints or articulations.

inhalent siphon
The projecting tube in many bivalves through which water is drawn into the animal's shell, providing it with oxygen and food particles. *See also* exhalent siphon.

internal mould
A fossil mineral impression of the inside of a now-vanished original shell or other part of an organism.

invertebrate
Any animal without a backbone, i.e. other than a vertebrate.

keystone predator
A predator whose removal would cause a drastic change to its ecosystem – for example, by allowing its former prey to proliferate.

leaflet
A single division of a leaf, in the case of a leaf that is divided into several parts. *See also* palmate, pinna, pinnate, pinnule.

lepidosaurs
The group of reptiles that includes the present-day lizards and snakes, plus various extinct groups such as plesiosaurs. *Compare* archosaurs.

lepospondyls
A diverse group of small-bodied extinct amphibians widespread in the Carboniferous and Permian periods. They included both aquatic and terrestrial forms. *See also* temnospondyls.

lichens
Composite living things made up of a fungus and an alga growing in close association. There are many types, typically growing as crusts or bushy outgrowths on rocks, tree branches, etc.

lignophytes
Vascular plants that produce specialized woody tissue. They include seed plants and some extinct relatives that are often grouped together as progymnosperms. *See also* progymnosperms.

lineage
Any given branch on an evolutionary family tree. A lineage consists of a species or group plus its forebears, traced back to an original ancestor. *See also* clade.

liverworts
A group of small, low-growing land or freshwater plants, often growing in the form of a flat sheet, but sometimes with lobe-like leaves. They lack vascular tissue and were traditionally classified with two other groups of "non-vascular" plants, mosses and hornworts, as "bryophytes". *See also* bryophytes, vascular plants.

living fossil
Refers either to an individual species that has hardly changed over millions of years, or a living representative of a taxonomic group that is otherwise extinct.

lobe-fins
See sarcopterygians.

lobopods
A group of soft-bodied arthropod-like crawling invertebrate animals, superficially resembling caterpillars, that includes the living *Peripatus*, plus many extinct forms dating back to the Cambrian and beyond.

lycophytes
A group of vascular plants that includes the lycopods and zosterophylls. Lycopods include the clubmosses of the present day, plus many extinct forms, including giant forest trees in the Carboniferous period. Lycophytes differ from all other vascular plants (euphyllophytes) in their very simple, scale-like leaves. *See also* euphyllophytes, vascular plants, zosterophylls.

mammals
Warm-blooded vertebrates that suckle their young on milk and whose skin is typically covered in hair or fur. Mammals evolved from synapsid ancestors during the Triassic period. *See also* synapsids.

mandible
The lower jaw of a vertebrate, or (plural) the biting mouthparts of an arthropod.

mantle
In molluscs, the tissue, originally forming the upper surface of the animal, that secretes the mollusc's shell. *See also* molluscs.

marsupials
A group of mammals in which offspring are born in a relatively undeveloped state; they often continue their growth within an external pouch on the mother. Includes kangaroos, opossums, and many other living and extinct forms.

matrix
The material on which an organism rests or is embedded.

medullosans
An extinct group of seed plants with fern-like leaves and often large, fleshy seeds.

Mesozoic
The second era of the Phanerozoic eon, dating from the beginning of the Triassic period to the end of the Cretaceous.

metacarpals
The bones occurring between the wrist and the fingers in the human hand; also, the equivalent bones in other species.

metamorphic rock
Any rock that has been transformed underground by natural heat or pressure into a new form. For example, marble is metamorphized limestone.

metazoa
The group that includes all multicellular animals except for sponges, which are considered to have a simpler form of body organization.

microfossil
A microscopic or very small fossil.

mid-ocean ridge
A submerged range of mountains running along any part of the deep-ocean floor, marking the place where new ocean crust is being created by molten rocks rising from deep in the Earth.

midrib
The central vein of a leaf (including its leaf-stalk).

Miocene
The first epoch of the Neogene period. It lasted from 23.03 to 5.332 mya.

molluscs
A major group (phylum) of invertebrate animals that includes the gastropods, bivalves, cephalopods, and other forms. Molluscs are soft-bodied and typically have hard shells, though some subgroups have lost their shells during evolution.

monocots
Also called monocotyledons, the group within the angiosperms that includes grasses, orchids, lilies, and palm trees, as well as many other species. They are characterized by having only one cotyledon (seed-leaf) within each seed. Most monocots have leaves with parallel veins. *See also* angiosperms.

monotremes
Egg-laying mammals, including the platypus and the echidnas (spiny anteaters). This egg-laying habit is thought to be the original mode of reproduction for mammals.

mosasaurs
A group of Cretaceous marine reptiles, thought to be close relatives of snakes.

mosses
A group of small, low-growing land or freshwater plants, often bushy or cushion-like in appearance, that lack vascular tissue. *See also* bryophytes, vascular plants.

multituberculates
An extinct group of mainly rodent-like early mammals found from the Jurassic to the Paleogene period.

mya
Million years ago.

myriapods
The group of many-legged arthropods that includes millipedes and centipedes.

natural selection
Evolutionary process by which the environment favours the fittest individuals in a given population, weeding out those that are less fit. ("Fitness" in this context means the sum total of the qualities that give an individual an increased chance of leaving descendants.) Since fitness is partly inherited, this should result in a population or species changing genetically over time and (other things being equal) becoming better adapted to their environment.

nautiloids
The group of cephalopod molluscs containing the living *Nautilus* plus many extinct forms. Their coiled external shell contains gas-filled chambers that lighten the body in the water.

Neolithic
The final phase of the stone age, occurring at different times in different parts of the world and characterized by the transition to an agricultural lifestyle and by new technologies such as pottery.

niche
Roughly, the ecological "role" that an animal or other living thing plays (for example "small night-active insect-eater living high up in trees").

noeggerathians
A group of extinct plants that have proved difficult to classify. They have been linked to horsetails and to ferns, but recent research suggests that they were more closely allied to progymnosperms.

nothosaurs
See sauropterygians.

notochord
See chordates.

occipital
Relating to the back of the head.

Oligocene
The third and final epoch of the Paleogene period. It lasted from 33.9 to 23.03 mya.

ooze
Sediment on the deep-sea floor that contains a large proportion of skeletons of dead microscopic planktonic life forms. *See also* plankton.

operculum
A term with various applications in zoology, all related to the original Latin meaning of "covering" or "lid". It can mean: the horny or calcareous disc used by many snails to shut themselves into their shells; a flap that covers the gills of bony fish and tadpoles.

oral surface
The body surface where the mouth is located. *See also* aboral surface.

ornithischians
One of the two main subgroups of dinosaurs, the name literally meaning "bird-hipped". Ornithischians include stegosaurs, iguanodons, armoured dinosaurs, and horned dinosaurs. *Compare* saurischians.

ossicles
Small bony or other hard parts collectively making up the skeletal support in some animals.

osteichthyans
The largest group of jawed fish, made up of two subgroups, the actinopterygians (ray-finned fish) and the sarcopterygians (lobe-finned fish). Also called bony fish.

ozone layer
A layer high in the atmosphere with a high concentration of ozone (a 3-atom form of

oxygen). Ozone absorbs harmful ultraviolet radiation that can damage living things.

palaeoclimatology
The study of past climates.

palaeoecology
The study of the ecology and environments of past geological time periods.

Paleocene
The first epoch of the Paleogene period and Cenozoic era. It lasted from 65.5 to 55.8 mya.

palmate
Of a leaf: divided into leaflets spreading out from the base of a leaf like the fingers of a hand.

paranthropine
Any member of the genus *Paranthropus* of extinct hominins, with fossils known from between 2.7 and 1.2 mya.

parareptiles
A group of extinct amniotes considered not to be true reptiles. They include mesosaurs and several other groups.

pedicle
The flexible stalk of a brachiopod that attaches it to its substrate.

pelvic fins
A pair of fins situated on the under surface of most fish.

Pentoxylales
An extinct group of Mesozoic seed plants, known from India, Australia, and New Zealand. They have leaves similar to those of some bennettitaleans. *See also* bennettitaleans.

period
A unit of geological time that is a division of an era – for example, the Jurassic period of the Mesozoic era.

perissodactyls
Odd-toed hoofed mammals, including horses, rhinoceroses, and tapirs. *See also* artiodactyls.

permineralization
A form of fossilization where minerals occupy the spaces between the hard parts of the original organism, but do not replace the hard parts themselves. *See also* petrifaction.

petrifaction
Also called petrification, a form of fossilization where the detailed structures of an original organism are replaced by minerals, sometimes in a way that preserves very fine details. *See also* permineralization.

Phanerozoic
The most recent large time-period (eon) of Earth's history. It stretches from the Cambrian period to the present.

photosynthesis
Process in plants, algae, and cyanobacteria whereby the Sun's energy, trapped by the green pigment chlorophyll, is used to build energy-containing food molecules from carbon dioxide and water.

phragmocone
The chambered part of a cephalopod shell.

phylum (pl. phyla)
The highest-level grouping in the traditional classification of the animal kingdom. Molluscs, arthropods, and chordates are examples of phyla.

phytosaurs
A group of crocodile-like aquatic reptiles that were common in the Triassic period.

pinna
In leaves: a leaflet or primary segment of a pinnate leaf. *See also* leaflet, pinnate.

pinnate
In leaves: being divided into leaflets arranged in regular rows on either side of the leaf's midrib. *See also* leaflet, midrib.

pinnule
In leaves: any of the ultimate leaflets of a bipinnately compound leaf. *See also* leaflet.

placental mammals
See eutherians.

placoderms
A group of extinct jawed fish, widespread in the Devonian period, whose skin was protected by bony "armour-plating".

placodonts
Extinct Triassic marine reptiles that fed on molluscs and other shellfish.

plankton
Any living species (plants, animals, or micro-organisms) that live in open water but cannot swim strongly (or cannot swim at all) and so drift with the currents. Most plankton are small, but some larger types such as jellyfish are also included.

plate tectonics
Phenomena relating to the fact that the Earth's crust and its underlying rocks are divided into rigid sections called tectonic plates that move relative to one another.

Pleistocene
The third epoch of the Neogene period. It lasted from 1.806 to 0.0117 mya.

plesiosaurs
see sauropterygians.

pleurae (sing. pleura)
In trilobites, the lateral parts of the thoracic segments.

Pliocene
The second epoch of the Neogene period. It lasted from 5.332 to 1.806 mya.

pliosaurs
See sauropterygians.

polyp
The body form of many cnidarians, including sea anemones and coral animals. Polyps are typically tubular, attached at their base, and with a single opening (mouth) at the top surrounded by tentacles. *See also* cnidarians.

primates
The group of mammals that includes monkeys, apes, and humans, plus various less advanced forms (prosimians). Typical characteristics include grasping hands and forward-facing eyes. *See also* prosimians.

proboscidians
The group to which elephants and mammoths belong, characterized by a long, flexible, grasping trunk. *See also* Afrotheria.

progymnosperms
Spore-producing extinct plants with woody tissue similar to that of present-day seed plants. *See also* lignophytes.

prokaryotes
Living things such as bacteria whose individual cells are smaller, simpler in structure, and lack a well-defined nucleus, compared with those of eukaryotes. *See also* eukaryotes, bacteria, Archaea.

prosimians
Primates that are consider less "advanced" than the monkeys and apes. Living forms include lemurs, galagos, and tarsiers.

protists
A wide grouping of often unrelated, mainly microscopic, eukaryotic organisms, traditionally classified as a single kingdom. Protists are mainly single-celled, and include both plant-like forms (algae) and animal-like forms (protozoans). They can be defined informally as including all eukaryotes that have not been assigned to another kingdom. *See also* eukaryotes, algae, protozoans.

protozoa
Animal-like single-celled organisms, mainly microscopic, that are common in virtually all habitats and may be free-living or parasitic. The thousands of species are in many different subgroups, not all closely related. *See also* protists.

pteridophytes
A miscellaneous group of vascular plants that reproduce by spores rather than by seeds. Living pteridophytes include ferns, clubmosses, and horsetails. These plants are of diverse evolutionary relationships.

pteridosperms
A miscellaneous group of seed plants that has been used in the past to include several very different kinds of plants, such as medullosans and Caytoniales. These plants are of diverse evolutionary relationships.

pterosaurs
Flying reptiles related to the dinosaurs, with wings formed by skin stretched over their front limbs. Originating in the Triassic, they went extinct at the end of the Cretaceous.

rachis
The central stalk or midrib in a leaf or a flower shoot. *See also* midrib.

radial symmetry
A form of body symmetry analogous to the spokes of a wheel, as in echinoids and starfish.

radiation (evolutionary)
The emergence of a large number of novel evolutionary forms as the result of new adaptations.

radio-dating
Measuring naturally occurring radioactivity to estimate the age of rock or other material.

radius (bone)
One of the two main bones of the forearm.

ratites
The group of flightless birds that includes the living ostriches, emus, rheas, and kiwis, together with related extinct forms.

rauisuchians
A group of dinosaur-like reptiles that were actually more closely related to crocodilians. Widespread in the Triassic before the rise of dinosaurs, they became extinct before the end of that period.

reducing conditions
Conditions low in oxygen, sometimes favourable to the preservation of fossil animal and vegetable material.

reptiles
(1) In traditional classification schemes, all amniotes other than birds and mammals were classified as reptiles. Present-day forms, including lizards, snakes, turtles, and crocodilians, can also be defined as cold-blooded, but this does not necessarily apply to all extinct forms such as dinosaurs, and so does not itself define a reptile. (2) Modern classifications, which lay more emphasis on shared ancestry, tend to define reptiles more narrowly, in particular excluding the synapsids (the group from which mammals descended). On the other hand, birds do now count as reptiles, strictly speaking, since they descended from dinosaurs. *See also* cladistics, synapsids.

reptilomorphs
The group of tetrapods that contains all the amniotes plus some reptile-like amphibians that are thought to be their nearest relatives.

respiration
(1) Breathing. (2) Also called cellular respiration, the biochemical processes within cells that break down food molecules to provide energy, usually by combining the molecules with oxygen.

rhizome
A creeping underground stem that helps plants to spread asexually, to overwinter, and/or to store food. *See also* asexual.

rhynchosaurs
A group of plant-eating reptiles related to the archosaurs that lived only in the Triassic period. *See also* archosaurs.

rift valley
A large block of land that has dropped vertically compared with the surrounding regions, due to plate tectonic activity.

RNA
Short for ribonucleic acid. It has a similar structure to DNA and has various essential roles in cells, but is not used to store the main genetic information of an organism, except in some viruses.

robust
In palaeontological contexts: larger and/or more strongly built.

rodents
A very diverse group of mammals characterized by front teeth (incisors) that are specialized for gnawing. They include mice, rats, squirrels, beavers, and porcupines, together with many other forms.

rostroconchs
A group of extinct Paleozoic molluscs, superficially similar to bivalves but not closely related. *See also* bivalves, molluscs.

rotifers
Microscopic multicelled invertebrates living in both sea water and fresh water and having a variety of lifestyles.

ruminants
A subgroup of the even-toed hoofed mammals (artiodactyls) whose members have a specialized four-chambered stomach that allows them to digest tough plant food. Ruminants include cattle, sheep, antelopes, deer, and giraffes, but not pigs or hippopotamuses.

sarcopterygians
A mainly extinct group of fish having fleshy, muscular bases to their paired front and back fins. Also called lobe-finned fish, they are first known from the Silurian period and include the present-day lungfish and coelacanth. *See also* osteichthyans, tetrapods.

saurischians
One of the two main divisions of dinosaurs, in turn made up of two major subgroups, the theropods and the sauropodomorphs. The term "saurischian" means "lizard-hipped". *Compare* ornithischians.

sauropodomorphs
The group of dinosaurs that includes the well-known sauropods, huge herbivores with long necks and long tails that were the largest land animals ever to have lived. Other sauropodomorphs were smaller, and some early species were bipedal.

sauropterygians
A diverse group of semi-aquatic and fully aquatic reptiles that existed from the Triassic until the end of the Cretaceous. Well known members include the long-necked plesiosaurs and the shorter-necked, large-headed pliosaurs. Early forms, such as some nothosaurs, still possessed legs, but in later sauropterygians the limbs were modified as flippers.

scyphozoans
The group of cnidarians that includes jellyfish. *See also* cnidarians.

sedentary
Of animals such as worms: habitually staying in one position. *See also* sessile.

sedimentary rock
Rock formed when small particles settle out from water or are deposited by the wind, volcanic activity, etc., before later hardening to rock.

seed ferns
A general term for several groups of extinct seed plants that have fern-like leaves but belong to the seed plant group. *See also* bennettitaleans, Caytoniales, medullosans.

seed plants
Plants that reproduce by seeds as distinct from simpler spores. *See also* seed fern, gymnosperm, angiosperm, spore.

septum (pl. septa)
A partition within the body of an animal.

sessile
Of an animal: attached permanently to a surface, especially without a stalk, and not able to move around. *See also* sedentary.

sexual dimorphism
Condition in which the males and females of a species differ obviously in appearance.

sexual selection
Type of natural selection in which a feature has developed solely because of its direct advantage in furthering mating and reproduction, even if it appears disadvantageous in other ways. A male peacock's tail is an example.

silica/silicate
The chemical compound silicon dioxide, which as the mineral quartz is the main component of sand. Silicates consist of silicon and oxygen in chemical combination with atoms of various metals.

speciation
The formation of new species.

species divergence
The increasing dissimilarity over time of two or more species derived from a common ancestor.

sphenophytes
A group of vascular plants, also known as horsetails, whose above-ground stems have a jointed appearance, with rings of small branches bearing tiny scale-like leaves. The few living species are outnumbered by the many extinct forms, some as big as medium-sized trees.

sponges
A large group (phylum) of marine invertebrates with a very simple structure that feed by creating currents through their bodies and filtering small particles from the water. They have no muscles or nerve cells. Many have skeletons built of small hard elements called spicules.

sporangium (pl. sporangia)
A structure that produces spores: found in plants other than seed plants, and also in fungi.

spore
A usually microscopic structure produced, often in large numbers, by plants (except seed plants), fungi, and many micro-organisms, from which a new individual can grow. Spores are usually spread by wind or water, and may be produced by sexual or asexual reproduction.

sporophore
A spore-bearing structure.

stomata (sing. stoma)
Small openings on the surfaces of most plants, especially on the underside of their leaves, that allow gases (expecially oxygen, carbon dioxide, and water vapour) to enter or leave the plant. *See also* guard cell.

stratigraphy
The geological study of the order and relative position of strata (layers of sedimentary rock) in Earth's crust.

stromatolites
Large, hard, dome-like structures, built up of thin layers, that are usually produced in shallow water by the action of many generations of cyanobacteria. Some of Earth's oldest fossils are of stromatolites.

subduction
The forcing down of oceanic crust belonging to one tectonic plate beneath another plate when two plates collide. *See also* plate tectonics.

suture
A rigid junction between hard parts of animals, such as shell or bone. In internal moulds of ammonoids, a suture line marks the junction between the internal dividing walls and the inside wall of the shell.

symbiosis
A close living relationship between two species, especially one in which both benefit.

synapsids
A major group (formerly referred to as "mammal-like reptiles") that branched off early in the evolution of amniote tetrapods, and eventually gave rise to the mammals. *See also* amniotes, teptrapods.

taxon
Any named group of organisms that forms part of a classification system.

taxonomy
The science of classifying organisms; also, a classificatory scheme resulting from the scientific study of a particular group.

teleosts
An evolutionarily advanced group of ray-finned bony fish (actinopterygians) that include the majority of living fish species.

temnospondyls
A diverse group of extinct amphibians widespread from the Carboniferous to the Triassic, and not finally becoming extinct till the Cretaceous. They included some massive crocodile-like forms. *See also* lepospondyls.

Tertiary
The first period of the Cenozoic era, lasting from 65.5 to 1.8 mya.

tetrapods
Vertebrates with four limbs, including amphibians, reptiles, birds, and mammals.

theca (pl. thecae)
In graptolites: the organic-walled tubes housing zooids. *See also* graptolites, zooids.

therapsids
An extinct group of land vertebrates that include the direct ancestors of mammals. A subgroup of synapsids. *See also* synapsids.

theropods
A major group of bipedal dinosaurs that includes well known top predators such as *Tyrannosaurus rex*, as well as many smaller species.

trace fossil
A fossil that preserves the activity of a living thing, rather than the living thing itself – for example, a dinosaur footprint.

trilobites
An enormously diverse group of extinct marine arthropods. They became extinct at the end of the Permian period.

type species
A species that defines the genus in which it is included, and to which the genus name originally assigned is permanently attached even if later reclassification occurs.

ulna
One of the two main bones of the forearm.

valve
In molluscs and brachiopods: a shell, especially a separate section of a shell when it comes in more than one part. *See also* bivalves.

vascular bundle
A longitudinal strand of tissues within the stem of a vascular plant that conducts water and food materials.

vascular plants
Plants that have specialized tissue for conducting water and nutrients between different parts of their bodies. Most land plants, except for bryophytes, are vascular plants. *See also* bryophytes, lycophytes, euphyllophytes.

ventral
Relating to the lower surface or belly of an animal. *See also* dorsal.

vertebrates
Animals with backbones, including fish, amphibians, reptiles, birds, and mammals. The backbone elements (vertebrae) are usually made of bone, but sometimes of cartilage. Vertebrates also share many other distinctive features, such as the possession of a skull (cranium) and the way their internal organs are arranged. *See also* chordates.

water vascular system
A system of hydraulically operated water-filled tubes that is characteristic of the body structure of echinoderms. It typically ends in small projecting "tube feet", although these may be used for food capture and other purposes as well as walking, depending on the type of animal. *See also* echinoderms.

zooid
An individual of a colonial animal, such as corals, graptolites, and bryozoans.

zosterophylls
An extinct group of primitive vascular plants included together with the clubmosses (lycopods) in the lycophytes. Zosterophylls are among the earliest land plants known. They had simple, usually leafless, stems. *See also* lycophytes.

- **Dinosauria**
  - **Saurischia**
    - **Herrerasauridae**
    - **Sauropodomorpha**
      - **Massospondylidae**
      - **Plateosauria**
      - **Sauropoda**
        - Vulcanodontidae
        - Mamenchisauridae
        - Cetiosauridae
        - Diplodocoidea
          - Rebbachisauridae
          - Dicraeosauridae
          - Diplodocidae
        - Macronaria
          - Brachiosauridae
          - Titanosauria
    - **Theropoda**
      - **Coelophysidae**
      - **Dilophosauridae**
      - **Ceratosauria**
        - Ceratosauridae
        - Abelisauroidea
          - Noasauridae
          - Abelisauridae
      - **Tetanurae**
        - Spinosauroidea
          - Megalosauridae
          - Spinosauridae
            - Baryonychinae
            - Spinosaurinae
        - Allosauroidea
          - Sinraptoridae
          - Carcharodontosauridae
        - Coelurosauria
          - Tyrannosauroidea
            - Tyrannosauridae
          - Ornithomimosauria
            - Ornithomimidae
          - Compsognathidae
          - Maniraptora
            - Alvarezsauridae
            - Therizinosauroidea
            - Oviraptorosauria
              - Caenagnathidae
              - Oviraptoridae
            - Deinonychosauria
              - Troodontidae
              - Dromaeosauridae
                - Unenlagiinae
                - Microraptorinae
                - Velociraptorinae
                - Dromaeosaurinae
            - Avialae
              - Scansoriopterygidae

- **Ornithischia**
  - **Heterodontosauridae**
  - **Thyreophora**
    - **Stegosauria**
      - Stegosauridae
    - **Ankylosauria**
      - Nodosauridae
      - Polacanthidae
      - Ankylosauridae
  - **Cerapoda**
    - **Ornithopoda**
      - Iguanodontia
        - Rhabdodontidae
        - Dryosauridae
        - Camptosauridae
        - Hadrosauridae
          - Hadrosaurinae
          - Lambeosaurinae
    - **Marginocephalia**
      - Pachycephalosauria
        - Pachycephalosauridae
      - Ceratopsia
        - Chaoyangosauridae
        - Leptoceratopsidae
        - Protoceratopsidae
        - Ceratopsidae
          - Chasmosaurinae
          - Centrosaurinae

Dinosaur List

These pages list and briefly describe all current, non-avian dinosaur genera. There are over 800. The table to the left shows the relationship between dinosaur groups, with Saurischia and Ornithischia as the first main divisions.

Abelisaurus Large theropod from the Cretaceous of Argentina, with a blunt, deep skull. One of the first recognized members of the theropod group Abelisauridae.

Abrictosaurus Jurassic heterodontosaurid from South Africa. Had serrated, fang-like teeth, and shorter arms than related forms such as *Heterodontosaurus*.

Abrosaurus Chinese Jurassic sauropod known from excellent skull material. Had a deep snout and enormous bony nostril openings. It may be a primitive Macronarian.

Acanthopholis British ankylosaur from the Cretaceous. Based on fragmentary remains including odd bones and armour fragments, which lack unique features and cannot be distinguished from other ankylosaur bones.

Achelousaurus Cretaceous ceratopsid from the USA, closely related to *Einiosaurus* and *Pachyrhinosaurus*. Its nasal horn curved forwards over the tip of its snout.

Achillesaurus Poorly known Argentinean alvarezsaurid (a relative of *Mononykus*) from the Cretaceous. Presumably had the enlarged thumb and short arms of its relatives.

Achillobator Large dromaeosaurid from the Cretaceous of Mongolia, similar to *Utahraptor*. Its pubic bones pointed downwards, rather than in the typical backwards manner.

Acrocanthosaurus Large allosauroid theropod from the Cretaceous of the USA. It had a huge skull and a prominent muscular ridge along the length of its back.

Adamantisaurus Poorly known South American titanosaur from the Cretaceous of Brazil, known only from tail vertebrae. It is named after the Adamantina Formation.

Adasaurus Small dromaeosaurid maniraptoran from the Cretaceous of Mongolia. Its hind limbs were powerful and the sickle claws on its toes unusually small.

Adeopapposaurus Jurassic sauropodomorph from Brazil, known from the remains of four individuals originally identified as *Massospondylus*.

Aegyptosaurus Egyptian sauropod from the Cretaceous, known from remains that were destroyed during World War II.

Aeolosaurus Titanosaur from the Cretaceous of Argentina. Two species have been named.

Aepisaurus French sauropod from the Cretaceous, known only from an arm bone that has been lost.

Aerosteon Argentinean allosauroid theropod from the Cretaceous, known from a partial skeleton. Similar to *Allosaurus* but with a surprising number of air-filled bones.

Afrovenator Spinosauroid theropod from the Cretaceous of Niger, closely related to *Eustreptospondylus*. It had a long skull and low, rounded horns in front of its eyes.

Agilisaurus Small, bipedal Jurassic ornithischian from China. Originally it was classified as an ornithopod, but it is now thought more primitive.

Agnosphitys Poorly known, primitive Triassic dinosaur from Britain. It may not be a dinosaur at all.

Agujaceratops Cretaceous ceratopsid from the USA, originally classified as a new species of *Chasmosaurus*. It had long brow horns and a long, erect frill.

Agustinia Unusual sauropod from the Cretaceous of Argentina. Tall spines were arranged across its back.

Alamosaurus Cretaceous sauropod from the USA. Typically thought to be the only Cretaceous North American titanosaur, and to be an invader from South America or Asia.

Alaskacephale Pachycephalosaur from the Cretaceous of Alaska, named for a skull fragment.

Albertaceratops Canadian ceratopsid from the Cretaceous, and the most primitive centrosaurine. Unlike other centrosaurines, it had long brow horns.

Albertonykus Canadian relative of *Mononykus* from the Cretaceous. Its remains include arm and leg bones from several individuals. It was less than one metre long.

Albertosaurus Well known tyrannosaurid from the Cretaceous of North America with triangular horns in front of its eyes and particularly short arms.

Alectrosaurus Medium-sized tyrannosaurid from the Cretaceous of Mongolia. It had slender hind limbs.

Aletopelta Cretaceous ankylosaur from the USA. Mosaic-like armour plates covered the back and a pair of tall spikes projected upwards from the shoulders.

Algoasaurus Sauropod from the Jurassic or Cretaceous of South Africa, known from a few bones and of uncertain affinity.

Alioramus Medium-sized tyrannosaurid from the Cretaceous of Mongolia. It had a long, low snout, the upper surface of which was decorated by a series of small horns.

Allosaurus Jurassic allosauroid theropod, first discovered in the USA but now known from Portugal. It had triangular horns in front of its eyes. Several species have been named.

Alocodon Portugese ornithischian from the Jurassic, known only from teeth.

Altirhinus Mongolian iguanodontian ornithopod from the Cretaceous with a massive, inflated nasal region. It had a small thumb-spike and was similar to *Iguanodon*.

Altispinax German theropod from the Cretaceous known only from a single tooth that has been lost. The name *Altispinax* was later given, incorrectly, to *Becklespinax*.

Alvarezsaurus Small, slender Argentinean theropod from the Cretaceous. The first member of the maniraptoran group Alvarezsauridae to be named.

Alwalkeria Poorly known primitive saurischian from the Triassic of India. A short gap separated the teeth at the snout tip from those further back.

Alxasaurus Medium-sized theropod from the Cretaceous of Mongolia. It had a long neck and its limbs and feet show that it was a relatively slow runner.

Amargasaurus Remarkable Argentinean diplodocoid sauropod from the Cretaceous, with long bony spikes that projected upwards and backwards from its neck bones. Closely related to *Dicraeosaurus* from Tanzania.

Amargatitanis Titanosaur from the Cretaceous of Argentina, known from vertebrae, limb bones, and a shoulder blade. Its shoulder blade was wide and flat and the head of its thigh bone was unusually large.

Amazonsaurus Poorly known Brazillian sauropod, and the first dinosaur named from the Amazon region. Probably a diplodocoid.

Ammosaurus Small primitive sauropodomorph from the Jurassic of the USA, thought by some to be *Anchisaurus*.

Ampelosaurus French titanosaur from the Cretaceous, known from numerous bones belonging to many specimens. It had several different kinds of armour on its back.

Amphicoelias Jurassic sauropod from the USA, generally thought to be a primitive diplodocoid similar to *Diplodocus*. One species may be the largest sauropod.

Amtosaurus Name given to ornithischian skull remains from the Cretaceous of Mongolia. These have conventionally been identified as belonging to an ankylosaurid, but this identification cannot be supported.

Amurosaurus Russian hadrosaurid from the Cretaceous, originally described from partial skull bones but now known from numerous remains.

Amygdalodon Jurassic sauropod from Argentina known from teeth, vertebrae, and hip and limb bones. Its oval teeth resemble British sauropod teeth, *Cardiodon*.

Anabisetia Small, bipedal ornithopod from the Cretaceous of Argentina, known from numerous remains of several individuals.

Anasazisaurus Deep-snouted hadrosaurid from the Cretaceous of the USA, regarded by many experts as identical to either *Gryposaurus* or *Kritosaurus*.

Anatotitan Large hadrosaurine hadrosaurid from the Cretaceous of North America. It had particularly long, duck-billed jaws. Extremely similar to *Edmontosaurus* and thought by some experts to be part of that genus.

Anchiceratops Rhino-sized ceratopsid from the Cretaceous of Canada. It had a long snout, long brow horns, and a long, rectangular frill.

Anchiornis Small, feathered maniraptoran from China with a raised second toe, known from excellent partial skeletons. Originally described as a bird close to *Archaeopteryx*.

Anchisaurus Small North American sauropodomorph from the Jurassic, with a pointed snout.

Andesaurus Primitive titanosaur from the Cretaceous of Argentina, originally regarded as a close relative of *Argentinosaurus*. Known from vertebrae and limb and hip bones.

Aniksosaurus Coelurosaur from the Cretaceous of Argentina of uncertain affinities. It had stout hand claws and relatively robust arm and hind limbs bones.

Animantarx Medium-sized nodosaurid ankylosaur from the Cretaceous of the USA, known from a single specimen that includes a partial skull. It had a domed skull roof and small horns behind its eyes.

Ankylosaurus Gigantic Cretaceous ankylosaurid ankylosaur from North America. It had a broad body covered in spikes and plates and a bony club at the end of its tail.

Anoplosaurus Name used for various ankylosaur remains from the Cretaceous of Britain. None of these (mostly vertebrae and armour plates) can be identified precisely.

Anserimimus Ornithomimid ornithomimosaur from the Cretaceous of Mongolia. Compared to its relatives, it had straight hand claws with flattened undersides.

Antarctopelta Ankylosaur from the Cretaceous of Antartica, probably a nodosaurid. Small horns projected sideways from above its eyes.

Antarctosaurus Gigantic sauropod, probably a titanosaur, from the Cretaceous of Argentina. Its lower jaws were rectangular when seen from above.

Antetonitrus Primitive sauropod from the Triassic of South Africa, closely related to *Lessemsaurus* from Argentina. Its hands were specialized for weight-bearing, but still had grasping features.

Antrodemus Jurassic theropod from the USA, named for fragmentary remains that cannot be identified with certainty but probably belonged to *Allosaurus*.

Apatodon Old name given to a partial, damaged vertebra from the Jurassic of the USA that has since been lost. This was originally misidentified as a fossil pig jaw.

Apatosaurus Jurassic diplodocoid sauropod with a long, whip-like tail. It had a broad neck and massive, muscular limbs. Several species are known, one was originally named *Brontosaurus*.

Appalachiosaurus North American tyrannosauroid from the Cretaceous, closely related to tyrannosaurids and differing from them only in minor details.

Aragosaurus Sauropod from the Cretaceous of Spain, known from vertebrae, hip bones, and hind limb bones.

Aralosaurus Hadrosaurine hadrosaurid from the Cretaceous of Kazakhstan. It had a deep snout and a large, hooked crest in front of its eyes, and would have looked similar to *Gryposaurus* from North America.

Archaeoceratops Small ceratopsian from the Cretaceous of China, known from two partial skeletons. It had slim limbs, a narrow, hooked beak, and was probably bipedal.

Archaeodontosaurus Jurassic sauropod from Madagascar, known only from its lower jaw. Its teeth were primitive compared to those of most other sauropods.

Archaeornithoides Enigmatic small theropod from the Cretaceous of Mongolia, known from a partial skull that seems to have been bitten by a Mesozoic mammal. Once suggested to be a hatchling tyrannosaurid.

Archaeornithomimus Ornithomimid ornithomimosaur from the Cretaceous of China, known from numerous bones and originally identified as a species of *Ornithomimus*. Its foot bones are more robust than those of related forms.

Argentinosaurus Gigantic Argentinean titanosaur from the Cretaceous, known from vertebrae and limb bones. It had peg-and-socket joints between its vertebrae, making it more primitive than most titanosaurs.

Argyrosaurus Poorly known, large titanosaur from the Cretaceous of Argentina, known from forelimb bones and vertebrae. Its hand bones were long while the arm bones were stout.

Aristosaurus Small sauropodomorph from the Jurassic of South Africa, named for a partial juvenile skeleton that includes jaw fragments. Generally thought to be the same animal as *Massospondylus*.

Aristosuchus Small theropod from the Cretaceous of Britain, named for vertebrae and hip bones. These are similar in shape to the corresponding bones of *Compsognathus*.

Arrhinoceratops Long-frilled, rhinocerous-sized chasmosaurine ceratopsid from the Cretaceous of Canada. It had a short, deep snout with a very short nose horn and long, forward-curving brow horns.

Arstanosaurus Poorly known hadrosaurid from the Cretaceous of Kazakhstan, named for a partial skull bone that cannot be distinguished from other hadrosaurids.

Asiaceratops Ceratopsian from the Cretaceous of Uzbekistan, named for skull fragments, teeth, and other bones. These cannot be distinguished with certainty from the remains of other ceratopsians.

Asiamericana Theropod from the Cretaceous of Uzbekistan, named for teeth originally thought to be from a spinosaurid or fish. They cannot be reliably distinguished from other theropod teeth but might belong to *Richardoestesia*.

Asiatosaurus Cretaceous sauropod from China, named for two large teeth that have similarities with the teeth of brachiosaurs and camarasaurs.

Astrodon Macronarian sauropod from the Cretaceous of the USA, named for small spatulate teeth. Often thought to be the same as *Pleurocoelus*.

Asylosaurus Small, primitive sauropodomorph from the Triassic of Britain, originally included within *Thecodontosaurus*, but said to differ from it in the shape of the upper arm bone. A lightly built, omnivorous biped or quadruped.

Atlantosaurus Jurassic sauropod from the USA, known from remains that cannot be identified precisely, and might belong to *Apatosaurus* or *Camarasaurus*.

Atlasaurus Peculiar but poorly known sauropod from the Jurassic of Morocco, with unusual tall vertebrae.

Atlascopcosaurus Small bipedal ornithopod from the Cretaceous of Australia, named for a skull bone and teeth. Usually thought to be similar to *Hypsilophodon*.

Atrociraptor Dromaeosaurid maniraptoran from the Cretaceous of Canada, known from a deep-snouted skull with strongly recurved teeth.

Aublysodon Small tyrannosaurid from the Cretaceous of North America, originally thought to be a peculiar form, but now mostly regarded as a juvenile *Tyrannosaurus*.

Aucasaurus Abelisaurid from the Cretaceous of Argentina, closely related to *Carnotaurus*. It had a short, deep head and peculiar straight arms with short fingers.

Auroraceratops Small ceratopsian from the Cretaceous of China. It had a short, wide snout and peculiar patches of wrinkled bone on its face and jaws.

Australodocus African diplodocid sauropod from the Jurassic of Tanzania, originally misidentified as *Barosaurus*.

Austroraptor Large maniraptoran from the Cretaceous of Argentina with a particularly long-snouted skull. Probably a close relative of *Unenlagia*.

Austrosaurus Poorly known Australian sauropod from the Cretaceous, known from vertebrae and limb bones from several individuals. Likely to be a titanosaur.

Avaceratops Small ceratopsid ceratopsia from the Cretaceous of the USA, often thought to be a centrosaurine. It had a deep snout, solid frill, and short, curved brow horns.

Aviatyrannis Small tyrannosauroid theropod from Portugal, named for a hip bone. Originally identified as *Stokesosaurus*.

Avimimus Small Mongolian oviraptorosaur with a deep, toothless skull. Like other oviraptorosaurs it would have been feathered like a bird.

Avipes Named for a partial foot from a small Triassic reptile from Germany, originally thought to be a theropod. The foot does not definitely belong to a dinosaur.

Bactrosaurus Hadrosaurid from the Cretaceous of China, known from many specimens. It is usually thought to have been a crestless, primitive hadrosaurid.

Bagaceratops Small ceratopsian from the Cretaceous of Mongolia, known from numerous specimens. It had a short frill and a low nasal horn.

Bagaraatan Medium-sized coelurosaur from the Cretaceous of Mongolia, with a deep lower jaw and slim hind limbs. Possibly a primitive tyrannosauroid.

Bahariasaurus Poorly known theropod from the Cretaceous of northern Africa, known from various fragments. Some of these might belong to *Deltadromeus* or *Carcharodontosaurus*.

Bainoceratops Cretaceous ceratopsian from Mongolia, similar in size to *Protoceratops*. Named for a partial vertebral column that is said by some to fall within the range of variation seen in *Protoceratops*.

Balochisaurus Cretaceous sauropod from Pakistan, based on various incomplete vertebrae. Said to be a titanosaur and a close relative of *Marisaurus*.

Bambiraptor Small dromaeosaurid maniraptoran from the Cretaceous of the USA, closely related to *Saurornitholestes* and *Atrociraptor*.

Barapasaurus Primitive Jurassic sauropod from India. It had particularly long limbs. Suggested by some experts to be closely related to *Cetiosaurus* from Britain.

Barosaurus Jurassic diplodocoid sauropod from the USA, related to *Diplodocus* but with an even longer, slimmer neck. It was probably able to reach particularly high into trees.

Barsboldia Mongolian hadrosaurid from the Cretaceous, known from vertebrae, ribs, and hip bones. It had tall spines on its vertebrae and is suggested to be a lambeosaurine.

Baryonyx Cretaceous spinosauroid theropod with a crocodile-like skull and powerful forelimbs. Originally discovered in Britain, it is also known from Spain, Germany, and Portugal.

Baurutitan Titanosaur from the Cretaceous of Brazil, known from tail vertebrae and some other remains.

Becklespinax British Cretaceous theropod known only from three vertebrae with tall, robust, bony spines. It may have had a low sail along its back and hips.

Beipiaosaurus Therizinosauroid from the Cretaceous of China, preserved with long quill-like "proto-feathers" all over its body. One of the most primitive therizinosauroids.

Bellusaurus Chinese sauropod from the Jurassic, known from juvenile skeletons. It has been identified as a titanosaur but also as a primitive macronarian close to *Jobaria*.

Berberosaurus Jurassic theropod from Morocco, originally suggested to be a primitive abelisaur. Found in close association with the sauropod *Tazoudasaurus*.

Betasuchus Cretaceous theropod from the Netherlands, named for a single partial thigh bone, and originally named *Megalosaurus*. It might belong to an abelisaurid.

Bienosaurus Thyreophoran from the Jurassic of China, originally suggested to be a close relative of *Scelidosaurus*. Known from a lower jaw and a partial skull roof.

Bihariosaurus Poorly known iguanodontian ornithopod from the Cretaceous of Romania, thought similar to *Camptosaurus* by some.

Bissektipelta Ankylosaurid from the Cretaceous of Uzbekistan, known only from a braincase and armour scutes.

Blikanasaurus Primitive sauropod from the Triassic of South Africa, known only from the hind limb bones of two specimens.

Bonatitan Small Cretaceous titanosaur from Argentina known from substantial remains that include skull bones. Its neck vertebrae were very long for a titanosaur.

Bonitasaura Unusual small titanosaur from the Cretaceous of Argentina. Toothless spaces on the sides of its jaw bones seem to have formed guillotine-like blades used in slicing vegetation.

Borealosaurus Titanosaur from the Cretaceous of China, known only from tail vertebrae. These are concave on their back surfaces like those of *Opisthocoelicaudia*.

Borogovia Troodontid maniraptoran from the Cretaceous of Mongolia, known from hind limb material. The large claws on its second toes are less strongly curved than they are in other troodontids.

Bothriospondylus Old name given to sauropod vertebrae from the Jurassic of Britain. These cannot be identified precisely and the name is rarely used.

Brachiosaurus Giant sauropod from the Jurassic of the USA, famous for its tall shoulders, long arms, and long neck.

Brachyceratops Ceratopsid ceratopsian from the Cretaceous of North America, initially known from the remains of several juveniles discovered together.

Brachylophosaurus Hadrosaurine hadrosaurid from the Cretaceous of North America. It had a deep snout and a sheet-like crest that projected backwards over the eyes.

Brachypodosaurus Controversial dinosaur from the Cretaceous of India, known only from an arm bone that might belong to an ankylosaur. Lacks features to distinguish it from the arm bones of other ankylosaurs.

Brachyrophus Jurassic ornithopod from the USA, generally regarded to be the same animal as *Camptosaurus* but originally identified as a sauropod.

Brachytrachelopan Small dicraeosaurid diplodocoid from the Jurassic of Argentina, known as an articulated partial skeleton. It had a short neck with very short vertebrae.

Bradycneme Maniraptoran known only from the end of a shin bone from the Cretaceous of Romania.

Breviceratops Cretaceous ceratopsian from Mongolia. Extremely similar to *Protoceratops* and originally included in this genus. Some think that it is the same as *Bagaceratops*.

Brohisaurus Sauropod from the Jurassic of Pakistan, named for limb fragments thought to belong to a titanosaur similar to Cretaceous forms *Khetranisaurus* and *Pakisaurus*.

Bruhathkayosaurus Poorly known sauropod from the Cretaceous of India, known from a gigantic tibia and from forelimb and hip bones, and vertebrae. This may have been one of the largest sauropods of all.

Bugenasaura Bipedal North American ornithopod from the Cretaceous, known from a partial skull, hind limb, and other bones originally thought to belong to *Thescelosaurus*.

Buitreraptor Dromaeosaurid maniraptoran from the Cretaceous of Argentina, known from a near-complete skeleton. It had a very long, low snout. Closely related to *Unenlagia*.

Byronosaurus Mongolian troodontid maniraptoran from the Cretaceous, with a long, slender snout, and numerous teeth.

Caenagnathasia Tiny oviraptorosaur from the Cretaceous of Uzbekistan, known only from a partial lower jaw. One of the smallest non-avian theropods.

Caenagnathus Oviraptorosaur from the Cretaceous of Canada, named for lower jaw bones once thought to be from a giant flightless bird. Often regarded today as the same animal as *Chirostenotes*.

Calamosaurus Small theropod from the Cretaceous of Britain, known only from two neck vertebrae. These might be from a *Dilong*-like tyrannosauroid.

Calamospondylus Enigmatic small theropod from the Cretaceous of Britain, known from a partial hip region that has since been lost. One of the earliest small theropods to be discovered.

Callovosaurus Iguanodontian ornithopod from the Jurassic of Britain, named for a single thigh bone, which might have belonged to a dryosaurid.

Camarasaurus Jurassic sauropod from the USA, with a relatively short, powerful neck and a short, deep skull. Known from many specimens belonging to several species, and is probably the best known of all sauropods.

Camelotia Large Triassic sauropodomorph from Britain, thought by some to be a relative of *Melanorosaurus* from South Africa. Known from hind limb bones and other remains.

Camposaurus Poorly known coelophysoid theropod from the Triassic of the USA, known from fragmentary hind limb bones.

Camptosaurus Iguanodontian ornithopod from the Jurassic of the USA, known from many specimens belonging to several species. It had a heavy body, stout hind limbs, and short arms. Its fingers were stout and it had a large, conical thumb claw.

Campylodoniscus Sauropod from the Cretaceous of Argentina, named for a single partial skull bone originally named *Campylodon*. The precise identity of this bone cannot be determined; some experts have suggested that it belonged to a titanosaur.

Carcharodontosaurus Giant African allosauroid theropod with a huge, deep skull lightened by massive bony openings. Two species are known, from northern Africa.

Cardiodon Enigmatic British sauropod from the Jurassic, known from a single tooth often assumed to have belonged to *Cetiosaurus*.

Carnotaurus Well known abelisaurid theropod from the Cretaceous of Argentina, famous for its short, deep snout, and horns. The horns were rounded and thick. The only known specimen includes skin impressions.

Caseosaurus Triassic fossil from the USA, suggested by some to be a herrerasaurid but by others to be a primitive saurischian of uncertain affinities. It may be the same animal as *Chindesaurus*. Some experts argue that it may not be a dinosaur at all.

Cathartesaura Rebbachisaurid diplodocoid from the Cretaceous of Argentina, known from vertebrae and shoulder, hip, and limb bones. It was medium-sized, and related to *Limaysaurus* and *Rebbachisaurus*.

Cathetosaurus Macronarian sauropod from the Jurassic of the USA, generally thought to be the same animal as *Camarasaurus*. Originally argued to differ from *Camarasaurus* in features of its vertebrae.

Caudipteryx Small Chinese oviraptorosaur from the Cretaceous, and one of the first non-avian dinosaurs to be discovered with definite feathers. It had a short tail and long hind limbs. Its skull was short and tall with teeth at the jaw tips alone.

Cedarosaurus Medium-sized macronarian sauropod from the Cretaceous of the USA, known from a partial skeleton. It had particularly long, slender forelimb bones, and was probably similar in appearance to *Brachiosaurus* and *Giraffatitan*.

Cedarpelta Large Cretaceous ankylosaur from the USA, known from two skulls and other material. Originally regarded as an early ankylosaurid, closely related to *Gobisaurus* and *Shamosaurus* from eastern Asia.

Cedrorestes Iguanodontian ornithopod from the Cretaceous of the USA, mostly known from hind limb and pelvic bones.

Centrosaurus Ceratopsid from the Cretaceous of North America. A short-frilled, rhino-sized horned dinosaur with a long nasal horn. Closely related to *Styracosaurus*.

Cerasinops Small Cretaceous ceratopsian from the USA. It had a short bony frill and two pairs of fang-like teeth at the front of the upper jaw. Suggested to be capable of walking bipedally, unlike other North American ceratopsians.

Ceratonykus Alvarezsaurid from the Cretaceous of Mongolia, related to *Mononykus* and *Shuvuuia*, and known from a near-complete skull and other remains. Like other alvarezsaurids, it may have been a specialized ant-eating theropod.

Ceratops Old name once used for ceratopsid remains – horns and a fragment from the back of the skull – from the Cretaceous of North America. This original material cannot be reliably distinguished from the remains of other ceratopsids.

Ceratosaurus Jurassic theropod from the USA, Portugal, and perhaps Tanzania, with a nasal horn and triangular horns over its eyes. It had particularly long teeth and small armour scutes along its backbone.

Cetiosauriscus British Jurassic sauropod known from a partial skeleton. Often thought to be a diplodocoid, it may in fact be related to *Mamenchisaurus* from China.

Cetiosaurus Jurassic sauropod from Britain, and the first sauropod to be known from good remains. Many species named, but some have since been shown to be diplodocoids, brachiosaurs, or other kinds of sauropod.

Changchunsaurus Poorly known small, bipedal ornithischian from the Cretaceous of China, suggested by its describers to be an ornithopod. It had a long, pointed snout and small bony lumps on the sides of its cheeks.

Chaoyangsaurus Small primitive ceratopsian from the Jurassic of China, known from a partial skull, neck vertebrae, and forelimb bones. It had two pairs of fang-like teeth at the tip of the upper jaw, and the edges of the upper jaw's beak were serrated.

Charonosaurus Gigantic lambeosaurine hadrosaurid from the Cretaceous of China, known from abundant remains. Closely related to *Parasaurolophus*.

Chasmosaurus Cretaceous ceratopsid from North America. Had a long, rectangular frill with large openings. Various species differed in the length of their snout and brow horns.

Chassternbergia Nodosaurid ankylosaur from the Cretaceous of North America. Thought by most experts to be the same animal as *Edmontonia*.

Chebsaurus Medium-sized sauropod from the Jurassic of Algeria, known from a partial skeleton that includes some skull material.

Chialingosaurus Chinese stegosaur from the Jurassic, said to have slender forelimb bones. Only known specimen is a juvenile.

Chiayusaurus Cretaceous sauropod from China, named for a single tooth that cannot be identified precisely. Sometimes thought to be the same as *Asiatosaurus*.

Chilantaisaurus Chinese theropod from the Cretaceous, known from fragmentary remains that might belong to a spinosauroid.

Chindesaurus Triassic saurischian from the USA, known from fragments. Suggested by some to be a herrerasaurid but generally thought to be a primitive saurischian.

Chinshakiangosaurus Jurassic sauropod from China, originally described from several bones, most of which are now missing. Its jaw bones suggest that it had fleshy cheeks.

Chirostenotes Large, North American oviraptorosaur from the Cretaceous, with long, slender hands, generally included in the group Caenagnathidae. Controversy remains as to whether it is the same as *Caenagnathus*.

Chondrosteosaurus Obscure British sauropod from the Cretaceous, known from neck vertebrae. Once identified as similar to *Camarasaurus*, it might have been a brachiosaur or titanosaur.

Chuandongocoelurus Jurassic theropod from China, known from fragmentary remains that have similarities with the bones of *Elaphrosaurus* from Tanzania.

Chuanjiesaurus Chinese sauropod from the Jurassic, thought by some to be a close relative of *Mamenchisaurus*. Its tail vertebrae are similar to those of *Bellusaurus*.

Chubutisaurus Argentinean sauropod from the Cretaceous, known from vertebrae and limb bones, and suggested by some experts to be a brachiosaur or titanosaur.

Chungkingosaurus Small Chinese stegosaur from the Jurassic, known from a partial skull and some armour fragments. Its hip bones allow it to be distinguished from other stegosaurs.

Cionodon Old name given to a partial hadrosaurid jaw from the Cretaceous of the USA. Cannot be identified precisely.

Citipati Mongolian oviraptorid from the Cretaceous, originally thought to be the same as *Oviraptor*. Some specimens are preserved on top of egg-filled nests.

Claosaurus Small hadrosaurid from the Cretaceous of the USA, known from a partial skeleton originally identified as a species of *Hadrosaurus*. Its skull is unknown but for one small jaw fragment.

Clasmodosaurus Sauropod from the Cretaceous of Argentina, named for teeth that cannot be reliably distinguished.

Coelophysis Triassic theropod known from hundreds of specimens from New Mexico, USA. Lightly built, with a long, shallow snout.

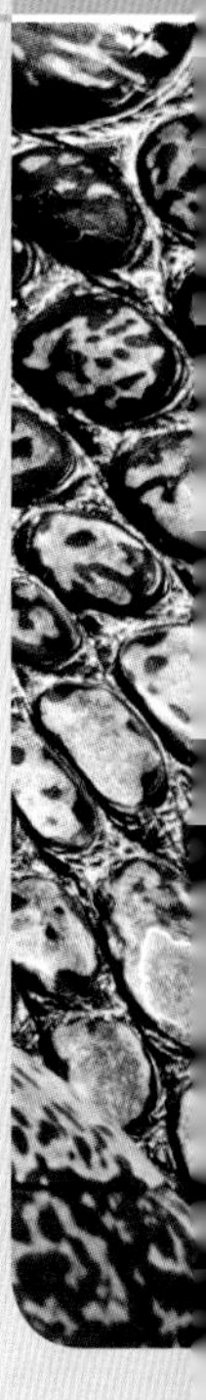

Coeluroides Small theropod from the Cretaceous of India, known from tail vertebrae once assumed to be from a coelurosaur, but now known to be from an abelisaur.

Coelurus Small coelurosaur from the Jurassic of the USA, often confused with *Ornitholestes*. Unlike related forms, it had the remnants of a fourth finger.

Colepiocephale Dome-headed pachycephalosaur from the Cretaceous of Canada, previously thought to be a species of *Stegoceras*. It lacked the bony shelf present at the back of the skull in *Stegoceras*.

Coloradisaurus Sauropodomorph from the Triassic of Argentina, originally named *Coloradia*. It had a short, triangular snout. Closely related to *Massospondylus*.

Compsognathus Small, slender Jurassic theropod known from Germany and France. It had long, slim hind limbs and a very long tail. It was once thought to possess only two fingers, but this is now uncertain.

Compsosuchus Poorly known Indian theropod known from neck bones. Recent work has shown that it is probably an abelisaur similar to *Masiakasaurus*.

Conchoraptor Mongolian oviraptorid from the Cretaceous with a deep, rounded snout. Closely related to *Ingenia* and *Khaan*.

Condorraptor Jurassic theropod from Argentina, known from an incomplete skeleton that includes vertebrae, hip, and hind limb bones. It was perhaps a spinosauroid, or a close relative of this group.

Corythosaurus North American lambeosaurine hadrosaurid from the Cretaceous, with a shallow snout and rounded, plate-like crest. Known from spectacular specimens, some of which have skin preserved.

Craspedodon Cretaceous ornithischian from Belgium, known only from teeth. Long thought to be from an ornithopod, they may in fact be from a ceratopsian.

Crataeomus Small Cretaceous ankylosaur from Austria, generally thought to be the same as *Struthiosaurus*.

Craterosaurus British stegosaur from the Cretaceous, known from a single vertebra originally identified as a lizard skull.

Creosaurus Large allosauroid theropod from the Jurassic of the USA, mostly thought today to be the same animal as *Allosaurus*.

Crichtonsaurus Medium-sized Jurassic ankylosaur from China, named after Michael Crichton, author of the book *Jurassic Park*. It is known from a partial lower jaw, vertebrae, and limb and shoulder bone fragments.

Cristatusaurus Cretaceous spinosaurid theropod from Niger, originally said to be *Baryonyx* but now thought by some to belong to *Suchomimus*. Known only from a snout tip.

Cryolophosaurus Antarctica's best dinosaur discovered to date, this is a Jurassic theropod with a bizarre crest. It was closely related to *Dilophosaurus*.

Cryptosaurus British ankylosaur from the Jurassic, long known as *Cryptodraco*. Named for a thigh bone, and suggested by some to be a polacanthid similar to *Hoplitosaurus*.

Cryptovolans Small dromaeosaurid maniraptoran from the Cretaceous of China with long, modern-type feathers on its limbs. Thought by many to be *Microraptor*.

Cumnoria British iguanodontian ornithopod from the Jurassic, often regarded as a species of *Camptosaurus* but differing from it in skull shape. Known from vertebrae, limb bones, and skull fragments.

Daanosaurus Macronarian sauropod from the Jurassic of China, suggested to be similar to *Bellusaurus*. The similarity might be due to both being known from juvenile remains.

Dacentrurus Large European Jurassic stegosaur known from England, Spain, and Portugal. Had long spikes on its hips and tail.

Dakotadon North American iguanodontian ornithopod from the Cretaceous, known from a partial skull originally said to be a species of *Iguanodon* but differing from it substantially.

Dandakosaurus Poorly known theropod from the Jurassic of India, known from a pelvic bone, probably from a ceratosaur.

Danubiosaurus Nodosaurid ankylosaur from the Cretaceous of Austria, widely thought to be the same as *Struthiosaurus*.

Dashanpusaurus Obscure Jurassic sauropod from China, known from vertebrae and hip and hind limb fragments. Several specimens have been identified.

Daspletosaurus North American tyrannosaurid from the Cretaceous, usually said to be more heavily built than its close relatives *Albertosaurus* and *Gorgosaurus*. Unlike *Tyrannosaurus* it had short triangular horns in front of its eyes.

Datousaurus Chinese sauropod from the Jurassic, known from two specimens. It has a stout, deep skull and robust, spatulate teeth, and was originally thought to be a close relative of *Camarasaurus*.

Daxiatitan Titanosaur from the Cretaceous of China, known from neck vertebrae and a thigh bone.

Deinocheirus Giant theropod from the Cretaceous of Mongolia, known only from its long arms and three-fingered hands. The broad, weakly curved claws do not seem to have been those of a dedicated predator.

Deinonychus Cretaceous dromaeosaurid from the USA. Like other members of the group, it had long hands and an enlarged, curved claw on each second toe. Thought by some researchers to have hunted in packs.

Deltadromeus Large theropod from the Cretaceous of Morocco with slender hind limbs. Originally identified as a primitive coelurosaur but now thought to be an abelisaur related to *Masiakasaurus* and *Velocisaurus*.

Denversaurus North American nodosaurid ankylosaur from the Cretaceous. Thought by most experts to belong to *Edmontonia*, but said to have a different skull shape.

Diclonius Old name once used for North American hadrosaurid teeth from the Cretaceous.

Dicraeosaurus African diplodocoid sauropod from the Jurassic of Tanzania. It had a comparatively short neck, and a tall bony ridge ran along its back. Closely related to *Amargasaurus* and *Brachytrachelopan*.

Didanodon Lambeosaurine hadrosaurid from the Cretaceous of Canada, named for teeth that are now generally thought to belong to *Lambeosaurus*.

Dilong Small, primitive tyrannosauroid from the Cretaceous of China. It had three-fingered hands and quill-like proto-feathers.

Dilophosaurus Medium-sized, slender theropod from the Jurassic of the USA, with a second species from China. Famous for its twinned, plate-like head crests.

Dimodosaurus Old name once used for sauropodomorph remains from the Triassic of Germany, probably the same as *Plateosaurus*.

Dinheirosaurus Portugese Jurassic sauropod, known from vertebrae, ribs, and limb bones originally thought to belong to *Lourinhasaurus*. Probably a diplodocoid.

Dinodocus Poorly known British sauropod from the Cretaceous, known from a limb and pelvic bones originally though to belong to a large plesiosaur.

Diplodocus Giant diplodocoid sauropod from the Jurassic of the USA, famous for its long neck. Its tail was strong and muscular, with a slender, whiplike tip.

Diplotomodon North American dinosaur from the Cretaceous, known from a single serrated tooth, which might belong to the tyrannosauroid *Dryptosaurus*.

Dolichosuchus Coelophysoid theropod from the Triassic of Germany, named for a single limb bone, which cannot be reliably distinguished from other coelophysoids.

Dollodon Iguanodontian ornithopod from the Cretaceous of Belgium, long confused with *Iguanodon*, and later with *Mantellisaurus*. It was a slender iguanodontian with long jaws.

Dongbeititan Chinese Cretaceous macronarian sauropod, closely related to brachiosaurs and titanosaurs. Known from a partial skeleton found in the same sediments as the feathered Chinese theropods.

Dongyangosaurus Chinese macronarian sauropod from the Cretaceous, known from distinctive back and tail vertebrae, which had unusual struts and hollows on their sides.

Draconyx Poorly known iguanodontian ornithopod from the Jurassic of Portugal. Known from various fragments. Its teeth and thigh bone are similar to those of *Camptosaurus*.

Dracopelta Portugese ankylosaur from the Jurassic, and one of the oldest members of its group. Less than three metres long. Known from ribs, vertebrae, and armour scutes.

Dracorex North American pachycephalosaur with a small skull-dome. Large bony spikes and knobs covered the sides of its dome and surface of its snout. It may be a growth stage of *Pachycephalosaurus*.

Dracovenator Large South African theropod from the Jurassic. Closely related to *Dilophosaurus* and *Cryolophosaurus*, but known from less complete remains.

Dravidosaurus Controversial Indian dinosaur long said to be the only stegosaur that survived late into the Cretaceous. Its fossils may not be from a dinosaur at all.

Drinker Small North American ornithischian from the Jurassic with coarsely serrated cheek teeth, a narrow beak, and fang-like teeth in the upper jaw tip.

Dromaeosauroides Medium-sized maniraptoran from the Cretaceous of Denmark, known from teeth similar to those of *Dromaeosaurus* from North America.

Dromaeosaurus Dromaeosaurid maniraptoran from the Cretaceous of North America. Its skull was deeper and stouter than that of relatives such as *Velociraptor*.

Dromiceiomimus North American ornithomimid from the Cretaceous, very similar to *Struthiomimus* and thought by some experts to be the same animal.

Dromicosaurus South African Jurassic sauropodomorph, named for vertebrae and limb bone fragments. Now generally thought to be the same as *Massospondylus*.

Dryosaurus Medium-sized iguanodontian ornithopod from the USA, and one of the best-known members of the group Dryosauridae. It had a short, deep skull, toothless beak, and short, slender forelimbs with small hands.

Dryptosauroides Indian theropod from the Cretaceous, known only from vertebrae. These can be identified as abelisaurid, but cannot be reliably distinguished from other members of the group.

Dryptosaurus Large Cretaceous theropod from North America, known from various fragments and recently argued to be a tyrannosauroid. Probably had longer arms than tyrannosaurid tyrannosauroids.

Dubreuillosaurus French megalosaurid theropod from the Jurassic, closely related to *Eustreptospondylus*. Its fossils come from sediments deposited near the sea.

Duriavenator Jurassic megalosaurid from Britain, long misidentified as a species of *Megalosaurus*. Known from skull bones that possess unique pits and other concavities.

Dyoplosaurus Cretaceous ankylosaurid ankylosaur from Canada, known from a tail with a bony club and from various other bones. Generally thought to be the same as *Euoplocephalus*.

Dysalotosaurus Medium-sized iguanodontian ornithopod from the Jurassic of Tanzania, often thought to be the same animal as *Dryosaurus* but differing in skull shape and some other details.

Dysganus Old name given to ornithischian teeth from the Cretaceous of North America. Several species were named and most belong to ceratopsids. These teeth cannot be reliably distinguished from other ceratopsids.

Dyslocosaurus Enigmatic North American sauropod that might be from the Jurassic or Cretaceous. Suggested by some to be a diplodocoid, and by others to be a titanosaur.

Dystrophaeus North American Jurassic sauropod, known from forelimb bones and a shoulder blade. Suggested to be a cetiosaurid or diplodocoid, its affinities remain unknown.

Echinodon Tiny British heterodontosaurid known from jaw and skull fragments. One of only a few members of the group known from the Cretaceous.

Edmarka Large Jurassic megalosaurid theropod from the USA, known from skull fragments and shoulder, hip, and rib bones. Probably the same animal as *Torvosaurus*.

Edmontonia Large nodosaurid ankylosaurid from the Cretaceous of North America. Large spikes projected forwards and outwards from its shoulders. Several species are known.

Edmontosaurus Large North American Cretaceous hadrosaurid, famous for its long, duck-billed skull, and hundreds of teeth.

Efraasia Medium-sized sauropodomorph from the Triassic of Germany, often confused with *Sellosaurus* and *Plateosaurus* and known from several specimens. It had a shorter neck and longer arms than these related forms.

Einiosaurus Centrosaurine ceratopsid from the Cretaceous of the USA, closely related to *Styracosaurus*. Its nasal horn was curved and pointed forwards, and two spikes projected backwards from the top of its frill.

Ekrixinatosaurus Large abelisaurid theropod from the Cretaceous of Argentina, known from a partial skeleton discovered after an explosion opened the rocks at its site.

Elaphrosaurus Jurassic Tanzanian ceratosaur, closely related to abelisaurids. It had a very long neck and long, slim legs, and was probably a fast runner. Its skull is unknown.

Elmisaurus Mongolian Cretaceous maniraptoran, known from partially fused foot bones, belonging to a caenagnathid oviraptorosaur. North American specimens have also been identified as *Elmisaurus*.

Elopteryx Poorly known Romanian Cretaceous theropod, known from fragments. May have been a flightless bird or troodontid.

Emausaurus Small primitive thyreophoran from the Jurassic of Germany with a broad skull, and armour on its back and sides.

Embasaurus Cretaceous theropod from Kazakhstan, known from a few vertebrae that are not thought to be precisely identifiable.

Eobrontosaurus Jurassic sauropod from the USA, originally thought to be a diplodocoid closely related to *Apatosaurus* but later thought to be a camarasaur.

Eocarcharia African carcharodontosaurid allosauroid from Niger, closely related to *Acrocanthosaurus* and with a similar-shaped skull. Large bony bosses decorated its skull.

Eoceratops Ceratopsid from the Cretaceous of Canada, once thought to be a distinct chasmosaurine but now generally thought to be the same as *Chasmosaurus*.

Eocursor Small, bipedal ornithischian from the Triassic of South Africa. Unlike most other ornithischians, its hands were long and capable of strong grasping.

Eolambia Iguanodontian ornithopod from the Cretaceous of the USA, originally identified as an early hadrosaurid but now thought to be a close relative of *Altirhinus*.

Eomamenchisaurus Chinese Jurassic sauropod, known from fragmentary remains including vertebrae, and hip and hind limb bones.

Eoraptor Small, bipedal predator from the Triassic of Argentina. It had short arms and both leaf-shaped and recurved teeth. It might be one of the most primitive of all dinosaurs.

Eotriceratops Giant Canadian horned dinosaur from the Cretaceous, similar to *Triceratops* but differing from it in the details of its frill, cheek region, and snout. Closely related to *Torosaurus*, *Nedoceratops*, and *Triceratops*.

Eotyrannus Tyrannosauroid theropod from the Cretaceous of Britain. It had long, three-fingered hands and thick, fused bones along the top of its snout that may have provided it with a particularly strong bite.

Epachthosaurus Argentinean titanosaur from the Cretaceous, known from vertebrae and pelvic fragments. Suggested to be a primitive titanosaur, and similar in some respects to the gigantic *Argentinosaurus*.

Epanterias North American allosauroid theropod from the Jurassic, generally thought to represent a large example of *Allosaurus*.

Epidendrosaurus Tiny Cretaceous theropod from China, known only from a juvenile specimen. It had very long hands and particularly long third fingers.

Epidexipteryx Close relative of *Epidendrosaurus* from the Cretaceous of China. It had a short, tall skull and four very long, strap-like structures at the end of its tail.

Equijubus Cretaceous iguanodontian ornithopod from China, closely related to hadrosaurids. It had a short, down-curved snout and particularly large bony openings at the back of the skull.

Erectopus Poorly known Cretaceous theropod from France, named from hind limb, hand, and skull bones. Recently argued to be an allosauroid.

Erketu Mongolian Cretaceous sauropod, and a close relative of *Euhelopus*. It had very long, slender neck bones, and its neck might have been twice as long as its body.

Erliansaurus Chinese therizinosauroid maniraptoran from the Cretaceous, named from beautifully preserved arm bones.

Erlikosaurus Mongolian therizinosauroid from the Cretaceous. It had leaf-shaped teeth, a broad body, and broad feet with large, narrow claws.

Eshanosaurus Controversial dinosaur from the Jurassic of China. Originally identified as a therizinosauroid, it is the oldest member of the group if this is true.

Eucamerotus British sauropod from the Cretaceous, known from vertebrae that have many similarities with those of the Jurassic *Brachiosaurus*.

Eucercosaurus Old name given to various ornithopod vertebrae from the Cretaceous of Britain, once thought to be from an ankylosaur.

Eucnemesaurus South African sauropodomorph from the Jurassic, previously confused with the bones of a predator and known as *Aliwalia rex*. Closely related to *Riojasaurus* from Argentina.

Euhelopus Chinese Cretaceous sauropod, with a particularly long neck and a short, deep skull with robust, spoon-shaped teeth. Once thought to be a close relative of *Mamenchisaurus*, it now seems to be a macronarian related to titanosaurs.

Euoplocephalus Cretaceous ankylosaur from North America, closely related to *Ankylosaurus* but not as large. Known from several exceptionally preserved specimens. Its feet only had three toes, and it had a large, bony tail-club.

Euronychodon Small maniraptoran from the Cretaceous of Portugal, known only from teeth sometimes identified as belonging to dromaeosaurids or troodontids.

Europasaurus Dwarf macronarian sauropod from the Cretaceous of Germany. It had an arched bony skull crest and, even when adult, was about the size of a cow.

Euskelosaurus Triassic sauropodomorph from South Africa, known from vertebrae and limb bones that cannot be reliably distinguished from those of the other sauropodomorphs known from the same rocks.

Eustreptospondylus Jurassic spinosauroid theropod from Britain, known from an excellent partial skeleton found in marine sediments. Closely related to *Afrovenator* from Niger and *Dubreuillosaurus* from France.

Fabrosaurus Name once used for a partial lower jaw from the Jurassic of Lesotho.

Falcarius Long-necked maniraptoran from the Cretaceous of the USA, probably one of the most primitive members of Therizinosauroidea. It had large teeth at the tips of its jaws.

Ferganasaurus Poorly known sauropod from the Cretaceous of Kyrgyzstan, known from vertebrate, pelvic, and hind limb bones. It had short hand bones and a stout pelvis.

Ferganocephale Cretaceous ornithischian from Kyrgyzstan, known only from teeth. These were originally suggested to belong to a pachycephalosaur but this is doubtful.

Frenguellisaurus Bipedal Triassic herrerasaurid from Argentina, generally thought to be the same as *Herrerasaurus*.

Fukuiraptor Medium-sized Japanese allosauroid theropod from the Cretaceous, originally misidentified as a dromaeosaurid because of its large, curved claws.

Fukuisaurus Iguanodontian ornithopod from the Cretaceous of Japan. It had a very deep lower jaw but was otherwise similar to *Iguanodon* in many features.

Fulgurotherium Australian ornithopod from the Cretaceous, typically thought to be closely related to *Hypsilophodon*.

Fusuisaurus Chinese sauropod from the Cretaceous. One of several recently named sauropods, showing that brachiosaurs and titanosaurs were common in China.

Futalognkosaurus Giant Cretaceous titanosaur from Argentina, over 30m (98ft) long. Closely related to *Mendozasaurus*.

Gallimimus Large ornithomimid ornithomimosaur from the Cretaceous of Mongolia, similar to *Struthiomimus* but differing in limb proportions and skull shape.

Galveosaurus Spanish sauropod from the Jurassic or Cretaceous. Thought by some to be a close relative of *Cetiosaurus*, and by others to be close to the gigantic *Turiasaurus*.

Gargoyleosaurus Small ankylosaur from the Jurassic of the USA; possibly an early ankylosaurid. It had a long, narrow snout and bony bumps and horns on its skull.

Garudimimus Toothless ornithomimosaur from the Cretaceous of Mongolia, often said (incorrectly) to have had horns. Its feet were broader and shorter than those of most other "ostrich" dinosaurs.

Gasosaurus Medium-sized theropod from the Jurassic of China, originally identified as a megalosaur.

Gasparinisaura Small, bipedal ornithopod from the Cretaceous of Argentina, known from several specimens. It had a short, pointed snout.

Gastonia Cretaceous ankylosaur from the USA, similar in some details to *Polacanthus* from Britain. Known from skulls and the remains of several partial skeletons.

Genusaurus Cretaceous theropod from France, known from hip and hind limb bones. Originally suggested to be a late-surviving coelophysoid, it may instead be an abelisaur.

Genyodectes Argentinean theropod from the Cretaceous, known from partial jaws and teeth. Suggested over the years to be a megalosaur, tyrannosaur, or abelisaur, it was recently argued to be a close relative of *Ceratosaurus*.

Geranosaurus Heterodontosaurid from the Jurassic of South Africa, known from jaw fragments. Its remains cannot definitely be distinguished from those of other heterodontosaurids.

Giganotosaurus Gigantic allosauroid theropod from the Cretaceous of Argentina, closely related to *Mapusaurus*. Rows of bony lumps decorated its deep snout. One of the biggest of known theropods, and similar in size to *Tyrannosaurus*.

Gigantoraptor Gigantic oviraptorosaur from the Cretaceous of Mongolia, almost 3m (9ft) tall. It had long, slender hind limbs and the longest arms of any oviraptorosaur.

Gigantosaurus Name once used for sauropod remains from the Jurassic of Britain. The remains cannot be identified precisely and the name is mostly ignored.

Gigantoscelus Sauropodomorph from the Triassic of South Africa, conventionally thought to be the same as *Euskelosaurus*.

Gigantspinosaurus Stegosaur from the Jurassic of China, possibly one of the most primitive members of the group. Large spikes with massive bases projected from its shoulders.

Gilmoreosaurus Hadrosaurid from the Cretaceous of Mongolia, known from skull bones and remains from several individuals. Perhaps the same as *Bactrosaurus*.

Giraffatitan Brachiosaur from the Jurassic of Tanzania, famous for its arched skull crest and long forelimbs. Conventionally included within *Brachiosaurus*.

Glacialisaurus Jurassic sauropodomorph from Antarctica, discovered close to *Cryolophosaurus* and probably closely related to *Massospondylus*. Known from a partial foot and other remains.

Glyptodontopelta Poorly known ankylosaur from the Cretaceous of the USA, named for armour scutes that have a distinctive shape.

Gobiceratops Ceratopsian from the Cretaceous of Mongolia, named for a tiny skull less than 4cm (1½in) long. Suggested to be a close relative of *Bagaceratops*, though it is likely a juvenile specimen of that dinosaur.

Gobisaurus Cretaceous ankylosaurid ankylosaur from Mongolia, known from a complete skull and other remains. It had large eye sockets and nostrils and a relatively narrow snout. Its skull lacks the armoured texture normally present in ankylosaurids.

Gobititan Titanosaur from the Cretaceous of China, known from vertebrae and hind limb bones. It seems to have had a longer tail than most other titanosaurs, and may have been a primitive member of the group.

Gojirasaurus Coelophysoid theropod from the Triassic of the USA, known from vertebrae, limb bones, and a tooth. Named after Gojira, the Japanese name for Godzilla.

Gondwanatitan Brazilian titanosaur from the Cretaceous, suggested by some experts to be the same as *Aeolosaurus*. Its tail bones have distinctive heart-shaped end surfaces.

Gongbusaurus Small bipedal ornithischian from the Jurassic of China, named for teeth that cannot be reliably distinguished from those of other ornithischians.

Gongxianosaurus Poorly known primitive sauropod from the Jurassic of China, known from many bones including a partial skull. Little information is available on it.

Gorgosaurus Large Cretaceous tyrannosaurid from North America, long confused with both *Albertosaurus* and *Daspletosaurus*. Its eye sockets are rounder than those of other tyrannosaurids.

Goyocephale Flat-headed pachycephalosaur from the Cretaceous of Mongolia, known from a partial skull and various bones. Usually regarded as one of the most primitive of all pachycephalosaurs.

Graciliceratops Small ceratopsian from the Cretaceous of Mongolia, originally identified as *Microceratops*. The margins of its bony frill were particularly slender.

Graciliraptor Small dromaeosaurid maniraptoran from the Cretaceous of China, known from a skull bone, tooth, limb bones, and vertebrae. Possibly a close relative of *Microraptor* and *Sinornithosaurus*.

Gravitholus Pachycephalosaur from the Cretaceous of Canada with a particularly thick, heavy skull roof and small braincase. Suggested by some to be *Stegoceras*.

Gresslyosaurus Triassic sauropodomorph from Germany, named for remains usually thought today to belong to *Plateosaurus*.

Gryponyx South African sauropodomorph from the Jurassic, generally thought to be the same as *Massospondylus*.

Gryposaurus Hadrosaurid from the Cretaceous of North America. Had a deep snout and a bony bump in front of its eyes. This gave it a "roman-nosed" appearance. Probably the same as *Kritosaurus*.

Guaibasaurus Bipedal primitive saurischian

from the Triassic of Brazil, suggested to be an early theropod. Known from hind limb bones, vertebrae, and shoulder and hip bones.

Guanlong Jurassic tyrannosauroid theropod from China, known from excellent specimens. A bizarre bony crest projected upwards and backwards from its skull.

Gyposaurus Primitive sauropodomorph from the Jurassic of South Africa, generally thought to be same as *Massospondylus*.

Hadrosaurus Hadrosaurid from the Cretaceous of the USA, one of the first North American dinosaurs to be named.

Hagryphus Oviraptorosaur from the Cretaceous of the USA, known from a long, slender hand, and wrist and partial arm bones. One of the largest oviraptorosaurs, probably similar to *Chirostenotes*.

Halticosaurus Triassic theropod from Germany named for incomplete remains that are generally thought to belong to the coelophysoid *Liliensternus*.

Hanssuesia Dome-skulled pachycephalosaur from the Cretaceous of North America, long included within *Stegoceras*. The front of its skull dome was wide and flattened and the bony shelf at the back of the dome was reduced.

Haplocanthosaurus Jurassic sauropod from the USA, known from remains that belong to two species. Skulls are unknown.

Harpymimus Ornithomimosaur from the Cretaceous of Mongolia. It had six conical teeth in its lower jaws. Unlike other ornithomimid dinosaurs, it still had a small first toe and its thumb was shorter than the other two fingers.

Heishansaurus Enigmatic ornithischian from the Cretaceous of China, known only from skull fragments first suggested to be from a pachycephalosaur but later thought to be from an ankylosaur.

Heptasteornis Romanian maniraptoran theropod from the Cretaceous, known from the incomplete end of a shin bone originally thought to be from a giant owl. Suggested to be an alvarezsaurid or oviraptorosaur.

Herrerasaurus Large, bipedal predator from the Triassic of Argentina. It had a deep skull with long teeth and large, curved claws on its inner three fingers. The best-known member of the saurischian group Herrerasauridae.

Hesperonychus Tiny dromaeosaurid from the Cretaceous of Canada, the first definite North American member of the microraptorine group. Known from claws and hip bones, and equipped with long feathers.

Hesperosaurus Jurassic stegosaur from the USA. Its skull was shorter and broader than that of *Stegosaurus*, and its plates were lower and longer. Regarded by some researchers as being part of the genus *Stegosaurus*.

Heterodontosaurus Best-known heterodontosaurid. A long-legged ornithischian with fang-like teeth near the front of its jaws, and long, grasping hands. From the Jurassic of South Africa.

Heterosaurus Poorly known Cretaceous ornithopod from France, known from fragmentary remains generally thought today to belong to *Iguanodon* or *Mantellisaurus*.

Hexinlusaurus Small bipedal ornithischian from the Jurassic of China, known from a partial skull and other remains. It possessed a concavity above and behind its eye sockets.

Heyuannia Small oviraptorid oviraptorosaur from the Cretaceous of China. It had a short, deep, crestless skull and a long neck.

Hierosaurus Ankylosaur from the Cretaceous of the USA, named for armour plates that cannot be reliably distinguished from those of other ankylosaurs.

Histriasaurus Poorly known sauropod from the Cretaceous of Croatia known from vertebrae. A member of the diplodocoid group Rebbachisauridae.

Homalocephale Mongolian Cretaceous pachycephalosaur. It had a flat skull roof with rows of small, bony lumps.

Hongshanosaurus Cretaceous ceratopsian from China, closely related to *Psittacosaurus* but with a narrower, shallower skull.

Hoplitosaurus Ankylosaur from the Cretaceous of the USA, named from vertebrae, limb bones, ribs, and other remains. Regarded by some as a close relative of *Polacanthus*.

Hoplosaurus Cretaceous ankylosaur from Australia, generally thought to be the same as *Struthiosaurus*.

Hortalotarsus Sauropodomorph from the Jurassic of South Africa, mostly thought to be the same animal as *Massospondylus*.

Huabeisaurus Large Chinese sauropod from the Cretaceous, known from teeth, massively built limb bones, and vertebrae. Probably a titanosaur.

Huanghetitan Cretaceous sauropod from China, closely related to brachiosaurs and titanosaurs. Its long ribs show that it had a particularly deep body.

Huaxiagnathus Small compsognathid theropod from the Cretaceous of China, similar to *Sinosauropteryx*. Known from an articulated, near-complete skeleton.

Huayangosaurus One of the most primitive stegosaurs. Some individuals had small horns above their eyes. From the Jurassic of China.

Hudiesaurus Sauropod from the Jurassic of China, named for a large vertebra. Forelimb bones and teeth from a second, smaller individual were thought to be from the same animal, but this cannot be proven.

Hulsanpes Small maniraptoran from the Cretaceous of Mongolia, sometimes suggested to be a dromaeosaurid. Known only from partial foot bones.

Hungarosaurus Medium-sized Cretaceous nodosaurid ankylosaur from Hungary, known from four specimens and hundreds of bones. It had bony bumps on the back of its skull and a large bony plate covered its hip region. Two spikes projected upwards from this plate.

Hylaeosaurus British ankylosaurid from the Cretaceous, perhaps closely related to *Polacanthus*. Long spikes projected from its neck and shoulder region. One of the earliest dinosaurs to be described.

Hypacrosaurus North American lambeosaurine hadrosaurid from the Cretaceous, closely related to *Corythosaurus*. It had a tall, rounded bony crest and tall bony spines on its vertebrae, which would have given it a noticeable ridge along its back.

Hypselosaurus Name given to numerous sauropod bones from the Cretaceous of France. Many eggs have been given the same name but their identity is uncertain.

Hypsibema Hadrosaurid from the Cretaceous of the USA, previously identified as a sauropod. Known from fragmentary remains and not considered distinguishable from other hadrosaurids.

Hypsilophodon Small, bipedal ornithopod from the Cretaceous of Britain, Spain, and Portugal. Had a narrow beak with small fangs in the upper jaw and leaf-shaped cheek teeth.

Hypsirophus Stegosaur from the Jurassic of the USA, known from incomplete remains that cannot be reliably distinguished from those of other stegosaurs. Usually thought the same animal as *Stegosaurus*.

Iguanodon Large iguanodontian ornithopod from the Cretaceous of Europe. The name is restricted to the large, heavily built species first discovered in Belgium. It had large, spike-like thumbs and a deep, massive skull.

Iliosuchus Poorly known small theropod from the Jurassic of Britain, found in the same rocks as *Megalosaurus*. Known only from hip bones and perhaps from a partial leg bone.

Ilokelesia Medium-sized Cretaceous abelisaur from Argentina, known from skull fragments, vertebrae, and finger and toe bones. It had bony shelves over its eyes.

Incisivosaurus Primitive oviraptorosaur from the Cretaceous of China, known from an excellent skull with projecting, rodent-like teeth at the jaw tips.

Indosaurus Large theropod from the Cretaceous of India, nowadays thought to be a abelisaurid. It had a thickened skull roof, but lacked the horns or bony bosses seen in some members of the group.

Indosuchus Abelisaurid from the Cretaceous of India, known from three partial braincases previously identified as those of ankylosaurs or tyrannosaurids. Probably similar to *Abelisaurus* from Argentina.

Ingenia Mongolian oviraptorid oviraptorosaur with a rounded, crestless skull and short fingers. The best-known member of the oviraptorid group Ingeniinae.

Inosaurus Poorly known Cretaceous theropod, named for vertebrae and a shin bone fragment from Niger.

Irritator Brazilian spinosaurid from the Cretaceous, known from an excellent skull that was originally modified to resemble that of a pterosaur. It had long, crocodile-like jaws and a bony crest on its forehead.

Isanosaurus Sauropod from the Triassic of Thailand; one of the earliest members of its group. Known from vertebrae, ribs, and bones from the shoulders and hips. It had robust thigh bones, and was a heavy-bodied quadruped.

Ischisaurus Predatory saurischian from the Triassic of Argentina, and almost certainly the same animal as *Herrerasaurus*.

Ischyrosaurus British macronarian sauropod from the Jurassic, known from a single upper arm bone that might belong to a brachiosaur.

Isisaurus Unusual titanosaur from the Cretaceous of India, originally identified as a species of *Titanosaurus* and known from numerous remains. It had short neck vetrebrae, tall shoulders, and slim limbs.

Iuticosaurus British Cretaceous sauropod, known only from isolated tail vertebrae originally thought to belong to *Titanosaurus*. These belong to a titanosaur but reveal no other important information.

Jainosaurus Indian titanosaur from the Cretaceous, known from a skull and other remains. Its limb bones are larger and more slender than those of most Indian titanosaurs.

Janenschia Sauropod from the Jurassic of Tanzania, named for limb bones originally thought to belong to *Tornieria*. Its stout arm bones suggest that it is a titanosaur, and one of the oldest members of this group.

Jaxartosaurus Hadrosaurid from the Cretaceous of Kazakhstan, named for a toothless jaw bone. This cannot be reliably distinguished from other hadrosaurid jaw bones so the name is not used today.

Jeholosaurus Small, bipedal ornithischian from the Cretaceous of China, known from good remains and originally thought to be an ornithopod. Now regarded as more primitive within the ornithischian family tree.

Jenghizkhan Name once used for a giant tyrannosaurid from the Cretaceous of Mongolia, now universally regarded as the same animal as *Tarbosaurus*.

Jiangjunosaurus Jurassic stegosaur from China, known from a partial skull, neck, and some plates. The small plates on its neck were diamond shaped, and the jaw curved strongly downwards at its tip.

Jiangshanosaurus Poorly known titanosaur from the Cretaceous of China, known from vertebrae, shoulder, hip bones, and a thigh bone fragment. Its shoulder blade was distinctive, with a large, wide base.

Jinfengopteryx Small maniraptoran theropod from the Jurassic of China with numerous closely packed teeth. Originally thought to be a bird, it is more likely to be a troodontid.

Jingshanosaurus Jurassic sauropodomorph from China, regarded by some experts as very similar to (and possibly the same as) *Yunnanosaurus*.

Jinzhousaurus Cretaceous iguanodontian ornithopod from China, known from excellent remains.

Jiutaisaurus Obscure sauropod from the Cretaceous of China, known from a series of articulated tail vertebrae that could not be identified to a specific group. A few features of the vertebrae were said to be unique.

Jobaria Sauropod from the Cretaceous of Niger, known from many excellent remains. It had a deep-snouted skull with massive nostril openings and was archaic for its time.

Jubbulpuria Small theropod from the Cretaceous of India, named for vertebrae originally thought to be from a coelurosaur, but now known to be from a small abelisaur similar to *Ligabueino* from Argentina.

Juravenator Small theropod from the Jurassic Solnhofen Limestone of Germany, known for a virtually complete, articulated skeleton that preserves skin impressions. Perhaps related to compsognathids.

Kaijiangosaurus Poorly known Jurassic theropod from China, named for vertebrae conventionally thought to be from a megalosaur.

Kakuru Australian Cretaceous theropod, named for the end of a shin bone that has been turned into the precious stone opal. When complete, the shin bone would have been slender and long.

Kangnasaurus Iguanodontian ornithopod from the Cretaceous of South Africa, named for a tooth. Additional remains, including vertebrae and hind limb bones, were later thought to belong to the same animal.

Karongasaurus Titanosaur from the Cretaceous of Malawi, known only from a lower jaw and some teeth.

Kelmayisaurus Chinese Cretaceous theropod, named for skull and lower jaw bones. The exact affinities of this theropod are unknown but it has usually been regarded as a megalosaur-like form.

Kentrosaurus Medium-sized African stegosaur from the Jurassic of Tanzania. It had long spikes on its tail and hip region, and long spikes projecting from its shoulders.

Kerberosaurus Hadrosaurine hadrosaurid

from the Cretaceous of Russia, known from abundant remains and thought to be closely related to *Prosaurolophus* and *Saurolophus*.

Khaan Oviraptorid oviraptorosaur from the Cretaceous of Mongolia, known from complete skeletons. Had a short, rounded snout without a crest, and horizontally aligned nostril openings. Similar to *Conchoraptor*.

Khetranisaurus Titanosaur from the Cretaceous of Pakistan, named for a single tail vertebra. Said to be from an animal closely related to *Pakisaurus* and *Sulaimanisaurus*.

Klamelisaurus Enigmatic Jurassic sauropod from China. It had a long neck composed of 16 vertebrae, and broad, spatulate teeth. Its affinities cannot yet be determined but it might have been similar to *Omeisaurus*.

Koparion Small Jurassic theropod from the USA, named for a single serrated tooth. Originally suggested to be a troodontid, though this is unlikely.

Kotasaurus Primitive sauropod from the Jurassic of India.

Koutalisaurus Hadrosaurid from the Cretaceous of Spain, named for an elongate lower jaw originally thought to belong to *Pararhabdodon*. The jaw has a particularly long toothless region between the beak at the front and the tooth batteries further back.

Kritosaurus Cretaceous hadrosaurine hadrosaurid from North America, very similar to *Gryposaurus* and regarded by some experts to be the same animal.

Kryptops Abelisaurid from the Cretaceous of Niger, known from skull material similar to that of *Rugops*. Had a blunt, deep snout and a bony sheet covered part of the snout.

Kulceratops Poorly known ceratopsian from the Cretaceous of Uzbekistan, named for a partial skull bone. This bone cannot be reliably distinguished from those of other ceratopsians and the name is rarely used.

Labocania Enigmatic theropod from the Cretaceous of the USA, named for partial skull roof bones and other remains. Originally suggested to have tyrannosaurid affinities, but some have suggested it may be an abelisaur.

Labrosaurus Old name for theropod remains from the Jurassic of the USA. These are generally thought to belong to *Allosaurus*, but their precise identity is unclear.

Laevisuchus Small theropod from the Cretaceous of India, named for vertebrae once thought to be from a coelurosaur but now thought to be from an abelisaur.

Lamaceratops Small ceratopsian from the Cretaceous of Mongolia with a short frill and very small, conical nasal horn. Suggested to be *Bagaceratops* or a close relative.

Lambeosaurus Large lambeosaurine hadrosaurid from the Cretaceous of North America with a rectangular head crest that projected forwards over the snout. A spike projected backwards from the crest in some individuals.

Lametasaurus Cretaceous dinosaur from India, named for various fragments that have been identified as belonging to titanosaurs, ankylosaurs, and abelisaurs.

Lamplughsaura Primitive sauropod from the Jurassic of India, known from substantial remains including skull bones. It had coarsely serrated teeth and a straight thumb claw.

Lanasaurus Jurassic heterodontosaurid from South Africa, named for partial jaw bones generally thought to be *Lycorhinus*.

Lanzhousaurus Large iguanodontian ornithopod from the Cretaceous of China, known from a partial skeleton. Its leaf-shaped cheek teeth were proportionally very large.

Laosaurus Jurassic ornithischian from the USA, known from fragmentary remains that cannot be specifically identified.

Laplatasaurus Cretaceous titanosaur from Argentina, known from tail vertebrae and limb bones that have been confused with remains identified as belonging to *Titanosaurus*.

Lapparentosaurus Jurassic sauropod from Madagascar known from the remains of several individuals that were originally thought to be from *Bothriospondylus*.

Leaellynasaura Small bipedal ornithopod from the Cretaceous of Australia, typically thought to be similar to *Hypsilophodon* but known from fragmentary remains.

Leipsanosaurus Ankylosaur from the Cretaceous of Austria, generally thought today to be the same animal as the dwarf nodosaurid *Struthiosaurus*.

Leptoceratops Ceratopsian from the Cretaceous of North America, known from several specimens. It had a proportionally large skull with a short bony frill and claws (not hooves) on its fingers and toes.

Leptospondylus Old name for Jurassic sauropodomorph remains from South Africa, generally thought to be *Massospondylus*.

Lesothosaurus Small bipedal ornithischian from the Jurassic of Lesotho. Typically thought to be the most primitive ornithischian, possibly closely related to the thyreophorans.

Lessemsaurus Primitive sauropod from the Triassic of Argentina, closely related to *Antetonitrus* from South Africa. Named for a series of articulated vertebrae.

Lexovisaurus Jurassic stegosaur from Britain, though also reported from France. Its remains cannot be reliably distinguished from those of other stegosaurs; the best specimen has been renamed *Loricatosaurus*.

Liaoceratops Small primitive ceratopsian from the Cretaceous of China, known from complete skulls. Had a short, rounded frill and three pairs of cylindrical teeth at the tip of the upper jaw. Possesses features previously thought unique to the psittacosaurids.

Liaoningosaurus Ankylosaur from the Cretaceous of China, known from the tiny skeleton of a juvenile less than 40cm (15½in) long. It had armour plates on its belly.

Ligabueino Tiny theropod from the Cretaceous of Argentina, known from vertebrae and hind limb and hip bones. Thought to be a small abelisaur.

Ligabuesaurus Large titanosaur from the Cretaceous of Argentina, known from vertebrae, limb bones, a skull bone, and some other remains. Had long forelimbs and was likely brachiosaur-like in its proportions.

Liliensternus Large Triassic coelophysoid theropod from Germany, known from remains belonging to two individuals originally included in *Halticosaurus*. Because it might be related to *Dilophosaurus*, it is sometimes shown with twin bony head crests.

Limaysaurus Argentinean diplodocoid sauropod from the Cretaceous, originally included in *Rayososaurus*.

Lirainosaurus Sauropod from the Cretaceous of Spain, known from limb bones, vertebrae, teeth, and other remains, belonging to several individuals. Generally thought to be a titanosaur, and one of the best-known members of the group from Europe.

Loncosaurus Obscure dinosaur from the Cretaceous of Argentina, named for a thigh bone fragment and tooth. The bone is from an ornithopod, but cannot be reliably distinguished from other species.

Longosaurus Coelophysoid theropod from the Triassic of the USA, named for a hip bone thought by one expert to be distinct from *Coelophysis*. Now thought to be *Coelophysis*.

Lophorhothon Hadrosaurid from the Cretaceous of the USA, known from a partial skull and other remains. It is probably a very primitive hadrosaurine.

Lophostropheus Coelophysoid theropod from the Triassic of France, based on vertebrae and hip bones originally identified as *Halticosaurus* and later as a species of *Liliensternus*. It had long, slender neck bones.

Loricatosaurus British stegosaur from the Jurassic, originally identified as a specimen of *Lexovisaurus*. It had long spikes on its tail.

Loricosaurus Name given to armour plates from the Cretaceous of Argentina, originally thought to be from an ankylosaur but now regarded as belonging to a titanosaurian sauropod, perhaps to *Saltasaurus*.

Losillasaurus Spanish sauropod from the Jurassic or Cretaceous, known from a partial skeleton. Originally suggested to be a diplodocoid, it may instead be related to *Cetiosauriscus* and *Mamenchisaurus*.

Lourinhanosaurus Jurassic allosauroid theropod from Portugal, known from a partial skeleton and thought to be a sinraptorid.

Lourinhasaurus Portugese sauropod from the Jurassic, known from remains belonging to several individuals. Originally misidentified as *Apatosaurus*, and later thought to be a camarasaur, its affinities are unknown.

Luanchuanraptor Dromaeosaurid maniraptoran from the Cretaceous of China, known from vertebrae, arm bones, skull bones, and other remains.

Lufengocephalus Sauropodomorph from the Jurassic of China. The name is not used today as it is generally thought to be the same animal as *Lufengosaurus*.

Lufengosaurus Medium-sized Chinese sauropodomorph from the Jurassic, known from the remains of several individuals. Closely related to *Massospondylus*. It had unusual bony lumps on its cheeks and snout.

Lukousaurus Poorly known Jurassic fossil from China, known from a partial skull. Small horns projected upwards from in front of the eyes. Suggested by some to be ceratosaur or coelurosaur, but it may not be a dinosaur.

Lurdusaurus Iguanodontian ornithopod from the Cretaceous of Niger, known from a very large, heavily built partial skeleton and skull. It had massively built arms with spike-like thumbs, and a very broad, rotund body.

Lusitanosaurus Ornithischian from the Jurassic of Portugal, named for a partial snout with teeth. Originally claimed to be a close relative of *Scelidosaurus*, it might be a primitive thyreophoran.

Lusotitan Macronarian sauropod from the Jurassic of Portugal, known from a partial skeleton and remains originally thought to be *Brachiosaurus*. Had long, slender arm bones.

Lycorhinus Heterodontosaurid from the Jurassic of South Africa, named for jaws originally thought to be from early relatives of mammals. Thought by many to include the remains once named *Lanasaurus*.

Macrodontophion Old name given to a tooth from the Jurassic of the Ukraine, often thought to be from a theropod. The tooth seems to have been lost, and may not have been from a dinosaur at all.

Macrogryphosaurus Relatively large, bipedal iguanodontian ornithopod from the Cretaceous of Argentina, with flange-like growths on its ribs. Suggested to be closely related to *Talenkaeun*.

Macrophalangia Cretaceous Canadian coelurosaur, named for a foot once thought to be from an ornithomimid but now known to be from a caenagnathid oviraptorosaur. Generally thought to be *Chirostenotes*.

Macrurosaurus British Cretaceous sauropod, known from vertebrae and poor remains. Some belong to a titanosaur, but they cannot be reliably distinguished.

Magnirostris Ceratopsian from the Cretaceous of Mongolia, known from a crushed skull with small brow and nasal horns. Said by some to be *Bagaceratops*.

Magnosaurus Poorly known British theropod from the Jurassic, suggested to be *Megalosaurus* or *Eustreptospondylus* but currently thought distinct.

Magyarosaurus Small titanosaur from the Cretaceous of Romania, known from numerous bones and armour scutes. It seems to have been an island-dwelling dwarf.

Mahakala Small dromaeosaurid maniraptoran (less than a metre long) from the Cretaceous of Mongolia. It had a rounded skull roof and very large eye sockets.

Maiasaura North American hadrosaurid from the Cretaceous, famous for being discovered with nests, eggs, and babies. Known from over 200 specimens. It had a bony crest just in front of its eyes and a shallow, down turned snout.

Majungasaurus Madagascan abelisaurid from the Cretaceous, known from spectacular remains. It had a large bony lump on its forehead and stout legs.

Majungatholus Name once used for a skull roof from the Cretaceous of Madagascar, originally thought to be from a pachycephalosaur but now generally thought to belong to the abelisaurid *Majungasaurus*.

Malarguesaurus Sauropod from the Cretaceous of Argentina, known from tail vertebrae, limb bone fragments, and ribs.

Malawisaurus African titanosaur from the Cretaceous of Malawi, known from a substantial amount of material. Its skull bones suggest that it had a short, deep snout.

Maleevosaurus Asian tyrannosaurid from the Cretaceous, first thought to be a distinctive dwarf form but now generally regarded as a juvenile *Tarbosaurus*.

Maleevus Mongolian ankylosaurid from the Cretaceous, named for a skull originally included in *Syrmosaurus*. Cannot be reliably distinguished from other ankylosaurids.

Mamenchisaurus Chinese sauropod from the Jurassic, famous for its incredibly long neck and known from many specimens. Numerous species have been named.

Mandschurosaurus Chinese hadrosaurid from the Cretaceous, known from fragments originally thought to belong to *Trachodon*.

Manospondylus Old name given to poorly preserved tail vertebrae from the Cretaceous of the USA, which probably belonged to *Tyrannosaurus*.

Mantellisaurus Iguanodontian ornithopod from the Cretaceous of Britain, long included in *Iguanodon* but with shorter arms and a shallower snout than that dinosaur.

Mapusaurus Large allosauroid from the Cretaceous of Argentina, closely related to *Giganotosaurus*.

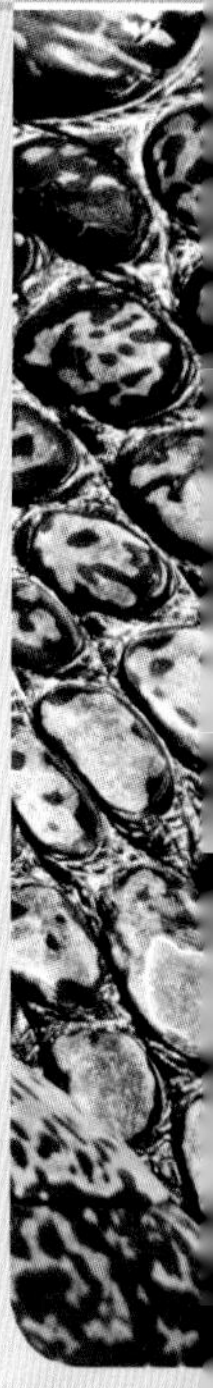

Marisaurus Titanosaur from the Cretaceous of Pakistan, named for partial tail vertebrae. A partial skull and some hind limb bones have also been assigned to it.

Marmarospondylus Old name used for a British sauropod vertebra from the Jurassic. The specimen had previously been identified as *Bothriospondylus*.

Marshosaurus Medium-sized theropod from the Jurassic of the USA, known from a hip bone and other fragments.

Masiakasaurus Small Madagascan abelisaur from the Cretaceous, known from excellent remains. It had a bizarre skull where the teeth at the jaw tips projected outwards and were longer than the back teeth.

Massospondylus Sauropodomorph from the Jurassic of South Africa, known from numerous specimens, some of which are embryos and juveniles. Specimens from South and North America have been identified as *Massospondylus*, but this is controversial.

Maxakalisaurus Cretaceous titanosaur from Brazil, known from vertebrae, ribs, limb bones, and various other bones. A partial snout bone shows that it had slim, pencil-like teeth.

Megalosaurus Spinosauroid theropod from the Jurassic of Britain, known from many bones that might have belonged to more than one animal. The first dinosaur to be named.

Megapnosaurus Coelophysoid theropod from the Jurassic of Zimbabwe and the USA, named for species previously labelled *Syntarsus*. Very similar to *Coelophysis* and regarded by most as the same animal.

Megaraptor Controversial large theropod from the Cretaceous of Argentina, originally thought to be a large coelurosaur with huge curved claws on its feet. The claws now seem to be from the hand.

Mei Small troodontid theropod from the Cretaceous of China, famously preserved in a curled-up sleeping posture and with its head tucked under its arm.

Melanorosaurus Large quadrupedal sauropodomorph from the Triassic of South Africa, close to the origin of sauropods. It had massive, heavy limb bones.

Mendozasaurus Titanosaur from the Cretaceous of Argentina, known from limb bones, armour plates, and other bones. Some of the armour plates from its tail were unusually large, and with a conical shape.

Metriacanthosaurus British theropod from the Jurassic, known from fragmentary remains but clearly similar to *Yangchuanosaurus* and *Sinraptor* from China.

Microceratus Small ceratopsian from the Cretaceous of China, long known as *Microceratops*. Cannot be reliably distinguished from other ceratopsians.

Microcoelus Old name given to Cretaceous titanosaur remains from Argentina. These remains are now thought to belong to the armoured titanosaur *Saltasaurus*.

Microhadrosaurus Name given to a juvenile hadrosaurid jaw from the Cretaceous of China, originally thought to be a pygmy *Edmontosaurus*-like form. The name is no longer used.

Micropachycephalosaurus Small ornithischian from the Cretaceous of China, originally identified as a pachycephalosaur. Its precise affinities cannot be determined.

Microraptor Tiny Cretaceous dromaeosaurid maniraptoran from China. It had long feathers on its arms and hind limbs and was probably capable of gliding.

Microvenator Small Cretaceous oviraptorosaur from the USA, known only from a single, juvenile specimen originally confused with the remains of *Deinonychus*. It had a deep, toothless lower jaw.

Minmi Small ankylosaur from the Cretaceous of Australia. Triangular armour plates projected from the sides of its tail and it had a deep snout and large eye sockets.

Minotaurasaurus Cretaceous ankylosaurid ankylosaur from Mongolia, known only from its skull. It had a square-shaped beak and long horns at the back of its skull.

Miragaia Portugese stegosaur from the Jurassic, closely related to *Dacentrurus*. It had a long neck formed of 17 vertebrae – more than were present in many sauropods.

Mirischia Small Cretaceous theropod known only from hip and hind limb bones discovered in Brazil. Perhaps similar to *Compsognathus*.

Mochlodon Iguanodontian ornithopod from the Cretaceous of Austria, named for vertebrae and a jaw bone similar to those of the rhabdodontid *Zalmoxes*.

Mongolosaurus Poorly known Asian sauropod from the Cretaceous of Mongolia, known only from vertebrae and teeth.

Monkonosaurus Poorly known Jurassic stegosaur from Tibet, known from its hips, a few vertebrae, and three armour plates. Its plates were similar to those of *Stegosaurus*.

Monoclonius Old name given to various ceratopsid remains from the Cretaceous of North America. Most of these remains in fact belong to *Centrosaurus*.

Monolophosaurus Medium-sized Chinese theropod from the Jurassic, regarded by some as an allosauroid. It had an unusual hollow crest that ran along the top of its skull.

Mononykus Best-known alvarezsaurid. A small, long-legged maniraptoran from the Cretaceous of Mongolia with a long neck and short, strong forelimbs.

Montanoceratops Ceratopsian from the Cretaceous of North America, closely related to *Leptoceratops*. It had a small nose horn and claws instead of hooves.

Morinosaurus French sauropod known only from a single tooth from the Jurassic. Though sometimes allied to *Pelorosaurus* from England, its true identity remains a mystery.

Morosaurus Jurassic macronarian sauropod from the USA, almost certainly the same animal as *Camarasaurus*.

Mussaurus Argentinean sauropodomorph from the Triassic, known from eggs, hatchlings, juveniles, and adults. In adult form, superficially similar to *Plateosaurus*.

Muttaburrasaurus Australian ornithopod from the Cretaceous, with a deep, bulbous snout region. Thick, strong cheek bones may have allowed it to bite tough plants.

Muyelensaurus Macronarian sauropod from the Cretaceous of Argentina, known from vertebrae, slender limb bones, and a partial skull. Closely related to *Rinconsaurus*.

Mymoorapelta Small Jurassic ankylosaur from the USA, thought by some to be a close relative of *Polacanthus*. A bony shield covered its hip region.

Naashoibitosaurus Cretaceous hadrosaurid from the USA with a deep snout and bony nasal crest.

Nanningosaurus Chinese hadrosaurid from the Cretaceous, suggested to be one of the most primitive lambeosaurines. Its remains include incomplete skull material.

Nanosaurus Small ornithischian from the Jurassic of the USA, named from remains that cannot be identified precisely. *Othnielosaurus* was originally included in *Nanosaurus*.

Nanotyrannus Small tyrannosaurid from the Cretaceous of North America, thought by some to be a unique dwarf but by others to be a juvenile *Tyrannosaurus*.

Nanshiungosaurus Cretaceous therizinosauroid from China, known from vertebrae and hips. Originally misidentified as a sauropod.

Nanyangosaurus Chinese iguanodontian ornithopod from the Cretaceous, closely related to hadrosaurids.

Nedcolbertia Small theropod from the Cretaceous of the USA, known from several individuals. Probably a primitive coelurosaur.

Nedoceratops Chasmosaurine ceratopsid from the Cretaceous of the USA, known only from a single skull. Very similar to *Triceratops* but with holes in its frill.

Neimongosaurus Cretaceous therizinosauroid from China known from a lower jaw, limb bones, and vertebrae. Probably very similar to *Segnosaurus*.

Nemegtomaia Cretaceous oviraptorid maniraptoran from Mongolia, originally named *Nemegtia*. It had a tall, hollow crest at the front of its skull and was similar to *Citipati*.

Nemegtosaurus Slender-toothed Mongolian titanosaur from the Cretaceous, known only from its skull. Originally thought to be a late-surviving diplodocoid.

Neosodon Old name given to a single large sauropod tooth from the Jurassic of France. It might have belonged to a sauropod similar to *Turiasaurus* from Spain.

Neovenator Cretaceous allosauroid from Britain, recently shown to be the most primitive member of Carcharodontosauridae. It had a curved snout with twin bony crests on the upper surface of its snout.

Neuquenraptor Dromaeosaurid maniraptoran from the Cretaceous of Argentina, probably closely related to *Unenlagia*.

Neuquensaurus Argentinean titanosaur from the Cretaceous, named for tail vertebrae and limb bones. Regarded by some experts as the same as *Saltasaurus*.

Nigersaurus Small Cretaceous diplodocoid sauropod from Niger. Its mouth was wider than the rest of its skull and it had many teeth.

Niobrarasaurus Cretaceous nodosaurid ankylosaur from the USA, originally known as *Hierosaurus*. Known from a partial skeleton discovered in marine sediments.

Nipponosaurus Hadrosaurid from the Cretaceous of Sakhalin Island, Russia. It had a plate-like skull and was similar to *Corythosaurus*.

Noasaurus Small South American theropod from Argentina, originally thought to be a dromaeosaurid-like animal with a raised toe claw. Now known to be an abelisaur.

Nodocephalosaurus North American ankylosaurid from the Cretaceous. Small rounded armour plates covered its skull.

Nodosaurus North American nodosaurid ankylosaur from the Cretaceous, often shown with armour bands covering its body. Like most other nodosaurids, it probably had long spikes on its neck and shoulders.

Nomingia Oviraptorosaur from the Cretaceous of Mongolia, known from hips, vertebrae, and leg bones. The vertebrae at the end of its tail were fused together, perhaps to support of a fan-like arrangement of feathers.

Nopcsaspondylus Cretaceous diplodocoid sauropod from Argentina, named for a distinctive vertebra that is currently lost.

Nothronychus Cretaceous therizinosauroid from the USA. It had long arms, a wide body, and broad feet. Some of its bones were originally misidentified as those of the ceratopsian *Zuniceratops*.

Notoceratops Poorly known dinosaur from the Cretaceous of Argentina, once thought to be the only South American ceratopsian. Known from fragments.

Notohypsilophodon Cretaceous ornithopod from Argentina, known from vertebrae and limb bones. Its affinities are uncertain and little is known about it.

Nqwebasaurus Coelurosaur from the Cretaceous of South Africa. It had a particularly stout thumb bone and might be related to the compsognathids.

Nuthetes British theropod from the Cretaceous known only from very small teeth and jaw fragments.

Ohmdenosaurus Poorly known primitive sauropod from the Jurassic of Germany, known only from ankle bones. Though sometimes regarded as a close relative of *Vulcanodon*, its affinities are uncertain.

Oligosaurus Old name given to Cretaceous ornithopod remains from Austria. Generally thought today to be the same as *Rhabdodon*.

Olorotitan Distinctive lambeosaurine hadrosaurid from the Cretaceous of China. It had a long neck and a large, fan-like crest that extended backwards over the skull.

Omeisaurus Jurassic sauropod from China with an incredibly long neck. Probably a close relative of *Mamenchisaurus*. Said to have a bony club at the end of its tail.

Onychosaurus Name given to ornithischian teeth from the Cretaceous of Romania, thought to belong to a *Stegosaurus*-like dinosaur. These are thought today to be rhabdodontid teeth, probably from *Zalmoxes*.

Opisthocoelicaudia Mongolian titanosaur from the Cretaceous. It had a broad, rounded body and stout, robust limbs. Its skull is unknown but it has sometimes been regarded as the same animal as *Nemegtosaurus*.

Oplosaurus British sauropod from the Cretaceous known only from a single tooth. This is large and spoon-shaped and has a heavily worn crown.

Orkoraptor Mysterious coelurosaur from the Cretaceous of Argentina, known from skull bones, teeth, vertebrae, and hind limb fragments. Its affinities are uncertain.

Ornatotholus North American pachycephalosaur from the Cretaceous, known from a partial skull roof that might belong to a young specimen of *Stegoceras*.

Ornithodesmus Poorly known maniraptoran, probably a dromaeosaurid, from the Cretaceous of Britain. Known only from the vertebrae between its hips.

Ornitholestes Small coelurosaur from the Jurassic of the USA, known from a near-complete skull and other good remains.

Ornithomerus Old name given to ornithischian limb bones from the Cretaceous of Austria, thought today to be *Rhabdodon*.

Ornithomimoides Indian theropod from the Cretaceous, named for vertebrae that seem to belong to an abelisaurid similar to *Majungasaurus*. Originally thought to be an ornithomimosaur-like theropod.

Ornithomimus North American ornithomimid ornithomimosaur from the Cretaceous with long, slender hands, a long neck, and a bird-like, toothless skull.

Ornithopsis Cretaceous macronarian sauropod from Britain, known only from poorly preserved vertebrae that reveal little.

Ornithotarsus Name once used for partial hind limb bones belonging to a Cretaceous hadrosaurid from the USA.

Orodromeus Small, bipedal ornithopod from the Cretaceous of the USA, known from several specimens. Traditionally thought similar to *Hypsilophodon*. Eggs thought to belong to it are actually from *Troodon*.

Orosaurus Large sauropodomorph from the Triassic of South Africa, thought to be the same as *Euskelosaurus* and *Plateosauravus*.

Orthogoniosaurus Cretaceous theropod from India, named for a fragmentary tooth.

Orthomerus Old name given to hadrosaurid remains from the Cretaceous of the Netherlands and Ukraine. Often thought to be the same as *Telmatosaurus*.

Oryctodromeus Small, bipedal ornithopod discovered inside a burrow. Various features of its skull and arm bones show that it was a good digger. Closely related to *Orodromeus* and *Zephyrosaurus*.

Othnielia Jurassic ornithischian from the USA, named for scrappy remains. A good skeleton once labelled *Othnielia* had to be renamed *Othnielosaurus*.

Othnielosaurus Small bipedal ornithischian. Probably a fast runner.

Otogosaurus Cretaceous sauropod from China, known from foot and hip bones.

Ouranosaurus Iguanodontian ornithopod from the Cretaceous of Niger, well known for having tall spines on its back. These may have supported a sail or hump.

Oviraptor Toothless Mongolian maniraptoran from the Cretaceous. Poorly known, but with a longer, lower skull than most other oviraptorids. Many specimens once labelled *Oviraptor* have now been renamed.

Ozraptor Australian theropod from the Jurassic, known only from an incomplete shin bone. It might be from an abelisaur.

Pachycephalosaurus Large North American pachycephalosaur, from the Cretaceous, with a thick, rounded skull dome.

Pachyrhinosaurus Large centrosaurine ceratopsid from the Cretaceous of North America. Had a massive, thickened bony boss on its snout and spikes on its frill. Closely related to *Achelousaurus*.

Pachysauriscus Name formerly used for sauropodomorph remains from the Triassic of Germany. These bones are generally considered today to belong to *Plateosaurus*.

Pachyspondylus Old name for Jurassic saurodomorph remains from South Africa, which might be *Massospondylus*.

Pakisaurus Titanosaur from the Cretaceous of Pakistan, named for poorly preserved, incomplete tail vertebrae.

Palaeopteryx Small theropod from the Jurassic of the USA, named for a leg bone suggested to be from a bird.

Palaeoscincus One of North America's earliest named dinosaurs. Named for an ornithischian tooth from the Cretaceous. It probably belonged to a nodosaurid such as *Edmontonia* or *Panoplosaurus*.

Paluxysaurus Cretaceous macronarian sauropod from the USA, known from the remains of four individuals. Perhaps closely related to *Brachiosaurus* and *Giraffatitan*.

Panoplosaurus North American nodosaurid ankylosaur from the Cretaceous. Similar to *Edmontonia* but with a shorter snout and wider, rounder skull. Seems to have lacked the long shoulder spikes of *Edmontonia*.

Panphagia Primitive sauropodomorph from the Triassic of Argentina, probably closely related to *Thecodontosaurus*. Similar to *Eoraptor*. Teeth suggest an omnivorous diet.

Pantydraco Small Triassic sauropodomorph from Britain, originally classified as a species of *Thecodontosaurus*. Lightly built.

Paralititan Giant titanosaur from the Cretaceous of Egypt, found in a mangrove environment and similar to *Argentinosaurus*.

Paranthodon African stegosaur from the Cretaceous of South Africa, known only from a partial snout originally identified as that of a non-dinosaurian reptile.

Pararhabdodon Hadrosaurid from the Cretaceous of Spain, known from numerous remains initially misidentified as belonging to a *Rhabdodon*-like iguanodontian.

Parasaurolophus Cretaceous lambeosaurine hadrosaurid from North American with a tube-like crest. This was curved in some species but almost straight in others.

Parksosaurus Small, bipedal ornithopod from the Cretaceous of Canada known from an incomplete skull and skeleton. Compared to the better-known *Hypsilophodon*, it had a longer snout, more teeth, and stouter hind limb bones.

Paronychodon North American theropod from the Cretaceous, known from a single, tiny, unserrated tooth. Other teeth have been discovered since and labelled *Paronychodon*. They probably belong to small maniraptorans.

Parrosaurus Hadrosaurid from the Cretaceous of the USA, known from vertebrae and skull bones that were originally labelled *Neosaurus*. Sometimes thought to be the same as *Hypsibema*.

Parvicursor Tiny Mongolian alvarezsaurid from the Cretaceous, with very long, slender hind limbs. Known from numerous vertebrae, hind limb bones, and other parts of the skeleton. Probably a specialized ant-eater.

Patagonykus Relatively large alvarezsaurid from the Cretaceous of Argentina, known from a partial skeleton that includes forelimb and hind limb bones. It had short, stout forelimbs with a massive thumb claw.

Patagosaurus Jurassic sauropod from Argentina, conventionally thought to be a close relative of *Cetiosaurus* from Britiain. Known from several individuals.

Pawpawsaurus Cretaceous nodosaurid ankylosaur from the USA, known from a skull and other bones. It had a long snout and broad cheeks, and bumpy armour on top of its skull. Suggested by some to be *Texasetes*.

Pectinodon Cretaceous maniraptoran from North America, known from teeth thought to belong to a troodontid; probably *Troodon*.

Pedopenna Small maniraptoran, closely related to birds, from the Jurassic or Cretaceous of China. Only known from a hind limb, with long feathers attached.

Peishansaurus Cretaceous ornithischian from China, named for a partial lower jaw and one tooth. Usually thought to be an ankylosaur, though its true affinities have yet to be determined.

Pelecanimimus Primitive ornithomimosaur (ostrich dinosaur) from the Cretaceous of Spain. Had a long, slender skull housing over 200 teeth – more than any other theropod.

Pellegrinisaurus Argentinean titanosaur from the Cretaceous, originally misidentified as *Epachthosaurus*. Known from vertebrae belonging to the back and tail. Some of the back vertebrae were particularly broad.

Peloroplites Cretaceous nodosaurid ankylosaur from the USA, known from a partial skull and other remains. Its lower jaw was heavily built, and its lower arm bones were straight compared to other nodosaurids.

Pelorosaurus British sauropod from the Cretaceous, named for a large, relatively straight arm bone and some tail bones. Perhaps similar to *Brachiosaurus*.

Penelopognathus Iguanodontian ornithopod from the Cretaceous of Mongolia, named for a lower jaw bone. This is relatively long and straight.

Pentaceratops Large to gigantic North American ceratopsid from the Cretaceous, known from excellent remains including complete skulls. It had a massive, erect frill and long brow horns.

Phaedrolosaurus Cretaceous theropod, named for a single tooth originally thought similar to that of *Deinonychus*.

Phuwiangosaurus Cretaceous sauropod from Thailand, known from numerous bones, some of which are from juveniles. It seems to be a titanosaur, perhaps closely related to *Nemegtosaurus*.

Phyllodon Portugese Jurassic ornithischian, named for a jaw fragment and numerous tiny teeth. Conventionally been regarded as a small ornithopod similar to *Hypsilophodon*.

Piatnitzkysaurus Jurassic theropod from Argentina known from two partial skeletons, both of which include skull material.

Pinacosaurus Mongolian ankylosaurid from the Cretaceous, known from numerous specimens and two species. Had a short, deep snout, horns at the back of the skull, and other openings around its nostrils.

Pisanosaurus Triassic ornithischian from Argentina, perhaps the most primitive of the group. Unlike other ornithischians, its pubic bones seem to have pointed forwards.

Pitekunsaurus Cretaceous titanosaur from Argentina, known from a partial skeleton that includes skull remains. Suggested to be a close relative of *Rinconsaurus*.

Piveteausaurus Jurassic theropod from France, known from a partial braincase usually thought to be from a theropod.

Planicoxa Cretaceous iguanodontian ornithopod from the USA known from pelvic and hind limb bones belonging to several individuals. Two species have been named.

Plateosauravus Large Triassic sauropodomorph from South Africa, originally included, incorrectly, in *Plateosaurus* and later *Euskelosaurus*. Named for limb and hip bones and incomplete vertebrae.

Plateosaurus One of the best-known primitive sauropodomorphs, a European Triassic animal known from many specimens and species. Several species are recognized, different in skull shape, size, and proportions.

Platyceratops Small ceratopsian from the Cretaceous of Mongolia known from a near-complete skull. It had a short, deep snout, a short frill, and a small, conical nasal horn. Thought by some to be *Bagaceratops*.

Pleurocoelus Macronarian sauropod from the Cretaceous of the USA, known from fragments. Usually thought to be a small, brachiosaur-like form the same as *Astrodon*.

Pleuropeltus Old name for skull bones, armour plates, and other fragments from the Cretaceous of Austria, once thought to be a turtle but now thought to be *Struthiosaurus*.

Podokesaurus Small Triassic coelophysoid theropod, known from numerous fragments now lost. Similar to *Coelophysis*.

Poekilopleuron Spinosauroid theropod from the Jurassic of France, closely related to *Megalosaurus*.

Polacanthoides Name once used for ankylosaur remains from the Cretaceous of Britain. Most of the bones belonged either to *Polacanthus* or *Hylaeosaurus*.

Polacanthus Cretaceous ankylosaur from Britain. A fused armour plate covered its hip region and triangular spikes projected from the sides of its tail. Often incorrectly thought to be the same animal as *Hylaeosaurus*.

Polyodontosaurus North American maniraptoran from the Cretaceous, named for a lower jaw originally thought to be from a large lizard but later identified as troodontid. Almost certainly the same as *Troodon*.

Polyonax Old name given to ceratopsid remains from the Cretaceous of North America. These likely belong to *Triceratops*.

Ponerosteus Name given to a single poorly preserved specimen from the Cretaceous of the Czech Republic. Thought to be the internal cast of a shin bone.

Pradhania Medium-sized sauropodomorph from the Triassic of India, known from fragmentary skull bones, vertebrae, and a partial hand. It had a prominent bony ridge on the inside surface of one of its skull bones.

Prenocephale Mongolian pachycephalosaur from the Cretaceous. It had a short face and tall, rounded skull dome. Large bony knobs grew on the back of its skull.

Prenoceratops Primitive ceratopsian from the Cretaceous of North America, known from several individuals found together. It had a long, low head, and nostril openings located further away from the snout tip than is usual. Closely related to *Leptoceratops*.

Priodontognathus British ankylosaur, known only from a single skull bone and teeth. Originally thought to belong to *Iguanodon*, may be from a nodosaurid. Uncertain whether it is Jurassic or Cretaceous in age.

Probactrosaurus Iguanodontian ornithopod from the Cretaceous of China with a superficially *Iguanodon*-like skull. Related to hadrosaurids. Three species are named.

Proceratosaurus Jurassic theropod from Britain, known only from its skull. Top of the skull is missing, but a crest ran along the top of the snout. One of the oldest coelurosaurs.

Procheneosaurus Cretaceous lambeosaurine hadrosaurid from North America. Probably a juvenile *Lambeosaurus*.

Procompsognathus Small Triassic theropod from Germany, known from a partial skeleton thought to be related to *Coelophysis*. A skull thought to belong to this species has been argued by some to be a crocodilian.

Prodeinodon Mongolian theropod from the Cretaceous, named for a single tooth that cannot be precisely identified. Other teeth and fragmentary limb bones have also been regarded as belonging to the same animal.

Prosaurolophus Large North American hadrosaurine hadrosaurid from the Cretaceous, known from remains assigned to two species. A blunt bony lump projected between its eyes and its head was large.

Protarchaeopteryx Chinese maniraptoran

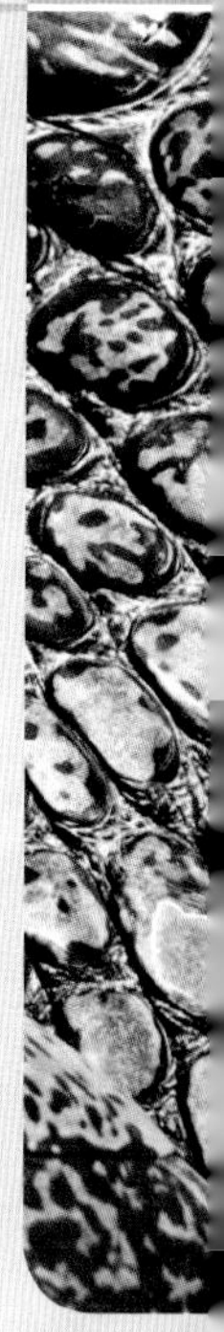

from the Cretaceous, preserved with feathers and thought to be similar to *Incisivosaurus*.

Protiguanodon Mongolian ceratopsian from the Cretaceous, widely agreed nowadays to be a species of *Psittacosaurus*.

Protoceratops Primitive ceratopsian from the Cretaceous of Mongolia. It had a deep snout, a narrow beak, and two pairs of fang-like teeth. Its frill was short and erect.

Protognathosaurus Chinese sauropod from the Jurassic, known from a partial lower jaw bone originally named *Protognathus*.

Protohadros North American iguanodontian ornithopod from the Cretaceous with a deep, scoop-like lower jaw and down-curved beak.

Psittacosaurus Primitive ceratopsian from the Cretaceous of eastern Asia. It had a short, deep skull with flaring cheeks. One specimen has long, quill-like filaments on its tail.

Pteropelyx Old name given to fragmentary hadrosaurid remains from the Cretaceous of the USA. Some of these remains might have belonged to *Corythosaurus*.

Pterospondylus Triassic coelophysoid theropod from Germany, named for a single vertebra usually thought to have belonged to *Procompsognathus*.

Puertasaurus Gigantic titanosaur from the Cretaceous of Argentina, known from gigantic, broad vertebrae.

Pukyongosaurus Poorly known Cretaceous sauropod from Korea, known from vertebrae, ribs, and limb bones.

Pycnonemosaurus Poorly known Brazilian theropod from the Cretaceous, known from vertebrae, limb bone fragments, and teeth. Probably an abelisaurid.

Pyroraptor French maniraptoran from the Cretaceous, known from fragmentary remains. Suggested to be a dromaeosaurid.

Qantassaurus Cretaceous ornithischian from Australia, named after the airline company Qantas. Probably a small, agile biped with short arms and usually imagined to have been a *Hypsilophodon*-like form.

Qingxiusaurus Cretaceous titanosaur from China, known from vertebrae and bones from the chest and forelimb. The spines on its vertebrae were distinctive in being paddle shaped. Its upper arm bones were broad with large muscle attachments.

Qinlingosaurus Asian sauropod from the Cretaceous, known from vertebrae and hip bones.

Quaesitosaurus Mongolian titanosaur from the Cretaceous, known only from a partial skull. Very similar to *Nemegtosaurus*.

Quilmesaurus Poorly known theropod named for partial leg bones from the Cretaceous of Argentina.

Rahonavis Small maniraptoran from the Cretaceous of Madagascar, known from forelimb and hind limb bones, and other remains. It had long arms and could probably fly or glide. A close relative of *Unenlagia*.

Rajasaurus Large abelisaurid from the Cretaceous of India, known from numerous bones of a single individual. Had massive hind limbs and a thick, bony lump on its forehead.

Rapator Mysterious Australian theropod from the Cretaceous, known only from a single hand bone.

Rapetosaurus Medium-sized Cretaceous titanosaur from Madagascar, known from excellent remains. Closely related to *Nemetosaurus* with slim, peg-like teeth.

Rayososaurus Diplodocoid sauropod from Argentina, originally identified as *Rebbachisaurus*. Like other rebbachisaurids, it had a distinctively shaped shoulder blade.

Rebbachisaurus African diplodocoid sauropod from the Cretaceous, first reported from Morocco, and the first rebbachisaurid to be named. Known from numerous vertebrae and limb bones.

Regnosaurus British thyreophoran known only from a fragmentary jaw from the Cretaceous. Suggested to belong to a stegosaur, though this is now doubtful.

Rhabdodon Medium-sized ornithopod from the Cretaceous of Romania, France, Austria, Spain, and Hungary. Broad-bodied herbivore with stout jaw bones. Related to *Zalmoxes*.

Rhadinosaurus Poorly known fossil from the Cretaceous of Austria, known from limb bones, vertebrae, and other fragments. It may not even be a dinosaur.

Rhodanosaurus French nodosaurid ankylosaur from the Cretaceous, known from remains that cannot be identified precisely. They might belong to *Struthiosaurus*.

Rhoetosaurus Australian sauropod from the Jurassic, known from vertebrae and various fragments. Long thought allied to *Cetiosaurus*, but may be a sauropod.

Richardoestesia Cretaceous theropod first named from teeth and jaws found in North America, but later reported from Europe and Asia. Its affinities are uncertain.

Rinchenia Oviraptorid oviraptorosaur from the Cretaceous of Mongolia, often wrongly labelled as *Oviraptor* in the past. It had a tall, rounded crest on the top of its skull.

Rinconsaurus Small Cretaceous titanosaur from Argentina, known from bones from three individuals. Unusual in that its tail vertebrae varied in their styles of articulation: some have convex front faces, others had convex back faces, and others were convex on both faces.

Riojasaurus Large, bulky Triassic sauropodomorph from Argentina, with stout limb bones.

Rocasaurus Titanosaur from the Cretaceous of Argentina, known from vertebrae and hind limb and pelvic bones.

Ruehleia German Triassic sauropodomorph, known from a near-complete headless skeleton. It had complicated wrist bones.

Rugops Abelisaurid theropod from the Cretaceous of Niger. It had a short, deep skull. Unusual texturing on the side of its face suggests a covering of horny tissue.

Ruyangosaurus Large Chinese sauropod from the Cretaceous, known from vertebrae and a shin bone more than a metre long.

Sahaliyania Cretaceous lambeosaurine hadrosaurid from China. Its lower jaws were down-curved and scoop-like and it had deep depressions on its skull roof.

Saichania Mongolian ankylosaurid ankylosaur from the Cretaceous, known from excellent remains. It had stout limb bones, and bulbous armour plates covered its short, deep skull. Curved horns projected from the back of the skull.

Saltasaurus Cretaceous titanosaur from Argentina, and one of the best known members of its group. Small armour scutes covered its back and sides.

Sanpasaurus Jurassic dinosaur from China, named from jumbled remains that belonged to both a sauropod and ornithopod. The name is not generally used today.

Santanaraptor Small Brazilian theropod from the Cretaceous, known from its hips and hind limb bones, and discovered with skin fragments. Possibly a tyrannosauroid.

Sarcolestes British ankylosaur from the Jurassic, known from an incomplete jaw.

Sarcosaurus Jurassic theropod from Britain, known from fragmentary remains. Probably a coelophysoid, though once thought to be a small relative of *Ceratosaurus*.

Saturnalia Small primitive sauropodomorph from the Triassic of Brazil. Probably both bipedal and quadrupedal, and omnivorous.

Saurolophus Cretaceous hadrosaurid known from North America and Mongolia. A solid bony crest projected backwards from the top of its skull.

Sauropelta Large nodosaurid ankylosaur from the Cretaceous of the USA. It had a long tail, a covering of small armour plates, and long spikes on its neck and shoulders.

Saurophaganax Large Jurassic allosauroid theropod from the USA, thought by some to be a giant species of *Allosaurus*.

Sauroplites Poorly known Cretaceous ankylosaur from China, named for armour scutes, ribs, and a possible hip bone. Thought by some to an ankylosaurid, and perhaps the same as *Shamosaurus*.

Sauroposeidon Giant brachiosaur from the Cretaceous of the USA, and a close relative of *Brachiosaurus*. Known from four long neck vertebrae. Its full size is unknown but it may have been larger than *Brachiosaurus*.

Saurornithoides Large troodontid maniraptoran from the Cretaceous of China, known from two species represented by partial skulls and skeletons. It had huge eye sockets and numerous, closely packed teeth.

Saurornitholestes North American dromaeosaurid from the Cretaceous, named for skull fragments and other remains. Usually thought to have been a relative of *Velociraptor*.

Scansoriopteryx Tiny Cretaceous theropod from China, similar to *Epidendrosaurus*. It had incredibly long hands with long third fingers. The only known specimen is a juvenile.

Scelidosaurus Primitive thyreophoran from the Jurassic of Britain. Rows of armour ran along its neck, body, and tail, and horns and spikes projected from the back of its head.

Scipionyx Cretaceous theropod from Italy, known only from the partial skeleton of a juvenile. Long teeth and large hands suggest that it was already able to care for itself.

Scolosaurus Ankylosaurid ankylosaur from the Cretaceous of Canada, named for an excellent near-complete skeleton generally thought to belong to *Euoplocephalus*.

Scutellosaurus Small thyreophoran from the Jurassic of the USA, and one of the most primitive members of its clade. It had a long tail and rows of armour on its back and sides.

Secernosaurus Poorly known Cretaceous hadrosaurid from Argentina, known from a braincase and various other bones. Its affinities within the group are uncertain. One of only a few South American hadrosaurids.

Segisaurus Small Triassic theropod from the USA, probably closely related to *Coelophysis*. Once incorrectly thought to have solid bones.

Segnosaurus Large, wide-bodied therizinosauroid maniraptoran from the Cretaceous of Mongolia. The chin region of its lower jaw curved downwards.

Seismosaurus Gigantic diplodocoid sauropod from the Jurassic of the USA. Now thought to be *Diplodocus*. Originally claimed to be the longest sauropod, has more recently been down-sized to about 30m (98ft).

Sellosaurus Sauropodomorph from the Triassic of Germany, almost certainly the same animal as *Plateosaurus*. It had a shallower snout than the best-known *Plateosaurus* species and a shorter neck.

Serendipaceratops Australian ornithischian from the Cretaceous, known from an arm bone. Suggested to be a small ceratopsian, though this is doubtful.

Shamosaurus Large Mongolian ankylosaurid ankylosaur from the Cretaceous, known from two skulls and a partial skeleton. Unlike many ankylosaurids, it had a narrow snout and relatively large eye sockets.

Shanag Small dromaeosaurid maniraptoran from the Cretaceous of Mongolia, known from a partial skull. It had a shallow, bird-like snout and may have been closely related to Chinese dromaeosaurids like *Microraptor*.

Shanshanosaurus Cretaceous tyrannosauroid from China, once thought to be an unusual, dwarf form but now generally regarded as a juvenile of *Tarbosaurus*.

Shantungosaurus Gigantic hadrosaurine hadrosaurid from the Cretaceous of China, one of the largest non-sauropod dinosaurs.

Shanxia Chinese Cretaceous ankylosaur, known from its broad skull. Triangular horns projected from the skull's rear edges.

Shanyangosaurus Poorly known Chinese theropod from the Cretaceous, known from vertebrae, a shoulder blade, and limb bones.

Shenzhousaurus Small primitive ornithomimosaur from the Cretaceous of China. It had a long, shallow snout, and conical teeth at the tips of the lower jaws.

Shidaisaurus Jurassic theropod from China, known from vertebrae, ribs, limbs, and limb girdles. It might be a spinosauroid.

Shixinggia Oviraptorosaur from the Cretaceous of China, known from hip, hind limb bones, and vertebrae.

Shuangmiaosaurus Cretaceous iguanodontian ornithopod from China, named for skull bones. It seems to be very close to the ancestry of the hadrosaurids.

Shunosaurus Jurassic sauropod from China, known from many specimens. It had a deep snout with large nostrils, and is unusal in having a bony club at the end of its tail.

Shuvuuia Mongolian alvarezsaurid from the Cretaceous. It had a bird-like skull, numerous tiny unserrated teeth, and huge eyes.

Siamosaurus Cretaceous theropod from Thailand, known from teeth thought to be from a spinosaurid.

Siamotyrannus Cretaceous theropod from Thailand, known from hips and vertebrae and originally identified as a tyrannosauroid. More likely to be an allosauroid.

Sigilmassasaurus North African theropod from the Cretaceous, named for neck vertebrae. Some experts think that these actually belong to *Carcharodontosaurus*.

Siluosaurus Chinese ornithischian from the Jurassic, known only from leaf-shaped teeth conventionally thought to be from a small, *Hypsilophodon*-like form.

Silvisaurus North American nodosaurid ankylosaur from the Cretaceous of the USA. It had a pear-shaped skull and long rows of spikes on its neck and shoulders.

Similicaudipteryx Chinese Cretaceous oviraptorosaur, very similar to *Caudipteryx*, and known from a single, jumbled skeleton.

Sinocalliopteryx Chinese theropod from the Cretaceous. Close relative of *Sinosauropteryx*, although much larger.

Stomach contents show that it ate other theropods.

Sinocoelurus Jurassic theropod from China, named for teeth that cannot be precisely identified. The teeth are curved and serrated, and typical for theropods.

Sinornithoides Slender, long-limbed troodontid maniraptoran from the Cretaceous of Mongolia. Only about 1m (3ft) long. One specimen is preserved curled up as if asleep.

Sinornithomimus Chinese ornithomimid ornithomimosaur from the Cretaceous. Several specimens were found together, seemingly in a herd.

Sinornithosaurus Dromaeosaurid maniraptoran from the Cretaceous of China, preserved with masses of feathers. Its skull bones had distinctive textured surfaces.

Sinosauropteryx Long-tailed compsognathid from the Cretaceous of China. It had large thumbs, a long tail, and was covered in quill-like proto-feathers.

Sinosaurus Poorly known theropod from the Triassic or Jurassic of China, named for skull bones, teeth and some partial skeletons. The skeletons seem to belong to sauropodomorphs but the affinities of the skull bones are uncertain.

Sinovenator Small maniraptoran from the Cretaceous of China, originally suggested to be a troodontid. It had a short snout with numerous closely packed teeth.

Sinraptor Chinese allosauroid theropod from the Jurassic, known from excellent remains. Closely related to *Yangchuanosaurus* but with a longer, shallow skull with small concavities on the sides of the facial bones.

Sinusonasus Troodontid maniraptoran from the Cretaceous of China, known from an articulated skeleton that only lacks the arms, shoulder girdles, and some vertebrae. It had relatively large teeth for a troodontid and thickened bones along the top of the snout.

Skorpiovenator Abelisaurid from the Cretaceous of Argentina, known from an excellent, articulated skeleton. It had a very deep, blunt face and lacked horns like those seen in its close relative, *Carnotaurus*.

Sonidosaurus Titanosaur from the Cretaceous of China, known from vertebrae, ribs, and a pelvis. One of several specimens showing that titanosaurs were abundant and diverse in Cretaceous Asia.

Sonorasaurus Macronarian sauropod from the Cretaceous of the USA, known from vertebrae, ribs, hind limb, and hip bones.

Sphaerotholus Dome-skulled pachycephalosaur from the Cretaceous of North America. Rows of small bony lumps decorated the back of its skull. Regarded by some experts as the same as *Prenocephale*.

Sphenospondylus Old name for iguanodontian ornithopod vertebrae from the Cretaceous of England, usually thought to have belonged to *Iguanodon* but in the old, inclusive use of the name.

Spinosaurus Gigantic spinosaurid theropod from the Cretaceous of northern Africa. It had a long, crocodile-like skull, conical teeth that lacked serrations, and tall bony spines on its vertebrae. Perhaps the largest theropod.

Spinostropheus African theropod from the Cretaceous of Niger, known from neck vertebrae and other remains that were originally identified as *Elaphrosaurus*. Closely related to ceratosaurs like the abelisaurs.

Staurikosaurus Bipedal Triassic saurischian from Brazil, closely related to the larger *Herrerasaurus*. It had a long, slender tail and its hips were primitive.

Stegoceras Small pachycephalosaur from the Cretaceous of North America. Adults had a thickened, rounded skull roof. A bony shelf jutted out from the back of the skull.

Stegopelta Nodosaurid ankylosaur from the Cretaceous of the USA, named for a partial skeleton. Perhaps the same as *Nodosaurus*.

Stegosaurides Poorly known thyreophoran from the Cretaceous of China, named for incomplete vertebrae and armour fragments. Usually regarded as an ankylosaur.

Stegosaurus Best known stegosaur, from the Jurassic of the USA and Portugal, but also thought by some to be present in China. It had large, diamond-shaped plates and four tail spikes at its tail tip.

Stenonychosaurus Troodontid maniraptoran from the Cretaceous of Canada, now generally thought to be *Troodon*.

Stenopelix Small ornithischian from the Cretaceous of Germany, known by the back part of a skeleton. Likely a pachycephalosaur.

Stenotholus Cretaceous pachycephalosaur from North America, named for a skull fragment now thought to be *Stygimoloch*.

Stephanosaurus Lambeosaurine hadrosaurid from the Cretaceous of North America, likely the same as *Lambeosaurus*.

Sterrholophus Old name given to ceratopsid remains from the Cretaceous of North America. These cannot be identified precisely but probably belong to *Triceratops*.

Stokesosaurus Tyrannosauroid theropod from the Jurassic of the USA and Britain, originally named for hip bones but now known from many regions of the skeleton.

Stormbergia Primitive ornithischian from the Jurassic of South Africa and Lesotho, known from several specimens. Some of these had previously been identified as *Lesothosaurus*.

Streptospondylus Old name originally given to theropod bones from the Jurassic of France that had been confused with those of marine crocodilians.

Struthiomimus North American ornithomimid ornithomimosaur from the Cretaceous. It had a particularly slender third finger. Very similar to *Ornithomimus*.

Struthiosaurus Small nodosaurid ankylosaur known from several species from various locations across Europe, including France, Austria, and Romania. Possibly an island-dwelling dwarf.

Stygimoloch North American Cretaceous pachycephalosaur, with a lower skull dome and longer spikes than *Pachycephalosaurus*. May be a juvenile of that dinosaur.

Stygivenator Cretaceous tyrannosaurid from North America, named for a partial snout and lower jaw. Almost certainly a juvenile *Tyrannosaurus*.

Styracosaurus Centrosaurine ceratopsid from the Cretaceous of North America. Long spikes projected from the back of its frill. Closely related to *Centrosaurus*.

Suchomimus Spinosaurid from the Cretaceous of Niger, known from numerous remains. It had a long, crocodile-like skull and was very similar to *Baryonyx*.

Suchosaurus Named for teeth from the Cretaceous of Britain, long thought to be from crocodilians but now regarded as from *Baryonyx* or a similar theropod.

Sulaimanisaurus Titanosaur from the Cretaceous of Pakistan, based on incomplete tail vertebrae said to possess unique features. Claimed to be from a sauropod closely related to *Pakisaurus* and *Khetranisaurus*.

Supersaurus Gigantic diplodocoid sauropod from the Jurassic of the USA, argued to be a close relative of *Apatosaurus*. Had a very long neck.

Suuwassea Jurassic diplodocoid sauropod from the USA, closely related to *Apatosaurus*. Medium-sized for this group.

Suzhousaurus Therizinosauroid maniraptoran from the Cretaceous of China. Like other members of the group, it had a broad belly and long, scythe-like hand claws.

Symphyrophus Old name given to vertebrae and other fragments from the Jurassic of the USA. Some might belong to crocodilians, but the dinosaur bones are now generally thought to belong to *Camptosaurus*.

Syrmosaurus Mongolian ankylosaurid from the Cretaceous, known from a fragmentary skull now thought to be *Pinacosaurus*.

Szechuanosaurus Name given to Jurassic theropod teeth from China. Partial skeletons belonging to an allosauroid have also been identified as *Szechuanosaurus*.

Talarurus Mongolian ankylosaurid ankylosaur from the Cretaceous, known from several specimens. It had a broad skull, and four toes on each hindfoot.

Talenkauen Bipedal Cretaceous iguanodontian ornithopod from Argentina, known from a partial articulated skeleton. It had a small skull and plate-like growths on its ribs.

Tangvayosaurus Sauropod from the Cretaceous of Laos, known from two partial skeletons. Probably a primitive titanosaur.

Tanius Poorly known Cretaceous hadrosaurid from China, known from a partial skull, vertebrae, limb bones, and other remains. Had tall spines on its vertebrae, stout upper arm bones, and lacked a crest.

Tanycolagreus Small Jurassic theropod from the USA, known from a good partial skeleton. Superficially similar to *Coelurus* but possibly close to Tyrannosauroidea.

Tanystrosuchus Old name given to tail vertebrae from the Triassic of Germany, sometimes suggested to be from *Coelophysis*.

Tarascosaurus European theropod from the Cretaceous of France, suggested to be an abelisaurid. Named for a partial thigh bone.

Tarbosaurus Giant Asian tyrannosaurid from the Cretaceous. Similar to *Tyrannosaurus* but with a slimmer, less heavily built skull, and perhaps more closely related to *Alioramus* and *Daspletosaurus* than to *Tyrannosaurus*.

Tarchia Giant Cretaceous ankylosaurid ankylosaur from Mongolia. It had bulbous armour plates on the top of its head and curved horns at the back.

Tastavinsaurus Cretaceous macronarian sauropod from Spain, known from numerous bones. It may be closely related to *Venenosaurus* from the USA.

Tatisaurus Small ornithischian from the Jurassic of China, known from a lower jaw suggested by some to be similar to *Scelidosaurus*. Its true affinities are uncertain.

Taveirosaurus Cretaceous ornithischian from Portugal, known from teeth. Suggested to be a pachycephalosaur or ankylosaur.

Tawasaurus Dinosaur from the Jurassic of China, named for a partial skull once thought to be from a bipedal ornithischian, but actually from a baby sauropodomorph.

Tazoudasaurus Relatively small primitive sauropod from the Jurassic of Morocco, known from bones belonging to several individuals. Shares features with *Vulcanodon*.

Technosaurus Small dinosaur from the Triassic of the USA, named for skull fragments. Originally thought to be a *Fabrosaurus*-like ornithischian, it was named for remains that belong to two dinosaurs: some are from a small ornithischian, but others are from a sauropodomorph.

Tehuelchesaurus Jurassic sauropod from Argentina, known from a partial skeleton that was preserved with skin impressions. Might have been similar to *Omeisaurus*.

Teinurosaurus French theropod from the Jurassic, named for a single tail vertebra that has since been lost.

Telmatosaurus Primitive hadrosaurid from the Cretaceous of Romania, known from numerous remains. It was small compared to most other hadrosaurids, and seems to have been an island-dwelling dwarf.

Tendaguria Unusual Jurassic sauropod from Tanzania, named for broad vertebrae that are very different from those of other sauropods.

Tenontosaurus Cretaceous iguanodontian ornithopod from the USA with a long tail. Limb proportions suggest that it was quadrupedal.

Texasetes Cretaceous ankylosaur from the USA, known from a skull fragment, limb bones, armour plates, and other bones. Probably a nodosaurid, and suggested by some to be the same as *Pawpawsaurus*.

Teyuwasu Triassic dinosaur from Brazil, named for partial hind limb bones. Its affinities are uncertain, it may not even be a dinosaur.

Thecocoelurus British dinosaur from the Cretaceous, known only from a single partial neck vertebra.

Thecodontosaurus Small, primitive sauropodomorph from the Triassic or Jurassic of Britain, known from numerous isolated bones. Probably an omnivore able to be bipedal or quadrupedal.

Thecospondylus Medium-sized British dinosaur from the Cretaceous, known only from a sandstone cast of its sacrum (the fused vertebrae present in between the hips).

Theiophytalia North American Jurassic ornithopod with horn-like lumps on its cheeks. The skull was wrongly identified as that of *Camptosaurus* for years.

Therizinosaurus Giant Asian therizinosauroid from the Cretaceous, best known from its giant arms. Its long, slender hand claws were over 70cm (28in) long. A hind limb may also belong to it.

Thescelosaurus Medium-sized bipedal Cretaceous ornithopod from North America, once thought (incorrectly) to have armour plates on its back. It had short arms, a narrow beak, and leaf-shaped cheek teeth.

Thespesius Hadrosaurid from the Cretaceous of the USA, known only from vertebrae and a toe bone. The name is no longer in use.

Tianchisaurus Ankylosaur from the Jurassic of China, originally known as *Jurassosaurus* after the move *Jurassic Park*. Cannot be reliably distinguished from other ankylosaurs.

Tianyulong Cretaceous heterodontosaurid from China, known from an excellent partial skeleton covered in hair-like and quill-like structures, like those in some theropods.

Tianzhenosaurus Medium-sized ankylosaurid ankylosaur from the Cretaceous of China. It is very similar to *Saichania* but may also be close to *Pinacosaurus*.

Tichosteus Jurassic ornithischian from the USA, known only from vertebrae. Often thought to be an ornithopod.

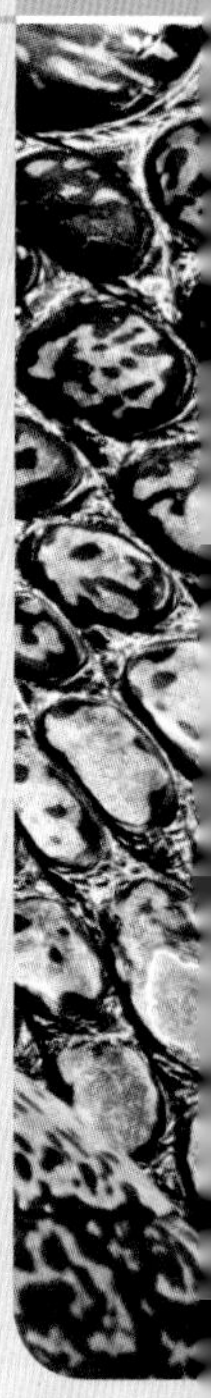

Tienshanosaurus Poorly known Jurassic sauropod from China.

Timimus Obscure Cretaceous theropod from Australia known only from two leg bones. It was originally identified as an ornithomimosaur, but this is no longer thought correct. It might be an abelisaur.

Titanosaurus Cretaceous sauropod named for tail bones from India. The same name was later applied to similar bones from all around the world. The name is no longer widely used.

Tochisaurus Mongolian troodontid maniraptoran from the Cretaceous, known only from its very long, slender foot bones.

Tornieria Diplodocoid sauropod from the Jurassic of Tanzania, long thought to be an African species of *Barosaurus*. It differs from *Barosaurus* and *Diplodocus* in having more heavily built hind limbs and tail vertebrae.

Torosaurus Giant Cretaceous chasmosaurine ceratopsid from North America, related to *Triceratops* but with a longer frill perforated by rounded openings. It had long brow horns and a short nose horn.

Torvosaurus Large Jurassic megalosaurid theropod from the USA and Portugal. It probably looked like a more heavily built version of *Megalosaurus*.

Triceratops Giant, short-frilled chasmosaurine ceratopsid from the Cretaceous of North America. Two species are currently recognized. They differ in snout length, body size, and other features.

Trigonosaurus Brazilian titanosaur from the Cretaceous, known from vertebrae of two individuals, and from a hip bone. It had particularly long neck vertebrae.

Trimucrodon Tiny Jurassic ornithischian from Portugal, known only from its leaf-shaped teeth.

Troodon Troodontid maniraptoran from the Cretaceous of North America, previously known as *Stenonychosaurus*. It had slender legs and a narrow skull.

Tsaagan Mongolian dromaeosaurid maniraptoran from the Cretaceous, similar to *Velociraptor*, but with a deeper, stouter skull.

Tsagantegia Cretaceous ankylosaurid ankylosaur from Mongolia, known only from its skull. In contrast to other ankylosaurids, the armour plates on the top of the skull did not form a distinct mosaic-like pattern.

Tsintaosaurus Bizarre Cretaceous hadrosaurid from China, famous for having a forward-pointing crest like a unicorn.

Tugulusaurus Small Cretaceous coelurosaur from China, known from a partial skeleton. Its tail vertebrae were broad, and its fingers and shin bones allow it to be distinguished from other theropods.

Tuojiangosaurus Jurassic stegosaur from China, known from a partial skeleton. Closely related to *Stegosaurus* but with longer, thinner plates and more spikes on its tail. The top of its skull was particularly broad.

Turanoceratops Ceratopsian from the Cretaceous of Asia, the only member of Ceratopsidae known from outside North America. Long brow horns but no nasal horn.

Turiasaurus Gigantic Jurassic or Cretaceous sauropod from Spain, perhaps as large or larger than *Argentinosaurus*. Known from a partial skull, forelimb, vertebrae, and other remains. Identified as part of a new sauropod group, Turiasauria.

Tylocephale Pachycephalosaur from the Cretaceous of Mongolia, known from an incomplete skull. This had a tall, rounded dome and a shelf jutting out from the back. Probably looked similar to *Stegoceras*.

Tyrannosaurus Giant tyrannosaurid theropod from the Cretaceous of North America, well known for its massively built, wide-cheeked skull and short, two-fingered forelimbs. The largest tyrannosaurid and one of the biggest theropods.

Tyrannotitan Large and heavily built allosauroid from the Cretaceous of Argentina, known from a partial skeleton that includes a lower jaw and bones from the cheek. Closely related to carcharodontosaurids.

Uberabatitan Titanosaur from the Cretaceous of Brazil, known from the remains of three individuals with neck, back, and tail vertebrae, and hip and limb bones.

Udanoceratops Ceratopsian from the Cretaceous of Uzbekistan, known from a partial skull and skeleton. Had a particularly short, deep snout. A low bump on the snout may be an incipient nasal horn.

Ugrosaurus Ceratopsid from the Cretaceous of the USA, known from a partial snout and frill. Originally thought to be unique in having a nasal boss instead of a horn, but now regarded by most as part of *Triceratops*.

Uintasaurus Jurassic sauropod from the USA, named for Uinta County in Utah. Widely thought to be the same as *Camarasaurus*.

Ultrasauros Name once used for a sauropod vertebra and other bones from the Jurassic of the USA. Originally thought to be from a gigantic brachiosaur, now regarded as belonging to the diplodocoid *Supersaurus*.

Ultrasaurus Name originally used for *Ultrasauros* but also given to a sauropod from the Cretaceous of Korea, represented by only an arm bone and vertebra.

Unaysaurus Medium-sized sauropodomorph from the Triassic of Brazil, known from a partial skeleton that includes a skull. Closely related to *Plateosaurus*.

Unenlagia Cretaceous maniraptoran from Argentina with bird-like arms. A close relative of *Rahonavis* and *Buitreraptor*: all form the dromaeosaurid group Unenlagiinae.

Unquillosaurus Argentinean theropod from the Cretaceous known only from one hip bone. This appears to be unique and has a distinctive shape. Suggested by some researchers to be a large maniraptoran.

Urbacodon Cretaceous troodontid maniraptoran from Uzbekistan, known from jaws and unserrated teeth. It had less teeth than troodontids like *Troodon* and was similar in size to small forms like *Sinornithoides*.

Utahraptor Giant dromaeosaurid from the Cretaceous of the USA, more than twice as large as *Deinonychus*. Originally thought to have gigantic hand claws.

Valdoraptor Cretaceous theropod from Britain known only from an incomplete foot, originally identified as that of an ankylosaur. Might have been an allosauroid.

Valdosaurus Bipedal iguanodontian ornithopod from the Cretaceous of Britain, closely related to *Dryosaurus*. Named for a thigh bone originally identified as belonging to a new *Dryosaurus* species.

Variraptor Cretaceous maniraptoran from France, known from vertebrae and other remains thought to belong to a dromaeosaurid. The remains were originally identified as belonging to *Elopteryx*.

Vectisaurus Cretaceous iguanodontian ornithopod from Britain, named for vertebrae and generally thought to be the same animal as *Mantellisaurus*. Regarded by some as a small, dryosaurid-like animal.

Velafrons Cretaceous hadrosaurid from Mexico, closely related to *Hypacrosaurus* and *Corythosaurus*, and also possessing a rounded, plate-shaped crest. A bony bar at the back of its skull was arch-shaped, whereas this bar was straight in related forms.

Velocipes Triassic fossil from Poland, named for a partial foot skeleton. Usually thought to belong to a theropod, it cannot be identified precisely and may not even be a dinosaur.

Velociraptor Cretaceous dromaeosaurid from Mongolia with a low snout that has a concave upper surface. Known from excellent specimens, one of which is preserved locked in combat with a *Protoceratops*.

Velocisaurus Small Argentinean theropod from the Cretaceous, known only from slender, elongate foot bones. Probably a noasaurid abelisaur related to *Masiakasaurus*.

Venenosaurus Cretaceous macronarian sauropod from the USA, possibly a primitive titanosaur. It may have been similar to *Brachiosaurus*, though about half the size. Known from vertebrae, limb bones, and ribs.

Vitakridrinda Cretaceous theropod from Pakistan, known from fragmentary limb bones, vertebrae, teeth, and possibly a snout.

Volkheimeria Jurassic sauropod from Argentina, known from a partial skeleton.

Vulcanodon Primitive sauropod from the Jurassic of Zimbabwe. It had unusual, nail-like toe claws, and its first toe was not massive and stout, as in more advanced sauropods.

Wakinosaurus Cretaceous theropod from Japan, named for a single tooth. This cannot be identified precisely and the name *Wakinosaurus* is not much used.

Walgettosuchus Theropod from the Cretaceous of Australia, named for a single tail vertebra.

Wannanosaurus Small pachycephalosaur from the Cretaceous of China, less than one metre long. Known from a partial skull roof, vertebrae, and limb bones. Its upper arm bone is strongly curved along its length

Wuerhosaurus Chinese stegosaur from the Cretaceous that has long, low plates. Two species are known. Thought by some to be part of the genus *Stegosaurus*.

Wulagasaurus Hadrosaurine hadrosaurid from the Cretaceous of China, possibly one of the most primitive members of the group. It had very long, slender jaws.

Xenoposeidon Unusual medium-sized British sauropod from the Cretaceous, known from a single, strikingly distinctive vertebra.

Xenotarsosaurus Theropod from the Cretaceous of Argentina, known from its hind limb. Usually thought to be an abelisaurid.

Xiaosaurus Small, bipedal ornithischian from the Jurassic of China, known from jaw remains, vertebrae, and limb bones. Originally thought to be similar to *Hypsilophodon*, it is probably not an ornithopod at all.

Xinjiangovenator Theropod from the Cretaceous of China, named for remains originally included in *Phaedrolosaurus*. Shares some hind limb features with *Bagaraatan*.

Xuanhanosaurus Jurassic theropod from China, known from vertebrae and stout forelimb bones. Often thought to be a megalosaur or close relative.

Xuanhuaceratops Primitive ceratopsian from the Jurassic of China, closely related to *Chaoyangsaurus* and known from remains belonging to several individuals. It had a short, deep face and only a single pair of fang-like teeth near the upper jaw's tip.

Yaleosaurus Primitive sauropodomorph from the Jurassic of the USA, thought today to be the same as *Anchisaurus*.

Yamaceratops Small primitive ceratopsian from the Cretaceous of Mongolia, known from a partial skull. It had a short bony frill and wide, flaring cheeks.

Yandusaurus Small, bipedal ornithischian from the Jurassic of China, often said to be an ornithopod but more recently shown to be outside of this group.

Yangchuanosaurus Large allosauroid theropod from the Jurassic of China. Closely related to *Sinraptor*, it had a short, deep skull and a tall bony ridge along its back.

Yaverlandia Controversial Cretaceous British dinosaur known only from a partial skull roof. Originally identified as a pachycephalosaur, it may be a theropod.

Yimenosaurus Poorly known Chinese sauropodomorph from the Jurassic, known from the remains of several individuals. It had a particularly short, tall skull.

Yinlong Small, primitive ceratopsian from China known from a near-complete skeleton. Had a short snout and a bony crest at the back of its head. Some of the skull bones had an unusual bumpy texture.

Yixianosaurus Mysterious Chinese maniraptoran theropod from the Cretaceous, known only from a pair of slender, feathered arms. Its affinities are unknown.

Yuanmousaurus Poorly known Chinese sauropod from the Jurassic. Known from incomplete remains but thought by some to be a close relative of *Euhelopus*.

Yunnanosaurus Chinese sauropodomorph from the Jurassic with slender, narrow teeth and a line of bumps along the top of its skull.

Zalmoxes Cretaceous iguanodontian ornithopod from Romania, known from numerous remains. It had a stout, broad skull, narrow beak, and bulky body.

Zapalasaurus Diplodocoid sauropod from the Cretaceous of Argentina, known from vertebrae and hip and hind limb bones.

Zapsalis Cretaceous theropod from North America, named for a single small, serrated tooth. Thought today to be *Paronychodon*.

Zephyrosaurus North American ornithopod from the Cretaceous, usually thought to be similar to *Hypsilophodon*. Had bony bumps on its cheeks and small, fang-like teeth.

Zhejiangosaurus Poorly known ankylosaur from the Cretaceous of China, suggested by some to be a nodosaurid. Known from vertebrae, hips, and hind limbs.

Zhongyuansaurus Cretaceous ankylosaur from China known from good remains,. Originally suggested to be a nodosaurid, may be an ankylosaurid close to *Shamosaurus*.

Zhuchengosaurus Giant, poorly known hadrosaurid ornithopod from the Cretaceous of China.

Zigongosaurus Long-necked Jurassic sauropod from China, thought by many to be the same as *Mamenchisaurus*.

Zizhongosaurus Poorly known primitive sauropod from the Jurassic of China, suggested to be a relative of *Barapasaurus*.

Zuniceratops Medium-sized, Cretaceous horned dinosaur from the USA, with a frill and long brow horns, but no nose horn.

Zupaysaurus Medium-sized Triassic theropod from Argentina, with two parallel crests on its snout, like *Dilophosaurus*.

Index

Page numbers in **bold** indicate the main reference, page numbers in *italics* an illustrated reference.

A

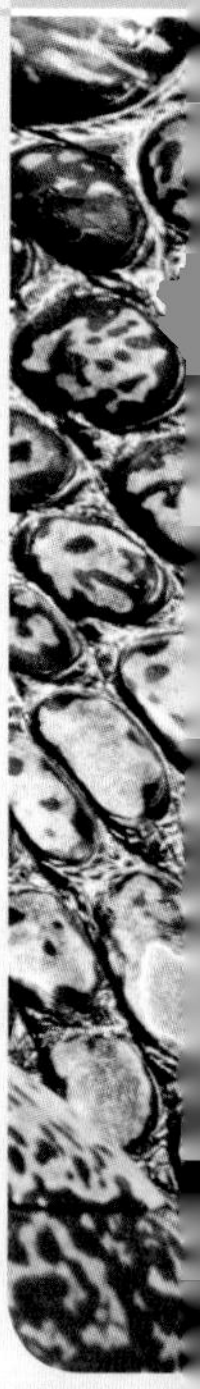

D

E

I

J

K

L

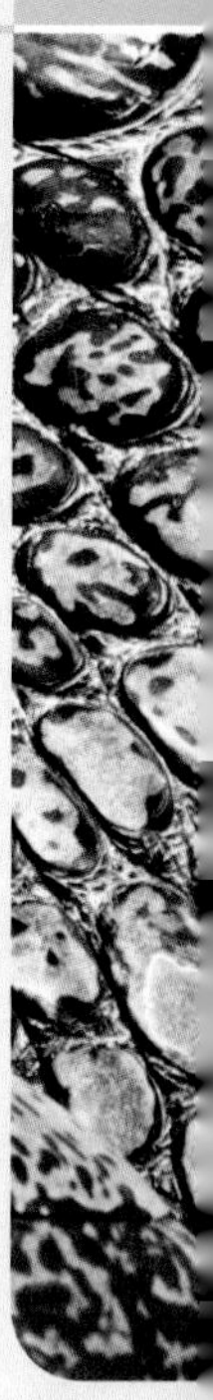

N

O

P

R

S

Acknowledgments

Dorling Kindersley would like to thank the following people for their help in the preparation of this book: Anushka Mody for additional design help; Regina Franke at DK Verlag and Riccie Janus for help with setting up photo shoots in Germany; Roger Jones for access to his fossil collection; Dr Charles Wellman at Sheffield University for plant spore images; and Steve Willis and Mel Fisher for colour work.

Many thanks go to the following for access to their fossil collections for photography: Else Marie Friis, Kamlesh Khullar, Steve McLoughlin and the rest of the team at Naturhistoriska riskmuseet, Stockholm; Doris Von Eiff and all the staff at Senckenberg Forschungsinstitute u. Naturmuseen, Frankfurt; Kirsten Andrews-Speed, Eliza Howlett, Monica Price, Derek Siveter, Malgosia Nowak-Kemp, and Tom Kemp at Oxford University Museum of Natural History; Bob Owens, Caroline Buttler, and John Cope at Amgueddfa Cymru – National Museum Wales.

Sources for the illustrations listed below are as follows: **p.23** Oxygen isotopes diagram: Simon Lamb and David Sington, *Earth Story*, p.149; **p.24** Rain, altitude, and leaf types diagram: Simon Lamb and David Sington, *Earth Story*, p.137; **p.29** House Sparrow size map: http://evolution.berkeley.edu/evosite/evo101/IVB1aExamples.shtml; **p.38** Dinosaur locomotion: http://www.nature.com/nature/journal/v415/n6871/full/415494a.html; **p.78** Non-vertebrate and vertebrate diagram: F. Harvey Pough, Christine M. Janis, John B. Heiser, *Vertebrate Life* – seventh edition, p.25; **p.92** Dermal plate development in fish: L.B. Tarlo, *The Downtonian Ostracoderm Corvaspis kingi Woodward, with notes on the development of dermal plates in the Heterostrachi*; **p.100** Tabulate corals diagram: http://faculty.cns.uni.edu/~groves/LabExercise09.pdf; **p.104** Rolled trilobite diagram: http://www.trilobites.info/enrollment.htm; **p.106** Jaw development, F. Harvey Pough, Christine M. Janis, John B. Heiser, *Vertebrate Life* – seventh edition, p.57; **p.112** Water conducting cells diagram: Wilson N. Stewart and Gar W. Rothwell, *Paleobotany and the Evolution of Plants*, p.86; **p.121** Elkinsia cross-section: Wilson N. Stewart and Gar W. Rothwell, *Paleobotany and the Evolution of Plants*; **p.123** *Heliophyllum* diagram: http://faculty.cns.uni.edu/~groves/LabExercise09.pdf; **p.164** *Falcatus*: after illustration in Fossil fishes of Bear Gulch, 2005 by Richard Lund and Eileen Grogan; **p.179** *Fenestella* artwork: http://www.kgs.ku.edu/Extension/fossils/bryozoan.html; **p.202** Ammonoid sutures diagram: http://faculty.cns.uni.edu/~groves/LabExercise09.pdf; **p.234** *Magellania*: E.N.K. Clarkson, Invertebrate Palaeontology and Evolution, p.159; **p.294** *Archaeanthus*, Wilson N. Stewart and Gar W. Rothwell, *Paleobotany and the Evolution of Plants*, p.448; **p.304** Feather evolution: http://www.nature.com/nature/journal/v420/n6913/fig_tab/nature01196_F5.html; **p.424** Scallop anatomy: http://www.fao.org/docrep/007/y5720e/y5720e07.htm; **p.425** *Magellania* pedicle: E.N.K. Clarkson, *Invertebrate Palaeontology and Evolution*.

Plate Tectonic and Paleogeographic Maps on pages 56–57, 66–67, 82–83, 96–97, 110–11, 142–43, 196–97, 224–25, 282–83, 360–61, 388–89, 416–17 by C.R. Scotese, © 2007, PALEOMAP Project (www.scotese.com)

The publisher would like to thank the following for their kind permission to reproduce their photographs:

(Key: a-above; b-below/bottom; c-centre; f-far; l-left; r-right; t-top)

6 Alamy Images: Louis Champion (tl). **Ardea:** Pat Morris (cra/Cambrian). **Corbis:** Michael Freeman (crb/Devonian); Layne Kennedy (cra/Archean); Kevin Schafer (br/Carboniferous). **Getty Images:** Art Wolfe (tc). **Photolibrary:** Iain Sarjeant (cra/Proterozoic). **Science Photo Library:** Hervé Conge, ISM (crb/Silurian); Paul Whitten (cr/Ordovician). **7 Alamy Images:** WaterFrame (clb/Paleogene). **Corbis:** Micha Pawlitzki / zefa (bl/Quaternary) (cla/Jurassic). **DK Images:** Natural History Museum, London (cla/Permian). **Getty Images:** Peter Chadwick / Gallo Images (tc); Ralph Lee Hopkins / National Geographic (cla/Triassic); Louie Psihoyos / Science Faction (cl/Cretaceous). **Photolibrary:** Tam C. Nguyen (clb/Neogene). **8–9 Corbis:** Michael S. Yamashita. **10–11 Alamy Images:** Louis Champion. **12 Anne Burgess:** (bl). **Corbis:** Hulton-Deutsch Collection (cb). **DK Images:** Satellite Imagemap Copyright © 1996-2003 Planetary Visions (br). **Science Photo Library:** Bernhard Edmaier (cl); Dirk Wiersma (clb). **12–13 Getty Images:** Ralph Lee Hopkins / National Geographic (tc). **13 Corbis:** Jonathan Blair (cr). **DK Images:** Natural History Museum, London (tr). **14 Corbis:** Bettmann / Mariner 10 (crb); NASA / CXC / GSFC / U. Hwang / EPA (cb). **Getty Images:** Stocktrek Images (cl). **NASA:** JPL (cr/Venus). **15 NASA:** (crb). **16 Alamy Images:** Blue Gum Pictures (b). **17 Getty Images:** Johnny Johnson / The Image Bank (tr). **Martina Menneken:** (bl). **Science Photo Library:** Michael Abbey (bc); Eye of Science (bc/archaea). **18 Corbis:** image100 (tr); David Pu'u (ca). **Bradley R. Hacker :** (br). **19 Corbis:** Roger Ressmeyer (br). **20 Corbis:** Image Plan (br). **Getty Images:** Astromujoff (ca). **20–21 Corbis:** Arctic-Images (tc). **21 Corbis:** Lloyd Cluff / Comet (cb); George D. Lepp (br/green broadbill); Jochen Schlenker / Robert Harding World Imagery (br/cockatoo); Jim Sugar / Comet (tr). **DK Images:** University Museum of Zoology, Cambridge (bl); Natural History Museum, London (clb) (bc). **22 Corbis:** Chinch Gryniewicz / Ecoscene (tr); Douglas Pearson / Flirt (b). **Paul F. Hoffman:** (ca). **23 Corbis:** Tony Wharton / Frank Lane Picture Agency (tl). **Science Photo Library:** (br); British Antarctic Survey (cl); Andrew Syred (tr). **25 Corbis:** Catherine Karnow (t); Michael T. Sedam (crb). **DK Images:** Natural History Museum, London (crb/fossil mammoth tooth). **26 Corbis:** Philip Gould (tr); Elisabeth Sauer / zefa (ca). **Science Photo Library:** M.I. Walker (cra). **27 Corbis:** Michael & Patricia Fogden (bl). **DK Images:** Robert L. Braun – modelmaker (crb); Natural History Museum, London (tl/bones). **Science Photo Library:** Michael W. Tweedie (cl). **28 Corbis:** Michael & Patricia Fogden (cla) (bc); Martin Harvey (c). **DK Images:** Oxford Scientific Films (cb). **Getty Images:** Bob Thomas / Popperfoto (ca). **Science Photo Library:** Photo Researchers (cra). **29 Corbis:** Stephen Frink (bl). **Getty Images:** Wally McNamee (br); Steve Winter / National Geographic (tr). **naturepl.com:** Dave Watts (cla). **30 Corbis:** DLILLC (crb/dolphin photo) (cra/dolphin photo); Jeffrey L. Rotman (cra/dolphin and calf). **32 Corbis:** Jonathan Blair (tr) (cra); Frans Lanting (b). **Science Photo Library:** Mark Pilkington / Geological Survey of Canada (ca). **33 Corbis:** Jonathan Blair (cb) (clb); Michael & Patricia Fogden (crb). **Science Photo Library:** George Bernard (ftr). **34 Corbis:** Jonathan Blair (cl); George H.H. Huey (cra); Vienna Report Agency (ca). **DK Images:** Natural History Museum, London (bc). **Royal Saskatchewan Museum:** (clb). **Science Photo Library:** Steve Gschmeissner (cr). **35 Corbis:** Jonathan Blair (bl); Layne Kennedy (ca). **DK Images:** Rainbow Forest Museum, Arizona (ca/petrified log); Natural History Museum, London (tl). **Derek Siveter:** (br/nymphon and haliestes). **36 DK Images:** Natural History Museum, London (bl). **Mark V. Erdmann:** (clb/living coelacanth). **38 Corbis:** Louie Psihoyos (r/main image); Visuals Unlimited (cla). **Oxford University Museum of Natural History:** (clb). **39 DK Images:** Natural History Museum, London (cl). **Getty Images:** J. Sneesby / B. Wilkins (br). **40–41 Corbis:** Louie Psihoyos. **42–45 Alamy Images:** Louis Champion. **45 Science Photo Library:** James King-Holmes (tc); Dr Ken Macdonald (cra). **46–47 Alamy Images:** Phil Degginger. **48–49 Getty Images:** Art Wolfe. **50–51 Corbis:** Layne Kennedy. **52 Martin Brasier:** (cr). **Getty Images:** Peter Hendrie / Photographer's Choice (cl); Carsten Peter / National Geographic (bl). **52–53 Corbis:** Layne Kennedy. **53 Martin Brasier:** (b/all 3 images). **naturepl.com:** Doug Perrine (t). **Science Photo Library:** B. Murton / Southampton Oceanography Centre (cl). **54–55 Photolibrary:** Iain Sarjeant (main). **56–63 Photolibrary:** Iain Sarjeant (sidebars). **56–57 Photolibrary:** Iain Sarjeant (b/background). **56–57** Plate Tectonic and Paleogeographic Map C.R. Scotese, © 2007, PALEOMAP Project (www.scotese.com). **56 Alamy Images:** Randy Green (tr). **Getty Images:** O. Louis Mazzatenta / National Geographic (cl). **57 Alamy Images:** Blue Gum Pictures (ca); David Wall (cra). **Corbis:** Jonathan Blair (bl); Kazuyoshi Nomachi (tl) (b/background). **58 Photolibrary:** Iain Sarjeant (t/microscopic lettering). **60 J. Gehling, South Australian Museum:** (clb). **Photolibrary:** Iain Sarjeant (t/invertebrates lettering). **61 © Leicester City Museums Service:** (tr). **63 J. Gehling, South Australian Museum:** (ca). **64–65 Ardea:** Pat Morris (main). **66–79 Ardea:** Pat Morris (sidebars). **66–67 Ardea:** Pat Morris (b/background). **66–67** Plate Tectonic and Paleogeographic Map C.R. Scotese, © 2007, PALEOMAP Project (www.scotese.com). **66 Corbis:** Yann Arthus-Bertrand (tr); David Muench (cl). **The Natural History Museum, London:** (br). **67 Alamy Images:** All Canada Photos (cra). **Getty Images:** O. Louis Mazzatenta / National Geographic (clb/Burgess Shale); Dr. Marli Miller / Visuals Unlimited (cla/evaporite). **The Natural History Museum, London:** (bc). **Science Photo Library:** Jonathan A. Meyers (tl) (b/background). **68 Ardea:** Pat Morris (t/microscopic lettering). **Martin Brasier:** (cb). **69 David Siveter (University of Leicester):** (br) (b/background). **70 Ardea:** Pat Morris (t/invertebrates lettering). **John Cope:** (cra). **Getty Images:** O. Louis Mazzatenta / National Geographic (cl). **71 Science Photo Library:** Alan Sirulnikoff (cra). **72 Getty Images:** O. Louis Mazzatenta / National Geographic (cr). **The Natural History Museum, London:** (tc). **74 Alamy Images:** All Canada Photos / T. Kitchin & V. Hurst (bl). **The Natural History Museum, London:** (cla). **77 Alamy Images:** Kevin Schafer (t). **US Geological Survey:** (br). **78 Ardea:** Pat Morris (t/vertebrates lettering). **Natural Visions:** (br). **80–81 Science Photo Library:** Paul Whitten (main). **82–93 Science Photo Library:** Paul Whitten (sidebars). **82–83** Plate Tectonic and Paleogeographic Map C.R. Scotese, © 2007, PALEOMAP Project (www.scotese.com). **82–83 Science Photo Library:** Paul Whitten (b/background). **82 Corbis:** Gary Braasch (cl). **Ken McNamara:** (tr). **83 Corbis:** Stephen Frink (tl/smaller image); Galen Rowell (cra). **Getty Images:** Jeff Rotman (tl). **84 DK Images:** Natural History Museum, London (cla) (b/background). **Science Photo Library:** Paul Whitten (t/invertebrates lettering) (b/background). **92 Science Photo Library:** Paul Whitten (t/vertebrates lettering). **94–95 Science Photo Library:** Hervé Conge, ISM (main). **95 DK Images:** Natural History Museum, London (cra). **96–107 Science Photo Library:** Hervé Conge, ISM (sidebars). **96–97 Science Photo Library:** Hervé Conge, ISM (b/background). **96–97** Plate Tectonic and Paleogeographic Map C.R. Scotese, © 2007, PALEOMAP Project (www.scotese.com). **96 Corbis:** Free Agents Limited (tr); Wilfried Krecichwost / zefa (cl). **97 Corbis:** Christophe Boisvieux (ca). **DK Images:** Natural History Museum, London (clb). **Getty Images:** Medioimages / Photodisc (tl). **Science Photo Library:** Sinclair Stammers (cra) (t/plants lettering). **98 Science Photo Library:** Hervé Conge, ISM (b/background). **99 DK Images:** Natural History Museum, London (t). **100 Alamy Images:** Maximilian Weinzierl (clb). **Science Photo Library:** Hervé Conge, ISM (t/invertebrates lettering). **101 Alamy Images:** David Bagnall (br). **Patrick L. Colin, Coral Reef Research Foundation:** (cra). **The Natural History Museum, London:** (crb). **104 Jan Hartmann:** (tc) (b/background). **106 Science Photo Library:** Hervé Conge, ISM (t/vertebrates lettering). **108–109 Corbis:** Michael Freeman (main). **110–139 Corbis:** Michael Freeman (sidebars). **110–111 Corbis:** Michael Freeman (b/background). **110–111** Plate Tectonic and Paleogeographic Map C.R. Scotese, © 2007, PALEOMAP Project (www.scotese.com). **110 DK Images:** Natural History Museum, London (clb) (cb). **Getty Images:** Image Source (tr); Dr. Marli Miller / Visuals Unlimited (cl). **Ken McNamara:** (clb/fish). **111 Alamy Images:** Phil Lyon / Sylvia Cordaiy Photo Library Ltd (cra). **DK Images:** Natural History Museum, London (clb). **Science Photo Library:** Sinclair Stammers (tl). **112 Corbis:** Michael Freeman (t/plants lettering). **Science Photo Library:** Hervé Conge, ISM (b/background). **112–113 Corbis:** Michael Freeman (b/background). **113 Science Photo Library:** Dr Jeremy Burgess (cla). **114 Courtesy of the Smithsonian Institution – Photo by Carol Hotton:** (cla). **115 Science Photo Library:** Sinclair Stammers (tr) (b/background). **122 Corbis:** Michael Freeman (t/invertebrates lettering). **From** *"The Biota of Early Terrestrial Ecosystems: The Rhynie Chert"* **(www.abdn.ac.uk/rhynie/intro.htm) , © University of Aberdeen:** (cra). **123 Corbis:** Jeffrey L. Rotman (cr). **127 DK Images:** Royal Museum of Scotland, Edinburgh (cla). **128 Corbis:** Michael Freeman (t/vertebrates

lettering); Stephen Frink (cb).
128–129 Corbis: Michael Freeman (b/ background).
129 AAP Image : Museum Victoria (crb). **DK Images:** Courtesy of the University Museum of Zoology, Cambridge, on loan from the Geological Museum, University of Copenhagen (fcla).
139 Alamy Images: Yves Marcoux / First Light (cl).
140–141 Corbis: Kevin Schafer (main).
142–168 Corbis: Kevin Schafer (t/plants lettering))sidebars).
142–143 Plate Tectonic and Paleogeographic Map C.R. Scotese, © 2007, PALEOMAP Project (www.scotese.com).
142–143 Corbis: Kevin Schafer (b/ background).
142 DK Images: Natural History Museum, London (bl) (br). **Getty Images:** Peter Essick / Aurora (cl); Travel Ink / Gallo Images (tr).
144 Corbis: Kevin Schafer (b/background). **DK Images:** Natural History Museum, London (cl).
145 Corbis: Kevin Schafer (tr).
149 Alamy Images: blickwinkel / Ziese (cra). **Corbis:** Ashley Cooper (tl). **DK Images:** Natural History Museum, London (cb) (bl). **Valdosta State University :** Paleothyris cast obtained from Robert Carroll of McGill University's Redpath Museum; displayed on http://.fossils.valdosta.edu (bl).
151 DK Images: Natural History Museum, London (cla).
154 Alamy Images: Steffen Hauser / botanikfoto (cl) (b/background). **Corbis:** Kevin Schafer (t/invertebrates lettering).
161 Alamy Images: The Natural History Museum (cla). **Collection of the Illinois State Museum:** (fcr).
162–163 Corbis: Kevin Schafer (b/ background).
163 Corbis: Anthony Bannister / Gallo Images (cra).
164 Eileen Grogan / Richard Lund, The Bear Gulch Project: Carnegie Museum of Natural History, St Joseph's University (bl) (tl).
165 Eileen Grogan / Richard Lund, The Bear Gulch Project: Carnegie Museum of Natural History, St Joseph's University (tc). **DK Images:** Natural History Museum, London (cl). **© Hunterian Museum & Art Gallery, University of Glasgow:** (cr)
166 Alamy Images: John Cancalosi (bl). **Roger Jones Collection:** (tc).
168 DK Images: Royal Museum of Scotland, Edinburgh (cl). **Craig Slawson:** Geoconservation UK (cb). **Valdosta State University :** Paleothyris cast obtained from Robert Carroll of McGill University's Redpath Museum; displayed on http://.fossils. valdosta.edu (b).
169 Getty Images: Alan Marsh / First Light (cra).
170–171 DK Images: Natural History Museum, London (main).
171 DK Images: Natural History Museum, London (crb).
172–193 DK Images: Natural History Museum, London.
172 DK Images: Natural History Museum, London (br). **Getty Images:** Anthony Boccaccio (cl). **Photolibrary:** Oxford Scientific (OSF) / Konrad Wothe (tr).
172–173 DK Images: Natural History Museum, London (b/background).
173 DK Images: Natural History Museum, London (bl). **Getty Images:** Michael Fay / National Geographic (tl). **Jon Ranson, NASA Goddard Space Flight Center:** (cra).
174 Corbis: James L. Amos (cr) (t/plants lettering). **DK Images:** Natural History Museum, London (b/background). **FLPA:** Krystyna Szulecka (cl).
178 Corbis: Jonathan Blair (clb) (b/ background) (t/invertebrates lettering). **DK Images:** Natural History Museum, London (cra).
182 DK Images: Natural History Museum, London (b/background).
184 Corbis: Jack Goldfarb/Design Pics (cla).
184–185 Alamy Images: WaterFrame (bc). **DK Images:** Natural History Museum, London (t).
185 DK Images: Natural History Museum, London (cra). **Getty Images:** Ken Lucas / Visuals Unlimited (br).
186 Corbis: Martin Schutt / DPA (cla). **DK Images:** University Museum of Zoology, Cambridge (clb). **Getty Images:** Hulton Archive (c).
188 DK Images: Natural History Museum, London (tl).
189 Corbis: Lester V. Bergman (ca). **Getty Images:** Ken Lucas / Visuals Unlimited (tr).
190 Corbis: W. Perry Conway (t).
191 DK Images: Natural History Museum, London (cr). **Henssen PalaeoWerkstatt , www.palaeowerkstatt.de:** (br).
194–195 Getty Images: Ralph Lee Hopkins / National Geographic (main).
195 DK Images: Natural History Museum, London.
196–221 Getty Images: Ralph Lee Hopkins / National Geographic (sidebands).
196–198 Getty Images: Ralph Lee Hopkins / National Geographic (b/background).
196–197 Plate Tectonic and Paleogeographic Map C.R. Scotese, © 2007, PALEOMAP Project (www.scotese.com).
196 Corbis: Rose Hartman (tr). **DK Images:** Natural History Museum, London (br) (bl) (clb) (crb). **Paul Olsen:** (cl).
197 Corbis: Fridmar Damm / zefa (cra); Martin B. Withers / Frank Lane Picture Agency (tl).
198 Corbis: George H.H. Huey (cl). **Getty Images:** Ralph Lee Hopkins / National Geographic (t/plants lettering).
199 DK Images: Natural History Museum, London (tr) (t).
200 DK Images: Natural History Museum, London (cb).
200–201 DK Images: Natural History Museum, London (c).
202 Getty Images: Ralph Lee Hopkins / National Geographic (invertebrates lettering) (b/background).
206–207 Getty Images: Ralph Lee Hopkins / National Geographic (b/background).
207 Museum of Texas Tech University: (cla).
208 Giuseppe Buono (http:// fossilspictures.wordpress.com/ and http:// paleonews.wordpress.com/): (cla). Dr Rainer Schoch / Staatliches Museum für Naturkunde: (tc).
209 DK Images: Royal Tyrrell Museum of Palaeontology, Alberta, Canada (br).
210 DK Images: Natural History Museum, London (bl) (c).
211 DK Images: Institute of Geology and Palaeontology, Tubingen, Germany (ca); Natural History Museum, London (tl).
212 Corbis: Jonathan Blair (cra).
213 Museum of Texas Tech University: (tr).
217 Corbis: Louie Psihoyos (bc/Bob Bakker).
219 DK Images: Institute of Geology and Palaeontology, Tubingen, Germany (clb) (cla).
221 DK Images: Natural History Museum, London (cr).
222–223 DK Images: Natural History Museum, London (main).
223 DK Images: Natural History Museum, London (crb).
224–279 DK Images: Natural History Museum, London (sidebands).
224–225 Plate Tectonic and Paleogeographic Map C.R. Scotese, © 2007, PALEOMAP Project (www.scotese.com).
224 Corbis: Tom Bean (tr) (clb/flower). **DK Images:** Natural History Museum, London (bl) (bc). **Getty Images:** Jeff Foott / Discovery Channel Images (cl). **Science Faction Images:** Louie Psihoyos (tr/dinosaur tracks).
224–226 DK Images: Natural History Museum, London (b/background).
225 Corbis: Theo Allofs (cra). **DK Images:** Natural History Museum, London (clb) (bl). **Getty Images:** Fumio Tomita / Sebun Photo (tl).
226 Corbis: David Spears / Clouds Hill Imaging Ltd. (cr). **DK Images:** Natural History Museum, London (t/plants lettering). **Science Photo Library:** Dee Breger (clb).
227 Getty Images: Kevin Schafer / Photographer's Choice (bc).
228 DK Images: Natural History Museum, London (bl).
230 DK Images: Natural History Museum, London (cla).
234 Corbis: Derek Hall / Frank Lane Picture Agency (cl) (b/background). **DK Images:** Natural History Museum, London (t/ invertebrates lettering).
237 DK Images: Natural History Museum, London (clb/shells).
243 Corbis: Douglas P. Wilson / Frank Lane Picture Agency (bl). **Photoshot:** Ken Griffiths / NHPA (cra).
244–245 DK Images: Natural History Museum, London (b/background).
247 Corbis: Kevin Schafer (crb).
250 Photolibrary: Rich Reid / National Geographic (bl).
251 Corbis: Naturfoto Honal (cla).
253 Science Photo Library: D.J.M. Donne (tr).
255 DK Images: Robert L. Braun – modelmaker (crb).
261 DK Images: Staatliches Museum für Naturkunde, Stuttgart (tc); American Museum of Natural History (tr).
262 Corbis: Bettmann (cr).
264 DK Images: Natural History Museum, London (br).
265 Corbis: Louie Psihoyos (bl); University of the Witwatersrand / EPA (crb).
266 The Natural History Museum, London: (crb).
267 DK Images: Natural History Museum, London (bl). **© Frank Luerweg, Uni Bonn:** (cra).
270 DK Images: Carnegie Museum of Natural History, Pittsburgh (cla).
274 DK Images: Leicester Museum (clb). **Martin Williams:** (cla).
274–275 DK Images: Natural History Museum, London (bc).
275 Martin Williams: (br).
278 Visuals Unlimited, Inc.: Ken Lucas (cra).
279 Visuals Unlimited, Inc.: Albert Copley (bl).
280–281 Getty Images: Louie Psihoyos / Science Faction (main).
Science Photo Library: Andrew Syred (tr).
282–285 Getty Images: Louie Psihoyos / Science Faction (b/background).
282–357 Getty Images: Louie Psihoyos / Science Faction.
282–283 Plate Tectonic and Paleogeographic Map C.R. Scotese, © 2007, PALEOMAP Project (www.scotese.com).
282 DK Images: Natural History Museum, London (clb) (br). **Getty Images:** Don Klumpp / Photographer's Choice (ftr); Panoramic Images (cl).
283 DK Images: Natural History Museum, London (clb) (bc) (bl). **Getty Images:** Travel Ink / Gallo Images (tl); Eric Van Den Brulle (cra).
284 Getty Images: Louie Psihoyos / Science Faction (t/plants lettering). **© Sharon Milito, Paleotrails Project:** (crb).
286 Alamy Images: Scenics & Science (bl). **Photolibrary:** Phototake Science (cla).
287 DK Images: Natural History Museum, London (cra).
289 DK Images: Royal Museum of Scotland, Edinburgh (cb); Natural History Museum, London (b/araucaria). **Barry Thomas:** (crb/ *Araucaria*).
290–291 Alamy Images: Bildarchiv Friedrichsmeier / FAN travelstock.
292 DK Images: Natural History Museum, London (tr).
293 Ardea: John Mason. **DK Images:** Natural History Museum, London .
294 DK Images: Natural History Museum, London (bl). **Getty Images:** Jonathan Blair / National Geographic (cr).
295 DK Images: Royal Tyrrell Museum of Palaeontology, Alberta, Canada (b) (b/ background).
296 Getty Images: Louie Psihoyos / Science Faction (t/invertebrates lettering).
301 Corbis: Lawson Wood (bl).
304 Getty Images: Louie Psihoyos / Science Faction (t/vertebrates lettering).
304–305 Getty Images: Louie Psihoyos / Science Faction (b/background).
305 DK Images: Natural History Museum, London (cla).
306 Alamy Images: Michael Patrick O'Neill (bl).
306–307 DK Images: Natural History Museum, London.
308 Alamy Images: Danita Delimont (crb). **Getty Images:** George Grall / National Geographic (crb/gar fish).
310 Corbis: Louie Psihoyos / Science Faction (cb).
312 Yale University Peabody Museum Of Natural History: (bl).
316 Bayerische Staatssammlung für Paläontologie und Geologie, München (cb).
318 DK Images: Natural History Museum, London (tr).
319 Alamy Images: Javier Etcheverry (cra/ giganotosaurus). **DK Images:** Natural History Museum, London (tl). **Ryan Somma:** (b).
321 DK Images: Robert L. Braun – modelmaker (tl); Natural History Museum, London (bl) (cra). **Getty Images:** Ira Block / National Geographic (tr).
324 Corbis: Paul Vicente / EPA (bl).
328–329 DK Images: Natural History Museum, London; Natural History Museum, London (ca).
330 DK Images: Royal Tyrrell Museum of Palaeontology, Alberta, Canada (cra).
331 Corbis: Louie Psihoyos / Science Faction (cla); Kevin Schafer (br). **DK Images:** Peabody Museum of Natural History, Yale University (tl); Luis Rey – modelmaker (fcra). **Getty Images:** Louie Psihoyos / Science Faction (cra).
332 Getty Images: Ira Block / National Geographic (bl); Spencer Platt (cla).
334 DK Images: Queensland Museum, Brisbane, Australia (clb).
335 DK Images: Royal Tyrrell Museum of Palaeontology, Alberta, Canada (tr).
338 DK Images: Natural History Museum, London (br) (tl). **Getty Images** (cl).
342 DK Images: Royal Tyrrell Museum of Palaeontology, Alberta, Canada (tr/skull) (c).
343 DK Images: Royal Tyrrell Museum of Palaeontology, Alberta, Canada (b).
345 DK Images: American Museum of Natural History (br).
350 DK Images: American Museum of Natural History (ca). **Getty Images:** Louie Psihoyos / Science Faction (cr).
351 DK Images: American Museum of Natural History.
354 DK Images: Royal Tyrrell Museum of Palaeontology, Alberta, Canada (bl/ skeleton and skull). **Getty Images:** O. Louis Mazzatenta / National Geographic (br).
355 DK Images: American Museum of Natural History (c). **Hai-lu You:** (tr).
356 Corbis: Peter Foley / EPA (tr); Joe McDonald (cra). **Jin Meng:** (ca).
357 DK Images: American Museum of Natural History (br). **Getty Images:** Nicole Duplaix / National Geographic (cla). **Steve Morton:** Museum Victoria / Monash University (tl).
358–359 Alamy Images: WaterFrame (main).
360–413 Alamy Images: WaterFrame (sidebands).
360–361 Plate Tectonic and Paleogeographic Map C.R. Scotese, © 2007, PALEOMAP Project (www.scotese.com).
360 Steve Brusatte and Nicole Lunning: (tr/Gubbio). **Corbis:** Momatiuk–Eastcott (tr). **DK Images:** Natural History Museum, London (bc) (fbr). **Getty Images:** Bill Hatcher / National Geographic (cl).
361 Alamy Images: JTB Photo Communications, Inc. (tl). **DK Images:** Natural History Museum, London (clb) (bl). **Getty Images:** Kim Westerskov (cra) (t/plants lettering).
362 Alamy Images: WaterFrame (b/ background). **Corbis:** image100 (clb). **Science Photo Library:** Maurice Nimmo (cra); Maria and Bruno Petriglia (ca).
367 Corbis: Frans Lanting (cl). **DK Images:** Natural History Museum, London (bl) (bc).
368 Alamy Images: WaterFrame (t/ invertebrates lettering). **SeaPics.com:** James D. Watt (clb).
372 Getty Images: Richard Herrmann / Visuals Unlimited (br) (t/vertebrates lettering).

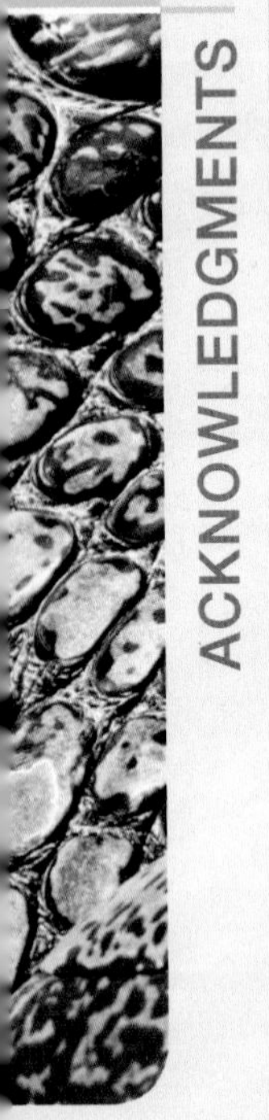

374 Alamy Images: WaterFrame (b/background). **Corbis:** Jonathan Blair (cl). **DK Images:** Natural History Museum, London (crb).
375 Corbis: Radius Images (clb); Visuals Unlimited (br).
377 Ray Carson –UF Photography: (crb). **DK Images:** Natural History Museum, London (ca) (br) (clb).
379 Ryan Somma: (br).
380 DK Images: Natural History Museum, London (cla).
381 Corbis: Bettmann (tl). **DK Images:** Natural History Museum, London (cla). **Getty Images:** Kevin Schafer / Visuals Unlimited (br). **Ryan Somma:** (crb).
382 Corbis: Michael & Patricia Fogden (cla). **DK Images:** Natural History Museum, London (cra). **Hessisches Landesmuseum Darmstadt:** Photo: Wolfgang Fuhrmannek. Hessisches Landesmuseum Darmstadt. (tc). **Dr Kent A. Sundell, owner – Douglas Fossils (www.douglasfossils.com):** (bc).
383 Corbis: Kevin Schafer (tr). **DK Images:** Natural History Museum, London (tl/skulls and foot).
384 Getty Images: Johnny Sundby / America 24-7 (tl).
385 DK Images: Natural History Museum, London (crb). **Getty Images:** Stan Honda / AFP (tl). **University of Nebraska State Museum:** with special thanks to Nancy Lindsley-Griffin and Ron Pike (cla).
386–387 Photolibrary: Tam C. Nguyen (main).
387 DK Images: Natural History Museum, London (crb).
388–389 Photolibrary: Tam C. Nguyen (b/background).
388–389 Plate Tectonic and Paleogeographic Map C.R. Scotese, © 2007, PALEOMAP Project (www.scotese.com).
388 Corbis: (cl); Wolfgang Deuter / zefa (tr). **DK Images:** Natural History Museum, London (br).
389 DK Images: Natural History Museum, London (bl) (br). **Getty Images:** Thomas Dressler / Gallo Images (cra/left image); Joseph Sohm – Visions of America / Photodisc (tl); Space Frontiers / Hulton Archive (cra).
390 Corbis: Mike Grandmaison (cra). **Getty Images:** Joseph Sohm – Visions of America / Stockbyte (clb) (t/plants lettering). **Photolibrary:** Tam C. Nguyen (b/background).
391 Corbis: David Muench (tc/taxodium).
394 DK Images: Natural History Museum, London (clb) (br/Palmoxylon).
395 DK Images: Natural History Museum, London (cra/porana) (b/background).
396 Photolibrary: Tam C. Nguyen (t/invertebrates lettering).
397 Getty Images: David Wrobel / Visuals Unlimited (cra).
398 Alamy Images: John T. Fowler (br).
402–403 Alamy Images: Phil Degginger.
404 DK Images: Natural History Museum, London (ca) (cb) (b/background). **Photolibrary:** Tam C. Nguyen (t/vertebrates lettering).
405 Bone Clones, Inc.: (bl). **Corbis:** Louie Psihoyos (cla). **DK Images:** Natural History Museum, London (tr) (clb).
406 DK Images: Natural History Museum, London (ca).
407 Nicholas D. Pyenson: Paleontologists from the San Diego Natural History Museum (cb). **Valley Anatomical Preparations, Inc.:** (bl).
408 Agate Fossil Beds National Monument: (tl). **Mira Images:** Phil Degginger c/o Mira.com (clb). **Ryan Somma:** (tc) (bc).
409 Ashfall Fossil Beds / University of Nebraska State Museum: (tr); Natural History Museum, London (br). **DK Images:** American Museum of Natural History (tl). **Photolibrary:** Steve Turner / Oxford Scientific (OSF) (fbr).
412 DK Images: Natural History Museum, London (br).
413 DK Images: Natural History Museum, London (cl) (bl) (clb) (t).
414–415 Corbis: Micha Pawlitzki/zefa (main).
416–439 Corbis: Micha Pawlitzki / zefa (sidebands).
416–418 Corbis: Micha Pawlitzki / zefa (b/background).
416–417 Plate Tectonic and Paleogeographic Map C.R. Scotese, © 2007, PALEOMAP Project (www.scotese.com).
416 DK Images: Natural History Museum, London (bl/both skulls on the left). **Getty Images:** William Albert Allard / National Geographic (tr); Vilem Bischof / AFP (br); Stocktrek Images (cl).
417 Corbis: Momatiuk–Eastcott (tl). **DK Images:** Natural History Museum, London (bc). **Getty Images:** PhotoLink / Photodisc (cra). **Science Photo Library:** Javier Trueba / MSF (bl).
418 Corbis: Craig Aurness (crb); Wayne Lawler / Ecoscene (cra) (b/background); Micha Pawlitzki / zefa (t/plants lettering). **Science Photo Library:** David Scharf (cl).
419 Alamy Images: Walter H. Hodge / Peter Arnold, Inc. (br).
420 Alamy Images: blickwinkel / Koenig (bc).
421 Photo Biopix.dk: J.C. Schou (cla). **Getty Images:** DEA / RANDOM / De Agostini Picture Library (bl). **Wikimedia Commons:** Kahuroa (cra).
422 Alamy Images: Marek Piotrowski (tr). **Corbis:** Stuart Westmorland (tc).
423 Science Photo Library: Eye of Science (clb).
424 Ardea: Steve Hopkin (clb) (b/background). **Corbis:** Micha Pawlitzki / zefa (t/invertebrates lettering).
426 The Natural History Museum, London: (cla).
427 Corbis: Ken Wilson / Papilio (bc).
428 Getty Images: David Wrobel / Visuals Unlimited (c).
430 Corbis: Micha Pawlitzki / zefa (vertebrates lettering) (b/background). **Getty Images:** Tom Walker / The Image Bank (crb); American Museum of Natural History (tr).
431 DK Images: Natural History Museum, London (crb) (tc). **Image courtesy of DigiMorph.org:** (bc).
432 DK Images: Natural History Museum, London (c/foot) (cr). **The Natural History Museum, London:** (b).
433 DK Images: Natural History Museum, London (cra) (tr). **Getty Images:** Theo Allofs / Photonica (br).
434 Corbis: Ted Soqui (crb). **DK Images:** Down House / Natural History Museum, London (cra). **photo by David K. Smith:** Arctodus simus specimen from The Mammoth Site, Hot Springs, South Dakota (clb).
435 DK Images: Natural History Museum, London (cr).
437 DK Images: National Museum of Wales (br).
438 DK Images: Natural History Museum, London (t/mastodon bones).
439 DK Images: Rough Guides / Simon Bracken (bl); Natural History Museum, London (tr) (br).
440–441 Getty Images: Peter Chadwick / Gallo Images.
442 DK Images: Natural History Museum, London (tr). **Getty Images:** Lluis Gene / AFP (cra) (cla).
442–479 Getty Images: Peter Chadwick / Gallo Images.
443 Corbis: Waltraud Grubitzsch / EPA (cl). **DK Images:** University College, London (tl). **Falling Pixel Ltd.:** Oleg Belobabov (bl).
444 Corbis: Wolfgang Kaehler (bl).
445 Alamy Images: Kolvenbach (tl). **The Natural History Museum, London:** (cla/homo habilis). **Science Photo Library:** Javier Trueba / MSF (cra) (cb).
446 Camera Press: Marc Deville / Gamma (br/Orrorin tugenensis fossil bones). **Getty Images:** Alain Beauvilain / AFP (cr).
447 Corbis: Natural Selection Ralph Curtin / Design Pics (cr).
448 Professor Michel Brunet: M.P.F.T (bc). **www.skullsunlimited.com:** (bl/& clb). **Stone Age Institute:** Dr. Sileshi Semaw (project director) (ca); Dr. Scott Simpson (project palaeontologist) (tr).
449 Alamy Images: Kolvenbach (br). **Getty Images:** Richard du Toit / Gallo Images (bl). **Science Photo Library:** John Reader (cr); Sinclair Stammers (tr).
450 Science Photo Library: John Reader (bl).
450–491 Getty Images: Peter Chadwick / Gallo Images (sidebars).
451 Getty Images: Vilem Bischof / AFP (bl). **Science Photo Library:** John Reader (r). **Paul Szpak:** (cl).
452 Alamy Images: Natural History Museum, London (br). **Science Photo Library:** John Reader (cb); Javier Trueba / MSF (clb).
453 Corbis: Brian A. Vikander (ca). **DK Images:** Natural History Museum, London (crb). **The Natural History Museum, London:** (cra). **Science Photo Library:** Des Bartlett (tr).
454 The Natural History Museum, London: (tr) (b).
455 Corbis: Ric Ergenbright (cla). **naturepl.com:** Karl Ammann (bl).
456 Alamy Images: John Warburton-Lee Photography (br).
457 Alamy Images: Marion Kaplan (br). **Getty Images:** Chris Johns / National Geographic (tr).
458 Getty Images: Kenneth Garrett / National Geographic (clb) (c); Paul Joynson Hicks / Gallo Images (b).
459 Alamy Images: Danita Delimont / Kenneth Garrett (cb) (bc). **Getty Images:** STR / AFP (tr). **The Natural History Museum, London:** (cla) (crb).
460 Copyright NNM, Leiden, The Netherlands: (br).
461 Alamy Images: Chris Howes / Wild Places Photography (br). **DK Images:** Natural History Museum, London (crb). **Getty Images:** Peter Chadwick / Gallo Images (cr).
462 Science Photo Library: Javier Trueba / MSF (cla) (bl) (clb).
463 The Natural History Museum, London: (fbr); The Boxgrove Project (br) (tl/insert). **Science Photo Library:** Javier Trueba / MSF (c) (bl) (clb) (tl).
464 Science Photo Library: John Reader (cl).
465 Bone Clones, Inc.: (r). **Science Photo Library:** Volker Steger (bl).
466 Getty Images: Alvis Upitis / The Image Bank (ca). **The Natural History Museum, London:** Janos Jurka (bl).
466–467 DK Images: Natural History Museum, London (cb).
467 Getty Images: Travel Ink / Gallo Images (cr) (tl/backed knife). **The Natural History Museum, London:** (tc).
468 Getty Images: Ira Block / National Geographic (bl); Azuhiro Nogi / AFP (cb); Anna Zieminski / AFP (br). **The Natural History Museum, London:** (tr).
469 The Natural History Museum, London: (r/skulls). **Science Photo Library:** Pascal Goetgheluck (l).
470 Science Photo Library: Alain Pol, ISM (bl).
471 Getty Images: Jim Watson / AFP (b).
472 akg-images: (br); Herbert Kraft (bl); Erich Lessing (bc). **Corbis:** Hanan Isachar (tr). **Science Photo Library:** Pascal Goetgheluck (ca).
473 Corbis: Michael Amendolia (cla); Ricardo Azoury (cra); Beawiharta / Reuters (ftr); Handout / Reuters (tr); Warren Morgan (crb). **DK Images:** Natural History Museum, London (bl).
474 Corbis: Gianni Dagli Orti (crb); Gustavo Tomsich (bl). **DK Images:** Natural History Museum, London (br). **The Natural History Museum, London:** (ca). **Science Photo Library:** RIA NOVOSTI (tr).
475 The Bridgeman Art Library: Musée des antiquités nationales, St Germain-en-Laye, France / Lauros / Giraudon (tc); Gianni Dagli Orti / The Picture Desk Limited (bl). **Corbis:** Gianni Dagli Orti (cla) (cr). **The Natural History Museum, London:** (br).
476 akg-images: Erich Lessing (cla). **French Ministry of Culture and Communication, Regional Direction for Cultural Affairs – Rhône-Alpes region – Regional department of archaeology:** (bl). **Getty Images:** AFP (br); Robert Frerck (cra).
477 Corbis: Gianni Dagli Orti (b). **Getty Images:** Sisse Brimberg / National Geographic (cla); Ralph Morse (tr).
478 The Trustees of the British Museum: (cl). **National Museum Of Denmark:** Lennart Larsen (r). **The Natural History Museum, London:** (bl).
479 Alamy Images: INTERFOTO Pressebildagentur (bl). **Ancient Art & Architecture Collection:** Ronald Sheridan (br) (ca). **Corbis:** Bettmann (cr). **Getty Images:** The Bridgeman Art Library (clb) (cb).
480–512 Getty Images: Art Wolfe (sidebars).